BASIC VIROLOGY

In Memoriam

Edward K. Wagner

(May 4, 1940 to January 21, 2006)

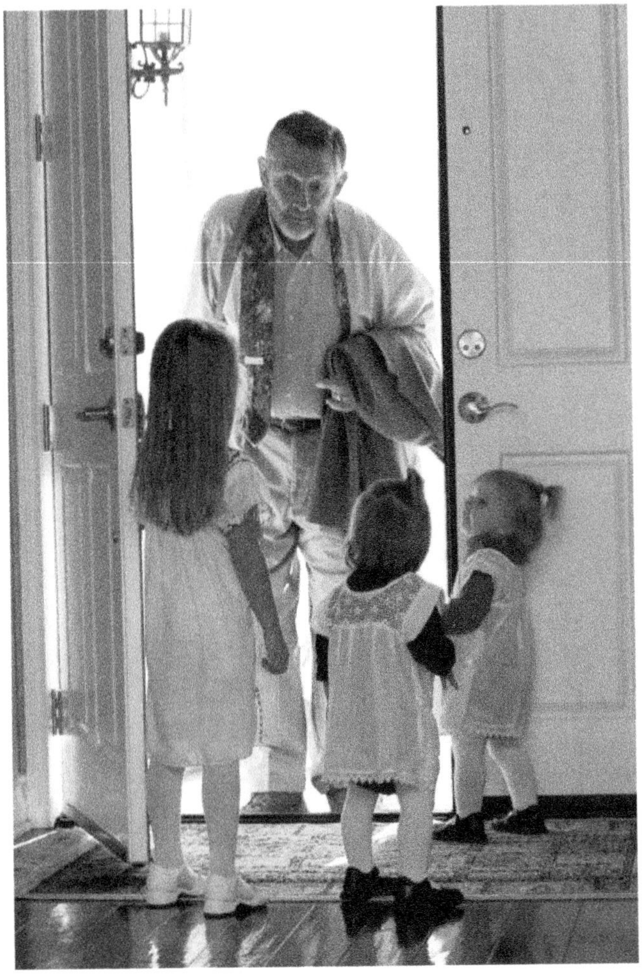

It was one of those telephone calls that you do not want to receive. Each of us, that weekend in late January, heard of the untimely passing of our colleague, co-author, collaborator, mentor, and friend, Ed Wagner. Ed will be remembered for his many contributions to the teaching of virology and for his research contributing to our understanding of the intricacies of the herpes-viruses. From his graduate work at MIT, through his postdoctoral research at the University of Chicago, and on to his professorship at the University of California, Irvine, Ed was a passionate champion for the most rigorous and critical thinking and the most dedicated teaching, setting a standard for the discipline of virology. Beyond the laboratory and the classroom, Ed loved life to the fullest, with his family and friends. The last time we were together as a writing team, in the fall of 2005, we all remember an intense day of work in a conference room at UCI, followed by an evening of touring some of Ed's favorite haunts in the Southern California coastal towns he called home. It is with those thoughts etched into our memories that we dedicate this edition of *Basic Virology* to Edward K. Wagner.

Basic Virology

Fourth Edition

Martinez "Marty" Hewlett
University of New Mexico
Taos, NM, USA

David Camerini
Center for Virus Research
University of California, Irvine
Irvine, CA, USA

David C. Bloom
Department of Molecular Genetics and Microbiology
University of Florida
Gainesville, FL, USA

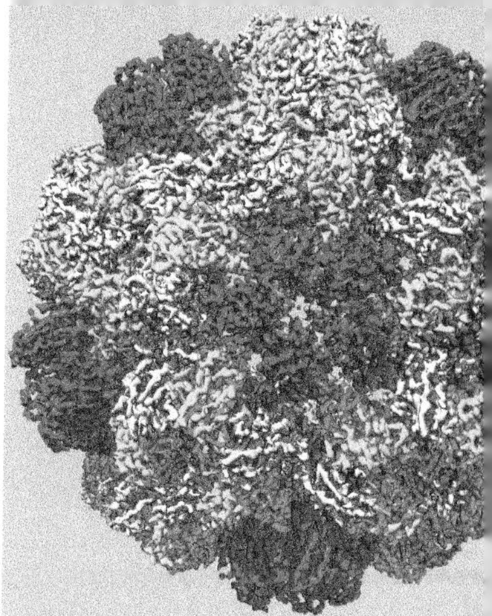

WILEY Blackwell

This fourth edition first published 2021

Edition History:

1e; (1999, Wiley-Blackwell) 2e; (2003, John Wiley & Sons), 3e; (2007, Wiley-Blackwell).

Registered Office
John Wiley & Sons, Inc., 111 River Street, Hoboken, NJ 07030, USA

Editorial Office
Boschstr. 12, 69469 Weinheim, Germany

For details of our global editorial offices, customer services, and more information about Wiley products visit us at www.wiley.com.

Wiley also publishes its books in a variety of electronic formats and by print-on-demand. Some content that appears in standard print versions of this book may not be available in other formats.

Library of Congress Cataloging-in-Publication Data
Names: Hewlett, Martinez "Marty", author. | Camerini, David, author. | Bloom, David (David C.), author.
Title: Basic virology / Marty Hewlett, David Camerini, David Bloom.
Description: Fourth edition. | Hoboken, NJ: Wiley-Blackwell, 2021. | Preceded by: Basic virology / Edward K. Wagner . . . [et al.]. 3rd ed. 2008. | Includes bibliographical references and index.
Identifiers: LCCN 2020026463 (print) | LCCN 2020026464 (ebook) | ISBN 9781119314059 (paperback) | ISBN 9781119314042 (adobe pdf) | ISBN 9781119314066 (epub)
Subjects: MESH: Virus Diseases–virology | Viruses–pathogenicity | Virus Replication | Genome, Viral
Classification: LCC QR360 (print) | LCC QR360 (ebook) | NLM WC 500 | DDC 579.2–dc23
LC record available at https://lccn.loc.gov/2020026463
LC ebook record available at https://lccn.loc.gov/2020026464

Cover Design: Wiley Blackwell
Cover Image: © Wah Chiu

Set in 10.5/12.5pt Adobe Garamond Pro by SPi Global, Pondicherry, India

SKY10031586_120321

Brief Contents

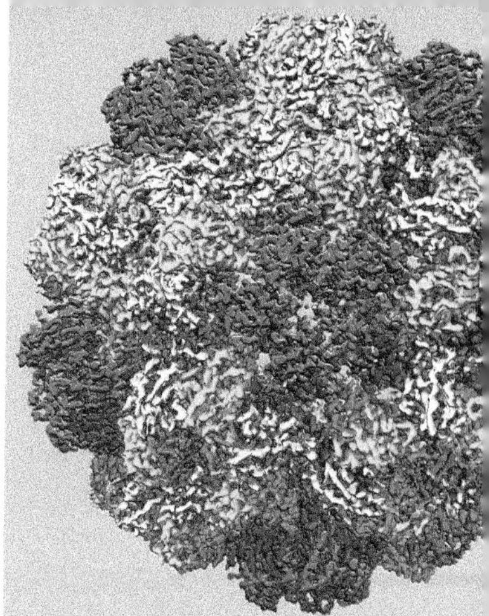

Contents

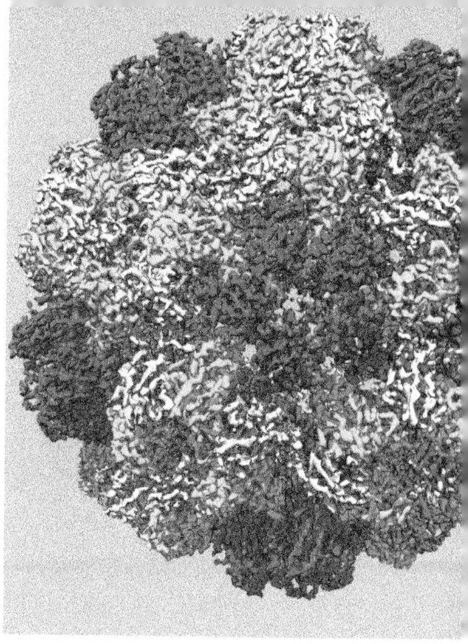

Preface to the First Edition

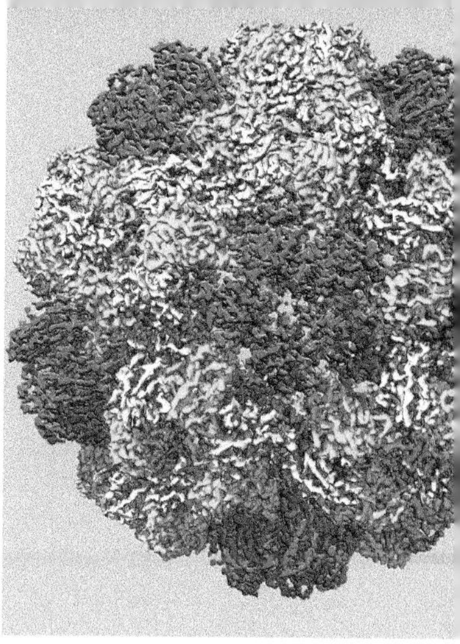

Viruses have historically flickered in and out of the public consciousness. In the eight years since we finished the first edition of *Basic Virology* much has happened, both in the world and in virology, to fan the flames of this awareness.

In this period we have seen the development of a vaccine to protect women against human papilloma virus type 16. This major advance could well lead to a drastic reduction in the occurrence of cervical cancer. In addition, viruses as gene delivery vectors have increased the prospect of targeted treatments for a number of genetic diseases. The heightened awareness and importance of the epidemiological potential of viruses, in both natural and man-caused outbreaks, have stimulated the search for both prophylactic and curative treatments.

However, the events of September 11, 2001, dramatically and tragically altered our perceptions. A new understanding of threat now pervades our public and private actions. In this new arena, viruses have taken center stage as the world prepares for the use of infectious agents such as smallpox in acts of bioterrorism.

Naturally occurring virological issues also continue to capture our attention. West Nile virus, originally limited to areas of North Africa and the Middle East, has utilized the modern transportation network to arrive in North America. Its rapid spread to virtually every state in the union has been both a public health nightmare and a vivid demonstration of the opportunism of infectious diseases. The continuing AIDS pandemic reminds us of the terrible cost of this opportunism. In addition, we are now faced with the very real prospect of the next pandemic strain of influenzas, perhaps derived from the avian H5N1 virus now circulating in wild and domestic birds.

It is against this backdrop of hope and concern that we have revised *Basic Virology*.

This book is based on more than 40 years in aggregate of undergraduate lectures on virology commencing in 1970 given by the coauthors (Wagner, Hewlett, Bloom, and Camerini) at the University of California, Irvine (UCI), the University of Arizona, and the University of Florida. The field of virology has matured and grown immensely during this time, but one of the major joys of teaching this subject continues to be the solid foundation it provides in topics running the gamut of the biological sciences. Concepts range from population dynamics and population ecology, through evolutionary biology and theory, to the most fundamental and detailed analyses of the biochemistry and molecular biology of gene expression and biological structures. Thus, teaching virology has been a learning tool for us as much as, or more than, it has been for our students.

Our courses are consistently heavily subscribed, and we credit that to the subject material, certainly not to any special performance tricks or instructional techniques. Participants have been mainly premedical students, but we have enjoyed the presence of other students bound for postgraduate studies, as well as a good number of those who are just trying to get their degree and get out of the "mill" and into the "grind."

At UCI, in particular, the course had a tremendous enrollment (approximately 250 students per year) in the past 5–8 years, and it has become very clear that the material is very challenging for a sizable minority studying it. While this is good, the course was expanded in time to five hours per week for a 10-week quarter to accommodate only those students truly interested in being challenged. Simply put, there is a lot of material to master, and mastery requires a solid working knowledge of basic biology and, most importantly, the desire to learn. This "experiment" has been very successful, and student satisfaction with the expanded course is, frankly, gratifying. To help students acquire such working knowledge, we have encouraged further reading. We have also included a good deal of reinforcement material to help students learn the basic skills of molecular biology and rudimentary aspects of immunology, pathology, and disease. Furthermore, we have incorporated numerous study and discussion questions at the end of chapters and sections to aid in discussion of salient points.

It is our hope that this book will serve as a useful text and source for many undergraduates interested in acquiring a solid foundation in virology and its relationship to modern biology. It is also hoped that the book may be of use to more advanced workers who want to make a quick foray into virology but who do not want to wade through the details present in more advanced works.

Preface to the Second Edition

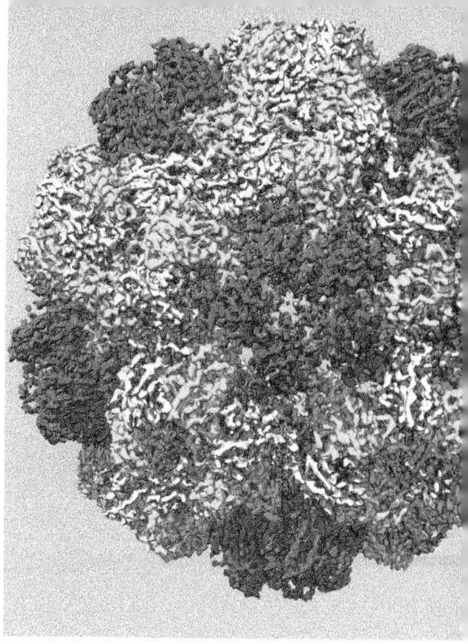

The text retains our organizational format. As before, Part I concerns the interactions of viruses and host populations, Part II is about the experimental details of virus infection, Part III discusses the tools used in the study of viruses, and Part IV is a detailed examination of families and groupings of viruses. We have found, in our own teaching and in comments from colleagues, that this has been a useful approach. We have also kept our emphasis on problem solving and on the provision of key references for further study.

What is new in the second edition has been driven by changes in virology and in the tools used to study viruses. Some of these changes and additions include

- a discussion of bioterrorism and the threat of viruses as weapons;
- updated information on emerging viruses such as West Nile, and their spread;
- the current state of HIV antiviral therapies;
- discussions of viral genomics in cases where sequencing has been completed;
- discussion of cutting-edge technologies, such as atomic force microscopy and DNA microarray analysis; and
- updated glossary and reference lists.

We have, throughout the revision, tried to give the most current understanding of the state of knowledge for a particular virus or viral process. We have been guided by a sense of what our students need in order to appreciate the complexity of the virological world and to come away from the experience with some practical tools for the next stages in their careers.

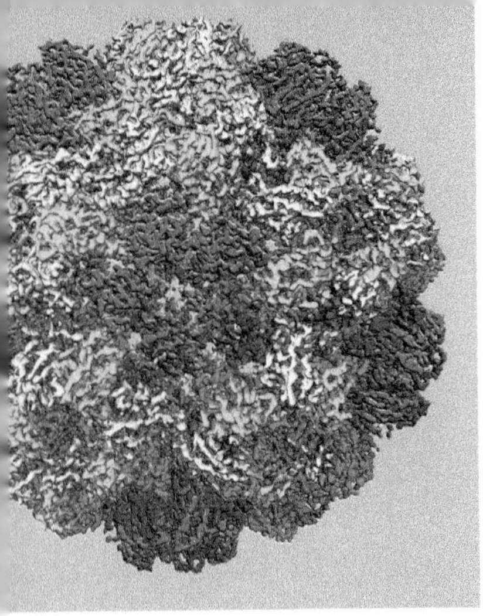

Preface to the Third Edition

It is with a true sense of our loss that the three of us sit in Irvine, California, Gainesville, Florida, and Taos, New Mexico, working toward completion of this edition. The absence of our friend and colleague, Ed Wagner, is all the more apparent as we write the preface to this latest edition of *Basic Virology*. In his spirit, we offer our colleagues and students this book that is our latest view of the field that Ed pursued with such passion and dedication.

In this new edition, we have attempted to bring the current state of our discipline into focus for students at the introductory and intermediate levels. To this end, we have done the job of providing the most current information, at this writing, for each of the subjects covered. We have also done some reorganization of the material. We have added three new chapters, in recognition of the importance of these areas to the study of viruses.

The book now includes a chapter devoted completely to HIV and the lentiviruses (Chapter 20), previously covered along with the retroviruses in general. Given that we continue to face the worldwide challenge of AIDS, we feel that this is an important emphasis.

You will also notice that this version now includes a Part V ("Viruses: New Approaches and New Problems"). This section begins with a consideration of the molecular tools used to study and manipulate viruses (Chapter 22), follows with coverage of viral pathogenesis at the molecular level (Chapter 23), and continues with a chapter dealing with viral genomics and bioinformatics (Chapter 24). We intend that these three chapters will give our students insight into the current threads of molecular and virological thinking. Part V concludes with our chapter on "Viruses and the Future" (Chapter 25), containing updated material on emerging viruses, including influenza, as well as viruses and nanotechnology.

A major change in this edition is the use of full-color illustrations. We welcome this effort from our publisher, Blackwell Science, and hope that you find this adds value and utility to our presentation.

In conjunction with the expanded coverage, the Glossary has been revised. In addition, all of the references, both text and web-based, have been reviewed and made current as of this writing.

Most of these changes were either finished or discussed in detail before Ed's untimely passing. As a result, we are proud to say that *Basic Virology*, Third Edition, bears the welcome imprint of the scientist/teacher who inspired the first one. We hope you agree and enjoy the fruits of this effort.

Marty Hewlett, Taos, NM, USA
Dave Bloom, Gainesville, FL, USA
David Camerini, Irvine, CA, USA

Preface to the Fourth Edition

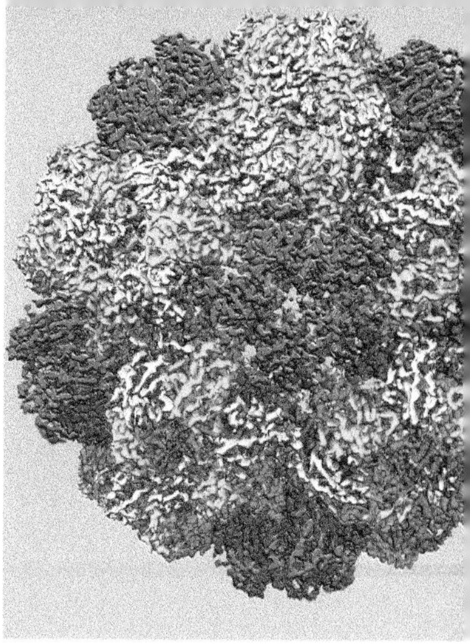

It seems like only yesterday that Ed and Marty spoke on the phone and said, in effect, "Let's do this thing," giving birth to *Basic Virology*. And here we are, completing the revisions for the fourth edition of what we hope will remain a useful and relevant textbook for the teaching of introductory virology to undergraduates.

As it has from its very beginnings, the field of virology is changing at an astounding pace, with newly recognized diseases and their viral causes being reported, accompanied by ever more sophisticated techniques for studying these entities that exist at the fringes of the living world.

In this latest edition we have attempted to capture some of this dynamism, while retaining the organization and pedagogical approach of the original. To that end we have added new and expanded discussions of such agents as Ebola virus, Zika virus, and H1N1 and H7N9 influenza virus, as well as the SARS-CoV-2/COVID-19 pandemic, with information that is current as of this writing. We have modified our presentation of techniques, removing some that are outdated (CoT curves, as an example), retaining the classics that have defined the field (pulse and pulse-chase labeling), and introducing the newest approaches that are opening new areas of investigation (CRISPR-Cas).

The organization of the book has been retained from the third edition, with 25 chapters divided into five parts, including the Case Studies, updated as necessary. We have tried to avoid textbook size creep by making judicious editorial choices. Figures have been changed as needed to reflect new information, with the addition of new graphics where necessary to complete new or expanded coverage.

We hope that you find this version of our work both useful and relevant in your teaching of our favorite topic... virology!

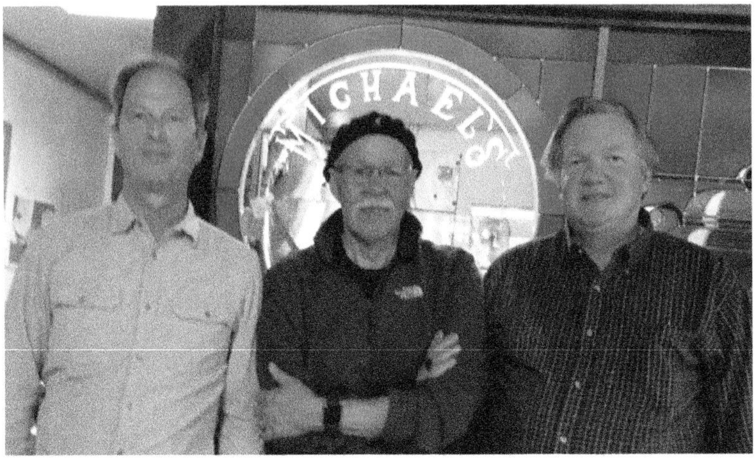

David Camerini, Martinez "Marty" Hewlett, and Dave Bloom: Michael's Kitchen and Bakery, Taos, New Mexico, March 2017.

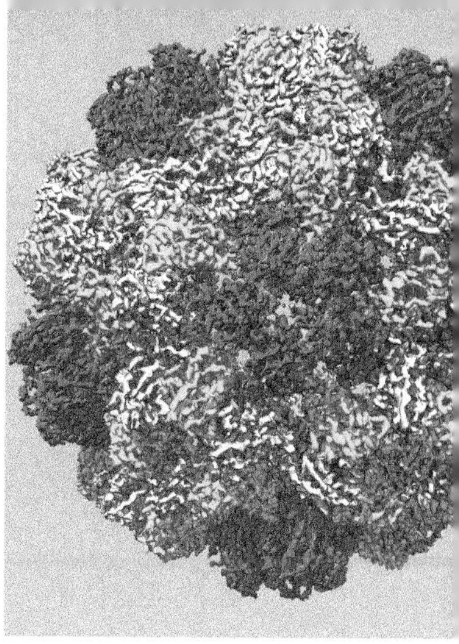

Acknowledgments

Even the most basic text cannot be solely the work of its author or authors; this is especially true for this one. We are extremely grateful to a large number of colleagues, students, and friends. They provided critical reading, essential information, experimental data, and figures, as well as other important help for all four of the editions of *Basic Virology*. This group includes the following scholars from other research centers: Wah Chiu, Stanford University; J. Brown, University of Virginia; J. B. Flannegan and R. Condit, University of Florida; J. Conway, National Institutes of Health; K. Fish and J. Nelson, Oregon Health Sciences University; D. W. Gibson, Johns Hopkins University; P. Ghazal, University of Edinburgh; H. Granzow, Friedrich-Loeffler-Institut, Insel Riems; C. Grose, University of Iowa; J. Hill, Louisiana State University Eye Center–New Orleans; S. Karst, University of Florida; J. Langland, Arizona State University; D. Leib, Dartmouth College; F. Murphy, University of California, Davis; S. Rabkin, Harvard University; S. Rice, University of Alberta–Edmonton; S. Silverstein, Columbia University; B. Sugden, University of Wisconsin; Gail Wertz, University of Alabama–Birmingham; and J. G. Stevens, University of California, Los Angeles. Colleagues at the University of California, Irvine who provided aid include R. Davis, S. Larson, A. McPherson, T. Osborne, R. Sandri-Goldin, D. Senear, B. Semler, S. Stewart, W. E. Robinson, I. Ruf, and L. Villarreal. Both current and former workers in Edward Wagner's laboratory did many experiments that aided in a number of illustrations; these people include J. S. Aguilar, K. Anderson, R. Costa, G. B. Devi-Rao, R. Frink, S. Goodart, J. Guzowski, L. E. Holland, P. Lieu, N. Pande, M. Petroski, M. Rice, J. Singh, J. Stringer, and Y.-F. Zhang. Colleagues of David Camerini that did experiments and helped make figures used in the fourth edition are Joseph J. Campo, Shailesh K. Choudhary, Arlo Randall, and Robert M. Scoggins.

We were aided in the writing of the second edition by comments from Robert Nevins (Milsap College), Sofie Foley (Napier University), David Glick (King's College), and David Fulford (Edinboro University of Pennsylvania).

We want to remember the many people who contributed to the physical process of putting the first edition of this book together. R. Spaete of the Aviron Corp carefully read every page of the manuscript and suggested many important minor and a couple of major changes. This was done purely in the spirit of friendship and collegiality. K. Christensen used her considerable expertise and incredible skill in working with us to generate the art. Not only did she do the drawings, but also she researched many of them to help provide missing details. Two undergraduates were invaluable to us. A. Azarian at University of California, Irvine made many useful suggestions on reading the manuscript from a student's perspective, and D. Natan, an MIT student who spent a summer in Edward Wagner's laboratory, did most of the Internet site searching, which was a great relief and time saver. Finally, J. Wagner carried out the very difficult task of copyediting the manuscript.

From the beginning, a number of people at Blackwell Science represented by Publisher N. Hill-Whilton demonstrated a commitment to a quality product. We especially thank Nathan Brown, Cee Brandson, and Rosie Hayden, who made great efforts to maintain effective communications and to expedite many of the very tedious aspects of this project. Blackwell Science directly contacted a number of virologists who also read and suggested useful modifications to this manuscript: Michael R. Roner, University of Texas, Arlington; Lloyd Turtinen, University of Wisconsin, Eau Claire; and Paul Wanda, Southern Illinois University.

All of these colleagues and friends represent the background of assistance we have received, leading to the preparation of this fourth edition. We would especially like to acknowledge Dr. Luis Villarreal and the Center for Virus Research at the University of California, Irvine for supporting our efforts in bringing this book to a timely completion.

Virology and Viral Disease

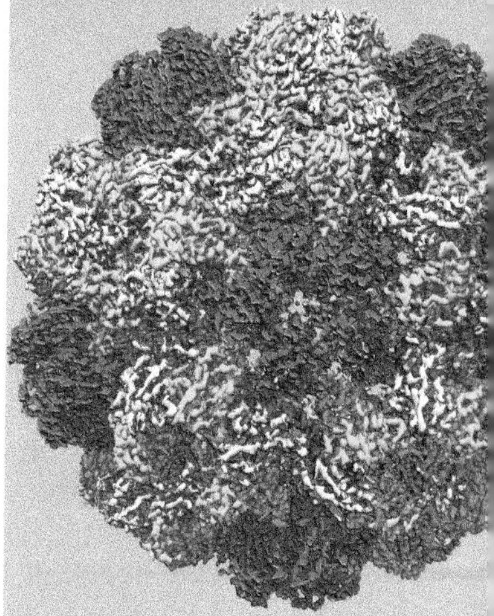

PART I

Introduction – The Impact of Viruses on Our View of Life

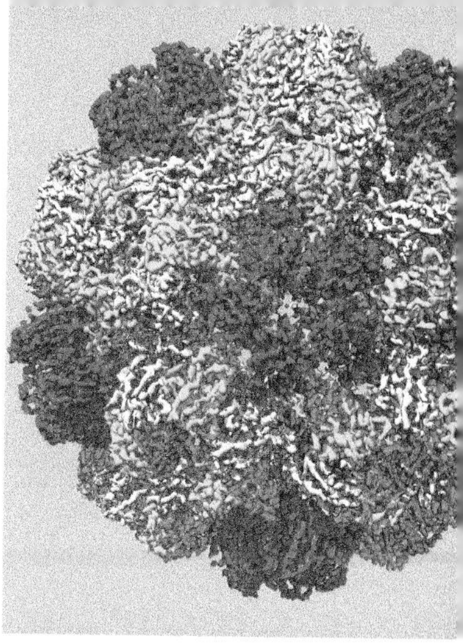

CHAPTER

1

* THE SCIENCE OF VIROLOGY
* The effect of virus infections on the host organism and populations – viral pathogenesis, virulence, and epidemiology
* The interaction between viruses and their hosts
* The history of virology
* Examples of the impact of viral disease on human history
* Examples of the evolutionary impact of the virus–host interaction
* The origin of viruses
* Viruses have a constructive as well as destructive impact on society
* Viruses are not the smallest self-replicating pathogens
* QUESTIONS FOR CHAPTER 1

THE SCIENCE OF VIROLOGY

The study of viruses has historically provided and continues to provide the basis for much of our most fundamental understanding of modern biology, genetics, and medicine. Virology has had an impact on the study of biological macromolecules, processes of cellular gene expression, mechanisms for generating genetic diversity, processes involved in the control of cell growth and development, aspects of molecular evolution, the mechanism of disease and response of the host to it, and the spread of disease in populations.

In essence, viruses are collections of genetic information directed toward one end: their own replication. They are the ultimate and prototypical example of "selfish genes." The viral genome contains the "blueprints" for virus replication enciphered in the genetic code, and must be decoded by the molecular machinery of the cell that it infects to gain this end. Viruses are thus obligate intracellular parasites dependent on the metabolic and genetic functions of living cells.

Basic Virology, Fourth Edition. Martinez "Marty" Hewlett, David Camerini, and David C. Bloom.

Given the essential simplicity of virus organization – a genome containing genes dedicated to self-replication surrounded by a protective protein shell – it has been argued that viruses are nonliving collections of biochemicals whose functions are derivative and separable from the cell. Yet this generalization does not stand up to the increasingly detailed information accumulating describing the nature of viral genes, the role of viral infections in evolutionary change, and the evolution of cellular function. A view of viruses as constituting a major subdivision of the **biosphere**, as ancient as and fully interactive and integrated with the three great branches of cellular life, becomes more strongly established with each investigational advance.

It is a major problem in the study of biology at a detailed molecular and functional level that almost no generalization is sacred, and the concept of viruses as simple parasitic collections of genes functioning to replicate themselves at the expense of the cell they attack does not hold up. Many generalizations will be made in the survey of the world of viruses introduced in this book; most if not all will be ultimately classified as being useful, but unreliable, tools for the full understanding and organization of information.

Even the size range of viral genomes, generalized to range from one or two genes to a few hundred at most (significantly less than those contained in the simplest free-living cells), cannot be supported by a close analysis of data. While it is true that the vast majority of viruses studied range in size from smaller than the smallest organelle to just smaller than the simplest cells capable of energy metabolism and protein synthesis, the mycoplasma and simple unicellular algae, the recently discovered mimivirus (distantly related to poxviruses such as smallpox or variola) contains nearly 1000 genes and is significantly larger than the smallest cells. With such caveats in mind, it is still appropriate to note that despite their limited size, viruses have evolved and appropriated a means of propagation and replication that ensures their survival in free-living organisms that are generally between 10 and 10 000 000 times their size and genetic complexity.

The effect of virus infections on the host organism and populations – viral pathogenesis, virulence, and epidemiology

Since a major motivating factor for the study of virology is that viruses cause disease of varying levels of severity in human populations and in the populations of plants and animals that support such populations, it is not particularly surprising that virus infections have historically been considered episodic interruptions of the wellbeing of a normally healthy host. This view was supported in some of the earliest studies on bacterial viruses, which were seen to cause the destruction of the host cell and general disruption of healthy, growing populations of the host bacteria. Despite this, it was seen with another type of bacterial virus that a persistent, **lysogenic** infection could ensue in the host population. In this case, stress to the lysogenic bacteria could release infectious virus long after the establishment of the initial infection.

These two modes of infection of host populations by viruses, which can be accurately modeled by mathematical methods developed for studying predator–prey relationships in animal and plant populations, are now understood to be general for virus–host interactions. Indeed, persistent infections with low or no levels of viral disease are universal in virus–host ecosystems that have evolved together for extended periods – it is only upon the introduction of a virus into a novel population that widespread disease and host **morbidity** occur.

While we can therefore consider severe virus-induced disease to be evidence of a recent introduction of the virus into the population in question, the accommodation of the one to the other is a very slow process requiring genetic changes in both virus and host, and it is by no means certain that the accommodation can occur without severe disruption of the host population – even its extinction. For this reason, the study of the replication and propagation of a given virus in a population is of critical importance to the body politic, especially in terms of formulating and implementing health policy. This is, of course, in addition to its importance to the scientific and medical communities.

The study of viral **pathogenesis** is broadly defined as the study of effects of viral infection on the host. The *pathogenicity* of a virus is defined as the sum total of the virus-encoded functions that contribute to virus propagation in the infected cell, in the host organism, and in the population. Pathogenicity is essentially the genetic ability of members of a given specific virus population (which can be considered to be genetically more or less equivalent) to cause a disease and spread through (**propagate** in) a population. Thus, a major factor in the pathogenicity of a given virus is its genetic makeup or **genotype**.

The basis for severity of the symptoms of a viral disease in an organism or a population is complex. It results from an intricate combination of expression of the viral genes controlling pathogenicity, physiological response of the infected individual to these pathogenic determinants, and response of the population to the presence of the virus propagating in it. Taken together, these factors determine or define the **virulence** of the virus and the disease it causes.

A basic factor contributing to virulence is the interaction among specific viral genes and the genetically encoded defenses of the infected individual. It is important to understand, however, that virulence is also affected by the general health and genetic makeup of the infected population, and in humans, by the societal and economic factors that affect the nature and extent of the response to the infection.

The distinction and gradation of meanings between the terms *pathogenesis* and *virulence* can be understood by considering the manifold factors involved in disease severity and spread exhibited in a human population subjected to infection with a disease-causing virus. Consider a virus whose genotype makes it highly efficient in causing a disease, the signs and symptoms of which are important in the spread between individuals – perhaps a respiratory infection with accompanying sneezing, coughing, and so on. This ideal or optimal virus will incorporate numerous, random genetic changes during its replication cycles as it spreads in an individual and in the population. Some viruses generated during the course of a disease may, then, contain genes that are not optimally efficient in causing symptoms. Such a virus is of reduced virulence, and in the extreme case, it might be a virus that has accumulated so many mutations in pathogenic genes that it can cause no disease at all (i.e., has mutated to an **avirulent** or **apathogenic** strain). While an avirulent virus may not cause a disease, its infection may well lead to complete or partial **immunity** against the most virulent genotypes in an infected individual. This is the basis of **vaccination**, which is described in Part II, Chapter 8. But the capacity to generate an immune response and the resulting generation of **herd immunity** also mean that as a virus infection proceeds in a population, either its virulence must change or the virus must genetically adapt to the changing host.

Other factors not fully correlated with the genetic makeup of a virus also contribute to variations in virulence of a pathogenic genotype. The same virus genotype infecting two **immunologically naive** individuals (i.e., individuals who have never been exposed to any form of the virus leading to an immune response) can cause very different outcomes. One individual might only have the mildest symptoms because of exposure to a small amount of virus, or infection via a suboptimal route, or a robust set of immune and other defense factors inherent in his or her genetic makeup. Another individual might have a very severe set of symptoms or even death if he or she receives a large inoculum, or has impaired immune defenses, or happens to be physically stressed due to malnutrition or other diseases.

Also, the same virus genotype might cause significantly different levels of disease within two more-or-less genetically equivalent populations that differ in economic and technological resources. This could happen because of differences in the ability of one society's support net to provide for effective medical treatment, or to provide for isolation of infected individuals, or to have available the most effective treatment protocols.

Taken in whole, the study of human infectious disease caused by viruses and other pathogens defines the field of **epidemiology** (in animals, it is termed **epizoology**). This field requires a good understanding of the nature of the disease under study and the types of medical and other remedies available to treat it and counter its spread, and some appreciation for the dynamics and particular nuances and peculiarities of the society or population in which the disease occurs.

The interaction between viruses and their hosts

The interaction between viruses (and other infectious agents) and their hosts is a dynamic one. As effective physiological responses to infectious disease have evolved in the organism and (more recently) have developed in society through application of biomedical research, viruses themselves respond by exploiting their naturally occurring genetic variation to accumulate and select mutations to become wholly or partially resistant to these responses. In extreme cases, such resistance will lead to periodic or episodic reemergence of a previously controlled disease – the most obvious example of this process is the periodic appearance of human influenza viruses causing disease.

The accelerating rate of human exploitation of the physical environment and the accelerating increase in agricultural populations afford some viruses new opportunities to "break out" and spread both old and novel diseases. Evidence of this is the ongoing **acquired immune deficiency syndrome (AIDS)** epidemic, as well as sporadic occurrences of viral diseases such as hemorrhagic fevers in Asia, Africa, and the southwestern United States. Investigation of the course of a viral disease, as well as societal responses to it, provides a ready means to study the role of social policies and social behavior of disease in general.

The recent worldwide spread of AIDS is an excellent example of the role played by economic factors and other aspects of human behavior in the origin of a disease. There is strong evidence to support the view that the causative agent, **human immunodeficiency virus (HIV)**, was introduced into the human population by an event fostered by agricultural encroachment of animal habitats in Equatorial Africa. This is an example of how economic need has accentuated risk.

HIV is not an efficient pathogen; it requires direct inoculation of infected blood or body fluids for spread. In the Euro-American world, the urban concentration of homosexual males with sexual habits favoring a high risk for venereal disease had a major role in spreading HIV and resulting AIDS throughout the male homosexual community. A partial overlap of this population with intravenous drug users and participants in the commercial sex industry resulted in spread of the virus and disease to other portions of urban populations. The result is that in Western Europe and North America, AIDS has been a double-edged sword threatening two disparate urban populations: the relatively affluent homosexual community and the impoverished heterosexual world of drug abusers – both highly concentrated urban populations. In the latter population, the use of commercial sex as a way of obtaining money resulted in further spread to other heterosexual communities, especially those of young, single men and women.

An additional factor is that the relatively solid medical and financial resources of a large subset of the "economic first world" resulted in wide use of whole blood transfusion and, more significantly, pooled blood fractions for therapeutic use. This led to the sudden appearance of AIDS in hemophiliacs and sporadically in recipients of massive transfusions due to intensive surgery. Luckily, the incidence of disease in these last risk populations has been reduced owing to effective measures for screening blood products.

Different societal factors resulted in a different distribution of HIV and AIDS in Equatorial Africa and Southeast Asia. In these areas of the world, the disease is almost exclusively found in heterosexual populations. This distribution of AIDS occurred because a relatively small concentration of urban commercial sex workers acted as the source of infection of working men living apart from their families. The periodic travel by men to their isolated village homes resulted in the virus being found with increasing frequency in isolated family units. Further spread resulted from infected women leaving brothels and prostitution to return to their villages to take up family life.

Another important factor in the spread of AIDS is technology. HIV could not have spread and posed the threat it now does in the world of a century ago. Generally lower population densities and lower concentrations of individuals at risk at that time would have precluded HIV from gaining a foothold in the population. Slower rates of communication and much more restricted travel and migration would have precluded rapid spread; also, the transmission of blood and blood products as therapeutic tools was unknown a century ago.

Of course, this dynamic interaction between pathogen and host is not confined to viruses; any pathogen exhibits it. The study and characterization of the genetic accommodations that viruses make, both to natural resistance generated in a population of susceptible hosts and to human-directed efforts at controlling the spread of viral disease, provide much insight into evolutionary processes and population dynamics. Indeed, many of the methodologies developed for the study of interactions between organisms and their environment can be applied to the interaction between pathogen and host.

The history of virology

The historic reason for the discovery and characterization of viruses, and a continuing major reason for their detailed study, involve the desire to understand and control the diseases and attending degrees of economic and individual distress caused by them. As studies progressed, it became clear that there were many other important reasons for the study of viruses and their replication.

Since viruses are parasitic on the molecular processes of gene expression and its regulation in the host cell, an understanding of viral genomes and virus replication provides basic information concerning cellular processes in general.

The whole development of molecular biology and molecular genetics is largely based on the deliberate choice of some insightful pioneers of "pure" biological research to study the replication and genetics of viruses that replicate in bacteria: the bacteriophages. (Such researchers include Max Delbrück, Salvadore Luria, Joshua Lederberg, Gunther Stent, Seymour Benzer, André Lwoff, François Jacob, Jacques Monod, and many others.)

The bacterial viruses (**bacteriophages**) were discovered through their ability to destroy human enteric bacteria such as *Escherichia coli*, but they had no clear relevance to human disease. It is only in retrospect that the grand unity of biological processes, from the simplest to the most complex, can be seen as mirrored in replication of viruses and the cells they infect.

The biological insights offered by the study of viruses have led to important developments in biomedical technology and promise to lead to even more dramatic developments and tools. For example, when infecting an individual, viruses target specific tissues. The resulting specific signs and symptoms, as already noted, define their pathogenicity. The normal human, like all vertebrates, can mount a defined and profound response to virus infections. This response often leads to partial or complete immunity to reinfection. The study of these processes was instrumental to gaining an increasingly clear understanding of the immune response and the precise molecular nature of cell–cell signaling pathways. It also provided therapeutic and preventive strategies against specific virus-caused disease. The study of virology has and will continue to provide strategies for the **palliative treatment** of metabolic and genetic diseases not only in humans, but also in other economically and aesthetically important animal and plant populations.

Examples of the impact of viral disease on human history

There is archeological evidence in Egyptian mummies and medical texts of readily identifiable viral infections, including genital papillomas (warts) and poliomyelitis. There are also somewhat imperfect historical records of viral disease affecting human populations in classical and medieval times. While the recent campaign to eradicate smallpox has been successful and the virus no longer exists in the human population (owing to the effectiveness of vaccines against it, the genetic stability of the virus, and a well-orchestrated political and social effort to carry out the eradication), the disease periodically wreaked havoc and had profound effects on human history over thousands of years. Smallpox epidemics during the Middle Ages and later in Europe resulted in significant population losses as well as major changes in the economic, religious, political, and social life of individuals. Although the effectiveness of vaccination strategies gradually led to decline of the disease in Europe and North America, smallpox continued to cause massive mortality and disruption in other parts of the world until after World War II. Despite smallpox being eradicated from the environment, the attack of September 11, 2001, on the World Trade Center in New York has led some government officials to be concerned that the high virulence of the virus and its mode of spread might make it an attractive agent for **bioterrorism**.

Other virus-mediated epidemics had equally major roles in human history. Much of the social, economic, and political chaos in native populations resulting from European conquests and expansion from the fifteenth through nineteenth centuries was mediated by introduction of infectious viral diseases such as measles. Significant fractions of the indigenous population of the Western Hemisphere died as a result of these diseases.

Potential for major social and political disruption of everyday life continues to this day. As discussed in later chapters of this book, the "Spanish" influenza (H1N1) of 1918–1919 killed tens of millions worldwide and, in conjunction with the effects of World War I, came very close to causing a major disruption of world civilization. Remarkable medical detective work using virus isolated from cadavers of victims of this disease frozen in Alaskan permafrost has led to recovery of the complete genomic sequence of the virus and reconstruction of the virus itself (some of the methods used will be outlined in Part V). While we may never know all the factors that caused it to be so deadly, it is clear that the virus was derived from birds and passed directly to humans. Further, a number of viral proteins have a role in its virulence. Ominously, there is no reason why another strain of influenza could not arise with a similar or more devastating aftermath or **sequela** – indeed, in the spring/summer of 2005, there was legitimate cause for concern because a new strain of avian influenza (H5N1) had been transmitted to humans. At the present time, human transmission of H5N1 influenza has not been confirmed, but further adaptation of this new virus to humans could lead to it establishing itself as a major killer in the near future.

In April 2009, a new version of H1N1 influenza (now called the 2009 H1N1 flu by the Centers for Disease Control and Prevention [CDC]) was identified in Veracruz, Mexico. Initially H1N1 became epidemic in Mexico, and, as the virus spread rapidly, on June 11, 2009, the World Health Organization declared a pandemic. Because this virus has the same surface markers (H1 and N1) as the infamous "Spanish flu" of nearly 100 years before, there were fears of the same kinds of morbidity and mortality as seen in the early years of the twentieth century. However, this virus turned out to be of no greater lethality than the other seasonal influenza strains currently circulating in the human population. We will discuss more details about this later in this book.

A number of infectious diseases could become established in the general population as a consequence of their becoming drug resistant or introduced as weapons of bioterrorism, or because of human disruption of natural ecosystems. As will be discussed in later chapters, a number of different viruses exhibiting different details of replication and spread could, potentially, be causative agents of such diseases.

Animal and plant pathogens are other potential sources of disruptive viral infections. Sporadic outbreaks of viral disease in domestic animals, such as vesicular stomatitis virus in cattle and avian influenza in chickens, result in significant economic and personal losses. Rabies in wild animal populations in the eastern United States has spread continually during the past half-century. The presence of this disease poses real threats to domestic animals and, through them occasionally, to humans. An example of an agricultural infection leading to severe economic disruption is the growing spread of the cadang-cadang viroid in coconut palms of the Philippine Islands and elsewhere in Oceania. The loss of coconut palms has led to serious financial hardship in local populations.

Examples of the evolutionary impact of the virus–host interaction

There is ample genetic evidence that the interaction between viruses and their hosts has a measurable impact on evolution of the host. Viruses provide environmental stresses to which organisms evolve responses. Also, it is possible that the ability of viruses to acquire and move genes between organisms provides a mechanism of gene transfer between lineages.

Development of the immune system, the cellular-based antiviral **interferon (IFN)** response, and many of the inflammatory and other responses that multicellular organisms can mount to ward off infection is the result of successful genetic adaptation to infection. In addition, virus infection may provide an important (and as yet underappreciated) basic mechanism to affect the evolutionary process in a direct way.

There is good circumstantial evidence that the specific origin of placental mammals is the result of an ancestral species being infected with an immunosuppressive proto-retrovirus. It is suggested that this immunosuppression permitted an immunological accommodation in the mother to the development of a genetically distinct individual in the placenta during a prolonged period of gestation!

Two current examples provide very strong evidence for the continued role of viruses in the evolution of animals and plants. Certain parasitic wasps lay their eggs in the caterpillars of other insects. As the wasp larvae develop, they devour the host, leaving the vital parts for last to ensure that the food supply stays fresh! Naturally, the host does not appreciate this attack and mounts an immune defense against the invader – especially at the earliest stages of the wasp's embryonic development. The wasps uninfected with a **polydnavirus** do not have a high success rate for their parasitism, and their larvae are often destroyed. The case is different when the same species of wasp is infected with a polydnavirus that is then maintained as a persistent genetic passenger in the ovaries and egg cells of the wasps. The polydnavirus inserted into the caterpillar along with the wasp egg induces a systemic, immunosuppressive infection so that the caterpillar cannot eliminate the embryonic tissue at an early stage of development! The virus maintains itself by persisting in the ovaries of the developing female wasps.

A further example of a virus's role in development of a symbiotic relationship between its host and another organism can be seen in replication of the ***Chlorella* viruses**. These viruses are found at concentrations as high as 4×10^4 infectious units/ml in freshwater throughout the United States, China, and probably elsewhere in the world. Such levels demonstrate that the virus is a very successful pathogen. Despite this success, the viruses can only infect free algae; they cannot infect the same algae when the algae exist semi-symbiotically with a species of paramecium. Thus, the algae cells that remain within their symbiotes are protected from infection, and it is a good guess that existence of the virus is a strong selective pressure toward establishing or stabilizing the symbiotic relationship.

The origin of viruses

In the last decade or so, molecular biologists have developed a number of powerful techniques to amplify and sequence the genome of any organism or virus of interest. The correlation between sequence data; classical physiological, biochemical, and morphological analyses; and the geological record has provided one of the triumphs of modern biology. We now know that the biosphere is made up of three domains, the **eubacteria** (bacteria), the **eukaryotes** (nucleated cells), and the **archaebacteria** – the latter only discovered through the **ribosomal RNA (rRNA)** sequence studies of Woese and his colleagues in the past 30 years or so. Further, analysis of genetic changes in conserved sequences of critical proteins as well as rRNA confirms that eukaryotes are more closely related to (and thus derived from) the ancestors of archaea than they are to eubacteria.

Carefully controlled statistical analysis of the frequency and numbers of base changes in genes encoding conserved enzymes and proteins mediating essential metabolic and other cellular processes can be used to both measure the degree of relatedness between greatly divergent organisms, and provide a sense of when in the evolutionary time scale they diverged from a common ancestor. This information can be used to generate a **phylogenetic tree**, which graphically displays such relationships. An example of such a tree showing the degree of divergence of some index species in the three domains is shown in Figure 1.1.

Although there is no geological record of viruses (they do not form fossils in any currently useful sense), analyses of the relationship between the amino acid sequences of viral and cellular proteins and of the nucleotide sequences of the genes encoding them provide ample genetic evidence that the association between viruses and their hosts is as ancient as the origin of the hosts themselves. Some viruses (e.g., retroviruses) integrate their genetic material into the cell they infect, and if this cell happens to be germ line, the viral genome (or its relict) can be maintained essentially forever. Analysis of the sequence relationship between various retroviruses found in mammalian genomes demonstrates integration of some types before major groups of mammals diverged.

While the geological record cannot provide evidence of when or how viruses originated, genetics offers some important clues. First, the vast majority of viruses do not encode genes for

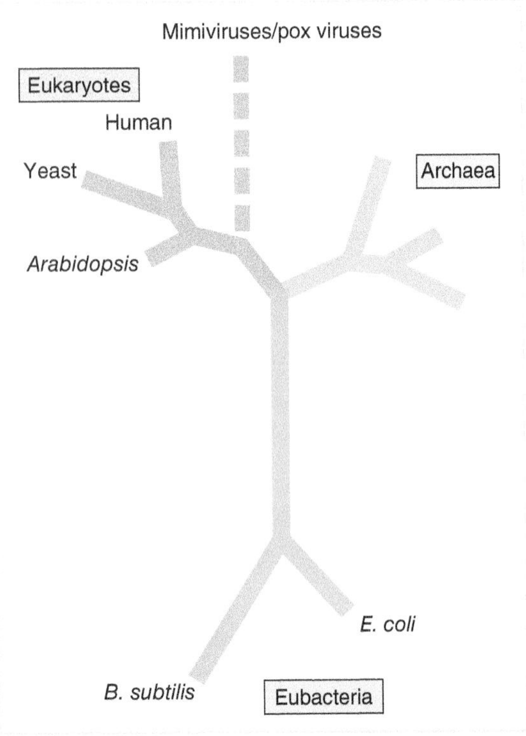

Figure 1.1 A phylogenetic tree of selected species from the three domains of life: Eukaryota (or Eukarya), Eubacteria, and Archaea. The tree is based upon statistical analysis of sequence variation in seven universally conserved protein sequences: arginyl-t-RNA synthetase, methionyl-t-RNA synthetase, tyrosyl-t-RNA synthetase, RNA pol II largest subunit, RNA pol II second largest subunit, PCNA, and 5′-3′ exonuclease. Source: Based upon Raoult, D., Audic, S., Robert, C., et al. (2004). The 1.2-megabase genome sequence of mimivirus. *Science* 306: 1344–1350.

ribosomal proteins or genetic evidence of relicts of such genes. Second, this same vast majority of viruses do not contain genetic evidence of ever having encoded enzymes involved in energy metabolism. This is convincing evidence that the viruses currently investigated did not evolve from free-living organisms. This finding distinctly contrasts with two eukaryotic organelles, the mitochondrion and the chloroplast, which are known to be derived from free-living organisms.

Genetics also demonstrates that a large number of virus-encoded enzymes and proteins have a common origin with cellular ones of similar or related function. For example, many viruses containing DNA as their genetic material have viral-encoded DNA polymerases that are related to all other DNA polymerase isolated from plants, animals, and archaea.

Statistical analysis of the divergence in three highly conserved regions of eukaryotic DNA polymerases suggests that the viral enzymes, including those from **herpesviruses** and from **poxviruses** and relatives (including mimiviruses), have existed as long as the three domains themselves. Indeed, convincing arguments exist that the viral enzymes are more similar to the ancestral form. This, in turn, implies that viruses or virus-like self-replicating entities (**replicons**) had a major role, if not the major role, in the origin of DNA-based genetics. The phylogenetic tree of relationships between two forms of eukaryotic DNA polymerase (alpha and delta), two forms of the enzyme found in archaebacteria, as well as those of three groups of large DNA viruses and some other DNA viruses infecting algae and protists is shown in Figure 1.2.

Another example of the close genetic interweaving of early cellular and viral life forms is seen in the sequence analysis of the reverse transcriptase enzyme encoded by retroviruses, which is absolutely required for converting retroviral genetic information contained in RNA to DNA. This enzyme is related to an important eukaryotic enzyme involved in reduplicating the telomeres of chromosomes upon cell division – an enzyme basic to the eukaryotic mode of genome replication. Reverse transcriptase is also found in cellular transposable genetic elements (**retrotransposons**), which are circular genetic elements that can move from one chromosomal location to another. Thus, the relationship between certain portions of the replication cycle of retroviruses and mechanisms of gene transposition and chromosome maintenance in cells is so intimately involved that it is impossible to say which occurred first.

A major complication to a complete and satisfying scheme for the origin of viruses is that a large proportion of viral genes have no known cellular counterparts, and viruses themselves may be a source of much of the genetic variation seen between different free-living organisms. In an extensive analysis of the relationship between groups of viral and cellular genes, L. P. Villarreal points out that the deduced size of the **Last Universal Common Ancestor (LUCA)** of eukaryotic and prokaryotic cells is on the order of 300 genes – no bigger than a large virus – and provides some very compelling arguments for viruses having provided some of the distinctive genetic elements that distinguish cells of the eukaryotic and prokaryotic kingdoms. In such a scheme, precursors to both viruses and cells originated in a pre-biotic environment hypothesized to provide the chemical origin of biochemical reactions leading to cellular life.

At the level explored here, it is probably not terribly useful to spend great efforts to be more definitive about virus origins beyond their functional relationship to the cell and organism they infect. The necessarily close mechanistic relationship between cellular machinery and the genetic manifestations of viruses infecting them makes viruses important biological entities, but it does not make them organisms. They do not grow, they do not metabolize small molecules for energy, and they only "live" when in the active process of infecting a cell and replicating in that cell. The study of these processes, then, must tell as much about the cell and the organism as it does about the virus. This makes the study of viruses of particular interest to biologists of every sort.

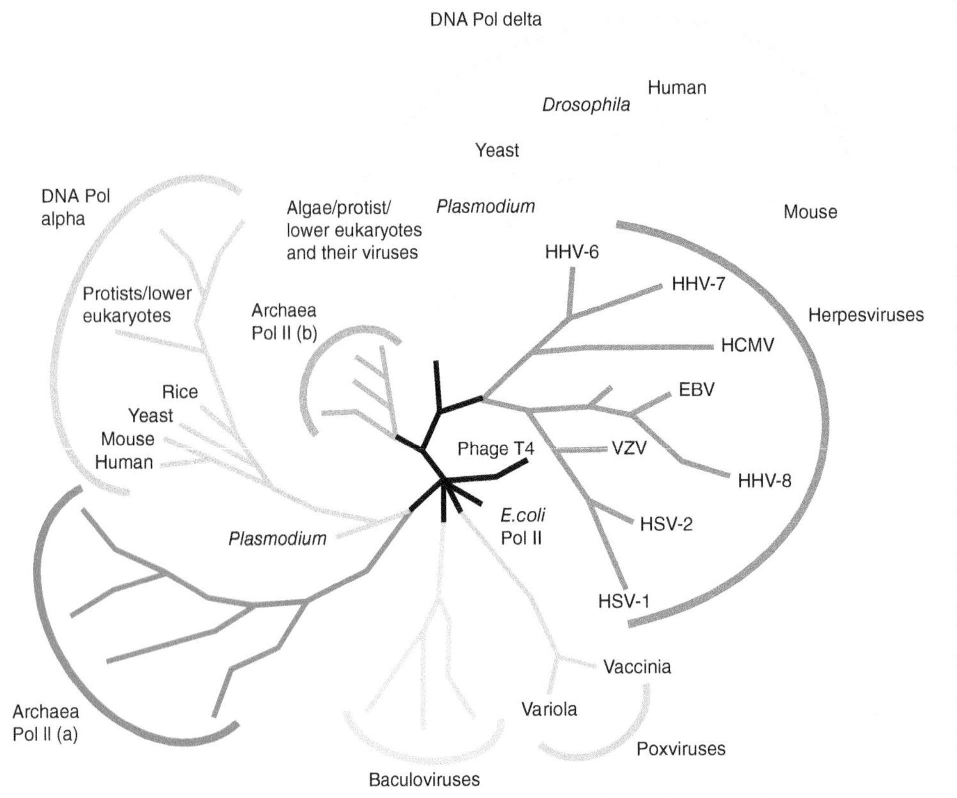

Figure 1.2 A phylogenetic tree of selected eukaryotic and archaeal species along with specific large DNA-containing viruses based upon sequence divergence in conserved regions of DNA polymerase genes. Source: Based upon Villarreal, L.P. and DeFilippis, V.R. (2000). A hypothesis for DNA viruses as the origin of eukaryotic replication proteins. *Journal of Virology* 74: 7079–7084.

Viruses have a constructive as well as destructive impact on society

Often the media and some politicians would have us believe that infectious diseases and viruses are unremitting evils, but to quote Sportin' Life in Gershwin's *Porgy and Bess*, this "ain't necessarily so." Without the impact of infectious disease, it is unlikely that our increasingly profound understanding of biology would have progressed as it has. As already noted, much of our understanding of the mechanisms of biological processes is based in part or in whole on research carried out on viruses. It is true that unvarnished human curiosity has provided an understanding of many of the basic patterns used to classify organisms and fostered Darwin's intellectual triumph in describing the basis for modern evolutionary theory in his *Origin of Species*. Still, focused investigation on the microscopic world of pathogens needed the spur of medical necessity. The great names of European microbiology of the nineteenth and early twentieth centuries – Pasteur, Koch, Ehrlich, Fleming, and their associates (who did much of the work with which their mentors are credited) – were all medical microbiologists. Most of the justification for today's burgeoning biotechnology industry and research establishment is medical or economic.

Today, we see the promise of adapting many of the basic biochemical processes encoded by viruses to our own ends. Exploitation of viral diseases of animal and plant pests may provide a useful and regulated means of controlling such pests. While the effect was only temporary and had some disastrous consequences in Europe, the introduction of **myxoma** virus – a pathogen of South American lagomorphs (rabbits and their relatives) – had a positive role in limiting the predations of European rabbits in Australia. Study of the adaptation dynamics of this disease to the rabbit population in Australia taught much about the co-adaptation of host and parasite.

The exquisite cellular specificity of virus infection is being adapted to generate biological tools for moving therapeutic and palliative genes into cells and organs of individuals with genetic and degenerative diseases. Modifications of virus-encoded proteins and the genetic manipulation of viral genomes are being exploited to provide new and (hopefully) highly specific **prophylactic** vaccines as well as other therapeutic agents. The list increases monthly.

Viruses are not the smallest self-replicating pathogens

Viruses are not the smallest or the simplest pathogens able to control their self-replication in a host cell – that distinction goes to **prions**. Despite this, the methodology for the study of viruses and the diseases they cause provides the basic methodology for the study of all subcellular pathogens.

By the most basic definition, viruses are composed of a genome and one or more proteins coating that genome. The genetic information for such a protein coat and other information required for the replication of the genome are encoded in that genome. There are genetic variants of viruses that have lost information either for one or more coat proteins or for replication of the genome. Such virus-derived entities are clearly related to a parental form with complete genetic information, and thus, the mutant forms are often termed **defective virus particles**.

Defective viruses require the coinfection of a **helper virus** for their replication; thus, they are parasitic on viruses. A prime example is hepatitis delta virus, which is completely dependent on coinfection with hepatitis B virus for its transmission.

The hepatitis delta virus has some properties in common with a group of RNA pathogens that infect plants and can replicate in them by still-unknown mechanisms. Such RNA molecules, called **viroids**, do not encode any protein, but can be transmitted between plants by mechanical means and can be pathogens of great economic impact.

Some pathogens appear to be entirely composed of protein. These entities, called prions, appear to be cellular proteins with an unusual folding pattern. When they interact with normally folded proteins of the same sort in neural tissue, they appear to be able to induce abnormal refolding of the normal protein. This abnormally folded protein interferes with neuronal cell function and leads to disease. While much research needs to be done on prions, it is clear that they can be transmitted with some degree of efficiency among hosts, and they are extremely difficult to inactivate. Prion diseases of sheep and cattle (scrapie and "mad cow" disease) recently had major economic impacts on British agriculture, and several prion diseases (**kuru** and **Creutzfeldt–Jacob disease [CJD]**) affect humans. Disturbingly, the inadvertent passage of sheep scrapie through cattle in England has apparently led to the generation of a new form of human disease similar to, but distinct from, CJD. Details of this are covered in Part IV, Chapter 15.

The existence of such pathogens provides further circumstantial evidence for the idea that viruses are ultimately derived from cells. It also provides support for the possibility that viruses had multiple origins in evolutionary time.

QUESTIONS FOR CHAPTER 1

1 Viruses are a part of the biosphere. However, there is active debate concerning whether they should be treated as living or nonliving.

(a) Briefly describe one feature of viruses that is *also found* in cell-based life forms.

(b) Briefly describe one feature of viruses that *distinguishes* them from cell-based life forms.

2 Why is it likely that viruses have not evolved from free-living organisms?

3 Give examples of infectious agents that are smaller self-replicating systems than viruses.

4 Ebola virus is a deadly (90% case-fatality rate for some strains) infectious agent. Most viruses, however, are not nearly as lethal. Given the nature of viruses, why would you expect this to be so?

5 Given that viruses are a part of the biosphere in which other organisms exist, what might be the kinds of selective pressure that viruses exert on evolution?

6 Viruses were originally discovered because of their size, relative to known bacterial cells. Tobacco mosaic virus was called a "filterable infectious agent" by this criterion. Why is size not a good defining feature for viruses? What is a better definition?

An Outline of Virus Replication and Viral Pathogenesis

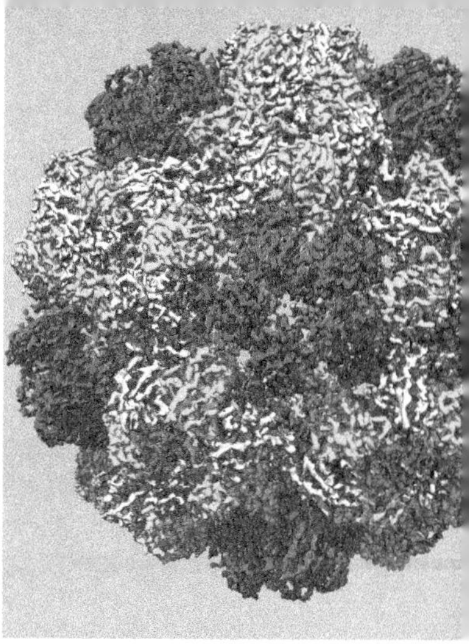

CHAPTER 2

VIRUS REPLICATION

Viruses must replicate in living cells. The most basic molecular requirement for virus replication is for a virus to induce either profound or subtle changes in the cell so that viral genes in the genome are replicated and viral proteins are expressed. This will result in the formation of new viruses – usually many more than the number of viruses infecting the cell in the first place. When reproducing, viruses use at least part of the cell's equipment for replication of viral nucleic acids and expression of viral genes. They also use the cell's protein synthetic machinery, and the cell's metabolic energy resources.

The dimensions and organization of "typical" animal, plant, and bacterial cells are shown in Figure 2.1. The size of a typical virus falls in the range between the diameters of a ribosome and of a centriolar filament. With most viruses, infection of a cell with a single virus particle will result in the synthesis of more than one (often by a factor of several powers of 10) infectious virus. Any infection that results in the production of more infectious virus at the end than at the start is classified as a **productive infection**. The actual number of infectious viruses produced in

Basic Virology, Fourth Edition. Martinez "Marty" Hewlett, David Camerini, and David C. Bloom.

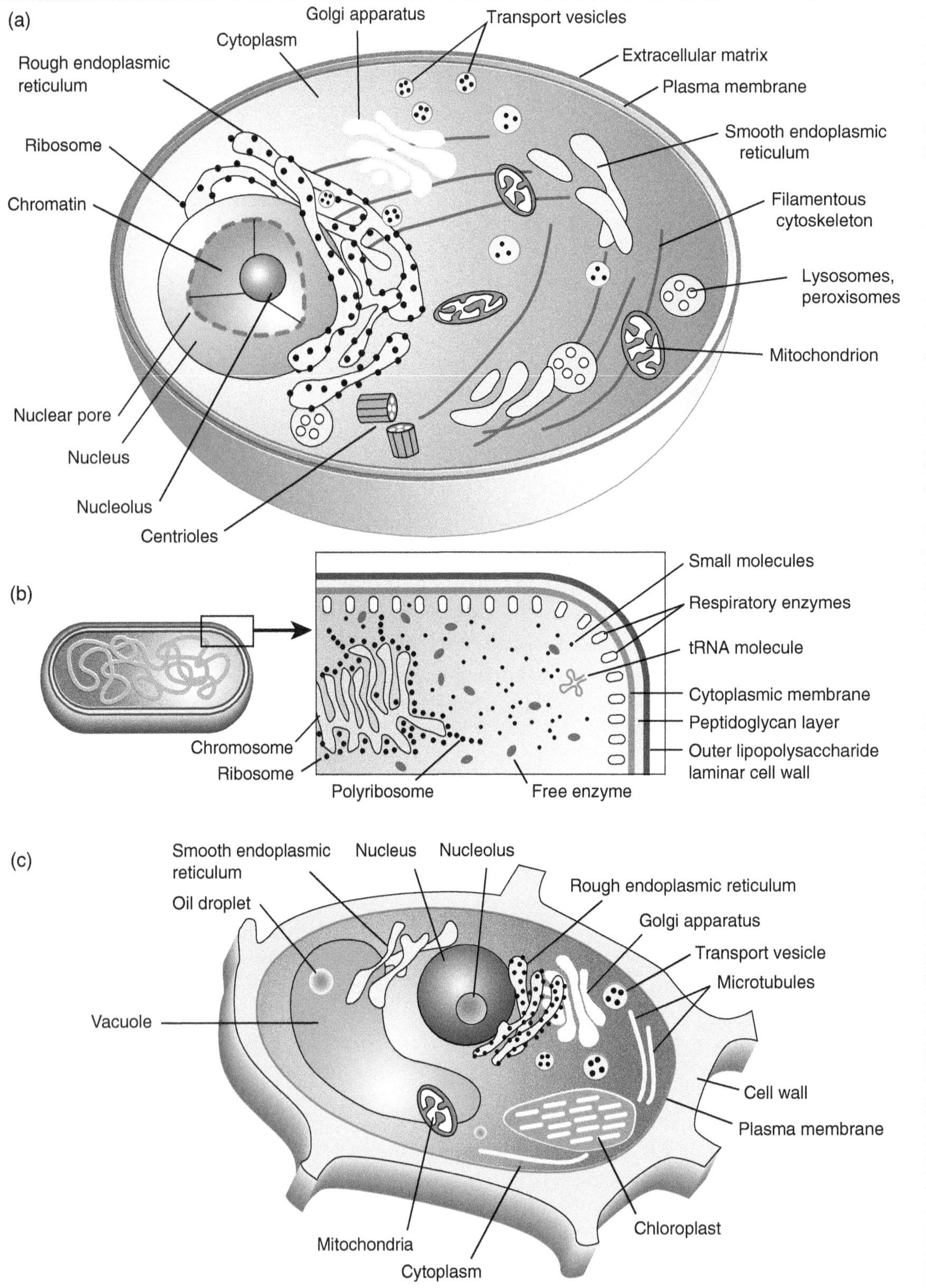

Figure 2.1 Dimensions and features of "typical" (a) animal, (b) plant, and (c) bacterial cells. The dimensions of plant and animal cells can vary widely, but an average diameter of around 50 µm (5×10^{-5} m) is a fair estimate. Bacterial cells also show great variation in size and shape, but the one shown here is *Escherichia coli*, the true "workhorse" of molecular biologists. Its length is approximately 5 µm. Based on these dimensions and shapes of the cells shown, the bacterial cell is approximately 1/500th of the volume of the eukaryotic cell shown. Virus particles also vary greatly in size and shape, but generally range from 25 to 200 nm ($0.25–2.00 \times 10^{-7}$ m).

an infected cell is called the **burst size**, and this number can range from less than 10 to over 10 000, depending on the type of cell infected, the nature of the virus, and many other factors.

Infections with many viruses completely convert the cell into a factory for replication of new viruses. Under certain circumstances and/or in particular cells, however, virus infection leads to a state of coexistence between the cell and infecting virus, which can persist for as long as the life of the host. This process can be a dynamic one in which there is a small amount of virus produced constantly, or it can be passive where the viral genome is carried as a "passenger" in the cell with little or no evidence of viral gene expression. Often in such a case, the virus induces some type of change in the cell so that the viral and cellular genomes are replicated in synchrony. Such coexistence usually results in accompanying changes to the protein composition of the cell's surface – the immune "signature" of the cell – and often there are functional changes as well. This process is called **lysogeny** in bacterial cells and **transformation** in animal and plant cells.

In animal cells, the process of transformation often results in altered growth properties of the cell and can result in the generation of cells that have some or many properties of **cancer cells**. There are instances, however, where the coexistence of a cell and an infecting virus leads to few or no detectable changes in the cell. For example, **herpes simplex virus (HSV)** can establish a **latent infection** in terminally differentiated sensory neurons. In such cells there is absolutely no evidence for expression of any viral protein at all. Periods of viral latency are interspersed with periods of **reactivation (recrudescence)** where virus replication is reestablished from the latently infected tissue for varying periods of time.

Some viral infections of plant cells also result in stable association between virus and cell. Indeed, the variegation of tulip colors, which led to economic booms in Holland during the sixteenth century, is the result of such associations. Many other examples of **mosaicism** resulting from persisting virus infections of floral or leaf tissue have been observed in plants. However, many specific details of the association are not as well characterized in plants as in animal and bacterial cells.

Stages of virus replication in the cell

Various patterns of replication as applied to specific viruses, as well as the effect of viral infections on the host cell and organism, are the subject of many of the following chapters in this book. The best way to begin to understand patterns of virus replication is to consider a simple general case: the **productive infection** cycle – this is shown schematically in Figure 2.2. A number of critical events are involved in this cycle. The basic pattern of replication is as follows:

1 **The virus specifically interacts with the host cell surface, and the viral genome is introduced into the cell.** This involves specific recognition between virus surface proteins and specific proteins on the cell surface (**receptors**) in animal and bacterial virus infections.

2 **Viral genes are expressed using host cell processes.** This viral gene expression results in synthesis of a few or many viral proteins involved in the replication process.

3 **Viral proteins modify the host cell and allow the viral genome to replicate using host and viral enzymes.** While this is a simple statement, the actual mechanisms by which viral enzymes and proteins can subvert a cell are manifold and complex. This is often the stage at which the cell is irreversibly modified and eventually killed. Much modern research in the molecular biology of virus replication is directed toward understanding these mechanisms.

4 **New viral coat proteins assemble into capsids, and viral genomes are included.** The process of assembly of new virions is relatively well understood for many viruses. The successful description of the process has resulted in a profound linkage of knowledge about the principles of macromolecular structures, the biochemistry of protein–protein and protein–nucleic acid interactions, and the thermodynamics of large macromolecule structures.

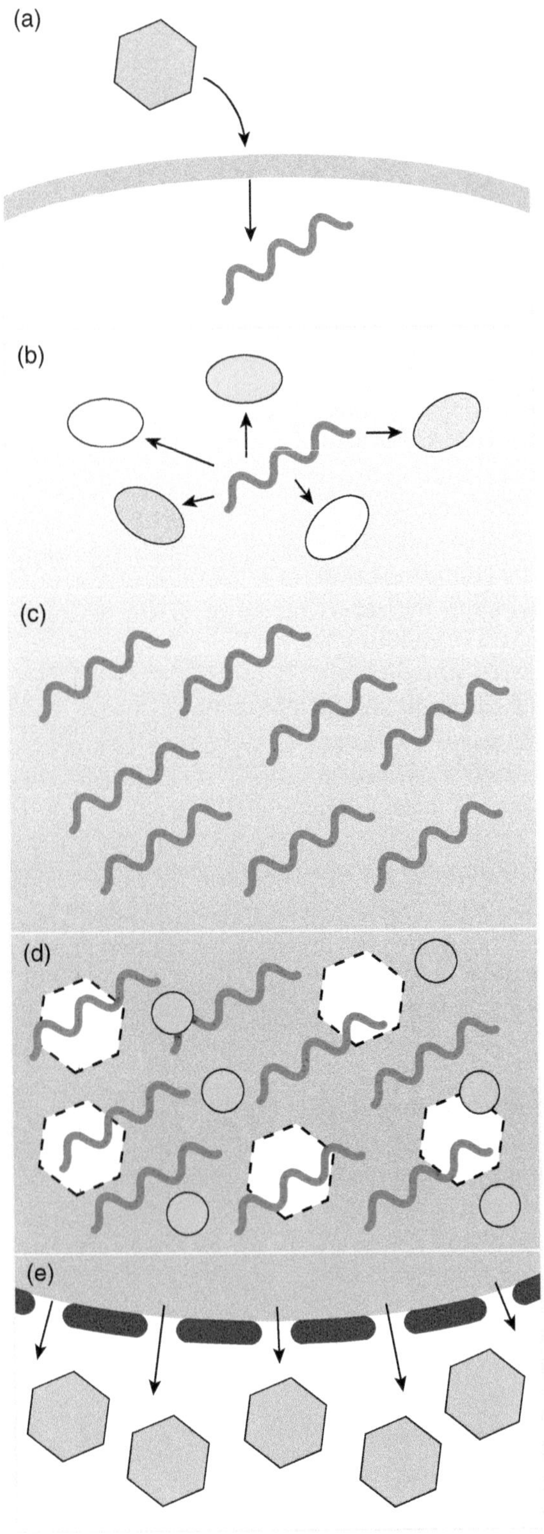

Figure 2.2 The virus replication cycle. Most generally, virus replication can be broken into the stages shown: (a) initial recognition between virus and cell and introduction of viral genetic material into the host cell, (b) virus gene expression and induction of virus-induced modification of host, allowing (c) virus genome replication. Following this, (d) virus-associated proteins are expressed, and (e) new virus is assembled and released, often resulting in cell death.

5 Virus is released where it can infect new cells and repeat the process. This is the basis of virus spread, whether from cell to cell or from individual to individual. Understanding the process of virus release requires knowledge of the biochemical interactions between cellular organelles and viral structures. Understanding the process of virus spread between members of a population requires knowledge of the principles of epidemiology and public health.

PATHOGENESIS OF VIRAL INFECTION

Most cells and organisms do not passively submit to virus infection. As noted in Chapter 1, the response of organisms to virus infection is a major feature of evolutionary change in its most general sense. As briefly noted, a complete understanding of pathogenesis requires knowledge of the sum total of genetic features a virus encodes that allows its efficient spread between individual hosts and within the general population of hosts. Thus, the term *pathogenesis* can be legitimately applied to virus infections of multicellular, unicellular, and bacterial hosts.

A major challenge for viruses infecting bacteria and other unicellular organisms is finding enough cells to replicate in without isolating themselves from other populations of similar cells. In other words, they must be able to "follow" the cells to places where the cells can flourish. If susceptible cells can isolate themselves from a pathogen, it is in their best interest to do so. Conversely, the virus, even when constrained to confine all its dynamic features of existence to the replication process per se, must successfully counter this challenge or it cannot survive.

In some cases, cells can mount a defense against virus infection. Most animal cells react to infection with many viruses by inducing a family of cellular proteins termed *interferons* that can interact with neighboring cells and induce those cells to become wholly or partially resistant to virus infection. Similarly, some viral infections of bacterial cells can result in a **bacterial restriction** response that limits viral replication. Of course, if the response is completely effective, the virus cannot replicate. In this situation, one cannot study the infection, and in the extreme situation, the virus would not survive.

Viruses that infect multicellular organisms face problems attendant with their need to be introduced into an animal to generate a physiological response fostering the virus's ability to spread to another organism (i.e., they must exhibit virulence). This process can follow different routes.

Disease is a common result of the infection, but many (if not most) viral infections result in no measurable disease symptoms – indeed, inapparent infections are often hallmarks of highly coevolved virus–host interactions. But inapparent or **asymptomatic** infections can be seen in the interaction between normally virulent viruses and a susceptible host as a result of many factors. A partial list includes the host's genetic makeup, host health, the degree of immunity to the pathogen in the host, and the random (**stochastic**) nature of the infective process.

Stages of virus-induced pathology

Pathogenesis can be divided into stages – from initial infection of the host to its eventual full or partial recovery, or its virus-induced death. A more-or-less typical course of infection in a vertebrate host is schematically diagrammed in Figure 2.3. Although individual cases differ, depending on the nature of the viral pathogen and the immune capacity of the host, a general pattern of infection would be as follows:

Initial infection leads to virus replication at the site of entry, and multiplication and spread into favored tissues. The time between the initial infection and the observation of clinical symptoms of disease defines the **incubation period**, which can be of variable length, depending on many factors.

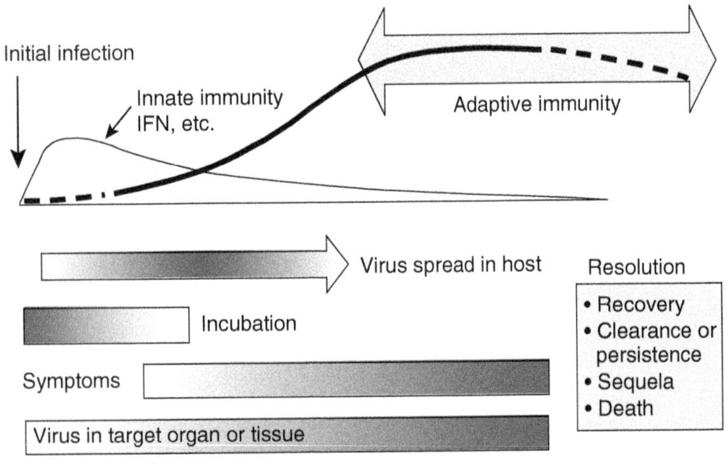

Figure 2.3 The pathogenesis of virus infection. Typically, infection is followed by an incubation period of variable length in which virus multiplies at the site of initial infection. Local and innate immunity, including the interferon response, counter infection from the earliest stages; and if these lead to clearing, disease never develops. During the incubation period, virus spreads to the target of infection (which may be the same site). The adaptive immune response becomes significant only after virus reaches high enough levels to efficiently interact with cells of the immune system; this usually requires virus attaining high levels or titers in the circulatory system. Virus replication in the target leads to symptoms of the disease in question and is often important in spread of the virus to others. Immunity reaches a maximum level only late in the infection process and remains high for a long period after resolution of the disease.

The host responds to the viral invasion by marshaling its defense forces, both local and systemic. The earliest defenses include expression of interferon and tissue inflammation. Ultimately the major component of this defense – **adaptive immunity** – comes into play. For disease to occur, the defenses must lag as the virus multiplies to high levels. At the same time, the virus invades favored sites of replication. Infection of these favored sites is often a major factor in the occurrence of disease symptoms and is often critical for the transmission to other organisms. As the host defenses mount, virus replication declines and there is recovery – perhaps with lasting damage and usually with immunity to a repeat infection. If an insufficient defense is mounted, the host will die.

Initial stages of infection – entry of the virus into the host

The source of the infectious virus is termed the **reservoir**, and virus entry into the host generally follows a specific pattern leading to its introduction at a specific site or region of the body. Epidemiologists working with human, animal, and plant diseases often use special terms to describe parts of this process. The actual means of infection between individuals is termed the *vector of transmission* or, more simply, the **vector**. This term is often used when referring to another organism, such as an arthropod, that serves as an intermediary in the spread of disease.

Many viruses must continually replicate to maintain themselves – this is especially true for viruses that are sensitive to desiccation and are spread between terrestrial organisms. For this reason, many virus reservoirs will be essentially dynamic; that is, the virus constantly must be replicating actively somewhere. In an infection with a virus with broad species specificity, the external reservoir could be a different population of animals. In some cases, the vector and the reservoir are the same – for example, in the transmission of rabies via the bite of a rabid animal. Also, some arthropod-borne viruses can replicate in the arthropod vector as well as in their primary vertebrate reservoir. In such a case the vector serves as a secondary reservoir, and this second round of virus multiplication increases the amount of pathogen available for spread into the next host.

Some reservoirs are not entirely dynamic. For example, some algal viruses exist in high levels in many bodies of freshwater. It has been reported that levels of some viruses can approach 10^7 per milliliter of seawater. Further, the only evidence for the presence of living organisms in some bodies of water in Antarctica is the presence of viruses in that water. Still, ultimately all viruses must be produced by an active infection somewhere, so in the end all reservoirs are, in some sense at least, dynamic.

Viruses (or their genomes) enter cells via the cooperative interaction between the host cell and the virus – this interaction requires a hydrated cell surface. Thus, initial virus infection and entry into the host cell must take place at locations where such cell surfaces are available, not, for example, at the desiccated surface layer of keratinized, dead epithelial cells of an animal's skin, or at the dry, waxy surface of a plant. In other words, virus must enter the organism at a site that is "wet" as a consequence of its anatomical function or must enter through a trauma-induced break in the surface. Figure 2.4 is a schematic representation of some modes of virus entry leading to human infection.

The incubation period and spread of virus through the host

Following infection, virus must be able to replicate at the site of initial infection in order for it to build up enough numbers to lead to the symptoms of disease. There are several reasons why this takes time. First, only a limited amount of virus can be introduced. This is true even with the most efficient vector. Second, cell-based innate immune responses occur immediately upon infection. The best example of these is the interferon response.

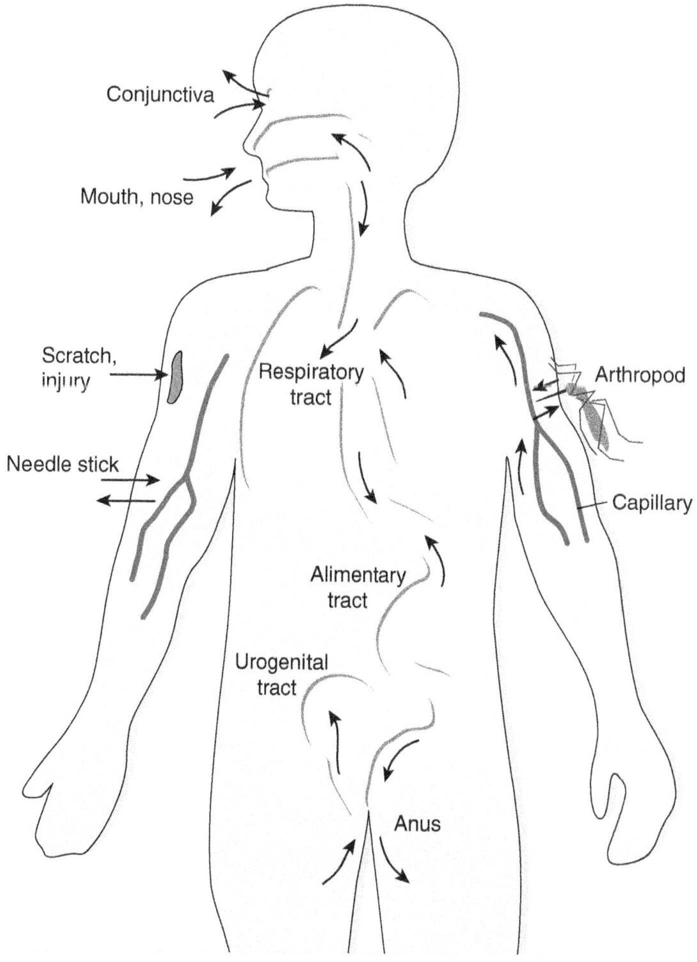

Figure 2.4 Sites of virus entry in a human. These or similar sites apply to other vertebrates. Source: Adapted from Mims, C.A. and White, D.O. (1984). *Viral Pathogenesis and Immunity*. Boston: Blackwell Science.

This "early" stage or incubation period of disease can last from only a few days to many years, depending on the specific virus. In fact, probably many virus infections go no further than this first stage, with clearance occurring without any awareness of the infection at all. Also, some virus infections lead only to replication localized at the site of original entry. In such a case, extensive virus spread need not occur, although some interaction with cells of the immune system must occur if the animal host is to mount an immune response.

Following entry, many types of viruses must move or be moved through the host to establish infection at a preferred site, the infection of which results in disease symptoms. This site, often referred to as the **target tissue** or **target organ**, is often (but not always) important in mediating the symptoms of disease, or the spread, or both.

There are several modes of virus spread in the host. Perhaps the most frequent mode utilized by viruses is through the circulatory system (**viremia**). A number of viruses can spread in the bloodstream either passively as free virus or adsorbed to the surface of cells that they do not infect, such as red blood cells. Direct entry of virus into the lymphatic circulatory system also can lead to viremia. Some viruses that replicate in the gut (such as poliovirus) can directly enter the lymphatic system via **Peyer's patches (gut-associated lymphoid tissue)** in the intestinal mucosa. Such patches of lymphoid tissue provide a route directly to lymphocytes without passage through the bloodstream. This provides a mode of generating an immune response to a localized infection. For example, poliovirus generally replicates in the intestinal mucosa and remains localized there until eliminated; the entry of virus into the lymphatic system via Peyer's patches leads to immunity. Virus invasion of gut-associated lymphoid tissue is thought to be one important route of entry for HIV spread by anal intercourse, as infectious virus can be isolated from seminal fluid of infected males.

Infection of lymphoid cells can also be a factor in the spread of infectious virus. HIV infects and replicates in **T lymphocytes** and macrophages, leading to the generation of active carrier cells that migrate to lymph nodes. This facilitates spread of the virus throughout the immune system. Many other viruses infect and replicate in one or another cells of the lymphatic system. Some of the viruses known to infect one or another of the three major cells found in lymphatic circulation are shown in Table 2.1.

While spread via the circulatory system is quite common, it is not the only mode of general dissemination of viruses from their site of entry and initial replication in animals. The nervous system provides the other major route of spread. Some **neurotropic viruses**, such as HSV and

Table 2.1 Some viruses that replicate in cells of the immune system.

Cells Infected	Virus
B lymphocytes	Epstein–Barr virus (herpesvirus)
	Some retroviruses
T lymphocytes	Human T-cell leukemia virus
	HIV
	Human herpesvirus 6
	Human herpesvirus 7
Monocytes	Measles virus
	Varicella-zoster virus (herpesvirus)
	HIV
	Parainfluenza virus
	Influenza virus
	Rubella (German measles) virus
	Cytomegalovirus (herpesvirus)

rabies virus, can spread from the peripheral nervous system directly into the central nervous system **(CNS)**. In the case of HSV, this is a common result of infection in laboratory mice; however, it is a relatively rare occurrence in humans, and is often correlated with an impairment or lack of normal development of the host's immune system. Thus, an initial acute infection of an infant at the time of birth or soon thereafter can lead to HSV encephalitis with high frequency.

Multiplication of virus to high levels – occurrence of disease symptoms

Viral replication at specific target tissues often defines **symptoms** of the disease. The nature of the target and the host response are of primary importance in establishing symptoms. The ability of a virus to replicate in a specific target tissue results from specific interactions between viral and cellular proteins. In other words, one or another viral protein can recognize specific molecular features that define those cells or tissues favored for virus replication. These virus-encoded proteins, thus, have a major role in specifying the virus's tissue **tropism**. Host factors, such as speed of immune response and inflammation, also play a major role. For example, a head cold results from infection and inflammation of the nasopharynx. Alternatively, liver malfunction due to inflammatory disease (*hepatitis*) could result from a viral infection in this critical organ.

One major factor in viral tropism is the distribution and occurrence of specific viral receptors on cells in the target tissue. The role of such receptors in the infection process is described in Part II, Chapter 6. For the purposes of the present discussion, it is enough to understand that there must be a specific and spatially close interaction between proteins at the surface of the virus and the surface of the cell's plasma membrane for the virus to be able to begin the infection process.

One example of the role of receptors in tissue tropism involves the poliovirus receptor, which is found on cells of the intestinal mucosa and in lymphatic tissue. A related molecule is also present on the surface of motor neurons, which means that neurotropic strains of poliovirus can invade, replicate in, and destroy these cells under certain conditions of infection. In another example, HIV readily infects T lymphocytes by recognizing the CD4 surface protein in association with a specific **chemokine** receptor that serves as a coreceptor. Rabies virus's ability to remain associated with nervous tissue probably is related to its use of the acetylcholine receptor present at nerve cell synapses. The ability of vaccinia virus (like the related smallpox virus) to replicate in epidermal cells is the result of its use of the epidermal growth factor receptor on such cells as its own receptor for attachment.

While tissue tropism is often understandable in terms of a specific viral receptor being present on the surface of susceptible cells, the story can be quite complicated in practice. This is the case for Epstein–Barr virus **(EBV)**, which is found in **B lymphocytes** in patients who have been infected with the virus. It is thought that primary infection of epithelial cells in the mucosa of the nasopharynx, followed by association with lymphocytes during development of the immune response, leads to infection of B cells that carry the EBV-specific receptor, CD21.

Even though the infection of target tissue is usually associated with the occurrence of symptoms of that viral infection, the target is not always connected with the spread of a virus subsequent to infection. For example, HIV infection can be readily spread from an infected individual long before any clinical symptoms of the disease (AIDS) are apparent. An individual can undergo a subclinical reactivation episode where there is virus in the saliva, but no fever blister can transmit HSV. Finally, paralytic polio is the result of a "dead-end" infection of motor neurons, and the resulting death of those neurons and paralysis have nothing to do with spread of the virus.

Later stages of infection – changes in the cell

Eventually, viral infection leads to distinct changes within the infected cell. Such alterations are termed **cytopathology**. These effects can most easily be observed in viral infected cells in culture. Viral-induced changes might be found in both the cytoplasm and the nucleus, depending, of course, on the virus in question. Table 2.2 lists some examples of such cytopathic outcomes of infection.

The later stages of infection – the immune response

Infections with virus do not necessarily lead to any or all symptoms of a disease. The severity of such symptoms is a function of the virus genotype, the amount of virus delivered to the host, and the host's general immune competence – the factors involved with virulence of the infection. The same virus in one individual can lead to an infection with such mild symptoms of disease that they are not recognized for what they are, while infection of another individual can lead to severe symptoms.

Generally, a virus infection results in an effective and lasting immune response. This is described in more detail in Part II, Chapter 7; briefly, the host's immune response (already activated by the presence of viral **antigens** at any and all sites where virus is replicating) reaches its highest level as clinical signs of the disease manifest.

A full immune response to virus infection requires the maturation of B and T lymphocytes. The maturation of lymphocytes results in the production of short-lived **effector T cells**, which kill cells expressing foreign antigens on their surfaces. Another class of effector T cells helps in the maturation of effector B cells for the secretion of antiviral antibodies. Such a process takes several days to a week after stimulation with significant levels of viral antigen. An important part of this immune response is the generation of long-lived memory lymphocytes to protect against future re-infection.

In addition to the host's immune response, which takes some time to develop, a number of nonspecific host responses to infection aid in limitation of the infection and contribute to virus clearing. Interferons quickly render sensitive cells resistant to virus infection. Therefore, their action limits or interferes with the ability of the virus to generate high yields of infectious material. Other responses include tissue inflammation, macrophage destruction of infected cells, and increases in body temperature, which can result in suboptimal conditions for virus infection.

Table 2.2 Some examples of viral cytopathic effect.

Cytopathic Effect	Features	Virus
Lysis	Lytic infection ultimately results in the loss of integrity of the plasma membrane of the cell.	Enteroviruses
Transformation	Cells lose their requirement for anchorage to a surface during growth. Also lost is contact inhibition. As a result, cells grow over each other and also grow as a suspension culture.	Oncornaviruses
Vacuolization	Proliferation of cytoplasmic vacuoles late during the infectious cycle	Flaviviruses
Cell fusion	Cell-cell fusion leading to the formation of syncytium and polykaryons	Paramyxoviruses
Inclusion bodies	Densely staining structures within the cytoplasm or nucleus of the cell; often indicates location of viral assembly	Various virus families
Apoptosis	Programmed cell death, characterized by nuclear events leading to chromosomal disaggregation	Lentiviruses

The later stages of infection – virus spread to the next individual

Virus exit is essentially the converse of virus entry at the start of the infection. Now, however, the infected individual is a reservoir of the continuing infection, and symptoms of the disease may have a role in its spread. Some examples should illustrate this simple concept. Infection with a mosquito-borne encephalitis virus results in high **titers** of virus in the victim's blood. At the same time, the infected individual's malaise and torpor make him or her an easy mark for a feeding mosquito. In **chickenpox** (caused by **herpes zoster virus**, also called **varicella zoster virus [VZV]**), rupture of virus-filled **vesicles** at the surface of the skin can lead to generation of viral aerosols that transmit the infection to others. Similarly, a respiratory disease–causing virus in the respiratory tract along with congestion can lead to sneezing, an effective way to spread an aerosol. A virus such as HIV in body fluids can be transmitted to others via contaminated needles or through unprotected sexual intercourse, especially anal intercourse. Herpesvirus in saliva can enter a new host through a small crack at the junction between the lip and the **epidermis**.

The later stages of infection – fate of the host

Following a viral or any infectious disease, the host recovers or dies. While many **acute infections** result in clearance of virus, this does not invariably happen. While infections with influenza virus, cold viruses, polioviruses, and poxviruses resolve with virus clearance, herpesvirus infections result in a lifelong latent infection. During the latent period, no infectious virus is present, but viral genomes are maintained in certain protected cells. Periodically, a (usually) milder recurrence of the disease (reactivation or recrudescence) takes place upon suitable stimulation.

In distinct contrast, *measles* infection resolves with loss of infectious virus, but a portion of the viral genome can be maintained in neural tissue. This is not a latent infection because the harboring cells can express viral antigens, which lead to lifelong immunity, but infectious virus can never be recovered.

Other lasting types of virus-induced damage can be much more difficult to establish without extensive epidemiological records. Chronic liver damage due to hepatitis B virus infection is a major factor in hepatic carcinoma. Persistent virus infections can lead to immune dysfunction. Virus infections may also result in the appearance of a disease or **syndrome** (a set of diagnostic signs and symptoms displayed by an affected individual) years later that has no obvious relation to the initial infection. It has been suggested that diseases such as diabetes mellitus, multiple sclerosis, and rheumatoid arthritis have viral **etiologies** (ultimate causative factors). Virus factors have also been implicated in instances of other diseases such as cancer and schizophrenia.

QUESTIONS FOR CHAPTER 2

1 A good general rule concerning the replication of RNA viruses is that they require what kind of molecular process?

2 What is the role of a vector in the transmission of a viral infection?

3 It is said that viruses appear to "violate the cell theory" ("cells only arise from preexisting cells"). To which phase of a virus life cycle (growth curve) does this refer? What is the explanation for this phase of the growth curve?

4 Viruses are called "obligate intracellular parasites." For which step of gene expression do *all* viruses completely depend on their host cell?

5 Viruses are said to "violate the cell theory," indicating that there are differences between viruses and cells. The following table lists several features of either viruses or cells or both. Indicate which of these features is true for viruses and which for cells. In each case, write a "Yes" if the feature is *true* or a "No" if the feature is *not true*.

Feature	Cells	Viruses
The genetic information may be RNA rather than DNA.		
New individuals arise by binary fission of the parent.		
Proteins are translated from messenger RNAs.		
New individuals assemble by spontaneous association of subunit structures.		

6 At this writing, the avian influenza H5N1 is transmitted from bird to bird, although it may, at some point, mutate to allow transmission from human to human. What feature of virus–host interaction does this characterize?

Virus Disease in Populations and Individual Animals

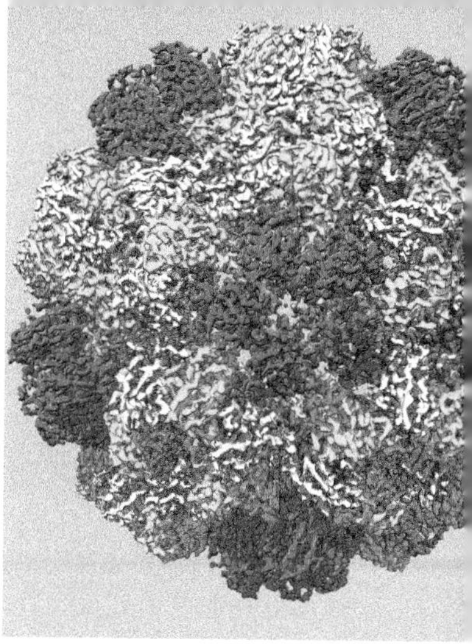

CHAPTER 3

THE NATURE OF VIRUS RESERVOIRS

Since viruses must replicate to survive, actively infected populations are the usual source of infection. Still, some viruses such as **poxviruses**, **noroviruses**, and **coronaviruses** have a high resistance to desiccation. In this case a contaminated object such as a desk, pen, book, contaminated clothing, or other inanimate object can be identified as the immediate reservoir. The persistence of some viruses in fecal material is also a potentially long-lasting, essentially passive reservoir of infection. Noroviruses, which are notorious for causing outbreaks of diarrhea on cruise ships and in nursing homes, are extremely difficult to eliminate from these facilities after an outbreak. Other viruses can even persist for long periods of time in arid environments. The last documented cases of smallpox in Somalia were apparently acquired from contaminated soil. Aerosols of infectious **hantavirus** and canine **parvovirus**

Basic Virology, Fourth Edition. Martinez "Marty" Hewlett, David Camerini, and David C. Bloom.

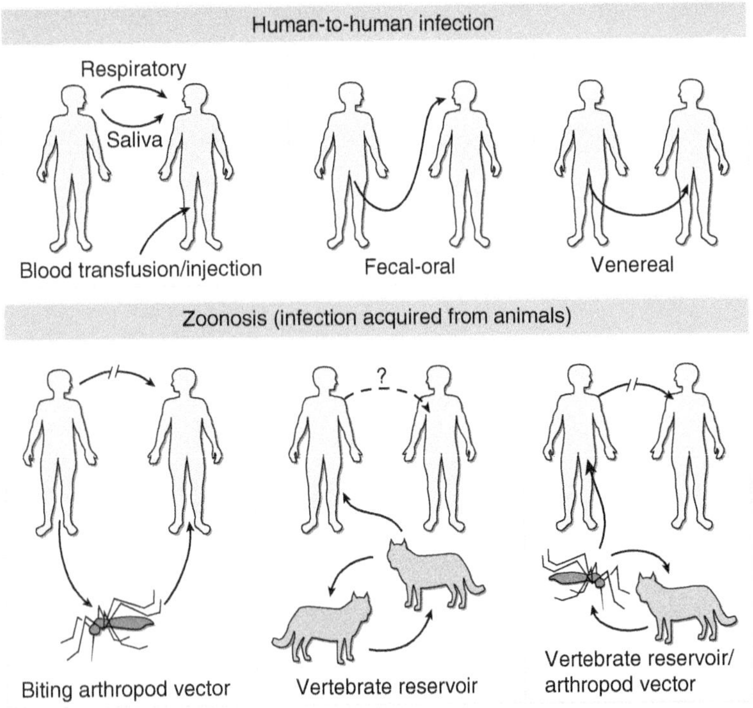

Figure 3.1 Some transmission routes of specific viruses from their source (reservoir) to humans. The mode of transmission (vector) is also shown. Source: Based on Mims, C.A. and White, D.O. (1984). *Viral Pathogenesis and Immunology.* Boston: Blackwell Science.

can be infectious for many months after secretion. Also, some viruses, especially hepatitis A virus, can be isolated from contaminated water sources for several days or even weeks after inoculation.

Even though infectious virus can sometimes be maintained for a time in the environment in a passive state, viruses are more commonly transmitted via contact with an infectious host. The two most usual reservoirs for human disease are other humans or other animals (**zoonosis**). Often this transmission requires direct contact between hosts, or may involve an intermediate vector such as a mosquito or tick; modes of spread of some human viruses are illustrated in Figure 3.1, and pathogenic viruses and their reservoirs discussed in this section are listed in Table 3.1.

Some viruses with human reservoirs

A significant number of human viruses leading to either mild or life-threatening disease are maintained solely in human populations. The list runs the gamut from colds caused mainly by rhinoviruses, warts caused by papillomaviruses, to AIDS caused by HIV. The mode of passage of viruses between humans (i.e., the vector) is intimately linked to human behavior. This behavior can be modified by the disease symptoms themselves. Thus, a respiratory infection leads to irritation of the airways resulting in coughing and sneezing, which spreads an aerosol of droplets containing virus. Herpes simplex virus 1 (HSV-1) is spread in saliva resulting in transfer of the virus to the recipient's oral epithelial cells; in contrast, the closely related varicella zoster virus (VZV, or chickenpox) is spread by inhalation of a virus-loaded aerosol or dust containing dead skin flakes. Warts are spread by direct physical contact between the virus-loaded source (another wart or dead skin from a wart) and layers of the skin below the keratinized epidermis exposed by small cuts or abrasions. Poliovirus and the related coxsackieviruses are spread by virus-containing

Table 3.1 Some pathogenic viruses, their vectors or routes of spread, and their hosts.

Virus	Vector/Route	Host	Disease
Poliovirus	Human-fecal contamination of water or food	Human	Enteric infection; in rare cases, CNS infection (poliomyelitis)
Western equine encephalitis	Mosquito	Horse	Viral encephalitis in the horse – occasional infection of human
La Crosse encephalitis	Mosquito	Squirrel, fox (reservoir); human	No obvious disease in squirrel or fox; viral encephalitis in human
Sin Nombre virus (hantavirus)	Deer mouse	Deer mouse, other rodents (reservoir); human	Hantavirus hemorrhagic respiratory distress syndrome
HIV	Direct injection of virus-infected body fluids into blood	Human	AIDS
Measles	Aerosol	Human	Skin rash, neurological involvement
Yellow fever	Mosquito	Tropical monkeys; human	Malaise, jaundice
Chikungunya	Mosquito	Human, primates	Mild to severe hemorrhagic and arthritic disease
Dengue fever	Mosquito	Human, primates	Mild to severe hemorrhagic disease
Ebola	Unknown vector, but nosocomial transmission	Fruit bat; humans and primates	Often fatal hemorrhagic fever
Hepatitis A	Fecal contamination of water or food	Human	Acute hepatitis
Hepatitis B	Direct injection of blood	Human	Chronic hepatitis, liver carcinoma
Hepatitis C	Direct injection of blood	Human	Acute and chronic hepatitis
Hepatitis delta	Blood, requires coinfection with hepatitis B	Human	Acute hepatitis
Hepatitis E	Fecal contamination of water or food	Human	Mild acute hepatitis, except often fatal to pregnant women
Norovirus	Fecal contamination of water or food	Human	Severe diarrhea
Rabies	Bite of infected animal	Vertebrates	Fatal encephalitis
Herpes simplex virus (HSV)	Saliva, other secretions	Human	Surface lesions followed by latency, rare encephalitis
Varicella zoster (VZV, or chickenpox)	Aerosol, infected skin cells (dander)	Human	Rash, shingles, latency
Epstein-Barr virus (EBV)	Saliva	Human	Infectious mononucleosis, latency
Influenza	Aerosol	Human, many vertebrates	Flu
Myxoma	Insect bite	Rabbits	Variable mortality, skin lesions
Rhinovirus	Aerosol	Human	Colds
Coronavirus (CoV)	Aerosol	Civet cat (SARS-CoV); camel (MERS-CoV); human	Colds, SARS, MERS, COVID-19
Rubella (German measles)	Aerosol	Human	Mild rash, severe neurological involvement in first-trimester fetus
Adenovirus	Aerosol, saliva	Human	Mild respiratory disease
Papillomavirus	Contact	Human	Benign warts, some venereally transmitted, some correlated with cervical carcinomas

Table 3.1 *Continued*

Virus	Vector/Route	Host	Disease
HTLV (human T-cell leukemia virus)	Injection of blood	Human	Leukemia
Tomato spotted wilt (bunyavirus)	Thrip	Broad range of plant species	Necrosis of plant tissue, destruction of crops
Cadang-cadang (viroid)	Physical transmission via pruning	Coconut palm	Coconut palm pathology
Prion (protein pathogen)	Ingestion or inoculation of prion protein	Human, other mammals have specific types, cross-species spread possible	Non-inflammatory encephalopathy
Plant rhabdoviruses	Leaf hoppers, aphids, plant hoppers	Broad range of plant species	Necrosis of plant tissue, destruction of crops
Zika virus	Mosquito	Human, primates	Mild to severe disease, microencephaly, Guillain-Barré syndrome

feces contaminating food, drinks, or fomites that are ingested; this is termed *oral-fecal spread*. In the case of HIV, body fluids – including blood, breast milk, serum, vaginal secretions, and seminal fluid – are sources of infection. The virus can be spread by passive inoculation of, for example, a contaminated hypodermic syringe, or by transfusion, breastfeeding, or sexual activity.

Some viruses with vertebrate reservoirs

While many human viral diseases are maintained in the human population itself, some important pathogens are maintained primarily in other vertebrates. A disease that is transmissible from other vertebrates to humans is termed a *zoonosis*. Rabies is a classic example of a zoonosis that affects humans only sporadically. Because humans rarely transmit the virus to other animals or other humans, infection of a human is essentially a dead end for the virus. The rabies virus, which is transmitted in saliva via a bite, is maintained in populations of wild animals, most generally carnivores. The long incubation period and other characteristics of the pathogenesis of rabies mean that an infected animal can move great distances and carry out many normal behavioral patterns prior to the onset of disease symptoms. These symptoms may include hypersensitivity to sound and light, and finally, hyperexcitability and frenzy. Except in rare instances of inhalation of aerosols, humans only acquire the disease upon being bitten by a rabid animal; however, the fact that the disease can be carried in domestic dogs and cats means that when unvaccinated pets interact with wild animal sources, the pets become potential vectors for transmission of the disease to humans. Vaccination of pets provides a generally reliable barrier.

Many viral zoonoses require the mediation of an arthropod vector for spread to humans. The role of the arthropod in the spread can be mechanical and passive in that it inoculates virus from a previous host into the current one without virus replication having occurred (a favored route with animal poxviruses), but the arthropod's role as a vector can be dynamic. For viruses with RNA genomes that are transmitted between hosts via arthropods (such as those responsible for yellow fever, a number of kinds of encephalitis, dengue fever, and Zika virus), virus replication in the vector provides a secondary reservoir and a means of virus amplification. This makes spread to a human host highly efficient, since even a small inoculation of the virus into the arthropod vector can result in a large increase in virus for transmission to the next host.

VIRUSES IN POPULATIONS

Most (but certainly not all) virus infections induce an effective and lasting immune response. Some of the basic features of this response are described in Part II, Chapters 7 and 8. An effective immune response means that local outbreaks of infection result in the formation of a population of resistant hosts – often termed **herd immunity**. This means that any virus that induces protective immunity must maintain itself either in another reservoir or by dynamically spreading in "waves" through the population at large. If enough members of the susceptible population become immune, virus cannot spread effectively and it becomes extinct. This herd immunity is a major factor in both gradual and abrupt acquisition of genetic alterations that create new serotypes of viruses that can escape immunity to the original strain.

Viral epidemiology in small and large populations

The occurrence of mild respiratory infections (such as a common cold) in isolated communities provides graphic examples of the process of virus extinction. For example, when scientists visit the Antarctic research stations at the beginning of the Antarctic summer, they bring in colds to infect the resident population. When scientists stop arriving with the onset of winter, the prevailing respiratory diseases run their course and disappear. Figure 3.2 charts a classic epidemiological study of respiratory illness in an isolated fishing and mining population on Spitzbergen Island, Norway, in the Arctic Ocean. Note that after the last contact with the "outside world," the incidence of such viral-borne respiratory infections rapidly declines to an undetectable level.

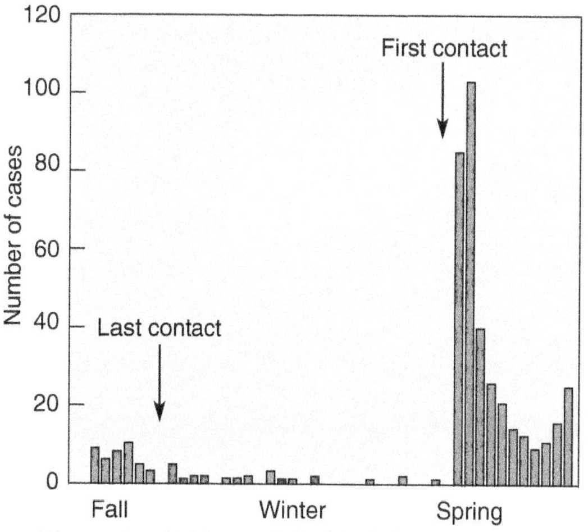

Figure 3.2 Occurrence of respiratory illness in an arctic community (Spitzbergen Island, Norway) that is isolated during the winter months. Following the last boat communication with the European mainland, the number of respiratory illnesses declines from a low number to almost nil. With the first boat arriving in the spring, new serotypes of respiratory viruses are communicated from the crew and passengers, and a "mini-epidemic" occurs. As the virus passes through the population, resistance builds and infections decline to a low level. Source: Based on data originally published by Paul, J.H. and Freese, H.L. (1933). An epidemiological and bacteriological study of the "common cold" in an isolated Arctic community (Spitsbergen [sic]). *American Journal of Hygiene* 7:517.

In large populations, the rate of virus spread greatly surpasses the limitations of the generation of herd immunity, and the introduction of a novel pathogenic virus leads to epidemic spread of disease. The outbreak of severe acute respiratory syndrome (SARS) in China in the early part of the 2000s and its spread to Canada provide an important case study of this process, as well as providing examples of effective and ineffective public health measures set up to deal with it. The SARS virus is a member of the coronavirus family and distantly related to one that causes mild colds in humans. The virus appears to have been maintained in wild animal populations in southeast China and was introduced into humans in Guangdong Province and the city of Guangzhou (Canton) through the custom of using such animals as dietary delicacies. While human infection is characterized by flu-like symptoms, the persistence, severity, and relatively high death rate suggested that this was a novel type of infection – a novel virulent form of influenza or an uncharacterized respiratory virus. Evidence suggests that the Chinese government, in hopes of avoiding loss of tourist and business travel revenues, suppressed news of this outbreak.

The disease was spread by a physician who had treated infected individuals in China and then traveled to Hong Kong on business – as the first identified source of infection, he was termed the **index case**. He contaminated the registration desk of the hotel in which he was registered, and this desk served as a source of infection for a number of tourists from other parts of the world (including Toronto, Canada) who happened to be staying in the same hotel. The disease spread to individuals in Hong Kong and was eventually described and quarantined there, but not before other infected individuals traveled back to Canada and, in lesser numbers, to the United States.

In Toronto, the index case of the local epidemic was a woman who infected her immediate family members upon returning from Hong Kong. She and one son subsequently died, but not before being admitted to the hospital where a physician treating them as well as other members of the hospital staff were infected. This illustrates a continuing conundrum of modern medicine – the concentration of individuals suffering from an infectious disease in a hospital can serve as a potent reservoir for the spread of that disease through the staff attending them and, subsequently, others. Such **nosocomial** infections are a major occupational hazard for hospital personnel as well as patients suffering other maladies, yet hospitals are obviously necessary for the treatment of the severely ill.

The Canadian public health authorities were reluctant to initiate stringent quarantines for infected individuals in the hospital where the first patients were housed, and the hospital served as a source of infected individuals who spread the disease to others both through family and social contacts and through contact in other workplaces. In contrast, in the United States, the infection was initiated somewhat later. By that time sufficient information concerning the disease, its spread, and its control led to rapid quarantine of SARS patients, especially among health workers. These control methods were successful in the United States and Europe, as well as in Hong Kong, and the virus never spread beyond the first intimate contacts.

A fictionalized sequence of events based on the Canadian SARS outbreak is shown schematically in Figure 3.3. Without the intervention of public health and other government agencies, the spread would continue through a susceptible population for an extended period of time. Further, it is clear that rapid recognition of symptoms and effective quarantine of affected individuals are key to stopping spread. In the case of SARS, the suppression of information concerning its appearance until it was potentially out of control in Asia could have led to a widespread epidemic there and in neighboring countries.

Many have suggested that only the lucky fact that the SARS coronavirus (SARS-CoV) is not particularly efficient at spreading between individuals saved us from a much more serious situation. Further, it has been argued that SARS provided a testing ground for public health response

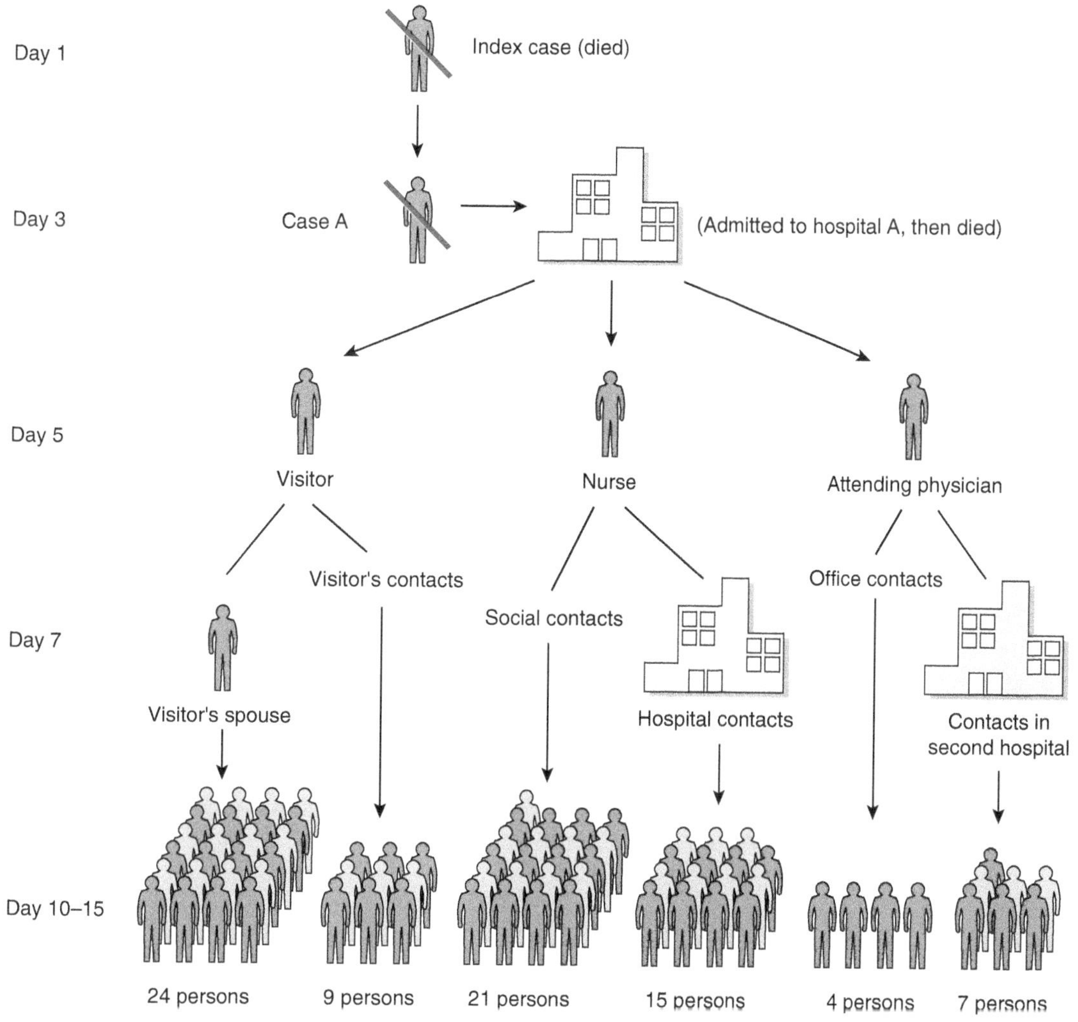

Figure 3.3 Fictionalized timeline of the spread of SARS virus following its introduction into Toronto, Canada, from Hong Kong in early 2003. Source: Data based on material presented on the CDC website (http://www.cdc.gov/ncidod/sars) and in the February 2004 issue of the *Journal of Emerging Infectious Diseases*, which was dedicated to studies on the SARS outbreak of late 2002–early 2003.

strategies, which worked reasonably well. Other examples of serious virus epidemics have not had as felicitous outcomes – for example, the HIV epidemic, which continues to grow and consume increasingly significant public health resources. Several other potentially lethal epidemics threaten the human population in the future, including a strain of **avian influenza**, and another coronavirus that is related to SARS-CoV, Middle East respiratory syndrome coronavirus (MERS-CoV), which caused a similar severe respiratory disease in the Middle East and is transmitted to humans from camels. Finally, the emergence of SARS-CoV-2 in 2019 with its more efficient human-human transmission (than SARS-CoV) resulted in the world-wide pandemic of COVID-19.

General features of these diseases and the viruses that cause them will be discussed in Parts III and IV, and an overview of the potential threats of virus disease in the future will be briefly addressed in Part V. It suffices here to note that the dynamics of virus spread are not the problem; rather, it is coordinating political, public health, medical, and scientific resources targeted at the control of infection in a timely and efficient manner that is and will continue to be major challenges.

Factors affecting the control of viral disease in populations

Generation of lasting immunity provides an effective means of controlling and even eradicating certain viral diseases. The antigenic stability of the smallpox virus and effective immunity against it allowed effective vaccination programs to eradicate the disease from the population. Polio and measles are current candidates for partial or total elimination from the population due to availability of effective vaccines. In addition, currently a program is underway to try to vaccinate wild populations of raccoons and other small carnivores against rabies with use of vaccine-laced bait. It is hoped that such an approach will reduce or eliminate the growing incidence of rabies in US wild animal populations. Of course, the reason for this solicitude has little to do with the animals involved; rather, it is to afford protection to domestic animals, and ultimately to humans.

Despite our considerable abilities, not all viral diseases can be readily controlled even under the most favorable economic and social conditions. Flu virus variants arise by genetic mixing of human and animal strains, and it is not practical to attempt a widespread vaccination campaign with so many variables. HIV remains associated with lymphatic tissue in infected individuals even when antiviral drugs effectively suppress virus replication. The intimate association of HIV with the immune system may make vaccination campaigns only partially effective. The ability of herpesviruses to establish latent infections and to reactivate suggests that a completely effective vaccine may be difficult if not impossible to generate.

A major obstacle to the control of viral and other infectious diseases in the human population as a whole is economic. It costs a lot of money to develop, produce, and deploy a vaccine. Many of the nations most at risk of deadly infectious disease outbreaks are financially unable to afford effective control measures, and pharmaceutical corporations involved with vaccine research and production are primarily interested in bottom-line profit. Perhaps more tragically, some nations at risk also lack the political will and insight to mount effective efforts to counter the spread of viral disease. Such problems constantly change character but are never ending.

ANIMAL MODELS TO STUDY VIRAL PATHOGENESIS

The great German clinical microbiologist Robert Koch formulated a set of rules for demonstrating that a specific microorganism is the causative agent of a specific disease. These rules are very much in force today. In essence, Koch's rules are as follows:

1 The same pathogen must be able to be cultured from every individual displaying the symptoms of the disease in question.

2 The pathogen must be cultivated in pure form.

3 The pathogen must be able to cause the disease in question when inoculated into a suitable host.

While these rules can be applied (with *caution*) to virus-mediated human diseases, it is clearly not ethical to inoculate a human host with an agent suspected to cause a serious or life-threatening disease (criterion 3). Regrettably, this ethical point has been missed more than once in the history of medicine. Examples of the excesses of uncontrolled human experimentation stand as a striking indictment of Nazi Germany, but excesses are not confined to totalitarian forms of government. The infamous Tuskegee syphilis studies are an example of a medical experiment gone wrong. These studies, ostensibly to evaluate new methods to treat syphilis, were carried out on a large group of infected African-American men in the rural southern United States by physicians of the US Public Health Service in the 1930s and 1940s. Even though effective treatments were known, a number of men were treated with **placebos** (essentially sugar pills) to serve as "controls" and to allow the physicians to accumulate data on progression of the disease in untreated individuals. Other examples of potentially life-threatening experiments in the United States with little effort to explain the dangers or potential benefits (the criteria for **informed consent**) using volunteer prisoners as test subjects are also well documented.

This discussion should not lead to the conclusion that it is never appropriate to use human subjects to study a disease or its therapy. Human experimentation (such as in clinical trials) is critical to ensuring treatment safety and effectiveness, but to do such studies in an ethical manner, the risks and benefits must be fully understood by all those involved.

One extremely effective way to obtain reliable data on the dynamics of disease and its course in an individual is to develop an accurate animal model. A researcher's need to experimentally test variable factors during infection in order to build a detailed molecular and physiological picture of the disease in question can only be accomplished with a well-chosen model. The lack of a suitable animal model for a viral disease is almost always a great impediment to understanding its disease processes and developing treatments.

Another important reason for using an animal model to study virus infection is that useful information can often be obtained with very defined and controlled experimental approaches. The ability of a virus to cause specific symptoms can be determined by careful control of the viral genotype and site of inoculation in the animal, followed by observation of the symptoms as they develop. The passage of a virus throughout the body during infection can be studied by dissection of specific organs, careful gross and microscopic observation, and simple measurement (**assay**) of virus levels in those organs. The host response to infection can be determined (in part, at least) by measuring the animal's production of antibodies and other immune factors directed against the infecting virus.

Much information concerning the interaction between a virus and its animal host can be obtained by using a combination of sophisticated molecular analyses on animals with defined genetic properties. Some examples are outlined in Parts III and IV. For example, transcription of a portion of the HSV genome in latently infected neurons can be observed by use of sophisticated methods to detect viral RNA in tissue in situ. Methods for introduction, mutation, and inactivation of specific genes controlling one or another aspect of the immune response can be introduced into mice (and potentially other animals) using methods described in Part V. These and many other techniques provide detail and richness to the "picture" of the virus–host interaction, and all are required for a full understanding of the interaction between virus and cell and virus and host. However, the basic outline of the course of viral infection in animals can be obtained by using the most simple and readily applied experimental tools: observation, dissection, and measurement of virus.

The use and sacrifice of animals raise significant ethical questions, and the suffering caused must be thoroughly considered in the design of appropriate experimental protocols. For example, an experimental study that establishes important aspects of a disease may be too devastating to repeat as a casual laboratory exercise. Appropriate treatment of animals, limiting pain and suffering and maximizing comfort to the animal, is a practical as well as ethical requirement for animal study.

It is very important to realize that animal models, by their very nature, can only approximate the course and nature of the disease occurring in humans. In order to carry out a valid and reproducible scientific study, the mode of infection, the amount of infecting virus, the age and sex of the animal, its nutritional background, and many other factors must be carefully and reproducibly controlled. Thus, any experimental animal model for a human disease is a compromise with the real world. For example, the amounts of virus inoculated into the animal and the site of inoculation (e.g., mouth, eye, **subcutaneous**, or **intracerebral**) must be the same for all test subjects, a situation very different from the "real" world. Also, the model disease in the animal may well be different, in whole or in part, from the actual disease seen in a human population. Genetic makeup of the animal (**inbred** or **outbred**, specific genetic markers present or absent), age, and sex of the infected host must be controlled to generate interpretable and reproducible results. Obviously, while certain diseases favor certain age groups, an infected population will evidence a wide range of variation in genetic and physical details.

Another complication is that the viral pathogen sometimes must be specifically adapted to the test animal. Virus directly isolated from an infected human may not cause disease in an animal unless it is first passaged in cells from that animal, or a mouse strain that is deficient in certain immune responses is used. In addition, safety considerations must be taken into account. Working with virus characterized by a very high mortality rate, such as Ebola virus, requires heroic and expensive precautions and containment facilities for study.

Despite very real problems with the use of animals to study virus-caused disease, it often is the only way to proceed. Careful and accurate clinical observations of infected individuals, animals, or plants provide many details concerning the course of viral infection. But only in a complete plant or animal model can the full course of disease and recovery as a function of controlled variations of infection and physiological state of the host be studied. This is true even when many aspects of virus infection can be studied in cultured cells and with cloned fragments of the viral genome.

The animal models for viral disease described in this and later chapters demonstrate some of the methods, successes, and limitations involved in the use of animals. Despite the problems associated with working with experimental animals, key basic data could not have been and cannot be obtained any other way.

A mouse model for studying poxvirus infection and spread

Many of the models developed for the study of viral pathogenesis involve the use of mice. These animals have an excellent immune system, can be infected with many viruses adapted from human diseases, and are relatively inexpensive to use. Frank Fenner's studies on the pathogenesis of mousepox carried out in the 1950s provided what is still a classic model for experimental study of viral pathogenesis.

Although smallpox virus is extinct in the wild, the recent realization that smallpox has been extensively studied as a weapon, and fears that it may be in the possession of terrorists, bring these classic studies into sharp focus. Further, other animal poxviruses such as monkeypox can infect humans, and human encroachment of tropical habitats has led to significant occurrence of this disease in tropical Africa. Another poxvirus, myxoma virus, is endemic in rabbit populations in South America, and was used in a temporarily successful attempt to control the ecological threat posed by the high rate of rabbit multiplication in Australia. While touted at the time as an example of successful biological control, numerous complications occurred with its use. Thus, this "experiment" is a valuable example of the benefits and problems involved with biological control.

In Fenner's classic study of mousepox pathogenesis, virus was introduced by subcutaneous injection of the footpad, and virus yields in various organs, antibody titer, and rash were scored. As noted, the basic experiment thus required only careful dissection of the infected animal, measurement of virus titers, and careful observation, but provided a detailed picture of the virus's pattern of spread within the animal. The patterns of virus spread and the occurrence of disease symptoms are illustrated in Figure 3.4.

Of course, the model is just that; it does not completely describe virus infection in the wild. An example of a significant deviation from one "natural" mode of infection is when poxvirus is transmitted as an aerosol, leading to primary infection in the lungs. This is a difficult infection route to standardize and is only rarely utilized. Also, examining experimental infection of animals in the laboratory ignores the dynamics of infection and the interactions between virus and the population. As a consequence, genetic changes in virus and the host, both of which are the result of the disease progressing in the wild, are ignored.

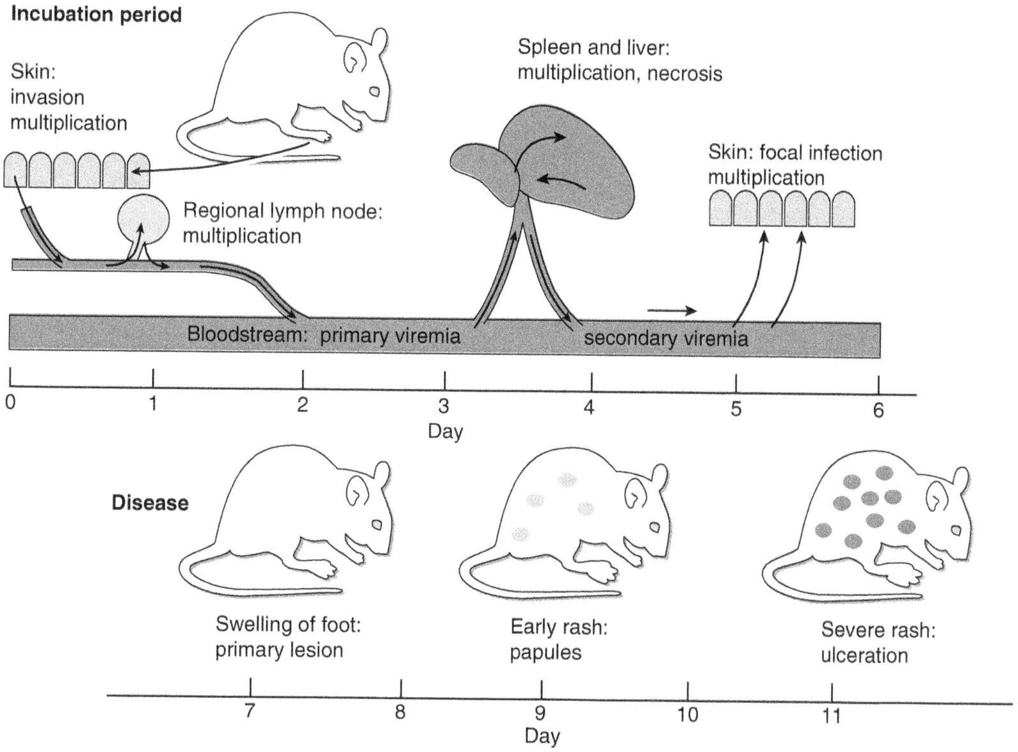

Figure 3.4 The course of experimental poxvirus infection in laboratory mice. Virus is inoculated at day 0 in the footpad of each member of a large group of genetically equivalent mice. Mice are observed daily, and antibody titers in their serum are measured. Selected individuals are then killed, and various organ systems assayed for the appearance and presence of virus. Note that symptoms of the disease (rash and swollen foot) only become noticeable after a week.

Rabies: where is the virus during its long incubation period?

Rabies and its transmission by the bite of infected animals to other animals and humans are well known in almost all human cultures. The disease and its transmission were carefully described in Arabic medical books dating to the Middle Ages, and there is evidence of the disease in classical times. One of the puzzles of rabies virus infection is the very long incubation period of the disease. This long period plays an important role in the mechanism of spread, and it is clear that animals (or humans) infected with the virus can be vaccinated weeks to months *after* infection and still mount an effective immune response.

The pathogenesis of rabies has been studied for over a century, and our current understanding is well founded in numerous careful studies made at varying levels of sensitivity using a number of approaches. An example of the use of immunological methods is shown in Figure 3.5. The basic course of infection starts with inoculation of virus at a wound caused by an infected animal, followed by limited virus replication at the site of primary infection. For the disease to develop, the virus must enter a neuron at a sensory nerve ending. These sensory nerve endings exist in all sites where the virus is known to enter an animal. Following this, the virus spreads passively to the nerve cell body in a dorsal root ganglion, where it replicates to a high level. Either this replicated virus, or other virus moving directly, passes into neurons of the cerebellum and cerebral cortex, where it replicates to high levels. Such replication leads to distinct behavioral changes associated with virus transmission. The virus also moves away from the central

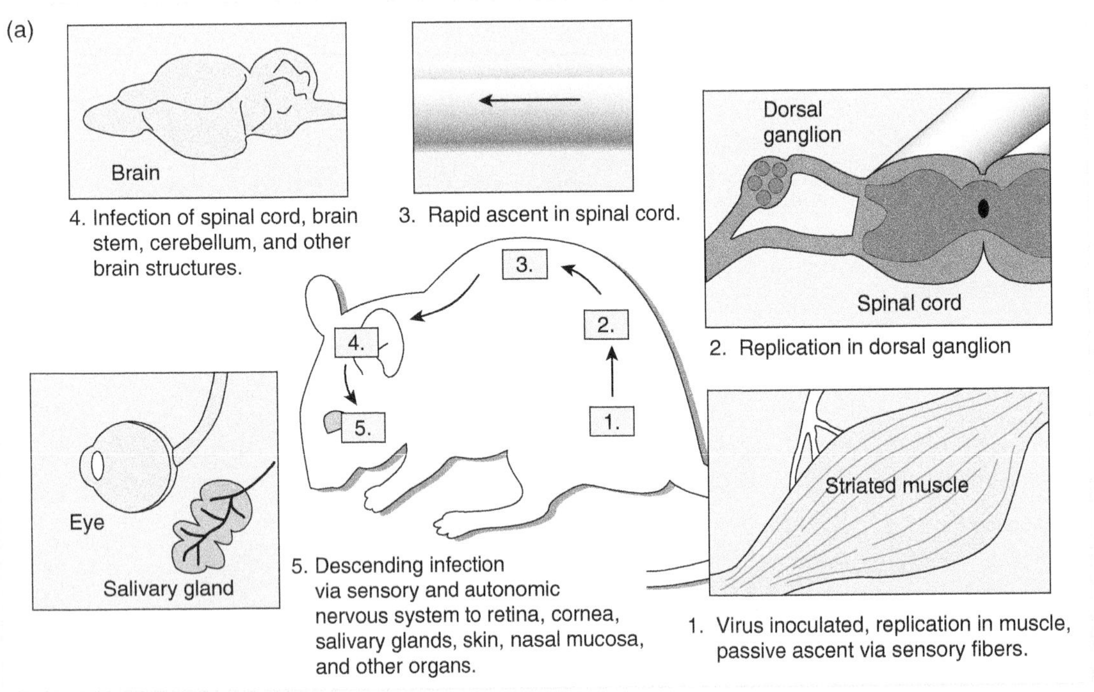

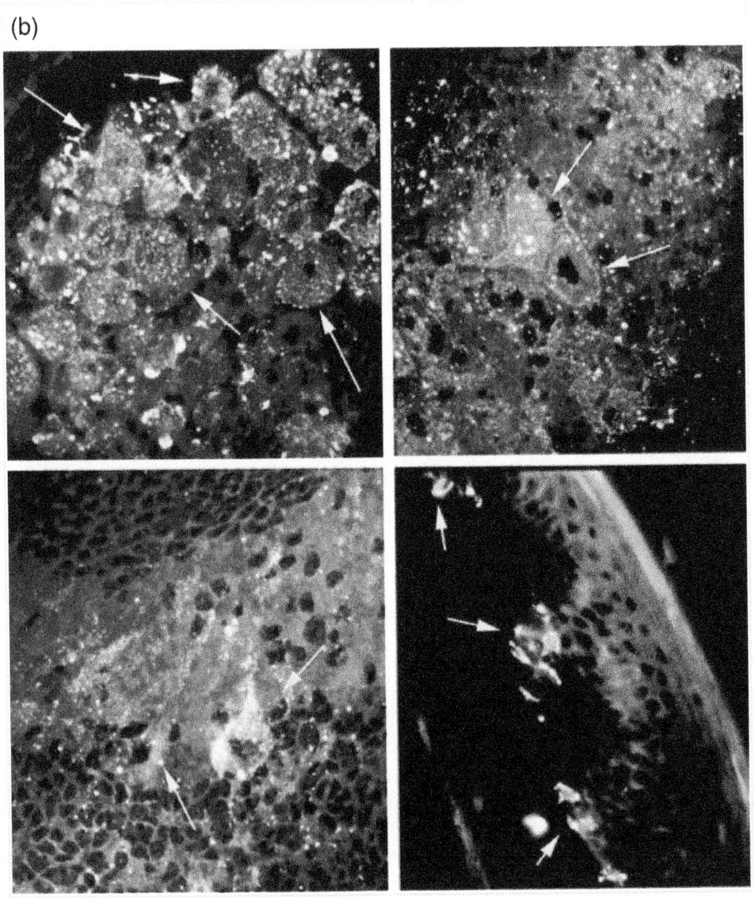

Figure 3.5 Visualization of rabies virus–infected neurons in experimentally infected animals. (a) A schematic representation of the pathogenesis of rabies in an experimentally infected laboratory animal. (b) Immunofluorescent detection of rabies virus proteins in neurons of infected animals. As described in Chapters 7 and 12, the ability of an antibody molecule to specifically combine with an antigenic protein can be visualized in the cell using the technique of immunofluorescence. The cell and the antibody bound to it are then visualized in the microscope under ultraviolet light, which causes the dye to fluoresce (a yellow-green color). The top left panel shows replication of rabies virus in a sensory nerve body in a dorsal root ganglion along the spine of an animal infected in the footpad. The bottom left panel shows the virus replicating in a neuron of the cerebellum, while the top right panel shows infected neurons in the cerebral medulla. Infection of the brain leads to the behavior changes so characteristic of rabies infections. Finally, the sensory nerve endings in the soft palate of a hamster infected with rabies virus at a peripheral site contain virus, as shown by the fluorescence in the bottom right panel. This virus can move to the saliva, where it can be spread to another animal. The arrows point to selected cells showing the variation in signal intensity that is typical of infections in tissues.

nervous system (CNS) to sensory neurons and salivary glands of the oral mucosa, where it replicates and is available for injection into another animal.

As early as 1887, CNS involvement was shown to result from direct spread of the virus from the site of infection into the CNS, as experimental animals that had their sciatic nerve severed prior to injection of the footpad with rabies virus did not develop the disease. The following experiment showed that the virus can remain localized at the site of infection for long periods of time: The footpad of several experimental animals was injected with virus at day 0 and then the inoculated foot was surgically removed from different groups at days 1, 2, 3, and so on after infection. Mice whose foot was removed as long as three weeks after infection survived without rabies, but once neurological symptoms appeared, the mice invariably died. Since removal of the foot saved the mice, it is clear that the virus remained localized there until it invaded the nervous system.

Finally, a similar experiment showed that rabies virus virulence for a specific host could be increased by multiple **virus passages** (rounds of virus replication) in that host. Virus isolated from a rabid wild animal takes as long as a week to 10 days to spread to the CNS of an experimentally infected laboratory animal. In contrast, isolation of virus from animals developing disease and re-inoculation into the footpad of new animals several times result in a virus stock that can invade the test animal's CNS in as little as 12–24 hours. Further, the virus stock that has been adapted to the laboratory animal is no longer able to efficiently cause disease in the original host. As described in Chapter 8, this is one way of isolating strains of virus that are avirulent for their natural host and have potential value as vaccines.

Herpes simplex virus latency

There are two closely related types of herpes simplex virus: type 1 (facial, HSV-1) and type 2 (genital, HSV-2). Both establish latent infections in humans, and reactivation from such infections is important to virus spread. Some details concerning latent infection by HSV are discussed in Chapter 18 in Part IV. Different animal models demonstrate both general similarities and specific differences. These differences illuminate a major limitation of many animal models for human disease: A model often only partially reflects the actual course of disease in the natural host – in this case, in humans.

Murine models

HSV infection in the eye or the footpad of mice can lead to a localized infection with spread of virus to the CNS and then to the brain. Although some animals die, as shown in Figure 3.6, survivors maintain a latent infection in sensory nerve ganglia. During this latent infection, no infectious virus can be recovered from nerve tissue, but if the nerve ganglia are **explanted** (dissected, dissociated, and maintained on a **feeder layer** of cultured cells), virus will eventually appear and begin to replicate. This observation demonstrates both that the viral genome is intact in the latently infected neuron, and that virus is not present in infectious form until something else occurs.

This model is quite useful for the study of genetic and other parameters during *establishment* and *maintenance* of a latent infection. For example, the sensory neurons can be isolated and viral DNA can be recovered. But since mice do not efficiently reactivate HSV, the physiological process of reactivation, where virus can be recovered at the site of initial infection, cannot be effectively studied in mice.

Rabbit models

Infection of rabbit eyes with HSV leads to localized infection and recovery. The rabbits maintain virus in their trigeminal ganglia, and viral DNA, virus, or both can be recovered using methods

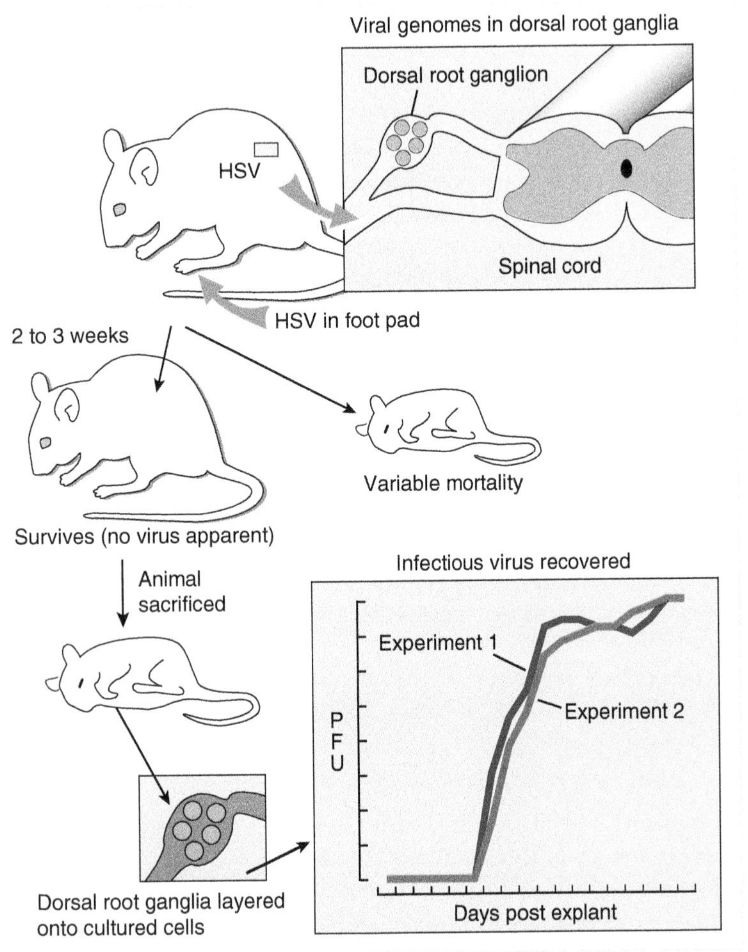

Figure 3.6 Analysis of the establishment and maintenance of latent HSV infections in mice. A number of mice are inoculated in the footpad, and following the symptoms of primary disease, which include foot swelling and minor hindquarter paralysis, many mice recover. Those that do not recover have infectious virus in their CNS. The mice that recover are latently infected and no infectious virus can be detected, even with high-sensitivity measurements of nervous and other tissue. HSV genomes, but not infectious virus, can be detected in nuclei of sensory nerve dorsal root ganglia. When these ganglia are cultured with other cells that serve both as an indicator of virus replication and as a feeder layer for the neurons (i.e., explanted), a significant number demonstrate evidence of virus infection and infectious virus can be recovered, as shown on the inset graph (two separate experiments are shown, with essentially the same results).

described for the murine model. Unlike mice, rabbits spontaneously reactivate HSV, and virus occasionally can be recovered from the rabbit's tear film. Further, this reactivation can be *induced* by **iontophoresis** of epinephrine with high frequency. Rabbits, because HSV can reactivate in them, are vital to the design of experiments to investigate induced reactivation, although they are more expensive to purchase and keep than mice.

Guinea pig models

Guinea pigs are favored experimental animals for the study of infection and disease because they are readily infected with many human pathogens. They are an important model for the study of HSV-2, which cannot be studied effectively in the murine and rabbit models just described.

Guinea pigs can be infected vaginally with inoculation of virus, and following a localized infection, latency can be established. As occurs in the murine and rabbit models, virus or viral DNA can be recovered from latently infected neurons (those enervating the vaginal area in this case). As in rabbits, latent infection in guinea pigs will spontaneously reactivate, and periodic examination can be used to measure reactivation rates. Unlike rabbits, however, guinea pig reactivation cannot be induced. Also, HSV-2 reactivates much more frequently than does HSV-1 in the guinea pig model; therefore, this model may be of some value in establishing the subtle genetic differences between these two types of viruses that manifest as a differential tropism for **mucosa**.

QUESTIONS FOR CHAPTER 3

1 In the case of rabies virus, how would you classify humans with respect to their role as a host?

2 What characteristics are shared by *all* hepatitis viruses?

3 Using the data presented in Table 3.1, answer the following questions:

(a) Which of the viruses in the table are vectored by mosquitoes?

(b) Which of the viruses in the table are transmitted in an aerosol?

(c) Which of the viruses in the table are transmitted by injection of blood?

(d) Which of the viruses in the table are neurotropic?

4 You are a viral epidemiologist studying the population of Spitzbergen Island off the coast of Norway (see Figure 3.2). Suppose that a team of scientists plans to visit this island by special boat during the Christmas holiday season. How might this visit change the pattern of respiratory infections you have been observing? What criteria must exist for this visit to have an effect on the pattern of viral respiratory illness on the island?

5 You have isolated two mutant strains of virus Z – mutant 1 and mutant 2. Neither strain can replicate when infected into cells, but either can be propagated in cell culture when coinfected with mutant virus 3. When you coinfect cultured cells with mutants 1 and 2 together, infection proceeds, but only mutant 1 and mutant 2 can be recovered from the infected cells. What is the best explanation of these results?

Patterns of Some Viral Diseases of Humans

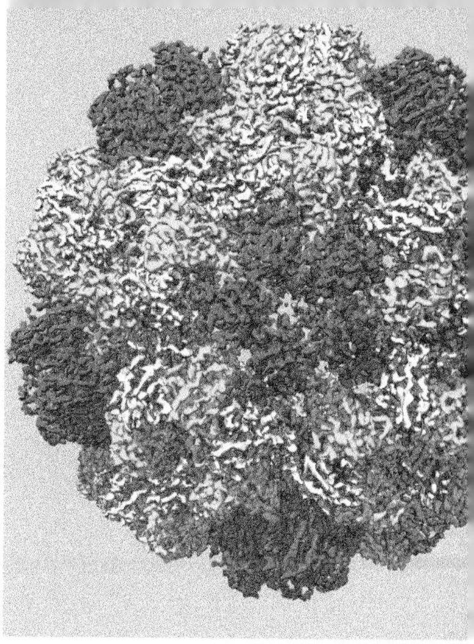

CHAPTER

4

- THE DYNAMICS OF HUMAN-VIRUS INTERACTIONS
- The stable association of viruses with their natural host places specific constraints on the nature of viral disease and mode of persistence
- Classification of human disease-causing viruses according to virus-host dynamics
  Viral diseases leading to persistence of the virus in the host are generally associated with viruses having long associations with human populations
  Viral diseases associated with acute, severe infection are suggestive of zoonoses
- PATTERNS OF SPECIFIC VIRAL DISEASES OF HUMANS
- Acute infections followed by virus clearing
  Colds and respiratory infections
  Influenza
  Variola
- Infection of an "accidental" target tissue leading to permanent damage despite efficient clearing
- Persistent viral infections
  Papilloma and polyomavirus infections
  Herpesvirus infections and latency
  Other complications arising from persistent infections
- Viral and subviral diseases with long incubation periods
  Rabies
  HIV/AIDS
  Prion diseases
- SOME VIRAL INFECTIONS TARGETING SPECIFIC ORGAN SYSTEMS
- Viral infections of nerve tissue

Basic Virology, Fourth Edition. Martinez "Marty" Hewlett, David Camerini, and David C. Bloom.

THE DYNAMICS OF HUMAN–VIRUS INTERACTIONS

We have seen that the process of infection and consequent disease is controlled by a number of factors ranging from the effect of specific genes controlling aspects of pathogenesis to more subjective factors that can be classified as important in overall virulence of the disease. The nature of the viral disease – or more accurately, the viral infection, its severity, the fate of the host, and the fate of the virus causing the disease – is important from a purely medical point of view. But, as importantly, the features of the dynamic interaction between virus and host provide important clues as to how long a particular virus has been associated with its host. Further, the nature of the interaction provides clues as to the evolutionary history of the host.

The stable association of viruses with their natural host places specific constraints on the nature of viral disease and mode of persistence

As noted in Chapter 3, viruses are maintained by active rounds of infection somewhere in their reservoir. We have seen that a virus infection leading to immunity to re-infection will lead to virus extinction in a small population once the pool of susceptible individuals is exhausted. Also, even in a large reservoir, if the virus infection directly or indirectly leads to death of a large enough number of individuals, the host population of the reservoir will crash and, in extreme cases, may become extinct. Clearly, a virus that can only replicate within that population will also become extinct. These limitations, which can be described with precision using the mathematics of population biology and epidemiology, lead to a number of evolutionary constraints on the dynamics of the virus–host interaction. Viruses whose infection leads to an acute disease followed by clearing and immunity need a large host population, while the outcome of the disease cannot be too lethal or the virus cannot be maintained. If infection results in mild or inapparent symptoms, however, there still must be efficient spread. This latter pattern of virus infection is a common feature of viruses with animal reservoirs that contain large populations, such as flocks of migratory birds or herds of ungulates. Since the agricultural/urban revolution starting about 10 000 years ago, which engendered rapid increases in our population, urbanized human populations fit these criteria also.

The early history of humans, however, was not one of large sedentary populations. Rather, our ancestors lived in small, nomadic groups organized along familial lines. This organization is similar in broad outline to that of predators such as wolves and large cats. In such a population, no virus that is cleared with resulting immunity can persist; therefore, only a virus that can establish a persistent infection of its host, with little or no diminution of the host's ability to survive and propagate, can persist. Furthermore, this persistence must allow for infectious virus to be present at opportune moments for infection of new, susceptible individuals (i.e., infants and occasional adults encountered from other groups). A number of viruses have replication strategies and genetic capacities to establish such infections, and it is striking that genetic analysis of such viruses demonstrates ancient associations with humans.

These two basic, non-exclusive strategies of virus replication are diagrammed in Figure 4.1. Of course, not all viruses are constrained by their **narrow host range** to infect just one species

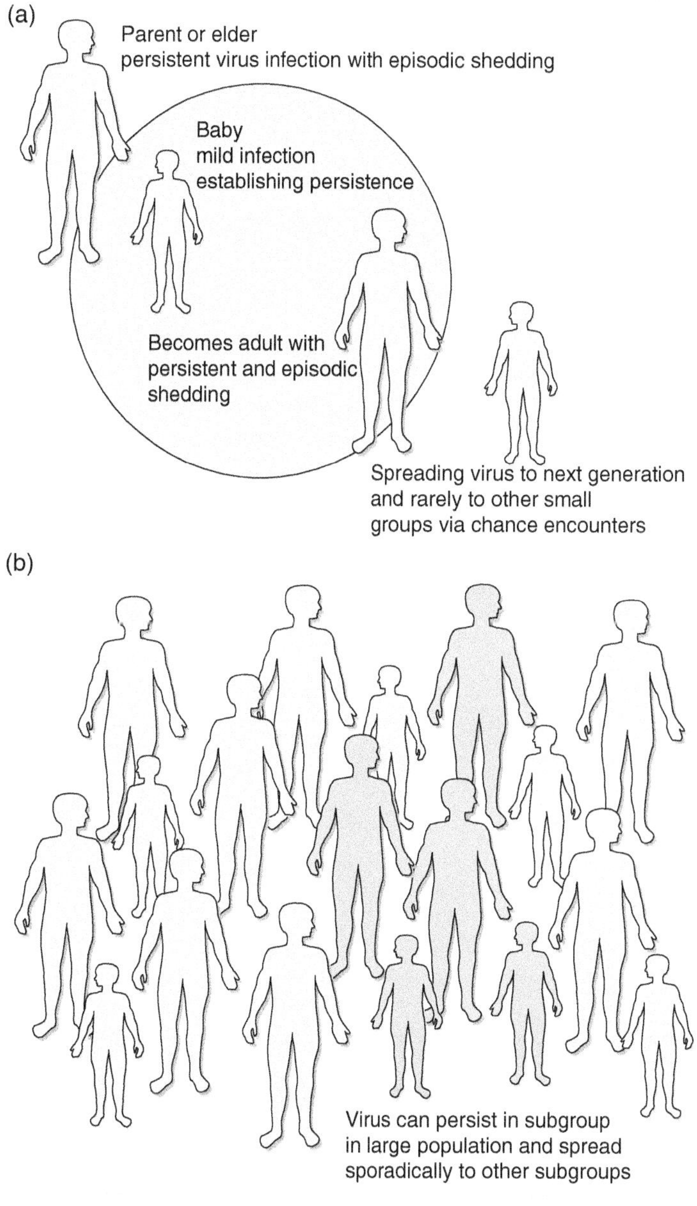

Figure 4.1 Virus maintenance in small and large populations. (a) In a small population, virus infection can only occur when there is an immunologically naive individual available. This requires a virus within such a population to be able to maintain itself in an infectious state in individuals long after they have been infected. A favored mode of infection would be from parent to child. Clearly, high mortality rates or severe disease symptoms would be selected against. (b) In a large population, there will be a large number of susceptible individuals appearing at the same time. This can result in local episodic infections of such individuals. The large size of the host population insures that some virus is available from actively infected adults at all times. While persistence is not excluded, it need not be strongly selected for, especially if the course of the acute phase of the disease is relatively long compared to the generation time of the population.

or type of host. Some, notably a number of viruses using RNA as their genetic material, have a **broad host range** and can readily jump from one species of host to another. With such a virus, the constraints on the mortality of the disease caused in the novel or ancillary host target population are counterbalanced by less severe disease and therefore persistence in a different species that the virus infects. It is not particularly surprising, then, that mortality rates of some diseases caused by zoonotic viruses are quite high.

Other factors further complicate the simple patterns of virus infection and persistence outlined here. A notable one is that if the period of time between the initial infection and the appearance of symptoms (the **incubation period**) is longer than the generation time of the host, constraints on mortality are lost. This is the case for rabies, which exhibits essentially a 100% mortality rate in infected carnivores – its natural host – but has a very long incubation period that allows reproduction even after infection. The lifespan of humans is so long that this might not seem to be a major factor in maintaining virus infections with high mortality rates, but the association between certain persistent human virus infections and the very much later appearance of tumors and immuno-pathologies is a consequence of a long incubation period between initial infection and ultimate pathology.

Classification of human disease-causing viruses according to virus–host dynamics

We can use the nature of the virus–host interaction to generate a simple classification of viruses, especially those of humans. These criteria are a useful aid for organizing detailed information concerning diseases with a viral **etiology** within the context of possible courses and outcomes. A number of specific examples are outlined in this chapter.

Some of the basic criteria that can be incorporated into such a scheme include:

1 Do the symptoms of the viral disease take a short or long time to develop after first encountering the virus?

2 Are the symptoms of initial infection relatively mild or severe?

3 Can the infected individual be expected to recover completely, or partially?

4 Does the virus stay associated with the victim following apparent recovery?

5 If the association is lasting, is the virus maintained in an infectious form either sporadically or constantly?

Viral diseases leading to persistence of the virus in the host are generally associated with viruses having long associations with human populations

Humans, like other animals, are subject to numerous viral infections mediated by viruses either maintained solely in the host population or in another population of animals with which humans interact. We have seen that two basic patterns occur in the course of virus replication in humans – persistent infections with incomplete virus clearing, and acute infections with efficient virus clearing upon recovery from the acute infection. It might be argued that persistent infections represent associations between virus and host that have stabilized over time, such that the viruses are maintained within the host population without a large negative effect. Conversely, acute infections may involve viruses that have recently moved from a different host. In the case of humans, such viruses may originate from zoonotic infections. While influenza A viruses and hantaviruses are examples that support this model, human rhinoviruses, among others, would not, since there are no known animal reservoirs. These virus groups and some of the symptoms caused by their infections of humans are listed in Table 4.1.

Table 4.1 Some viruses infecting humans.

Family	Genome	Primary Reservoir	How Long Associated with Humans	Virus Type	Acute Disease	Primary Infection	Mortality Rate	Persistent/ Latent?	Reactivation	Chronic Disease/ Complications
Herpesviridae	DNA	Humans	Ancient	HSV-1	Facial lesion	Epidermis	Nil	Yes	Frequent at site	Encephalitis (rare)
				HSV-2	Genital lesion	Epidermis	Nil	Yes	Frequent at site	Encephalitis (rare)
				VZV	Chickenpox	Epidermis	Nil	Yes	Once	Shingles/ disseminated infection upon immune suppression
				HCMV	Mononucleosis	Hematopoietic tissue	Nil	Yes	Asymptomatic/ infrequent?	Disseminated infection upon immune suppression/ retinitis
				EBV	Mononucleosis	Lymphoid tissue	Nil	Yes	Asymptomatic/ infrequent?	Lymphoma/ carcinoma
				HHV-6	Roseola	Lymphoid tissue	Nil	Yes	Asymptomatic/ infrequent?	?
				HHV-7	Roseola	Lymphoid tissue	Nil	Yes	Asymptomatic/ infrequent?	?
				HHV-8	?	Lymphoid tissue	Nil	Yes	Asymptomatic/ infrequent?	Kaposi's sarcoma
Polyomaviridae	DNA	Humans	Ancient	JC	None	Kidney/bladder	Nil	Yes	Infrequent (?) shedding	Encephalitis upon immune suppression
				BK	None	Kidney/bladder	Nil	Yes	Infrequent (?) shedding	Kidney infection
Papillomaviridae	DNA	Humans	Ancient	>60 types	Warts	Epidermis	Nil	Yes	Constant shedding at site	Cervical carcinoma (Types 6, 11, 16, and 18)
Adenoviridae	DNA	Humans	Ancient?	>12 types	Mild respiratory	Respiratory tract	Nil	Yes	Infrequent shedding?	?

Table 4.1 *Continued*

Family	Genome	Primary Reservoir	How Long Associated with Humans	Virus Type	Acute Disease	Primary Infection	Mortality Rate	Persistent/ Latent?	Reactivation	Chronic Disease/ Complications
Poxviridae	DNA	Humans	Recent	Variola	Smallpox	Epidermis	Moderate to high	No	N/A	None
Orthomyxoviridae	RNA	Birds, pigs	Sporadic/ current	Influenza A	Influenza	Respiratory tract	Usually low/ rarely high	No	N/A	None
		Humans	Sporadic/ current	Influenza B	Influenza	Respiratory tract	Nil			None
Coronaviridae	RNA	Humans	?	Human coronavirus	Cold	Nasopharynx	Nil	No?	N/A	None
		Palm civets	Current	SARS	Acute respiratory failure	Respiratory tract	Moderate to high	No	N/A	None
Picornaviridae	RNA	Humans	Recent?	Poliovirus	None to mild digestive upset	GI tract	Nil	No	N/A	Paralysis
		Humans	Recent?	Hepatitis A virus	Hepatitis	Liver	Low	No	N/A	Rare
		Humans	?	Rhinoviruses	Cold	Respiratory tract	Nil	No?	N/A	None
Hepeviridae	RNA	Human	?	Hepatitis E	Hepatitis; severe in newborns	Liver	Nil, except pregnant women; moderate in newborns	No	N/A	Rare
Flaviviridae	RNA	Birds	Sporadic/ current	West Nile	Encephalitis	Brain	Low	No	N/A	Neurological
		Primates	Sporadic/ current	Yellow fever	Encephalitis	Brain	Moderate	No	N/A	Neurological
		Human	?	Hepatitis C virus	Hepatitis	Liver	Low	Occasional	Chronic hepatitis with no virus shedding	Liver failure/ carcinoma
		Human	Sporadic/ current	Dengue	Fever/joint pain	Leukocytes	Low	No	N/A	Hemorrhagic fever
		Human/ primate	Sporadic/ current	Zika	Fever/joint pain/ rash	?	Low	No?	N/A	Neurological, teratogen
Rhabdovirus	RNA	Carnivores	Sporadic/ current	Rabies	Encephalitis	Brain	100%	No	N/A	N/A

Togavirus	RNA	Horses	Sporadic/current	Equine encephalitis virus	Encephalitis	Brain	Low	No	N/A	Neurological
		Human?	?	Rubella virus (German measles)	Rash	Skin/developing nerve tissue	Nil in adults; severe neurological symptoms in developing fetus	No	N/A	Fetal infection
Paramyxovirus	RNA	Carnivore?	Recent	Measles	Rash	Respiratory tract	Low/moderate	Yes/no	Chronic virus antigen present/no infectious virus	SSPE
		Human	?	Mumps	Glandular inflammation	Respiratory tract	Nil	No	N/A	Infertility
		?	?	Respiratory syncytial virus	Mild respiratory in adults	Nasopharynx	Nil in adults	Yes	Virus shed from nasopharynx	Infections of newborns
Hepatitis delta	RNA	Human	?	Hepatitis D	Hepatitis	Liver	Usually low, but infection with hepatitis B leads to acute liver failure	Yes	Virus antigen present and infectious in blood	Liver failure
Bunyavirus	RNA	Mosquitoes	?	Lacrosse encephalitis virus	Encephalitis	CNS	Low; more severe for children	No	N/A	None
		Rodents	?	Hantavirus	Severe respiratory failure	Respiratory tract	Moderate, especially in young adults	No	N/A	Respiratory failure
Hepadnavirus	RNA/DNA	Human	?	Hepatitis B	Hepatitis	Liver	Low	Yes	Virus antigen present and infectious in blood	Liver failure/carcinoma
Retrovirus/RNA tumor virus	RNA/DNA	Humans	Ancient	Human T-cell leukemia	None	Lymphoid tissue	Nil?	Yes	Infectious virus shed?	Lymphoma/paraparesis
Retrovirus/lentivirus	RNA/DNA	Chimpanzees/mangabeys	1930 ± 20 years/?	HIV-1/HIV-2	None/flu-like or mononucleosis-like	Lymphoid tissue	100%	Yes	Infectious virus shed	Immunodeficiency

A number of important human viruses are either asymptomatic or cause relatively mild symptoms of primary infection, which is followed by a stable association between virus and host that lasts as long as the latter lives. During this more or less stable association, some viruses are constantly shed, while persistent infections with others lead to loss of any detectable virus. In the latter case, various types of stress to the host can lead to viral **recrudescence** (reappearance of infectious virus along with mild or no symptoms) with the potential for virus spread to other individuals.

Notable persistent viruses include the human herpesviruses, human papilloma and related polyoma viruses, human retroviruses, and two completely unrelated viruses – hepatitis B and hepatitis C viruses – whose chronic infection leads to severe liver damage. The details of the replication of these viruses will be covered in later chapters in Part IV; what is important to consider now is that the persistent mode of virus–host interaction is well suited to small groups of wandering related individuals with only occasional contact with other such groups. Virus persistence insures that virus can survive without infecting new individuals for the long life of the human host, yet virus recrudescence insures that there will be opportunities for virus spread at opportune times.

A detailed analysis of the genes of these viruses, especially those originally appropriated from the host, using procedures outlined in Chapter 1 demonstrates both their long association with humans and the considerable evidence for their replicative isolation – again expected for an infectious agent constrained to maintain itself in a small, relatively isolated environment.

Viral diseases associated with acute, severe infection are suggestive of zoonoses

While a pattern of virus infection leading to persistence of infectious virus correlates with the ancient association between humans and specific groups of viruses, those viruses characterized by an acute, severe infection with variable or high mortality rates suggest a virus that either has a primary reservoir other than humans (a zoonosis) or has recently "jumped" from such a reservoir to establish and maintain itself in the human population. Influenza A virus is a typical example of such an acute disease-causing virus. The primary reservoir for this virus is birds, but its broad host range allows it to establish secondary reservoirs in pigs as well as humans. Not all strains of the virus introduced into the human population are marked by high mortality rates, but occasionally an avian- or swine-adapted strain with high mortality rates can establish itself in the human population with severe and even devastating results. The great worldwide epidemic of influenza (H1N1) in 1918–1919 (also called a **pandemic**) was the result of a direct transmission of the virus from birds to humans, with subsequent mutation to allow human-to-human spread. There are great concerns that recently described avian flu outbreaks (i.e., H5N1), which originated in eastern China, could do the same.

Numerous other examples of zoonotic viruses associated with severe human disease exist. The 2002–2003 SARS (severe acute respiratory syndrome) outbreak was the result of a coronavirus established in animals (probably the bat and/or palm civet) adapting to human spread. The epidemic of **Ebola virus** disease, which killed over 11 000 people in West Africa in 2013–2016, whose natural reservoir was probably in bats is another example; others are the sporadic appearance of West Nile virus harbored in avian populations, as well as various insect-borne encephalitis viruses.

The most devastating virus introduced into the human population from a zoonosis in recent time is the human immunodeficiency virus (HIV), which is the causative agent of AIDS. There are two types of HIV, which differ in sequence by 25%. Both types of virus have been shown to be closely related to retroviruses carried asymptomatically in nonhuman primates in Africa, and they clearly entered the human population through contact with such animals, probably in their use as a delicacy in local cooking. The virus is spread through sexual contact and intravenous drug usage, and, as described in a later chapter, human behavior has favored its establishment as a formidable threat to public health, especially in the developing world.

We can infer that other viruses characterized by their spread as sporadic epidemics with high mortality rates in the human population have had relatively recent histories in human populations. A notable example is smallpox (variola) virus, which is completely adapted to humans, but which requires a large population for its maintenance and spread due to its lack of persistence and its engendering of a strong and lasting immunity. Since related poxviruses are found in many animals maintaining large populations, introduction into humans after the development of agriculture and urbanization can be inferred. The same can be said for polio, measles, Ebola, and other viral diseases, which spread by acute infection and leave an immune population cleared of virus in their path. Indeed, Jared Diamond has made a strong case for such diseases being a critical feature in the development of modern, urban society in his book *Guns, Germs, and Steel: The Fates of Human Societies.*

PATTERNS OF SPECIFIC VIRAL DISEASES OF HUMANS

While classification of human viruses according to the history of their association with us as host is very useful from an epidemiological and evolutionary point of view, a classification of these viruses by the nature of the disease they cause and its **sequelae** is more useful from a medical perspective. Some important viral diseases are classified in this way throughout this section.

Acute infections followed by virus clearing

Colds and respiratory infections

Cold viruses (rhinoviruses, adenoviruses, and coronaviruses) are spread as aerosols. Infection is localized within the nasopharynx, and recovery involves immunity against that specific virus **serotype**. The vast array of different cold viruses and serotypes ensures that there will always be another one to infect individuals. Although generally these types of respiratory diseases are mild, infection of an immune-compromised host or a person having complications due to another disease or advanced age can lead to major problems.

Influenza

The epidemiology of **influenza** is an excellent model for the study of virus spread within a population. While symptoms can be severe, in part due to host factors, the virus infection is localized, and the virus is efficiently cleared from the host. Flu viruses have evolved unique mechanisms to ensure constant generation of genetic variants, and the constant appearance of new influenza virus serotypes leads to periodic epidemics of the disease. Some of these mechanisms are described in detail in Chapter 15, Part IV. The respiratory distress caused by most strains of flu virus is not particularly life-threatening for healthy individuals, but poses a serious problem for older people and individuals with immune system or respiratory deficiencies. Some strains of the virus cause more severe symptoms with accompanying complications than others. At least one strain, the Spanish strain of 1918–1919 (H1N1), caused a worldwide epidemic with extremely high mortality rates in the years immediately following World War I.

Variola

The disease caused by infection with **smallpox (variola)** virus is an example of a much more severe disease than flu, with correspondingly higher mortality rates. There are (or were) two forms of the disease: *variola major* and *variola minor*. These differed in severity of symptoms and death rates. Death rates for variola major approached 20%, and during the Middle Ages in Europe, it reached levels of 80% or higher in isolated communities. Virus spread was generally by inhalation of virus aerosols formed by drying exudate from infected individuals. Variola virus is unusually resistant to inactivation by desiccation, and examples of transmission from contaminated material as long as several years after active infection were common.

The disease involves dissemination of virus throughout the host and infection of the skin. Indeed, the pathogenesis of mouse pox described in Chapter 3 provides a fairly accurate model of smallpox pathogenesis. The virus encodes **growth factors** that were originally derived from cellular genes. These growth factors induce localized proliferation at sites of infection in the skin, which results in development of the characteristic pox (see Chapter 18, Part IV).

Infection of an "accidental" target tissue leading to permanent damage despite efficient clearing

As outlined in Figure 4.2, some viruses can target and damage an organ or organ system in such a way that recovery from infection does not lead to the infected individual regaining full health despite generation of good immunity. A well-understood example is paralytic poliomyelitis. Poliovirus is a small enteric virus with an RNA genome (a **picornavirus**), and most infections (caused by ingestion of fecal contamination from an infected individual) are localized to the small intestine. Infections are often asymptomatic, but can lead to mild enteritis and diarrhea. The virus is introduced into the immune system by interaction with lymphatic tissue in the gut, and an effective immune response is mounted, leading to protection against reinfection.

Infection with poliovirus, however, can also lead to paralytic polio. The cellular surface protein to which the virus must bind for cellular entry (CD155) is found only on cells of the small intestine and on motor neurons. In rare instances, infection with a specific genotype that displays marked tropism for (a propensity to infect) neurons (i.e., a neurovirulent strain) leads to a situation where virus infects motor neurons and destroys them. In such a situation, destruction of the neurons leads to paralysis.

It should be noted that paralysis resulting from neuronal infection does not aid the virus's spread among individuals; this paralytic outcome is a "dead end." Perhaps ironically, the paralytic complications of poliovirus infections have had negative selective advantages, since if such a dramatic outcome did not occur, there would have been no interest in developing a vaccine against poliovirus infection!

A variation on the theme of accidental destruction of neuronal targets by an otherwise relatively benign course of acute virus infection can be seen in **rubella**. This disease (also called **German measles**), which is caused by an RNA virus, is a mild (often asymptomatic) infection resulting in a slight rash. Although infection is mild in an immunocompetent individual, the virus has a strong tropism for replicating and differentiating neural tissue. Therefore, women in the first trimester of pregnancy who are infected with rubella have a very high probability of having an infant with severe neurological damage from congenital rubella syndrome. Rubella is rare in the United States due to high acceptance of the vaccine, but vaccination of women who are planning to become pregnant is an effective method of preventing such damage during localized rubella epidemics in other parts of the world.

Persistent viral infections

While persistent viral infections often indicate a long history of coevolution between virus and host, the lack of serious consequences to the vast majority of those infected does not mean that

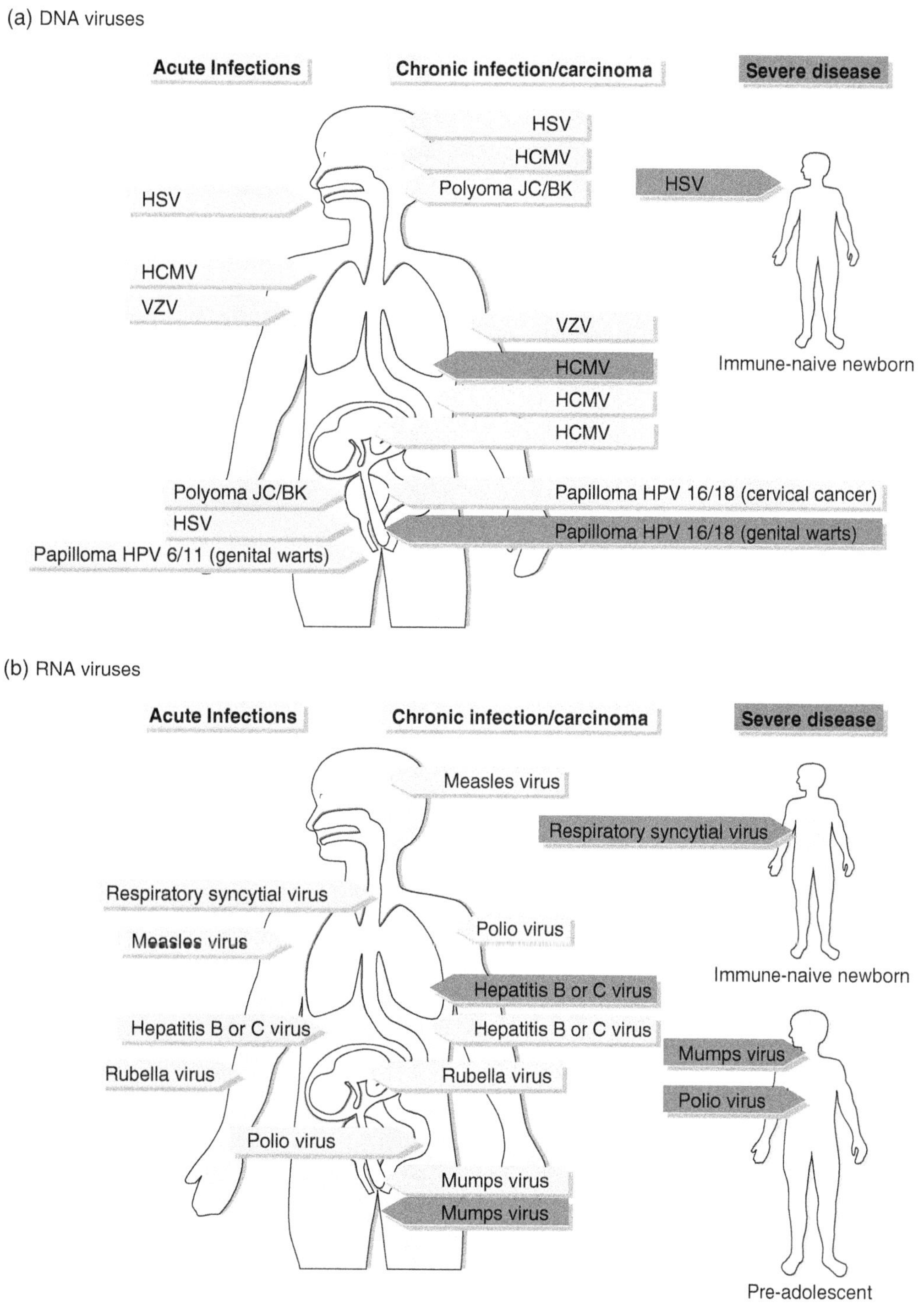

Figure 4.2 Examples of virus infection of specific organs or organ systems. (a) DNA genome viruses. (b) RNA genome viruses. Blue labels indicate acute infections, while pink labels indicate infections that result in either chronic disease states or carcinomas. Red labels indicate acute infections that can result in severe disease.

debilitating or lethal consequences are not possible. This is especially the case in situations where the immune system of the infected individual is compromised or has not yet developed. Some examples of persistent infections and the complications that can arise from these infections are shown in Figure 4.2.

Papilloma and polyomavirus infections

Some persistent infections are characterized by chronic, low-level replication of virus in tissues that are constantly being regenerated so that damaged cells are eliminated as a matter of course. An excellent example described in more detail in Chapter 16, Part IV, is the persistent growth and differentiation of **keratinized tissue** in a wart caused by a **papillomavirus**. In such infections, virus replication closely correlates with the cell's differentiation state, and the virus can express genes that delay the normal programmed death (**apoptosis**) of such cells in order to lengthen the time available for replication.

The distantly related polyoma viruses, the BK and JC viruses, induce chronic infections of kidney tissue. Such infections are usually asymptomatic and are only characterized by virus shedding in the urine; however, in immunosuppressed individuals, infections of the brain and other organs can be seen. Thus, these persistent viruses have a role in the morbidity of late-stage AIDS and in persons undergoing immunosuppression for organ transplants.

Herpesvirus infections and latency

As detailed in Chapter 17, Part IV, hallmarks of herpesvirus infections are an initial acute infection followed by apparent recovery where viral genomes are maintained in the absence of infectious virus production in specific tissue. Latency is characterized by episodic reactivation (recrudescence) with ensuing (usually) milder symptoms of the original acute infection. Example viruses include herpes simplex virus (HSV), Epstein–Barr virus (EBV), and varicella zoster virus (VZV).

In a latent infection, the viral genome is maintained in a specific cell type and does not actively replicate. HSV maintains latent infections in sensory neurons, whereas EBV maintains itself in B lymphocytes. Latent infections often require the expression of specific virus genes that function to ensure the survival of the viral genome or to mediate the reactivation process.

Reactivation requires active participation of the host. Immunity, which normally shields the body against re-infection, must temporarily decline. Such a decline can be triggered by the host's reaction to physical or psychological stress. HSV reactivation often correlates with a host stressed by fatigue or anxiety. VZV reactivation leads to **shingles**, a very painful recrudescence throughout the sensory nerve net serving as the site of the latent virus. Unlike HSV reactivation, VZV reactivation results in destruction of the nerve ganglia and is associated with a generalized decline in immunity associated with aging.

Effective immunity is vital for controlling and maintaining herpesvirus latency and localizing its sites of replication. Newborns not protected by maternal immunity are subject to profound disseminated HSV infections of their central nervous system (CNS; see the "Viral Infections of Nerve Tissue" section) if they encounter the virus, for instance by infection from a mother carrying a primary acute infection. Disseminated human cytomegalovirus (CMV) infections are a major cause of death in individuals undergoing immunosuppressive treatment to facilitate organ transplants. Furthermore, CMV infections of the eye are a leading cause of blindness in patients with advanced immune deficiencies due to infection by HIV. Also, primary CMV infection of a pregnant woman is a leading cause of neurological abnormalities in developing fetuses.

Other complications arising from persistent infections

Persistent infections caused by some viruses can (rarely) lead to a **neoplasm** (a cancerous growth) due to continual tissue damage resulting in mutation of cellular genes controlling cell division (i.e., **oncogenes** and **tumor suppressor genes**). Examples include infections with slow-transforming retroviruses such as **human T-cell leukemia virus (HTLV),** chronic hepatitis B and hepatitis C virus infections of the liver, certain genital papilloma virus infections, and EBV infections. The latter require the additional action of auxiliary cancer-causing factors (*co-carcinogens*).

Autoimmune diseases such as **multiple sclerosis** (MS) are thought by many investigators to result from an abnormal immune response to viral protein antigens continually present in the body due to a persistent infection. Such persistent infections need not result in the reappearance of infectious virus. For example, infection with measles virus usually leads to rash and recovery, although portions of viral genomes and antigens persist in certain tissues, including neural tissue. The mechanism of this persistence is not fully understood, but it is clear that virus maturation is blocked in such cells that bear viral genomes, and viral antigens are present in reduced amounts on the cell surface. The presence of antigen leads to lifelong immunity to measles, but it can result in immune complications where the host's immune system destroys otherwise healthy neuronal tissue-bearing measles antigens.

The fatal disease of **subacute sclerosing panencephalitis (SSPE),** which is a rare complication in children occurring a few years after a measles infection, is a result of such an autoimmune response. SSPE is a rare outcome of measles infection, but other severe sequelae of measles are common. One of the most frequent is damage to eyesight. The virus replicates in the host and infects surface epithelium, resulting in characteristic rash and lesions in the mouth, on the tongue, and on the eye's conjunctiva. Virus infection of the conjunctiva can clear, but movement of eye muscles in response to light, or in the process of reading, can lead to further infection of eye musculature, leading to permanent damage, which is why individuals infected with measles should be protected from light and kept from using their eyes as much as possible.

Viral and subviral diseases with long incubation periods

Most virus-induced diseases have low or only moderate mortality rates. Obviously, if a virus's mortality rate is too high, infection will kill off all the hosts so rapidly that a potential pool of susceptible individuals is lost. Exceptions to this rule do occur, however. Introduction of viral disease into a virgin population (perhaps due to intrusion into a novel ecosystem) can lead to high mortality. Prime examples are the spread of smallpox in Europe during the Middle Ages, and the destruction of native populations in the Western Hemisphere by the introduction of measles during the era of European expansion. Another exception to the low-mortality rule comes about as a manifestation of infection with a virus that has an unusually long incubation period between the time of infection and the onset of symptoms of disease.

Rabies

Some viral diseases have very high mortality rates despite being well established in a population. With rabies, for example, injection of virus via the saliva of an animal bearing active disease leads to unapparent early infection, followed by a long incubation period of two to eight weeks or more depending on the species and individual. During this time, the infected animal is a walking "time bomb." The symptoms of disease (irritability, frenzy, and salivation) are all important parts of the way the virus is spread among individuals. The very long incubation period allows animals bearing the disease to carry on normal activities, even breed, before the symptoms almost inevitably presaging death appear. A hypothetical viral infection that might lead to these physiological and behavioral changes but that resulted in a quick death could not be spread in such a way.

HIV/AIDS

AIDS, which is characterized by an acute infection by HIV and a latent period in which HIV can be transmitted, followed by severe disease, is an example of a "new" zoonotic viral disease. In humans, virus spread is often the result of behavioral patterns of infected individuals during HIV's long latent period. This pattern of spread makes it unlikely that there is any selective pressure over time toward amelioration of the late severe symptoms of AIDS.

Prion diseases

We have noted in Chapter 1 that while prions are not viruses, many of the principles developed for the study of viral diseases can be applied to study of the pathology of prion-associated diseases. The prion-caused **encephalopathies** are, perhaps, the extreme example of an infectious disease with a long incubation period. Periods ranging from 10 to 30 years between the time of exposure and onset of symptoms have been documented. Prion-induced encephalopathy does not lead to any detectable immune response or inflammation, probably because the prion is a host protein and the CNS is isolated from normal immune surveillance. The course of the disease is marked by a slow, progressive deterioration of brain tissue. Only when this deterioration is significant enough to lead to behavioral changes can the disease be discerned and diagnosed. No treatment or vaccination strategy is available at this time for human prion diseases, but a partly successful vaccine to prevent prion-mediated chronic wasting disease of cervids (deer and elk) has been developed.

SOME VIRAL INFECTIONS TARGETING SPECIFIC ORGAN SYSTEMS

While all the organ systems of the vertebrate host have important or vital functions in the organism's life, several play such critical roles that their disruption leads to serious consequences or death. Among these are the CNS with its influence on all aspects of behavior both innate and learned, the circulatory system, the immune system, and the liver. Virus infections of these systems are often life-threatening to the infected individual, and the tissue damage resulting from infection can lead to permanent illness or death. For example, destruction of CD4+ T cells of the immune system by HIV is the major symptom of AIDS and leads in >90% of untreated cases to death from **opportunistic infections** and **neoplasms**. Other viruses can cause as devastating a disease as HIV, but most viral infections are not as often fatal. A consideration of some CNS and liver virus infections provides some interesting examples of both destructive and limited disease courses.

The different patterns of sequelae following infection of a common target organ are also important demonstrations of several features of virus infection and pathogenesis.

First, specific tissue or cell tropism is a result of highly specific interactions between a given virus and the cell type it infects. Depending on the type of cell infected, the severity of symptoms, and the nature of the damage caused by the infection, different outcomes of infection are evident.

Second, persistent infection is a complex process. It is, in part, the result of virus interacting with and modulating the host's immune system. Often, persistence involves the virus adapting to a continuing association with the target cell itself.

Third, classifying viruses by the diseases they cause is not a particularly useful exercise when trying to understand relationships among viruses.

Fourth, and finally, viruses spread by very different routes can target the same organ. The movement of virus within the host is as important as the initial port of entry for the virus.

Viral infections of nerve tissue

The vertebrate nerve net can be readily divided into peripheral and central portions. The peripheral portion functions to move impulses to and from the brain through connecting circuits in the spinal cord. Viral infections of nerve tissue can be divided into infections of specific groups of neurons: neurons of the spinal cord (**myelitis**), the covering of the brain (**meningitis**), and neurons of the brain and brain stem itself (**encephalitis**).

The brain and CNS have a privileged position in the body and are protected by a physical and physiological barrier from the rest of the body and potentially harmful circulating pathogens. This barrier, often referred to as the **blood–brain barrier**, serves as an effective but incomplete barrier to pathogens. Viruses that migrate through neurons can breach it and traverse synapses between peripheral and central neurons, by physical destruction of tissue due to an active infection, by direct invasion via olfactory neurons (which are not isolated from the CNS), or by other less well-characterized mechanisms. Certainly, invasion of the CNS by pathogens is not all that rare since a specific set of cells in the CNS, the microglial cells, function in manners analogous or identical to macrophages in other tissues.

Many viruses can infect nerve tissue, and while some such infections are dead ends, other viruses specifically target nerve tissue. Viruses that do infect nerve tissue tend to favor one or another portion, and whereas the discrimination is not complete, many viruses, such as **enteroviruses** and genital HSV (HSV-2), tend to be causative agents of meningitis; while others, such as rabies and facial HSV (HSV-1), are almost always associated with encephalitis. Viral or **aseptic meningitis** tends in general to be less life-threatening than the majority of viral infections associated with encephalitis, but all are serious and can lead to debilitating diseases.

While many viral infections of the brain can have grave consequences, such consequences are not always the case. Some viral infections of the CNS have reasonably benign prognoses if proper symptomatic care is provided to the afflicted individual. Viruses that target the brain can be broken into several operational groupings, depending on the nature of brain involvement and whether it and associated tissue are a primary or secondary ("accidental") target.

Examples of viral encephalitis with grave prognosis

Rabies

Once the symptoms of disease become apparent, rabies virus infections are almost always fatal. The virus targets salivary tissue in the head and neck in order to provide itself with an efficient medium for transmission to other animals. Involvement of the CNS and brain is eventually widespread, with ensuing neuronal dysfunction. Prior to this, however, the involvement is only with specific cells that lead to alterations in the afflicted animal's behavior and ability to deal with sensory stimuli. During this period, which is often preceded by a **prodromal** period of altered behavioral patterns, the animal can be induced to an aggressive biting frenzy by loud sounds or by the appearance of other animals. This course is the "furious form" of the disease. This behavioral change is most marked for carnivores such as dogs, cats, and raccoons, but can be observed in other infected animals such as squirrels and porcupines. The behavioral changes obviously have a marked impact on transmission of the virus, as the frenzied animal bite is often the instrument of spread.

Despite its association with frenzy (the name *rabies* is derived from the Sanskrit term for doing violence), not all rabies infections lead to the furious form. There is another form of the disease (often termed "dumb") in which the afflicted animal becomes progressively more torpid and withdrawn, eventually lapsing into a coma and death.

The disease's long incubation period between the time of initial inoculation and final death is a very important factor, both in spread of the virus and in its being able to persist in wild populations, but there is also evidence that some animals can be carriers of the disease for long periods with no obvious, overt symptoms. While there are (extremely) rare examples of apparent recovery from the disease even after symptoms appear, generally one can consider the development of the symptoms of rabies as tantamount to a death sentence.

Herpes encephalitis

Encephalitis induced by HSV infection is the result of a physiological accident of some sort. Normally, HSV's involvement with neurons of the CNS and brain is highly restricted, although viral genomes can be detected at autopsy in brain neurons of humans who have died of other causes.

HSV encephalitis occurs only very rarely, but can be a result of either primary infection or an aberrant reactivation. Exactly what features of viral infection or reactivation lead to encephalitis are unknown, but a lack of effective immunity appears to be a major factor. Certainly, there is a much higher risk of invasive HSV encephalitis in neonates and infants with primary HSV infection prior to full development of their own immune defenses. If diagnosed during early clinical manifestations of disease, HSV encephalitis can be treated effectively with antiviral drugs (see Chapter 8, Part II). But within a short period of time (a few days at most), infection leads to massive necrotic destruction of brain tissue, coma, and death.

Although clinical isolates of HSV are often high in neurovirulence and neuroinvasive indices when they are tested in laboratory animals, there is no evidence that the virus recovered from patients with herpes encephalitis is any more virulent than those isolated from the more common, localized facial or genital infections. Moreover, there has never been any confirmed epidemiological pattern to the occurrence of herpes encephalitis that would suggest a specific strain of virus as a causative factor.

Viral encephalitis with favorable prognosis for recovery

Many of the viruses that cause encephalitis have RNA genomes and are carried by arthropod vectors from zoonoses, and human involvement is often incidental. Such viruses are often termed **arboviruses**, although this is an imprecise classification that includes two groups of viruses not closely related by other criteria.

The symptoms of encephalitis in wild animals can be difficult to measure, but several equine encephalitis viruses are known to cause serious disease in horses. Often the symptoms of viral encephalitis in humans are drowsiness, mild malaise, and sometimes coma. These mosquito-borne encephalitis viruses do not usually directly invade neural tissue itself, but rather infect supporting tissue. The host response to this infection and resulting inflammation lead to the observed neurological symptoms.

Since tissue at the periphery of neural tissue is the primary target for such encephalitis virus infections, the infection can be resolved and complete recovery will ensue, provided that the host's immune defenses work properly. During the disease's symptomatic period, lethargy and malaise of infected individuals make them vulnerable to other environmental hazards, including infection with other pathogens. But provided these risks are avoided by means of proper care, the disease generally resolves.

While humans are often accidental targets for encephalitis viruses, it is not clear that symptoms of the disease in humans have any major role in virus spread. As with all arthropod-borne diseases, transmission is by arthropod ingestion of blood-associated virus found during the viremic stage of animal infection, and the behavioral effects are incidental. Still, it may be that the lethargy manifested during active disease makes infected animals more easily bitten by arthropods, and perhaps this is a factor in natural transmission.

Viral infections of the liver (viral hepatitis)

Diseases of the liver hold a special place in many types of medicine, both because of the important physiological role of this organ and because all circulating blood and lymph pass through the liver frequently. A number of different and unrelated viruses target the liver; these are collectively known as *hepatitis viruses*. All hepatitis viruses cause liver damage that can be devastating to the infected host. Liver failure due to hepatitis virus infections is a major reason for liver transplantation. Furthermore, a number of these viruses establish persistent carrier states in which virus is present for many months or years following infection. Currently, there are five reasonably well-characterized human hepatitis viruses: A, B, C, delta (D), and E. The severity of the disease caused and the sequelae vary with each.

Hepatitis A

This virus is classified as a picornavirus, related to poliovirus. Hepatitis A virus (HAV) is spread by contaminated water or food, and causes a potentially severe but controllable loss of liver function and general malaise. Proper medical care will generally result in full recovery of liver function and full clearance of virus from the host, with effective immunity against reinfection. A relatively effective HAV vaccine is available for individuals at risk of infection, including those who live in, or travel to, endemic regions.

Hepatitis B

Hepatitis B virus is related to but clearly distinct from retroviruses. Unlike the situation with HAV, the B virus is spread mainly through blood, either during sexual activity or during other blood contamination events (sharing of needles, for instance), and primary infection is sometimes followed by persistent viremia and liver damage. Hepatitis B infection is a special risk to medical personnel owing to the possibility of transmission by needle stick from contaminated blood, and the virus is endemic among intravenous drug users, commercial sex workers, and their customers. The disease is also endemic in Southeast Asia, where the virus can be spread from mother to infant by birth trauma.

Hepatitis B virus infection can lead to acute disease with attendant liver failure or can be asymptomatic. In many cases, virus is completely cleared, leading to full or partial recovery of liver function. Unfortunately, 5–10% of infected individuals go on to become asymptomatic chronic carriers of the virus. Indeed, chronic hepatitis B infections are a leading factor in human liver cancer. A third form of the hepatitis B virus infection (**fulminant infection**) is marked by rapid onset of extensive liver damage and often death. An effective vaccine against hepatitis B virus is now widely used to prevent infection.

Hepatitis C

Hepatitis C virus is a member of the Flaviviridae, as are several other important human pathogens including dengue virus, Zika virus, and West Nile virus. The virus is transmitted by contaminated blood and blood products, and it is thought to cause as much as 25% of acute viral hepatitis worldwide. There is no current evidence of its being efficiently spread by arthropod vectors, but this possibility cannot be ruled out. Unlike those infected with HAV, a significant proportion of victims do not mount an effective immune response to the infection and have chronic infection that can last for many years with resulting accumulated liver damage and carcinoma. New anti–hepatitis C virus medicines introduced since 2011 have greatly improved the prognosis for infected people, but no vaccine currently exists.

Hepatitis delta (D)

Hepatitis delta (D) virus is a defective virus that cannot replicate without the aid of another virus, the hepatitis B virus. Rates of hepatitis D and B virus coinfection vary widely throughout the world. Hepatitis D and B virus coinfection in the same individual, however, leads to a higher incidence of chronic liver disease than does infection with hepatitis B virus alone. Hepatitis D virus infection can be prevented by use of the hepatitis B vaccine, since hepatitis D virus replication is dependent on the presence of hepatitis B virus.

Hepatitis E

Like HAV, hepatitis E virus is spread in the developing world by contaminated water and possibly by food. It is found throughout the world and has caused significant epidemics in India and Russia through problems with drinking water. In the developed world, hepatitis E virus is spread from animal reservoirs, primarily domestic pigs. The disease caused by this virus is usually mild, but can have high mortality rates in pregnant women. Recovery from acute infection is generally complete. Chronic infection occurs in immune-compromised individuals.

QUESTIONS FOR CHAPTER 4

1 The disease SSPE is a complication that may follow infection with measles virus. Discuss the possible mechanisms occurring during development of this rare disease.

2 What features of pathogenesis are shared by measles virus, varicella zoster virus, and variola virus?

3 What are some of the unique features of infection by rabies virus?

4 What features distinguish an acute from a persistent infection?

5 Distinguish encephalitis produced by herpesvirus from that resulting from infection with an arbovirus such as La Crosse encephalitis virus.

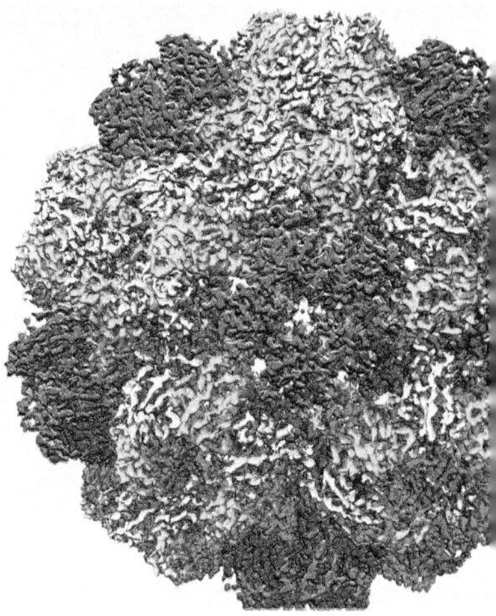

Problems

PART I

1 This part described the various patterns of viral infection that can be observed, among them acute, persistent, and latent. What common features may exist among these three types of infection? What are the distinguishing characteristics of each of these three types of infection?

2 The five hepatitis viruses have the same tissue tropism (the liver), and yet each is in a different virus family. One of them (hepatitis D or the delta agent) is actually a defective virus, sometimes called a subviral entity.

(a) In the table below, indicate the mode of transmission of each of these agents:

Agent	Transmitted by
Hepatitis A virus	
Hepatitis B virus	
Hepatitis C virus	
Hepatitis D (delta) agent	
Hepatitis E virus	

(b) What functions of the liver may allow all of these agents to have a common tissue tropism, despite their differing modes of transmission?

3 As part of a larger project, you have been given five unknown viruses to characterize. Your job is to determine, given the tools at your disposal, the host range and tissue tropism of these unknown viruses. You will be using two kinds of cells: human and mouse. In each case, you have a cell line that grows continuously in culture and is therefore representative of the organism, but not of a particular tissue (human: HeLa cells; mouse: L cells). In addition, you have cells that are derived from and still representative of specific tissues: muscle or neurons. For each virus, you have an assay system that indicates if the virus attaches to ("+") or does not attach to ("–") a particular type of cell. Using the data in the table below, determine, if possible, the host range and tissue tropism of each unknown virus.

	Human			Mouse		
Virus	HeLa	Muscle	Neuron	L	Muscle	Neuron
#1	+	–	–	–	–	–
#2	+	+	–	+	+	–
#3	–	–	–	+	+	+
#4	–	–	–	–	–	–
#5	+	–	+	–	–	–

Here is the report form you will send back with your results. Indicate with a check mark (✓) what your conclusions are for each of the unknown viruses.

		Virus				
		#1	#2	#3	#4	#5
Host range	Human					
	Mouse					
	Both					
	Neither					
Tissue tropism	Muscle					
	Neuron					
	No tropism					
	Cannot be determined from data					

Additional Reading for Part I

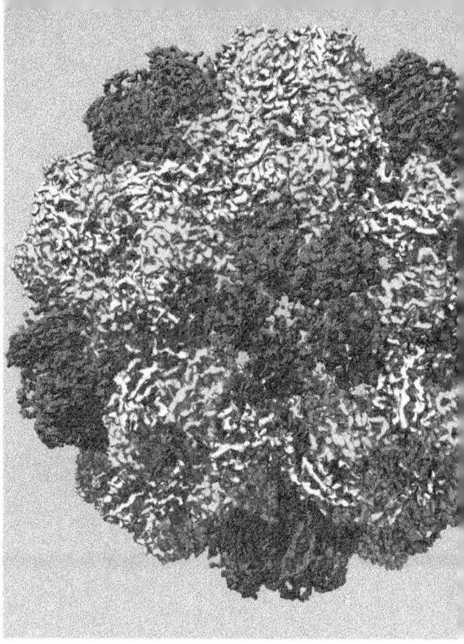

Note: See Resource Center for relevant websites.

Ahmed, R., Morrison, L.A., and Knipe, D.M. (1995). Persistence of viruses. In: Virology, 3e (eds. B.N. Fields and D.M. Knipe), 219–250. New York: Raven Press.

Baer, G.M. and Tordo, N. (1994). Rabies virus. In: Encyclopedia of Virology (eds. R.G. Webster and A. Granoff), 1180–1185. New York: Academic Press.

Barry, J.M. (2004). The Great Influenza: The Epic Story of the Greatest Plague in History. New York: Viking.

Baum, S.G. (2004). Acute viral meningitis and encephalitis. In: Infectious Diseases, 3e (eds. S.L. Gorbach, J.G. Bartlett and N.R. Blacklow), 1286–1291. Philadelphia: W.B. Saunders.

DeFilippis, V.R. and Villarreal, L.P. (1999). An introduction to the evolutionary ecology of viruses. In: Viral Ecology (ed. C.J. Hurst), 125–208. New York: Wiley.

Diamond, J. (1999). Guns, Germs, and Steel: The Fates of Human Societies. New York: W.W. Norton.

Domingo, E., Webster, R.G., and Holland, J.J. (eds.) (1999). Origin and Evolution of Viruses. San Diego, CA: Academic Press.

Fan, H., Conner, R.F., and Villarreal, L.P. (1989). The Biology of AIDS. Boston: Jones and Bartlett.

Fenner, F.J., Gibbs, E.P.J., Murphy, F.A. et al. (1993). Veterinary Virology, 2e, chaps. 4, 6, 7, 8, and 9. San Diego, CA: Academic Press.

Haase, A.T. (1997). Methods in viral pathogenesis: tissues and organs. In: Viral Pathogenesis (ed. N. Nathanson), 465–482. Philadelphia: Lippincott-Raven.

Koff, R.S. (2004). Hepatitis C. In: Infectious Diseases (eds. S.L. Gorbach, J.G. Bartlett and N.R. Blacklow), 2072–2074. Philadelphia: W.B. Saunders.

Koff, R.S. (2004). Hepatitis E. In: Infectious Diseases (eds. S.L. Gorbach, J.G. Bartlett and N.R. Blacklow), 2074–2076. Philadelphia: W.B. Saunders.

Koff, R.S., Hepatitis, B., and hepatitis, D. (1998). Infectious Diseases (eds. S.L. Gorbach, J.G. Bartlett and N.R. Blacklow) chap. 91. Philadelphia: W.B. Saunders.

Kolata, G. (2001). Flu: The Story of the Great Influenza Pandemic. New York: Touchstone.

Lemon, S.M. (1998). Type A viral hepatitis. In: Infectious Diseases (eds. S.L. Gorbach, J.G. Bartlett and N.R. Blacklow) chap. 90. Philadelphia: W.B. Saunders.

McNeill, W. (1996). Patterns of disease emergence in history. In: Emerging Viruses (ed. Morse), 29–36. New York: Oxford University Press.

Morse, S.S. (1996). Examining the origins of emerging viruses. In: Emerging Viruses (ed. S.S. Morse), 10–28. New York: Oxford University Press.

Nathanson, N. (2001). Epidemiology. In: Virology, 4e (eds. B.N. Fields and D.M. Knipe) chap. 14. New York: Raven Press.

Nathanson, N. and Tyler, K.L. (1997). Entry dissemination, shedding, and transmission of viruses. In: Viral Pathogenesis (ed. N. Nathanson) chap. 2. Philadelphia: Lippincott-Raven.

Oldstone, M.B.A. (1998). Viruses, Plagues, and History. New York: Oxford University Press.

Porterfield, J.S. and Htraavik, T. (1994). Encephalitis viruses. In: Encyclopedia of Virology (eds. R.G. Webster and A. Granoff), 362–371. New York: Academic Press.

Preston, R. (1994). The Hot Zone. New York: Random House.

Prusiner, S.B., Telling, G., Cohen, G., and DeArmond, S. (1996). Prion diseases of humans and animals. Seminars in Virology 7: 159–174.

Rotbart, H.A. (1991). Viral meningitis and the aseptic meningitis syndrome. In: Infections of the Central Nervous System (eds. W.M. Scheld, R.J. Whitley and D.T. Durack), 19. New York: Raven Press.

Scheld, W.M., Armstrong, D., and Hughes, J.M. (eds.) (1998). Emerging Infections, vol. 1 and 2. Washington, DC: ASM Press.

Shope, R.E. (1994). Rabies-like viruses. In: Encyclopedia of Virology (eds. R.G. Webster and A. Granoff). New York: Academic Press.

Shope, R.E. and Evans, A.S. (1996). Assessing geographic and transport factors and recognition of new viruses. In: Emerging Viruses (ed. S.S. Morse), 109–119. New York: Oxford University Press.

Smith, A.L. and Barthold, S.W. (1997). Methods in viral pathogenesis: animals. In: Viral Pathogenesis (ed. N. Nathanson), 483–506. Philadelphia: Lippincott-Raven.

Villarreal, L.P. (1997). On viruses, sex, and motherhood. Journal of Virology 71: 859–865.

Villarreal, L.P. (2004). Viruses and the Evolution of Life. Washington, DC: ASM Press.

Woese, C.R., Kandler, O., and Wheelis, M.L. (1990). Towards a natural system of organisms: proposal for the domains Archaea, Bacteria, and Eucarya. Proceedings of the National Academy of Sciences 87: 4576–4579.

Basic Properties of Viruses and Virus–Cell Interaction

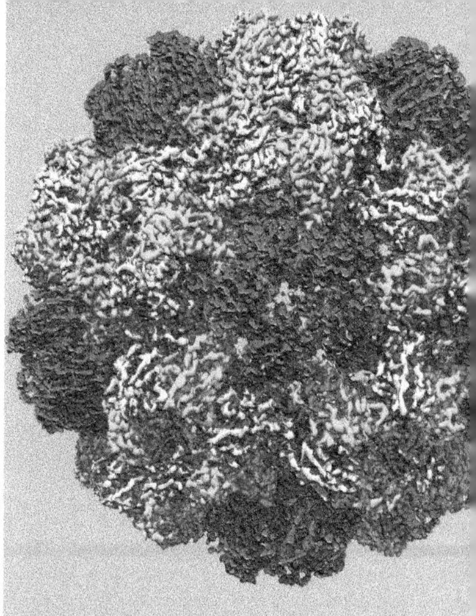

PART II

- Virus Structure and Classification
  - The Features of a Virus
  - Classification Schemes
  - The Virosphere
  - The Human Virome
- The Beginning and End of the Virus Replication Cycle
  - Outline of the Virus Replication Cycle
  - Viral Entry
  - Late Events in Viral Infection: Capsid Assembly and Virion Release
- The Innate Immune Response: Early Defense Against Pathogens
  - Host Cell-Based Defenses Against Virus Replication
  - The Adaptive Immune Response and the Lymphatic System
  - Control and Dysfunction of Immunity
  - Measurement of the Immune Reaction
- Strategies to Protect Against and Combat Viral Infection
  - Vaccination – Induction of Immunity to Prevent Virus Infection
  - Eukaryotic Cell-Based Defenses Against Virus Replication
  - Antiviral Drugs
  - Bacterial Antiviral Systems – Restriction Endonucleases
- Problems for Part II
- Additional Reading for Part II

Virus Structure and Classification

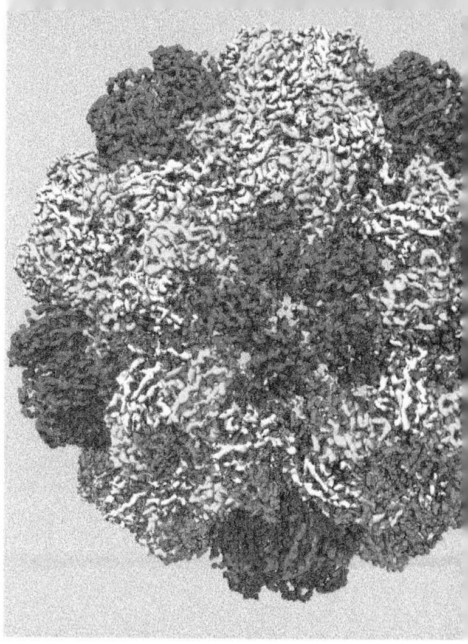

CHAPTER

5

- THE FEATURES OF A VIRUS
- Viral genomes
- Viral capsids
- Viral envelopes
- CLASSIFICATION SCHEMES
- The Baltimore scheme of virus classification
- Disease-based classification schemes for viruses
- THE VIROSPHERE
- THE HUMAN VIROME
- QUESTIONS FOR CHAPTER 5

THE FEATURES OF A VIRUS

Viruses are small compared to the wavelength of visible light; indeed, while the largest virus can be discerned in a good light microscope, the vast majority of viruses can only be visualized in detail using an electron microscope. A size scale with some important landmarks is shown in Figure 5.1.

Virus particles are composed of a nucleic acid **genome** or core, which is the genetic material of the virus, surrounded by a **capsid** made up of virus-encoded proteins. Viral genetic material encodes the **structural proteins** of the capsid and other viral proteins essential for other functions in initiating virus replication. The entire structure of the virus (the genome, the capsid, and – where present – the envelope) makes up the **virion** or virus particle. The exterior of this virion contains proteins that interact with specific proteins on the surface of the cell in which the virus replicates. The schematic structures of some well-characterized viruses are shown in Figure 5.2.

To date, more than 5000 different genotypes of viruses have been identified, and it is estimated that there may be as many as 10^6 in a kilogram of marine sediment. The National Center for Biological Information (NCBI) database contains more than 8000 complete viral genomes as February 2019. Although perhaps not as overwhelming, the number of different types of viruses associated with terrestrial plants and animals is very high, and, of course,

Basic Virology, Fourth Edition. Martinez "Marty" Hewlett, David Camerini, and David C. Bloom.

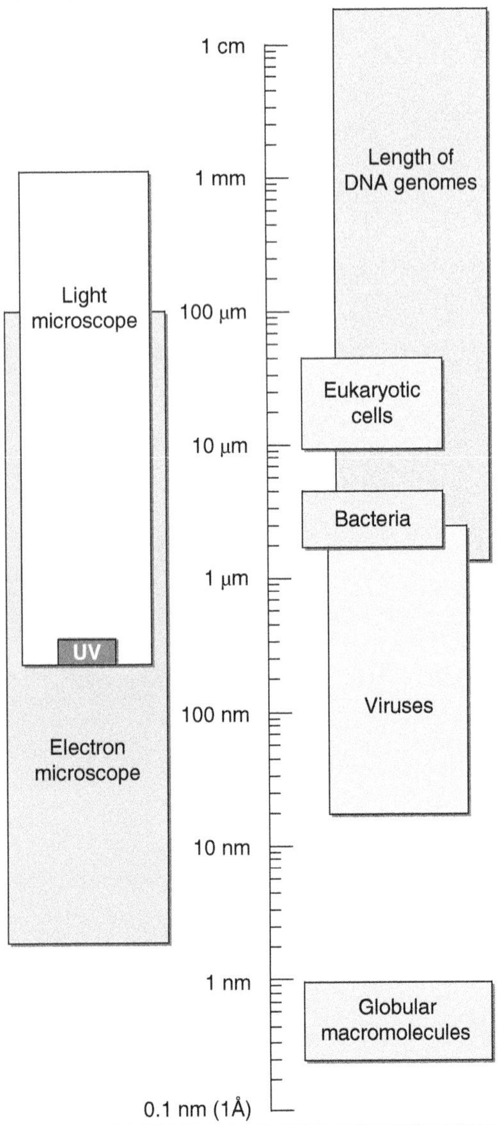

Figure 5.1 A scale of dimensions for biologists. The wavelength of a photon or other subatomic particle is a measure of its energy and its resolving power. An object with dimensions smaller than the wavelength of a photon cannot interact with it, and thus is invisible to it. The dimensions of some important biological features of the natural world are shown. Note that the wavelength of ultraviolet (UV) light is between 400 and 280 nm; objects smaller than that, such as viruses and macromolecules, cannot be seen in visible or UV light. The electron microscope can accelerate electrons to high energies; the resulting short wavelengths can resolve viruses and biological molecules. Note that the length of DNA is a measure of its information content, but since DNA is essentially "one-dimensional," it cannot be resolved by light.

bacteria and protists all have their own populations of associated viruses. Further, there are a very large number of subviral entities, which depend on viruses themselves for replication – these are subviral infectious agents and plant satellite nucleic acid elements that share at least some features of their replication strategies with viruses. And, finally, as we have noted briefly in Part I, there are infectious proteins (prions), which also can be studied using the techniques of virology.

The development of self-consistent classification schemes for this plethora of entities is a major challenge for virologists. Good classification schemes have a major role in helping organize the growing flood of detailed genetic and molecular information concerning viruses and their genes. Further, a valid classification scheme provides an important framework for understanding the different ways that viruses can utilize cellular and their own genes in maintaining themselves within the biosphere. Finally, valid classifications provide useful guides to our understanding of the origins of various virus groups, and the relationships between viruses in the same group and divergent groups.

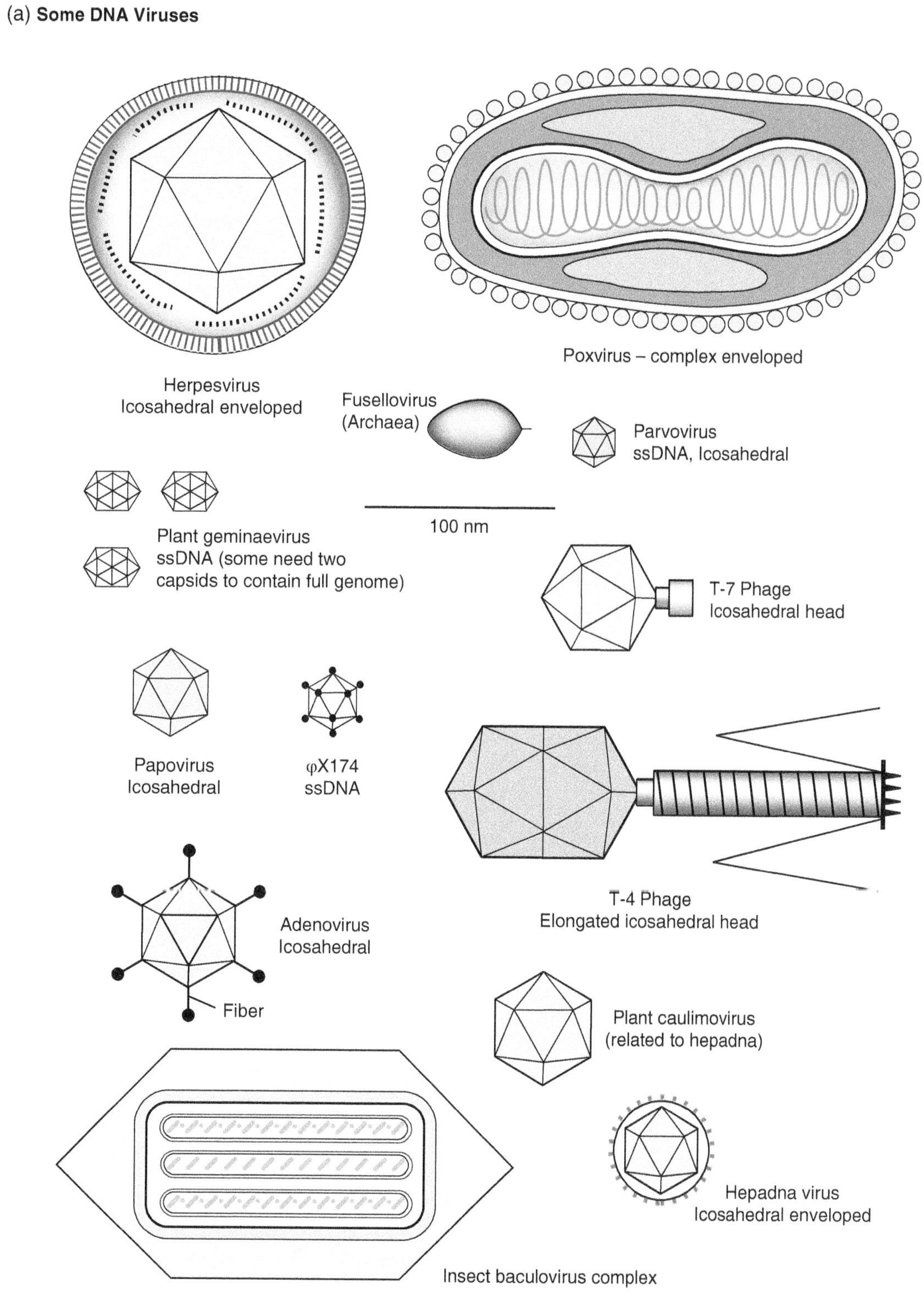

Figure 5.2 The structure and relative sizes of a number of (a) DNA and (b) RNA viruses. The largest viruses shown have dimensions approaching 300–400 nm and can be just resolved as refractile points in a high-quality ultraviolet-light microscope. The smallest dimensions of viruses shown here are on the order of 25 nm. Classifications of viruses based on the type of nucleic acid serving as the genome and the shape of the capsid are described in the text. ss: Single stranded; ds: double stranded.

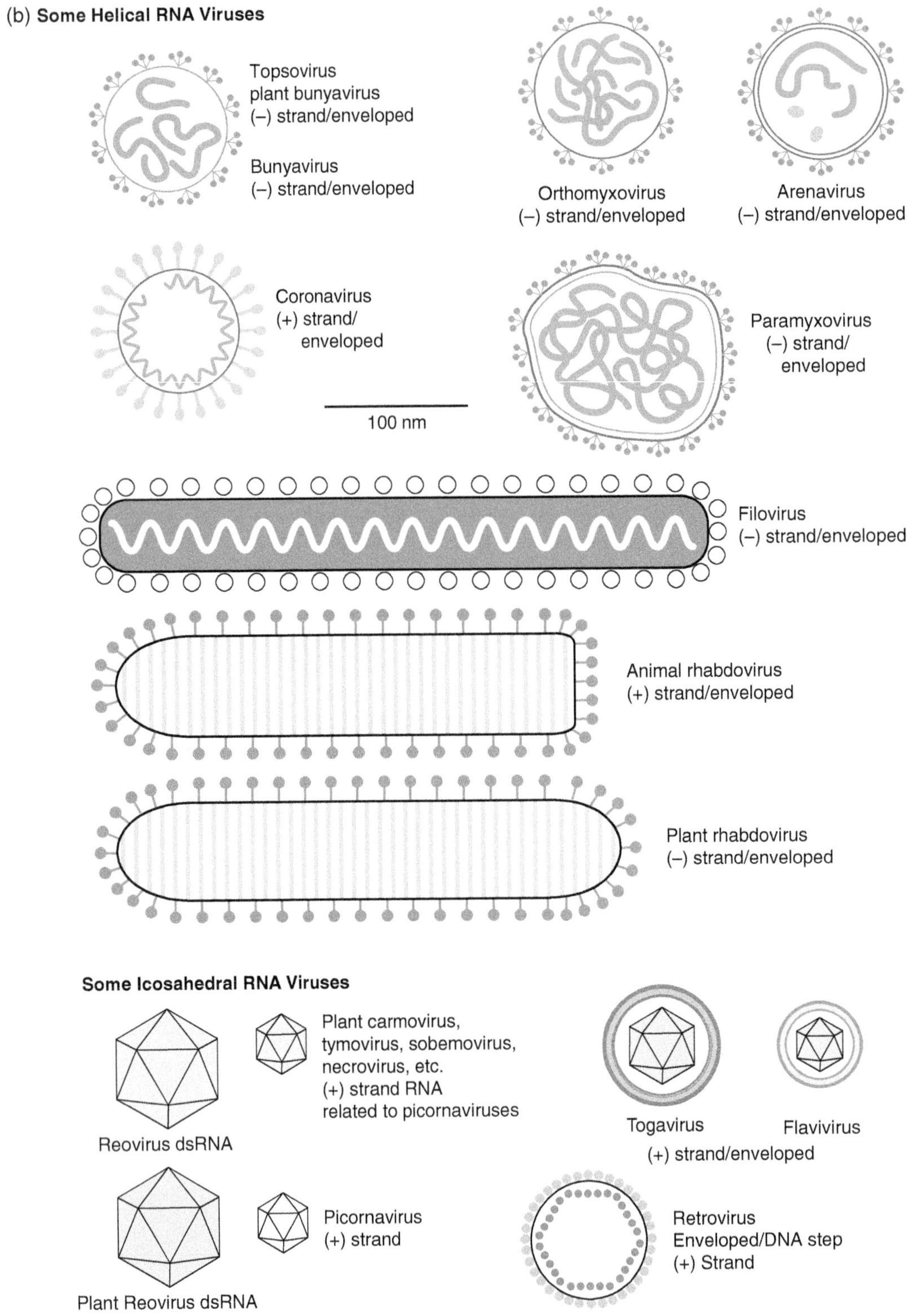

Figure 5.2 *Continued*

The International Committee on Taxonomy of Viruses (ICTV) was created at an international conference on microbiology in Moscow in 1966 in order to develop a single, universal taxonomic scheme for all the viruses infecting animals (vertebrates, invertebrates, and protozoa), plants (higher plants and algae), fungi, bacteria, and archaea. Its membership is made up of distinguished virologists throughout the world, and it has issued periodic reports describing its progress, and its problems, as well as databases containing the properties of viruses and appropriate computer-based tools for utilizing such databases. One of the notable achievements of this group and the community of virologists that it represents is the recognition of a limited number of viral features that can be used for classification; these include the nature of the viral genome, the presence of an envelope, and the morphology of the virus particle. The classification scheme uses the designation of "family," even though these phylogenetic terms do not strictly apply in the case of viruses. Table 5.1 lists the families of this scheme, in alphabetical order, as of their 2018 report.

Table 5.1 Classification of viruses according to the ICTV.

Family	Nature of the genome	Presence of an envelope	Morphology	Genome configuration	Genome size (kbp or kb)	Host
Abyssoviridae	ssRNA			+		
Ackermannviridae	dsDNA	–	Icosahedral tailed	1 linear	155	Bacteria
Adenoviridae	dsDNA	–	Isometric	1 linear	28–45	Vertebrates
Alloherpesviridae	dsDNA	+	Isometric	1 linear	134–248	Vertebrates
Alphaflexiviridae	ssRNA	–	Filamentous	1 + segment	7–9	Plants
Alphasatellitidae	ssDNA	–	N/A	1 linear	1.3–1.4	Plants
Alphatetraviridae	ssRNA	–	Isometric	1 linear	6.5	Invertebrates
Alvernaviridae	ssRNA	–	Isometric	1 linear	4.4	Dinoflagellates
Amalgaviridae	dsRNA	–	Isometric	1 linear	3.5	Plants
Ampullaviridae	dsDNA	+	Bottle-shaped	1 linear	23.8	Archaea
Anelloviridae	ssDNA	–	Isometric	1 linear	3–4	Vertebrates
Arenaviridae	NssRNA	+	Spherical	2 ± segments	11	Vertebrates
Arteriviridae	ssRNA	+	Isometric	1 + segment	13–16	Vertebrates
Artoviridae	ssRNA	+	Spherical	1 – linear	12.3	Invertebrates and vertebrates
Ascoviridae	dsDNA	+	Reniform	1 linear	100–180	Invertebrates
Asfarviridae	dsDNA	+	Spherical	1 circular	170–190	Vertebrates
Aspiviridae	ssRNA	–	Filamentous	3-4 - linear	11.3–12.5	Plants
Astroviridae	ssRNA	–	Isometric	1 + segment	7–8	Vertebrates
Avsunviroidae	ssRNA	N/A	N/A	1 + circular	0.25	Plant viroids
Bacilladnaviridae	ssDNA	–	Isometric	1 circular	5–6	Diatoms
Baculoviridae	dsDNA	+	Bacilliform	1 circular supercoiled	80–180	Invertebrates
Barnaviridae	ssRNA	–	Bacilliform	1 + segment	4	Fungi
Belpaoviridae	ssRNA (RT)	N/A	N/A	1 linear	?	Invertebrates
Benyviridae	ssRNA	–	Rod-shaped	4/5 + segments	14–16	Plants

Table 5.1 *Continued*

Family	Nature of the genome	Presence of an envelope	Morphology	Genome configuration	Genome size (kbp or kb)	Host
Betaflexiviridae	ssRNA	–	Filamentous	1 linear	6.5–9	Plants, fungi
Bicaudaviridae	dsDNA	+	Lemon-shaped	1 circular	62	Archaea
Bidnaviridae	ssDNA	–	Isometric	2 linear	6 and 6.5	Invertebrates
Birnaviridae	dsRNA	–	Isometric	2 segments	6	Vertebrates and invertebrates
Bornaviridae	NssRNA	+	Spherical	1 – segment	6	Vertebrates
Bromoviridae	ssRNA	–	Isometric	3 + segments	8–9	Plants
Caliciviridae	ssRNA	–	Isometric	1 + segment	7–8	Vertebrates
Carmotetraviridae	ssRNA	–	Isometric	1 linear	6.1	Invertebrates
Caulimoviridae	dsDNA-RT	–	Isometric, bacilliform	1 circular	8	Plants
Chrysoviridae	dsRNA	–	Isometric	4 linear	2.9–3.6	Fungi
Chuviridae	ssRNA	?	?	? (negative sense)	?	Invertebrates
Circoviridae	ssDNA	–	Isometric	1 circular	2	Vertebrates
Clavaviridae	dsDNA		Bacilliform	1 circular	5.3	Archaea
Closteroviridae	ssRNA	–	Filamentous	1/2 + segments	15–19	Plants
Coronaviridae	ssRNA	+	Isometric	1 + segment	27–31	Vertebrates
Corticoviridae	dsDNA	–	Isometric	1 circular supercoiled	9	Bacteria
Cruliviridae	ssRNA	?	?	? (negative sense)	?	Invertebrates
Cystoviridae	dsRNA	+	Spherical	3 segments	13	Bacteria
Deltaflexiviridae	ssRNA	?	?	1 + sense	6–8	Fungi, plants
Dicistroviridase	ssRNA	–	Isometric	1 linear	8.5–10.2	Invertebrates
Endornaviridae	dsRNA	N/A	No true capsid	1 linear	14	Plants
Euroniviridae	ssRNA	?	?	?	?	?
Filoviridae	NssRNA	+	Bacilliform	1 – segment	19	Vertebrates
Fimoviridae	ssRNA	+	Spherical	4 – sense segments	12	Plants
Flaviviridae	ssRNA	+	Isometric	1 + segment	10–12	Vertebrates
Fuselloviridae	dsDNA	+	Lemon-shaped	1 circular supercoiled	15	Archaea
Gammaflexiviridae	ssRNA	–	Filamentous	1 linear	6.8	Plants
Geminiviridae	ssDNA	–	Isometric	1 or 2 circular	3–6	Plants
Genomoviridae	ssDNA	–	Isometric	1 circular +/–	2.17	Mammals, birds, fungi
Globuloviridae	dsDNA	+	Spherical	1 circular	20–30	Archaea
Guttaviridae	dsDNA	+	Ovoid	1 circular	20	Archaea
Hantaviridae	ssRNA	+	Spherical	3 linear negative sense	11–20	Humans, rodents
Hepadnaviridae	dsDNA-RT	+	Spherical	1 circular	3	Vertebrates

Table 5.1 *Continued*

Family	Nature of the genome	Presence of an envelope	Morphology	Genome configuration	Genome size (kbp or kb)	Host
Hepeviridae	ssRNA	−	Isometric	1 linear	7.2	Vertebrates
Herpesviridae	dsDNA	+	Isometric	1 linear	125–240	Vertebrates
Hypoviridae	dsRNA	−	Pleomorphic	1 segment	12	Fungi
Hytrosaviridae	dsDNA	+	Filamentous	1 circular	120–190	Insects
Iflaviridae	ssRNA	−	Isometric	1 linear	8.8–9.7	Invertebrates
Inoviridae	ssDNA	−	Filamentous	1 + circular	7–9	Bacteria, mycoplasmas
Iridoviridae	dsDNA	−	Isometric	1 linear	140–383	Vertebrates and invertebrates
Lavidaviridae	dsDNA	−	Isometric	1 circular	17–30	Protists
Leviviridae	ssRNA	−	Isometric	1 + segment	3–4	Bacteria
Lipothrixviridae	dsDNA	+	Rod-shaped	1 linear	16	Archaea
Lispiviridae	ssRNA	+ (?)	Spherical (?)	1, linear	12	Arachnids
Luteoviridae	ssRNA	−	Isometric	1 + segment	6	Plants
Malacoherpesviridae	dsDNA	+	Spherical	1 linear	150	Mollusks
Marnaviridae	ssRNA	−	Isometric	1 linear	8.6	Seaweed
Marseilleviridae	dsDNA	−	Isometric	1 circular	368	Amoeba
Medioniviridae	ssRNA	?	?	?	?	Tunicates
Megabirnaviridae	dsRNA	−	Isometric	Linear, segmented	7	Fungi
Mesoniviridae	ssRNA	+	Spherical	1 linear	20	Vertebrates
Metaviridae	ssRNA	−	RT-spherical	1 + segment	4–10	Fungi, plants, invertebrates
Microviridae	ssDNA	−	Isometric	1 + circular	4–6	Bacteria, spiroplasmas
Mimiviridae	dsDNA	−	Isometric	1 linear	1200	Amoeba
Mymonaviridae	ssRNA	+	Filamentous	1 linear	10	Fungi
Myoviridae	dsDNA	−	Tailed phage	1 linear	39–169	Bacteria, archaea
Nairoviridae	ssRNA	+	Spherical	3 linear negative sense	11–20	Vertebrates, arthropods
Nanoviridae	ssDNA	−	Isometric	6–9 circular	6–9	Plants
Narnaviridae	ssRNA	−	RNP complex	1 + segment	2–3	Fungi
Nimaviridae	dsDNA	+	Ovoid	1 circular	293	Crustaceans
Nodaviridae	ssRNA	−	Isometric	2 + segments	4–5	Vertebrates and invertebrates
Nudiviridae	dsDNA	+	Rod-shaped	1 circular	96–231	Invertebrates
Nyamiviridae	ssRNA	+	Spherical	1 linear	11.6	Birds, invertebrates
Orthomyxoviridae	NssRNA	+	Pleomorphic	6–8 − segments	10–15	Vertebrates
Papillomaviridae	dsDNA	−	Isometric	1 circular	7–8	Vertebrates
Paramyxoviridae	NssRNA	+	Pleomorphic	1 − segment	15	Vertebrates
Partitiviridae	dsRNA	−	Isometric	2 segments	4–6	Plants, fungi

Table 5.1 *Continued*

Family	Nature of the genome	Presence of an envelope	Morphology	Genome configuration	Genome size (kbp or kb)	Host
Parvoviridae	ssDNA	–	Isometric	1 +/– circular	4–6	Vertebrates and invertebrates
Peribunyaviridae	ssRNA	+	Spherical	3, linear, negative sense	11–20	Humans, rodents, arthropods
Permutotetraviridae	ssRNA	–	Isometric	1 linear	5.6	Invertebrates
Phasmaviridae	ssRNA	+	Spherical	3, linear, negative sense	10.3	Insects
Phenuiviridae	ssRNA	+	Spherical	3, linear, negative sense	11–20	Ruminants, camels, humans, mosquitos
Phycodnaviridae	dsDNA	–	Isometric	1 linear	160–380	Algae
Picobirnaviridae	dsRNA	–	Isometric	2 linear segments	2.5 + 1.7	Mammals
Picornaviridae	ssRNA	–	Isometric	1 + segment	7–8	Vertebrates
Plasmaviridae	dsDNA	+	Pleomorphic	1 circular	12	Mycoplasmas
Pleolipoviridae	ssDNA or dsDNA	+	Pleomorphic	1 circular or linear	7–16	Archaea
Pneumoviridae	ssRNA	+	Spherical	1 linear	15	Vertebrates
Podoviridae	dsDNA	–	Tailed phage	1 linear	19–44	Bacteria
Polycipiviridae	ssRNA	–	Isometric	1 linear + sense	11	Insects
Polydnaviridae	dsDNA	+	Rod, fusiform	Multiple supercoiled	150–250	Invertebrates
Polyomaviridae	dsDNA	–	Isometric	1 circular	5	Vertebrates
Portogloboviridae	dsDNA	–	Isometric	1 circular	20	Sulfolobus S38A archaea
Pospiviroidae	ssRNA	N/A	N/A	1 circular	0.24–0.4	Plants
Potyviridae	ssRNA	–	Filamentous	1/2 + segments	8–12	Plants
Poxviridae	dsDNA	+	Pleomorphic	1 linear	130–375	Vertebrates and invertebrates
Pseudoviridae	ssRNA	–	RT-spherical	1 + segment	5–8	Fungi, plants, invertebrates
Qinviridae	ssRNA	?	?	?	?	?
Quadriviridae	dsRNA	+	Spherical	Linear, 4 segments	16.8	Fungi
Reoviridae	dsRNA	–	Isometric	10–12 segments	19–32	Vertebrates and invertebrates, plants
Retroviridae	ssRNA	–	RT + spherical	1 dimer + segment	7–12	Vertebrates
Rhabdoviridae	NssRNA	+	Bullet-shaped	1 – segment	11–15	Vertebrates, plants
Roniviridae	ssRNA	+	Bacilliform	1 linear	26	Crustaceans
Rudiviridae	dsDNA	+	Rod-shaped	1 linear	33–36	Archaea
Sarthroviridae	ssRNA	–	Isometric	1 linear	0.9	Crustaceans

Table 5.1 *Continued*

Family	Nature of the genome	Presence of an envelope	Morphology	Genome configuration	Genome size (kbp or kb)	Host
Secoviridae	ssRNA	–	Isometric	Linear/ segmented	24	Plants
Siphoviridae	dsDNA	–	Tailed phage	1 linear	22–121	Bacteria, archaea
Smacoviridae	ssDNA	–	Isometric	1 circular	2.3–2.8	Vertebrates (?)
Sphaerolipoviridae	dsDNA	–	Isometric	1 circular	16–19	Bacteria, archaea
Spiraviridae	ssDNA	–	Cylindrical	1 circular	25	Archaea
Sunviridae	ssRNA	?	?	?	17	Vertebrates
Tectiviridae	dsDNA	–	Isometric	1 linear	15	Bacteria
Tobaniviridae	ssRNA	+	Spherical	1 + segment	28	Vertebrates
Togaviridae	ssRNA	+	Isometric	1 + segment	10–12	Vertebrates
Tolecusatellitidae	ssDNA	N/A	N/A	1	0.7–1.35	Plants
Tombusviridae	ssRNA	–	Isometric	1/2 + segments	4–5	Plants
Totiviridae	dsRNA	–	Isometric	1 segment	5–7	Fungi, protozoa
Tristromaviridae	dsDNA	+	Rod-shaped	1 linear	15.9	Archaea
Turriviridae	dsDNA	–	Isometric	1 circular	17.6	Archaea
Tymoviridae	ssRNA	–	Isometric	1 linear	6.5–7	Plants
Virgaviridae	ssRNA	–	Rod-shaped	Linear, segmented, or nonsegmented	3.3–6.5	Plants
Wupedeviridae	ssRNA	?	?	?	?	Insects
Xinmoviridae	ssRNA	?	?	1 linear, – sense	12	Mosquitoes
Yueviridae	ssRNA	?	?	– sense	?	?

+ sense: Positive-sense; – sense: negative-sense; dsRNA: double-stranded RNA; N/A: not applicable; NssRNA: negative-sense single-stranded RNA; RNP: ribonucleoprotein; RT: reverse transcriptase; ssRNA: single-stranded RNA.

Viral genomes

The nucleic acid core can be DNA for some types of viruses, RNA for others. This genetic material may be single or double stranded and may be linear or circular, but it is always the same for any given type of virus. The type of genetic material (i.e., whether DNA or RNA) is an important factor in the classification of any given virus into groups. Thus, although all free-living cells utilize only double-stranded DNA (dsDNA) as genetic material, some viruses can utilize other types of nucleic acid.

Viruses that contain DNA as genetic material and utilize the infected cell's nucleus as the site of genome replication share many common patterns of gene expression and genome replication along with similar processes occurring in the host cell.

The viruses that use RNA as their genetic material have devised some way to replicate such material, since the cell does not have machinery for RNA-directed RNA replication. The replication of RNA viruses requires expression of specific enzymes that are not present in the uninfected host cell.

Although virus genes encode the proteins required for replication of the viral genome and these proteins have similarities to cellular proteins with roughly analogous functions, viral and cellular proteins are not identical. Viral replication proteins are enzymes involved both in nucleic acid replication and in the expression and regulation of viral genetic information. Viruses also encode enzymes and proteins involved in modifying the cell in which the virus replicates, in order to optimize the cell for virus replication.

Viral capsids

The capsid is a complex structure made up of many identical subunits of viral protein – often termed a **capsomer**. The capsid functions to provide a protein shell in which the chemically labile viral genome can be maintained in a stable environment. The association of capsids with genomes is a complex process, but it must result in an energetically stable structure. While viruses can assume a range of shapes, some quite complex, given the dimensions of virus structure and the constraints of the capsomer's structural parameters, a very large number assume one of two regular shapes. The first is the **helix**, in which the capsomers associate with helical nucleic acid as a **nucleoprotein** – these can be either stiff or flexible depending upon the properties of the capsid proteins themselves. The other highly regular shape is the **icosahedron**, in which the capsomers form a regular solid structure enfolding the viral genome. Despite the frequency of such regular shapes, many viruses have more complex and/or less regular appearances; these include spindle, kidney, lemon, and lozenge shapes. Further, some viruses can assume different shapes depending upon the nature of the cells in which they mature, and some groups of viruses – notably the poxviruses – are distinguished by having a number of different shapes characterizing specific members of the group. Arrangement of the capsid around its viral genetic material is unique for each type of virus. The general properties of this arrangement define the shape of the capsid and its **symmetry**, and since each type of virus has a unique shape and structural arrangement based upon the precise nature of the capsids proteins and how they interact, capsid shape is a fundamental criterion in the classification of viruses.

The technique of **x-ray crystallography** has been applied fruitfully to the study of capsid structures of some smaller icosahedral viruses, and structural solutions for human rhinovirus, poliovirus, foot and mouth disease virus, and canine parvovirus are available. In addition, the structures of a number of plant viruses have been determined. Since the method requires the ability to crystallize the subject material, it is not certain that it can be directly applied to larger, more complex viruses. Still, the structures of specific protein components of some viruses – such as the membrane-associated hemagglutinin of influenza virus – have been determined.

The x-ray crystallographic structure of *Desmodium* yellow mottle virus – a pathogen of beans – is shown in Figure 5.3, to illustrate the basic features of icosahedral symmetry. The icosahedral shell has a shape similar to a soccer ball, and the 12 vertices of this regular solid are arranged in a relatively simple pattern at centers of fivefold axes of symmetry. Each edge of the solid contains a twofold axis of symmetry, and the center of each of the 20 faces of the solid defines a threefold axis of symmetry. While a solid icosahedron can be visualized as composed of folded sheets, the virion structure is made up of repeating protein capsomers that are arrayed to fit the symmetry's requirements. It is important to see that the peptide chains themselves have their own distinct morphology, and it is their arrangement that makes up individual capsomers. The overall capsid structure reflects the next level of structure. Morphology of the individual capsomers can be ignored without altering the basic pattern of their arrangement. Further detail is shown in Figure 5.4, where the assembly of the single capsomer protein into two subunits of the capsid, a **penton** or a **hexon**, is shown.

Twelve pentons and 20 hexons assemble to form the capsid itself. The core of the capsid is filled with the viral genome, in this case RNA. This RNA is also arranged very precisely, with the bulk forming helical stretches and the regions coming in close contact with the inner surface of the capsid shell, forming open loops.

Viral envelopes

A naked capsid defines the outer extent of bacterial, plant, and many animal viruses, but other types of viruses have a more complex structure in which the capsid is surrounded by a lipid **envelope**. This envelope is made up of a lipid bilayer that is derived from the cell in which the

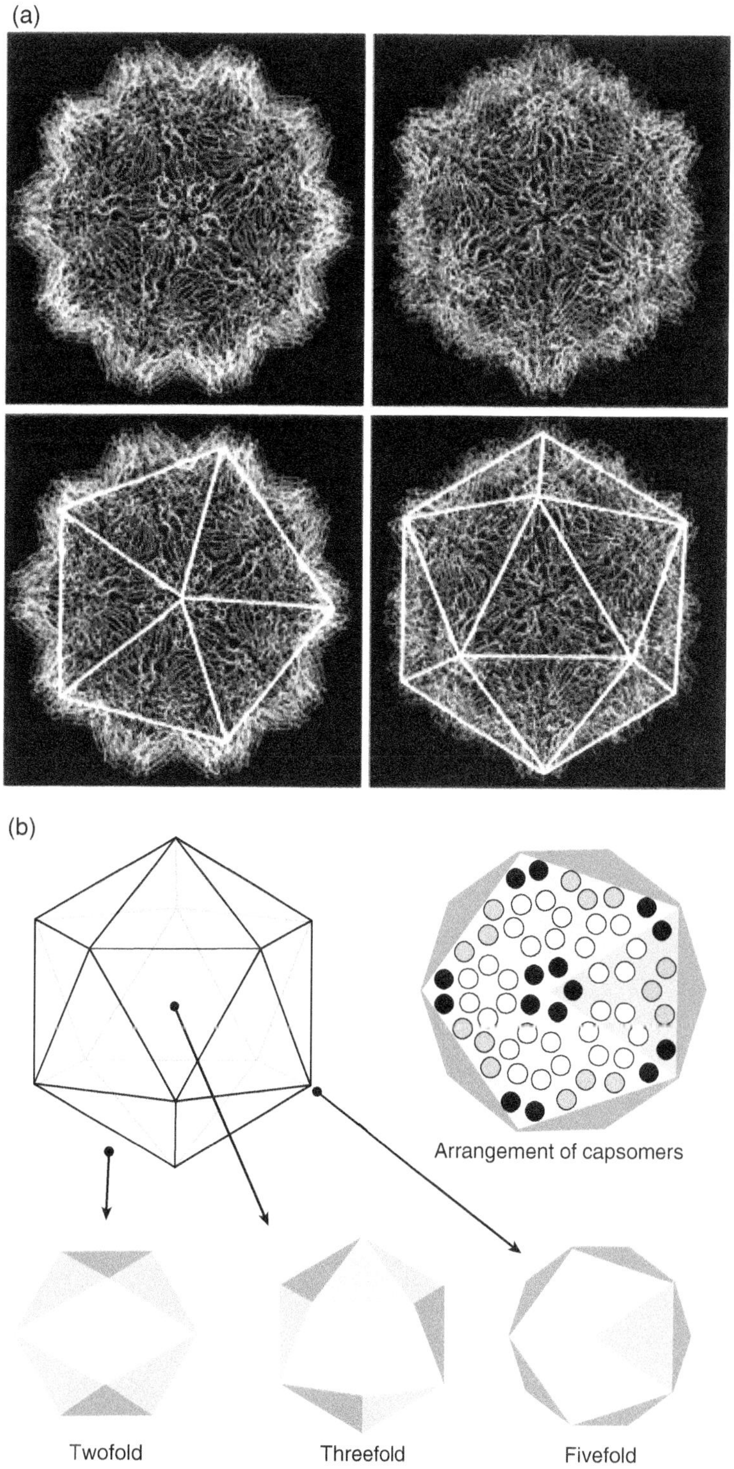

Figure 5.3 Crystallographic structure of a simple icosahedral virus. (a) The structure of *Desmodium* yellow mottle virus as determined by x-ray crystallography to 2.7-Å resolution. This virus is a member of the tymovirus group and consists of a single positive-strand RNA genome about 6300 nucleotides long. The virion is 25–30 nm in diameter and is made up of 180 copies of a single capsid protein that self-associates in two basic ways: in groups of 5 to form the 12 pentons, and in groups of six to form the 20 hexamers. Two views are shown: Panels at left are looking down at a fivefold axis of symmetry, and the right-hand panels look at the threefold and twofold axes. Note that the individual capsomers arrange themselves in groups of five at vertices of the icosahedra, and in groups of six on the icosahedral faces. Since there are 12 vertices and 20 faces, this yields the 180 capsomers that make up the structure. The axes are outlined in the lower panels. Source: Courtesy of S. Larson and A. McPherson, University of California, Irvine. (b) Schematic diagram of the vertices and faces of a regular icosahedron showing the axes of symmetry. Arrangements of the capsomers described in (a) are also shown.

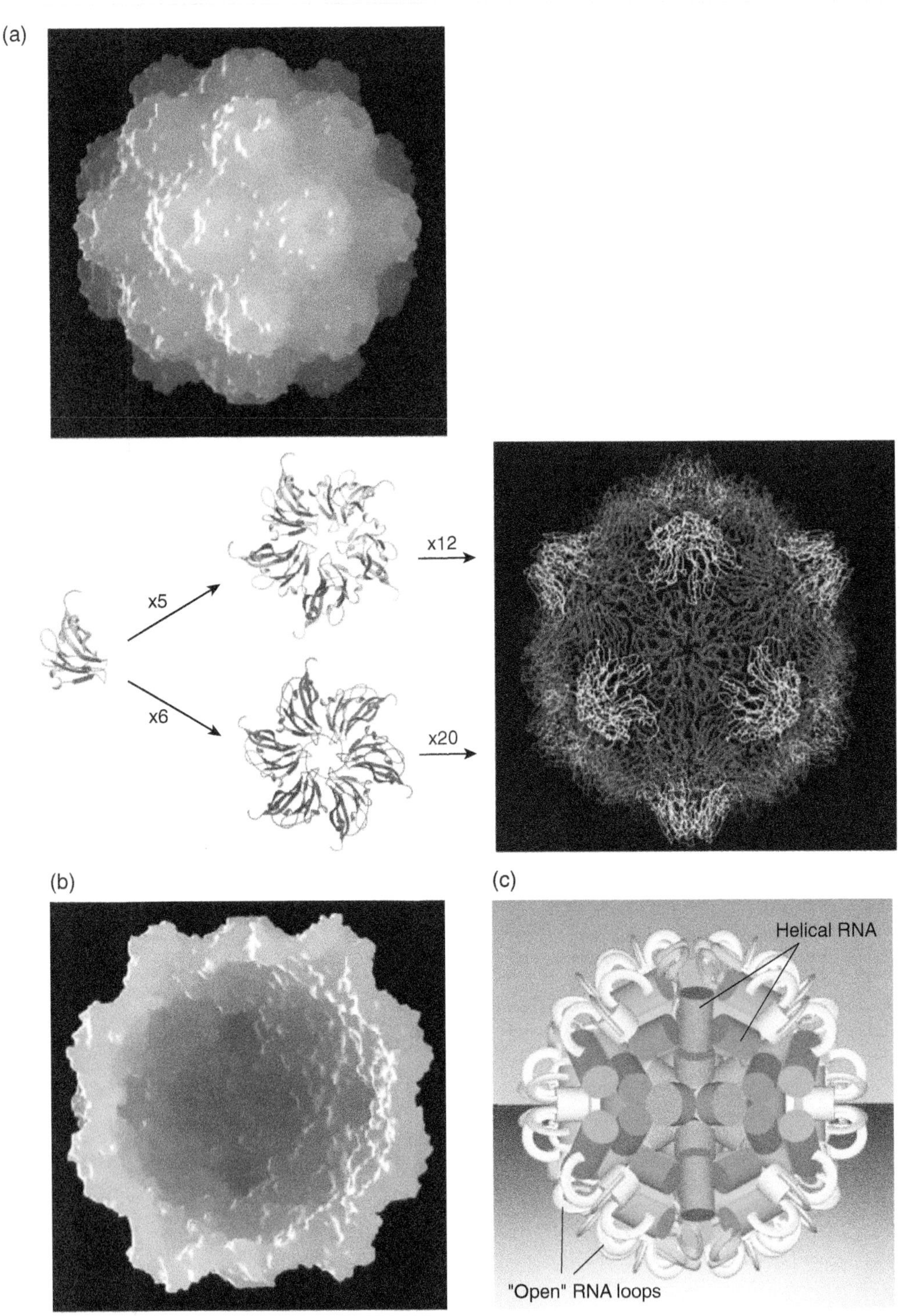

Figure 5.4 The structure of a simple icosahedral virus. (a) A space-filling model of the capsid of *Desmodium* yellow mottle virus as determined by x-ray crystallography to 2.7-Å resolution. (b) The assembly of the single capsid protein into 12 pentons and 20 hexons to form the capsid. Source: Courtesy of S. Larson and A. McPherson, University of California, Irvine. (c) The structure of the RNA genome inside the capsid as determined by x-ray crystallography.

virus replicates and from virus-encoded membrane-associated proteins. The presence or absence of a lipid envelope (described as enveloped or naked, respectively) is another important defining property of different groups of animal viruses.

The shape of a given type of virus is determined by the shape of the virus capsid and really does not depend on whether or not the virus is enveloped. This is because for most viruses, the lipid envelope is **amorphous** and deforms readily upon preparation for visualization using the electron microscope.

CLASSIFICATION SCHEMES

As we have noted, since it is not clear that all viruses have a common origin, a true Linnaean classification is not possible, but a logical classification is invaluable for understanding the detailed properties of individual viruses and how to generalize them. Schemes dependent on basic properties of the virus, as well as specific features of their replication cycle, afford a useful set of parameters for keeping track of the many different types of viruses. A good strategy for remembering the basics of virus classification is to keep track of the following:

1 What kind of genome is in the capsid: Is it DNA or RNA? Is it single stranded or double stranded? Is the genome circular or linear, composed of a single piece or segmented?

2 How is the protein arranged around the nucleic acid; that is, what are the symmetry and dimensions of the viral capsid?

3 Are there other components of the virion?

(a) Is there an envelope?

(b) Are there enzymes in the virion required for initiation of infection or maturation of the virion?

Note that this very basic scheme does not ask what type of cell the virus infects. There are clear similarities between some viruses whether they infect plants, animals, or bacteria. Despite this, however, it is clear that basic molecular processes are somewhat different between the Archaea, Eubacteria, and Eukaryota kingdoms; further, among eukaryotes, it is increasingly clear that there are significant differences in detail between certain processes in plants and animals. For this reason, viruses infecting different members of these kingdoms must make different accommodations to the molecular genetic environment in which they replicate. Thus, the nature of the host is an important criterion in a complete classification scheme.

Note also that there is no consideration of the disease caused by a virus in the classification strategy. Related viruses can cause very different diseases. For example, poliovirus and hepatitis A virus are clearly related, yet the diseases caused are quite different. Another more extreme example is a virus with structural and molecular similarities to rabies virus that infects *Drosophila* and causes sensitivity to carbon dioxide!

The Baltimore scheme of virus classification

Knowledge of the particulars of a virus's structure and the basic features of its replication can be used in a number of ways to build a general classification of viruses. In 1971, David Baltimore suggested a scheme for virus classification based on the way in which a virus produces messenger RNA (mRNA) during infection. The logic of this consideration is that in order to replicate, all viruses *must* express mRNA for translation into protein, but how they do this is determined by the type of genome utilized by the virus. In this system, viruses with RNA genomes whose genome is the same sense as mRNA are called **positive-sense (+ sense) RNA viruses**, while viruses whose genome is the opposite (**complementary**) sense of mRNA are called **negative-sense (− sense) RNA viruses**. Viruses with double-stranded genomes obviously have both senses of the nucleic acid.

The Baltimore classification has been used to varying degrees as a way of classifying viruses and is currently used mainly with reference to the RNA genome viruses, where positive- and negative-sense viruses are grouped together in discussions of their gene expression features. This classification scheme is not complete, however. Retroviruses, which are positive sense but utilize DNA in their replication cycle, are not specifically classified. Still, the scheme provides a fundamental means of grouping a large number of viruses into a manageable classification.

A more general classification based on a combination of the Baltimore scheme and the three basic criteria listed above is shown in Table 5.2. When compared to the listing of viruses in Table 5.1, it is clear that this scheme is not complete; for example, viruses with complex morphology are not well represented. More importantly, subtle distinctions such as the actual genetic relatedness of the proteins involved in viral genome replication are not taken into account. Indeed, only those viruses that have been characterized in some detail, and whose infection has some medical or economic impact upon humans, have been included; if a virus is not a human pathogen or if its occurrence has no obvious economic impact, it has been ignored. While the scheme can be expanded to include all known viruses, it then loses the value of relative simplicity.

Disease-based classification schemes for viruses

While molecular principles of classification are of obvious importance to molecular biologists and molecular epidemiologists, other schemes have a significant amount of value to medical and public health professionals. The importance of insects in the spread of many viral diseases has led to many viruses being classified as arthropod-borne viruses, or **arboviruses**. Interestingly, many of these viruses have general or specific similarities, although many arthropod-borne viruses are not part of this classification. The relationships between two groups of RNA viruses that are classified as arboviruses are described in some detail in Part IV, Chapter 13.

Viruses can also be classified by the nature of the diseases they cause, and a number of closely or distantly related viruses can cause diseases with similar features. For example, two herpesviruses, Epstein–Barr virus (EBV) and human cytomegalovirus (HCMV), cause infectious mononucleosis, and the exact cause of a given clinical case cannot be fully determined without virological tests. Of course, completely unrelated viruses can cause similar diseases too. Still, disease-based classification systems are of value in choosing potential candidates for the **etiology** of a disease. A general grouping of some viruses by similarities of the diseases caused or organ systems infected was presented in Table 3.1.

THE VIROSPHERE

The ICTV published their tenth report in 2017. This report is available online (https://talk.ictvonline.org/ictv-reports/ictv_online_report), and the current (2019) Master Species List database of viruses can be downloaded at the ICTV website (http://ictvonline.org). The current database contains 4958 different virus species arranged in 846 genera, 143 families, and 14 orders. They are listed by virus families in Table 5.1. While this is a notable achievement, it is not a complete one – the pace of discovery of new viruses and characterization of the genes they encode ensures that the number will change. Further, it is increasingly evident that the very nature of virus replication and association with their hosts leads to complications not found in classification schemes for cell-based life. The rate of genetic change in viruses can be great due to the rapidity and frequency of genome replication with the

Table 5.2 The Baltimore classification scheme for viruses.

RNA-containing viruses

I. Single-stranded RNA viruses
  A. Positive sense (virion RNA like cellular mRNA)
    1. Nonenveloped
      a. Icosahedral
        i. Picornavirus[a] (poliovirus,[a] hepatitis A virus,[a] rhinovirus[a])
        ii. Caliciviruses
        iii. Plant virus relatives of picornaviruses
        iv. MS2 bacteriophage[a]
    2. Enveloped
      a. Icosahedral
        i. Togaviruses[a] (rubella,[a] equine encephalitis, Sindbis[a])
        ii. Flaviviruses[a] (yellow fever,[a] dengue fever, Zika virus)
      b. Helical
        i. Coronavirus[a] (SARS-CoV, MERS-CoV)
  B. Positive sense but requires RNA to be converted to DNA via a virion-associated enzyme (reverse transcriptase)
    1. Enveloped
      a. Retroviruses[a]
        i. Oncornaviruses[a] (Rous sarcoma virus)
        ii. Lentiviruses[a] (HIV)
  C. Negative-sense RNA (opposite polarity to cellular mRNA, requires a virion-associated enzyme to begin replication cycle)
    1. Enveloped
      a. Helical
        i. Mononegaviruses[a] (rabies,[a] vesicular stomatitis virus,[a] Ebola virus[a])
        ii. Segmented genome (orthomyxovirus–influenza,[a] bunyavirus-hantavirus,[a] arenavirus[a])
II. Double-stranded RNA viruses
      a. Nonenveloped
    1. Icosahedral (reovirus,[a] rotavirus[a])
III. Single-stranded DNA viruses
      a. Nonenveloped
    1. Icosahedral
      a. Parvoviruses[a] (canine distemper, adeno-associated virus[a])
      b. Bacteriophage ΦX174[a]
IV. Double-stranded DNA viruses
  A. Nuclear replication
    1. Nonenveloped
      a. Icosahedral
        i. Small circular DNA genome (papovaviruses–SV40,[a] polyomaviruses,[a] papillomaviruses[a])
        ii. "Medium"-sized, complex morphology, linear DNA (adenovirus[a])
    2. Enveloped - nuclear replicating
Icosahedral
        i. Herpesviruses[a] (linear DNA)
        ii. Hepadnavirus[a] (virion encapsidates RNA that is converted to DNA by reverse transcriptase)
  B. Cytoplasmic replication
    1. Icosahedral
      a. Iridovirus
    2. Complex symmetry
      a. Poxvirus[a]
  C. Bacterial viruses
    1. Icosahedral with tail
      a. T-series bacteriophages[a]
      b. Bacteriophage λ[a]

[a] Discussed in text. MERS-CoV: Middle East respiratory syndrome coronavirus; SARS-CoV: severe acute respiratory syndrome coronavirus.

associated opportunity for error. Viruses can also, however, exchange genetic elements with their hosts and any other genomes present in the same milieu in which the virus is replicating. Such an occurrence can lead to the creation of a new virus species in which some of its genes are derived from one lineage and some from another – clearly, its classification will be complicated.

The best generalization that can be made concerning virus classification is that it depends on analysis of a number of features, and the importance of such features may vary depending upon the use being made of the classification. A classification scheme that combines the Baltimore basis along with the nature of the host and detailed genetic characterization of critical viral proteins can generate a global view of viruses as a **virosphere**, such as that shown in Figure 5.5.

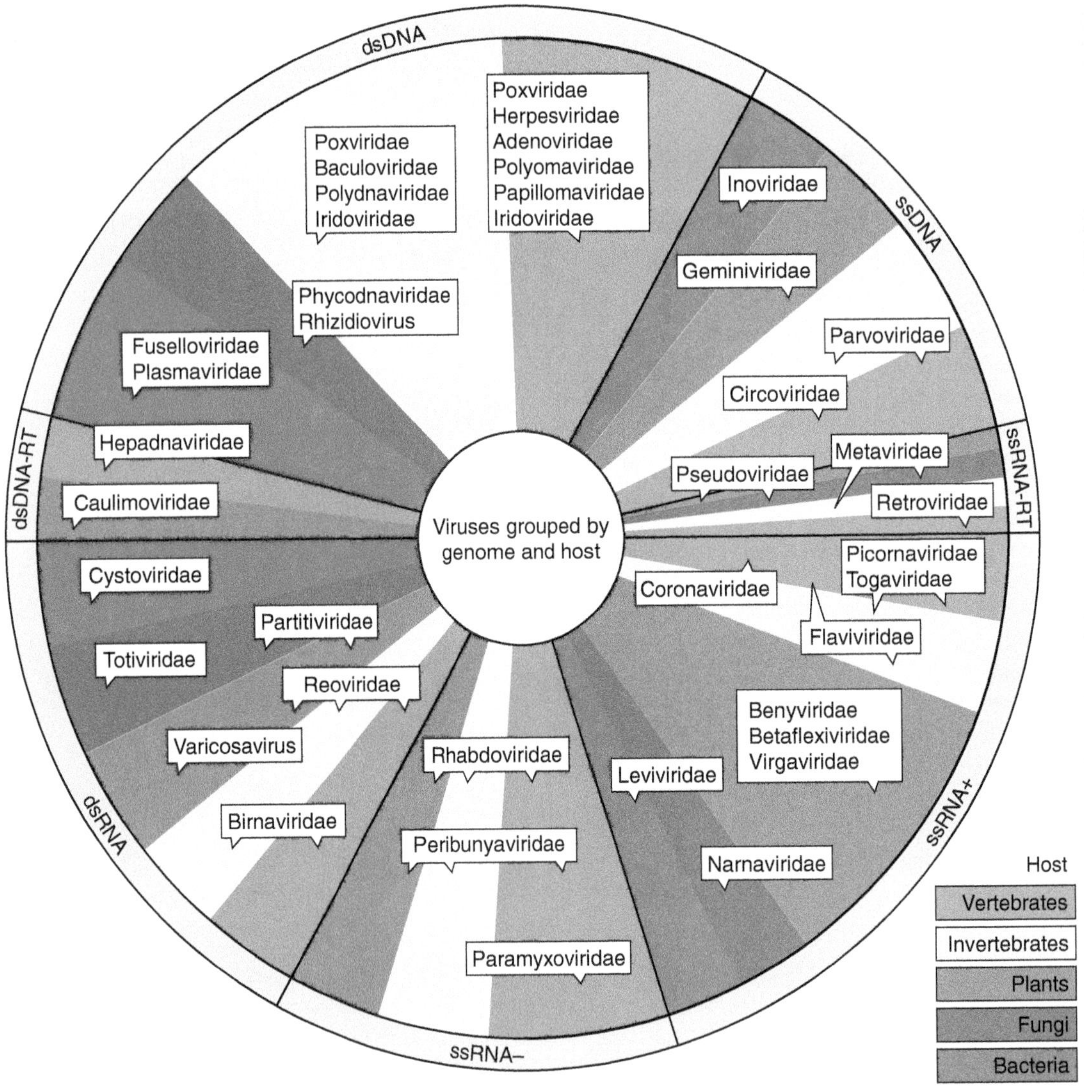

Figure 5.5 The virosphere. Classification of a major portion of the currently known families of viruses (*-viridae*) using criteria defined by the International Committee on Taxonomy of Viruses (ICTV). Major groupings are based on the nature of the viral genome and the nature of the host.

The features of viruses discussed in this chapter provide the basis for this comprehensive classification scheme, but they are not complete – for example, diseases caused by viruses cannot be readily listed. Further, relationships between virus families will often transcend the nature of the host – this would require a third dimension to the scheme (appropriate to the concept of a sphere). Since the concept of species in biology has always been a problem, it is no surprise that relationships between viruses pose a number of specific and profound problems. For more distantly related groups, the problems grow. Still, throughout this confusion, virus families made up of related species or types are clear, and it is possible to group some major virus families into superfamilies – we will see that this can be done for the Herpesviridae and certain bacteriophages. As a rationalization, it is useful to consider virus families and larger groupings as **polythetic** – a group whose members always have several properties in common, although no single common attribute is present in all of its members. As a result, no single property can be used as a defining property of a polythetic group on the basis that it is universally present in all the members and absent in the members of other groups. For viruses, it is impossible to use any one discriminating character for distinguishing related groups and families, because of the inherent variability of the members.

THE HUMAN VIROME

Recently, high-throughput sequencing (Chapter 11) coupled with metagenetic analysis (Chapter 22) have expanded our understanding of the number and types of viruses that are a normal part of the flora of the human body. The so-called human **virome** is in the process of being mapped for the blood, the gut, and various other locations that are a part of human anatomy, similar to work being done with bacterial species and the human microbiome. Since the techniques that lead to an understanding of the virome will be discussed later in this book, a detailed discussion of these explorations will be covered at that time. It suffices to say that the relationship we have with our normal viruses and their importance to our physiology are only partially understood.

QUESTIONS FOR CHAPTER 5

1 One structural form used in building virus particles is based on the icosahedron. Describe, either in words or in a diagram, the organization (number of capsomers, etc.) of the simplest virus particle of this form.

2 If a virus has a negative-sense RNA genome, what enzymatic activity (if any) will be found as part of the virion, *and* what will be the first step in expression of the viral genome?

3 List three properties of a virus that might be used as criteria for classification (taxonomy).

4 What is the basis of the Baltimore classification scheme?

5 What are some examples of virus structural proteins? What are some examples of proteins that have enzymatic activity included as part of a virus structure?

The Beginning and End of the Virus Replication Cycle

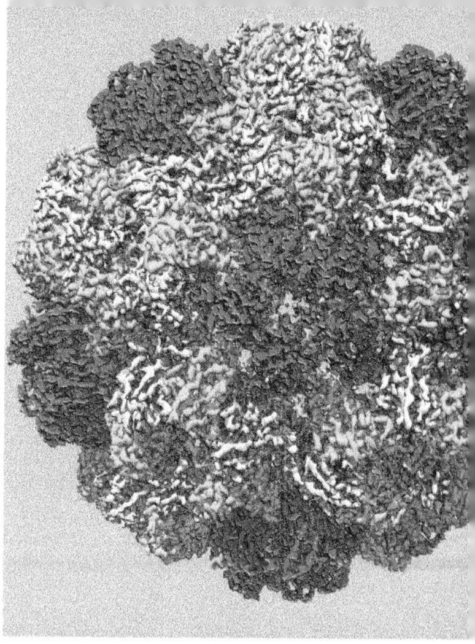

CHAPTER 6

OUTLINE OF THE VIRUS REPLICATION CYCLE

All viruses share the same basic replication cycle, but the time involved depends on a number of factors, including the size and genetic complexity of the virus itself as well as the nature of the host cell. As outlined briefly in Part I, the basic replication process involves the following steps:
1 Virus receptor recognition, attachment, and entry into the cell. Viruses must be able to utilize specific features of the host cell in which they will replicate to introduce their genome into that cell and ensure its being transported by cellular functions to where the virus replication cycle can continue. This requires either inducing the cell to engulf the whole virus particle in some specific way or, in the case of many bacterial viruses, injecting the viral genome into the host cell.
2 Viral gene expression and genome replication. Viral genes must be decoded from nucleic acid and translated into viral protein. This requires generation of messenger RNA (mRNA).

Basic Virology, Fourth Edition. Martinez "Marty" Hewlett, David Camerini, and David C. Bloom.

Different types of genomes obviously will require different mechanisms. One of the functions of viral gene expression is to allow the cell to carry out viral genome replication. It should be clear that the process for DNA viruses is different from that for RNA viruses.
3 Viral capsid formation and virion assembly. At the time that viral genomes are replicated, viral capsid proteins must be present to form virus structures. Often another stage of viral gene expression is required, and virion assembly may require scaffolding proteins (viral proteins that are needed to form the capsid structure, but are not part of the capsid structure). Following the formation of capsids, the virus must be released. Such release would involve an enveloped virus obtaining a membrane envelope.

Within this general pattern, there is a wealth of variation and difference in detail. Consider virus entry: While there is no known instance of a plant virus utilizing a specific cellular receptor for its entry, all animal and bacterial viruses do. The viral entry process for some bacterial viruses is an extremely complex one involving biochemical reactions between components of the virus capsid proteins to achieve injection of the viral genome.

There also is a lot of variation in the details of the virus release step. Here, most variations are seen among viruses being released from eukaryotic cells. In some infections, virus release results in cell death (a *cytocidal* infection). Such cell death might or might not involve cell lysis (**cytolysis**), depending on the virus. An infection leading to cell lysis is termed a cytolytic infection. Other changes to the cell (**cytopathology**) may also occur. Cytopathic effects due to viral infection can be used to measure the biological activity of many viruses.

Despite this type of variability, the process of capsid maturation and assembly is generally determined by the structural features of the virus in question. Thus, icosahedral bacterial viruses mature following steps that are quite similar to those characterized for herpesviruses (herpes simplex viruses [HSVs]). Again, helical plant, animal, and bacterial viruses all assemble in much the same way.

VIRAL ENTRY

Animal virus entry into cells – the role of the cellular receptor

Animal viruses must enter the cell in an appropriate manner through a complex plasma membrane composed of a lipid bilayer in which membrane-associated proteins "float" in the upper or lower surface (Figure 6.1). Some integral membrane proteins form pores (*channels*) in the membrane for transport of ions and small molecules. Other proteins project from the cell's surface and are modified by the addition of sugar residues (glycosylation). Such **glycoproteins** serve many functions, including mediating immunity, cellular recognition, cell signaling, and cell adhesion.

Virus infection requires interaction between specific proteins on the surface of the virion and specific proteins on the cell's surface – the **receptor** for that particular virus. It should be kept in mind that the physiological functions of a cell surface protein utilized as a virus receptor really are for purposes other than viral attachment and entry; some identified viral receptors and their known functions are shown in Table 6.1. The term *receptor* is just a way of defining the protein by the effect that is being studied – in this case, entry of a virus into a cell.

The type and distribution of receptors utilized by a given virus determine (in large part) both its ability to recognize and enter specific differentiated cells (its **tissue tropism**) as well as the particular animal species it favors (the virus' **host range**). For example, CD4 and certain chemokine receptors (usually CCR5 or CXCR4) on some T lymphocytes that are involved in the immune response are recognized by HIV to allow an infection of these lymphocytes. The

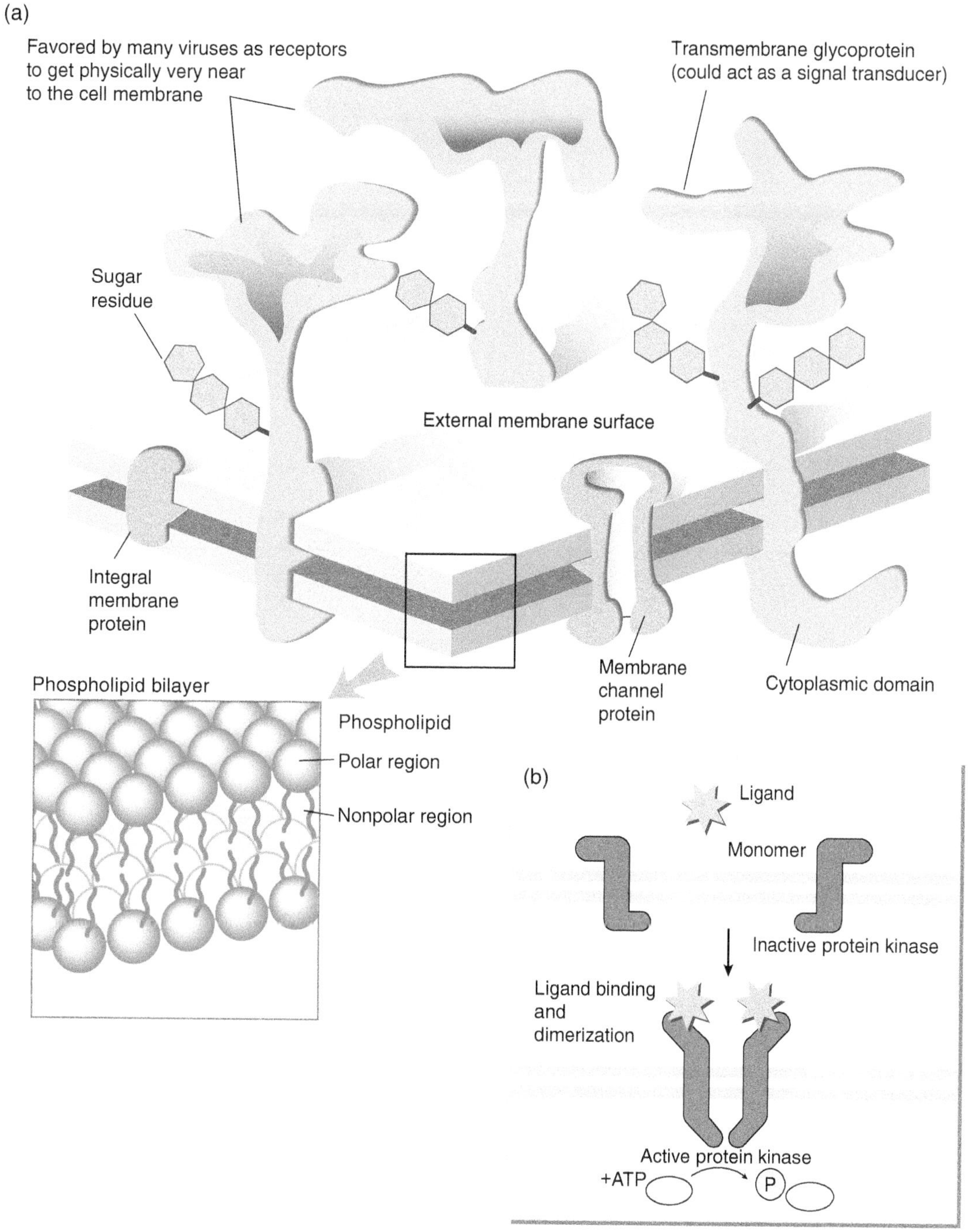

Figure 6.1 The surface of a "typical" animal cell. The lipid bilayer plasma membrane is penetrated by cell surface proteins of various functions. The proteins that extend from the surface (mainly glycoproteins) can be utilized by different viruses as "tether points" or "anchors" for bringing the virus close enough to the cell surface to initiate the entry process. This interaction between a cell surface protein serving as a virus receptor and the virus itself is highly specific between proteins. Integral membrane proteins, such as those mediating transport of small molecules and ions across the plasma membrane, tend not to project as far into the extracellular matrix and can be utilized by retroviruses, especially as receptors. Some viral receptors are listed in Table 6.1.

Table 6.1 Some cellular receptors for selected animal viruses.

Name	Cellular Function	Virus Receptor For
ICAM-1	Intracellular adhesion	Poliovirus
CD4	T-lymphocyte functional marker	HIV
MHC-I	Antigen presentation	Togavirus, SV40
MHC-II	Antigen presentation/stimulation of B-cell differentiation	Visnavirus (lentivirus)
Fibronectin	Integrin	Echovirus (picornavirus)
Cationic amino acid transporter	Amino acid transport	Murine leukemia virus (oncornavirus)
LDL receptor	Intracellular signaling receptor	Subgroup A avian leukosis virus (oncornavirus)
Acetylcholine receptor	Neuronal impulse transducer	Rabies virus
EGF	Growth factor	Vaccinia virus
CR2/CD21	Complement receptor	Epstein-Barr virus
HVEM	Tumor necrosis factor receptor family	Herpes simplex virus
Sialic acid	Ubiquitous component of extracellular glycosylated proteins	Influenza virus, reovirus, coronavirus

virus has evolved to recognize CD4 and **CCR5** or **CXCR4** and subvert their functions. Poliovirus utilizes an interaction with a major **intercellular adhesion molecule (ICAM)** in its infection. The slow progression of rabies virus up the neural net into the central nervous system (CNS) is accomplished by its use of acetylcholine receptors as its port of entry into neurons. These receptors are concentrated at the synapses between neurons, and thus, the virus can "jump" from neuron to neuron without causing destruction of the neuron. This pattern of progression minimizes tissue damage and inflammation resulting in virus "leakage" into the host's circulatory system with ensuing immune response. Finally, sialic acid residues are enzymatically added as modifications to the glycoproteins of secretory cells, especially of the nasopharynx and respiratory system. Influenza and some other respiratory viruses use these sialic acid residues to specifically target such host cells.

An important factor in the tissue tropism of a given virus is the physical availability of the receptor for interaction with the targeting virus. Poliovirus infects only primates because only primates express the appropriate ICAM utilized as the poliovirus receptor. Further, however, it can attach to and penetrate only specific cells of the small intestine's lining and motor neurons despite the fact that poliovirus-specific ICAMs are present on many other primate cells. In these refractory cells, however, other membrane proteins on the surface apparently mask the receptor. Conversely, if the gene expressing the poliovirus receptor is expressed in a non-primate cell such as those of a mouse using appropriate molecular genetic techniques, the virus can and does initiate a productive infection.

There is another very important factor in entry-mediated tissue tropism in virus infections. Many viruses utilize other proteins on the surface of cells as co-receptors in addition to the major receptor. In the case of HIV, an important co-receptor is one of a group of surface chemokine receptors. There must be a molecular interaction between both the CD4 receptor and the co-receptor for efficient HIV infection. With HIV, the co-receptor also determines tissue tropism. In addition to CD4, macrophages and some T cells express CCR5, which allows HIV variants that recognize this protein to show a marked tropism for these cells. Alternatively, some T lymphocytes express CD4 and a second HIV co-receptor,

CXCR4; some strains of HIV show a marked tropism for these cells. Finally, some HIV strains can utilize both co-receptors. Thus, a given virus may utilize a major receptor protein, but require the presence of one or several other proteins in addition. If a certain cell possesses the major receptor but not the co-receptor, infection cannot occur or occurs with impaired efficiency so that cell and tissue tropism are altered.

It is also important to understand that some viruses exhibit alternative methods of initiating infection in a cell neighboring the one initially infected via the receptor-mediated route. For example, infection of a cell may lead to membrane changes that allow fusion with a neighboring cell or cells. Then virus can pass freely into the cytoplasm of the uninfected cell without having to pass the plasma membrane; this is a well-established feature of infections with some strains of HSV that cause the formation of large groups of fused cells or **syncytia**. This and other aspects of virus-induced modifications to the infected cell are discussed in Chapter 10, Part III.

The contact between cells allowing virus spread need not be complete fusion. The close interaction between dendritic cells and other cells of the immune system in induction of the immune response, which is described in Chapter 7, may facilitate the passage of viruses that were taken up but not destroyed. This is clearly an important feature in the pathogenesis of HIV.

The virus itself may possess a surface protein involved in recognition and receptor-mediated entry that is dispensable under certain conditions. An excellent example is the situation with HSV-1 mutants that lack glycoprotein C (gC) on their envelope. As described in somewhat more detail in Part IV (Chapter 17), this glycoprotein interacts with heparan sulfate on the surface of the cell to allow it to come into close proximity of the ultimate receptor. Mutant viruses lacking gC demonstrate significant alterations in the details of their infection and pathogenesis in laboratory animals, but they replicate with excellent efficiency in many cultured cells in the laboratory. Here the culturing and frequent passage of the cells lead to alterations in the cell surface, so that gC-negative viruses can "find" their receptors with little difficulty.

Viruses may also inefficiently use other proteins on the surface to infect cells that do not bear the efficient receptor protein. Provided conditions are optimized, these proteins can substitute for the efficient receptor. This substitution is one reason why some viruses can be induced to infect certain cells in culture even though they do not possess the ideal receptor. An example is the ability of SV40 virus to inefficiently infect certain murine and hamster cells in culture. Such infections can be observed with ease in the laboratory, and there is good suggestive evidence that such atypical infections can occur with some frequency under natural conditions. The emergence of new infectious viruses in the environment is often associated with the appearance of a virus infecting a host previously unaffected by it. The emergence of novel infectious viruses is discussed in Part V, Chapter 25.

Some such occurrences can be inferred to result from an inappropriate infection followed by the novel virus adapting to utilize a previously unrecognized receptor. A rare inappropriate infection of an animal virus into a human with subsequent changes in the genetic properties of the virus was suggested to explain the relatively sudden appearance of HIV in the human community. Another example of such an occurrence may explain the sudden appearance of the H5N1 avian influenza virus that continues to spread worldwide. While the virus has been transmitted from birds to humans in a limited number of cases, it has not yet, at this writing, mutated to allow efficient human-to-human transmission.

Mechanisms of entry of nonenveloped viruses

Nonenveloped virus particles must be incorporated into the cell via a process called **translocation** across the lipid bilayer. This process is one in which the capsid or a cell-modified capsid physically crosses the cell plasma membrane. There are at least four mechanisms that result in

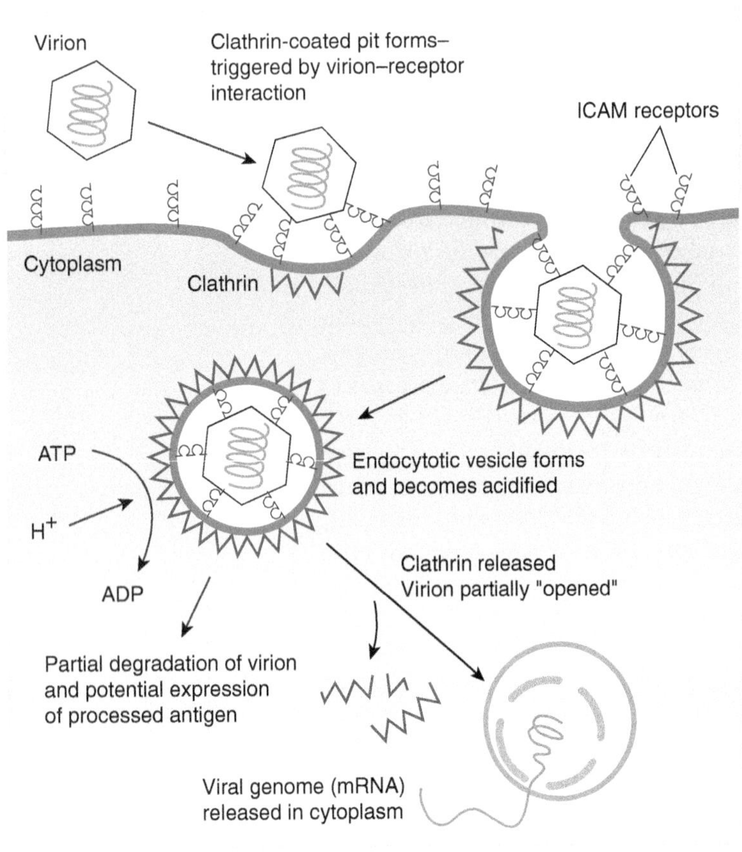

Figure 6.2 Schematic of receptor-mediated endocytosis utilized by rhinovirus for entry into the host cell. The endocytotic vesicle forms as a consequence of close association between the rhinovirus–receptor complex and the plasma membrane.

virus translocation across the membrane: **clathrin-mediated endocytosis**, **caveolae-mediated endocytosis**, **lipid-raft-mediate endocytosis**, and **macropinocytosis**. Each of these processes is endocytotic, in that they result in the formation of endosomal vesicles containing extracellular material (including the attached virus particles) that move into the cytoplasm of the cell. They are differentiated by the nature of the cellular components that line the endocytotic vesicle and mediate its formation. The clathrin-mediated pathway involving receptor binding is illustrated for rhinovirus in Figure 6.2. The acidic environment of the endocytotic vesicle causes specific changes to the rhinovirus capsid so that the internal genome (positive-sense RNA) is released into the cytoplasm, where it can be translated and begin gene expression.

A nuclear replicating nonenveloped virus, such as the papovavirus, SV40, begins entry in a similar fashion, but the interaction between viral capsid proteins and the vesicle, along with other **intracellular trafficking proteins**, allows the modified virion to be transported to the nuclear membrane. Once there, the viral genome is released, and viral DNA interacts with cellular transcription factors to begin gene expression. Because specific genetic alterations (**mutations**) in the SV40 capsid protein will interfere with this transport, it is known that the virus controls the process.

Entry of enveloped viruses

Enveloped viruses interact with cell receptors via the action of membrane-associated viral glycoproteins that project beyond the viral envelope. The viral glycoproteins are glycosylated with sugars in the cell Golgi apparatus during viral maturation. The process is similar to that carried out by the cell on its own glycoproteins.

Virus entry can involve **fusion** of the viral membrane at the cell's surface, or it can involve receptor-mediated endocytosis. These two processes are shown in schematic form in Figure 6.3a. As with nonenveloped viruses, the acidic pH of the endocytotic vesicle can lead to modifications of the viral envelope so that fusion between it and the vesicle's membrane can occur. The process of envelope fusion is shown schematically in Figure 6.3b; essentially, the association between membrane proteins in the viral and cellular envelopes leads to "clearance" of an area on the viral and cellular surfaces, then juxtaposition of these naked membranes leads to their fusion. Fusion of the pseudorabies virus (a close relative of HSV) with the plasma membrane of the cultured cells being infected is shown in the electron micrographs of Figure 6.3c.

The fusion interaction between the viral and plasma or vesicular membrane can be a simple one between one viral glycoprotein and one cellular receptor, or it can be a complex cascade of linked protein interactions. For example, with a herpesvirus such as HSV, five or six viral glycoproteins first bring the virus near the cell, and then allow entry, which requires interaction with a specific cellular surface receptor. The first interaction appears to be an association between viral glycoproteins and sulfated sugar molecules (polyglycans) like heparan sulfate, which is found attached to many surface proteins of the cell. Only then can the virion be brought close enough to the plasma membrane to allow interaction with the actual receptor.

Once the viral capsid is inside the cytoplasm, specific interactions can take place between components of the capsid and microfilaments and other proteins involved in cellular trafficking, leading to transport of the viral capsid to a location in the host cell suitable for the infection process to continue. This process is similar in very broad outline to the intracellular transport of a nonenveloped virus, and is outlined in Figure 6.3d.

Entry of virus into plant cells

A plant cell's special architecture, namely the presence of a rigid and fairly thick cell wall, presents a unique challenge for virus entry. Initial entry into the plant cell must take advantage of some break in integrity of the cell wall. Apparently, when the virus enters such a break and becomes situated in close proximity to the plant cell's plasma membrane, it can enter the cell without interaction with specific receptors.

Breaks or lesions in the plant cell's wall are most often produced by organisms that feed on the plant or by mechanical means. Above ground, invertebrates, such as aphids, leafhoppers, white flies, and thrips, are known vectors for a number of plant viruses. Nematodes, which feed on the root system of the plant, are another source of viral infection. In some cases, the virus is transferred from the invertebrate to the plant without growing in the vector. This is the case for *Geminivirus* transmission by white flies. Alternatively, viruses may replicate in both their invertebrate and plant hosts. This is seen with tomato spotted wilt virus (a plant bunyavirus) and its thrip vector. In either case, the viruses gain entry to cytoplasm of the plant host cell after the insect has begun to feed on plant tissue.

Mechanical damage to the plant's cell wall also can be a means of entry for plant viruses. This approach is used most often in experimental settings when the leaf surface is scratched or abraded prior to inoculation with a virus suspension. This also may happen in nature as a result of agricultural applications, such as harvesting. Brome mosaic virus, transmitted by beetles, can also gain entry into the plant during cutting operations.

Once inside the plant cell cytoplasm, viruses are uncoated and gene expression begins following patterns similar to those described for animal viruses. Passage of progeny virus from the initial site of infection to new host cells takes place through cell-to-cell connections called

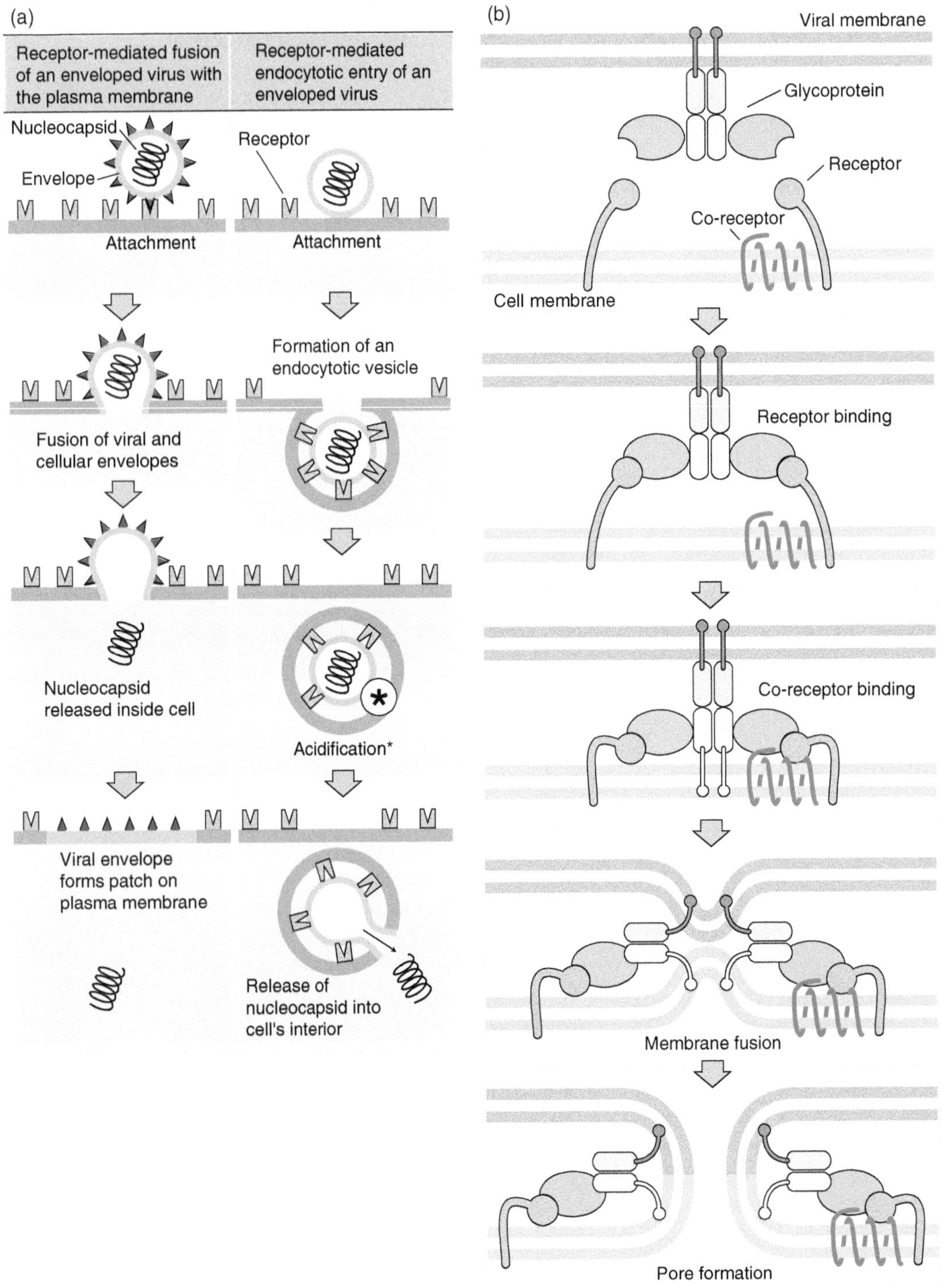

Figure 6.3 (a) The two basic modes of entry of an enveloped animal virus into the host cell. Membrane-associated viral glycoproteins either can interact with cellular receptors to initiate a fusion between the viral membrane and the cell plasma membrane, or can induce endocytosis. The fate of the input virus membrane differs in the two processes. (b) High-resolution schematic of the process of membrane fusion. The interaction between viral and cellular membrane-associated proteins results in the "clearance" of an area of the two lipid bilayers so that they can become closely juxtaposed, leading to fusion. (c) The fusion of pseudorabies virus with the plasma membrane of an infected cultured cell is shown in this series of electron micrographs (the bars represent 150 nm). Although each electron micrograph represents a single event "frozen in time," a logical progression from the initial association between viral envelope glycoproteins and the cellular receptor on the plasma membrane through the fusion event is shown. The final micrograph contains colloidal gold particles bound to antibodies against the viral envelope glycoproteins (dense dots). With them, the envelope can be seen clearly to remain at the surface of the infected cell. Source: Reprinted with the kind permission of the American Society for Microbiology from Granzow, H., Weiland, F., Jöns, A., et al. (1997). Ultrastructural analysis of the replication cycle of pseudorabies virus in cell culture: a reassessment. *Journal of Virology* 71: 2072–2082.) (d) The association of the viral capsid with the intracellular transport machinery following membrane fusion. This process leads to the virion and associated viral genome being transported to its appropriate location inside the cell to initiate the next step of the infection process – the expression of viral genes.

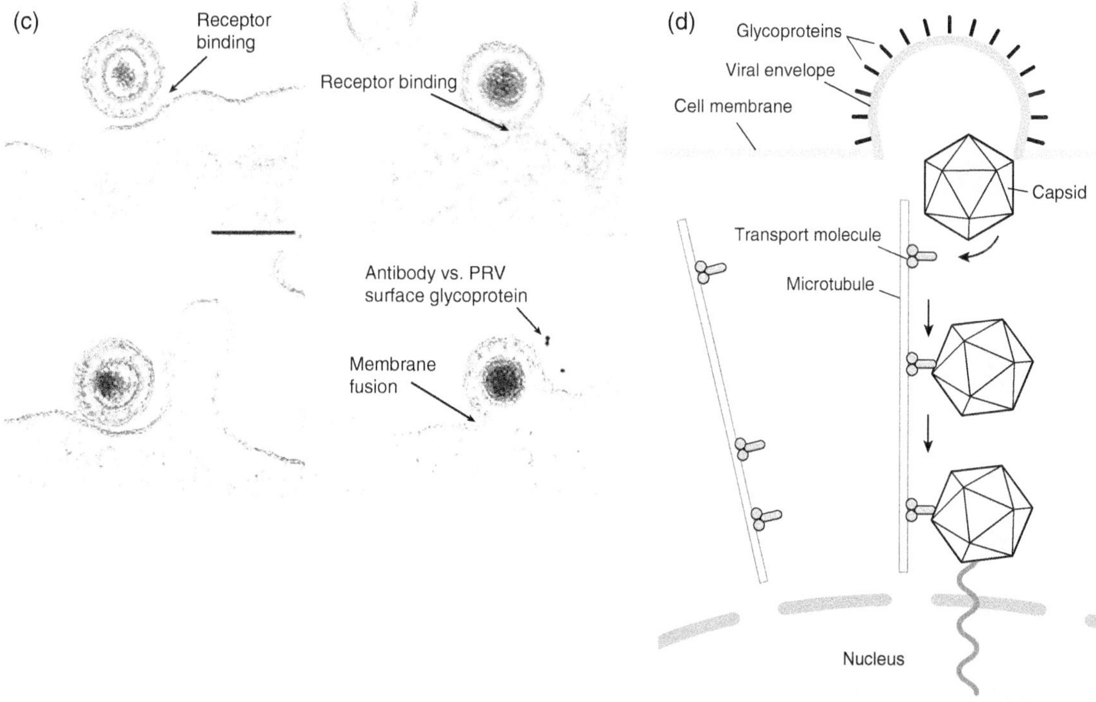

Figure 6.3 *Continued*

plasmodesmata and through the plant's circulatory system, the phloem. For this reason, most plant virus infections end up as systemic infections of the whole organism; thus, a single lesion and virus entry can result in virus lesions appearing throughout the plant.

The injection of bacteriophage DNA into *Eschcrichia coli*

Bacteriophages must interact with a receptor on the bacterial cell surface to successfully initiate replication. The outer surface of a prokaryotic cell presents a set of features to the external environment that includes structural materials (glycoproteins and lipopolysaccharides), transport machinery (amino acid or sugar transport complexes), and cell-to-cell interaction apparatus – the F or **sex pilus**. Sex pili are used by the bacteria in conjugation and exchange of genetic material with other bacteria of the opposite "sex." Attachment of the phage to host cells may employ any one of these structures, depending on the particular virus. Some features utilized by bacteriophages replicating in *E. coli* are shown in Table 6.2.

In some cases, attachment of phage to the host cell involves a physical rearrangement of the virus particle. For example, attachment of bacteriophage T4 to the surface of susceptible *E. coli* cells occurs in two steps, which are shown in Figure 6.4. First, there is a relatively weak interaction between the tips of the phage tail fibers and lipopolysaccharide residues on the surface of the cell's outer membrane. This triggers a second, stronger, and irreversible interaction. In this, tail pins on the baseplate of the virion interact with structures in the outer membrane itself, requiring a change in conformation of the tail fibers. This ultimately results in compression of the phage tail's contractile sheath and injection of phage DNA into the host cell. In this process, the phage tail tube penetrates the cell wall, but phage DNA must still cross the inner cell membrane. This last step is carried out with the help of a viral gene product called a **pilot protein**.

Table 6.2 Some *E. coli* bacteriophage receptors.

Virus	Structure	Normal Function
T2	OmpF	Porin protein
	Lipopolysaccharide	Outer membrane structure
T4	OmpC	Porin protein
	Lipopolysaccharide	Outer membrane structure
T6	Tsx	Nucleoside transport protein
T1 and T5	TonA	Ferrichrome transport
	LamB	Maltose transport protein
MS2	F pilus	Conjugation

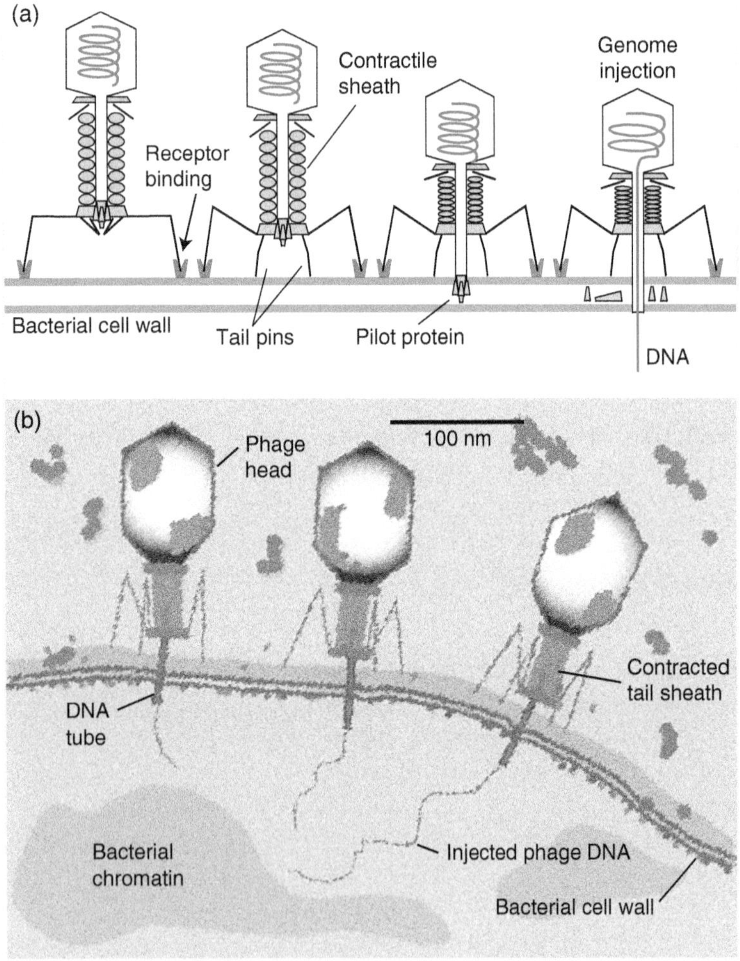

Figure 6.4 Entry of T4 bacteriophage DNA into an *E. coli* cell. Initial attachment is between the fibers to the ompC lipopolysaccharide receptor on the bacterial cell wall (a). The binding of protein pins on the baseplate to the cell wall leads to contraction of the tail fibers and sheath proteins, leading to insertion of the tail tube through the cell wall. As shown in the electron micrograph (b), phage pilot protein (arrow) allows the highly charged viral DNA genome to penetrate the bacterial plasma membrane and enter the cell. Phage DNA can be seen as shadowy lines emanating from the tail tube. Source: From Dimmock, N.J. and Primrose, S.B. (1994). *Introduction to Modern Virology*, 4e. Boston: Blackwell Science.

With some other phages, the interaction between virion and cell results in no immediate alterations to the phage structure, for instance in attachment of bacteriophage λ to its receptor LamB. Again, the attachment of MS2 bacteriophage to the F pilus does not result in changes to the virus structure. Since cells with a pilus structure (the product of an F-plasmid) are called *male*, MS2 and similar phages are sometimes termed *male specific*.

The actual amount of bacteriophage that enters the host cell is quite variable. In the case of tailed phage, only phage DNA and certain accessory proteins enter the host cell. For a nontailed phage such as MS2, however, the entire phage particle enters the cell and is uncoated in the cytoplasm.

Nonspecific methods of introducing viral genomes into cells

Clearly, the process of infection of a cell by a virus essentially involves the efficient insertion of the viral genome into an appropriate location within the cell so that viral genes can be expressed. The fact that viruses can be internalized into plant cells without the benefit of receptors suggests that other methods for the introduction of viral genomes can take place, if only rarely. In the laboratory, for example, cells can be made permeable by chemical or physical methods so that they can take up quite large particles. Appropriate treatment of cells and addition of high concentrations of virus particles can lead to virus uptake. The process will be inefficient, and most virus particles may be destroyed. Despite this, it is often possible to initiate productive infection in a few cells if enough virus particles are taken up so that an intact viral genome or two can get to the appropriate portion of the cell to initiate infection.

A similarly inefficient and nonspecific process called **transfection** is often used to introduce viral genomes (especially DNA genomes) into cells. Isolated genomes can be aggregated into the proper-sized particles by precipitation into aggregates using calcium phosphate ($Ca_3(PO_4)_2$), and cells can be treated to readily incorporate the aggregates. Alternatively, viral genomes can be concentrated inside lipid vesicles called *liposomes* in solution, and these can be readily assimilated by cells that have been specifically treated with mild detergents so that their plasma membrane can fuse with the liposome. Several other methods of introducing DNA, RNA, or even proteins into cells have also been effectively exploited. Transfection of plants can be efficiently carried out by literally "shooting" microscopic pellets of plastic coated with the appropriate macromolecular concoction using an air blast. Macromolecules can also be introduced into cells using electric fields. In all these cases the process is inefficient, but only a few viral genomes presented to the proper intracellular location are sufficient to induce a productive infection.

An example of the use of transfection to examine the properties of a viral protein is illustrated in Figure 6.5. Here, cells were transfected with a fragment of DNA containing the gene for the varicella zoster virus (herpes zoster virus) glycoprotein, gL. This gene is controlled by a promoter that can be expressed by transcriptional machinery of the uninfected cell (see Chapter 13). The three micrographs shown in Figure 6.5b were taken just after, 12 hours after, and 24 hours after transfection. Cells were incubated with a fluorescent antibody against gL (see Chapter 12). The expression of this protein in the cytoplasm is quite evident at the later times.

LATE EVENTS IN VIRAL INFECTION: CAPSID ASSEMBLY AND VIRION RELEASE

Assembly of helical capsids

The capsids of helical viruses must assemble around the genome. This process is relatively well studied in tobacco mosaic virus (TMV) of plants. As noted, the basic process appears to be similar for all helical viruses. This similarity depends on the fact that single- or double-stranded RNA (or DNA, for that matter) can readily form a helical structure when associated with the proper type of protein.

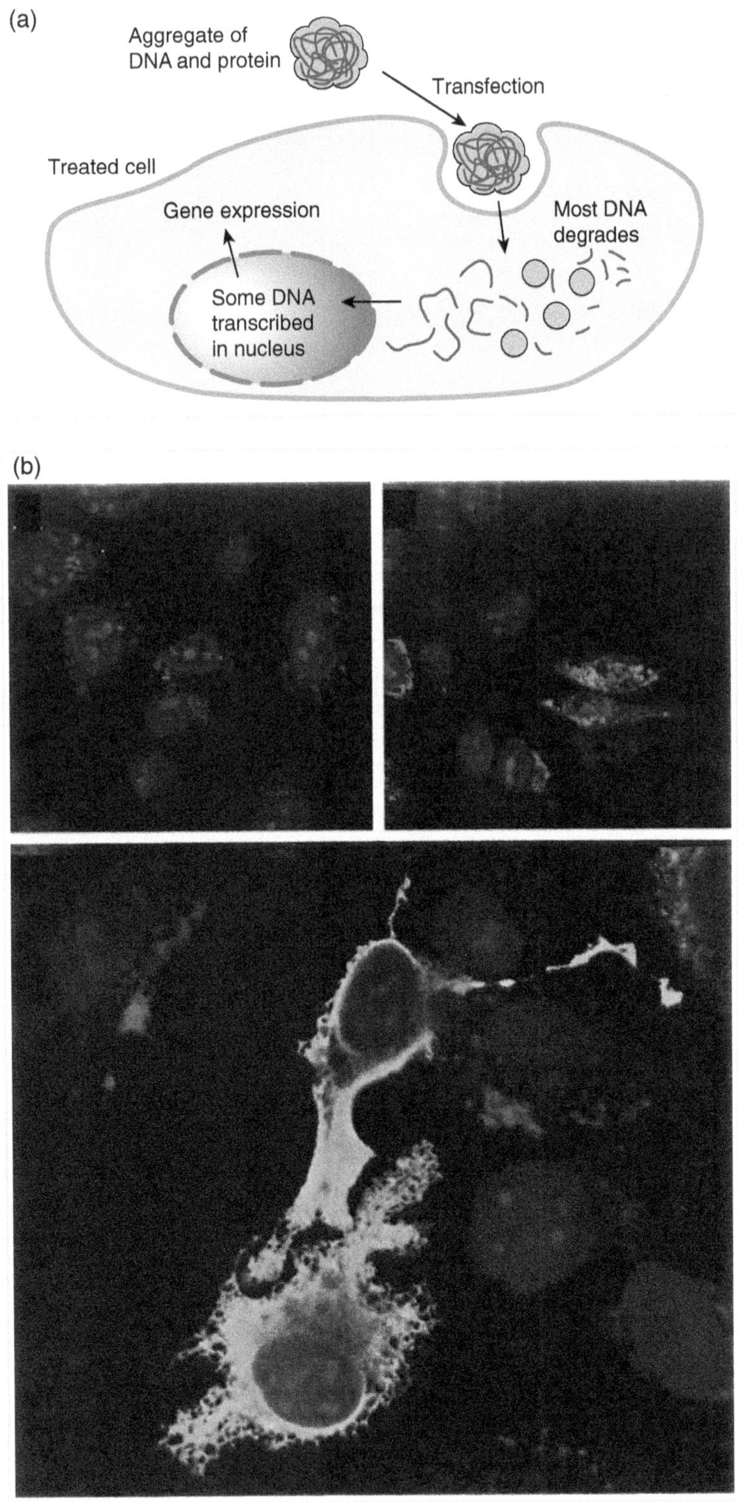

Figure 6.5 Expression of a varicella zoster virus protein following transfection of a cell with the viral gene under the control of a promoter that is active in the uninfected cell. (a) The basic process. The cell membrane is treated with agents that allow it to readily take up large aggregates of protein and nucleic acids by phagocytosis. The transfecting DNA is caused to form aggregates with the use of calcium phosphate ($Ca_3(PO_4)_2$) and then mixed with cells that have been appropriately treated. While most of the DNA taken up by the cell is degraded, some gets to the nucleus by nonspecific cellular transport of macromolecules, and this DNA can be transcribed and any genes within it expressed as proteins. (b) An actual experiment. Cells were made permeable and then transfected with DNA containing the varicella zoster virus glycoprotein L gene. The protein encoded in this gene was expressed following its transcription into mRNA (see Chapter 13). Cells were treated with fluorescent antibody reactive with the glycoprotein at (clockwise from the top left) 0, 12, and 24 hours after infection. The expression of the glycoprotein in the cytoplasm is clearly evident from the green fluorescence. (See Chapter 12 for a description of the method.) Source: Courtesy of C. Grose, University of Iowa.

The assembly of the helical capsid and RNA genome of TMV is shown in Figure 6.6. Capsomers self-assemble to form disks, and the disks formed by the capsomers initially interact with a specific sequence in the genome called *pac* (for **packaging signal**). Interaction with the RNA itself converts the disk into a "lock washer" conformation, and subsequent capsomer assemblies then thread onto the growing helical array to form the complete capsid. Note that, for TMV, the RNA forms the equivalent of a "screw," which penetrates the disk assembly of capsomers. This penetration allows translocation to a helical arrangement that grows by continued association with the genomic RNA.

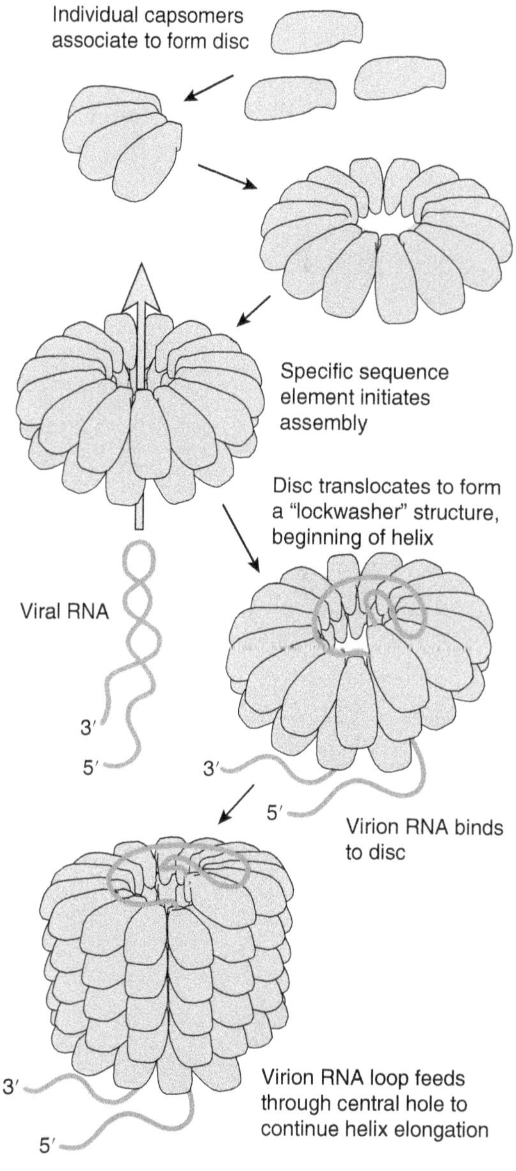

Figure 6.6 Assembly of the helical tobacco mosaic virus (TMV). Steps in the preassembly of the capsomer disk, insertion of viral RNA, and the translational "screwlike" helix assembly process with sequential addition of more capsomers are shown. Source: Adapted from Dimmock, N.J. and Primrose, S.B. (1994). *Introduction to Modern Virology,* 4e. Boston: Blackwell Science.

Assembly of icosahedral capsids

In the majority of cases studied in detail, maturation of the icosahedral capsid from an immature *procapsid* to final state involves specific proteolytic cleavage of one or several capsid proteins that were assembled into the immature virus particle. This cleavage results in subtle changes in structure or increased capsid stability, and often accompanies inclusion of the viral genome. These cleavage steps, accomplished by virus-encoded proteins called *maturational* proteases, are quite limited – only a few discrete peptide bonds are hydrolyzed. Thus, a fairly good general rule has the assembly of icosahedral capsids involving both preassembly of procapsids and specific covalent modifications of the virion proteins by proteolytic processing. The high specificity of maturational proteases and the fact that they are encoded by the viral genome make them attractive targets for antiviral therapy; protease inhibitors of HIV have been found to have great therapeutic value (see Chapter 8).

Some of the general models for assembly of an icosahedral capsid were based on early studies on poliovirus, a small RNA-containing virus. One characteristic of poliovirus infection in the laboratory is the formation of empty capsids. Thus, it is clear that the viral capsomers can self-assemble. This observation was interpreted as indicating that empty capsids assemble *before* the genome enters the virion. Notably, however, some recent studies on the assembly of poliovirus and related viruses suggested that the procapsid assembles directly around the viral RNA, and empty capsids are a nonfunctional by-product of the assembly process. Despite this, empty capsids can form a stable structure spontaneously, even if they are not intermediates in virion assembly.

With larger icosahedral viruses, the process of capsid assembly is complex, with scaffolding proteins forming a "mold" or pattern for the final capsid. In either case, capsid assembly occurs *before* entry of the viral genome into the capsid, and one of the hallmarks of icosahedral virus maturation is the generation of empty capsids.

Assembly of the head of bacteriophage P22 is shown in Figure 6.7 as an illustration of this process. The process is quite similar to the assembly of HSV capsids. Note that the pilot proteins, which are important for injection of the genome (see Figure 6.4), may

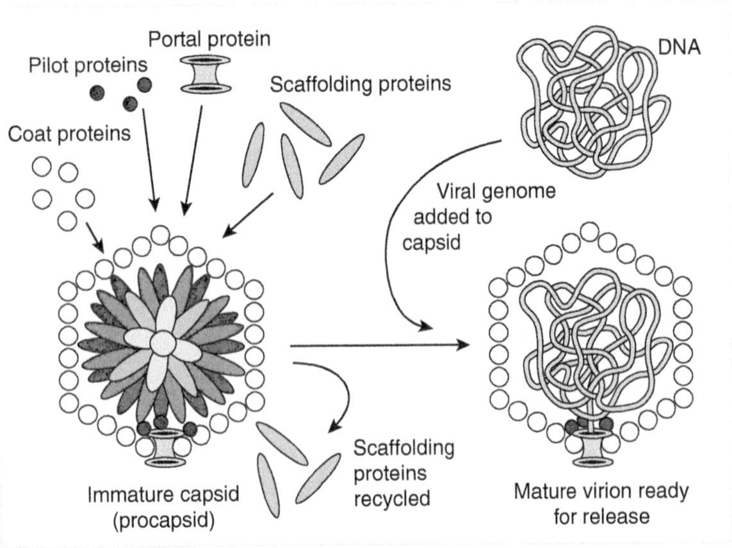

Figure 6.7 Assembly of the phage P22 capsid and maturation by insertion of viral genomic DNA. Individual capsomer subunits preassemble into a procapsid around scaffolding protein. This latter protein is recycled with phage P22 but can be proteolytically removed with a maturational protease with other icosahedral viruses. The empty head then associates with viral genomes. Genome insertion requires both energy and a conformational change in the procapsid.

also help the capsid proteins assemble. The scaffolding proteins can recycle and function in the assembly of more than one capsid. Also note that the term *pilot protein* here has a completely different meaning than when used in the T-even bacteriophage infection discussed previously.

Retrovirus proteases "activate" virion-associated enzymes during the final stages of virion maturation following release from the infected cell. These retrovirus proteases form part of the virion's structural protein. Antiviral drugs targeting the HIV protease have shown significant therapeutic benefit, and other viral proteases are targets for drug development because they are specific to the virus encoding them. This is discussed in more detail in Part IV, Chapter 20.

Generation of the virion envelope and egress of the enveloped virion

The lipid bilayer of the membrane envelope of the viruses that bear them is derived from the infected cell. Few (if any) viral genes directed toward lipid biosynthesis or membrane assembly are yet identified. While the lipid bilayer is entirely cellular, the envelope is made virus specific by the insertion of one or several virus-encoded membrane proteins that are synthesized during the replication cycle.

Some of the patterns of envelopment at the plasma membrane for viruses that assemble in the cytoplasm are shown in Figure 6.8. Viral glycoproteins, originally synthesized at the rough **endoplasmic reticulum** and then processed through the **Golgi apparatus**, arrive at

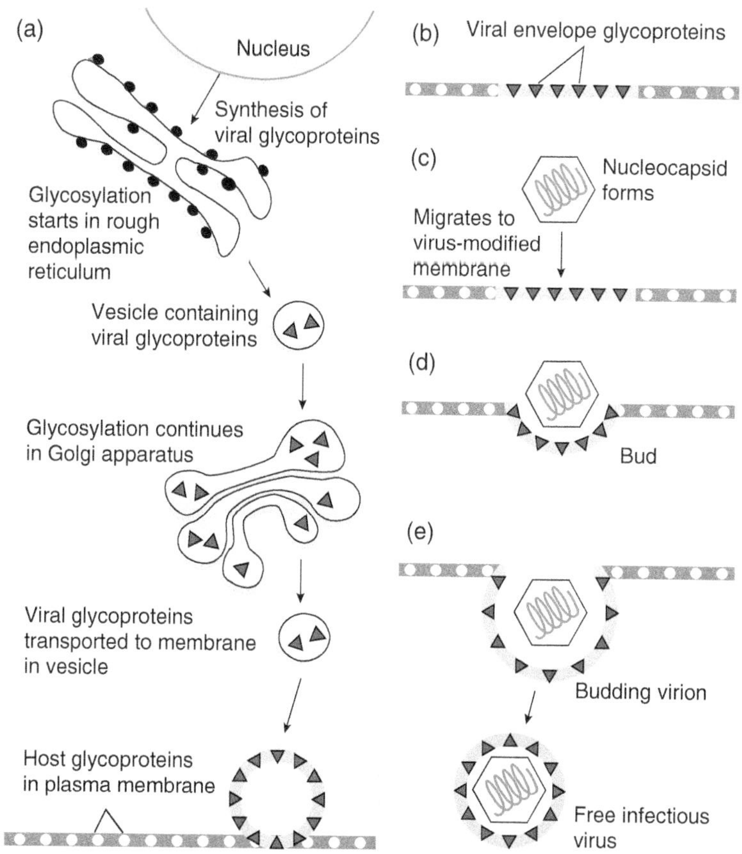

Figure 6.8 Insertion of glycoproteins into the cell's membrane structures and formation of the viral envelope. The formation of viral glycoproteins on the rough endoplasmic reticulum parallels that of cellular glycoproteins except that viral mRNA is translated (a). Full glycosylation takes place in the Golgi bodies, and viral glycoproteins are incorporated into transport vesicles for movement to the cell membrane where they are inserted (b). At the same time (c), viral capsids assembly and then associate with virus-modified membranes. This can involve the interaction with virus-encoded matrix proteins that serve as "adapters." Budding takes place (d, e) as a function of the interaction between viral capsid and matrix proteins and the modified cellular envelope containing viral glycoproteins.

the site of budding with their carboxy termini in the cytoplasm and their amino termini on the outside of the cell. At both sites, enveloped viruses recruit cellular proteins needed to pinch off the cytoplasmic membrane stalk connecting the budding enveloped particle to the cell surface.

Many viruses, including retroviruses, bud from the surface of the infected cell. The use of atomic force microscopy (outlined in Part III, Chapter 9) has provided dramatic visualization of this process as shown in Figure 6.9. Virions associate through capsid interactions with modified cytoplasmic membranes, leading to the formation of "blebs" on the surface of the plasma membrane. This process then continues until a bud is formed, which then extends outward and breaks off, forming a complete enveloped virion. The final stage of budding requires the action of one of the three cellular protein complexes primarily involved in cytoplasmic vesicle formation. These protein complexes normally function to pinch off budding vesicles from the parental membrane and carry out the same role in the final budding of the virion.

For viruses budding at other subcellular locations (such as the bunyaviruses, which bud into the Golgi itself; or HSVs, which bud from the nuclear membrane and then into **exocytotic vesicles**), a similar process occurs. In each case, the viral glycoproteins contain trafficking signals that direct the protein to its destination, using host cell machinery for this purpose. The plasma membrane of many cells in organized tissue is asymmetrical, and some viruses have evolved to utilize this asymmetry. Thus, certain viruses (e.g., influenza viruses) bud from the **apical surface** of such cells, while others (e.g., vesicular stomatitis virus) bud from the **basolateral surface.** Using elegant recombinant DNA techniques to produce hybrid versions of the relevant proteins, the trafficking signals in these cases were shown to reside in the amino terminal portion of the viral glycoprotein.

Specific details of envelope formation and virion release are complex for nuclear replicating enveloped viruses exemplified by the HSVs. As outlined here, capsid formation takes place in the nucleus, and full capsids presumably associate with **tegument** (matrix) proteins near the nuclear membrane that has become modified by inclusion of viral glycoproteins glycosylated in the cellular Golgi apparatus. Mettenleiter and colleagues have provided persuasive evidence using electron microscopy and defined viral mutants whose formation of extracellular infectious virus involves two cycles of envelopment. A specific viral glycoprotein is incorporated into the inner nuclear membrane, and viral capsids bud into the lumen between the inner and outer nuclear membranes. This enveloped "pre-virion" then infects the cytoplasm through fusion with the outer nuclear membrane, resulting in the loss of this pre-envelope. Subsequently, the capsids acquire their mature envelope by budding into exocytotic vesicles, and enveloped virus is transported to the cell surface for release. The process is very elegantly shown in the electron micrographs of the exocytosis of pseudorabies virus included in Figure 6.10. This process will be described in more detail in Chapter 17 (Part IV), where HSV replication is described.

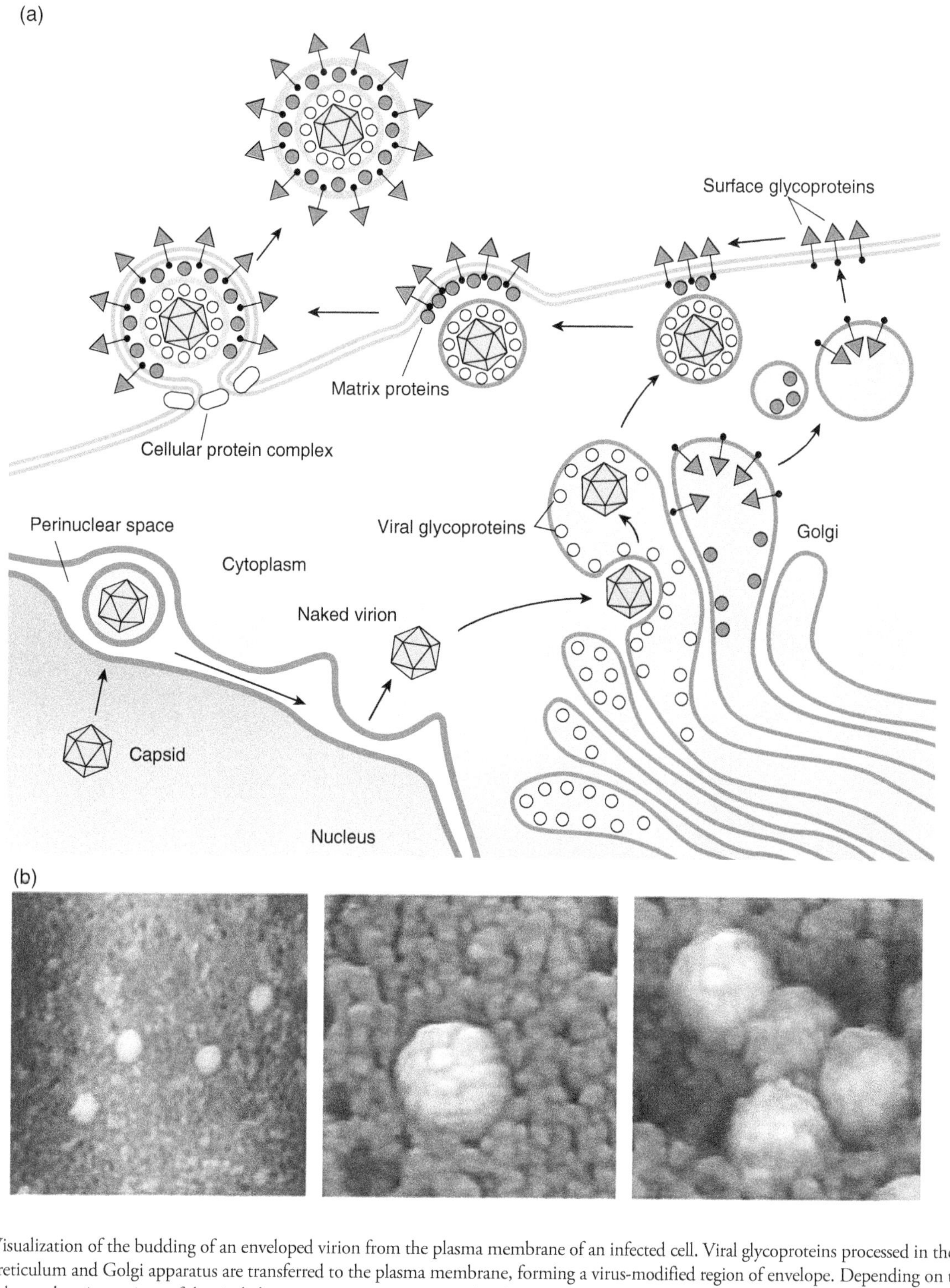

Figure 6.9 Visualization of the budding of an enveloped virion from the plasma membrane of an infected cell. Viral glycoproteins processed in the endoplasmic reticulum and Golgi apparatus are transferred to the plasma membrane, forming a virus-modified region of envelope. Depending on the virus, the C-terminal cytoplasmic portions of the viral glycoproteins may associate with other viral proteins of the matrix. The modified region of the plasma membrane can specifically associate with mature virions assembled inside the infected cell. This association leads to budding and release of mature enveloped virions. The appearance of enveloped murine leukemia virus (a retrovirus) at the surface of an infected cell, as visualized by atomic force microscopy, is also shown. Here the background plasma membrane of the cell has a slightly different appearance due to differences in the membrane-associate proteins present, and the budding of the virus at the surface forming enveloped virions is apparent. Source: Courtesy of Yuri G. Kuznetsov and Alex McPherson, University of California, Irvine.

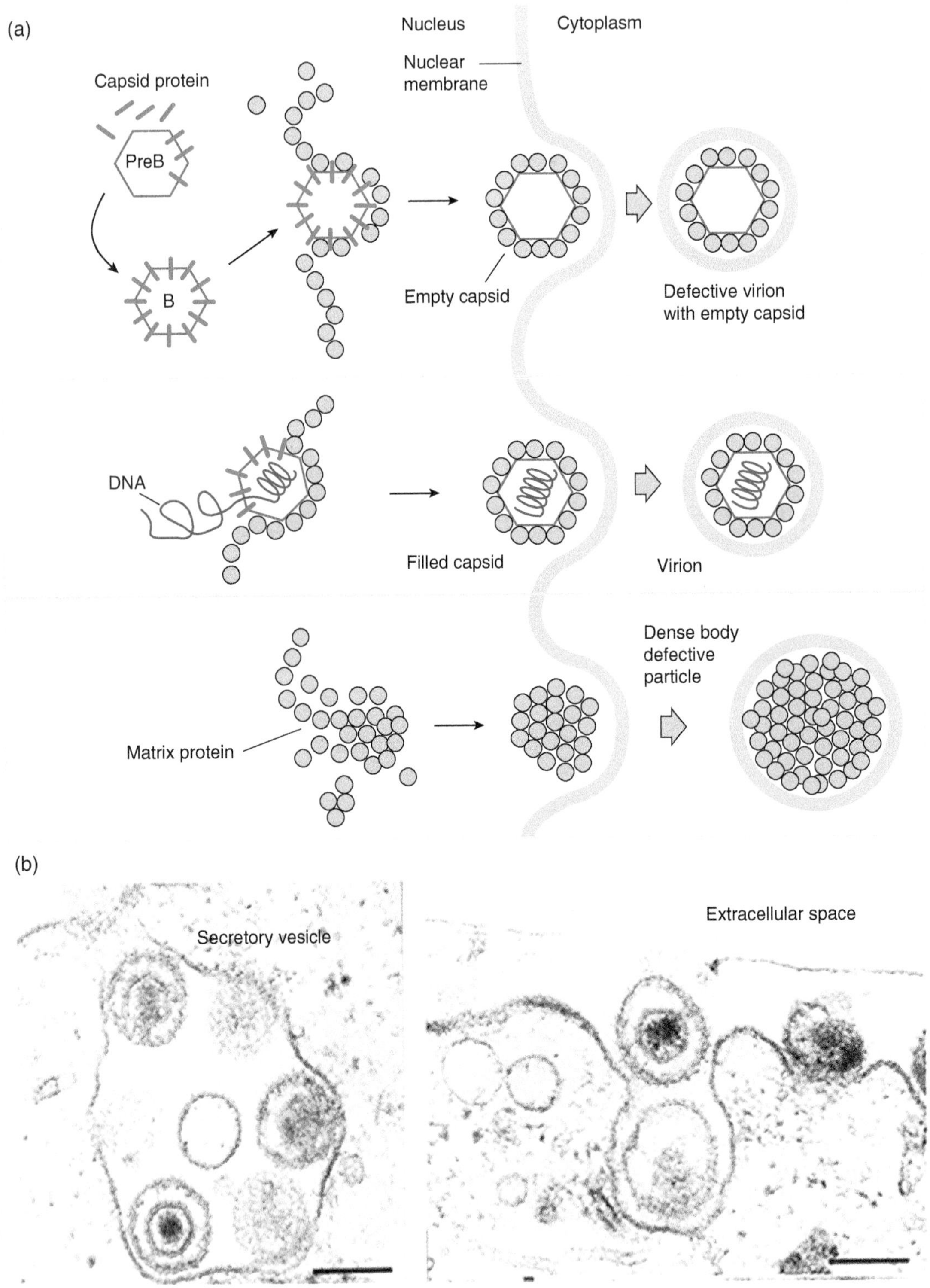

Figure 6.10 The envelopment and egress of herpesvirus. Electron micrographs of exocytosis of pseudorabies virus in the cytoplasm of the infected cell; release of enveloped virions is clearly shown. The bars represent 150 nm. Source: Reprinted with the kind permission of the American Society for Microbiology from Granzow, H., Weiland, F., Jöns, A., et al. (1997). Ultrastructural analysis of the replication cycle of pseudorabies virus in cell culture: a reassessment. *Journal of Virology* 71: 2072–2082.

QUESTIONS FOR CHAPTER 6

1 Briefly describe the two modes that enveloped viruses use for entry into their host cells.

2 How do nonenveloped viruses enter their host cells? Describe in detail one example.

3 How do plant viruses enter their host cells? What feature of the plant cell's architecture dictates these modes of entry?

4 Describe how the T-even bacteriophage attaches and enters the host cells. Which part of the virus particle enters the cell?

5 Simple virus capsids are found in two types of structural arrangements: helical and icosahedral. What are the key features in the assembly of these two kinds of particles?

6 How do enveloped viruses acquire their membranes during their maturation in animal cells?

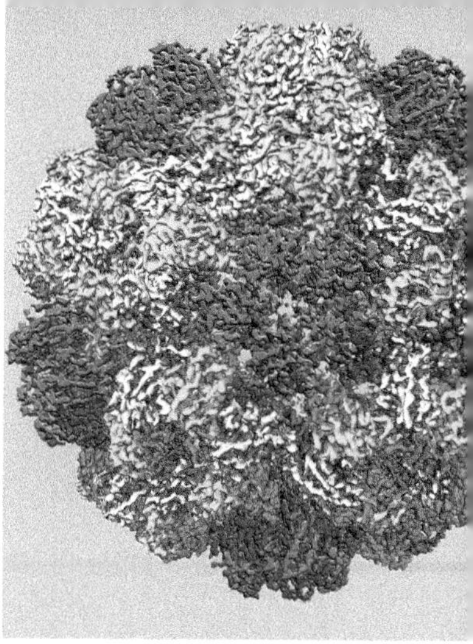

The Innate Immune Response: Early Defense Against Pathogens

CHAPTER

7

Basic Virology, Fourth Edition. Martinez "Marty" Hewlett, David Camerini, and David C. Bloom.

The immune system that protects the body from invading pathogens is composed of two different parts: the **intrinsic/innate immune response** and the **adaptive immune response**. The **intrinsic immune response** involves defense systems that are always present in cells and can immediately counteract viruses, like apoptosis, RNA silencing, and autophagy pathway proteins or RNAs. The **innate immune response** is a generalized response that is turned on after the cell "senses" certain proteins or molecules that are found on, or produced by, bacteria, viruses, or fungi. This response is the earliest antipathogen defense and results in rather nonspecific intracellular signaling responses to block the infection as well as extracellular inflammatory responses. In addition, the intrinsic and innate responses help to signal the body to the presence of an invading pathogen, and help promote the more specific and potent adaptive immune response, which involves B- and T-cell responses and must be activated.

HOST CELL-BASED DEFENSES AGAINST VIRUS REPLICATION

When a virus infects an immunologically naive host, one might expect that initially the odds are in favor of the virus. After all, most viruses have relatively short replication cycles, resulting in the rapid release of hundreds of new virions from a single cell. While the body immediately starts mounting specific antibody and cellular immune responses, it takes time for enough virus-specific B and T cells to accumulate in high enough numbers (even locally) to destroy infected cells, and to prevent the infection from spreading to other host cells and tissues. The innate immune response, one of the most primitive and ancient arms of the immune system, plays a critical role in slowing the spread of virus at very early times after infection. This innate response buys the host the critical time it needs to develop the more specific adaptive immune response to control the infection (see Figure 7.1).

Toll-like receptors

Elements of the innate immune response were first identified in *Drosophila* mutants (*Toll*) that were observed to be especially susceptible to fungal infections. These mutants lacked a key protein that has since been shown to be involved in cell-signaling response that promotes a nonspecific antifungal response. Years later a mouse strain that was particularly susceptible to gram-negative bacterial infections was found to lack a receptor protein that was related to the *Drosophila* protein. In normal mice, when this receptor protein encounters lipopolysaccharide (LPS), it initiates a cell-signaling response that causes a nonspecific inflammatory process. This inflammatory process alters the local cellular environment in a manner that slows down bacterial growth, until an antigen-specific immune response has been mounted. The LPS-specific mouse receptor was termed **"toll-like" receptor (TLR)** after the name of the original *Drosophila* mutant.

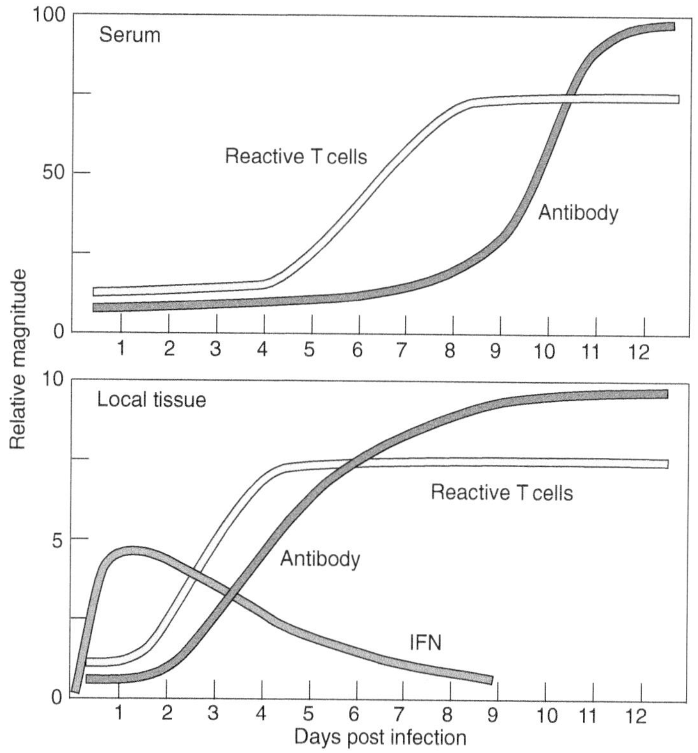

Figure 7.1 Schematic representation showing differences in the intensity and time of appearance of local versus systemic immunity against a typical virus infection in mice. IFN: Interferon. Source: Courtesy of D. C. Bloom.

It is now known that vertebrates possess a number of different TLRs (at least 10 in humans) that bind to different types of molecules that are associated with bacterial, fungal, viral, and protozoan pathogens. TLR3 and TLR9 play a particularly important role in the innate antiviral defense; for example, TLR3 recognizes and binds to double-stranded RNA (dsRNA), which is formed during the replication cycles of many viruses (RNA and DNA viruses alike).

Once TLRs are activated, they activate adaptor proteins that in turn induce pro-inflammatory cytokines. Each of the TLRs induces specific pathways through the activation of different adaptor proteins. For example, TLR3 (activated by dsRNA) induces type 1 interferons, described in Section 7.1.3. Interestingly, in addition to the antiviral effects mediated by the interferon and inflammatory responses induced by the specific TLRs, the cytokines induced by the innate immune response also play a key role in helping to directly activate and augment the development of specific cellular and humoral immune responses. For example, in response to LPS activation, TRL4 produces cytokines that specifically activate helper T cells (also called $CD4^+$ cells) that have been stimulated by specific antigens. Moreover, the type of cytokines produced by TLR innate responses is now believed to be an important determinant of the type of helper-T response that is mediated in response to a given pathogen. As discussed in this chapter, the selectivity of the helper-T response plays a critical role in dictating whether an antiviral response is primarily humoral or cellular, a process that no doubt has helped drive viral evolution.

Defensins

Another component of innate immunity is mediated by cellular proteins known collectively as **defensins**. These small proteins (composed of 30–50 amino acids) are secreted by a number of cells of the respiratory and gastrointestinal systems and bind to many pathogens, including bacteria, fungi, and some viruses. Such binding enhances elimination of bacteria and fungi. Defensins have been shown to interfere with the entry of influenza virus and HIV into human cells by cross-linking cellular membrane proteins. This cross-linking blocks the virus-induced clearance of cellular membrane proteins from the region juxtaposed to the viral envelope, thus blocking membrane fusion (see Figure 6.3b for an illustration of the normal process of membrane fusion).

Like TLRs, defensins also enhance the stimulation of both cellular and humoral immunity. Defensins also appear to play a role in helping generate immune responses to tumor cells in the host. This further illustrates the complex interactions between the different arms of the immune system and the continuing evolution of components of these arms toward cooperative interactions.

Interferon

The ability of cells to produce interferon (IFN) provides another important rapid response. The cells capable of such a response contain a complex set of gene products that can be induced in direct response to virus attack and that render neighboring cells more resistant to virus replication. IFN has a large number of biological effects, including the following:

- Inhibition of virus replication in IFN-treated cells (target cells)
- Inhibition of growth of target cells
- Activation of macrophages, natural killer cells, and cytotoxic T lymphocytes
- Induction of **major histocompatibility complex class I (MHC-I)** and **major histocompatibility complex class II (MHC-II)** antigens and Fc receptors
- Induction of a large number of other host intrinsic response effectors
- Induction of fever

It was shown in the late 1950s that culture media isolated from fibroblasts infected with certain viruses contained a substance or substances that would render uninfected cells more resistant to infection with similar viruses (i.e., the infected cells produced a substance that interfered with subsequent infection). Classic protein fractionation methods demonstrated that this substance – IFN – is actually a group of proteins, all very stable to acid pH and all able to function at very high dilutions so that only a few molecules interacting with a target cell render that cell resistant to viral infection.

There are two basic interferons, types I and II. Type I IFNs are stable at acid pH and heat. All are distinct and are encoded by separate cellular genes, but all have the same general size and have roughly similar effects. The two major type I IFNs are IFN-α, expressed by leukocytes, and IFN-β, expressed by fibroblasts. There are at least three others in this class. There is only one type II IFN, IFN-γ, expressed mainly by T lymphocytes. Type I IFNs are most active against virus infections, while IFN-γ modulates the immune response and appears to have some antitumor activity. All IFNs are very species specific; therefore, human IFN is active in human cells, mouse IFN in mouse cells, and so on.

The characterization of IFN followed by cloning and expressing IFN genes resulted in a lot of excitement concerning its potential use as an antiviral and anticancer drug. The promise has yet to be fully realized; it is now known that IFN proteins are very toxic to cells, and methods for their efficient delivery to regions of the body where it would be therapeutic have yet to be perfected. Thus, although it is clear that the IFN response has a role in natural recovery from virus infection and disease, its complete therapeutic potential is yet to be fully exploited.

Induction of interferon

IFN induction takes place in the infected cell in response to viral products. A major inducer is dsRNA, which is generated in infections by many RNA and DNA viruses. In addition, some viruses (e.g., orthoreoviruses) use dsRNA as their genetic material. A single molecule of dsRNA can induce IFN in a cell under the appropriate conditions. A cellular protein, **RIG-1**, binds dsRNA to domains in its C-terminal region and serves as the cellular detector of such RNA. When bound with dsRNA, the RIG-1 protein activates a number of cellular transcription factors, which act together to induce expression of type I IFNs. Interestingly, another antiviral protein expressed by mitochondria (**mitochondrial antiviral protein [MAV]**) mediates the effect of RIG-1. The activation of a receptor protein by its binding to the appropriate signaling ligand, leading to interaction and activation of further proteins leading to a transcriptional or other cellular response, is a common feature of cellular **signaling cascades**, and will be discussed briefly in Part III, Chapter 13.

Because IFN is expressed from cellular genes, only cells that are relatively intact and functioning when dsRNA is present will express it. The requirement for continuing cell function is one reason why viruses that replicate slowly are good IFN inducers. When a virus capable of rapid replication and quick host-cell shutoff initiates an infection under optimal conditions, little IFN is generally induced.

The antiviral state

IFN inducers cause the cell in which they are present to synthesize IFN. This protein is secreted and interacts with neighboring cells to put them in an antiviral state in which **antiviral effector molecules (AVEMs)** are expressed. Cells that have been induced by IFN express new membrane-associated surface proteins, have altered glycosylation patterns, produce enzymes that are activated by dsRNA to degrade messenger RNA (mRNA), and inhibit protein synthesis by ribosome modification. These effects are outlined in Figure 7.2. In the antiviral state, therefore, the cell is primed to trigger a number of responses to virus infection. As in the case of IFN induction, viral dsRNA acts as the trigger of these responses.

To date, expression of more than 300 cellular genes has been demonstrated to be induced or enhanced by IFN – many of these are involved in the establishment of the antiviral state. One protein – Mx – appears solely directed against influenza virus infections, although it also has activity against vesicular stomatitis virus (VSV). Some of these proteins that serve as AVEMs are listed in Table 7.1. Different mechanisms are involved in the different cellular responses to virus infection. Changes to the cell surface may make it more difficult for viruses to attach and penetrate. When presented with dsRNA, the antiviral cell activates **2′,5′-oligoA synthetase** – enzymatic activity that is induced by IFN, which produces an unusual oligonucleotide, 2′,5′-oligoA. This, in turn, activates a latent mRNA endonuclease (RNAse-L). Finally, this endonuclease rapidly degrades all mRNA (viral and cellular) in the cell. The IFN-primed cell also expresses a **dsRNA-dependent protein kinase (PKR)** that causes modifications resulting in partial inactivation of the translational initiation factor eIF2 in the presence of dsRNA. This makes the cell a poor producer of virus proteins, and thus an inefficient producer of new infectious virus, since all molecular processes are inhibited.

The action of IFN on cells is not always beneficial. Since IFN also acts as a negative growth regulator (the basis of its activity against tumor cells), its presence can interrupt the function of differentiated cells and tissues. Also, one cellular response to virus infection is the induction of a number of cellular genes that lead to programmed cell death (apoptosis); this process is outlined in more detail in Chapter 10, Part III. Such cell death is good for the host, since the reduction of virus replication is well worth the loss of a few cells, but in some cases IFN can block the induction of apoptosis and, thus, actually protect virus-infected cells! Further, IFN causes tissue inflammation and high fevers.

The toxic effects of the IFN response are alleviated by its being carefully balanced and controlled, so that it is maintained only as long as needed. The amount of IFN produced by any given infected cell is very small, so that only the cells within the immediate vicinity are affected and converted to the antiviral state. If the cells are not infected, they may eventually recover and resume their normal processes.

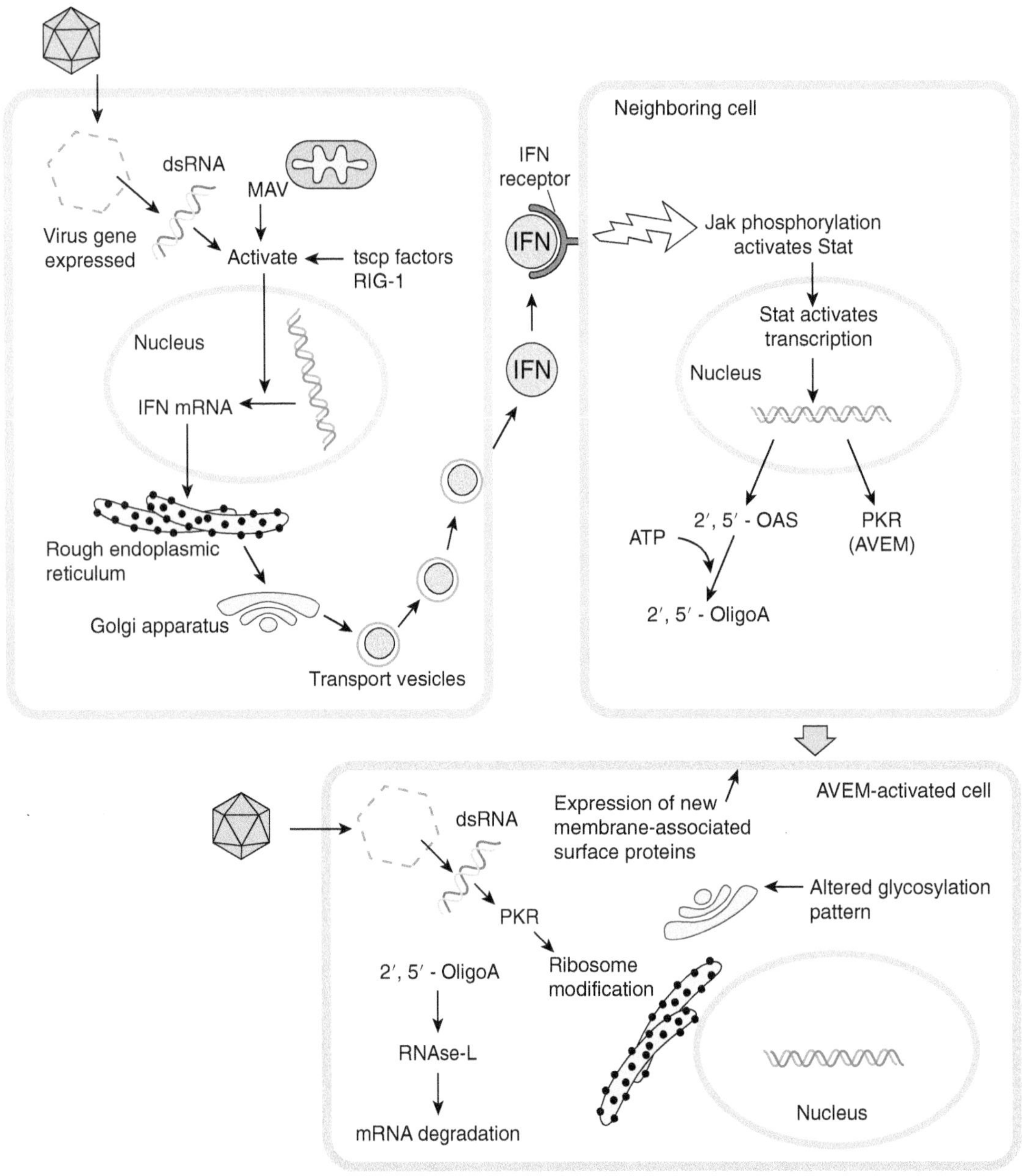

Figure 7.2 The cascade of events leading to expression of interferon (IFN) and induction of the antiviral state in neighboring cells. The interferon inducer (dsRNA) produced during virus infection leads to an infected cell secreting small numbers of the IFN proteins, which are extremely stable glycoproteins. These interact with neighboring cells to induce the antiviral state, in which a number of antiviral effector molecules (AVEMs) are expressed and can be triggered by the presence of dsRNA to alter the cell to markedly reduce the yield of infectious virus. dsRNA: Double-stranded RNA; PKR: dsRNA-dependent protein kinase; 2′,5′-OAS: 2′,5′-oligoA synthetase.

Measurement of interferon activity

IFN activity is measured in a number of ways because there are so many different types and different effects. An easy and rapid method in virology is the *plaque reduction assay*. This method is quite sensitive; it has been claimed that as few as 10 molecules of IFN can be detected with its careful use. *Plaque assays* are described in detail in Chapter 10, but in essence the process is as follows: Duplicate cell cultures are set up (see Chapter 9), and one culture is treated with IFN

Table 7.1 Some antiviral proteins induced or activated by interferon.

Protein	Function
2′,5′-Oligoadenylate synthetase	Activates latent RNAse-L
dsRNA-dependent protein kinase (PKR)	Phosphorylates eIF2
RNAse-L	mRNA degradation
IFN-1, IFN-2	Transcriptional regulation
MHC-I	Antigen presentation
Mx	Specific blockage of influenza (and vesicular stomatitis virus) entry

for several hours to allow the potential antiviral state to develop. Both are then infected with the same number of infectious units of indicator virus (often VSV since it is so sensitive to IFN). The IFN-treated cells will produce fewer and smaller centers of virus infection (*plaques*) than will the untreated control. Serial dilutions of the original sample can be made until the effect is no longer seen, and a measure such as *median effective dose* (**ED_{50}**) can be calculated. The ED_{50} is that dilution in which the number of plaques is reduced by 50% or plaques are 50% smaller than untreated ones. This reduction can be related to units of IFN activity and to the number of IFN molecules present.

Other cellular defenses against viral infection

Small RNA-based defenses

Discoveries starting in the early 1990s have demonstrated that small RNA molecules with double-stranded regions have a number of important roles in regulating eukaryotic cellular processes and protecting against pathogens beyond the induction of interferons. This is briefly described in Chapter 8; here, it suffices to note that there are pathways in eukaryotic cells for processing small RNA molecules encoded in the cell's genome into 22-base-pair dsRNA molecules (microRNAs or miRNAs). These miRNAs then bind to specific viral or cellular mRNA molecules, leading to their degradation.

Cells have a similar way of dealing with dsRNA occurring in transcripts, such as those produced by viral infections. These can be processed into 28-base-pair dsRNA molecules called small interfering RNAs (siRNAs). Such siRNAs interfere with the translation of mRNAs containing homologous sequences, also by inducing the degradation of those mRNAs. Thus, infection of a plant cell with a virus will lead to the spread of these to neighboring and more distant cells, resulting in resistance to viral spread. The presence of plant virus genes, which act to counter the function of plant siRNAs, demonstrates the extent of this system in the plant kingdom.

It is not yet clear just how extensive the roles of miRNA and siRNA are in protecting animal cells, but evidence of the importance of these molecules in virus infections and the antiviral response is growing. There are over 2000 miRNA-encoding sequences in the human genome, and recently miRNAs have been identified in a number of DNA viruses including human herpesviruses and SV40. While it is often difficult to identify the target of a given miRNA, an SV40 miRNA blocks part of cellular control of its replication cycle. In contrast, it has also been found that hepatitis C virus utilizes a small RNA species of human liver cells to increase the efficiency of translation of its mRNA. All available evidence suggests that siRNAs and miRNAs act as a kind of innate immune response directed against viral RNA motifs. There is also evidence that viruses may have co-opted miRNAs that target cellular defense mechanisms as a means of evading host responses.

THE ADAPTIVE IMMUNE RESPONSE AND THE LYMPHATIC SYSTEM

The human lymphatic system shown in Figure 7.3 is part of the general circulatory system and plays a critical role in developing the immune response to the presence of foreign proteins in the body. When any protein that is not part of the vast protein repertoire making up the vertebrate host is presented to the immune system by an **antigen-presenting cell (APC)**, both B-cell immunity (**humoral immunity**) and T-cell immunity (**cell-mediated immunity**) are mobilized. Such a foreign protein is usually termed an *antigen* and can be derived from an invading pathogen (virus, bacteria, or parasite), or it can be a novel cellular protein expressed as a result of abnormal growth properties of the cells – a **tumor antigen**. In general, an antigen that is not part of a host's normal protein composition can be recognized by the host's immune system as foreign and can become a target of the immune response.

Lymphocytes are produced, differentiate, and mature in certain specialized tissues, including bone marrow, spleen, and thymus. They circulate throughout the body in the circulatory and lymphatic systems, and can migrate between cellular junctions into tissue in response to infection. They are most concentrated in **lymph nodes**, where stimulation to provide a **systemic immune response** often begins. B cells produce **antibodies** that are secreted proteins able to bind specifically to the antigenic determinants on proteins. Activated T cells have antigen-binding sites on their surfaces and, upon encountering cells expressing foreign antigens (such as virus-infected cells), interact with

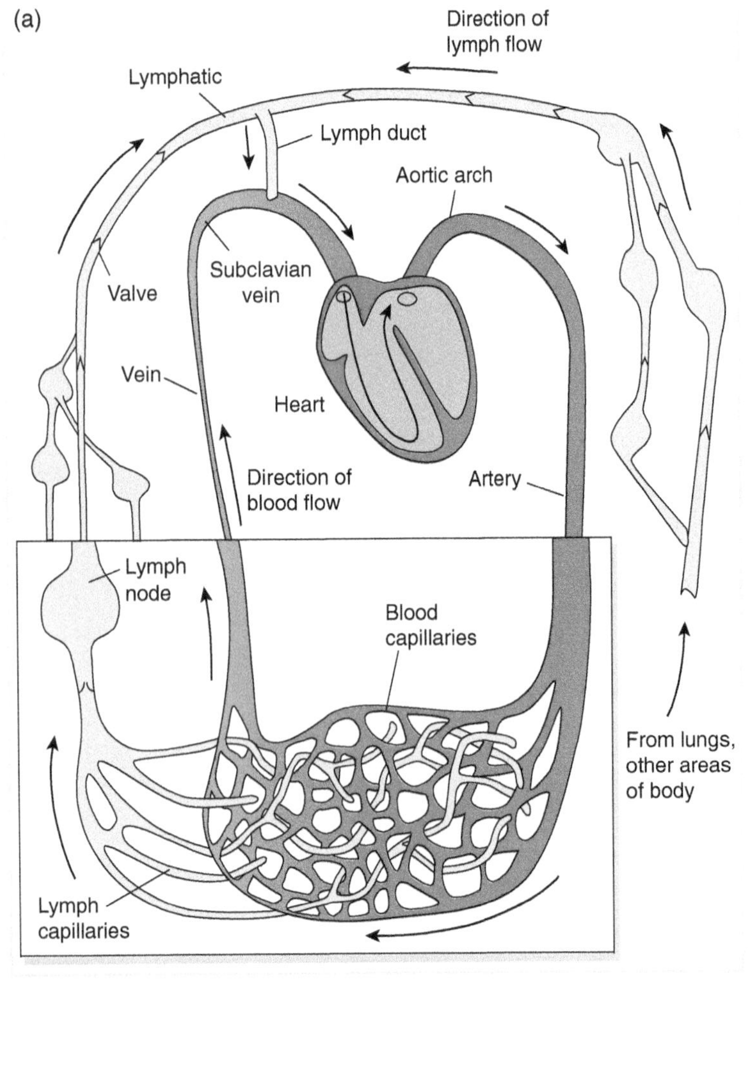

Figure 7.3 The human lymphatic system. The lymphatic system is the principal organ of the immune system. (a) The relationship between the lymphatic circulation and that of the blood. (b) Some of the important components of the lymphatic system as related to the immune response.

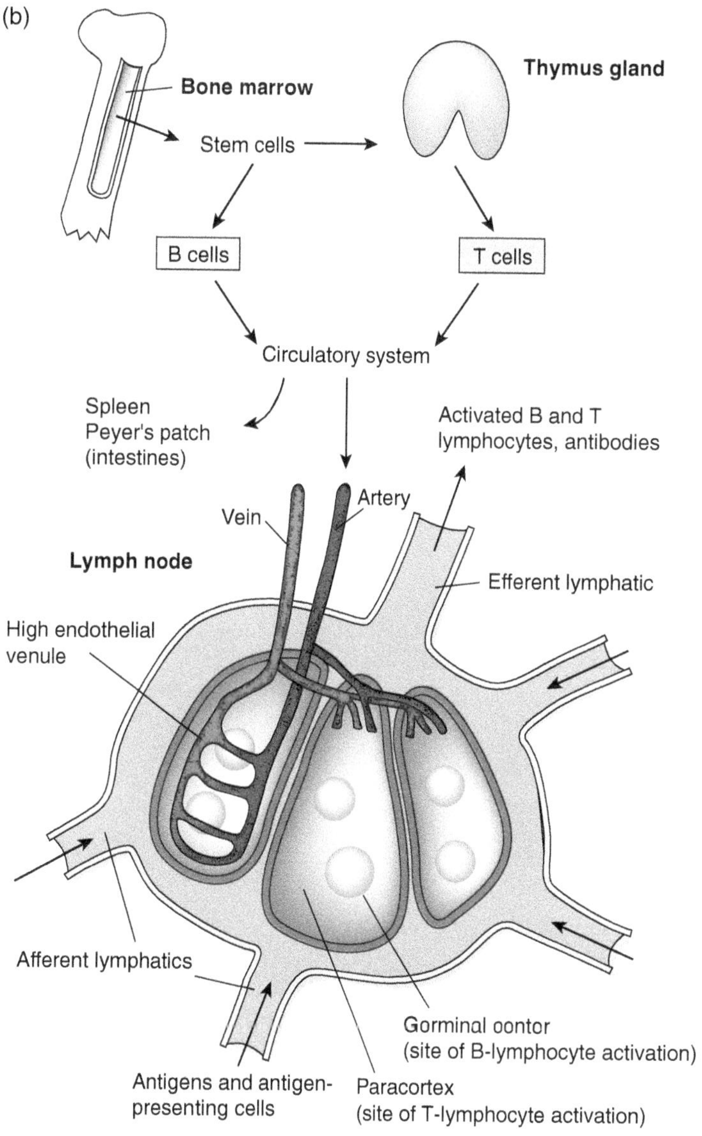

Figure 7.3 *Continued*

them, resulting in lysis of the infected target cells. Certain T cells (helper T cells) can also function in helping to promote the development of B-cell and cytolytic $CD8^+$ T-cell immunity.

Together, these two arms of the adaptive immune system interact to allow the host to detect and destroy or render noninfectious (inactivate or neutralize) both free virus and virus-infected cells that display viral proteins at their surface. A general outline of the interaction between an antigenic pathogen and the adaptive immune system is shown in Figure 7.4.

Two pathways of helper T response: the fork in the road

An early event in the development of a specific immune response is the presentation of virus-specific antigens to cells of the adaptive immune response. These include helper and regulatory T cells, cytotoxic T cells, and B cells. A key event in this process is defined by the presentation of antigens to $CD4^+$ helper T cells. Following this antigen presentation, some of the cells will

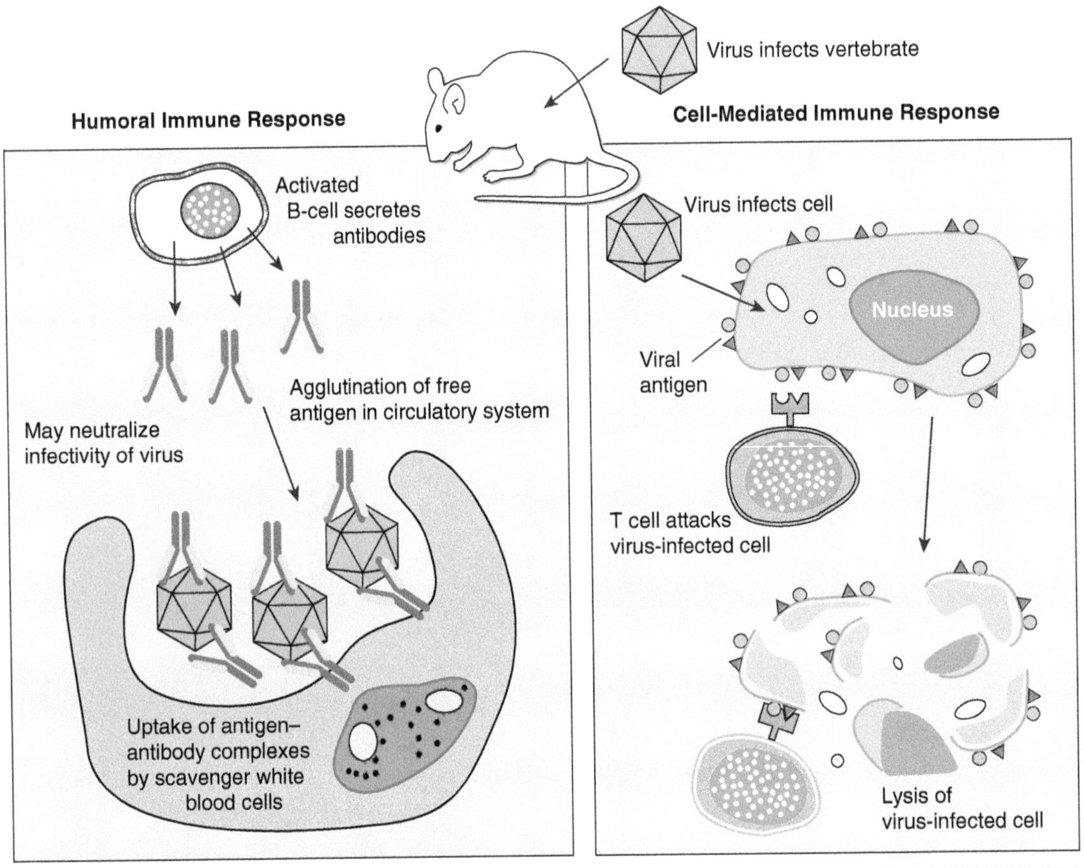

Figure 7.4 T and B cells in immunity. T lymphocytes play the central coordinating role in evoking the immune response. Upon activation by interaction with a specific antigenic determinant with which they can interact, they proliferate and carry out the functions shown. B cells reactive with specific antigens require reactive T cells for their maturation. Upon maturation, they secrete antibody proteins that bind to antigenic determinants.

differentiate into **helper T 1 (Th1) cells**, and some will differentiate into **helper T 2 (Th2) cells**. Th1 cells primarily secrete IFN-γ which mediates the differentiation of $CD8^+$ T cells to generate a cytolytic cellular immune response. Conversely, Th2 cells produce primarily IL-4, IL-5, IL-10, and IL-13, which promote the differentiation of B cells with resulting humoral antibody response.

A number of the TLRs discussed so far in this chapter have been shown to have roles in promoting either a Th1 or Th2 response. In this sense the innate immune response serves to not only buy time for the development of a specific immune response against a viral pathogen, but also play a role in guiding the development of the type and degree of response that is elicited. This is only one of a large number of examples of the interleaving of various components of the immune system into a functional whole.

The immunological structure of a protein

In any protein, certain clusters of amino acids (usually between 10 and 12) are able to interact with the appropriate antigen-recognizing T cells or antibody-producing B cells to lead to proliferation of those cells. These clusters are called antigenic determinants (**epitopes**). B-cell reactive epitopes are usually **hydrophilic** and, thus, hydrated. A viral protein can have none, a few, or many antigenic determinants, depending on its protein structure, amino acid sequence, sequence relation to cellular proteins, degree of glycosylation, and other factors. Two proteins can share some of the same or closely related determinants, and the closer the relation between the proteins, the greater the shared ones. This is why closely related viral serotypes share a high degree of immunological reactivity. A schematic representation of epitope types present in proteins is shown in Figure 7.5.

Epitopes are often composed of a specific sequence of amino acids. With such an epitope, denaturation of the antigenic protein will have little or no effect on its properties or how it is

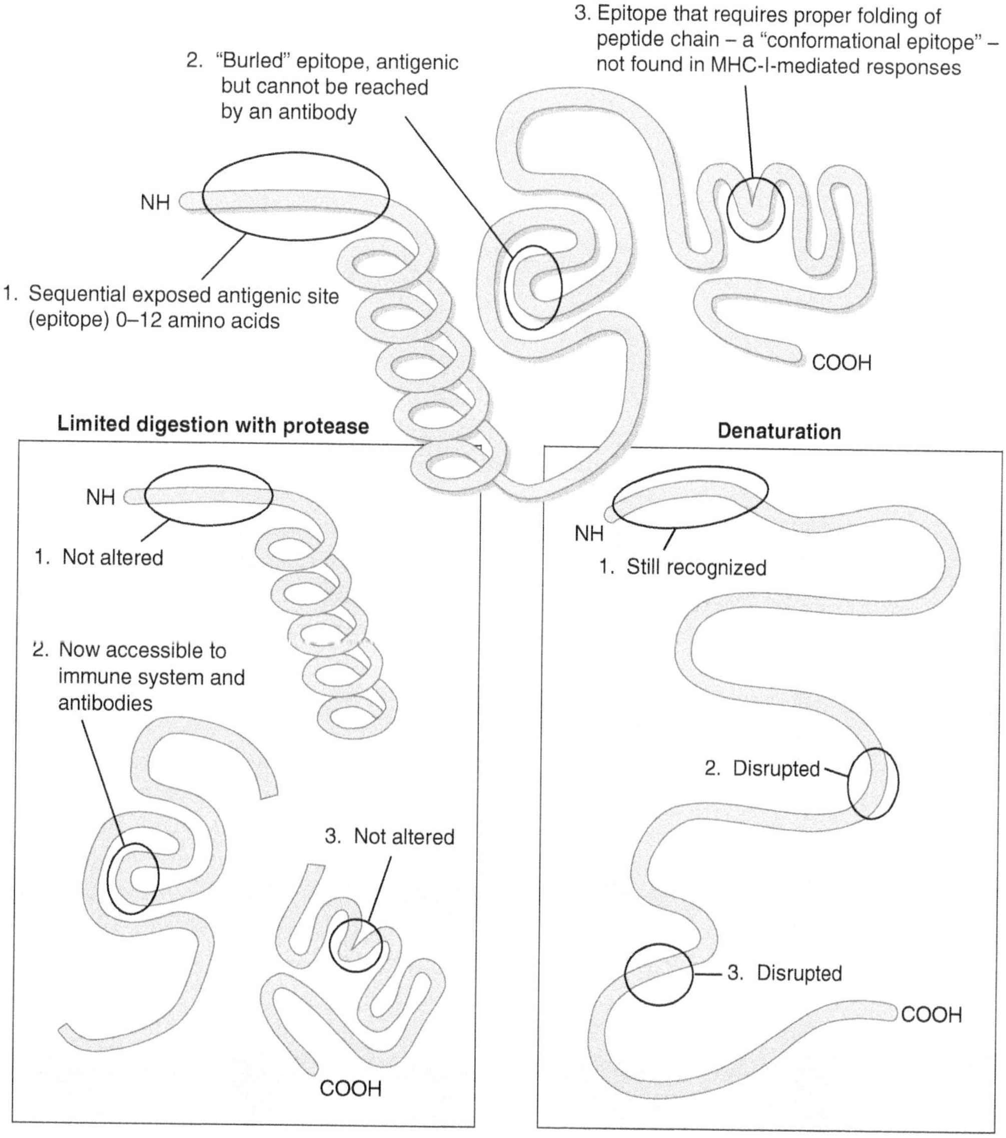

Figure 7.5 The antigenic structure of a protein. Specific groups of amino acids (usually hydrated) serve as specific antigenic determinants, or epitopes in an antigenic protein. Some of these are insensitive to the protein's physical structure; others require a specific conformation for presentation.

presented to the immune system. Such determinants expressed in a protein in either its native or its denatured state are called *sequential epitopes*.

Epitopes can also be sensitive to the structure of the protein region where they occur. For example, they could be made of amino acids that have been brought near each other by protein folding or conformation. These are **conformational epitopes**, which are sensitive to denaturation (disruption) of the protein structure.

Either sequential or conformational epitopes can be in the interior of a protein where they are not normally seen by the humoral immune system. These are *buried determinants*. Many of these are sequential and can be exposed by denaturation of the protein. A buried conformational determinant could be exposed by proper limited degradation of the protein, or by denaturation of the protein followed by its being refolded in a form that exposed the epitope.

Role of the antigen-presenting cell in initiation of the immune response

Any protein and many other macromolecules can be antigenic, but antigens must be "processed" and then presented at the surface of the cell bearing them (**antigen-presenting cell**) in the proper context to be able to evoke an immune response. This context is as a complex with one of two closely related heterodimeric cell surface glycoproteins, the **major histocompatibility** proteins. These MHC glycoproteins ensure that only antigen-presenting cells (e.g., macrophages and **dendritic cells**) from the same organism can present antigens to the immune system.

There are two basic pathways through which cells present antigens (see Figure 7.6). The first, which is a function of nearly all cells, is the presentation of endogenously expressed antigens on the surface via MHC-I (Figure 7.6a). As proteins are being synthesized, portions are complexed with a group of cellular proteins named **ubiquitins**, which target the proteins to proteolytic vesicles (**proteosomes**) where they are partially degraded into epitope-sized peptides. These peptides are then moved via transporter proteins (**TAPs**) into the Golgi apparatus, where the peptides associate with newly synthesized MHC-I glycoproteins and are presented on the surface of the cell. These MHC-I complexes serve as targets for surveying $CD8^+$ T cells, and if reactive, the cells bearing the antigen are destroyed. In this way, the immune system surveys all cells for the synthesis of foreign or abnormal proteins. This endogenous antigen presentation is important in the early immune detection of viral-infected cells, and is clearly a major factor in local immunity.

The establishment of systemic immunity and immune memory requires a relatively large population of freely circulating, relatively short-lived effector T cells that can recognize the antigen in question. This primarily occurs via the activity of long-lived specialized dendritic cells that were formed in the bone marrow and migrate to the epithelium where they remain. Dendritic cells and certain other cells of the immune system are often termed professional antigen-presenting cells, because of this primary role in evoking systemic immunity. Antigenic proteins or complexes are recognized in manners that are not fully understood, and are internalized and partially digested by receptor-mediated endocytosis. Fragments of antigens containing epitopes are re-expressed on the cell surface in the presence of MHC-II proteins. The antigenic fragment and the MHC molecules together form a surface structure that can be recognized by $CD4^+$ T and certain B cells in lymph nodes to begin the amplification of cells able to recognize the antigen – this is shown schematically in Figure 7.6b. MHC-II-mediated antigen presentation occurs in lymph nodes. Because antigen concentration must reach a high enough level to evoke the immune response, the process takes time and occurs only following a lag after initial infection and early replication of the virus. This delay is important in virus infections – such as herpes simplex virus (HSV) infections – where virus can invade sensory neurons and establish latent infections before a powerful immune response is achieved. Indeed, HSV, like some other viruses, can actually interfere with the MHC-I-mediated early presentation of its antigenic proteins at the surface of the infected cell by the action of a specific viral protein expressed immediately following infection.

Some viruses (notably HIV) can survive internalization by dendritic cells, and their presentation to T cells leads to infection of lymphocytes. HIV can replicate in $CD4^+$ lymphocytes, and eventually replication of the virus in infected lymphatic cells leads to destruction of the immune system.

As the T and B cells able to interact with the presented epitope continue to proliferate, immature B cells with surface receptors that can bind to antigen also internalize and process the antigen. These B cells provide an alternative mechanism for presenting antigen in the lymph nodes.

The internalization and processing of antigens are clearly of paramount importance to the ability to generate effective immunity. Nevertheless, in addition to the generation of sequential determinants, the host can generate immune responses to complex conformational epitopes, such as portions of dimeric and multimeric proteins found at the surface of the virus. Indeed, the host preferentially mounts strong antibody responses to the surface proteins of viruses. Part of the reason for these responses is that such proteins are present in large amounts and are at the

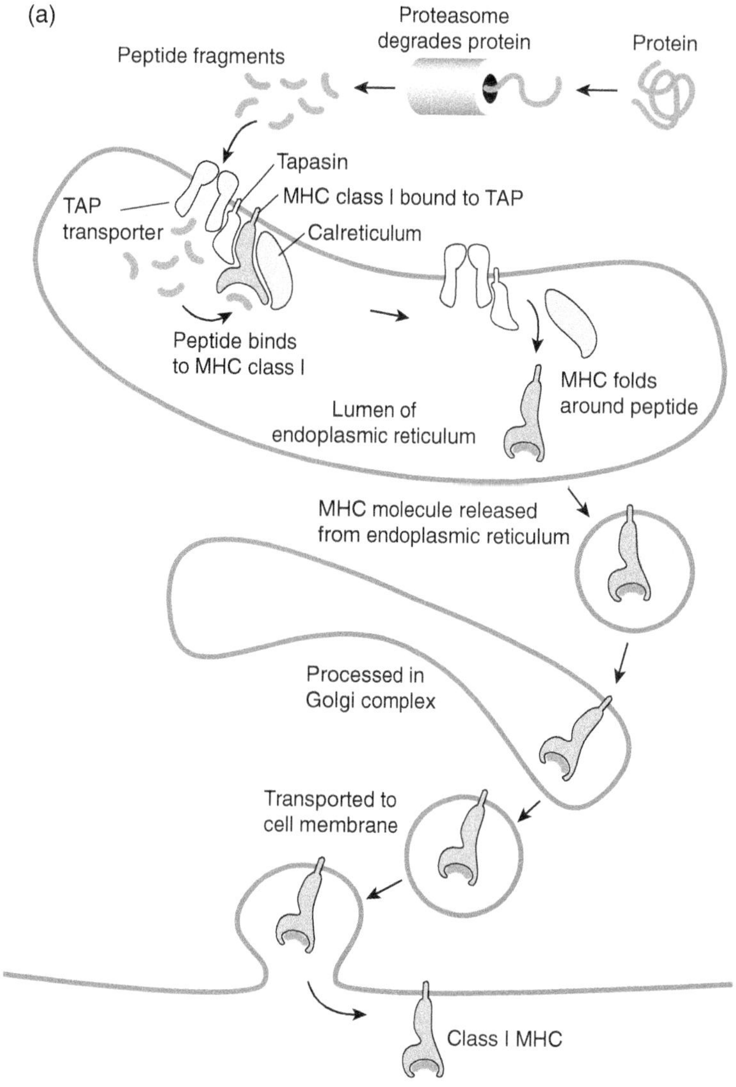

Figure 7.6 The processing of a foreign antigen and stimulation of the immune response. As described in the text, an antigenic protein can only stimulate the immune response when it is processed by a macrophage and then presented to cells of the immune system in lymph nodes in the presence of histocompatibility antigens. The processing is relatively rapid and involves partial degradation of the antigenic protein and expression of antigenic portions on the surface of the antigen-presenting cell. (a) MHC-I antigen processing and presentation. (b) MHC-II processing and presentation.
ER: Endoplasmic reticulum.

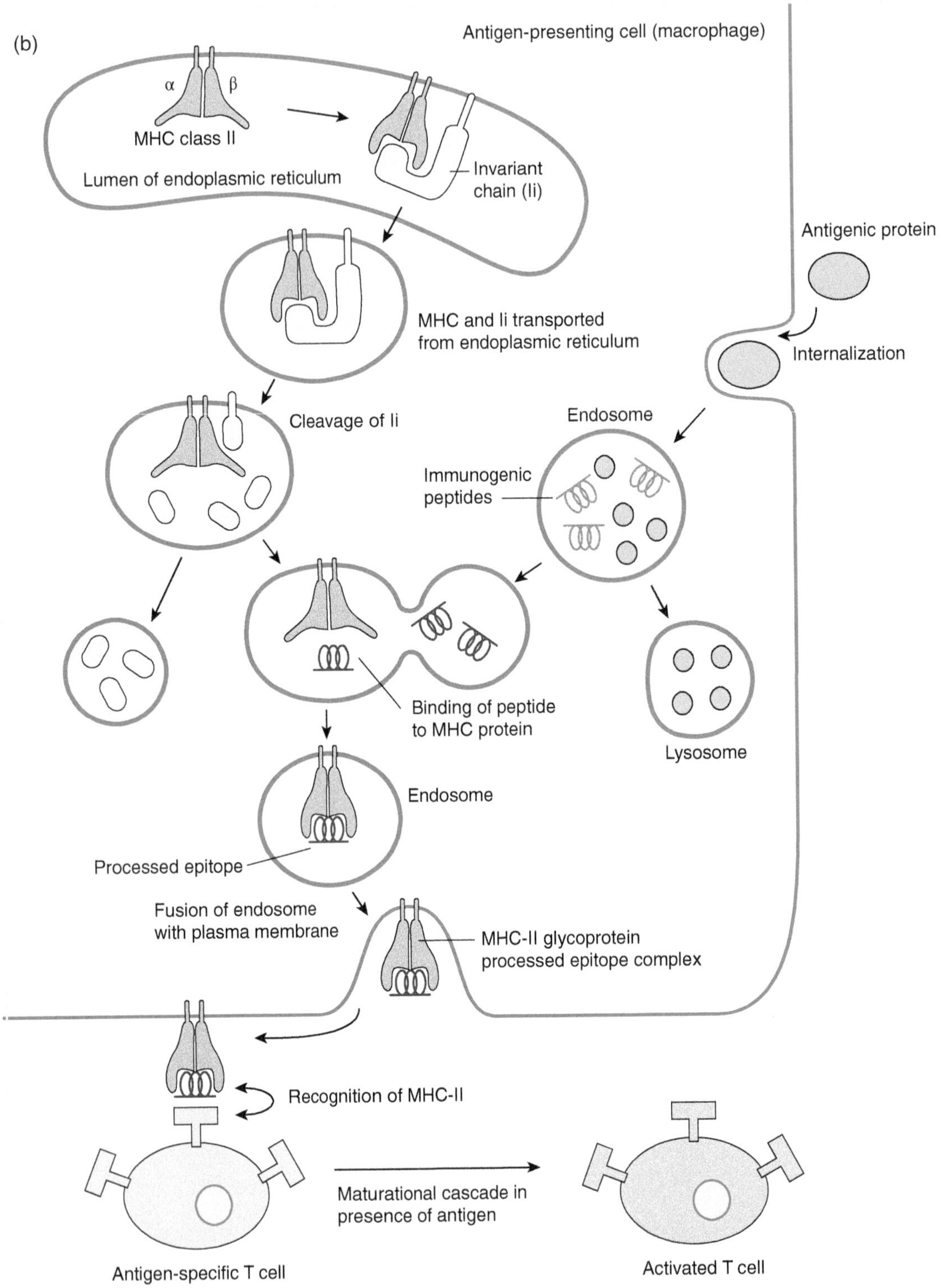

Figure 7.6 *Continued*

"interface" between the infection and the host antigenic response. Other factors are also involved, including structural features of the proteins, inherent resistance to extensive degradation, and the ability of surface antibody (immunoglobulin-G [IgG]) on immature B cells to recognize native (unprocessed) antigens.

Clonal selection of immune-reactive lymphocytes

When antigens are presented to immune cells that can recognize them, those T and B cells are stimulated to proliferate. As shown in Figure 7.7, the process of **clonal selection** takes place because each specific antibody-producing B cell and each specific epitope-recognizing T cell are derived from a single reactive cell (i.e., clones of that cell). This process takes place mainly in the lymph nodes because of the high concentration of cell populations that must interact. The ability to generate clones of antibody-producing B cells in the laboratory has provided an extremely important tool for studying the functional structure and relationships between various cellular and viral proteins. Some basic techniques using such material are outlined in Chapter 12, Part III.

As they are stimulated by the presence of a specific epitope that they recognize, B cells divide and differentiate (mature). Fully differentiated B cells secrete soluble antibodies. One class of effector T cells ($CD4^+$, or helper T, cells) mediates the maturation of B cells. Another class, $CD8^+$ cytotoxic T cells, attacks and destroys cells with foreign antigens on them, such as virus-infected cells. A third class of T cells (regulatory or T_{reg}) suppresses the immune response toward the end of the "crisis," when immunity is at a high level and antigen levels begin to decline.

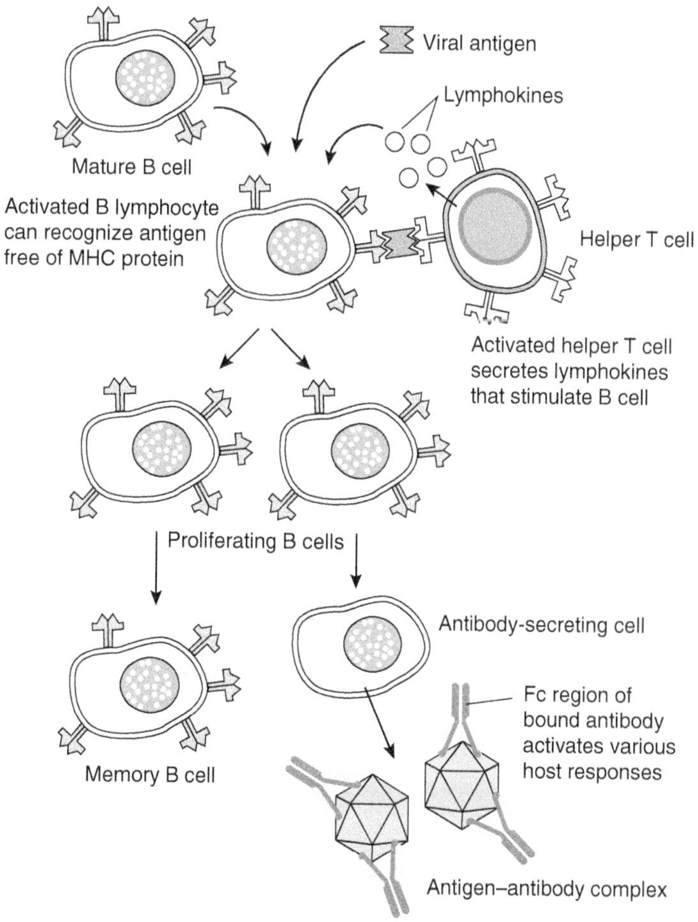

Figure 7.7 The clonal selection of B lymphocytes. Only the B lymphocytes reactive with a specific epitope can be stimulated to mature by the action of a helper T lymphocyte. Specific mature B cells secrete specific types of antibody molecules, but the same epitope will result in only the stimulation and maturation of B-cell clones reactive with it.

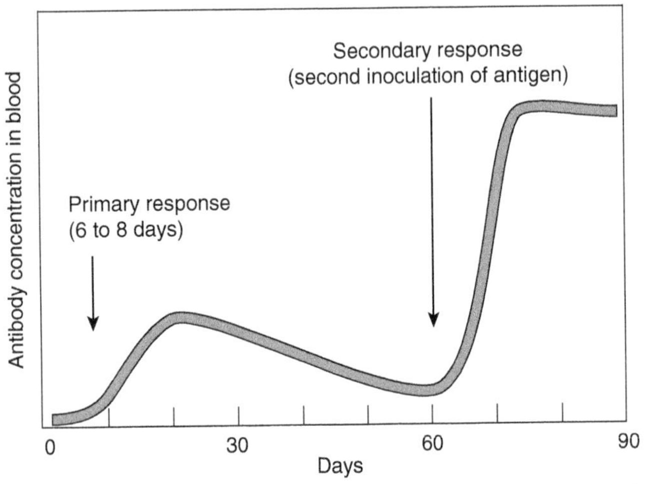

Figure 7.8 Immune memory. The first exposure to an antigen results in the primary response, which occurs after a week or so. During this time, maturation of immune-reactive cells is taking place. Once the primary response occurs, antibody and reactive T and B cells decline to a low level. Upon restimulation with the same antigen, the memory lymphocytes are rapidly mobilized and a more intense and more rapid immune response follows.

Immune memory

The immune system "remembers" the antigenic response and can rapidly respond to re-exposure to the antigen. Long-lived memory T and B cells mediate immune memory. Such memory cells reside mainly in lymph nodes and circulate in the blood and lymph. As antigen persists, the cells that respond to it continue to proliferate. While most have a finite lifetime and then undergo apoptosis, memory cells do not function in dealing with the antigen, but rather are long-lived and remain ready to respond to a second infection with the same or closely related pathogen. A second stimulation results in rapid interaction of the antigen with such memory cells and a secondary (remembered) immune response that is more rapid and more extensive than the first or primary response. The effect of immune memory on the strength and speed of the immune response is shown in Figure 7.8.

Complement-mediated cell lysis

Although T cells have a primary role in the destruction of cells bearing foreign antigens, B cells can also destroy antigen-bearing cells by use of the complement system, which leads to *complement-mediated cell lysis*. This system works because cells with antibodies bound to them trigger a cascade of interactions with serum complement proteins that leads to destruction of the cell; this process is outlined in Figure 7.9.

CONTROL AND DYSFUNCTION OF IMMUNITY

The T and B cells with antigenic recognition sites having the highest affinity for a given epitope are stimulated most efficiently. As general levels of antigens fall late in infection and during recovery, lower levels of high-affinity interactions can continue to stimulate immunity. Thus, the nature of the immune response changes with time after infection. A recovering patient will generally have higher-affinity and more specific antibodies than will an individual early in the course of a disease.

A population of $CD4^+$, $CD25^+$ regulatory T cells mature very late in the immune response and shut down immunity. These cells are important to the regulation of immunity. If they do not function properly, hyperimmune responses such as allergic reactions may occur.

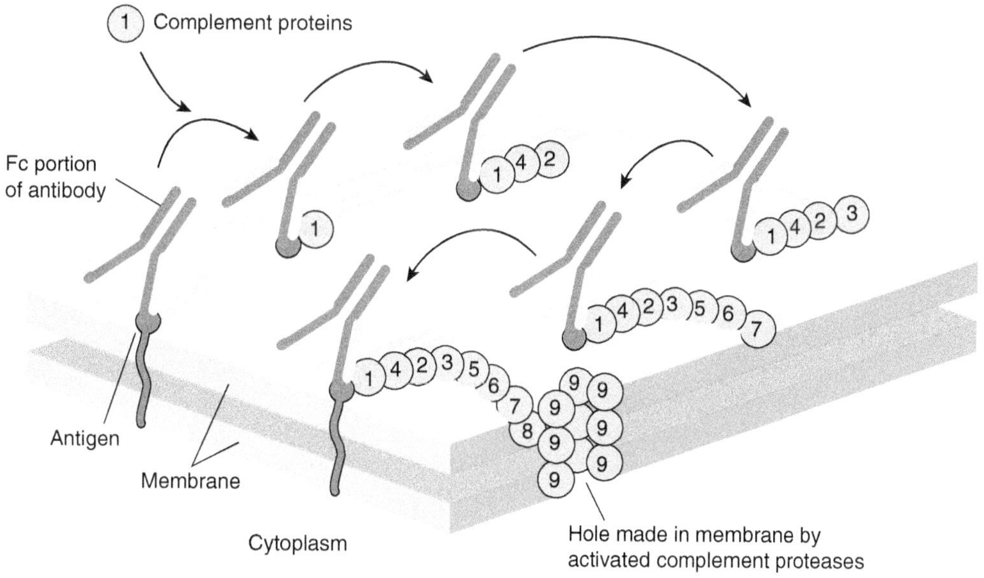

Figure 7.9 The maturational cascade of serum complement proteins upon binding to an antigen–antibody complex on the surface of a cell. The Fc region, a portion of the antibody molecule that is not involved in binding to the epitope of the antigen, specifically triggers this cascade.

If they function too well, inadequate immunity may result. Several autoimmune diseases are caused by a lack of regulatory T cells, which normally comprise 1–3% of the total population of $CD4^+$ T cells.

Other types of immune pathologies include autoimmune diseases where the immune system destroys seemingly healthy tissue in the body. This can be due to the immune system attempting to destroy cells that express viral antigens but are otherwise healthy.

An example of an autoimmune pathology due to viral infection and persistent presentation of antigen is subacute sclerosing panencephalitis (SSPE), which is a pathological response to persistence of measles virus antigen in neural tissue. This was briefly described in Chapter 4. Some other autoimmune diseases, such as multiple sclerosis, are thought to be caused by a previous virus infection and apparent recovery. It has been suggested that a previous infection with a virus (perhaps years before) can lead to immune pathology – in this case, demyelination of neurons. The exact mechanism of such pathology is not known, but a process termed **molecular mimicry**, where a specific epitope of the pathogen bears similarity to one in the host tissue, is suspected. Here, during the course of a normal immune response against the invading pathogen, normal tissue is also now recognized as foreign. This is known to be the mechanism for the role of group A *Streptococcus* in rheumatic fever, where the robust immune response to the bacterial epitope leads to problems because of similarity to an epitope found in a protein in heart tissue. This mechanism has not been proved for multiple sclerosis, and indeed, such cases require very careful statistical evaluation of long-term medical records to demonstrate correlations.

Specific viral responses to host immunity

The immune response is an effective one, and it plays a constant role in selection against viruses that do not mount an efficient infection. Despite the effectiveness of the immune response, it is clear that many virus infections survive and thrive in the setting of the host's immune capacity. Indeed, the great majority of nuclear-replicating DNA viruses establish long-lasting associations with their hosts. Clearly, they are able to deal with host attempts to clear the infection. A major factor in virus survival is the fact that viruses mount many effective counter-responses to the immune response. Some of these are essentially passive, while others involve virus-mediated blockage of specific portions of the immune response.

Passive evasion of immunity – antigenic drift

All animal viruses occur in antigenically distinct forms or serotypes. The number of forms varies with the type of virus. For example, there is only one strain of measles virus, three major serotypes of poliovirus, more than 40 for adenovirus, and as many as 100 for papillomaviruses. A serotype is stable and may be confined to a specific geographic location, and prior infection with one serotype of a specific virus will lead to no or only partial protection from reinfection with another.

Because RNA-directed RNA replication has no built-in enzymatic error-correction mechanism, in contrast to DNA replication, RNA viruses are generally more susceptible to the generation of mutations leading to serotype formation than are DNA viruses. This process is often termed **antigenic drift**; such drift is probably responsible for the large number of serotypes of rhinoviruses (more than 100), and is clearly responsible for the drift in influenza virus serotypes.

This mechanism for drift is countered by other factors that tend to favor antigenic "conservatism." For example, many RNA viruses (e.g., measles and poliovirus) do not exhibit large numbers of serotypes, and even where there is extensive drift, as with influenza, the internal proteins are antigenically relatively stable.

One factor in stabilizing protein sequences even when they are encoded by highly mutable RNA sequences is that important functional constraints on the amino acid sequence of viral proteins are imposed by enzymatic or precise structural functions. Such constraints do not operate with the same lack of tolerance for variation in the external glycoproteins of enveloped viruses.

Passive evasion of immunity – internal sanctuaries for infectious virus

Some viruses can evade the immune response of the host by establishing persistent or latent infections in tissue that is not subject to extensive immune surveillance. A classic example is the ability of HSV to establish latent infection in nondividing sensory neurons. Another example is the ability of respiratory syncytial virus to replicate at low levels in the mucous membranes of the nasopharynx, where secretory antibodies provide protection against invasion by the virus but cannot clear it. The highly localized replication of papillomaviruses, such as those causing skin warts, is another example of virus infection in a localized area that is removed from intense immune surveillance.

Passive evasion of immunity – immune tolerance

The immune system of fetuses and neonates is immature. This is an important strategy in the survival of the fetus as it develops in an antigenically distinct individual: its mother. Fetal and neonatal infections with viruses that normally cause generally mild infections in an immune-competent individual can be devastating in neonates. Rubella causes severe developmental abnormalities of the nervous system when it infects a developing fetus, and the fact that the virus does not evoke lasting immunity in adults means that it is a threat even to a mother who has been infected previously. A primary or reactivating HSV infection of the mother at the time of birth can lead to neonatal encephalitis with grave prognosis, and neonatal and uterine infections with cytomegalovirus are strongly linked to neurologically based developmental disorders. Active HIV replication at the time of delivery is also the major mechanism of mother-to-child transmission of this virus.

At least one group of viruses, the arenaviruses, utilizes the ability to selectively accommodate themselves to the developing immunity of the neonate. These viruses, of which lymphocytic choriomeningitis virus (LCMV) is the best-characterized laboratory model, persist in populations of rodents and are transmitted to newborns from the infected mother. The mouse develops relatively normally with persistent viremia and shows an impaired immune response to LCMV. The tolerant mouse has circulating antibody that is reactive with the virus but cannot neutralize it. Further, there is a lack of T-cell responsiveness to the virus. If an immune-competent adult mouse is infected with LCMV, a robust immune response is mounted, but the infection is usually fatal (see Chapter 23, Part V).

The mechanism for establishing immune tolerance is complex; it involves selection of specific viral genotypes with the ability to infect macrophages and some other cells of the immune system during the early stages of infection of the infant. This infection results in suppression of specific immunity against the virus.

Interestingly, the virus that is spread between individuals has tropism for neural tissue. These neurotropic and lymphotropic viruses differ only in a single amino acid in both the viral glycoprotein and the viral polymerase. The two variants are generated by random periodic mutations during replication of the resident virus in the animal, and while the neurotropic variant has little effect in the immune-tolerant animal, it causes severe disease in an uninfected adult. Similar patterns of infection are seen with other arenaviruses, several of which – including Lassa fever virus – are pathogenic for humans.

Active evasion of immunity – immunosuppression

Infections with a number of viruses lead to a transitory or permanent suppression of one or several branches of host immunity. Infectious mononucleosis caused by primary infection with Epstein–Barr virus (EBV) is a self-limiting generalized infection characterized by a relatively large induction of regulatory T lymphocytes. This not only results in the virus being able to maintain its infection effectively, but also results in the individual who has the infection being more susceptible to other infections. Some retroviruses, especially HIV, are able to specifically inhibit T-cell proliferation by the expression of suppressor proteins. Further, the continued destruction of T lymphocytes by HIV replication eventually leads to profound loss of immune competence: AIDS.

The polydnaviruses of certain wasps illustrate an evolutionary adaptation between virus and host based on the virus's ability to actively suppress immunity. This virus (mentioned in Chapter 1) is maintained as a persistent genetic passenger in the ovaries and egg cells of parasitic wasps. These wasps lay eggs in caterpillars of another insect species, and the developing larvae feed on the caterpillar as they develop. The polydnavirus inserted into the caterpillar along with the wasp egg induces a systemic, immunosuppressive infection so that the caterpillar cannot eliminate the embryonic tissue at an early stage of development. If wasps without such viruses inject eggs into the caterpillar host, there is a significant reduction in larval survival.

Active evasion of immunity – blockage of MHC antigen presentation

Adenovirus, HIV, and HSV specifically inhibit MHC-I antigen presentation. In each case, a specific virus protein that mediates this blockage is expressed. While it is apparent that the slowly replicating adenovirus will greatly benefit from its ability to interfere with host immunity, it requires a moment of reflection to see the importance of the blockage of MHC-I antigen presentation by HSV, which replicates very rapidly and efficiently in the cells it infects. Here, it is likely that the value is found in the earliest stages of reactivation from latent infection, where small amounts of virus must be able to initiate infection in a host that has a powerful immune memory biased against HSV replication. Similarly, one of the earliest genes expressed by HIV encodes a protein (Nef), which downregulates MHC-I expression, thus evading cytotoxic T-cell responses.

Consequences of immune suppression to virus infections

While some viruses are able to either mildly or profoundly suppress immunity during the course of infection and pathogenesis, immune suppression is also an important tool in certain medical conditions. Examples include the need to suppress host cell–mediated immunity prior to organ or tissue transplantation. Immune suppression also results from some types of intravenous drug abuse.

Major complications from immune suppression are reactivating herpesvirus infections such as varicella zoster (chickenpox) and cytomegalovirus infection. Of course, the same problems can occur when the immune system is disrupted by viral infections such as with HIV. A potentially more critical complication of significant populations of individuals evidencing immune suppression results from their serving as potential selective reservoirs for the development of antigenic and drug-resistant strains of pathogens. For example, the current increase in appearance of antibiotic-resistant tuberculosis is linked definitively to a combination of incomplete drug therapy, HIV infection, and drug-induced immunosuppression in critical urban, prison, and Third World populations.

MEASUREMENT OF THE IMMUNE REACTION

Measurement of cell-mediated (T-cell) immunity

T-cell proliferation assay

Measurement of cell-mediated T-cell immunity requires incubation of immune lymphocytes with a target cell and then measurement of a specific T-cell response: the lysis of the target cell. For measurement of T cell–mediated cell lysis, the release of radioactive chromium from target cells is a convenient method. Target cells are incubated under conditions such that they incorporate the radioactive metal. The cells are rinsed so that the only radioactivity is inside the cells. Thus, the radioactivity will sediment to the bottom of a centrifuge tube under low-gravity force (low speeds). In the presence of reactive killer T cells, the target cells are lysed and the "hot" chromium enters the solution and cannot be sedimented under low speeds.

A numerical assessment of the number of reactive lymphocytes can also be carried out by measuring cell replication as a response to a specific antigen. White blood cells are incubated with antigen and a radioactive nucleoside precursor to DNA. As T lymphocytes proliferate in response to antigen, they will incorporate this radioactive precursor. A measure of the incorporation of radioactivity in comparison to a control culture can be made and expressed as a lymphocyte **stimulation index**.

Another method for measuring T-cell immunity is to incubate antigen-bearing cells with lymphocytes. Reactive T lymphocytes will form *rosettes* around the antigen-bearing cell, and these can be observed and counted in the microscope.

Tetramer assay

Another powerful tool to detect and quantify T cells that specifically recognize a viral antigen is called a **tetramer assay**. This assay employs a synthetic "tetramer" of four MHC molecules that has been specifically engineered to contain a portion of a viral antigen. In this way the tetramer resembles how the MHC molecules of a virus-infected cell would present processed viral peptides on the cell surface to specific T cells. For this assay, these synthetic tetramers are labeled with a fluorescent molecule. When these tetramers are incubated with a sample of blood containing T cells, they will bind to only those T cells that recognized that viral antigen. These cells can then be detected on a fluorescence activated cell sorter, which can count the number of T cells that are present in the blood that have expanded to specifically recognize the virus protein.

Measurement of antiviral antibody

Antibody molecules are secreted glycoproteins that have the capacity to recognize and combine with specific portions of viral or other proteins foreign to the host. As described in Chapter 12, Part III, antibody molecules have a very specific structure in which the antigen-combining sites, which comprise variable amino acid sequences, are at one location on the antibody molecules, while a region of fixed amino acid sequence is found at another location. This constant region (**Fc region**) has a major function in mediating secretion of the antibody molecule from the B lymphocyte expressing it. Another major function of the F_C region is to serve as a signal to cells and other specific cellular proteins that the molecule bound to the antigen is, indeed, an antibody.

Enzyme-linked immunosorbent assays (ELISAs)

A number of methods to measure antibody reactions involve use of the antibody molecule's Fc region as a "handle." Extremely sensitive methods known collectively as enzyme-linked immunosorbent assays **(ELISAs)** use enzymes that can process a colorless substrate into a colored product bound to the Fc region of an antibody molecule. When the antibody is bound to an antigen, the enzyme affixed to the Fc region will also be bound. If the antigen–antibody complex is then incubated with appropriate substrates for the bound enzyme, the generation of color can be used as a measure of the antibody present. Examples of the method are outlined in Figure 7.10.

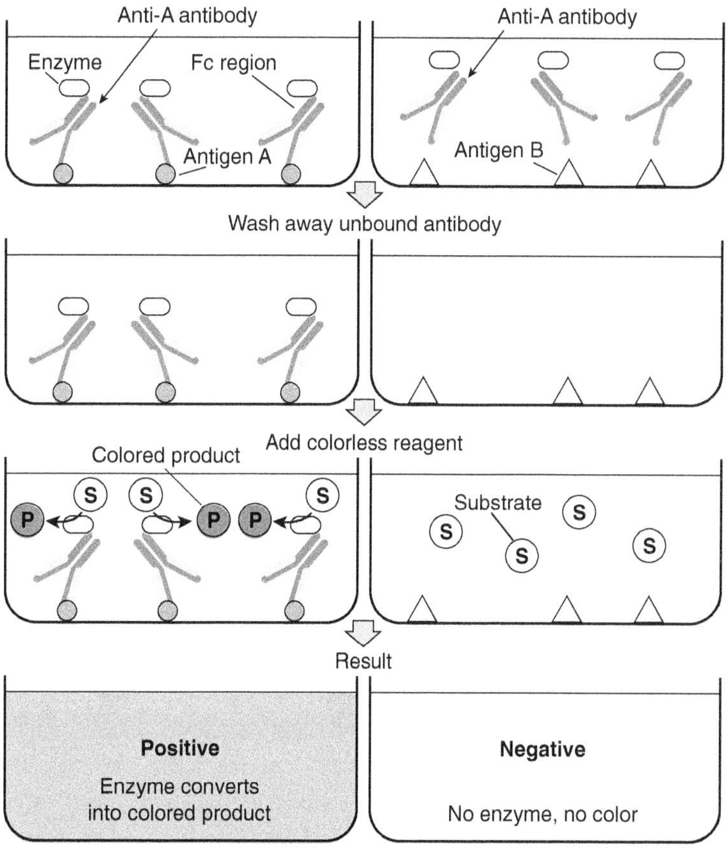

Figure 7.10 An enzyme-linked immunosorbent assay (ELISA): the method of using a color reaction mediated by an enzyme bound to the Fc region of the antibody molecule. P: Colored product; S: substrate.

ELISAs are of tremendous value for rapid diagnosis, and have great commercial significance. For example, if an antigenic peptide is bound to an insoluble matrix such as a flexible plastic strip onto which dry reagents are included and this strip is dipped into a plasma preparation that contains antibody against the peptide, a color will develop. Even if the amount of antibody is very low, incubation for a long enough period will generate some color as long as the enzyme used is relatively stable. The method is quite adaptable to quantitative as well as qualitative analysis, and can be adapted for use with automated equipment. A number of kits are currently commercially available where a small sample of body fluid that might contain either a virus or an antibody of interest can be spotted and dried. The kit is then sent to a laboratory where it can be quantitatively analyzed.

The use of lasers and microtechnology developed in the electronics industry promises to provide even more revolutionary changes to our ability to detect extremely small amounts of viral antigens or antibodies in test material. A microchip can be synthesized with a huge number of different potential antigens bound to it, and this can be incubated with unknown antibody and then subjected to either an ELISA or another method to generate a fluorescent signal where an antigen–antibody complex is formed. This can be rapidly scanned with a laser beam and fluorescent microscope or, alternatively, in a solid-state detection device. Such methods make it potentially possible to screen a given serum sample for the presence of antibodies directed against all or nearly all identified pathogenic agents in a few hours! Detection of such antibodies indicates current or previous exposure to the corresponding pathogen.

Neutralization tests

Some ways to measure the reaction between specific antibody molecules and an antigen involve the loss of specific functions by the target virus. Many antibodies will block the ability of a virus to initiate an infection in a cultured cell, and thus block the formation of a center of infection or *virus plaque*. Plaque assays are described in Chapter 10, Part III, and the inhibition of plaque formation is termed an *infectivity neutralization* or the *neutralization of a virus*. Here, a target virus with a known titer is incubated with test antibody dilutions. The more concentrated and specific the antibody, the more the initial antibody solution can be diluted and still block viral infectivity (and thus formation of plaques). Neutralization is illustrated schematically in Figure 7.11.

Inhibition of hemagglutination

Some methods for the measurement of antibody against viruses are based on the ability of the antibody to block some property of the virus. For example, it has been known since the first part of this century that many enveloped viruses will stick to red blood cells and cause them to agglutinate. This property of **hemagglutination** can be used as a crude measure of viral particle concentration in solution, as described in Chapter 9, Part III.

Many antibodies against enveloped viruses will inhibit virus-mediated agglutination of red blood cells, and this *hemagglutination inhibition (HI)* test can be used to measure antibody levels. The basic method was worked out long before a detailed understanding of the immune response was available, but it is based on the fact that many antibody molecules bind to the surface of viruses and physically mask them. If a virus that can cause hemagglutination is preincubated with an antibody to it, the virus will be coated with antibody and will not be able to stick to the red blood cells. This happens because the surface of the virus particle is relatively small, and once a protein molecule is stuck to it, that protein will block access to portions of the surface. If enough antibody sticks, the whole surface is obscured.

An experiment utilizing inhibition of hemagglutination (the HI test) is shown in Figure 7.12. All that is required to measure a patient's immune response is a standard virus stock and blood serum. The basic procedure is as follows: Standard samples of red blood cells (e.g., guinea pig or chicken red

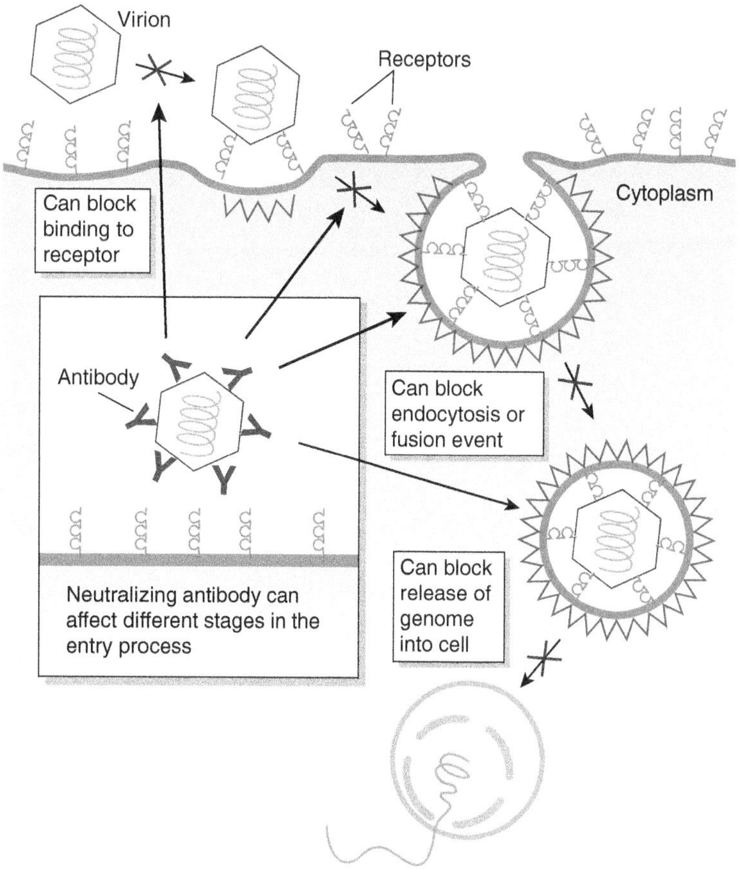

Figure 7.11 Antibody neutralization of virus infectivity. Specific types of antibody molecules, called *neutralizing antibodies*, can bind to surface proteins of the virus and block one or another aspect of the early events of virus–cell recognition or effective internalization of the virus.

blood cells for influenza virus) are mixed with a known amount of virus stock and different dilutions of an unknown antibody, which could be in a patient's serum. After a suitable period of time, the solution is gently shaken and subjected to low-speed centrifugation. If the red blood cells are agglutinated, the cells make a jelly-like clump and cannot sediment. Agglutination is characterized by a diffuse red or salmon-pink solution. If the red blood cells do not agglutinate because of sequestration of virus by the antibody, the cells pellet, forming a red "button" at the bottom of the tube. The beauty of using HI is not accuracy; it is relative speed, ease, and low cost of performance, which are very important in small clinical laboratories, especially in developing countries.

Complement fixation

Serum complement is made up of a number of soluble proteins that are able to stick to cells bearing antibody–antigen complexes. As this binding occurs, the complement proteins undergo structural changes, and finally the last protein bound is activated to become a protease, which then lyses the cell. The ability of complement to bind to antibody–antigen complexes at the Fc region of the antibody is termed fixation because, once bound, the complement is no longer free in solution. This property can be used as a relatively simple and inexpensive method to measure antibody–antigen reactions, called *complement fixation (CF)* titration.

In a CF assay, sheep red blood cells are used to make an antibody against their surface proteins, often in a horse, goat, or other large animal. The red blood cells are then "standardized" so that when a specific amount of antibody is added to them and the mix is incubated

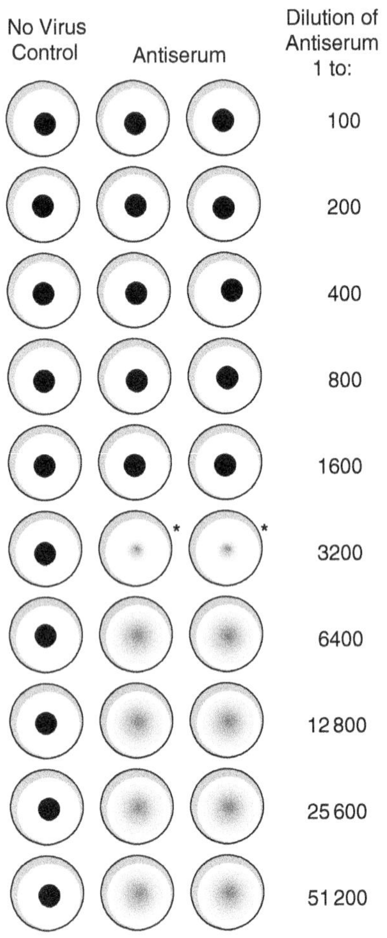

Figure 7.12 The hemagglutination inhibition assay for measuring antibody against a virus in serum. The assay is carried out by mixing constant amounts of a known hemagglutinating virus with serial dilutions of serum; then, the virus–serum mixture is added to red blood cells. Low dilutions of serum result in sequestering the virus so that it is not available for hemagglutination, and red blood cells in the wells pellet to the bottom under low centrifugal fields. Higher dilutions of the antiserum dilute the antibody concentration to a point where enough virus remains to cause a positive hemagglutinin reaction. If there were more antibody in the serum, a higher dilution would be required to accomplish this. Thus, the hemagglutination inhibition titer of the serum is a measure of how far it can be diluted and still block the hemagglutinin reaction. This is a measure of antibody concentration. In the example shown, a 1:3200 dilution of the original sample (asterisks) was the last one in which agglutination was inhibited. This is the endpoint of the antiserum dilution. Since a 1:3200 dilution was the endpoint, there were 3200 hemagglutination inhibition units in the original stock. Source: Based on a figure in Dimmock, N.J. and Primrose, S.B. (1994). *Introduction to Modern Virology*, 4e. Boston: Blackwell Science.

with guinea pig complement, the red blood cells lyse. Lysis of the red blood cells is readily assayed because when a solution of lysed red blood cells is centrifuged at low speed, the solution will stay red because there are no cells to take the hemoglobin to the bottom of the tube to form a pellet.

After the red blood cells, anti–red blood cell serum, and complement are standardized, they can be stored for relatively long periods in the cold. When they are used to assay an antibody–antigen reaction, the following process is carried out. Serial dilutions of either a solution of antibody of unknown strength and a fixed amount of known virus, *or* a solution with an unknown amount of virus and a fixed amount of known antibody, are incubated together. Then they are mixed with a known amount of guinea pig complement. If an antibody–antigen complex has formed, the complement will be *fixed* (i.e., bound) by it. If not, the complement will stay in solution. If there is an intermediate level of complex, then some complement will be fixed and some will be free.

Following incubation of the unknown antibody–antigen mix with the known amount of complement, the whole "mess" is incubated with standard amounts of red blood cells and anti–red blood cell antibody. If all the complement is fixed, there will be no lysis of the red blood cells. If some is fixed, there will be partial lysis of the red blood cells. If none is fixed, there will be complete lysis. Measurement of the degree of lysis (by measuring the amount of red color in solution following low-speed centrifugation) can be used to measure the amount of unknown antibody–antigen reaction and provides the CF titer.

Like HI, this method is not extremely precise or sensitive, but it is cheap and fast, and requires few expensive pieces of equipment. It is an ideal method for getting quick results in small laboratories or those with limited resources. It is still used in all modern hospitals.

QUESTIONS FOR CHAPTER 7

1 Which of the following statements is/are true?
(a) The only region in the body where a virus-infected cell can interact with T cells is in the lymph nodes.
(b) Virion surface proteins tend to elicit a stronger immune response during the course of natural infection than do internal components of the virion.
(c) Epitope-containing antigens must be digested to single amino acids and reassembled at the surface in the presence of histocompatibility antigens in order to provoke immunity.

2 Why are soluble antibodies (the products of the humoral response) good antiviral agents?

3 What are the roles of the following cells in the vertebrate immune response?
(a) B cells
(b) Helper T cells ($CD4^+$)
(c) Cytotoxic T cells ($CD8^+$)

4 What protein structural features are involved in the antigenic nature of epitopes?

5 What steps occur in the immune response following the primary infection of a vertebrate by a virus?

6 Assume you know that for a particular non-enveloped virus, gene A codes for a transcriptional activator, gene B for an origin binding protein, and gene C for a capsid protein. Following a normal infection in an animal, what would most likely generate a neutralizing antibody?

7 Interferons (IFNs)-α and -β are expressed in response to a virus infection and are released from the cell in which they are produced. IFNs induce an antiviral state in other neighboring cells.
(a) Which cellular process is inactivated when IFN-treated cells are infected with a virus?
(b) One arm of the IFN-induced antiviral state is the synthesis of 2′,5′-oligoA in response to viral infection. In one sentence or a simple diagram, what is the effect of this on the cell?
(c) Another arm of the IFN-induced antiviral state is activation of the protein kinase in response to viral infection. In one sentence or a simple diagram, what is the effect of this on the cell?
(d) All cells contain the genes for IFNs. IFN synthesis is stimulated by virus infection. Would you expect a cell that has been *treated with IFN* to synthesize IFN in response to a viral infection? Explain your answer.

8 The IFN response is one of the two major defense systems of animals in reaction to virus infection. The table below lists several activities that are associated with this response. Indicate which, if any of them, might be readily observed in cells before or after IFN treatment, with or without virus infection.

	Uninfected		Virus-Infected	
Activity	Normal Cells	IFN-Treated Cells	Normal Cells	IFN-Treated Cells
mRNA for IFN found in the cells				
mRNA for 2′,5′-oligo A synthase found in the cells				
2′,5′-oligo A found in the cells				
Inactive protein kinase found in the cells				
Receptor for IFN found on the surface of the cells				

Strategies to Protect Against and Combat Viral Infection

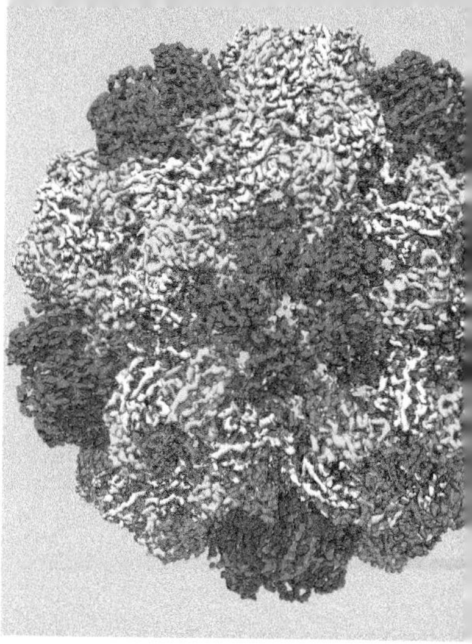

CHAPTER

8

Basic Virology, Fourth Edition. Martinez "Marty" Hewlett, David Camerini, and David C. Bloom.

Viruses and viral disease have coexisted with the bacterial and metazoan hosts in which they replicate since those hosts appeared in the biological universe. While this coexistence is a dynamic process, we humans can envision an "ideal world" where viral disease is controlled, if not eliminated, and its effects minimized. Despite all our knowledge of the biological world, it is clear that there are just two ways to deal with virus-induced disease: prevention and treatment.

Prevention of viral infection can be accomplished by application of public health measures to eliminate the spread of the virus or control its transmission, or it can be accomplished by making sure that there are no susceptible individuals available for the virus to infect. The latter can be accomplished by inducing immunity to infection. The specific application of appropriate antiviral drugs also can have a role in preventing virus infection. Treatment of virus infection may utilize methods to encourage the body's own highly evolved antiviral mechanisms to deploy before virus infection leads to serious damage. Additionally, treatment can be mediated by specific antiviral agents designed to block any stage of virus replication in the host.

VACCINATION – INDUCTION OF IMMUNITY TO PREVENT VIRUS INFECTION

Antiviral vaccines

Almost everyone has heard the term *vaccination* and, in fact, has been given a vaccine, whether it be for poliovirus, measles virus, or mumps virus. Just what is a vaccine? How is it prepared and administered? And is it possible to create one for every viral infection of significance?

The term *vaccinate* against a virus means to administer, as a single or multiple dose, a nonpathogenic antigen (intact virus or virion subunit) to an animal or human such that the immune system of the individual responds by producing antibodies (humoral immunity) and, in some cases, cell-mediated immunity directed against one, several, or all viral antigens. The successfully vaccinated individual retains an immunologic memory of the event that prevents subsequent viral infection and/or disease. The mechanism of such immunity formation was described in Chapter 7.

Smallpox and the history of vaccination

For more than 2000 years, the scourge of smallpox affected human populations. The virus (member of the Poxviridae family, *Orthopoxvirus* genus) appears to have originated in Asia and made its way into Africa, the Middle East, and the Western world by 800–1000 CE. The virus was brought to the New World by Spanish and other European explorers and colonizers, and hundreds of thousands of indigenous people in North and South America died as a result. In some cases, such as in the Caribbean, all native populations were wiped out. The process repeated itself with other infectious diseases such as measles both in the New World and in the islands of Oceania in the eighteenth and nineteenth centuries. Some of the processes involved in such spread of novel diseases in populations were briefly described in Part I, especially Chapters 2 and 3. Variola major, the more serious form of smallpox, had a case fatality rate of 30–40%. In contrast, variola minor, a less severe form of the disease, killed only about 1–5% of those infected.

Differences in disease severity were attributed to slight genetic differences between strains of smallpox virus only in the late nineteenth century. Despite this, perspicacious observers noted that survivors of the disease were immune for life, and those who contracted it late in a local epidemic had a higher chance of survival. This was exploited in the technique called **variolation**, which was developed in China and introduced into Europe from the Middle East in the early eighteenth century. Lady Mary Wortley Montagu, the wife of the British ambassador to Turkey, saw to it that her children underwent variolation, despite prejudices of those who argued that it would not work on Caucasians. Her success was responsible for introduction of the technique into England in 1718. In this (rather heroic) technique, an uninfected person, usually a child, would be exposed to scabs or crusts that formed on the skin of a patient recovering from a natural infection. This method often resulted in inducing disease with mortality rates well below 1% and lifelong immunity. We now know that this method inadvertently exploited the fact that virus in such a healing lesion will tend to be partially inactivated by the patient's own immune response as well as by partial desiccation.

Even though variolation was often successful, the failure rate (number of deaths from the technique) made it a dangerous practice. Still, this was a common preventative method used in many parts of China, the Middle East, and Africa well into the early parts of the twentieth century. In England, Edward Jenner, a country physician working in Gloucestershire, was experimenting with variolation when he learned from his patients – who were milkmaids – that those infected with a disease called cowpox would subsequently be immune to smallpox. Jenner had the insight to exploit this method as a relatively safe way to protect against the scourge of smallpox. As a result, he began experiments to purposely infect his patients with cowpox virus, giving them a mild, asymptomatic disease and subsequent protection against infection with smallpox. Jenner named the method *vaccination* from the Latin word for "cow," *vacca*.

The success of Jenner's technique led to the rapid spread of prophylactic vaccination against infection with smallpox, but the success was largely confined to the developed West until after World War II. Ultimately, the success of vaccination against smallpox culminated in the announcement by the World Health Organization that smallpox has been eradicated from the planet. The last naturally occurring case of smallpox in the world was in October 1977, in a man in Somalia. He died, and it was determined that he contracted the virus from an aerosol of desiccated contaminated material that had been improperly disposed of during an earlier epidemic!

The only existing stocks of smallpox virus are at the Centers for Disease Control and Prevention in Atlanta and at the Russian State Research Center of Virology and Biotechnology (Vector) at Koltsovo. By international agreement, these stocks were to be destroyed on June 30, 1999, thus making the virus extinct. However, disagreements over the advisability of this delayed the planned destruction. This situation changed dramatically on September 11, 2001. After the terrorist attacks on the World Trade Center and the Pentagon, the United States moved into a much different position with regard to the threat posed by potential biological agents that could be used against the population. In fact, the store of smallpox vaccine ready for use was found to be much smaller than needed. As a result, a spate of research was begun on smallpox and preventative measures, including both vaccine production as well as potential therapeutic modalities. The US stores of smallpox virus have not been destroyed and are again being tapped for experimental purposes, using the highest levels of containment.

Despite Jenner's success, little was understood about the dynamics of vaccine production or the reasons for generating avirulent variants of infectious agents, until the germ theory of disease was well established in the latter half of the nineteenth century. Notably, in 1885, Louis Pasteur produced the first effective vaccine for rabies virus, utilizing the technique of culturing the virus in a non-natural host using laboratory methods of infection. In the case of rabies, Pasteur injected virus isolated from a rabid dog directly into the brain of rabbits, and found that as the virus was maintained in this way, it became **attenuated** in its ability to infect dogs, but more virulent in its ability to cause the disease in rabbits. Considering how dangerous the disease of rabies is and the fact that it can be transferred to humans by needle stick, this method of generating avirulent virus was, indeed, heroic. Current practices take advantage of much more complete understanding of culturing methods as well as better precautions against accidental infection. Still, the generation of a vaccine against a human pathogen can be risky and is a potential hazard to laboratory workers.

How a vaccine is produced

Vaccines are produced either by changing the nature of a disease-causing virus so that immunity is evoked without accompanying disease or by using a component of such a virus to evoke an immune response in the absence of viral infection. The current vaccines available for human use

Table 8.1 Some human viral vaccines.

Virus	Vaccine Type	Route of Administration
Polio	Inactivated (Salk)	Intramuscular
Polio	Live, attenuated (Sabin)	Oral
Measles	Live, attenuated	Subcutaneous
Mumps	Live, attenuated	Subcutaneous
Rubella	Live, attenuated	Subcutaneous
Rabies	Inactivated	Intramuscular
Influenza	Inactivated	Intramuscular
Influenza	Live, attenuated	Nasal
Influenza	Subunit (surface hemagglutinin trimer)	
Yellow fever	Live, attenuated	Subcutaneous
Varicella zoster (chickenpox)	Live, attenuated	Subcutaneous
Rotavirus	Live, attenuated	Oral
Hepatitis A	Inactivated	Intramuscular
Hepatitis B	Subunit (surface antigen)	Intramuscular
Tick-borne encephalitis	Inactivated	Intramuscular
Japanese encephalitis	Inactivated	Subcutaneous
Smallpox (variola)	Live, attenuated (vaccinia)	Subcutaneous
Human papilloma viruses	Pseudovirion (coat protein assembly)	Intramuscular

include those shown in Table 8.1. General methods for vaccine production are described in the following sections of this chapter.

Live-virus vaccines

If a live virus is to be administered to elicit a protective immune response (a **live-virus vaccine**), it must be avirulent and cause either a mild disease or no disease at all. Jenner's vaccine is an example of such an avirulent virus, although it is not a typical one. The original vaccine against smallpox began as cowpox virus, but the modern vaccine utilizes a virus called *vaccinia*, which is related to cowpox and horsepox viruses and is less closely related to smallpox (variola) virus. It is not known precisely how vaccinia came to be cultured as a vaccine strain virus or when it became the laboratory entity that it now is. It was likely derived from cowpox following repeated passage in cow and horse skin. This method of propagation was used for many years prior to the advent of cell culture. Vaccinia is quite unlike other attenuated viruses used as vaccines in that it was not derived from virus that it protects against.

Typically, Pasteur's approach for attenuating virulence is used for production of live virus vaccines. Vaccine strains are produced in an empirical fashion by serial passage of a virulent strain of the virus in nonhuman cell culture multiple times. Intermediate passages are tested for virulence in appropriate animals, including primates. The process of attenuation introduces a number of point mutations into the viral genome, essentially mutating functions not required for replication but rather for pathogenesis. This technique was used to produce the Sabin strains of oral vaccine directed against the three serotypes of poliovirus present at that time.

Serial passage is a blind procedure, and the results cannot be predicted. As more information accumulates about the genetic basis of virus–host interactions and virulence, specific mutants can be produced, either as deletions of regions of the genome or as site-specific changes, such that the properties of the putative vaccine can be customized.

One great advantage of live-virus vaccines is that since an actual infection takes place, both humoral and cell-mediated immune responses are stimulated. As a result, immunity develops after one or at most three exposures and usually lasts many years. A disadvantage may be the occasional reversion of virus to virulence. This can take place either by the occurrence of **back mutation** as the vaccine virus replicates in the individual being immunized, or possibly by a recombination taking place between the genome of a virus in the individual and the vaccine strain. Reversion to virulence by back mutation is a problem with the Sabin type 3 poliovirus vaccine, and virulent virus can be isolated with high frequency from the feces of individuals who have been immunized with the vaccine. While this should not be a problem with a population enjoying good waste-treatment facilities, it could pose a significant problem in mass vaccinations in countries with inadequate public health facilities.

Live-virus vaccines also have other potential problems. A major one is that they must be carefully handled and preserved with refrigeration, which makes their use in the field somewhat difficult, especially in parts of the world where reliable sources of electrical power are lacking. This problem can be partially alleviated by freeze-drying (**lyophilizing**) if the virus is stable to such treatment, but rehydration will require reliable sources of sterile water among other things. In addition, there is always the risk of an unknown pathogen being present and undetected in the vaccine stock. As techniques of assay for adventitious contamination become more sensitive and sophisticated, this latter problem becomes less worrisome, but still exists.

Killed-virus vaccines

Even though smallpox and rabies vaccines were attenuated viruses, most of the successful vaccines produced in the first part of the twentieth century utilized inactivated virus. An inactivated virus for a vaccine is generated from stocks of the virulent strain of the virus grown in cultured cells (or animals). This potentially virulent virus is then made noninfectious (inactivated) by chemical treatment. Originally, formaldehyde (formalin) treatment was used to inactivate virus; the original and highly successful Salk poliovirus vaccine was a formalin-inactivated preparation of the three virus serotypes. Despite its wide use in early vaccines, formalin is difficult to remove and therefore has the danger of residual toxicity. More recently, betapropiolactone is the chemical of choice to inactivate virus because residual amounts of the reagent can be readily hydrolyzed to nontoxic products.

An advantage of the killed-virus vaccines is absence of the virus's capacity to revert to virulence, since there is no virus replication during immunization. Further, killed-virus vaccines can be stored more cheaply than can live-virus vaccines. These advantages are balanced against the fact that the vaccine must be injected, multiple rounds of immunization are generally required, and vaccination does not result in cell-mediated immunity, because an active infection does not occur. This latter complication also means that immunity is usually nowhere near as prolonged as it is with a live-virus vaccine. Another unforeseen complication arose from the fact that earliest preparations of the Salk strain of poliovirus grown in monkey cells were contaminated with simian virus 40 (SV40) – a virus of monkeys that can cause tumors in laboratory animals. The conditions for inactivation of poliovirus did not inactivate SV40, and those receiving the first preparations of Salk vaccine were inoculated with the monkey virus. Luckily, this has not led to any known sequelae to date, but antibodies to the virus can still be detected in people who were vaccinated. This accident had some major political and social repercussions, and led to rumors that the HIV-caused AIDS epidemic in Africa was the result of contaminated polio vaccine stocks. Scientists have carefully and exhaustively checked the original stocks (luckily preserved) for the presence of HIV sequences, and all were negative.

Recombinant virus vaccines

It is possible to use the process of **genetic recombination** to introduce the genes for proteins inducing protective immunity into the genome of another virus, which itself might be avirulent. For example, the capsid protein gene of hepatitis B virus, known to produce protective immunity, could be introduced into the vaccinia virus genome. The methods and general principles behind the generation of such *recombinant virus* are detailed in Chapter 22, Part V. The genes introduced either may replace genes not required for replication of the carrier virus when it is used as a vaccine, or they could be added to the viral genome. Such a recombinant virus could then be used to vaccinate an individual, leading to generation of immunity against the proteins in question. Since the carrier virus is able to replicate, it is able to generate a full repertoire of immune responses against the immunizing protein or proteins. Further, the carrier may be extensively modified to ensure that it was absolutely avirulent. Possible candidates for such carrier viruses include members of the poxviruses, the herpesviruses, and the adenoviruses.

Recombinant viruses are currently being tested for use as vaccines. There are two theoretical problems with the use of recombinant virus vaccines. First, it is not clear that the same level of immunity or repertoire of immune responses can be evoked from the expression of a "passenger" protein. Second, once a good carrier virus is produced, its use in a vaccine would provoke immunity against itself. This would preclude use of the same carrier virus for another vaccine at a later time. Thorough testing will resolve the first problem, and if a truly effective vaccine were made against an important pathogen, the second problem could be readily ignored.

Several vaccines against human diseases based on recombinant viruses are available or in development, including the dengue virus vaccine, Dengvaxia, which is an attenuated yellow fever virus bearing the membrane and envelope proteins of dengue virus. Several such vaccines are also available for animals. For example, an effective vaccine against the virus causing Newcastle disease of chickens was generated using recombinant cucumber mosaic virus of plants. In this case, the problem of induced immunity against the carrier virus is not a problem.

Capsid and subunit vaccines

Since the desired immune response is most often directed against a critical surface capsid or envelope protein of a pathogenic virus, this protein by itself could be used as a vaccine if it were properly presented to the immune system of the vaccine recipient. A subunit vaccine can be prepared by purification of the protein subunit from the viral particle, or by recombinant DNA cloning and expression of the viral protein in a suitable host cell, either bacterial or yeast. Some of the general procedures for utilizing either approach are described in Part III.

Direct administration of a protein will not induce a cell-mediated response the way a live-virus vaccine would. Still, the advantages of a subunit vaccine include the lack of any potential infectivity, either mild in the case of the attenuated strains or severe in the case of the virulent strains or revertants. In addition, subunit vaccines may serve when the virus in question is extremely virulent or when it cannot be grown conveniently in culture.

There are several important general problems with the use of subunit vaccines that may not be amenable to easy solution. Still, the speed with which they can be produced makes them very attractive candidates for specific uses. Several subunit vaccines are currently available using the hepatitis B virus surface antigen obtained by expression of a cloned gene in yeast cells. These vaccines are in common use in this country and throughout the world. In clinical trials, use of these vaccines reduced the incidence of liver disease and primary liver cancer markedly.

A whole capsid vaccine against four serotypes of human papilloma virus (HPV6, 11, 16, and 18) known to be associated with cervical carcinomas was approved by the Food and Drug Administration (FDA) for use in the United States in 2006. The vaccine is produced in a way that allows the capsid protein to assemble into empty virus-like structures known as

pseudovirions, since they do not contain the viral genome. Clinical trials with this vaccine (Gardasil) have shown that its use leads to a 100% reduction in HPV infections by these serotypes in uninfected women as compared to an unvaccinated control population and >90% reduction of genital neoplasias caused by any type of HPV – this is an extremely promising result. In 2014 an improved version of the vaccine that includes nine serotypes of HPV (HPV6, 11, 16, 18, 31, 33, 45, 52, and 58), Gardasil 9, was approved for use in the United States.

DNA and RNA vaccines

A more novel approach toward the production of an effective vaccine is to use a fragment of DNA encoding a protein known to confer protective immunity as a vaccine – this is termed a **DNA vaccine**. The idea behind a DNA vaccine is that if antigen-presenting cells can take up the DNA by a process of "natural" transfection as outlined in Chapter 6 and express the antigenic proteins, protection could be fostered without the need of inactivated or attenuated virus. Further, methods for the delivery and storage of such a vector is potentially cost-effective. While it may seem surprising, considering the inefficiency of the transfection process, DNA-based vaccines have been effective against herpes simplex virus (HSV), Zika virus, and several other viruses in animal tests.

A DNA vaccine effective in horses against West Nile encephalitis virus has been licensed by the US Department of Agriculture (USDA). Clinical trials of a DNA vaccine against Zika virus started in 2016 and are ongoing. To date, however, human tests have been rather ambiguous with a major problem being difficulties in getting high antibody titers without **adjuvants**. Adjuvants are compounds added to antigens being prepared for introduction into the host, which increase inflammation leading to heightened infiltration with cells of the immune system. Such inflammation is usually quite painful, however, and their general use is forbidden in humans and discouraged in animals.

RNA vaccines have been used experimentally against many viruses including influenza and Zika viruses. In 2020, however, two SARS-CoV-2 spike mRNA vaccines were rapidly developed and deployed worldwide to combat the pandemic of COVID-19. These vaccines were shown in clinical trials to be 90% to 95% effective.

Edible vaccines

Another approach is to express the antigenic protein or antigenic portion of a protein in a form that could be ingested and still generate protective immunity. Obviously, such an antigen must be able to survive the digestive system and be assimilated by antigen-presenting cells. To date, human clinical trials with transgenic potatoes have been carried out, and pre-clinical studies in which antigenic peptides were incorporated into tomatoes, spinach, lettuce, bananas, and other crop plants are underway in humans or animals so that these foods could be made available to provide protection against one or another major human or animal disease. This might be especially important to controlling infectious disease in developing nations.

Problems with vaccine production and use

The great success of so many vaccines, including those against smallpox, measles, mumps, rubella, polio, and rabies, has led to a serious commitment by the World Health Organization and other public health agencies to develop and distribute vaccines for protection against a variety of viral diseases, especially those affecting children. The Expanded Program on Immunization (EPI) of the World Health Organization, which started in 1976, targeted six childhood diseases for global immunization, two of which are viral: poliomyelitis and measles. In recent years many more vaccines have been included in the World Health Organization EPI recommendations, including several more vaccines against viral diseases.

Two of the major problems that arise to subvert such strategies are genetic instability and heat sensitivity of the vaccines. As mentioned in the "Live-Virus Vaccines" section, in certain cases, such as the type 3 Sabin strain of the oral poliovirus vaccine, revertants that are virulent can occur. Such instabilities can lead to vaccine-associated cases of the disease that is the target of the vaccination. These instabilities may be overcome with the use of recombinant vaccinia virus constructs, where the only gene expressed from the virulent virus is that of the surface antigen used to stimulate the immune response.

A serious problem with administering vaccines in the developing world is the need for refrigeration of some of the preparations. The requirement for a "cold chain" from the site of manufacture to the site of the vaccine's use is critical to efficacy of the immunization. As a result, a good deal of development has gone into two areas, one mechanical and one biological. Portable refrigerators and adequate cold packaging are constantly being redesigned. Accompanying this is the search for vaccine constructs that can withstand ambient temperatures during shipping and delivery. The development of heat-stable and yet highly immunogenic vaccines is a high priority for the World Health Organization and other organizations working to save children from the ravages of these diseases. The campaign for the eradication of poliovirus has made major advances. Only a few areas in the world still report reservoirs of wild viruses, but political opposition to vaccination is problematic in some of these regions.

The most serious problems are socioeconomic, and these may well persist – all efforts of scientists and medical researchers to the contrary. Public distrust of public health measures and vaccination campaigns can be a major problem. An example is the false, but persistent, myth linking use of the measles, mumps, and rubella (MMR) vaccine with autism in children, which led to a public reaction against measles vaccination in the United Kingdom, leading to an increase in the incidence of this life-threatening disease. While this myth originated from a fraudulent study that was withdrawn by the journal that published it, and has been repeatedly shown to be baseless, distrust has persisted. Another example of public mistrust of vaccines is seen in political reactions against polio vaccination in parts of Asia and Africa. This problem is ultimately the result of irresponsible political stances by national leaders, but was fostered by a general mistrust of science and technology in poorly educated populations. This latter problem is persistent and is not aided by the occasional lapses of ethical behavior on the part of large pharmaceutical companies and members of the scientific community.

The expense and financial liabilities involved in producing an effective vaccine are also problems. There are many expensive steps between discovery and characterization of a viral disease, to production and use of a truly effective vaccine. Such expense will only be borne by for-profit corporations provided they can get a return on their investments. While governments also may be able to cover the costs of vaccine production and application, it is clear that those ultimately supporting such efforts, the taxpayers, must be able to see the need for this expense. This requires education, information, and above all, good will. These items can be either plentiful or in short supply, depending on the historical and political background of the disease in question. Clearly, no general solution to such problems can be envisioned. Each disease will need to be dealt with as it occurs. Results inevitably will show both great success and great instances of lost opportunities.

EUKARYOTIC CELL-BASED DEFENSES AGAINST VIRUS REPLICATION

Cellular defenses against viral infection

Small RNA-based defenses

Discoveries starting in the early 1990s have demonstrated that small RNA molecules with double-stranded regions have a number of important roles in regulating eukaryotic cellular processes and protecting against pathogens beyond the induction of interferons (IFNs). This is briefly described in Chapter 7; here it suffices to note that there are pathways in eukaryotic cells for processing small RNA molecules encoded in the cell's genome into 22-base-pair double-stranded RNA molecules (microRNAs or miRNAs). These miRNAs then bind to specific viral or cellular messenger RNA (mRNA) molecules, leading to their degradation.

Cells have a similar way of dealing with double-stranded RNA occurring in transcripts, such as those produced in viral infections. These can be processed into 19–21-base-pair double-stranded RNA molecules, with two-base 3′ extensions at both ends, called small interfering RNAs (siRNAs). Such siRNAs interfere with the translation of mRNAs containing homologous sequences also by inducing the degradation of those mRNAs. Thus, infection of a plant cell with a virus will lead to the spread of these to neighboring and more distant cells, resulting in resistance to viral spread. The presence of plant virus genes, which act to counter the function of plant siRNAs, demonstrates the extent of this system in the plant kingdom.

It is not yet clear just how extensive the roles of miRNA and siRNA are in protecting animal cells, but evidence of the importance of these molecules in virus infections and the antiviral response is growing. There are almost 2000 miRNA-encoding sequences in the human genome, and over 500 miRNAs have been identified in a number of DNA viruses and retroviruses including human herpesviruses, polyomaviruses, and HIV. While it is often difficult to identify the target of a given miRNA, an SV40 miRNA blocks part of cellular control of its replication cycle. On the other hand, it has also been found that hepatitis C virus utilizes a small RNA species of human liver cells to increase the efficiency of translation of its mRNA. All available evidence suggests that siRNAs and miRNAs act as a kind of innate immune response directed against viral RNA motifs. There is also evidence that viruses may have co-opted miRNAs that target cellular defense mechanisms as a means of evading host responses.

Cellular factors that restrict retrovirus replication

Another form of cell-based antiviral activity can be seen in responses to retrovirus infection by mammalian cells. A group of cytidine deaminases termed APOBECs recognize newly synthesized retroviral DNA generated by reverse transcriptase and deaminate cytidines in that DNA to yield uracils. This leads to hypermutation and inactivation of the virus or to degradation of the altered DNA strand. This process is so effective that HIV has a specific viral gene directed against APOBEC activity!

A family of proteins containing a tripartite motif (TRIM proteins) block retroviral replication at the stage of complementary DNA (cDNA) synthesis. The best studied of these is TRIM5, which blocks retroviral cDNA synthesis by inducing premature disassembly of the reverse transcription complex. Another human protein, SAMHD1, also blocks retroviral cDNA synthesis by hydrolyzing deoxynucleotide triphosphates (dNTPs) to deoxynucleosides and free triphosphates.

Another human protein called tetherin blocks budding of many enveloped viruses, including retroviruses. Tetherin physically tethers budding virions to the cell by inserting into both cellular and viral membranes. HIV and other human viruses have proteins that block tetherin activity.

ANTIVIRAL DRUGS

All drugs effective against pathogenic microorganisms must target some feature of the pathogen's replication in the host that can be efficiently inhibited without unduly harming the host. Some drugs are effective when given to an individual before he or she is exposed or for a short time after exposure. Such prophylactic use is only effective and practical in large populations under specific circumstances; for example, military personnel prior to entering a biological hazard zone or as a precaution prior to unprotected sexual activity.

Despite the value of some prophylactic drugs, the most practically useful drugs are ones that can effectively interrupt infection and cure disease at any stage. The dramatic effectiveness of penicillin in treating numerous bacterial infections after World War II has proved a model for such drugs, but the earliest specific antibacterial drugs were made up of complex organic molecules containing mercury that Ehrlich utilized to combat syphilis at the end of the nineteenth century. He termed these "magic bullets" and developed them to reduce the toxicity of mercury, whose use as an antisyphilitic agent was known to be effective since the Renaissance in Europe. Perhaps not surprisingly, Ehrlich's success was marred by the anger of some moralists who argued that the disease was a punishment for sin!

The problem of therapeutic drug toxicity is a continuing one. Many effective inhibitors of metabolic processes, even if more or less specific for the pathogen, will have undesirable side effects in the person being dosed. The general ratio of benefit of a drug to its undesirable side effects is termed the **therapeutic index**. Determination of a drug's therapeutic index requires extensive animal testing and extensive documentation, and is a major factor in the expense involved in developing effective pharmaceuticals for any purpose.

Targeting antiviral drugs to specific features of the virus replication cycle

Given the fact that viruses are obligate intracellular parasites, it is easy to understand why a chemotherapeutic approach to halting or slowing a viral infection is difficult to achieve. Unlike bacterial cells, which are free-living, viruses utilize the host cell environment for much of their life cycle. Therefore, chemical agents that inhibit both virus and host functions are not a good choice for therapy.

The preferred strategy has been to identify the viral functions that differ significantly from or are not found within the host and are therefore unique. For each virus of clinical interest, a good deal of effort has been expended on understanding the virus's life cycle and attempting to develop drugs that can specifically block critical steps in this cycle. Table 8.2 lists targeted stages in the virus life cycle along with examples of existing or proposed agents that could block the cycle with

Table 8.2 Some targets for antiviral drugs.

Step in Virus Life Cycle Targeted	Molecular Target of Inhibitor	Example
Virus attachment	Viral surface protein-cellular receptor interaction	Maraviroc
Viral entry	Viral membrane protein	Fuzeon, amantadine
DNA virus genome replication	Viral DNA polymerase	Acyclovir, ganciclovir
RNA virus genome replication	Viral RNA replicase	Ribavirin, sofosbuvir
Retrovirus - reverse transcription	Reverse transcriptase	3TC, FTC, efavirenz, nevirapine
Retrovirus - integration	Integrase	Dolutegravir, bictegravir
Viral transcriptional regulation	HIV *tat*	Pre-clinical compounds
Viral mRNA posttranscriptional processing (splicing)	HIV *rev*	Pre-clinical compounds
Virion assembly	Viral protease	Ritonavir, indinavir
Virion assembly	Capsid protein-protein interactions, budding	Oseltamivir, protease inhibitors
Virion assembly	Assembly process localization	Ledipasvir

some measure of specificity. With each of these, the problem of resistant mutants always arises, leading to limitation of the drug's usefulness.

Acyclovir and the herpesviruses

The development of acycloguanosine (acG) in the late 1970s for use in herpesvirus infections marked a great advance in the chemotherapy of viral infections. This compound, prescribed under the name acyclovir, is the first of the nucleoside analogues that are chain-terminating inhibitors. The structure is shown in Figure 8.1. When the triphosphorylated form of acycloguanosine is incorporated into a growing DNA chain in place of guanosine, no further elongation can take place because of the missing 3′ OH.

The specificity of acyclovir for herpesvirus-infected cells results from two events. First, after the nucleoside is transported into the cell, it must be triphosphorylated to be utilized as a substrate for DNA replication. The first step in this process, the conversion of acG to the monophosphate (acGMP), requires the presence of the herpesvirus-encoded thymidine kinase (TK), whose activity is to phosphorylate deoxynucleosides. Following this, a cellular enzyme is able to add the next two phosphates, producing the triphosphate acGTP. This acGTP inhibits the synthesis of viral genomes by acting as a substrate for herpesvirus DNA polymerase. When this happens, the DNA chain is terminated – no additional bases can be added because of the missing 3′-OH group. The drug will inhibit the viral enzyme about 10 times more efficiently than it will the cellular DNA polymerases.

As a result of the requirement for herpesvirus TK and the inhibition and chain termination of herpesvirus DNA synthesis, acyclovir is highly specific for herpes-infected cells and is nontoxic to uninfected cells. Acyclovir has been used successfully in both topical and internal applications with both HSV type 1 and HSV type 2. While both types of HSV readily mutate to resistance in the laboratory, in both cases the mutant viruses do not replicate well in humans, and cessation of drug treatment results in the rapid appearance of wild-type (wt) virus with its accompanying drug sensitivity. This, and the low toxicity of the drug, have made acyclovir the most successful targeted antiviral drug yet produced.

Chemical modification of acG's structure has resulted in ganciclovir (9-[1,3-dihydroxy-2-propoxy] methylguanine) (Figure 8.1). This drug has the same properties as acG, except that it is specific for cells infected with cytomegaloviruses. Unfortunately, this drug has a severe toxicity when given intravenously and must be used with caution.

Blocking influenza virus entry and virus maturation

Type A influenza viruses enter their host cells by means of the receptor-mediated endocytotic pathway. In this process, the viral hemagglutinin molecules in the membrane of the particles undergo a conformational change when the pH of the endocytic vesicle is lowered to around five after fusion of the vesicle with an acidic endosome. At this lower pH, the viral membrane undergoes fusion with the vesicle membrane, and viral nucleocapsids enter the cell cytoplasm (see Chapter 6).

Two compounds were developed that interfere with the ability of the cell to change pH within influenza A virus–modified vesicles – amantadine and rimantadine. The first one introduced, amantadine, was approved for use in the United States in 1966. Both drugs are basic primary amines, and can prevent the acidification that is essential for completion of viral entry. These drugs were used to treat influenza until newer drugs with fewer side effects were developed. Moreover, most strains of influenza virus in circulation since the year 2000 are resistant to amantadine and rimantadine.

Starting in 1999, three new drugs that inhibit the influenza virus neuraminidase were approved. Oseltamivir was the first of these drugs to be introduced, followed by zanamivir and peramivir. All three drugs inhibit release of influenza virions from cells, which is dependent on

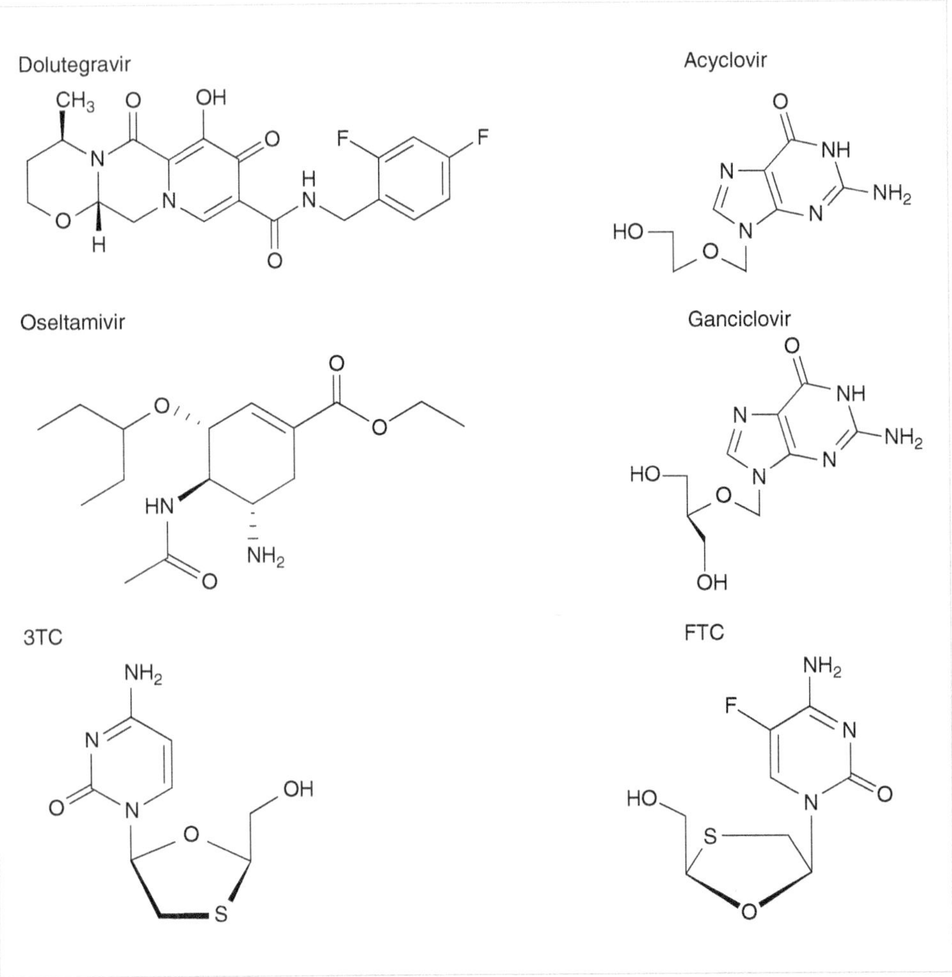

Figure 8.1 The structure of some currently effective antiviral drugs.

the activity of the neuraminidase enzyme on the viral surface. Neuraminidase normally cleaves off terminal neuraminic acid residues on the carbohydrate portions of cell surface proteins so that newly budding virions do not adhere to them via viral surface hemagglutinin molecules. Oseltamivir, the only oral medicine of the three neuraminidase inhibitors, is currently recommended for patients who are at risk for severe symptoms, including children less than two years old and adults over age 65. As of the 2018–2019 flu season, viral resistance to neuraminidase inhibitors is rare.

Chemotherapeutic approaches for HIV

When it was discovered that the viral agent that causes AIDS is, in fact, a retrovirus, the immediately obvious goal was the development of a drug that could specifically inhibit the unique viral replicative enzyme of the retroviruses: reverse transcriptase. A drug that had been developed as an antitumor agent was found to inhibit this enzyme: 3′-azido-2′3′-dideoxythymidine, commonly called azidothymidine or AZT.

Like acG, this drug, when transported into the cell and phosphorylated, can be utilized by the HIV polymerase to produce a chain termination because of the missing 3′ OH. Although the drug exhibits a good specificity for HIV reverse transcriptase compared with cellular DNA polymerases *in vitro*, toxic effects are still seen when the drug is administered to patients. Most importantly, because of the high mutability of HIV replication (see Part IV, Chapter 20), the development of AZT-resistant mutants occurs rapidly.

Other nucleoside/nucleotide analogues have been produced for therapeutic use. Notable are tenofovir and emtricitabine, which are often used in combination in a single pill called Truvada. Non-nucleoside analogue reverse transcriptase inhibitors have also been developed and approved for use in patients. These drugs are highly effective and less costly than other anti-HIV drugs, but viral resistance develops quickly so they can only be used for short periods alone or for longer periods in combination with other drugs (see the "Multiple-Drug Therapies to Reduce or Eliminate Mutation to Drug Resistance" section).

A major advance in the chemotherapeutic treatment of HIV infection was the production and use of the class of drugs known as *protease inhibitors*. Retroviruses, as well as many other viral families, require proteolytic processing of initial translation products so that the active viral proteins can be made. For HIV (like all retroviruses), this is carried out by a viral-encoded protease. The drugs act by inhibiting HIV protease; as a result, the posttranslational processing of viral products as well as the final proteolytic steps required during viral assembly are blocked (see Chapter 19).

Two approved drugs against HIV block viral entry. One is a fusion inhibitor named fuzeon. It is a peptide that has the same sequence as one of the alphahelical regions of the HIV envelope protein, gp41. For the gp41 protein to function to promote viral–cell membrane fusion in HIV infection, this region must associate with another helical region in the same protein, and the presence of fuzeon inhibits this pairing. The second HIV entry inhibitor is maraviroc, an inhibitor of HIV binding to CCR5 (see Chapter 19).

The newest category of drugs used to combat HIV infection are inhibitors of the viral integrase enzyme that mediates integration of HIV cDNA into the host genome. There are now three such drugs, raltegravir, elvitegravir, and dolutegravir, approved by the FDA in 2007, 2012, and 2013 respectively. These drugs are not as potent inhibitors of HIV replication as the protease inhibitors, but they have fewer serious side effects so they have gained wide use.

Multiple-drug therapies to reduce or eliminate mutation to drug resistance

The most promising therapy against HIV now being used involves the use of multiple drugs. The original protocol required the simultaneous administration of AZT, another nucleoside analogue such as ddC, and a protease inhibitor. Initial results with this cocktail were quite impressive. Clinical observations of AIDS patients showed reversal of symptoms and rebound of levels of CD4 cells. Viral loads decreased, and circulating virus all but disappeared. With the wide application of these therapies in the United States, most cities reported a decrease in deaths from AIDS by the end of 1997. Currently, the most popular combinations include an integrase inhibitor or a protease inhibitor in combination with two nucleoside or nucleotide analog reverse transcriptase inhibitors. Combinations of several drugs into single doses and other combinations are also available for use; in all of these cases, rapid and sustained reduction in viral load is the objective.

This exciting picture must be tempered by words of caution. First, these therapies are quite complicated and expensive. They cannot be readily applied to developing nations and to individuals at risk in this and other developed countries who do not have the financial resources required for the treatment, unless the medicines are provided free or at cost. Moreover, if dosages are skipped or missed, there is the great danger of developing resistant mutants that may destroy progress made by the patient. This fear was underlined by the finding that even after long periods of treatment, HIV genomes still exist in resting memory T lymphocytes and can be recovered as infectious virus if drug is removed. Therefore, therapy must be continued for the rest of the patient's life, often with adverse long-term effects.

Other approaches

The goal of developing methods for specifically targeting virus replication is so important that other methods are being actively pursued. One approach is *precise targeting*. The toxicity of many antiviral drugs is exacerbated by the fact that the drug must be presented to the whole body, thus affecting tissue that is free of virus. Localized HSV reactivation can be effectively treated with iodouridinedeoxyriboside (IUdR) by local application to the lips or genital area, even though this drug is relatively toxic when taken internally. Presently, research is directed toward the development of protocols that combine methods for ensuring the delivery of small amounts of even highly toxic drugs only to virus-infected tissue.

A second promising approach is the generation of short oligonucleotide polymers that have sequences complementary to specific portions of viral mRNA molecules. Such **antisense oligonucleotides** can be designed to specifically inhibit the translation of an important viral gene product with little or no attendant toxicity, mimicking the role of cellular siRNAs described in this chapter. Some antisense drugs are already being clinically tested.

Finally, the class of proteins called defensins (described in Chapter 7) have great promise in interfering with viral attachment to the surface of cells by the formation of networks between these small cellular proteins and viral glycoproteins. One type of defensin, the Retrocyclin 2 (RC2), effectively blocks the ability of HIV, HSV, and influenza to enter cells. The remarkable ability of this group of host peptides to block viral infection suggests that defensins, either given in large doses or as more stable derivatives, will be useful as antiviral drugs.

BACTERIAL ANTIVIRAL SYSTEMS – RESTRICTION ENDONUCLEASES

Bacterial cells do not have the ability to produce antibodies or IFN as do animal cells. However, they have evolved mechanisms through which viral infections can be aborted, or at least limited. **Bacterial restriction** is the most well-studied type of antiviral defense. The discovery of bacterial restriction systems not only led to a basic understanding of bacterial-viral interactions but also provided one of the most critical sets of tools used in modern molecular biology and biotechnology: **restriction endonucleases.**

Bacterial cells can "mark" their own DNA for identification by the covalent addition of methyl groups to critical bases within the nucleic acid. For example, adenosine residues can be enzymatically converted to 5-methyl adenosine by transfer of a methyl group from *S*-adenosylmethionine, catalyzed by bacterial enzymes called *DNA methylases*. These modifications are made at specific sites within the DNA. These sites are specific sequences of four, five, six, seven, or eight nucleotides; such sequences often display a dyad symmetry (GAATTC, for example) for *Eco*RI, methylase, and the resulting hexanucleotide (GA5-MeATTC) blocks cleavage of DNA at this site for *Eco*RI, one of the first restriction endonucleases characterized. Note that the sequence reads the same on both DNA strands; that is, it is a **palindromic sequence**.

Any DNA entering cell cytoplasm that does not have the host bacteria's specific modifications at these sites, such as GAATTC, will be cleaved with a **restriction enzyme** that can recognize the unmodified sequence. Thus, the system functions to restrict the growth of a virus whose genome has found its way into the cell. In effect, the host cell can recognize its own DNA as well as foreign viral DNA and destroy the invader before viral gene expression begins.

There are some cells in which a viral genome will be able to avoid the restriction enzymes for one of a number of reasons (perhaps the concentration of enzyme is too low to act quickly enough). These cells will produce progeny virus particles whose DNA is modified (methylated) in the same pattern as the host's DNA since host enzymes will work on this DNA as it is replicated. As a result, these progeny particles will be able to grow productively on cells of this particular restriction modification type. Thus, a balance is achieved in a population of cells between lytic replication of the virus with subsequent destruction of the host and complete inhibition of virus growth.

Later sections will explain that the exquisite specificity of restriction endonucleases (of which more than 500 are now known) makes them extremely valuable tools for manipulating DNA molecules. They can be used to cut genomes into specifically sized pieces, and are vital to the isolation and direct manipulation of individual DNA pieces containing genes of particular interest. The Nobel Prize was awarded in 1978 to W. Arber, H. Smith, and D. Nathans for their characterization of restriction endonucleases, and it is fair to say that this single discovery is probably the most directly seminal in the development of modern molecular genetics and recombinant DNA technology.

CRISPR/cas systems

More recently a mechanism of adaptive immunity to viruses was discovered in archaea and bacteria. Bacteria were found to have clustered, regularly interspersed short palindromic repeats **(CRISPR)** in the DNA sequences of their genomes as well as CRISPR-associated (**Cas**) genes nearby. Recent work has shown that Cas genes include endonucleases that cleave invading viral DNA into short fragments. These fragments are incorporated as spacers between palindromic repeats in the CRISPR region or regions of the bacterial or archaeal genomes. Other Cas gene products use CRISPR transcripts as a guide to find and destroy the genomes of invading viruses!

QUESTIONS FOR CHAPTER 8

1 Describe how bacterial restriction enzymes can cleave bacteriophage DNA as a part of a host defense mechanism.

2 What are some of the problems that arise in considering vaccination strategies for viral diseases?

Problems

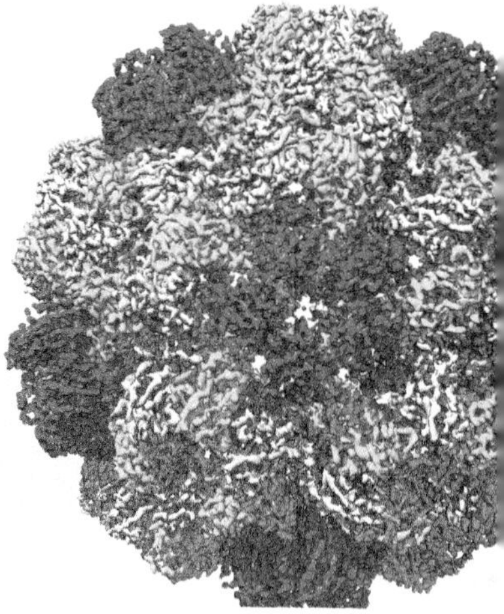

PART II

1 The table below shows the properties of the genomes of three different viruses. The data were obtained as follows: Nuclease sensitivity was measured by the ability of deoxyribonuclease (DNase) or ribonuclease (RNase) to destroy the genome (a "+" means sensitivity). The ability of the genome to act as mRNA was tested by incubating it in a cell-free system. If amino acids were incorporated into protein, the data are shown as a "+." Finally, the virus particles were tested for the presence of a virion polymerase. If an enzyme was present, the data show whether it could polymerize deoxynucleotide triphosphates (dNTPs) or nucleoside triphosphates (NTPs).

	Genome Properties				
	Nuclease Sensitivity?		Can Genome Be an mRNA?	Virion Polymerase?	
Virus	DNase	RNase		With dNTPs	With NTPs
#1	–	+	+	–	–
#2	–	+	–	–	+
#3	–	+	+	+	–

For each virus, indicate the strategy of the genome, using the Baltimore classification. What is the nature of the product of the virion polymerase when present?

2 Interferons are synthesized by cells in response to many different viral infections. The common result of the interferon-induced antiviral state is the cessation of protein synthesis. Predict the effect of the following treatments of *the indicated cell on protein synthesis in that cell.* (Assume, for the purpose of this question, that the virus does not inhibit cellular protein synthesis as a result of the infection.)

	Viral Infection of Cell	Insertion of dsRNA into Cell
Normal cell		
Interferon-treated cell		

3 You wish to produce a subunit vaccine for a nonenveloped *positive-sense RNA virus* that will stimulate the production of neutralizing antibodies in the person receiving it. Indicate which of the following viral proteins would be a logical candidate for such a subunit vaccine, and state a brief justification.

(a) Viral capsomer protein
(b) Viral protein that is bound to the RNA genome inside of the virion
(c) Viral RNA polymerase

4 Each year in late winter a behavioral "disorder" engulfs the people of New Orleans, Louisiana, reaching a climax on the day before Ash Wednesday. Together with virologists at Louisiana State University, you have isolated a virus from the affected people that you suspect is responsible for this condition. You have named the new isolate Mardi Gras virus (MGV). You have found a convenient host cell in which to grow MGV in the laboratory. The following table lists some of the properties of MGV you have discovered.

Initial Data for Mardi Gras Virus	
Experiment	**Observation**
Physical nature of the virion	Electron microscopy (EM) reveals 100-nm particles; shape indicates presence of envelope with visible surface projections; ether destroys particle integrity.
Chemical nature of viral genome	Digested with RNase; degraded by alkali; resistant to DNase.
Informational nature of viral genome	Genome cannot be translated in cell-free protein synthetic system.
Enzymatic analysis of virion	With NTP precursors: catalyzes RNA synthesis; with dNTP precursors: no reaction.
Biological analysis of virus	HeLa cells (human): attachment and penetration (observed by EM) and progeny virus produced; AGMK cells (simian): attachment and penetration (observed by EM) but *no* progeny virus produced.

(a) What would you predict to be the effect of treatment with ether or other lipid solvents on the infectivity of MGV?
(b) To which Baltimore class would you assign MGV? Give two reasons for this classification, based on the data in the table.
(c) Based on the data in the table, would you say that MGV is a human or a simian virus? Justify your answer briefly with reference to the data.

5 If a virus has a negative-sense RNA genome, what enzymatic activity (if any) will be found as part of the virion structure *and* what will be the first step in expression of the viral genome?

6 Influenza viruses gain entry into their host cells by attachment to *N*-acetylneuraminic acid residues on the cell surface, followed by receptor-mediated endocytosis. Predict what effect the treatments shown in the table below will have on (i) the attachment of an influenza virus to a susceptible host cell, *and* (ii) the subsequent uncoating of influenza virus in the same cell. Use a "+" to indicate that the event will take place or a "−" to indicate that it will not take place. Note: In each case it is assumed that the events would be occurring in the same cell that has undergone the treatment.

Treatment	Attachment?	Uncoating?
Treatment of the host cell with neuraminidase		
Treatment of the host cell with NH_4Cl, which prevents lowering of the lysosomal pH		
Treatment of the host cell with actinomycin D, which prevents synthesis of mRNA		

7 Cells produce mRNA by transcription of their DNA genomes. In contrast, single-stranded RNA genome viruses have three different strategies with respect to viral mRNA production. Briefly describe the production of mRNA for each of the following viruses.

(a) Poliovirus
(b) Vesicular stomatitis virus
(c) Rous sarcoma virus

8 Infection of a human with influenza virus can trigger both host defense systems: the interferon response and the immune response. In the table below, indicate with a "Yes" or a "No" regarding which of the events is characteristic of which defense system (either, both, or neither).

Event	The Interferon Response	The Immune Response
Both host and viral mRNA are degraded in the cell after infection.		
A fragment of viral protein in complex with major histocompatibility complex type I is displayed on the surface of the infected cell.		
The virally infected cell dies.		
Capped mRNA is no longer translated in the infected cell.		

Additional Reading for Part II

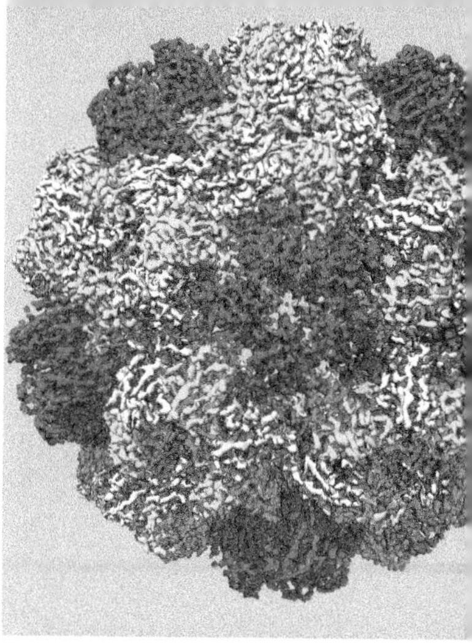

Note: See Resource Center for other relevant websites.

Brandenburg, O., Magnus, C., Regoes, R., and Trkola, A. (2015). The HIV-1 entry process: a stoichiometric view. Trends in Microbiology 23: 763–774.

Calendar, R. (ed.) (2005). The Bacteriophages, 2e. Oxford: Oxford University Press.

Coffin, J., Hughes, S., and Varmus, H. (1999). Retroviruses, chaps. 3, 12. Cold Spring Harbor, NY: Cold Spring Harbor Laboratory Press.

Fensterl, V., Chattopadhyay, S., and Sen, G. (2015). No love lost between viruses and interferons. Annual Review of Virology 2: 549–572.

Flint, S.J., Enquist, L.W., Krug, R.M. et al. (2015). Molecular Biology, vol. 1 of Principles of Virology, 4e, chaps. 4, 5, 13. Washington, DC: ASM Press.

Granzow, H., Weiland, F., Jöns, A. et al. (1997). Ultrastructural analysis of the replication cycle of pseudorabies virus in cell culture: a reassessment. Journal of Virology 71: 2072–2082.

HHMI BioInteractive. (N.d.). Viral Life Cycle. https://www.hhmi.org/biointeractive/viral-life-cycle

Iwasaki, A. and Pillai, P. (2014). Innate immunity to influenza virus infection. Nature Reviews Immunology 14: 315–328.

Johnson, J. and Rueckert, R. (1997). Packaging and release of the viral genome. In: Structural Biology of Viruses (eds. W. Chiu, R.M. Burnett and R.L. Garcea), 269–287. New York: Oxford University Press.

Katze, M., Korth, M., and Law, L. (2016). Viral Pathogenesis, 3e, chaps. 4, 5, 19, 20 (ed. N. Nathanson). New York: Academic Press.

Knipe, D. (2013). Fields Virology, 6e, chaps. 3, 4, 8, 9, 13, 14 (ed. P. Howley). Philadelphia: Lippincott Williams.

Maginnis, M. (2018). Virus–receptor interactions: the key to cellular invasion. Journal of Molecular Biology 430: 2590–2611.

Mahy, B. and Van Regenmortel, M. (2008). Encyclopedia of Virology, 3e. New York: Academic Press.

Morgridge Institute for Research, Outreach Experiences. (N.d.). VirusStructure. https://morgridge.org/wp-content/uploads/Virus-Structure.pdf

Murphy, K. and Weaver, C. (2016). Immunobiology, 9e, chaps. 2, 3, 9, 10. New York: Garland.

Oldstone, M. and Fujinami, R. (2008). Immunopathology. In: Encyclopedia of Virology, 3e (eds. B. Mahy and M. Van Regenmortel), 78–83. New York: Academic Press.

Saragovi, H., Sauvé, G., and Greene, M. (1999). Viral receptors. In: Encyclopedia of Virology, 2e (eds. A. Granoff and R.C. Webster), 1926–1928. New York: Academic Press.

Snyder, L. and Champness, W. (2002). Recombinant DNA techniques and cloning bacterial genes: the biological role of restriction modification systems; types of restriction modification systems. In: Molecular Genetics of Bacteria, 2e, chap. 13. Washington, DC: ASM Press.

Whitley, R. (2003). Antiviral therapy. In: Infectious Diseases, 3e (eds. S.L. Gorbach, J.G. Bartlett and N.R. Blacklow), chap. 32. Philadelphia: Lippincott.

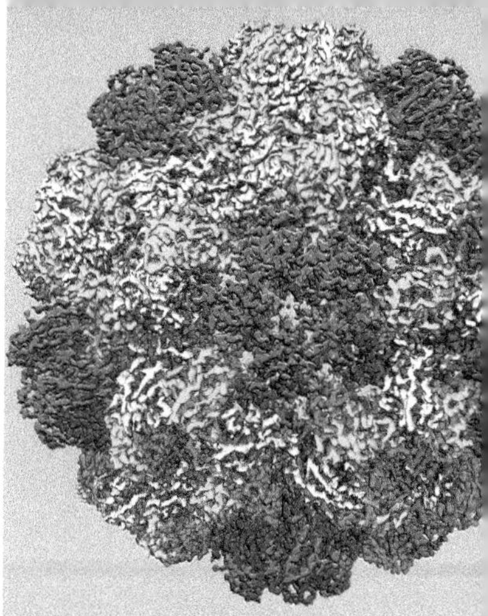

PART III

Working with Virus

Visualization and Enumeration of Virus Particles

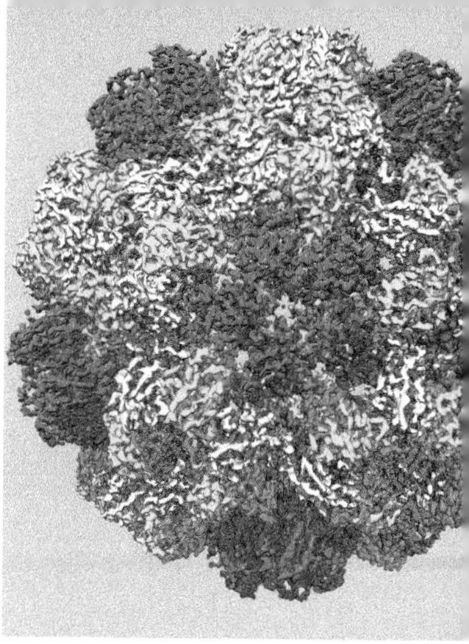

CHAPTER 9

- USING THE ELECTRON MICROSCOPE TO STUDY AND COUNT VIRUSES
- Counting (enumeration) of virions with the electron microscope
- ATOMIC FORCE MICROSCOPY – A RAPID AND SENSITIVE METHOD FOR VISUALIZATION OF VIRUSES AND INFECTED CELLS, POTENTIALLY IN REAL TIME
- INDIRECT METHODS FOR "COUNTING" VIRUS PARTICLES
- QUESTIONS FOR CHAPTER 9

Most viruses are submicroscopic physical particles, and while the largest can be discerned in an ultraviolet (UV)-light microscope, detailed visualization requires other methods. The development of physical and chemical methods for the study of viral structural properties and their unique shapes and sizes provides an important impetus for applying these techniques to the study of biological processes in general.

An investigator must know how many virus particles are in a sample, and what the sample's relationship is to the biological properties of the virus (measured in other ways), in order to carry out a meaningful physical study of virus particles.

The ability to count viruses ultimately depends on the ability to visualize them, and this requires special techniques that were not available until just prior to World War II. Notable among these is the **electron microscope (EM),** whose design required a sophisticated knowledge of modern particle physics and modern electrical and mechanical engineering. The EM has allowed scientists to instrumentally peer into the cell and biological processes, and much of the progress taken for granted in molecular biology and medicine would have been impossible without it.

USING THE ELECTRON MICROSCOPE TO STUDY AND COUNT VIRUSES

The dimensions of viruses are below the resolving power of visible light, so their visualization requires the shorter wavelengths available with the EM. The EM (schematically shown in Figure 9.1) accelerates electrons to high energy and magnetically focuses them. High energy

Basic Virology, Fourth Edition. Martinez "Marty" Hewlett, David Camerini, and David C. Bloom.

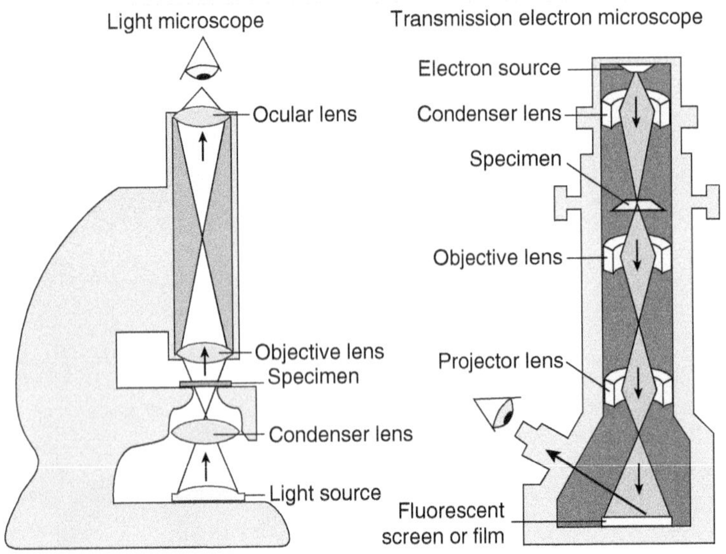

Figure 9.1 A schematic comparison of light and electron microscopes. The principles behind the focusing of the image are similar except that magnetic fields must be used to focus electrons. The higher energy of the electrons accelerated through high voltage produces very short wavelengths with resulting high resolving power.

gives the electrons a short wavelength, one that is much "smaller" than the virus particles. In fact, the EM can visualize DNA, RNA, and large proteins.

Despite the value of the EM's high resolving power, the energy needed to attain short wavelengths poses a problem. High-speed (short-wavelength) electrons are quite penetrating, and most biological subjects are transparent to them. Thus, in order to visualize viruses, they are generally either *stained* or *coated* with a heavy metal such as platinum or osmium. This coating or staining is done in such a way that the basic arrangements of the proteins and structure of the virus are not destroyed. The particles then are visualized by passing electrons through the specimen and observing it on a fluorescent screen or captured by a digital camera and displayed on a monitor. Areas where electrons do not pass because of the heavy metal appear dark on the screen, but appear white (light) in prints because they are photographed in negative.

The physics of electron acceleration and focusing mean that specimens must be observed in a vacuum; therefore, the sample must be completely dry and fixed. For this reason, the EM picture is only a representation of structure because subtle effects of protein hydration on the arrangement of the polypeptide chains, for example, may be altered or lost by preparation for the EM. Further, sample preparation means that the EM cannot visualize objects in motion but only "frozen" in time. The "snapshots" of virus entry, egress, or alterations to the infected cell therefore must be interpreted with caution. One never knows whether the observed virus is biologically functional (able to replicate) or whether the process seen is exactly the one leading to biological effects. This point is important to remember when interpreting the EM views of virus entry into and egress from cells such as those presented in Part II, Chapter 6.

The process of "shadowing" a virus particle with heavy metal is shown in Figure 9.2. Such a shadow-cast can provide exquisite detail of the geometry of the virus. Much of the early development of shadowing and visualization methods of viruses was carried out by Robley C. Williams at the University of California, Berkeley.

Even richer detail can be obtained with the use of subtle staining procedures where the heavy metal is linked to protein molecules. Other types of shadowing, such as carbon shadowing, can also increase detail. Application of computer image enhancement can provide further striking increases in apparent resolution and resolve features that are obscured in conventional EM. Many examples of such detail can be seen in references cited in the "Additional Reading" sections of this book.

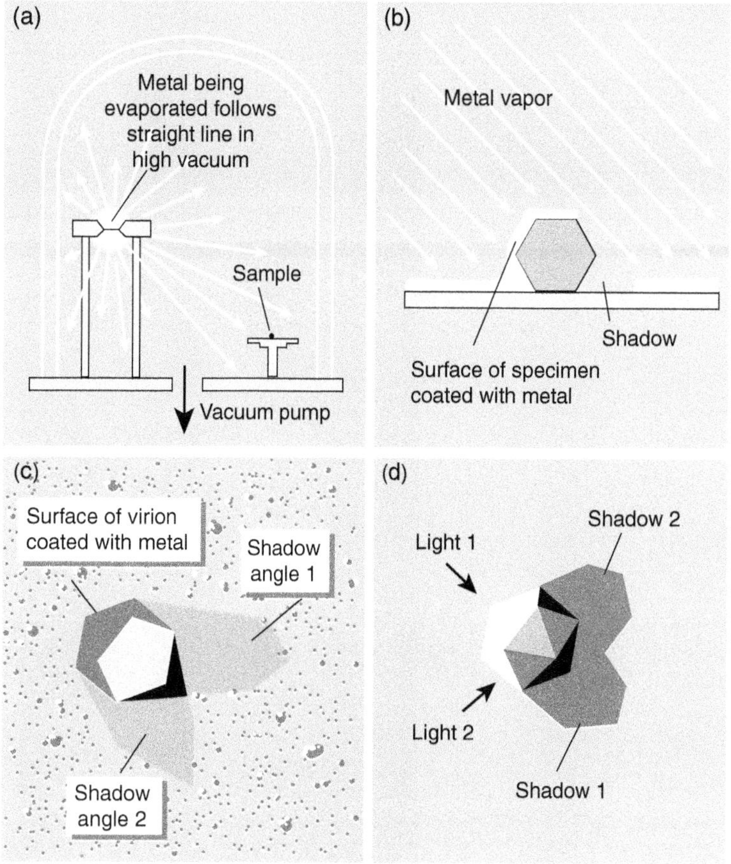

Figure 9.2 Shadowing specimens for viewing in the electron microscope. (a) A sample of heavy metal is vaporized in a vacuum chamber. This vapor travels in straight lines from the source and forms a layer on all surfaces in its path. (b) Any object in the path will cast a shadow on the grid on which it is supported. (c) A double-shadowed virus in the electron microscope. (Source: Drawn from photographs originally made by Robley C. Williams.) (d) An icosahedral model is placed in two light beams to show the equivalence of the shadows. This equivalence occurs because metal particles in vapor travel in straight lines, as does light. (Source: Drawn from photographs originally made by Robley C. Williams.)

To avoid the problems of structural deformation of particles that result during preparation for conventional EM, especially with enveloped viruses, a technique called **cryo-electron microscopy** was developed. This method employs no stains or heavy metal shadowing and therefore results in greater preservation of the particles. Instead, virus particles are rapidly frozen on the EM grid such that they are captured in a thin film of vitreous ice (ice in which large crystals cannot form). Within this glasslike matrix, the particles are hydrated in what is likely a more normal state, as opposed to the metal ion–stained and dried specimens of conventional EM. Since no stains are used, the frozen-hydrated particles are imaged by taking advantage of the difference between the electron density of protein or lipid and that of the surrounding water matrix. To prevent unwanted changes, the specimens are viewed in a microscope equipped with a cold stage to maintain the ice structure under vacuum, and data are collected at a very low dose of electrons to reduce damage from the intensity of the beam. The images observed can be enhanced by computer methods similar to those applied to the resolution of X-ray diffraction information. The herpes simplex virus (HSV) capsid image shown in Figure 9.3a was produced by these techniques. The structure of mimivirus with its huge 1.2 Mb genome has also been resolved using this method. The HSV capsid, like that of the distantly related poxviruses, has two lipid envelopes surrounding the inner virion and genome. These membranes, in turn, are encased in an outer protein shell (Figure 9.3b).

Counting (enumeration) of virions with the electron microscope

Since virus particles can be purified and visualized, they can be counted. Such counting does not tell how many of the particles are infectious (biologically active), but a count of particles in a

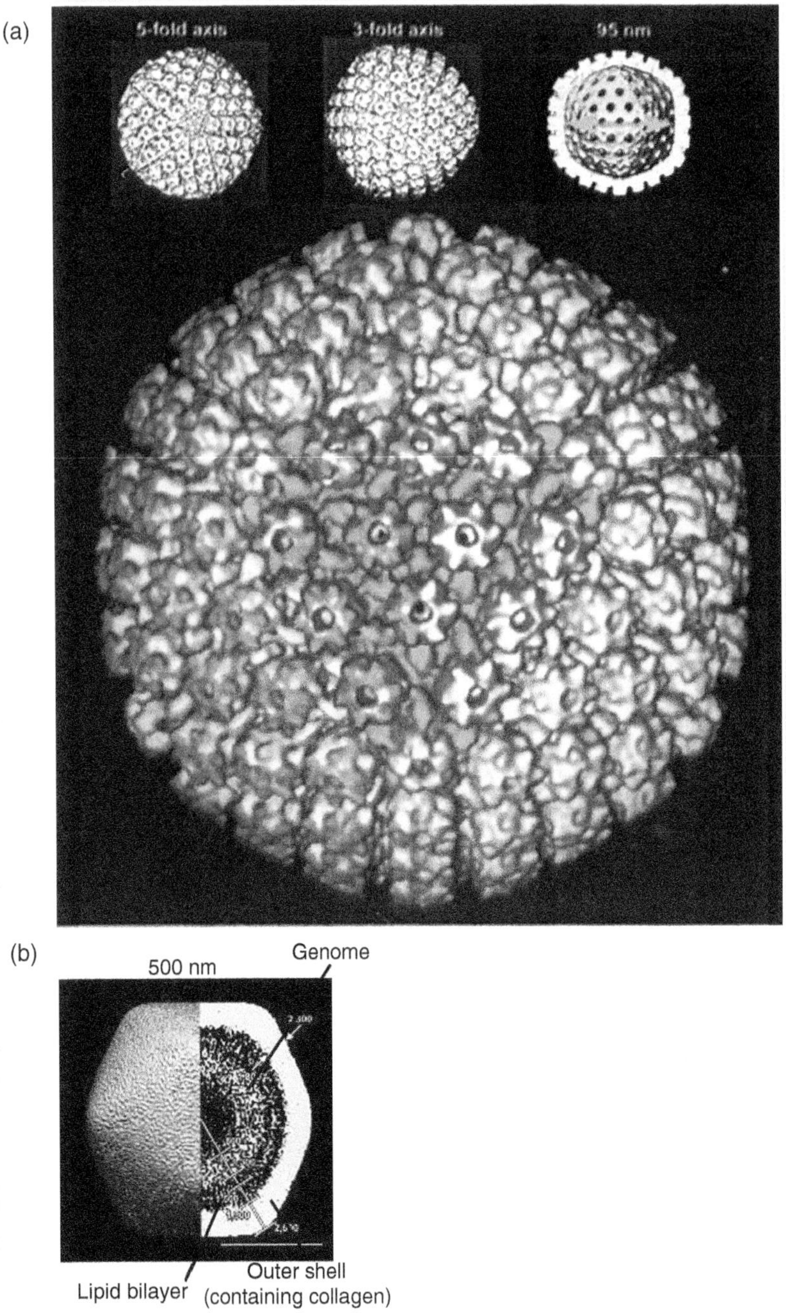

Figure 9.3 Computer-enhanced three-dimensional reconstruction of viral capsids using cryo-electron microscopy. The reconstructions are computed from electron micrographs of capsids preserved by freezing. For this type of electron microscopy, the samples are frozen and irradiated at liquid N_2 temperature with a very low flux of electrons to minimize damage. Information from many individual micrographs of particles is then combined in the computer to produce a reconstruction with a resolution higher than that of any single micrograph. (a) The HSV-1 capsid is reconstructed from 60 to 80 images, which provides a resolution on the order of 2.5 nm, but many more can be combined. Views showing the three major axes of symmetry and a cross section are shown at top. The bottom figure is a false color rendering of the information. One triangular face of the icosahedral capsid is shown in color. Pentons are orange, hexons red, and triplexes green. VP26, a small protein (molecular weight 12 000) associated with the hexons, is coded in blue. One VP26 molecule is bound to each VP5 molecule in each hexon. No VP26 is present in pentons. More detail concerning herpesvirus capsid structure can be found in Part IV, Chapter 18. See Plate 4 for color image. (Photographs courtesy of J. C. Brown and James Conway.) (b) A schematic rendering of the cryo-electron microscopic structure of the Mimivirus genome reconstructed to 7.5 nm resolution. (Source: Based on work of Xiao, C., Chipman, P.R., Battisti, A.J., et al. (2005). Cryo-electron microscopy of the giant mimivirus. *Journal of Molecular Biology* 353(3): 493–496.)

solution free of contaminating cellular material is very useful. The ratio of the number of virions to the number of infectious viruses is called the **particle-to-PFU ratio** (*PFU* stands for plaque-forming unit). Once the total number of particles is known in a solution, the measurement of total nucleic acid (genomes) allows calculation of the amount of genome per particle, and thus a measure of genome size. Again, the particle number can be used to tell the absolute amount of protein per capsid, and this (along with knowing the molar ratios of different capsid proteins determined by methods discussed in Chapter 11) allows one to work out details of the virus structure. Finally, the ability to count virus particles can be very useful for diagnostic and other medical purposes.

All counts require visualization, but once it is known that a certain number of virus particles contains a given amount of enzyme (i.e., reverse transcriptase for a retrovirus), or interacts with a certain number of test red blood cells (hemagglutination [HA]), or contains a given amount of DNA or RNA using quantitative polymerase chain reaction (PCR), then measure of these latter parameters can be related to particle number.

Counting of particles is simple in theory. For example, if one could be sure that each EM field contained virus from a specific volume of solution, one could readily calculate particle number. All that is required is knowledge of the fraction of the original sample being utilized for visualization. This fraction is a function of the volume of the observed sample as well as any dilution steps used in preparing the sample.

For example, if there were 30 particles in an average microscopic field and the volume of solution visualized corresponded to 10^{-4} ml of the original virus suspension, then that original suspension could be estimated to contain 3×10^5 ($30/10^{-4}$) particles per milliliter. However, this is not a particularly accurate way of measuring particle concentrations. The problem comes from the fact that despite the basic simplicity of the approach, it is difficult to achieve careful dilution and even spread of virus in the field of view of the EM, and many artifacts can arise.

Some uncertainties can be minimized by addition of a known amount of some standard in the original suspension, such as latex beads of uniform size. Then the number of both beads and particles can be counted in the EM field. The ratio of these, and knowledge of the number of beads used to make the solution, allow calculation of the number of particles in the original suspension. Since it is easy to add a known number of beads from a standard solution, the process can be applied to a series of different virus preparations.

ATOMIC FORCE MICROSCOPY – A RAPID AND SENSITIVE METHOD FOR VISUALIZATION OF VIRUSES AND INFECTED CELLS, POTENTIALLY IN REAL TIME

While the electron microscope can provide three-dimensional structural information, it is merely an averaging technique – that is, highly detailed structures are based upon the entire population of particles observed. This is an inherent limitation of even the highest resolution techniques available for studying molecular structures – **X-ray crystallography**. Further, these methods require extensive sample preparation and fixing, and subtle information regarding the characteristics of the individual particles and structures in a population of viruses as well as dynamic changes can only be inferred by painstaking statistical analysis, and then only with caution. Thus, grandly symmetrical and apparently perfect models of larger viruses derived from X-ray crystallography and cryo-electron microscopy may be somewhat deceptive, and not entirely representative of the entire population. Further, as has been discussed in Chapter 6, details about virus–cell interactions are often open to multiple interpretations.

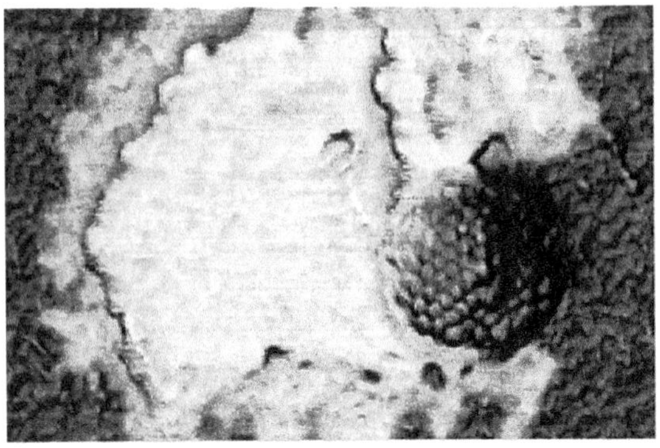

Figure 9.4 Atomic force microscopy was used to visualize the protein capsid of a herpes simplex virus on a glass substrate. The virus has lost most of its membrane, which forms the large mass to the left. Barely perceptible protein molecules are embedded in the membrane, which is folded back upon itself in places. The capsid is characterized by an icosahedral distribution of protein units, which enclose the double-stranded DNA genome.

A rather bizarre feature of molecules interacting at extremely close (quantum-scale) distances is that electrons can "tunnel" between atoms, producing a small but measurable force between them. This quantum force has been utilized in the technique of **atomic force microscopy (AFM)**, where a molecule-scale probe is held at a constant tunneling force over the surface of a cell, subcellular component, or virus so that as the probe is moved over the sample, a "contour map" of the surface can be generated. This method requires little or no sample preparation, and, in theory at least, could be done on living cells to provide animated real-time analyses of changes in cellular surface structure as virus infection proceeds. It introduces an effective complement to the techniques above. Most importantly, it can be used to examine the architecture of a single virus particle, or a collection of distinct individuals, and this may be carried out at a resolution very near that of cryo-electron microscopy. This method has been used for imaging capsid structures of viruses in crystals as well as viruses interacting with cells. An example of an AFM view of an HSV capsid is shown in Figure 9.4.

INDIRECT METHODS FOR "COUNTING" VIRUS PARTICLES

Once the number of virus particles in standard solution is known, this information can be correlated with other readily measurable properties of the virion. For example, the amount of virus causing agglutination can be related to particle number. As discussed in Chapter 7 in Part II, many enveloped viruses can agglutinate red blood cells, and this property can be used as a measure of virus particles because it takes a certain number to coat the red blood cells to cause agglutination. Under standard conditions for the assay, the number of influenza virions is about 10^4 virus particles per hemagglutination unit *(HA unit)*. An HA unit is just enough virus to cause agglutination of the standard sample. (Details of an HA unit definition can be found in many medical laboratory protocols.)

QUESTIONS FOR CHAPTER 9

1 The data in the table below show the results of attempting to infect three different cell lines with La Crosse encephalitis virus (LAC). With electron microscopy, observations were made to detect virus particles on the surface of the cells and virus particles present in endocytotic vesicles (endosomes) inside the cell. A "+" indicates that the virus was present in the majority (>80%) of the cells observed, whereas "+/−" indicates that the virus was present in only a few (<5%) of the cells observed. In addition, the *average* yield of virus per cell was measured. Using these data, answer the following questions about these cell lines.

	Data for La Crosse Encephalitis Virus		
Cell Line	Virus on Surface	Virus in Endosomes	Virus Yield per Cell
HeLa	+	+/−	5
CEF	−	−	0
BHK21	+	+	200

(a) Which cell lines are susceptible to infection by LAC? Why?
(b) Which cell lines appear to be permissive for LAC infection? Why?
(c) Propose a hypothesis to explain the data for HeLa cells compared to BHK-21 (baby hamster kidney-21) cells. How can you explain the difference in average yield per cell? How would you test your hypothesis?

2 You isolate virus particles and resuspend them in 2 ml of a buffered solution containing a total of 6×10^9 latex beads. After doing laborious and careful dilutions, shadowing, and other things necessary for electron microscopic examination, you view a number of equal fields and determine that you have three beads for every nine virions. What is the approximate number of virions present in each milliliter of your beginning stock solution?

3 What features of the electron microscope make it an excellent tool for examining virus particles?

4 How would you determine the number of virion particles in Question 2 that are actually infectious (particle-to-PFU ratio)?

Replicating and Measuring Biological Activity of Viruses

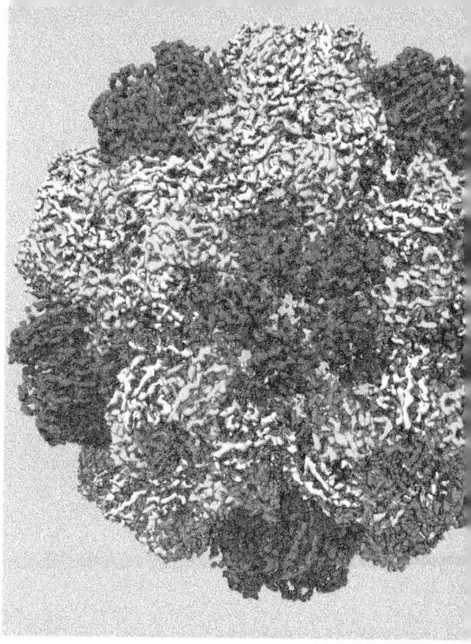

CHAPTER

10

Basic Virology, Fourth Edition. Martinez "Marty" Hewlett, David Camerini, and David C. Bloom.

CELL CULTURE TECHNIQUES

Growing and maintaining cells in the laboratory is an absolute necessity for any molecular biological investigation. Because viruses must replicate within the cell they infect, their study is greatly enhanced by the ability to maintain cultures of the cells in which the viruses grow most conveniently for the study at hand. Ultimately, cell culture involves taking a representative sample of cells from their natural setting, characterizing them to a sufficient degree so that their basic growth properties and any specific functional properties are known, and then keeping them in continuous or semicontinuous culture so that they are in ready supply. Depending on the type of virus being studied, and the specific property of that virus of interest, this task can be either routine or daunting.

Maintenance of bacterial cells

The study of bacterial viruses provided the model for the study of all viruses because it was convenient to replicate such viruses in easily grown bacterial cell cultures. Some bacterial cells are exceedingly difficult to grow in culture and have very slow generation times. But standard laboratory cultures of the most commonly used prokaryotic cells, such as *Escherichia coli* (*E. coli*), can be grown on simple, defined media consisting only of an energy and carbon source (usually glucose) and inorganic nitrogen, phosphorus, and sulfate sources such as NH_4Cl, $MgSO_4$, and phosphate buffers. Such ability to grow on media containing only sugar and inorganic molecules is called **prototrophy** and allows full knowledge of all the ultimate sources of biological reactions. More rapid growth is attained with a broth of yeast or beef extract, possibly supplemented with required inorganic materials.

Bacterial cells can be grown in liquid culture, where densities of 10^8 cells/ml are reached during the exponential (i.e., most rapid) phase of growth. Bacterial cells can also be grown on solid or semisolid surfaces, allowing formation of colonies or **clones** where all cells are the descendants of one single cell. The most common material used for this type of growth is agar, poured as a thin slab into glass or plastic Petri dishes.

Such plates are used extensively for *plaque assays* of bacterial viruses. Plaque assays take advantage of the fact that virus replication results in cell lysis and thus a center of virus infection will be devoid of cells. Techniques of plaque assays are described in more detail later in this chapter.

Plant cell cultures

Most plant viruses can be conveniently studied by infection of their intact hosts, which are usually not difficult to grow and maintain. This method allows basic analysis of many plant virus features. Indeed, early structural study of plant viruses was at a level fully equivalent to studies of bacterial viruses.

Molecular biological studies lagged until recently, however, due to a lack of reliable plant cell culture systems. Plant cell culture techniques have not developed as rapidly as those for animal cells, because plant cell architecture makes the manipulation of cells in culture (which is such a boon to the study of animal and bacterial cells) very difficult and often nearly impossible. These technical problems have resulted in plant cell culture not having a major impact on the development of plant virology. Plant cells without their cell walls can be cultured as protoplasts, and this has provided great impetus to the study of plant molecular biology; but as yet, little virology has been done with such systems.

Culture of animal and human cells

Maintenance of cells in culture

To maintain cells in culture, culture medium approximating blood plasma must be used. This medium contains salts similar to those found in plasma; most amino acids (since animal cells cannot synthesize many of these); vitamins; glucose as an energy source; buffers (usually carbon dioxide/sodium bicarbonate) to prevent lactic acid (resulting from glucose fermentation) from making the medium too acidic; and, most importantly, blood serum, which is usually obtained from calves or horses. This serum contains many growth factors (e.g., proteins) that the cell needs for growth. As noted, antibiotics also are included to preclude microbial contamination.

In addition to the uncertainties of exact culture requirements, the same type of cell (e.g., a fibroblast or skin cell) can have strikingly different properties depending on its species of origin, the age of the donor animal, state of the cell, and the specific culture history. Thus, each cell culture has its own pedigree and peculiarities.

A general method for obtaining mouse mammary epithelial cells is shown in Figure 10.1. Similar methods are used to generate cultures of many primary and tumor cells. Cells are usually grown in standard-sized culture dishes with specific areas. Popular sizes range from 25 cm^2 to 150 cm^2, depending on the number of cells needed.

Types of cells

Ultimately all animal cells are derived from living tissue; however, some – such as HeLa cells – have been in culture for so long (about 70 years) that they have lost all resemblance to the tissue from which they were isolated. Such **continuous cell lines** are very useful in that they

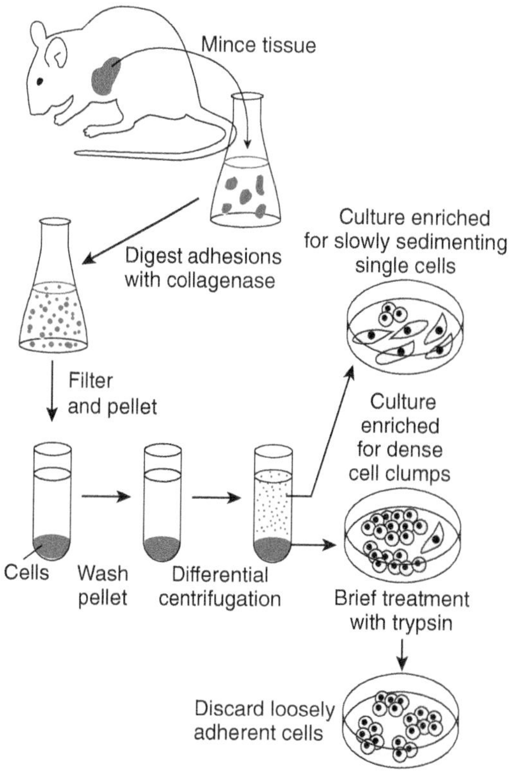

Figure 10.1 Generating a primary cell culture. Tissue is surgically removed from an anesthetized animal, and then minced and homogenized. Addition of collagenase breaks down extracellular collagen, but the enzyme does not attack intact cells. The cells are purified by filtration through a coarse mesh to remove large fragments, and then concentrated by deposition under a mild centrifugal field in a low-speed centrifuge. The pelleted cells are washed in various buffered media containing serum, and then can be subjected to differential low-speed centrifugation to partially separate cell types based on sedimentation rates (a function of cell size and density). Various fractions are plated onto culture dishes in the presence of a culture medium containing essential amino acids, vitamins, antibiotics, and serum. Cells grow as loose clumps that can be dispersed with mild trypsin treatment, and individual cell types then can be cultured.

grow rapidly to provide large amounts of virus for the study of some basic aspects of virus replication. They are not good, however, for studying the relatively subtle effects of virus infection on cell growth and control. Continuous cell lines are also not appropriate for the study of differentiated cell function.

Continuous cell lines generally have the following properties:

1 They have fragmented and reduplicated chromosomes; that is, they are **aneuploid**.

2 They are able to grow in suspension and in relatively low concentrations of serum, and can overgrow each other; they display no response to neighboring cells.

3 They are essentially immortal; if periodically diluted and fed with appropriate nutrients, they will continue replicating.

4 They generally do not display properties of differentiated cells and do not respond to modulators of cell growth or function.

5 If introduced into an animal (even one of the same species from which they were originally isolated), they will not grow and will be eliminated by the animal's immune system.

At the opposite extreme of laboratory cell type are **primary cells**. Primary cells are most conveniently isolated from embryonic (fetal) tissue or from newborn animals or tissue. Cells isolated from older animals tend to be difficult to culture and have a much shorter life in culture before they cease to divide (**senesce**) or die.

While the very act of culturing cells leads to rapid changes in the subtle properties of living cells, the earliest stages of culturing primary cells are very nearly identical to those in the tissue from which they derive. Although almost any type of replicating cell can be cultured if the tissue containing it is properly isolated, the more rapidly growing and replicating cells, such as fibroblasts, will outgrow other cells in a mixed-tissue source. For that reason, isolation of primary cells from whole embryos generally produces cultures of fibroblasts.

Most primary cells have the following properties:

1 They have normal chromosome numbers and shape.

2 They require high serum concentrations containing numerous growth factors.

3 They cannot divide or even survive for long unless they are maintained in contact with a solid surface.

4 They are subject to **contact inhibition** of growth and of cell movement. Contact inhibition means that when they touch other cells in a culture plate, they stop growing and stop moving. Thus, a given area of culture plate will allow cells to grow to a specific number. During contact inhibition, the cells are healthy and metabolize energy. When they are diluted (passaged) and placed into a new culture dish, they will begin to grow and divide again.

5 They have a finite lifetime measured in divisions. Normally, fibroblasts can divide 20–30 times after isolation and then the cells begin to senesce and die. Recent experimental evidence suggests that this finite lifetime is due in part to programmed loss of chromosome end regions (**telomers**) at each cell replication. When enough chromosomal DNA is lost, the cells begin to die.

6 They display all properties of differentiated cells and respond to modulators of cell growth or function.

7 If introduced into an animal of the same species from which the cells were originally isolated, they may survive but will not produce tumors.

This list of properties of primary cells is, of necessity, an idealized one. Some of the properties listed may not apply to a given type of cell isolated from an individual. For example, lymphocytes isolated from the small amount of blood found in the umbilical cord of a newborn will survive but not replicate when maintained in suspension. In contrast, lymphocytes cultured from many adolescents and adults who have had infectious mononucleosis not only will survive but also will divide for relatively long periods of time. Even though these immortalized lymphocytes are ostensibly normal, they maintain copies of the genome of the Epstein–Barr herpesvirus, and the expression of certain viral genes contained therein leads to these unusual properties. This type of transformed lymphocyte is not a tumor cell, but it clearly demonstrates some similar properties.

Loss of contact inhibition of growth and immortalization of primary cells

Immortalized B lymphocytes are but one example of cells available in the laboratory that have properties intermediate between the two extremes of continuous cell lines and primary cells. These cell types have undergone transformation and have at least some of the properties of tumor cells. Culturing primary cells for long periods can generate transformed cells. During the time in culture, there is a random accumulation of mutations that alter a critical number of growth control genes encoded by the cell. At a critical point during cultivation (the actual point will depend upon the cell type – it is usually 12–15 generations with fibroblasts), cells suffer from the cumulative effects of aging (senescence) where nearly all enter a crisis period and undergo apoptosis and die.

Senescence is a consequence of the defective replication of chromosomal DNA, which is linear and (as discussed in Chapter 13) cannot completely replicate itself at the ends. Thus, each round of DNA replication results in the loss of the critical telomeric sequences at the end of chromosomes. Normally, the telomeres bind to a number of specific cellular proteins, which mask the chromosomal ends that are structurally equivalent to double-stranded DNA breaks. When the ends are unmasked, a number of important cellular defenses are activated that initiate the apoptotic pathway.

As will be outlined in the discussion of carcinogenesis in Part V, a major factor in carcinogenesis is the abrogation of the normal apoptotic response to double-stranded DNA breaks. These breaks, if not properly repaired, can be mutagenic and can alter the function of numerous growth control and developmental genes, leading to uncontrolled cell growth. In cultured cells undergoing crisis, a very few may be able to protect the ends of critical chromosomes by a combination of inappropriate recombination events along with the activation of the telomerase enzyme normally active in early development that regenerates telomeric sequences lost during chromosome replication. Thus, cells surviving crisis share certain features of tumor cells. These *immortalized* cells become predominant and relatively rapidly overgrow the culture. Such cells eventually can be used to generate continuous lines.

Cells with the properties of transformed cells also can be isolated from tumors in an animal. Different tumor cells in an animal display one or several of the same transformation levels from normal cells that can be observed with the culture of primary cells. This is an important clue to the nature of the cellular events leading to cancers. It is important to be aware, however, that different tumor cells can display widely different deviations from normal growth properties of the cells from which they derive. Some tumor cells, especially those isolated early in the course of cancer development, display very few differences from normal cells – perhaps only the loss of contact inhibition of growth. Others, especially long after the cancer occurred, have many additional changes.

The process of change from primary cells to continuous-line cells and the relationship between these cells and tumors in the animal of origin are shown in schematic form in Figure 10.2. This process of change is a convincing experimental demonstration that the cellular changes in an organism from normal to cancerous involve multiple steps. The changes multiply as mutations of specific growth control and regulatory genes in the cells alter cell function and the cell's ability to respond to normal signals in the animal, limiting cell growth and function.

While it is not uncommon to generate an immortalized cell line by lengthy **passage** or other mutagenic processes, it is important to be aware that many tumor cells and some transformed

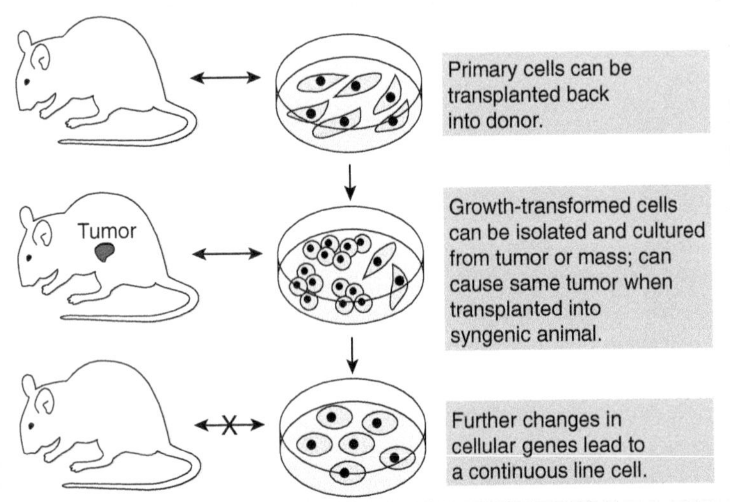

Figure 10.2 The progression of cells in culture from primary to transformed to continuous lines, and their relationship to tissues in the originating animal.

cells that have lost contact inhibition of growth still have a finite lifetime. The genes controlling lifespan and response to contact inhibition are not identical and can be mutated together or separately.

One of the most fruitful aspects of the study of some viruses is that they can cause transformation of normal primary cells into cancer or tumor cells. Such virus-transformed cells, when reintroduced into animals, can cause tumors. Since this transformation requires a specific interaction between viral and cellular genes or gene products, the study of the process has led to much current understanding of carcinogenesis and the nature of cancer cells.

THE OUTCOME OF VIRUS INFECTION IN CELLS

Fate of the virus

When a virus infects a cell, its genome enters that cell. Regardless of whether the virus capsid remains at the surface of the host, as in bacteriophage infection, or is internalized, it is modified and then disrupted. If an infecting virus is isolated after attachment and penetration, the virus is no longer stable and cannot initiate a new infection. Thus, following the initial steps of virus–cell interaction, the only way that infectious virus can be isolated is to either block its entry into the cell, or wait for progeny virus to be formed.

In some types of infection (generally called a **nonproductive infection**), new infectious virus is not produced. This type of infection is also termed an **abortive infection** because it does not proceed to completion of the replication process. Abortive infections can result from the virus infecting a **nonpermissive cell** (i.e., one that, for some reason, does not have the proper machinery for virus replication). A nonproductive infection could also be the result of infection with a virus that has some defective gene product interfering with replication. A general rule of thumb in differentiating types of infection is the following:

1 Productive infection: More virus out than in.

2 Abortive infection: No virus out; virus cannot replicate.

When an abortive infection occurs, the viral genome may be destroyed or it can be internalized. In the latter instance, one or several viral genes might be expressed. This situation could result in the cell's expression of viral antigens at its surface or elsewhere in the cell. Given the proper immune reagents (briefly described in Chapter 12), such antigens can be detected and studied.

From this explanation, it is clear that an abortive infection can have profound effects on the host cell, and perhaps ultimately on the organism. For example, the continuous presence of noninfectious measles virus in brain tissue can lead to severe complications (see Chapter 4, Part I). Also, many DNA viruses that cause cellular transformation do so only under conditions of abortive infection. Understanding the reason why a virus infection is abortive can be very important to understanding and describing the course of virus replication and the effects of virus infection on the host. Some questions important to characterizing abortive infections are the following:

1 Is the virus genome lost?
2 Is part of the genome maintained and expressed?
3 If the genome is maintained, is it physically integrated into the host genome, or is it maintained as a separate "mini-chromosome" or **episome**?

Other types of infection fall between the extremes of productive and abortive. For example, cells can be poor hosts for replication of a specific virus but not strictly nonpermissive for virus replication; often, such a cell is called **semipermissive**. Clearly, there is no real strict point at which a cell is permissive or semipermissive for virus replication; the terms are relative.

Other cell-based impediments to virus replication exist. Viruses can have mutations that are lethal only under certain conditions (conditional lethal mutations) such as high temperature (temperature-sensitive mutations). Dynamic situations can occur in which virus is slowly released over time at low levels. Such a situation can define a persistent, inapparent, or chronic infection. Under some conditions, such an infection in a cell culture or in an animal can lead to episodes of high levels of virus production with obvious cell destruction or disease. These episodic occurrences can alternate with periods in which virus is difficult to detect and the host (or cell culture) appears relatively healthy.

Under certain conditions of infection, many viruses will produce incomplete viral particles, and these particles may be able to infect other cells. Such particles are termed *defective virus particles*, and can be produced by a variety of mechanisms – often involving inefficient steps in virus maturation taking place very late in the infection cycle when the host cell is rapidly deteriorating due to virus-induced changes. The generation of empty capsids of cytomegalovirus as well as enveloped dense bodies made up of tegument proteins shown schematically in Chapter 6 is a good example of such an occurrence.

In addition to the formation of defective particles due to the packaging of empty capsids, viruses can randomly produce partial genomes during their replication. If these partial genomes contain a packaging signal, they can be encapsidated and form a specific class of defective particles. An infection of a cell with one of these particles will be abortive, since the genome is not complete.

Interestingly, the simultaneous infection of such a defective particle with an infectious one can lead to interference, which is a result of the smaller fragmentary genome being able to reproduce more copies in a given time than the complete genome. This is purely a mass effect. The shorter molecule can undergo more rounds of initiation and completion of replication per unit of time, but the result is that the yield of infectious particles will be reduced. For this reason, defective virus particles of this type are classified as defective interfering particles. Their presence in a virus stock can be a headache to a researcher trying to get a high yield of virus, but defective particles can be used to deliver genes to cells under certain instances. The use of viruses to deliver genes is briefly discussed in Part V, Chapter 22.

Finally, it should be recalled that herpesviruses and lentiviruses such as HIV-1 (as well as some other viruses) can remain as latent infections in which the viral genome is maintained in the cell or in some cells of the host but no infectious virus or viral structural proteins are detectable (see Chapters 17 and 20).

Fate of the cell following virus infection

Cell-mediated maintenance of the intra- and intercellular environment

As discussed in this chapter, long periods of passage of cells in culture as well as mutations in the genetic information carried by cells can alter their growth properties. Such changes can take place within the animal, leading to formation of a tumor, but usually this does not happen. This is because the vertebrate body and the cells comprising it have a number of "check points" that respond to genetic alterations of individual cells. This is a major function of major histocompatibility complex type I (MHC-I)-mediated antigen presentation. When an abnormal epitope (from a genetically damaged protein that would normally not be expressed) is presented at the surface of the cell, it triggers the destruction of the cell by a cytotoxic T lymphocyte (CTL). This interaction leads to the death of the cell by the apoptotic pathway. As noted in Chapter 8, Part II, apoptosis is a consequence of the action of specific cellular genes that lead to a phased shutdown of cellular functions and cell death. The process has a protective function in the body by inducing the death and elimination of highly differentiated cells no longer needed (such as effector cells of the immune system), aged cells, as well as cells with mutations in genes that normally function to limit cell division. It is important to understand that the apoptotic pathway leads to cell death *without* release of cellular contents to the immune system and resulting inflammation and potential pathology; rather, it is a highly regulated process designed only to eliminate those cells that are no longer of value in the tissue in question. The apoptotic pathway should be contrasted with the other major route of cell death, *necrosis*, where the swelling and bursting of the cell targeted for death lead to inflammation in order to stimulate the immune response. The two processes are schematically outlined and contrasted in Figure 10.3.

Obviously, it is of value for a virus replicating in a cell to ensure that the cell is maintained for a sufficient length of time to ensure an appropriate yield of virus, while at the same time limiting immune responses to the infection. Conversely, it is to the benefit of the cell and the organism comprised of such cells to mount a controlled immune response as rapidly as possible

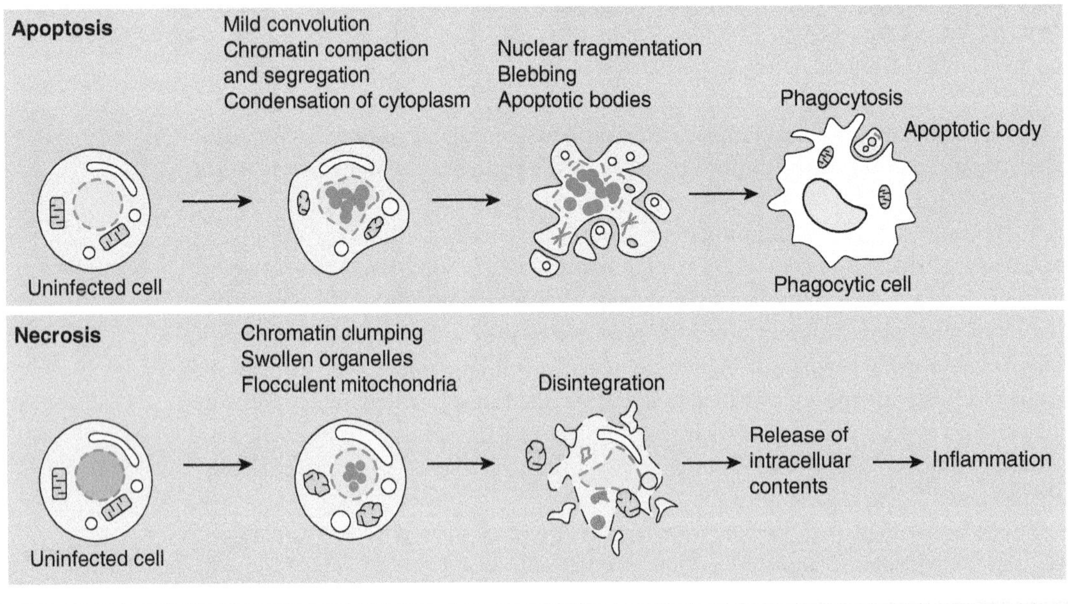

Figure 10.3 Apoptosis versus necrosis in cell death.

and eliminate infected tissue. It is the tension between these two processes that leads to evolutionary change in both virus and host, and the manifestations of both processes lead to macroscopic and microscopic changes in virus-infected cells that define cytopathology.

Virus-mediated cytopathology – changes in the physical appearance of cells

Some basic types of virus-induced changes to the host cell (cytopathology) result in changes that are readily observable by eye or with the aid of a low-power microscope. All cytopathology requires some specific interaction between viral gene products and the cell. Even the cell lysis induced by poliovirus or bacteriophage infection, in which the cell "explodes," is the result of very specific modifications to the cell's plasma membrane and lysosomes induced by specific poliovirus gene products. Less dramatic, but still clearly observable changes to the cell include the formation of cytoplasmic inclusion bodies (which is diagnostic for poxvirus infections), generation of nuclear inclusion bodies seen with herpesvirus infections, and alterations in chromosomes.

Cytopathology need not involve cell death. Virus-induced alterations in cell morphology, growth, and lifespan are all types of cytopathology. Even very subtle changes, such as a virus-induced change in the expression of a protein or appearance of a new macromolecule, are cytopathic changes, as long as they can be observed with some reproducible technique.

A major type of cytopathology involves changes to the cell surface due to expression there of viral proteins. Among other things, this can lead to the following:

1 Altered antigenicity: The altered cell will stimulate the immune system to generate antibodies to react with viral proteins or previously masked cellular proteins.

2 Hemagglutination or hemadsorption: Certain viral proteins will stick to red blood cells and cause these cells to stick together.

3 Cell fusion: Changes to the cell membrane can allow formation of large masses of fused cells or syncytia. Specific virus gene products are responsible for this. Such fusion induced by the Sindbis virus (a togavirus) can be used to generate somatic cell hybrids.

Another major type of cytopathology involves changes in cell morphology. An example is demonstrated in Figure 10.4; a herpes simplex virus 1 (HSV-1) infection is shown disrupting the cytoskeleton of the host cell, thereby changing the cell's shape. In this example, viral infection led to dissociation of the actin fibers, but not degradation of the actin. This very specific biochemical change in the actin subunits results in profound changes to the cell's morphology.

Virus-mediated cytopathology – changes in the biochemical properties of cells

Virus infection leads to specific changes in biochemical processes of the cell. Some viruses, such as HSV and poliovirus, specifically inhibit cellular protein synthesis. The mechanism for such

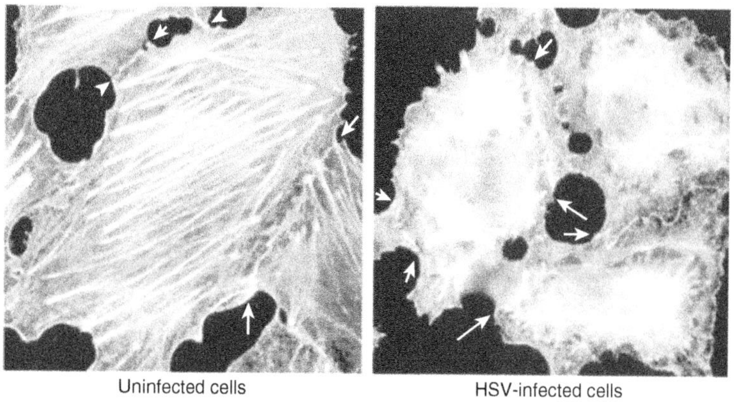

Figure 10.4 HSV-induced changes in the properties of actin microfilaments of a cultured monkey fibroblast. The cell was stained with a fluorescent dye that reacts with actin fibers so that they can be visualized in ultraviolet light. This technique is similar to immunofluorescence microscopy, which is discussed in Chapter 12. The left panel shows parallel arrangement of the microfibrils in the uninfected cell, while HSV infection (right panel) results in disassociation of the fibrils and diffusion of the actin throughout the cytoplasm. At the same time, the cell loses its spindle-shaped morphology and becomes rounded. The arrows indicate junctions between cells that are also rich in actin fibrils and are not disrupted by HSV infection at this time. Source: Courtesy of Stephen Rice.

inhibition is complex and differs for different viruses. Viral infection also can lead to specific inhibition of cellular messenger RNA (mRNA) synthesis. Gross inhibition of cellular macromolecular metabolism will lead to cell death. However, there are complex and multifaceted effects of virus infection on cell function resulting from subtle changes in cellular functions that do not result in cell death, but favor virus production.

A striking example of the ability of certain DNA tumor viruses to prevent cell death long enough to allow efficient virus replication is found in viral inhibition of apoptosis. Another very important consequence of infection is, as discussed earlier in the chapter, changes in the growth properties and lifespan of virus-infected cells. The growth rate, total number, and lifespan of differentiated cells are tightly controlled through the auspices of specialized **tumor suppressor genes**, so named because they block the formation of tumors. The interactions between viral genes and tumor suppressor genes are generally well understood in the replication of papovaviruses and adenoviruses, and are described in Chapter 16, Part IV. For the purposes of this discussion, it suffices to note that DNA-tumor viruses inhibit the tumor suppressor genes as a method to "activate" the cell for their own replication. The induction of apoptosis would interfere with the cell's ability to support virus replication. The mechanism of transformation varies between different tumor viruses, but in many cases specific virus-induced inhibition of apoptosis as well as inactivation of cellular genes actively inhibiting cell division are both important factors.

Another major effect of virus infection is interaction between the infected cell and the host's immune system. As briefly outlined in Chapter 8 and more specifically in chapters describing specific viruses (Part IV), many viruses contain genes that function to specifically inhibit the production of interferon in the infected cell. Further, certain viruses, such as HSV, can specifically inhibit MHC-I-mediated antigen presentation at the early stages of infection. Although eventually the cell will express viral antigens as infection proceeds, this early inhibition of antigen processing can provide the virus with a vital head start in its infection.

Virus infection of cells can lead to a number of specific cellular responses that involve the expression of new cellular genes, or the increase in expression of some cellular genes. The interferon response described in Chapter 7 is a good example of this. Several techniques of modern molecular biology allow very precise identification of cellular genes induced by virus infection.

One method is termed **differential display analysis** and requires the use of groups of oligonucleotide primers, retrovirus reverse transcriptase, and the **polymerase chain reaction (PCR)** to generate and amplify complementary DNA copies of cellular transcripts. By comparing the amplification patterns of products isolated from uninfected and infected cells, increases or decreases in levels of specific cellular gene transcription can be determined.

Other methods used involve microchip technology in which numerous (up to 64 000) oligonucleotide probes specific for various cellular genes are bound to a very small **microarray** and hybridized with PCR-amplified complementary DNA **(cDNA)** samples made from mRNA isolated from uninfected and infected cells and labeled with different-colored fluorescent dye. Comparison of the patterns of light emission when the microchip is scanned with a laser beam leads to identification of changes in cellular transcription. The general methodology for microchip analysis and PCR is discussed in Chapters 11 and 12.

MEASUREMENT OF THE BIOLOGICAL ACTIVITY OF VIRUSES

Quantitative measure of infectious centers

Plaque assays

Cytopathic effects on the host cell by the great majority of viruses cause observable damage or changes to the cells in which they are replicating. Even if cells are not killed or lysed, the alteration of cells in a local area due to a localized virus infection can readily be observed as a plaque or **focus of infection**. With proper dilutions and conditions, such a localized infection can be the result of infection with a single biologically active virus. A virus particle able to initiate a productive infection is termed a **plaque-forming unit (PFU)**.

The process of plaque formation is easy to envision. The first infected cell releases many viruses. If the viruses (big compared to even the most complex molecules in the growth medium) are kept from wide diffusion, they will remain in the vicinity of the original infected cells and will infect neighboring cells. This process is repeated a number of times. As long as the virus–cell interaction is kept localized (often by making the cell culture medium into which the virus is released quite viscous), the area of cytopathology resulting from a single infection of a single PFU will remain localized to the area surrounding the initially infected cell. The resulting plaque can be readily observed and counted a few days after infection. Some examples of plaques on cultured cells are shown in Figure 10.5.

Cell cultures are not the only way to obtain infectious centers. Infection of plant virus on a leaf of a susceptible plant, along with some type of mechanical abrasion to initiate the infection (perhaps rubbing the leaf with carborundum powder), results in formation of visible centers of infection. Examples are also shown in Figure 10.5.

The **chorioallantoic membrane** of developing chick or duck embryos can be used (and indeed, must be used) for assays of certain viruses. In this assay, fertilized eggs are incubated for two weeks, and then carefully opened to expose the membrane (the embryo is below this within the egg). Virus suspension is then placed on the membrane, the egg resealed, and virus pocks or plaques allowed to develop (Figure 10.6).

Generation of transformed cell foci

The same titration principles can be used to measure other biological effects of virus infection. Under certain conditions, some DNA viruses can transform cells so that normal growth control is altered. As outlined earlier in this chapter, transformed cells have a different morphology and tend to overgrow normal cells to form clumps of proliferating cells. Each infectious event, even if it is abortive and does not produce new virus, will result in the formation of transformed cell clumps, called a *focus of transformation*. An example of a focus of transformed cells is shown in Figure 10.7. The changes in cell morphology that form a focus (a type of cytopathic effect of

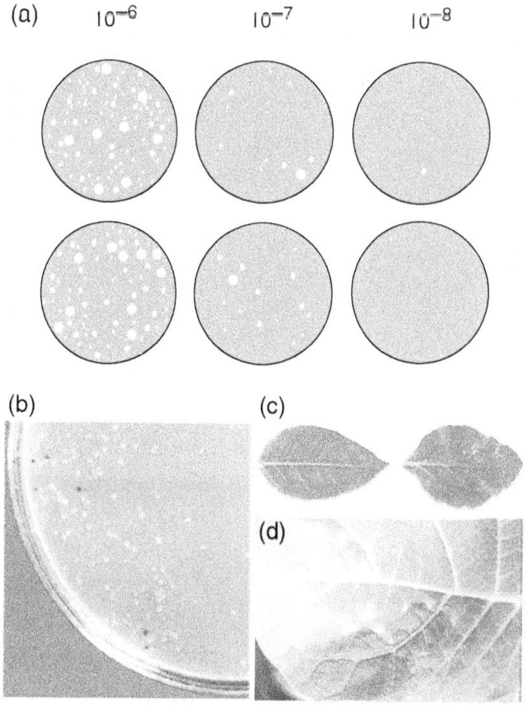

Figure 10.5 Visualization of virus plaques. Under appropriate conditions, virus infection can be localized to the vicinity of the originally infected cells. If a limited number of infectious units of virus (PFUs) are incubated in a culture dish or on tissue in which virus can cause a cytopathic effect, virus plaques can be visualized. (a) A continuous line of monkey cells (Vero cells) was grown in the six-well culture dish. When the cells reached confluence, they were infected in duplicate with a series of 10-fold dilutions of a HSV-1 stock. After 48 hours, the cells were partially dehydrated (fixed) with ethanol and stained. Areas of cell death show as white plaques, each representing a single infectious event with the input virus solution. Source: Courtesy of J. Langland. (b) A portion of the surface of a Petri dish containing agar with bacterial nutrient medium. A "lawn" of *E. coli* was grown on the plate's surface, and this layer of cells was infected with a solution containing a genetically "engineered" version of bacteriophage λ that can be used to clone inserted genes. (See Part V, Chapter 22, for some general details.) Bacteriophages that contain an inserted gene form clear plaques due to inactivation of an indicator gene (β-galactosidase), and viruses without the insert form dark-colored plaques. (c) Assay of tobacco mosaic virus (TMV). The left panel shows leaves of a resistant (bottom) and a susceptible (top) plant that have been infected with small amounts of virus. The right panel shows a higher magnification of plaque development.

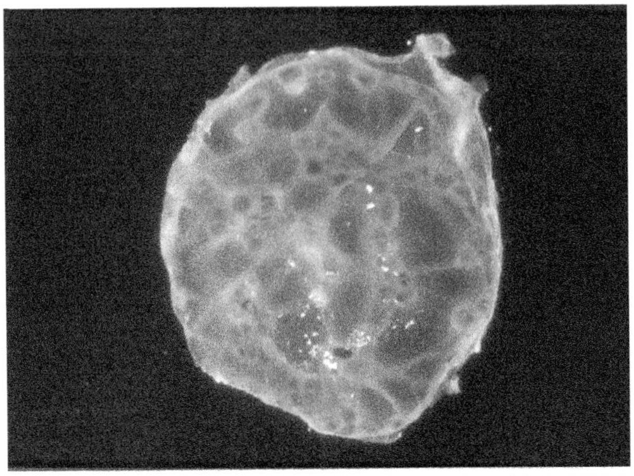

Figure 10.6 Rabbitpox virus pocks on the chorioallantoic membrane (CAM) of an embryonated chicken egg. The amount of infectious virions present in stocks of some viruses was historically determined by assaying for pock production on the CAM of embryonated chicken eggs. This is similar to plaque assays on cell monolayers. Shown are the classic red hemorrhagic pocks of rabbitpox virus on the CAM of embryonated eggs that were inoculated with rabbitpox viruses after 11 days of incubation. The image shows the pocks at 72 hours after inoculation. Source: Photograph courtesy of D. Bloom.

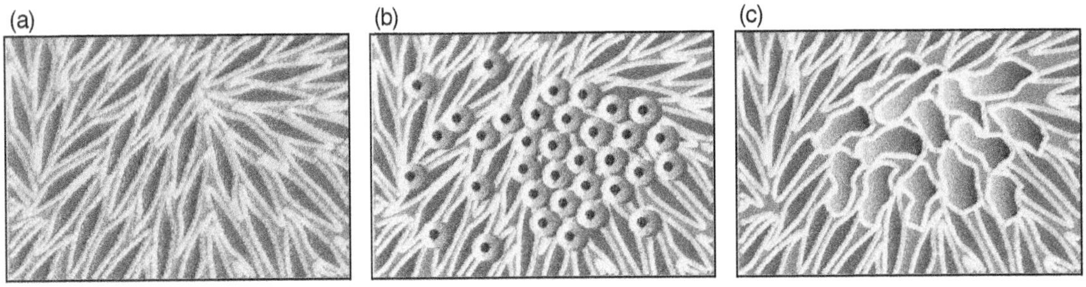

Figure 10.7 Some representative morphologies of rat fibroblast cells (F-111) infected with different transforming viruses. The top panel shows normal cells with their characteristic parallel orientation. A focus of transformed cells generated by infection with Rous sarcoma virus (an oncornavirus) is shown in the middle panel. Note the rounded morphology and density of these cells. The bottom panel shows the subtle difference in morphology when normal F-111 cells are infected with SV40 virus, for which they are nonpermissive. Source: Based on portions of a photograph in Benjamin, J., and Vogt, P.K. (1991). Cell transformation in viruses. In: *Fundamental Virology* (eds. B.N. Fields and D.M. Knipe), 2e. New York: Raven Press, Chapter 13.

transformation) are clearly evident. The number of focus-forming units can be counted just as with PFUs, but here one is counting the spread of transformed cells, not the spread of virus.

Use of virus titers to quantitatively control infection conditions

There are two important definitions relating to infectious virus particles or PFUs. The **particle-to-PFU ratio** measures just that: the proportion of total number of virus particles to infectious particles of virus. To obtain the ratio, one must count the total virus particles and do an assay for biologically functional ones (discussed in more detail in Chapter 9).

Some types of viruses (bacteriophages, and under very special circumstances poliovirus, for example) have particle-to-PFU ratios approaching 1. Preparations of viruses such as adenovirus and HSV typically have ratios of 10–100, but the best ratios for influenza A virus are on the order of 10^3. This high particle-to-PFU ratio is unusual but is inherent in the way that flu virus virions are formed. Particle-to-PFU ratios can vary depending on the specifics of the particular infection and virus, and each virus type has a characteristic optimum value that tells something about the efficiency of encapsidation and release of infectious virus from infected cells.

A second quantitative measure of conditions of virus infection is the **multiplicity of infection (MOI)**, which is simply the average number of PFUs per cell utilized in the original infection. An MOI of 1 means 1 PFU per cell, so if 10^6 cells were infected at an MOI of 1, one would need to add 10^6 PFUs of virus. It is important to note that an MOI can vary from zero to a very high number, depending on the concentration of virus in the original stock, the type of experimental problem being studied, and so on. MOI measures an average value; statistical analysis that demonstrates the number of PFUs interacting with one individual cell can vary over a wide range when a culture is infected at MOIs greater than 0.1 or so.

Examples of plaque assays

In the plaque assays shown in Figure 10.5a, the cell culture dishes contained 10^6 cells. Thus, the MOI used to generate the average of 40 plaques seen in the 10^7 dilution shown was calculated as follows: PFU/cell = $40/10^6 = 4 \times 10^{-5}$. One should be able to see that where the plaques could be readily counted, the MOI must be quite low, and any cell initiating a focus of infection or plaque must have been infected with only 1 PFU. This is a simple demonstration of the fact that with normal (wild-type [*wt*]) animal viruses, only one viral genome delivered to the right place in the cell is sufficient to carry out the whole infection. Indeed, a very high MOI may actually inhibit the replication process because particle-to-PFU ratios may increase rapidly with a high MOI. One way this can happen is by the generation of defective interfering particles, as outlined earlier in this chapter.

To do a plaque (or focus) assay, serial dilutions of a virus stock are made and aliquots of each dilution are added to a culture dish. The plaques are allowed to develop and then are counted. Simple arithmetic yields the original number of PFUs in the solution.

Figure 10.5a shows an example of HSV plaques developed on Vero cells. Serial 10-fold dilutions were added to separate plates in duplicate. After adsorption, the cells were rinsed and covered with a special overlay medium that inhibits virus spread beyond neighboring cells. Following incubation for 48 hours at 34 °C, the cells were rinsed, fixed, and stained. The clear areas are plaques. The average number of 40 plaques in the 10^7 dilution means that about that number of PFUs was added to each plate at that dilution of virus.

Another example is shown in Figure 10.8. Here a 100-ml stock of HSV was diluted and infectious units were measured by plaque assay, as shown in Table 10.1. One can readily calculate that the original stock was about 6×10^7 PFU/ml or 6×10^9 total units of infectious virus (PFU). The following formula is useful to make the calculation:

$$V_f = V_o / D$$

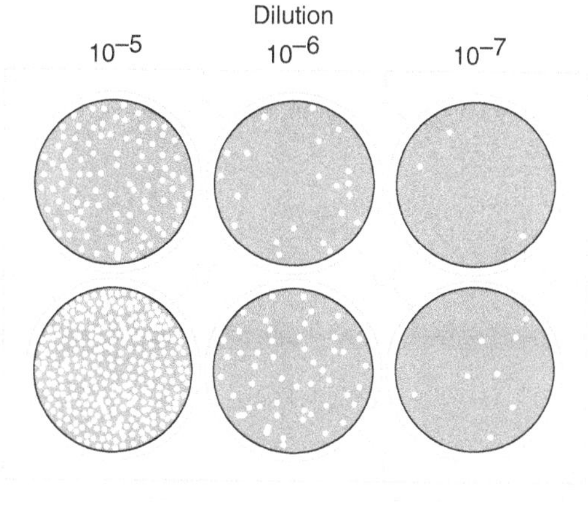

Figure 10.8 Serial 10-fold dilutions of HSV to determine the titer of virus in a stock solution. The details of the infection are as described in the legend to Figure 10.4a, and the calculation of the titer is shown in Table 10.1.

Table 10.1 An example of a set of dilutions for a plaque assay.

Operation	Dilution of Stock	Plaques per Dish
0.01 ml of stock diluted into 10 ml of buffer	10^3	Too many to count
1 ml of above diluted into 10 ml of buffer	10^4	Too many to count
1 ml of above diluted into 10 ml of buffer	10^5	500–1000 (estimated)
1 ml of above diluted into 10 ml of buffer	10^6	(20 + 100) / 2 = 60
1 ml of above diluted into 10 ml of buffer	10^7	(3 + 8) / 2 = 5.5
1 ml of above diluted into 10 ml of buffer	10^8	0
1 ml of above diluted into 10 ml of buffer	10^9	0

where V_f is the final concentration of PFU (units/ml), V_o is the original concentration, and D is the dilution factor.

Also note in this example that the number of plaques counted in two plates infected with the same amount of diluted stock varies quite a bit. Some of this variation is due to experimental error, but there also is an inherent statistical variation because one cannot be sure that the same amount of virus particles will be in a small volume at a given time. This type of variation is inherent when working with samples that contain a small number of particles.

Statistical analysis of infection

The statistics of chance mean that at low and moderate MOI values, the actual PFU number infecting any one cell will vary widely. For example, at an MOI of 2, a significant number of cells will *see* no *virus*, and a larger number for MOI will *get* 1 PFU. Some cells will get 3 PFUs, and some others (a smaller number) will get 4, 5, or more PFUs. The proportion (or probability) of any given cell being infected with any specific number of PFUs can be calculated using a statistical method originally developed for analyzing gambling results. This is the **Poisson analysis**, which describes the distribution of positive results in a low number of trials as

$$P_i = \left(m^i e^{-m}\right) / i!$$

where P_i is the probability that a cell will be infected with exactly i number of virus and m is the MOI (average number of PFUs added per cell). Using this equation, one can always calculate the probability of a cell being uninfected, and thus, the number of uninfected cells (if you know the number of cells in the sample). Since $m^0 = 1$ and $0! \equiv 1$:

$$P_0 = e^{-m}$$

For the MOI of 2 mentioned previously, the proportion (probability) of cells getting i number of PFUs is

$$P_0 = e^{-2} = 0.135$$

$$P_1 = 2e^{-2} = 0.27$$

$$P_2 = 2^2 e^{-2} / 2 = 0.27$$

$$P_3 = 2^3 e^{-2} / 6 = 0.18$$

$$P_{4\text{ or more}} = 1.0 - \left(P_{0-3}\right) = 0.14$$

One gets the last number ($P_{4\text{ or more}}$) from the fact that the total probability of a cell being infected with no PFUs or any number of PFUs must be 1.0.

This fact can be used in another way. For example, what MOI is needed to ensure that at least 99% of cells in a culture are infected?

$$P_0 = 1 - 0.99 = 0.01 = e^{-m}$$

Thus,

$$\ln(0.01) = -m \text{ or } 2.3\log(0.01) = -m$$

So m (MOI) must be at least 4.6 PFU/cell.

Dilution endpoint methods

If a virus stock is diluted far enough, and then a small measured sample (an **aliquot**) is taken, chances are that there will be no infectious virus present. The virus has not been destroyed, just diluted so much that its concentration is well below, say, 1 PFU/ml so that in any 1 ml, there is no virus.

Because virus stocks can be diluted so much that any given aliquot will usually have no PFUs, one can measure infection by dilution instead of by titration. This type of endpoint dilution method is often called a **quantal assay** because it is a statistical analysis, not a quantitative one. In this type of assay, a given number of subjects (animals, cell culture wells, etc.) must be infected with increasing dilutions of virus and then scored for illness, death, or cytopathicity.

In a quantal assay, localizing plaques is not necessary. By plotting log dilution versus percentage of infected subjects, one can estimate a virus dilution that results in half the aliquots in that dilution containing virus and half not. In an assay of a disease in animals, this endpoint is called **ID_{50} (median infectious dose)**, or **LD_{50} (median lethal dose)**. For measurement of gross cytopathology in tissue culture wells, it might be called **$TCID_{50}$ (median tissue culture infectious dose)**. The ED_{50} assay described to measure interferon activity in Chapter 7 is another example of a quantal assay.

The relation between dilution endpoint and infectious units of virus

Quantal endpoints are simply a measure of dilution of infectious virus, but they relate to the average number of PFUs in the aliquot. An example of a quantal assay is shown in Figure 10.9. An HSV stock was diluted as shown, and equal aliquots were added to individual wells of 48-well culture plates. Evidence of virus infection (cytopathic effect [CPE]) is shown by the black wells. For the titration, one can construct a table such as Table 10.2, and from the tabulated data, one can make the graph shown in Figure 10.10.

In Figure 10.10's graph, one can estimate that a dilution at which 50% of the wells would be infected is about 4×10^3; therefore, the $TCID_{50}$ was 4×10^3 in the original sample. More accurate measures of the ID_{50} of a virus stock can be obtained by using statistical methods such as the method of Reed and Muench, which is described in a variety of basic statistical texts.

Although ID_{50} is a measure of dilution, an ID_{50} *unit* is directly related to PFUs; 1 ID_{50} unit measures a dilution required to ensure that 50% of the aliquots in that dilution have infectious virus in them. This will only occur if there are 0.7 PFUs (average) per aliquot, or 7 PFUs in 10 ml in the above example.

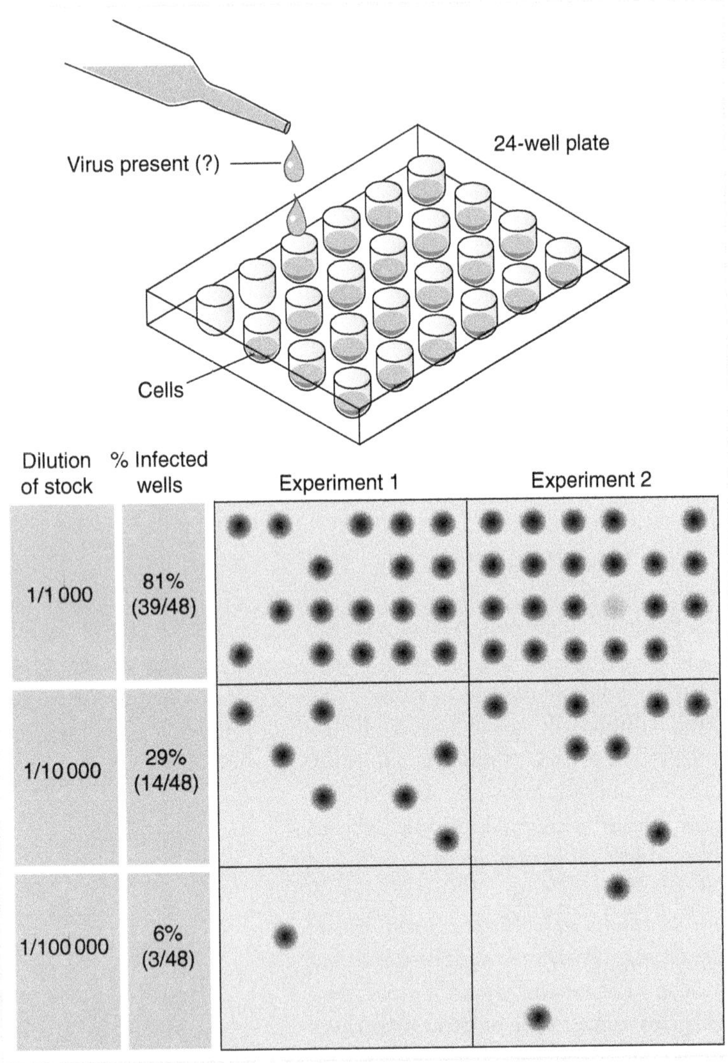

Figure 10.9 Quantal (endpoint dilution) assay of HSV in tissue culture wells. Replicate cultures of rabbit skin fibroblasts were grown to a density of about 5×10^4 cells per well of a 24-well tissue culture plate. Aliquots of the indicated stock virus dilutions were pipetted into the cultures, and the plate was incubated for 48 hours and then developed with a stain that indicates black for virus-infected cells. Any well that received at least 1 PFU of virus stained black (two separate experiments are shown). The percentage of positive (infected) wells is shown at each dilution.

Table 10.2 An example of a quantal assay for virus infectivity.

Sample Dilution	Log Dilution	No. of Infected Wells	Total No. of Wells	% Infected
None	0	100	100	100
1/1000	3	39	48	81
1/10 000	4	14	48	29
1/100 000	5	3	48	6
1/1 000 000	6	1	100	0

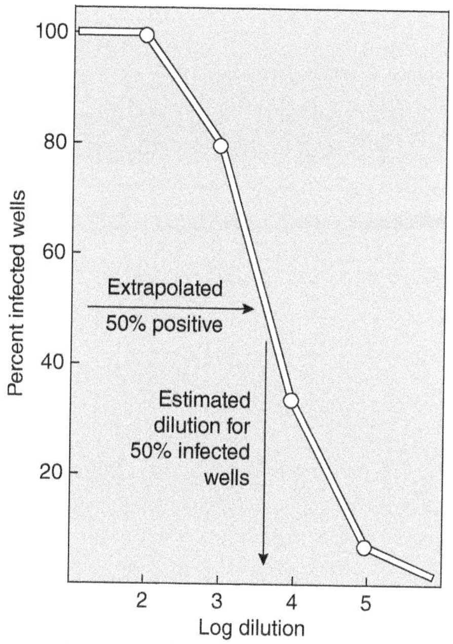

Figure 10.10 Graphic analysis of the data from Figure 10.8. The percentage of infected wells as a function of dilution is shown on a semilogarithmic plot. The dilution at which 50% of the wells would be infected (the $TCID_{50}$) can be estimated by graphic interpolation.

This finding follows from certain rough arithmetical considerations: If a certain X number of PFUs per milliliter in the original concentration was diluted by a factor D so that each animal or tissue culture well has a 50% probability of being infected with a PFU, then the final concentration of virus defines a type of MOI (call it m) where the probability of a positive infection is 50%. This value (m) should have the dimensions of units of infectivity in a standard volume (here 1 ml). Then:

$$P_0 = 0.5 = e^{-m} = 0.7\,\text{PFU} / \text{ml}.$$

QUESTIONS FOR CHAPTER 10

1 You have diluted a 1-ml sample of virus stock by taking 100 µl from the stock solution and adding to it 0.9 ml of buffer. You then take 10 µl of this dilution and dilute it into 1 ml. You then infect two plates that contain 10^5 cells each with 100 µl. One plate had 25 plaques, while the other had 29 plaques. What was the titer in the original stock?

2 One milliliter of bacterial culture at 5×10^8 cells/ml is infected with 10^9 phages. After sufficient time for more than 99% adsorption, phage antiserum is added to inactivate all unadsorbed phage. Cells from this culture are mixed with indicator cells in soft agar, and plaques are allowed to form. If 200 cells from the culture are put in a Petri dish, how many plaques would you expect to find?

3 You have a series of culture dishes that contain "lawns" of HeLa cells (human cells). You plan to infect these cells with poliovirus type 1 under a variety of conditions. You will measure the ability of the virus to form plaques on these cells. In the table below, predict which of the conditions will result in plaque formation by poliovirus type 1 on HeLa cells. Indicate your answer with a "Yes" or a "No" in the table.

Experiment	Virus Added	Treatment of Cells	Plaques?
Negative control	None	–	No
Positive control	Poliovirus type 1	–	Yes
A	Poliovirus type 1	Cells treated with interferon	
B	Poliovirus type 1	Antibody against rhinovirus added	
C	Poliovirus type 1	Antibody against poliovirus type 2 added	
D	Poliovirus type 1	Antibody against poliovirus type 1 added	

4 You have performed a plaque assay on a stock of bacteriophage T4. Your results show an average of 400 plaques when you assay 0.1 ml of a dilution prepared by mixing 1 part of the original virus solution with 999 999 parts of buffer.

(a) What is the titer of the original stock of bacteriophage?

(b) What volume of this stock would you have to use to infect a 10-ml culture of *E. coli* containing 4×10^6 cells/ml, such that the multiplicity of infection will be 10?

5 A stock of poliovirus is measured by plaque assay on a "lawn" of HeLa cells. When 0.1 ml of a 10^5 dilution of this stock is plated, an average of 200 plaques are observed.

(a) What is the titer of this stock?

(b) If 0.1 ml of this stock is used to infect 10.0 ml of HeLa cells containing 10^5 cells/ml, what is the MOI in this case?

6 Using the Poisson distribution, calculate the proportion (probability) of cells infected with the indicated number of PFUs, given the MOI shown in the table.

	Proportion (Probability) of Cells Infected With		
MOI	0 PFU	1 PFU	≥2 PFUs
0.01			
0.1			
1			
10			

7 Virus particles are very carefully isolated from an infected cell stock. You use this material to infect a culture of 10^6 cells with an MOI of 7 PFUs/cell. What is the maximum percentage of cells that *could* be productively infected?

8 You apply a virus stock solution containing 3×10^6 virus particles to 3×10^5 cells. What is the MOI for this infection?

Physical and Chemical Manipulation of the Structural Components of Viruses

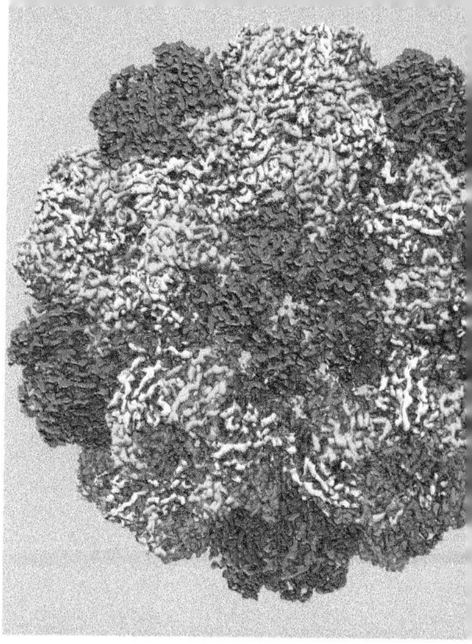

CHAPTER

11

VIRAL STRUCTURAL PROTEINS

Although viruses are nucleic acid genomes surrounded by a capsid (and sometimes by membrane-associated viral proteins), a large number of other virus-encoded nucleic acids and proteins are expressed during infection of the host cell and eventual formation of new virus

Basic Virology, Fourth Edition. Martinez "Marty" Hewlett, David Camerini, and David C. Bloom.

particles. If these nucleic acids and proteins do *not* end up in the structure of the virus itself, they are termed **nonstructural**. Thus, proteins involved in, for example, replication of herpesvirus DNA during its infection are nonstructural proteins. Indeed, during replication of a DNA virus, all the viral messenger RNA (mRNA) expressed and encoding viral proteins will be nonstructural components because this mRNA remains in the host cell when new virus particles form and exit the cell.

It is conceptually simple to differentiate structural and nonstructural proteins. Any protein found in purified virions (complete virus particles, i.e., genomes, capsid proteins, and any envelope and membrane-associated proteins *in* the virion) is structural. If a protein is viral encoded but not found in the virion, it is nonstructural. In practice, this differentiation can be somewhat difficult owing to problems with isolation of absolutely pure virus. Some enveloped viruses are almost impossible to isolate completely free of infected cellular debris or extracellular proteins. Many viruses have the irritating ability to include small amounts of cellular and viral material in their maturation that is not necessary for virus viability or replication.

The ability to isolate pure (or nearly pure) viral structural and nonstructural components is very important in research and medicine. Some of the uses for such material are as follows:

1 Sources of antigen for preparation of pure immunological reagents such as monoclonal or polyclonal monospecific antibodies, as well as prophylactic vaccines.

2 Enzymes that can be studied to develop specific antiviral drugs targeted against specific features of an enzyme's mechanism of action.

3 Pure "genes" encoding specific proteins that can be selectively modified to determine (i) how modifications in either DNA (or RNA) sequences that control expression of a specific mRNA affect such expression, or (ii) how modifications to specific amino acid codons within the gene affect activity of the encoded protein.

4 Proteins that can be modified and adapted for use in biotechnology and genetic engineering.

5 Proteins for structural and assembly studies.

6 Regulatory proteins with defined effects on the host cell, so that the mechanism of the interaction between such viral proteins and host cell regulatory pathways can be studied.

7 Nucleic acid "probes" that can be used to identify cellular genes that have similar nucleic acid sequences and thus can be inferred to have similar functions. They can also be used to monitor the virus load in patients following chemotherapy.

Isolation of structural proteins of the virus

A large number of techniques are available for fractionation of biological molecules and subcellular particles according to their size, density, or charge. **Buoyant density** differences are useful in fractionating enveloped viruses. Each subcellular particle has differences in buoyant density in aqueous solution. Those with large membrane components are "lighter" than those composed of only proteins and nucleic acids.

Virus particles also can be separated from cellular components of different density. This is accomplished by generating an equilibrium density gradient of sucrose or other material in an ultracentrifugal field. Virus particles will "band" or "float" at a specific location within the gradient corresponding to their equilibrium buoyant density (1.18 g/cm^3 in the example shown in Figure 11.1).

This position represents a balance of forces on the particle: the buoyant force trying to cause the particle to float, and the centrifugal force working to cause the particle to sediment lower in the gradient.

Size fractionation is widely used, especially for nonenveloped viruses. For subcellular particles, organelles, and virions, differential sedimentation under a centrifugal field (**rate zonal**

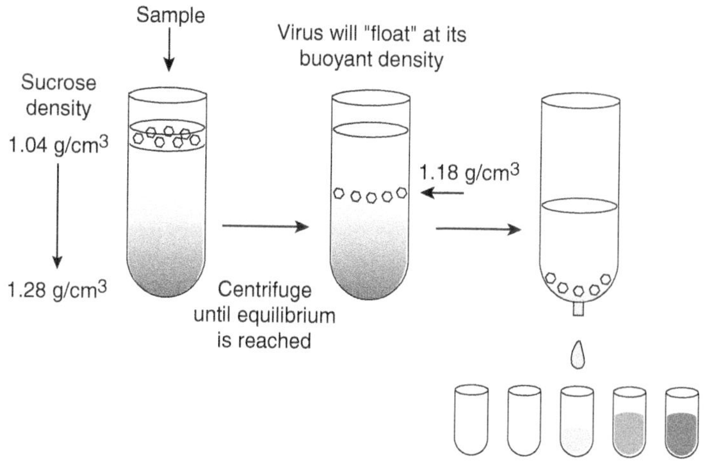

Figure 11.1 Equilibrium density gradient centrifugation of virus-infected cell components to isolate virus particles. A preformed sucrose density gradient is layered with a solution of infected cell material and subjected to centrifugation at high *g* force at 4 °C for several days. Virus particles sediment downward until they reach a layer with a density equivalent to their own. At this density, the virus particles will "float," and careful handling of the gradient in a clear plastic tube will reveal a turbid band of virions that can be removed. In this figure, the virus was collected by careful drop-wise fractionation of the gradient through a hole in the tube bottom into small tubes. The presence of virus in the appropriate fractions could be confirmed by plaque assay.

centrifugation) allows rapid fractionation and purification. In essence, one takes advantage of the difference in size of these components in the centrifugal field where the largest (the ones with the greatest sedimentation coefficient) will sediment most rapidly or under the least force. The practical aspects of such differential centrifugation can be complex. The basic approach is readily seen in Figure 11.2.

Since most viruses are smaller than mitochondria and larger than ribosomes, further fractionation could be obtained by taking the 100 000 g supernatant material and carrying out further differential centrifugation or more careful size fractionation.

Size fractionation of viral structural proteins

Once pure virus is obtained, gentle disruption of the virions with mild detergents or appropriate salt treatments can lead to disruption of the particle and solubilization of the components. Proteins and nucleic acids can be separated from each other by a variety of extraction or differential degradation regimens. For example, small amounts of nuclease could be used to digest nucleic acid into nucleotides, or proteases could be used to digest proteins. These macromolecular components then can be separated according to size or charge, or a combination of both.

It can be shown using physical chemical analysis that the sedimentation rate of a macromolecule is a function of its molecular size and its hydrodynamic volume. Thus, a globular macromolecule (such as most proteins) will migrate at a different rate than an extended (linear) macromolecule of the same size. Further, the same parameters apply to the rate of migration of a similarly charged macromolecule of equivalent shape when subjected to an electrical field, provided the molecules are suspended in a medium of high viscosity that discourages diffusion, such as an acrylamide gel. This is the principle of gel electrophoresis.

In electrophoresis, the rate of migration is *inversely proportional* to sedimentation rate (**s value**). Two macromolecules of equivalent hydrodynamic shape and unit charge will migrate, so that the molecule with the larger molecular size will migrate more slowly than the smaller molecule.

These principles are incorporated into a very powerful technique for the size fractionation of proteins. It involves mild denaturation (disruption) of protein structure with the detergent sodium dodecyl sulfate (SDS), which associates with denatured protein to give it a uniform net negative charge. Such proteins can then be size fractionated by electrophoresis on acrylamide

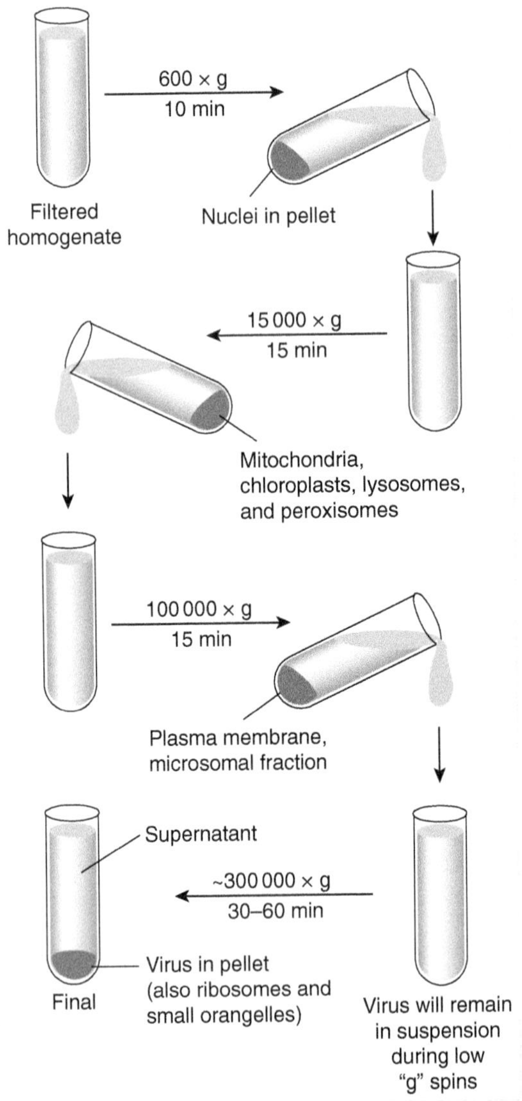

Figure 11.2 Differential centrifugation to purify virions. Infected cells are homogenized and then subjected to varying steps of centrifugation at increasing *g* forces. At low speeds, large cellular components pellet and can be removed. At the proper speed, viral particles sediment to the bottom of the tube.

gels where the *larger* proteins move more slowly through the gel network, and smaller proteins migrate more rapidly. If the procedure is properly done, such a gel provides good fractionation of viral structural proteins according to size.

Such gels can be stained with color reagents that provide a quantitative measure of the amount of protein of each size, as the color reaction is based on reactions with amino acids in the proteins. A small protein has fewer amino acids per polypeptide than a large one; therefore, a sample of, say, 1000 small protein molecules will stain less intensely than will a sample of 1000 larger protein molecules.

A hypothetical example of protein size fractionation and a method of estimating molar ratios are shown in Figure 11.3, where the fractionation of protein mixtures in a denaturing SDS-containing gel is represented. In this experiment, a solution of an equimolar mixture of four proteins of significantly different sizes (i.e., different number of amino acids in the peptide chain) was fractionated in lane 2. Another sample of three proteins of different sizes in variable amounts (with the smallest protein being present in higher molar concentration than the mid-sized one, and both present in higher concentration than the largest) was fractionated in lane 1.

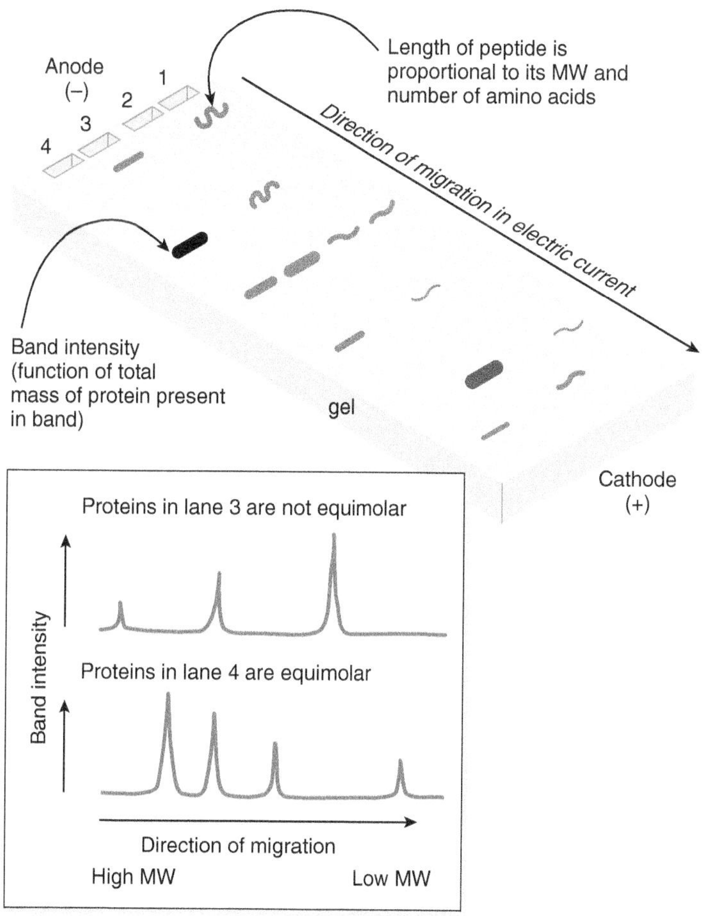

Figure 11.3 Denaturing gel electrophoresis of proteins. If proteins are gently denatured in a detergent solution such as sodium dodecyl sulfate (SDS), they will assume globular shapes and a net negative charge due to interaction with the detergent molecules. The proteins then can be fractionated by size on acrylamide gels. The proteins migrate in specific bands, and the amount of mass in each band can be determined with a color reaction that measures protein mass. The intensity of banding is a function of the *total* amount of amino acids (a direct correlate with the total mass) in the band, *not* the number of protein molecules per se. MW: Molecular weight.

The staining pattern of the gel is shown in lanes 3 and 4, where the staining intensity is represented by band thickness.

The pattern of staining intensity shown in lane 3 makes it clear that the proteins are not present in equimolar amounts. Since staining intensity of the most rapidly migrating band is greater than that of the midsized and large bands, there must be more amino acids in the band of small protein. This can only happen if there are more *copies* of the small-protein chains. The staining pattern of lane 4 shows a monotonically decreasing intensity of staining with size. Although a precise measure would be required, the band intensity appears to be (roughly, at least) *proportional to protein size*. This is the result expected for an equimolar mixture of proteins of different size, as one small-protein polypeptide chain will have fewer amino acids than a single-peptide chain of a larger protein.

Determining the stoichiometry of capsid proteins

The molar ratio of different structural proteins can be determined for a given virion or component of the virion (such as the capsid of an enveloped virus). This is possible because the relative amount of each protein can be measured by staining intensity or by other means, and because each capsid will yield only the number of capsomer copies present in it when isolated. Full stoichiometric analysis of the capsid's protein composition also requires knowledge of how many capsids are being analyzed. While the ratio of capsid proteins will be constant for different preparations, the absolute amount of protein must be related to the number of capsids to determine how many copies of each protein are present in each capsid.

There are important caveats to the application of this analysis. The most important is that the preparation of virions or capsids must be homogeneous. If a preparation is made up of partial capsids, or truncated helical capsids, the analysis will not be valid. Second, except for a few small enveloped viruses, such as togaviruses and flaviviruses, the number of glycoproteins in the envelope is not stoichiometric. One virion may be enveloped with a virus-modified cellular membrane that has significantly more or less of one glycoprotein than another.

The poliovirus capsid – a virion with equimolar capsid proteins

It is relatively easy to determine that the poliovirus capsid is made up of just four proteins, and that the four capsid proteins (VP1, VP2, VP3, and VP4) are present in equimolar amounts in the capsid. Groups of five copies of each protein are arranged at each of the 12 vertices of the icosahedral capsid (see Chapters 5 and 15). If the proteins are uniformly labeled with radioactive amino acids, more radioactivity will be in each large polypeptide chain than in each small one. A gel fractionation of the radiolabeled proteins extracted from purified capsids of poliovirus is shown in Figure 11.4.

There is much less radioactivity in the small VP4 band than in the larger protein bands; however, comparison of the bands' molecular weight (MW) with the amount of radioactivity in each reveals the equal numbers of protein molecules. Quantitative analysis of the results of a

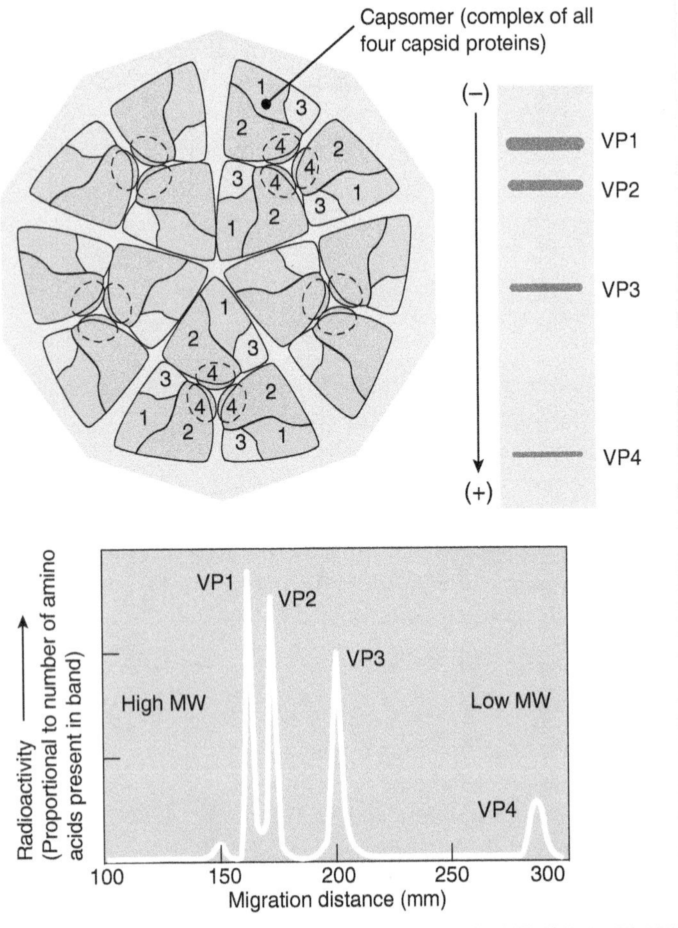

Figure 11.4 Electrophoretic fractionation of the capsid proteins isolated from purified poliovirus virions. The icosahedral capsid is made up of 60 capsomers, each containing one copy of each of the four viral proteins. The arrangement is shown schematically. Proteins from purified virions were solubilized in buffer and loaded onto a denaturing SDS-containing acrylamide gel. According to their size, viral proteins migrate in denaturing gel electrophoresis, and the amount of total mass in each band can be measured. The ratio of band intensity demonstrates that all four proteins are present in equimolar amounts. MW: Molecular weight.

Table 11.1 Gel fractionation of the poliovirus's four capsid proteins.

Protein	Molecular Weight	Radioactivity (cpm)
VP1	33 521	563 153
VP2	29 985	515 742
VP3	26 410	437 743
VP4	7385	124 806

Table 11.2 Protein composition of the HSV-1 capsid.

Gene	Molecular Weight	Copies per Capsid	Location in Capsid
UL19	149 075	960	Capsomers
UL38	50 260	375	Triplexes
UL26	45 000	87	Inside capsid
UL18	34 268	572	Triplexes
UL26.5	26 618	47	Inside capsid
UL35	12 095	952	Capsomer tips

similar gel fractionation is shown in Table 11.1. Note that the ratio of sizes of VP1 to VP4, for example, is 4.5, while the ratio of radioactivity between them is also 4.5.

Analysis of viral capsids that do not contain equimolar numbers of proteins

Most viruses that encode more than a few proteins in their genomes (e.g., adenovirus and herpesviruses) have capsids that contain proteins in vastly different molar amounts. The adenovirus capsid is shown in Figure 11.5.

A number of proteins within the capsid are not visible in the figure; these include core proteins and hexon-associated proteins. An example of an SDS gel fractionation for adenovirus is also shown in Figure 11.5. The penton base protein, which is only found at the 12 vertices of the icosahedral capsid, is present in much *smaller* molar amounts (i.e., *fewer copies per capsid*) than is the hexon protein. Conversely, the 24 000-Da core protein is present in many more copies per capsid than is the hexon protein. This conclusion comes from the fact that the core protein is considerably smaller than the hexon protein, yet it stains to an equivalent density, while the large penton base protein stains only faintly. Similarly, the capsid of herpesviruses contains proteins in varying molar amounts. The number of copies of the six herpes simplex virus (HSV) capsid proteins is tabulated in Table 11.2.

CHARACTERIZING VIRAL GENOMES

Isolation of purified virions provides a primary source of viral genomes. Isolating viral genomes from purified virions is relatively simple. All that is necessary is a mild disruption of the capsid proteins, and the nucleic acid can be isolated by phenol extraction. A famous electron micrograph of a partially disrupted capsid of bacteriophage T4 with its DNA genome extruded is shown in Figure 11.6.

An accurate determination of the viral genome's nature and molecular size is one of the first things that must be done when working with a newly isolated virus. Such information is important in establishing a basic idea of the virus's genetic complexity. This information, taken together with general characteristics of the virion (i.e., enveloped or not; icosahedral, helical, or complex

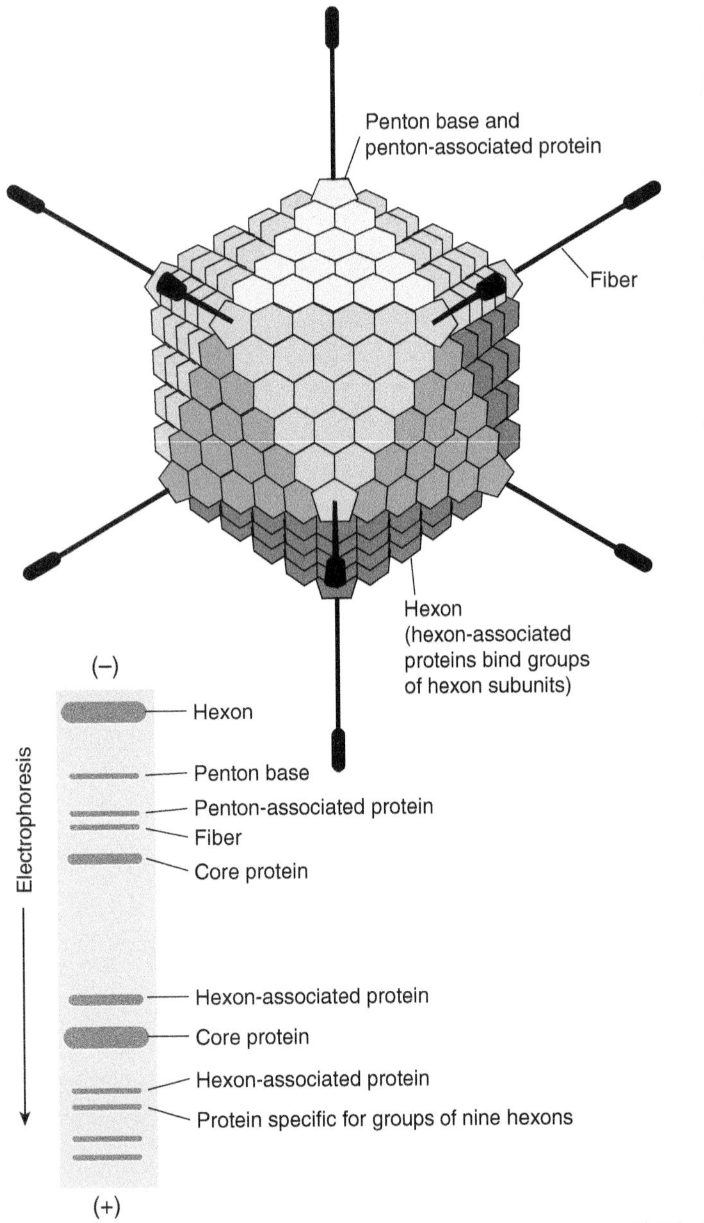

Figure 11.5 Electrophoretic fractionation of the capsid proteins isolated from purified adenovirus virions. This complex virion contains many different structural proteins that can be fractionated by denaturing gel electrophoresis. The different band intensities do not correlate with protein size. This result demonstrates that the structural proteins are not present in equimolar amounts.

shape), can be used to make a preliminary assignment of the relationship between the new virus and known virus families using criteria outlined in Part II, Chapter 5. Ultimately, of course, a full determination of nucleotide sequence of the viral genome will provide information as to the number and specific amino acid sequences of the proteins it encodes, as well as a precise measure of its degree of relatedness to other viruses.

Sequence analysis of viral genomes

The determination of a DNA virus genome sequence provides the ultimate physical description. While there are methods for sequencing RNA molecules, these methods are not applicable to determining the sequence of extremely large molecules such as those that are the genomes of

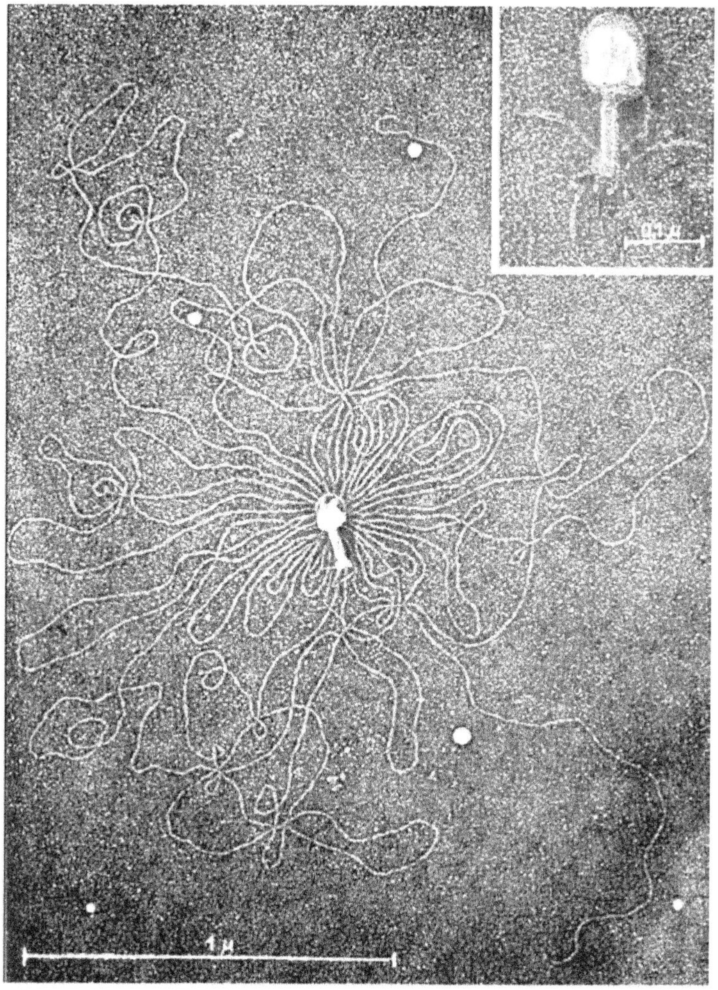

Figure 11.6 The famous Kleinschmidt electron micrograph of phage T4 DNA extruded from the capsid. Before this photograph was made, there was controversy about whether the viral genome was a single piece of DNA or multiple pieces – the fragility of large DNA molecules made them difficult to isolate without shearing. Kleinschmidt took purified bacteriophages and very carefully exposed them to low osmotic pressure. Under the proper conditions, viral DNA was gently released from the capsid and visualized in the electron microscope. Note the presence of two ends, showing that the DNA is linear. Source: Kleinschmidt, A.K., Lang, D.J., Jacherts, D., and Zahn, R.K. (1961). Darstellung und Längenmessungen des Gesamten Des oxyribonucleinsäure-1 haltes von T2-Bakteriophagen. *Biochimica et Biophysica Acta* 61: 857–864. Reprinted with permission from Elsevier.

RNA viruses. However, this problem is readily overcome in the study of RNA virus genomes because RNA can be conveniently converted to DNA using appropriate oligodeoxyribonucleotide primers and retrovirus reverse transcriptase. Enzymatic details of the conversion of RNA to complementary DNA (cDNA) and then double-stranded DNA (dsDNA) are outlined in Chapter 19 in Part IV.

Chemical methods for cleaving DNA at specific bases were originally described by Russian biochemists and perfected for use in DNA sequence analysis by Alan Maxam and Walter Gilbert. This method has largely been replaced by enzymatic methods. DNA sequence analysis requires only a few things: (i) DNA, (ii) a method for priming and extending the DNA and incorporating a labeled nucleotide(s), (iii) a method for detecting the terminated/labeled nucleotides, and (iv) a method either of separating the labeled fragments by size to identify the nucleotide that was labeled, or of identifying the labeled nucleotide addition as it is being extended.

All the necessary requirements are readily met with the repertoire of techniques available to molecular biologists. Labeling the fragments can be accomplished easily by use of one of a number of enzymatic methods to incorporate a nucleotide labeled with a fluorescent-tagged nucleotide derivative. Separation of deoxyribonucleotides by the technique of **capillary electrophoresis** using a polymer has provided high enough resolution to allow the separation of fragments ranging from around 10 to greater than 1000 bases.

Sanger sequencing

If a small amount of a dideoxynucleoside triphosphate (which causes chain termination due to lack of a 3′-OH group) is added to the primed synthesis reaction (where the deoxynucleoside triphosphates are in excess), the synthesis of the new DNA strand will terminate wherever the dideoxynucleotide is incorporated. The fact that strand synthesis can only proceed from the primer provides a convenient method for generating overlapping, nested sets of oligonucleotides complementary to any DNA sequence 5′ of the primer in question.

The enzymatic method was originally perfected by Sanger and collaborators, and has been modified in many ways. For example, and as described a bit later, the method has been automated so that analysis can be carried out and directly entered into computer databases with little human interfacing. The rapid progress made by the Human Genome Project, as well as the increasingly frequent publication of sequences of the entire genomes of free-living organisms, are due to the ease and speed of enzymatic methods. Indeed, even though it took several years to determine the complete sequence of HSV-1 (152 000 base pairs) over two decades ago, the same problem can now be solved in days! Complete sequence analysis of any virus of interest can be carried out essentially as soon as the virus is isolated and the genome purified.

To generate overlapping oligonucleotides with the same 5′ end, all that is needed is a primer sequence that will anneal to a region that is located 3′ to the sequence of interest. This is often a region in the vector used to clone the DNA in the first place. Annealing of the primer, which can be either labeled with a radioactive or fluorescent marker, or unlabeled, is followed by enzymatic synthesis of the complementary strand of the DNA template in the presence of a labeled base or bases. After synthesis is allowed to proceed for a short time to ensure the formation of highly labeled material, the reaction is broken into four aliquots, and a small amount of a single di-deoxy-base-triphosphate is added to generate oligonucleotides with random stops at a given base.

This is shown in Figure 11.7a and below for T (remember, lowercase nucleotides signify the complementary base on the antiparallel strand, and DNAY is the region of DNA to which the labeled primer, dnay*, binds):

5′-DNAX-ATACCGATCGTG-DNAY-3′
tagcac-dnay*-5′
5′-DNAX-ATACCGATCGTG-DNAY-3′
tggctagcac-dnay*-5′
5′-DNAX-ATACCGATCGTG-DNAY-3′
tatggctagcac-dnay*-5′
and so on.

Figure 11.7b shows application of this sequence analysis to compare the sequence of a wild-type and mutant virus.

Automated sequencing takes advantage of the fact that laser light of a given wavelength can excite specific dye molecules to fluoresce at specific frequencies. Different dye molecules fluorescing at different wavelengths can be chemically linked to each of the four di-deoxy-base-triphosphates in the reaction mixes described above. These can be used all together in the polymerase reaction to generate nested products terminating at every base in the sequence. This mixture is then loaded onto a capillary electrophoresis apparatus and subjected to a high voltage. The shortest fragments will, of course, migrate most rapidly through the capillary and past a laser-activated detector, where the presence of the terminating, dye-containing fragment will fluoresce at a wavelength characteristic of the terminating deoxynucleotide. A computer is used to record the order of appearance of the various colored signal peaks. An example of this methodology is shown in Figure 11.7c.

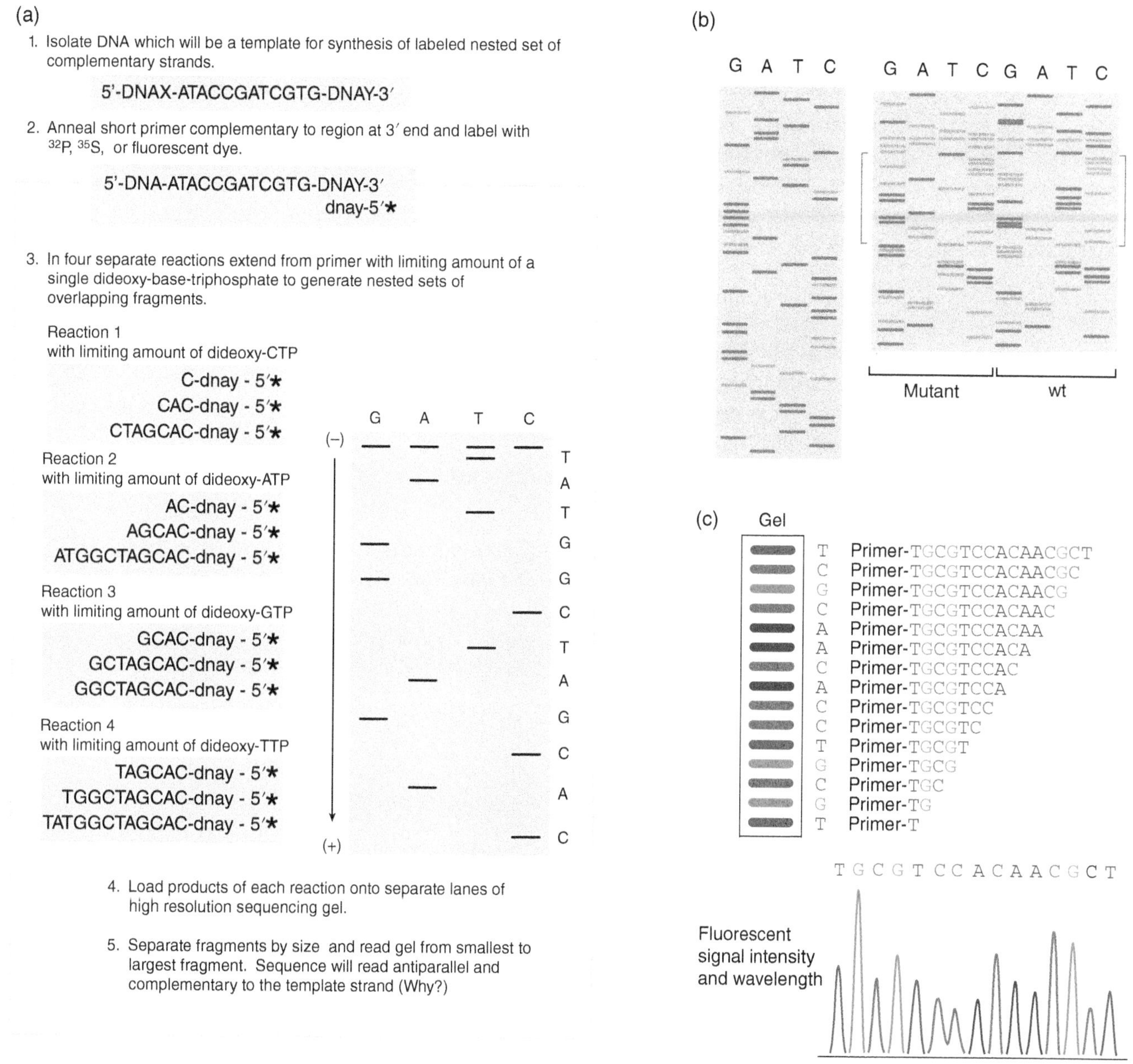

Figure 11.7 Enzymatic sequencing of DNA. The generation of overlapping oligonucleotide sets complementary to a template strand of DNA for sequence analysis was developed by Sanger and colleagues and is described in the text. (a) An outline of the basic method. One major advantage of the method is that it can be used to generate very long sequences with reactions using a single primer site. (b) For example, the gel on the left shows the sequence of a cloned fragment of HSV-1 DNA and the plasmid it is cloned into about 100 bases 3′ of the primer site. The sequence can be read as follows:

5′-ACGTC$_2$T$_2$A$_2$GCTAG$_2$C$_2$G$_2$C$_2$TCGC$_2$ATCG$_2$AG$_5$C$_2$TAGT$_2$CGA$_2$TAGCTA-3′

The right gel shows a comparative analysis of the sequence of a wild-type and mutant promoter region for an HSV-1 capsid protein mRNA. This region is about 300 bases 3′ of the location of the sequencing primer and shows that high resolution is still readily obtainable as long as the reaction products are fractionated under proper conditions, which in this case are long fractionation times under denaturing conditions. The regions of the two sequences that are different are indicated; the sequences read as follows:

Wild type: 5′-TCACAGGGTTGTCTGGGCCCCTGC-3′

Mutant: 5′-TCACAGGACCGGCTGACCGCCTGC-3′

Just above (i.e., 3′ of) this region is an example of a typical experimental artifact of this type of sequencing: a spot where there is termination in all reactions due to a structural feature of the sequence in question. Note that the sequence again can be read accurately beyond this point. (c) Automated DNA sequencing. In a typical sequencing reaction, each of the dideoxynucleotides is labeled with a specific dye that fluoresces to emit a given wavelength of light. Sensors measure the wavelength as the overlapping fragments of DNA in the reaction mixture are separated by electrophoresis and move past the detecting site. The results are recorded and stored in a database for later sequence interpretation.

High-throughput equencing (HTS)

Recent advances in sequencing technology are rapidly replacing Sanger sequencing for all but diagnostic and verification work in the lab, with highly automated and high-throughput sequence analyses methods that are often referred to collectively as **high-throughput sequencing (HTS)** or, formerly, as next-generation sequencing (NGS). These techniques are similar to the Sanger approach in that they utilize primers and incorporate fluorescent nucleotides as the DNA is being extended. The differences are: (i) The primer binding site is actually on an adaptor, or small dsDNA linker that is ligated on the ends of all of the molecules to be sequenced to create a **sequencing library**; (ii) the primers are extended on a glass chip with grooves or lanes called a flow cell; and (iii) using microfluidics and a sophisticated fluorescent detection system, the bases of the sequence are read in real time after each round of nucleotide addition. This technology allows hundreds of millions of DNA molecules to be read in a single lane and as such requires sophisticated bioinformatics analyses to interpret the sequence results. The process of HTS is diagrammed in Figure 11.8. This technique has allowed the rapid sequencing of thousands of viruses in the past few years.

The polymerase chain reaction – detection and characterization of extremely small quantities of viral genomes or transcripts

The ability to characterize, work with, and control many viruses is limited by the fact that they are present in very small quantities in a given cell, tissue, or host. The use of a fluorescent stain such as ethidium bromide allows the ready detection of 100 ng or less of dsDNA. For a

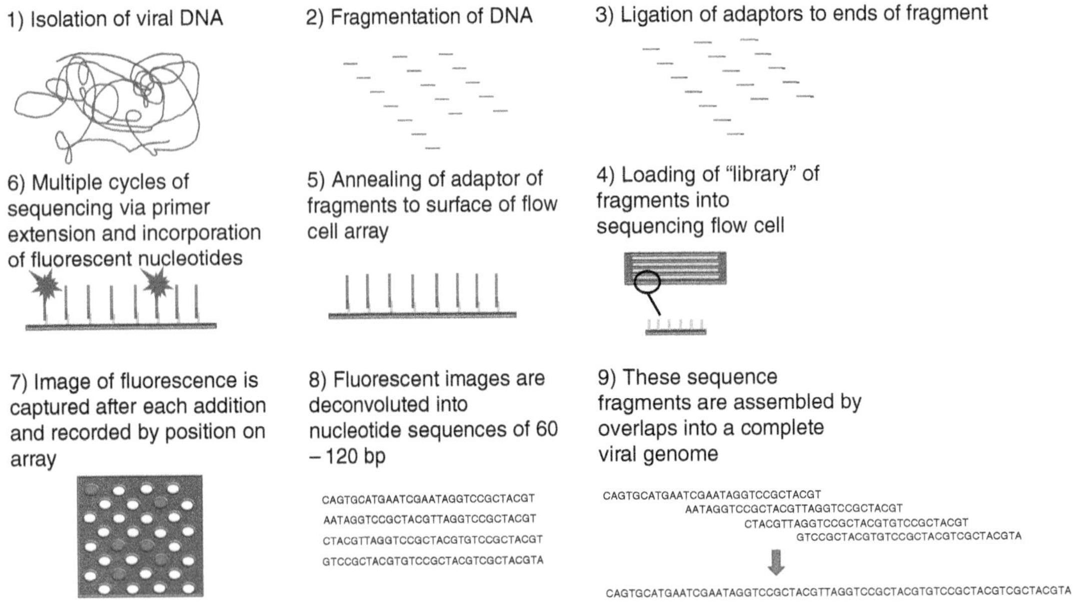

Figure 11.8 High-throughput sequencing (HTS) of DNA. Several different chemistries or platforms of HTS are currently in use. The one illustrated here is the Illuminia platform. While the chemistry is similar to Sanger sequencing in that it uses a polymerase to extend a primer and incorporate fluorescent nucleotides, it involves many more steps and a complex (and expensive) sequencing machine. The steps involved include generation of a "library" of fragmented DNA (generally prepared by sonication) with adapters ligated to the ends. These adapters have a region that will anneal to a complementary oligonucleotide on the sequencing flow cell. In addition, the adapters have unique nucleotide "codes" that allow the sequencer to know if sequencing products are from the same fragments. The flow cell is a microfluidic chip that has several "lanes," which are channels that the library is loaded into. The bottom (flow cell array) of the flow cell is where the library fragments will anneal. After the library is annealed to the flow cell array surface, the cell is placed in the sequencer, and cycles of fluorescent nucleotides and polymerase are rinsed through the flow cell. After each cycle, the sequencer takes an image of the fluorescence in order to "map" which positions on the array had a base added (corresponding to whether a G, A, C, or T was added in that cycle). After multiple cycles, the thousands of images that were taken are processed by the machine to convert the patterns of fluorescence into nucleotide sequence. At that point, the many millions of individual sequences or "reads" are analyzed and assembled into a full genome sequence.

viral genome of, for example, 30 000 base pairs, this works out to be approximately 5×10^{11} molecules. Radioactive labeling can greatly increase the sensitivity of detection, but it is not always possible to specifically label the DNA fragment of interest in the tissue being studied.

The problem of visualizing and manipulating extremely small quantities of DNA was overcome in large part by developments of the *polymerase chain reaction (PCR)* initiated and commercialized by scientists at the Cetus Corporation in the mid-1980s.

The principle, illustrated in Figure 11.9a, is quite simple. Consider a fragment of dsDNA present as even a single copy in a cell or animal. If this DNA is denatured and short oligonucleotide primers can be found to anneal to the opposite strands at positions not too far away from each other (e.g., within a thousand bases or so), a strand of cDNA can be synthesized using DNA polymerase. The new product will be double stranded in the presence of the nonprimed denatured DNA.

Now, if the newly synthesized dsDNA is itself denatured, and the priming and DNA synthesis step is repeated, this short stretch of DNA will be amplified as compared to the strands of DNA that did not bind primer. This process can be repeated many times in a chain reaction to amplify the desired strand of DNA to useful amounts.

To work properly, the oligonucleotide primers must be long enough to be highly specific, but short enough to allow frequent priming. The appropriate length works out to be about 20–30 bases. The technology for synthesis of 20- to 30-base oligodeoxynucleotides is well established and can be chemically performed relatively inexpensively. Indeed, numerous large and small biotechnology companies make oligonucleotides commercially.

Also important is the ability to do the reaction, denaturation, and reannealing in a single tube many times over. This is accomplished by using the heat-stable DNA polymerases isolated from organisms such as *Thermophilis aquaticus (Taq)*, which live in hot springs, and the use of computer-controlled thermal cyclers that can repeat the annealing, synthesis, and denaturation steps rapidly and repeatedly over one to four hours.

In practice, the method can be used to detect the presence of extremely small amounts (less than a single copy/cell) of a known viral genome by selection of appropriate primer pairs based on the knowledge of the sequence of the genome. An example of the use of PCR to detect HSV genomes is illustrated in Figure 11.9b.

PCR can also be used to look for the presence of genes related to a known gene. Such detection is based on the assumption that regions of a DNA sequence encoding a gene related to the one in hand will contain some stretches of identical or highly homologous sequences in their genomes. Detection can be accomplished by amplifying the DNA in question with a series of potential primer sets. If one or several of these yield products of a size within the range of those seen with the known gene, these products can be isolated and sequenced. If necessary, this can be done after the amplified fragment or fragments of interest are cloned using methods outlined in Chapter 22 (Part V).

Real-time PCR for precise quantitative measures of viral dna

In addition to its value for detecting vanishingly small amounts of viral genomes, PCR can also be used to make extremely precise quantitative measures of the amounts of viral genomes or transcripts present in different tissues, or under different conditions of infection. The amount of product formed in the PCR reaction is a function of a number of factors, but the most critical is the amount of target sequence available to begin the reaction in the first place. This follows from the dynamics of the rate of product formation, where there will be a limited period of time when there is an exponential rate of accumulation of the product as the number of cycles increases until the available number of primers falls to a lower level where the rate of product formations becomes linear and finally reaches a plateau. Since with standard PCR methods, the reaction is carried out for a set number of cycles, the amount of product formed will reflect the amount of target originally present only if the synthesis of PCR product is still in the linear range at the time of the last PCR cycle. Since total product formation is the endpoint, very rare sequences may only be amplified to a low level, moderately abundant sequences amplified to a level more or less proportional to their initial concentrations, and more abundant sequences will have reached a plateau in product formation relatively early in the course of the amplification. Thus, quantitative estimates of the amount of material present in the original sample can be difficult.

One way to overcome these reciprocity problems is to use a series of dilutions of the original sample for the amplification along with appropriate standards, as shown in Figure 11.9b where a series of dilutions of a fragment of HSV DNA corresponding to the copy numbers shown were

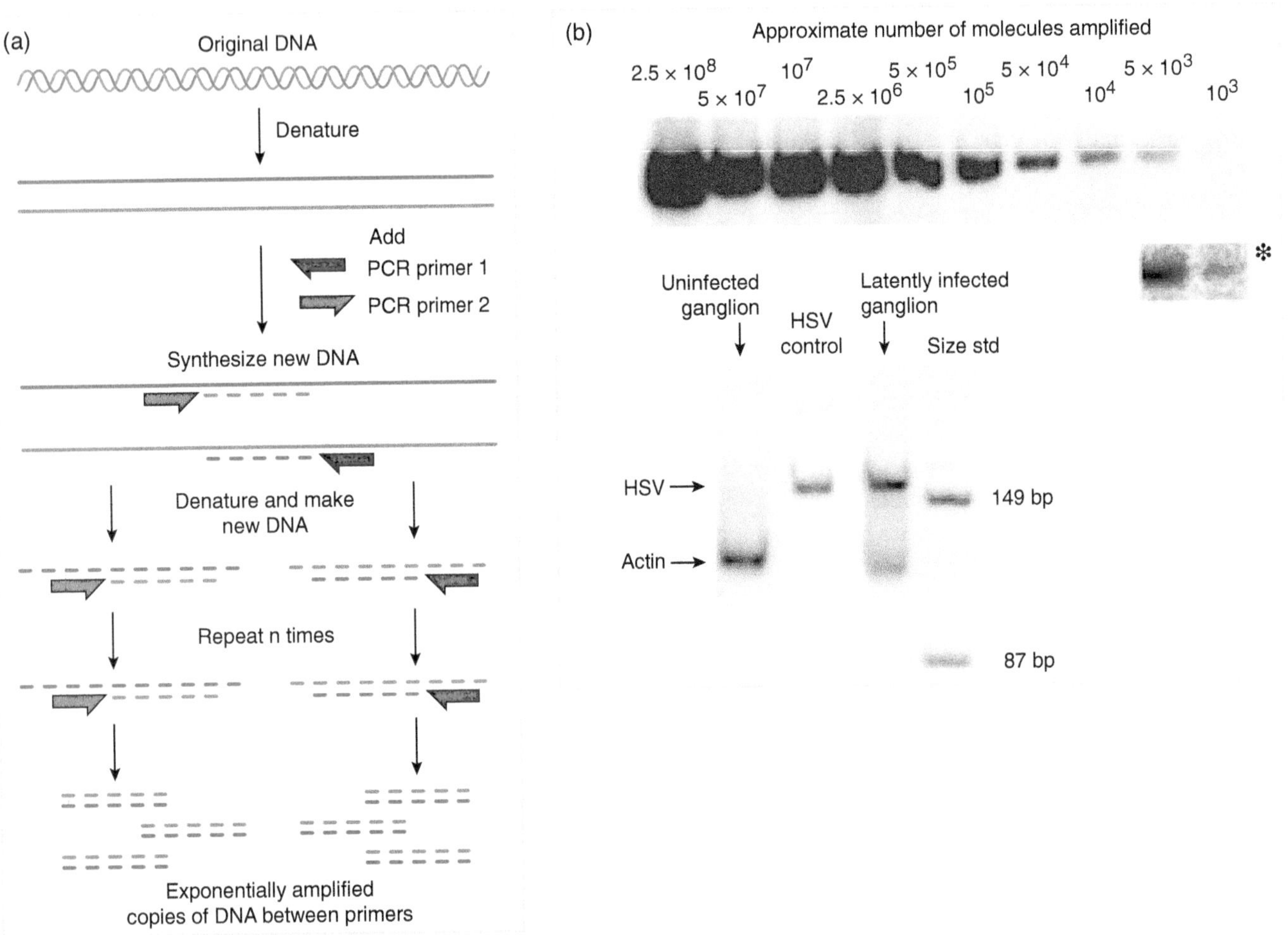

Figure 11.9 Amplification of DNA with the polymerase chain reaction (PCR). (a) The basic method requires specific primer sets that can anneal to opposite strands of the DNA of interest at sites relatively close to each other. After denaturation, the primers are annealed, and DNA is then synthesized from them. All other DNA in the sample will not serve as a template. Following synthesis, the reaction products are denatured, and more primer is annealed and the process repeated for a number of cycles. The use of heat-stable DNA polymerase allows the reaction to be cycled many times in the same tube. A single copy of a DNA segment of interest could be amplified to 10^9 copies in 30 cycles of amplification. Can you demonstrate this mathematically? (b) The amplified DNA products from a segment of HSV DNA. A total of 1 μg of nonspecific DNA was added to each of a series of tubes, and viral DNA corresponding to the copy numbers shown was added. Following this, primers, heat-stable DNA polymerase, and nucleoside triphosphates were added, and 30 cycles of amplification were carried out in an automated machine. The reaction products were fractionated on a denaturing gel and visualized by autoradiography. The lower gel shows the results of amplification under identical conditions of DNA isolated from two rabbit trigeminal ganglia. One was taken from a control rabbit, and the other was taken from a rabbit that had been infected in the eye with HSV followed by establishment of a latent infection. The use of rabbits to establish virus latency is shown in Figure 3.5. Amplified DNA from each sample was fractionated in the lanes shown; in addition to the amplification products, a sample with PCR-amplified HSV DNA as a standard (std) as well as some size markers were fractionated.

carried out and subjected to PCR amplification. The gel shown was used to fractionate the reaction products that were made radioactive by the addition of a small amount of radiolabeled nucleoside triphosphate to the reaction mix. An amplified signal from 1000 copies of the genome provided a detectable signal with a short exposure of the gel to x-ray film.

The technique of **real-time PCR** provides a much more reliable and precise method of quantitatively measuring the products of PCR reactions. This is accomplished by measuring the formation of the PCR products continually throughout all cycles of annealing and chain elongation. This is done by using primers that contain a fluorescent marker that is only detectable upon the formation of the amplified product. Such primers usually have a fluorescent tag (a **fluor**) that is quenched either by the secondary structure of the primer or by a second ligand (the *quench*) attached to the primer. When the primer is not annealed to a DNA product, illumination of the reaction mix with a laser or other suitable light source will yield no fluorescence, but when the primer is annealed to the target or the amplified strand of DNA, it then is able to generate a signal that can be quantitatively measured upon illumination. Different fluors, each fluorescing at a specific wavelength, can be incorporated into different primers so that the rate of formation of several products can be simultaneously measured in the same reaction mix. The quantitative analysis of the human globin gene in peripheral blood macrophage DNA is shown in Figure 11.10.

Nucleic acid from as little as a single cell can be subjected to PCR. The quantitative measure of viral genomes as a function of disease state or state of infection is vital for understanding the replication of HIV and its pathogenesis leading to AIDS. In the laboratory, PCR has also been very useful in studying the latent phase of infection of herpesviruses. Depending upon the details of infection and the exact strain of virus used, it has been determined that a typical latently infected neuron in an experimentally infected rabbit might harbor between 10 and 100 viral genomes.

PCR detection of RNA

PCR also can be used to detect viral RNA (either genomes or transcripts) present in very low amounts. Detection is accomplished by generating a cDNA copy of the RNA by use of retrovirus reverse transcriptase, followed by PCR amplification using a known primer set. If oligodeoxythymidine is used as a primer, it will anneal to the polyA tails of mRNA for the generation of cDNA. If the correct primers are used, PCR can detect vanishingly small numbers of transcripts. An example of such a use in the analysis of HSV gene expression during reactivation is shown in Figure 17.10.

The very high sensitivity of PCR, along with the ability to sequence the amplified products of PCR, also can be applied to determining splicing patterns of RNA expressed in cells. The application to analysis of viral transcription is briefly outlined in Chapter 13, and is illustrated in Figure 13.7b.

PCR as an epidemiological tool

Finally, PCR is invaluable for epidemiology and forensics. For example, it was used to amplify traces of influenza virus genomes still present in frozen cadavers of victims of the 1918–1920 influenza pandemic. Study of the sequence of such material has allowed scientists to establish some relationships between that virus and modern strains. Its use in forensics is somewhat outside the scope of this text, but it should be clear that the ability to amplify traces of DNA along with rapid sequencing methodology allow the identification of any genome present in more than a very few copies, from viral to human.

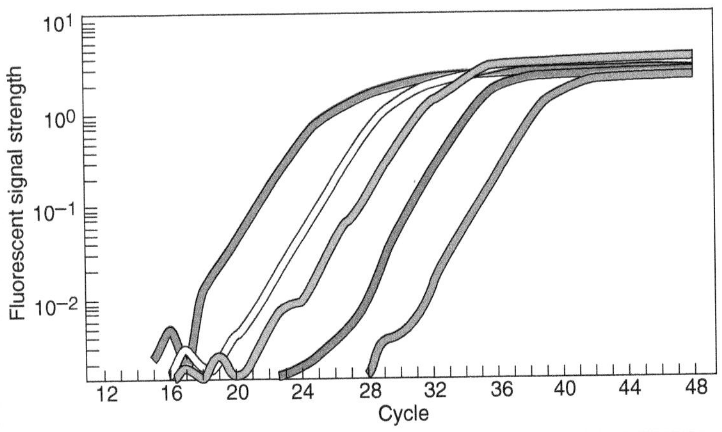

Figure 11.10 Real-time PCR amplification of globin DNA in blood macrophages. Fivefold dilutions of DNA from these cells were subjected to multiple cycles of PCR amplification under conditions where the amplified DNA can be measured by measuring fluorescence. As the dilutions increase, the range of cycles in which the amplified signal is logarithmic and, thus, the quantitative measure of the numbers of genes present increase.

QUESTIONS FOR CHAPTER 11

1 You have encountered a virus named hotvirus with three capsid proteins, E, K, and W. After gel fractionation of a purified stock of pure viral capsids that were uniformly radiolabeled with radioactive amino acids, you obtain the following results:

Protein	Molecular Weight	Radioactivity (cpm)
E	5280	29 348
K	18 795	101 185
W	10 776	122 674

What are the best values for the ratios of the proteins E to K to W?

2 Your laboratory has isolated a number of possible enteric viruses from samples of contaminated water. You have grown these viruses in appropriate cell cultures and have labeled the proteins with ^{35}S-methionine. You have purified virus particles from these cultures and separated the capsid proteins by SDS polyacrylamide gel electrophoresis. Below is an autoradiogram of this experiment with poliovirus type 1 (PV1) included as a control:

(a) Which of these isolates is potentially a virus identical to or very closely related to poliovirus type 1?

(b) Which of these isolates may be another member of the same family as poliovirus type 1?

(c) From which family of viruses might isolate B come? (Note: You will probably have to do some searching in Chapter 14 to find properties of enteric viruses in order to answer this question.)

3 How is SDS polyacrylamide gel electrophoresis used for the analysis of proteins? What is the basis for this technique?

4 Besides the molar ratio of the proteins, what would you need to know to determine the amount of specific proteins per capsid in a particular virus?

5 While analyzing the structural proteins of a pure stock of adenovirus by SDS polyacrylamide gel electrophoresis, you find, among others, two bands of equal intensity that migrate at 30 000 and 60 000 Da, respectively. What conclusion can you draw from this observation?

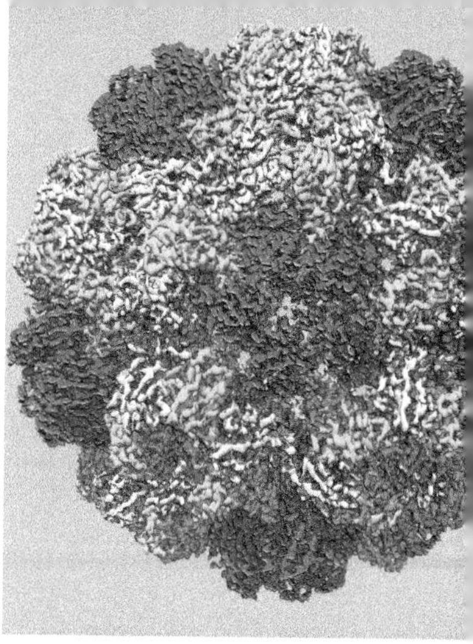

Characterization of Viral Products Expressed in the Infected Cell

CHAPTER

12

CHARACTERIZATION OF VIRAL PROTEINS IN THE INFECTED CELL

All viral proteins are synthesized in the infected cell; however, the amount and nature of these proteins, and the mRNAs encoding them, change with time following infection. The synthesis of nonstructural proteins generally occurs prior to the synthesis of viral structural proteins because nonstructural proteins include viral enzymes that function to modify the cell for virus replication, viral genome replication enzymes, and viral regulatory proteins, and all these must be expressed and function prior to the assembly of progeny virions. Thus, these nonstructural proteins have many important functions and are important to study. For example, the enzymes involved in the replication of HSV DNA during infection are good targets for chemotherapeutic drugs because they can be specifically inhibited with little effect on cellular DNA replication enzymes (see Chapter 8).

Basic Virology, Fourth Edition. Martinez "Marty" Hewlett, David Camerini, and David C. Bloom.

Pulse labeling of viral proteins at different times following infection

Study of the time of synthesis and nature of viral proteins in the infected cell requires the ability to distinguish virus-encoded proteins in a background of cellular ones, and to fractionate such viral proteins away from cellular components of the infected cell. Given the large amount of mass of the biological macromolecules contained in the cell, the process of viral protein or nucleic acid purification can be difficult and requires technical ingenuity.

Although the detection of viral proteins against the background of cellular material is difficult, the task is made somewhat more tractable in many virus infections because the infection leads to a partial or total shutoff of host cell mRNA or protein synthesis while viral proteins and mRNA are synthesized at high rates. This means that if radioactive amino acids are added to infected cells to serve as precursors to protein synthesis, they will be preferentially incorporated into viral products. In such a situation, the addition of radioactive precursors for a short period at a specific time after infection (a **pulse** of radioactive precursors), followed by isolation of total cellular material, will yield a mix of both viral and cellular material, but only the viral material will have incorporated significant amounts of radioactivity. Thus, size fractionation of the proteins in the infected cell provides a biochemical "snapshot" of whichever proteins are being synthesized at the time of labeling.

It is very important to remember that virus infection often leads to increased expression of some host cell proteins as part of its defenses (see Chapter 10). Therefore, the profile of proteins synthesized in a cell infected even with a virus that is extremely efficient in inhibiting host functions will not necessarily contain only viral products. Also, infections by some very important viruses do not result in efficient shutoff of host protein synthesis – in such a case, the proteins labeled in a pulse will be a mixture of cellular and viral proteins.

Examples of pulse-labeling experiments following infections with some viruses that do shut off host protein synthesis are shown in Figure 12.1. For the left panel, radiolabeled amino acids were added to poliovirus-infected cells at the time after infection shown in hours, and then proteins were fractionated. Many of the bands of radioactivity seen by exposing the gel to x-ray film are the result of the expression of viral proteins. Some of the more notable ones are indicated, as are some cellular proteins.

Several features of this pattern of pulse labeling are readily apparent. First, the amount of the capsid protein VP2 does not appear equimolar with that of VP1 and VP3, as was seen in the fractionation of proteins found in the mature capsid shown in Figure 11.4. The reason for this is that VP2 is derived from the processing of VP0, and therefore, some of the radioactivity that would be in the peak of VP2 is actually in the VP0 band.

Another feature is that the viral proteins indicated are in the same relative proportions at all times measured. As described in Chapter 14, poliovirus infection is characterized by the expression of only one mRNA molecule, and all proteins are derived from a large precursor that cannot be seen in this gel. However, portions of precursor proteins such as 3CD are clearly seen.

A third feature of the gel can be seen in examination of the cellular proteins labeled after infection. Although the synthesis of some is clearly shut off, the synthesis of others persists. This is an example of the fact that some cellular genes continue to be expressed (or can be induced) following infection.

The effect of an HSV-1 infection on total protein synthesis in infected cells is shown in the right panel of Figure 12.1. It is evident that the pattern of labeled viral proteins changes

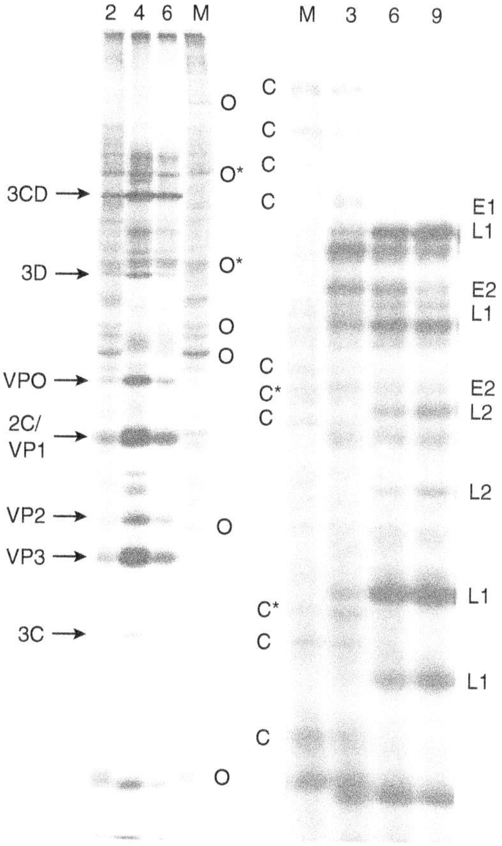

Figure 12.1 Changes in the proteins synthesized in virus-infected cells with time after infection. The left panel shows an experiment in which HeLa cells were infected with the Sabin (vaccine) strain of poliovirus, and labeled with ^{35}S-labeled methionine for two-hour pulses at the times (hours post-infection) shown at the top of the gel. Protein was isolated and then fractionated on a denaturing gel, and radioactive proteins were localized by autoradiography (exposure to x-ray film). The capsid proteins (VP0–VP3) are indicated, as are several nonstructural poliovirus-encoded proteins (2C, 3C, 3D, and 3CD). Some cellular proteins whose synthesis is shut off following infection are shown with the letter "O," while a couple whose synthesis continues are indicated by "O*." Source: Photograph courtesy of S. Stewart and B. Semler. The right panel shows a similar experiment carried out by labeling HSV-1-infected Vero cells for 30-minute periods at the times shown after infection. Some cellular proteins that are rapidly shut off are indicated with "C." "C*" marks proteins that do not appear to be shut off or whose synthesis increases for a period following infection. Viral proteins synthesized early after infection are indicated by "E." Note that there are at least two subsets, E1 and E2, which differ in the length of time that their synthesis continues. Similarly, there are at least two subsets of late proteins ("L"); some are clearly synthesized at the earliest times, while others are only synthesized later. Source: Photograph courtesy of S. Silverstein. In both panels, mock-infected cells (M) show the patterns of proteins synthesized in uninfected cells.

markedly with time. Some viral proteins synthesized at three hours following infection are no longer synthesized at later times. Conversely, some proteins are only labeled at later times after infection.

As described in Chapter 17, there are several reasons why the synthesis of some viral proteins readily detectable at one time after infection is not seen at other times. The basic reason for the temporal change in the patterns of expressed HSV proteins is that certain viral mRNAs are only expressed during a given window of time during infection; if the mRNAs are expressed at the earliest times, their synthesis declines at later times. The high constant rate of mRNA degradation in the cell (*mRNA turnover*) ensures that once the mRNA encoding a given protein is no longer synthesized, synthesis of that protein declines rapidly. This provides a ready means for the virus to control the timing and amount of protein synthesized at any given time.

A more modern way of analyzing the viral and cellular proteins expressed during viral infection is by high-pressure liquid chromatography (HPLC) followed by mass spectrometry, sometimes called proteomics. In this technique all the proteins of a virus-infected cell are digested with a proteolytic enzyme, separated by HPLC, and analyzed by mass spectrometry. The results can be compared to an identical proteomic analysis of uninfected cells to identify the expression pattern of viral and cellular proteins during viral infection.

Use of immune reagents for study of viral proteins

The immune response to viral infection in a vertebrate host is a complex process that was briefly outlined in Chapter 7, Part II. One of the major parts of this immune response is generation of antibody molecules, which are secreted glycoproteins with the capacity to recognize and combine with specific portions of viral or other proteins foreign to the host. The high degree of antibody molecule specificity, as well as the relative ease in obtaining them from immune animal serum, makes them important reagents in molecular biology. Antibody molecules isolated from the blood serum of animals following antigenic stimulation are made up of different molecules with different levels of affinity for different epitopes in the antigen. A mixture of various antibodies against a given antigen isolated from an animal is often termed an **antiserum** against that protein, organism, or virus in question. Although such antisera can react with many proteins, if care is used in purifying the antigen used to generate it in the animal, the immune serum will be specific for the antigen presented. Such an immune serum is **polyclonal**, as it is derived from many individual clones of antibody-secreting cells.

Working with antibodies

The structure of antibody molecules Antibody molecules have a very specific structure that is often described as a "wine glass" or "Y" shape. They are made up of two light and two heavy chains, and the two antigen-combining sites (made up of both heavy and light chains) are at the top of the wine glass or Y (the Fab region). Antibody molecules directed against different antigens have different amino acid sequences in these variable regions, which form the antigen-combining sites.

The stem of the wine glass or Y (the Fc region) is made up of a constant amino acid sequence for all antibody molecules of a given class, no matter what the antigen with which they react is. This region serves as a signal to the cell that an antibody molecule is there. It is important to the immune reaction and can be used both diagnostically and in the laboratory. An antibody molecule is shown diagrammatically in Figure 12.2.

Monoclonal antibodies The immune response is a result of proliferation of many different B- and T-cell types responsive to various antigenic determinants presented by the pathogen or by the antigen. Thus, each immature B cell stimulated by a specific epitope was stimulated into dividing into many daughter cells, all with identical genomes and all secreting identical antibody molecules. Such a clone of cells is short-lived in the body, but specific manipulations can be made to immortalize a single B cell so that a culture of clonally derived B cells, all secreting antibody molecules with identical sequence, can be isolated. The antibodies expressed by such a cell line are monoclonal antibodies and have many important uses in diagnostics, therapies, and research.

The original method for generation of monoclonal antibodies involves a number of steps that are outlined in Figure 12.3. These steps include immunizing the animal that is to be the source of the B cells (often a mouse), isolation of lymphocytes from the animal's spleen, transformation of cells to immortalize them, screening of specific populations, and selection of immortal cells that produce antibodies. Individual B cells that secrete only one antibody molecule reactive with only one determinant can be cloned by fusion of a mature B-lymphocyte population (each secreting a specific – and different – antibody) with immortal myeloma cells (tumor cells derived from lymphocytes that do not produce any antibody molecules).

If myeloma and B cells are induced to fuse with a very mild detergent, the cell culture contains short-lived parental B cells that will die, immortal myeloma cells, and fused cells. The key to the value of the method is that these fused cells (hybridoma cells) are also immortal. The job now is to get rid of the unfused cells, then screen the hybridoma cells for their ability to produce the desired antibody.

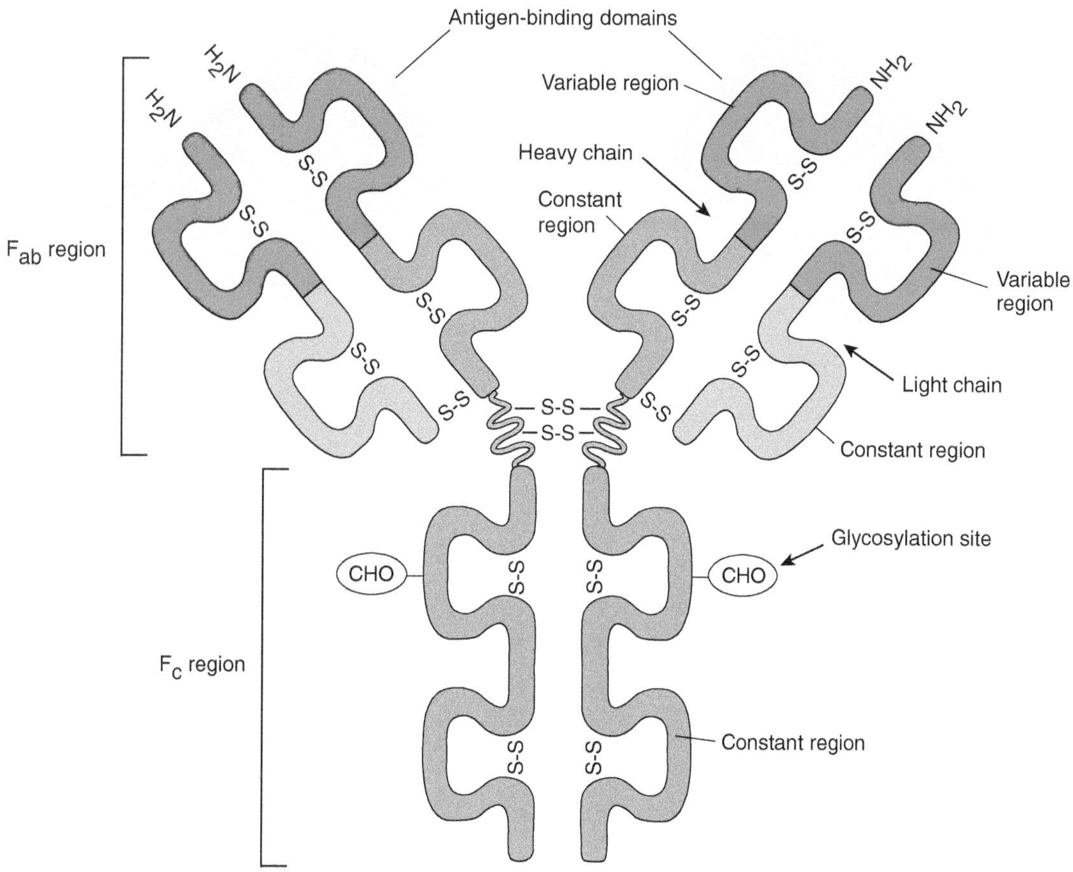

Figure 12.2 The structure of an antibody molecule, IgG. This molecule is made up of four chains: two heavy and two light. The antigen-combining domains are at the N-terminal of the four chains and are made up of variable amino acid sequences: a specific sequence for each specific antibody molecule. The C-terminal region has a constant amino acid sequence no matter what the antibody's specificity. This is the Fc region.

Getting rid of unfused B cells is no trick because they have a very short lifetime in culture and will die in a few days. Myeloma cells, however, offer a different problem because they are immortal and will continue to replicate, but they can be eliminated by using a mutant myeloma cell line that can be selected against. A convenient method uses a myeloma line that has been mutated so that it does not express hypoxanthine-guanine phosphoribosyltransferase (HGPRT negative), an essential enzyme in the biosynthesis of nucleotides. The advantage of this mutant is that since the parental myeloma cells cannot synthesize nucleotides, they need to get the nucleotides from the medium using a salvage pathway. This salvage pathway can be blocked with the drug aminopterin, which blocks the myeloma cell's ability to pick up nucleosides from the outside medium.

To understand this, remember that the hybridoma cells are not just derived from myeloma; they also have the genetic background of B cells, and the B cells are HGPRT positive. This means that adding aminopterin to the mixture of hybridoma and myeloma cells will result in the death of only the myeloma cells. The fused hybridoma cells will grow. The mixed hybridoma then can be screened by taking individual cells, growing clones from them, and testing the produced antibody for its ability to react with the antigen of interest.

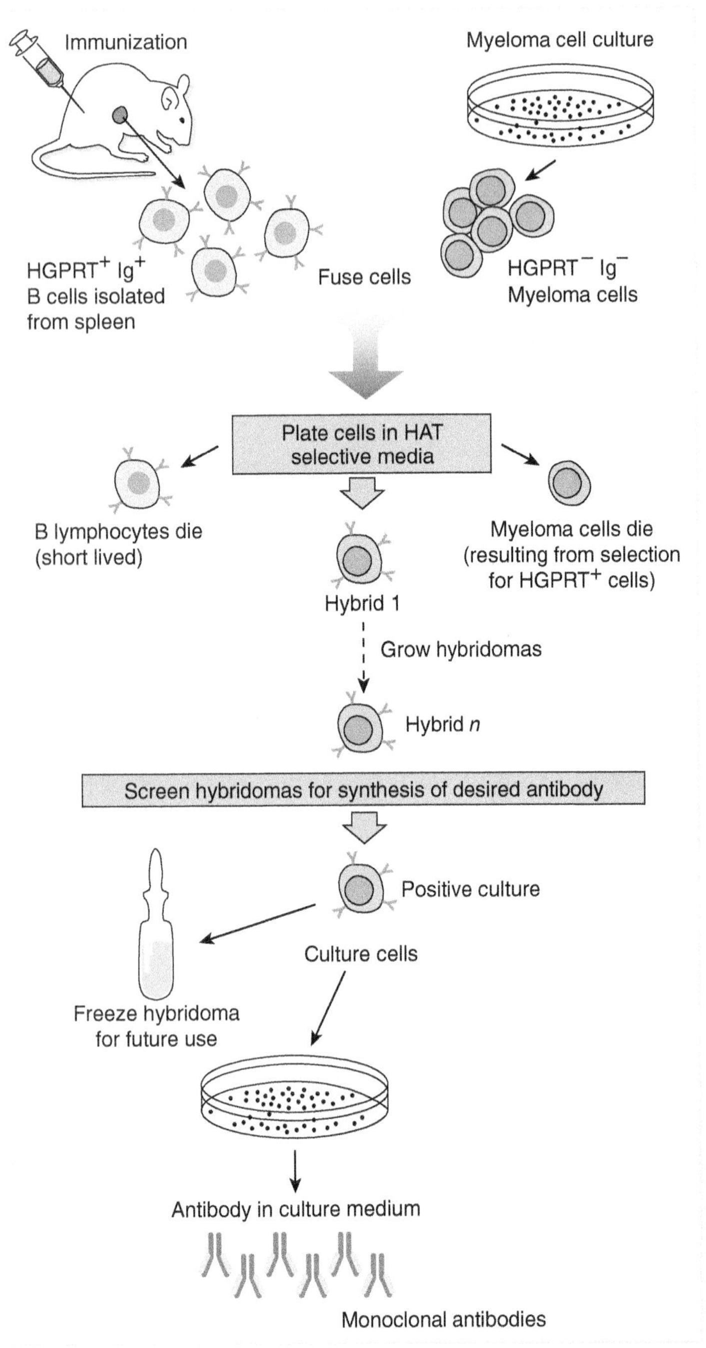

Figure 12.3 Generation of monoclonal antibodies by making hybridoma cells by fusion of mouse immune B lymphocytes and myeloma cells that are not able to grow in selective (HAT) medium. Antibody-secreting clones are screened by testing with an antigen. Once the hybridoma cell line is made, it can be stored frozen, and then grown in culture to produce pure monoclonal antibody. HAT: Hypoxanthine, aminopterin, and thymidine; HGPRT: hypoxanthine–guanine phosphoribosyltransferase.

Monoclonal antibodies are very useful for precise diagnosis of specific viral infections, as even closely related viruses will encode some proteins with different antigenic determinants. Each different determinant will react with only a specific monoclonal antibody generated against it. The monoclonal antibodies are also valuable tools for localizing viral proteins within the infected cell or animal, and as reagents to isolate and analyze specific viral proteins for study.

Monoclonal antibodies can also be made by a newer, molecular genetic technique. The variable regions of the immunoglobulin genes of an immunized animal or a virally infected person can be isolated from their B cells, used to reconstruct complete immunoglobulin genes, and then transferred to myeloma cells to create immortal monoclonal antibody–producing cells.

Detection of viral proteins using immunofluorescence

Many methods that measure antibody reactions involve use of the antibody molecule's Fc region as a "handle." Figure 12.4 shows some examples using a fluorescent dye either attached directly to the antibody (direct) or attached to a second antibody that is reacted against the Fc region of the first (indirect). Methods using **immunofluorescence** are very important to localize viral antigens inside infected cells, and to generate easily measurable immune reactions.

There are several micrographs in this book of infected and uninfected cells in which antigens of interest are located with fluorescent antibodies. A notable series is shown in Figure 3.5, where the passage of rabies virus through an infected animal was traced. Immunofluorescence can also be used with two (and even three) antibodies if each is tagged with a different chromophore. Two- and three-color immunofluorescence can provide a tremendous amount of information about the co-localization of proteins and other antigens of interest. The availability of lasers and prisms (or mirrors) that can differentially allow the passage of one wavelength of light while excluding others is used in **confocal microscopy** to allow the precise measure of the cellular distribution of viral and other antigens.

Although there are many variations on the method, confocal microscopy depends on the ability of a laser light source to be so coherent that it can be focused to a single focal plane within a cell. This, along with the use of appropriate prisms or filters and fluorescent dyes, allows one to visualize only the fluorescence emanating from a single plane within the cell. Since fluorescent radiation, of physical necessity, must be emitted at a wavelength longer than the incident radiation, the light path in a microscope can be used for both illumination and viewing.

The technique is shown schematically in Figure 12.5a, and an example of the type of data that can be obtained is shown in Figure 12.5b. For the studies shown in Figure 12.5b, cells were infected with human cytomegalovirus (CMV), a herpesvirus with a very long replication period, and then the expression of two proteins that localize to different parts of the cell was examined. The first protein, IE72, was detected with an antibody that was tagged with Texas red, which fluoresces red under illumination with the appropriate laser beam. This protein is synthesized in the cytoplasm,

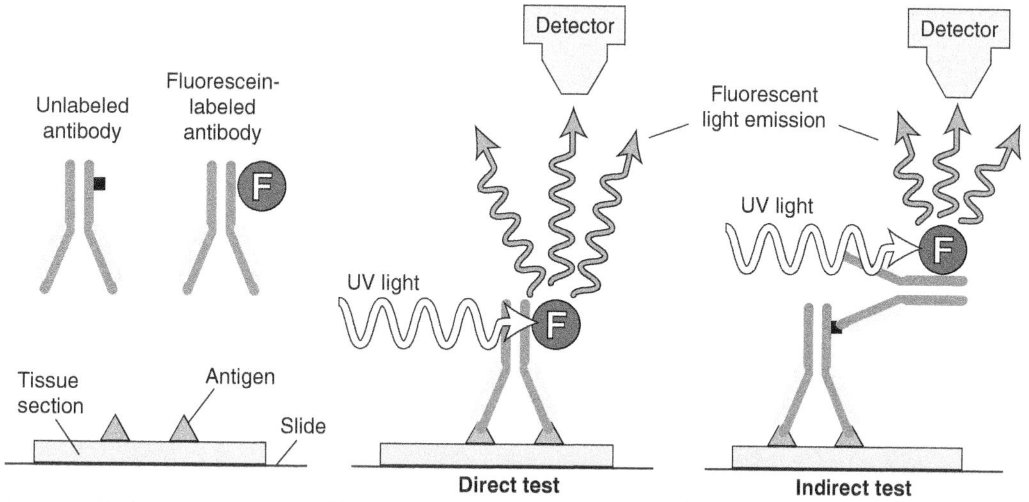

Figure 12.4 Outline of immunofluorescence as a means of detecting and localizing an antibody–antigen complex. The antibody specific against the antigen is allowed to react. If it has a fluorescent tag on its Fc region, it can be seen directly when illuminated with ultraviolet light since the tag emits visible light. For indirect immunofluorescence microscopy, a second antibody reactive with the Fc region of the first is used, and this antibody has the fluorescent tag. This method allows the same tagged antibody preparation to be used with a number of different antibodies of differing specificities.

(a)

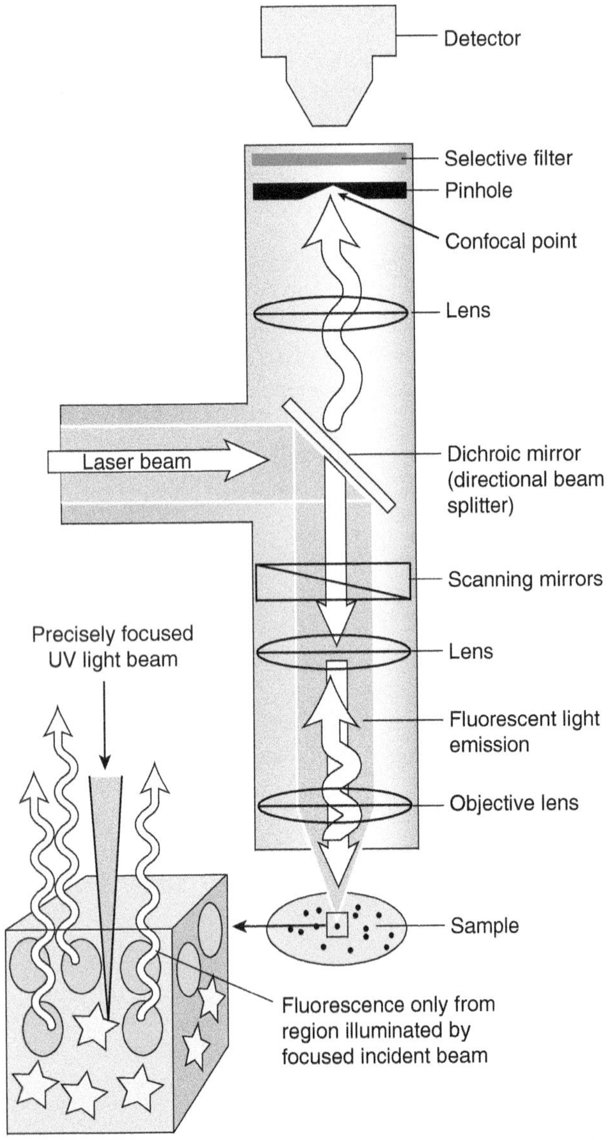

Figure 12.5 Confocal microscopy to detect co-localization of antigens. (a) The use of a laser beam and a specific filter to separate the incident laser light from the fluorescence that travels on the same light path. The ability to precisely focus the laser beam onto a single plane in the microscopic field allows one to observe fluorescence from proteins only in that plane. (b) Top: Confocal microscopic visualization of two human cytomegalovirus (HCMV) proteins, IE72 (red) and pp65 (green). Primary aortic endothelial cells were infected with a strain of HCMV isolated from a human patient. This high-magnification view of a cell shows nuclear and cytoplasmic staining of the two HCMV proteins at eight days following infection. Source: Photograph courtesy of K. Fish and J. Nelson. Middle row: A series of three photographs of the identical field viewed with three different filters to localize two specific proteins to the same region. The left panel shows the association of varicella zoster virus (VZV) glycoprotein E (gE), tagged with a green fluorescent antibody, with the surface of an infected cell. This glycoprotein was expressed in transfected cells. The middle panel shows the localization of the red fluorescence due to the transferrin receptor in the same cell, and the right panel shows that both fluorescent signals are located in the same sites on the cell, indicated by the yellow color, seen when a filter that allows both colors to pass is used for viewing. Bottom row: Same as middle row except that the interior of the cell is visualized. Source: Photographs courtesy of C. Grose.

but quickly migrates to the nucleus, where it remains and serves as a regulatory protein controlling expression of other CMV genes. The second protein, which fluoresces green due to a fluorescein isothiocyanate (FITC) tag, is pp65. This protein functions in the cytoplasm and is expressed later than IE72. The separation of the two proteins is clearly seen in the close view.

The lower photographs in Figure 12.5b demonstrate that another herpesvirus glycoprotein, varicella zoster virus (VZV) glycoprotein E (gE), localizes to the same region of the cell as does the transferrin receptor. This latter cellular protein is internalized into endocytotic vesicles of cells that are induced to take up iron borne by the carrier cellular protein transferrin. The fact that the VZV gE protein, which is expressed in transfected cells, co-localizes with the cellular receptor suggests that VZV may be internalized by endocytosis also. The specific glycoprotein for the virus (gE) was identified with green fluorescent FITC-tagged antibody, while the transferrin receptor was shown with Texas red–tagged fluorescent antibody. It is clearly evident that when both antibodies are observed, they are in the same precise location at the surface of the cell,

(b)

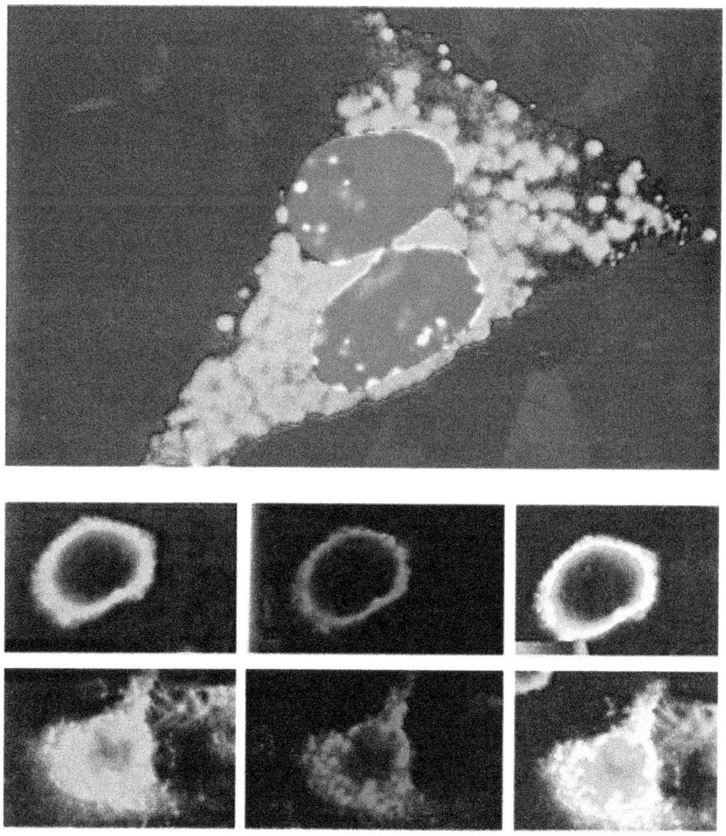

Figure 12.5 *Continued*

as indicated by the color of the fluorescent light in the middle right panel being yellow, which is a mix of the two colors. Co-localization of VZV gE and transferrin within the cell is similarly shown by the yellow fluorescent color in the bottom right panel.

Another widely used immunofluorescent method is flow cytometry, whereby thousands of cells can be analyzed per second. In this technique fluorescently labeled antibodies are bound to cells in a test tube and then run through a flow cytometer, which shines laser light on the flowing cells and collects the fluorescence associated with each individual cell. In this manner, complex mixtures of cells, such as human white blood cells (leukocytes), can be characterized using several different colored fluorescent molecules. An example of this is shown in Figure 12.6 in which human thymocytes were characterized with fluorescently tagged monoclonal antibodies reactive with CD4 and CD8 cell surface glycoproteins following infection with two clones of HIV-1. It is evident that both CCR5 tropic (R5 HIV-1 P1) and CXCR4 tropic HIV-1 (X4 HIV-1) have depleted CD4-positive thymocytes, although X4 HIV-1 did so more markedly. See Chapter 20 for a more complete description of HIV tropism.

Related methods for detecting antibodies bound to antigens

Other tags, such as enzymes, also can be bound to the Fc region of an antibody molecule. Enzyme-linked immunosorbent assays (ELISAs) were discussed in Chapter 7, Part II. A somewhat involved method is use of the enzyme peroxidase as a "tag" or indicator enzyme. Peroxidase will oxidize a soluble reagent containing a heavy metal, which then leads to precipitation of that metal near the antibody–antigen complex site. The precipitated metal can be observed in the

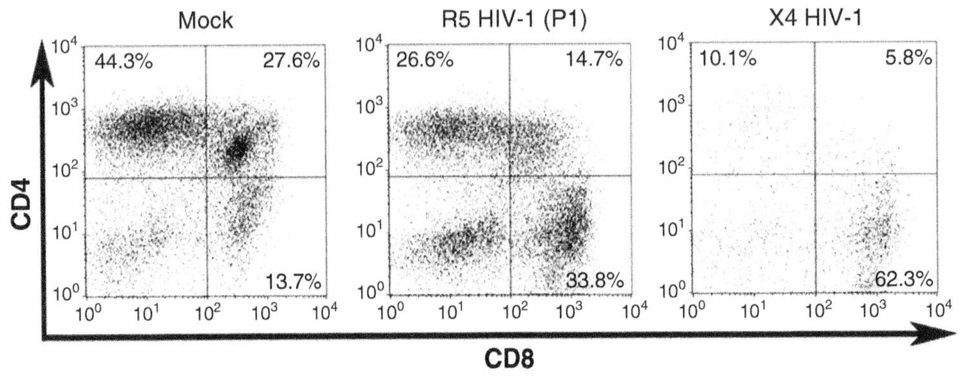

Figure 12.6 Human thymocytes in organ culture were mock infected or infected with a CCR5-tropic patient isolate of HIV-1 or the CXCR4-tropic infectious molecular HIV-1 clone, NL4–3. Twelve days later, cells were harvested and incubated with monoclonal antibodies directed to the major T-cell subset surface markers, CD4 and CD8, using two different colored fluorescent molecules. Data were collected by flow cytometry.

microscope (or electron microscope) to localize the immune reaction site. Individual antibody molecules bound to antigen can also be localized in the electron microscope using colloidal gold particles bound to the Fc region. These particles are so small as to have little effect on solubility of the antibody. An example of this technique is shown in Figure 6.3.

Use of bacterial staphylococcus a and streptococcus g proteins to detect and isolate antibody–antigen complexes Pathogenic staphylococci and streptococci express Fc-binding proteins on their surface to bind and inactivate antibody molecules by forcing them to face away from the bacterial cell. This reaction is quite useful in the laboratory, and the A protein of *Staphylococcus aureus* (staph protein A) and the G protein of group C streptococci (strep protein G) are commercially available for use as specific reagents to detect the presence of the Fc regions on human, rabbit, and mouse immunoglobulin G (IgG) molecules. An example is shown in Figure 12.7a.

In Figure 12.7a, all the proteins from a virus-infected cell were fractionated and blotted (immobilized) onto a membrane to which they tightly stick. This type of protein **transfer blot** is called a **Western blot** for a rather amusing reason. In the late 1960s and early 1970s, a scientist named Edward Southern developed a quantitative method for transferring gels of DNA fragments produced by restriction endonuclease digestion onto nitrocellulose filter paper. Such DNA transfer blots have ever since been called **Southern blots**. Subsequently, RNA transfer technology was developed and such blots were named **Northern blots** both to distinguish them from DNA blots and to establish similarity of the process. Protein blots were then named *Western blots* for comparable reasons.

In the example shown, the membrane and transferred proteins were incubated with antibodies to viral proteins. This version of the technique is therefore also called an immunoblot. These antibodies stick only to those proteins that they "recognize." The blot was rinsed and incubated with ^{35}S-methionine-labeled staph A or strep G protein. This protein reacts with the antibody's Fc region, and the area of immune complex is revealed.

Immunoaffinity chromatography Two variations on methods utilizing the binding of Fc regions to antibody molecules are frequently used to isolate specific proteins. Some methods using the affinity of staph protein A are shown in Figure 12.7b and 12.7c. In Figure 12.7b, an antibody against a protein in a complex mix is incubated with the protein mixture, and then passed through a Sepharose column (a high-molecular-weight polysaccharide) to which the Fc-binding protein was chemically bound. All antibody molecules bind to the column, and any proteins that are bound to the antibody molecules also stick. All other proteins are washed off the column and discarded. Finally, the protein is eluted from the antibody, which is itself bound to the column via the Fc-binding region, using conditions that will not disturb the antibody's binding to the column, and the protein can be recovered in pure form. In Figure 12.7c, the antibody is first bound to the column. It is then allowed to react with antigen as the protein mix flows slowly through the column. It can be eluted later, after unwanted proteins are thoroughly rinsed away.

An example of one use of this method, to characterize an HSV mutant that does not express a specific glycoprotein (glycoprotein C), is shown in Figure 12.8. Here, a polyclonal antibody against viral envelope proteins was prepared by immunizing rabbits. This antibody was allowed to bind to ^{35}S-labeled membrane proteins synthesized after infection with a wild-type and a gC^{-} mutant of HSV. The total protein mix and the purified envelope proteins that bound to the antibody preparation and were retained on the staph protein A column, as shown in Figure 12.7b, were then fractionated on a gel and exposed to x-ray film. Absence of the protein in the mutant virus is quite evident.

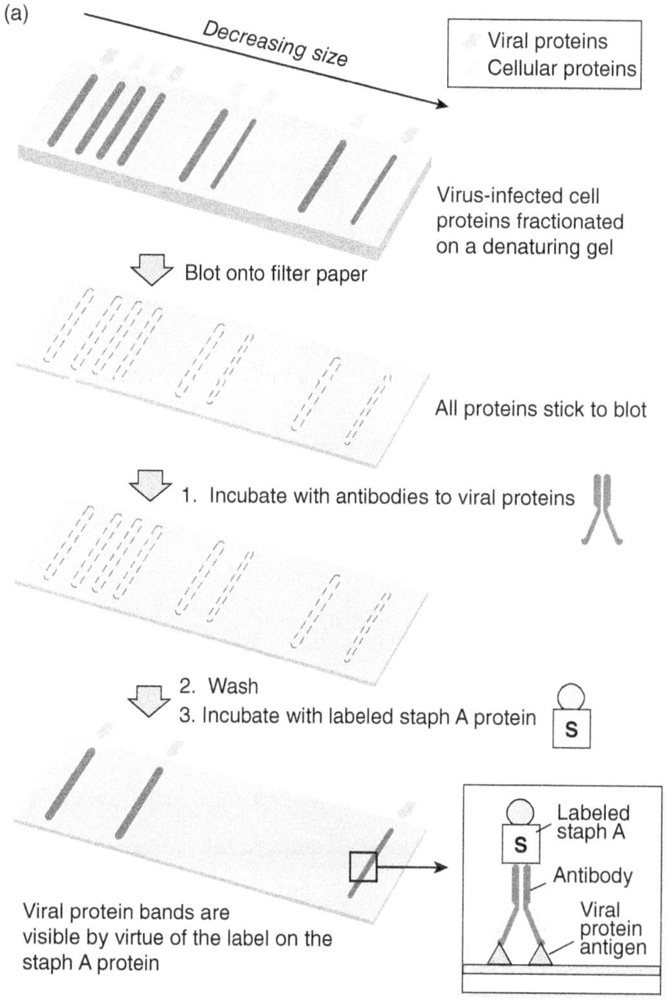

Figure 12.7 Detection and isolation of proteins reactive with a specific antibody by use of immunoaffinity chromatography. (a) Western blot. A mixture of viral and cellular proteins from an infected cell extract was fractionated on a sodium dodecyl sulfate gel, and the proteins blotted onto a membrane filter. The filter was then reacted with a specific antibody and washed, and then the antibody located by using radiolabeled staph protein A. (b) The antibody and antigen mixture is incubated so that specific interaction occurs. This is followed by passing the whole mix through a column with staph protein A bound to the column matrix (Sepharose). All antibody molecules bind through their Fc regions, and any antigen bound to them can be eluted with a gentle denaturation rinse that does not cause the staph protein A–Fc binding to be disrupted. (c) A similar approach in which the antibody first is bound to the column matrix, and the proteins are washed over the column for binding. Both methods provide essentially equivalent results.

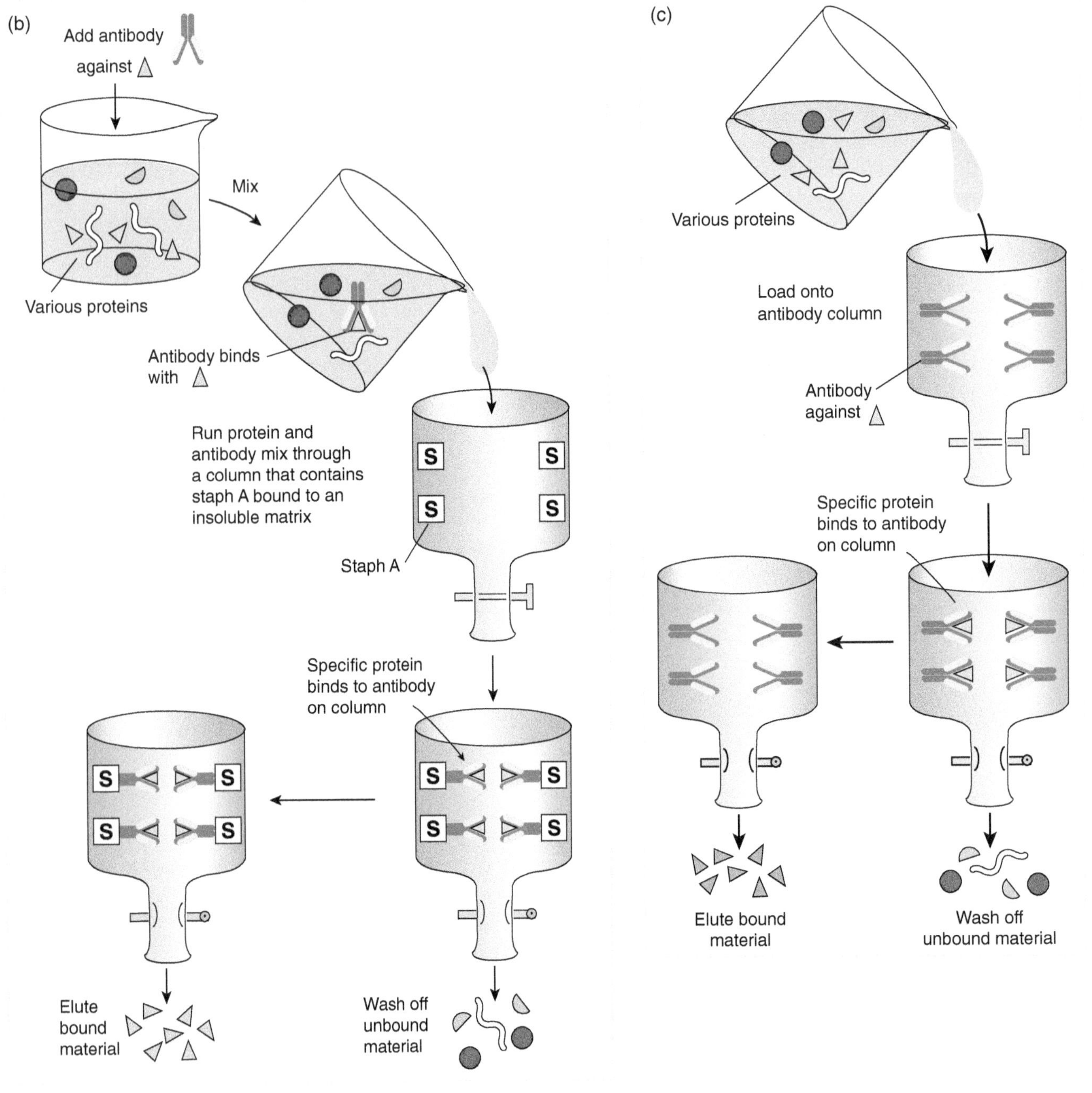

Figure 12.7 *Continued*

These same methods can be used with antibodies against the Fc region of antibodies from a different animal. Use of such antibody-binding methods provides another degree of specificity (just as did its use in immunofluorescence) and allows purification of even very small quantities of protein in a mix.

Another method of detecting and characterizing antibody responses to viral infection or viral vaccination is with viral protein microarrays. By printing a small amount of each viral protein, protein fragment, or epitope on a membrane-coated slide, investigators can discover which viral proteins are targeted by the antibody response. This may vary with severity of infection or disease outcome, and may be useful in the development of viral vaccines or diagnostics.

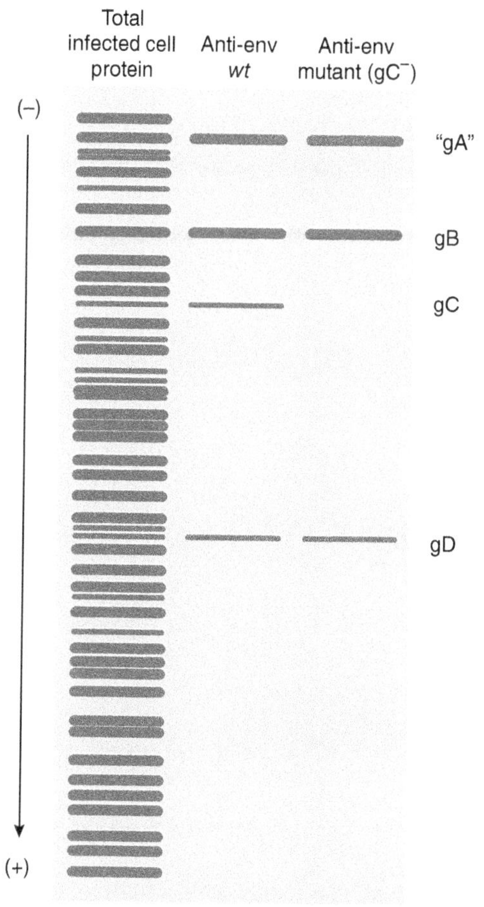

Figure 12.8 Use of immunoaffinity chromatography to isolate HSV envelope proteins from infected cells. Total infected cell protein was labeled by incubation with radioactive amino acids. The protein then was mixed with a polyclonal antibody specific for viral envelope proteins. The reactive proteins were isolated as described in Figure 12.7b and fractionated on a denaturing gel. The third column shows the results of a similar experiment where a virus unable to express glycoprotein C was used. wt: Wild type.

Figure 12.9 shows the reactivity of antibodies in saliva of people recently infected with HIV compared to people with long-term HIV infection and HIV-negative individuals on a protein microarray containing over 180 proteins, protein fragments, and epitopes from 10 types or subtypes of HIV. A heat map indicates antibody reactivity; green indicates no reaction, while black indicates antibody binding, and red shows stronger binding.

DETECTING AND CHARACTERIZING VIRAL NUCLEIC ACIDS IN INFECTED CELLS

Detecting the synthesis of viral genomes

Detection of viral DNA synthesized in an infected cell requires some method to separate viral material from the large background of cellular DNA. Viral DNA can be specifically detected using the polymerase chain reaction (PCR) with short (i.e., 20-base) oligonucleotide primers that hybridize specifically to viral DNA, as described in Chapter 11 and shown in Figure 11.9. Moreover, detection of viral DNA by PCR can be done quantitatively by use of a machine that can detect newly synthesized viral DNA at each step of the PCR. This is sometimes called "real-time" PCR since the product is detected as it is formed. Therefore, viral DNA synthesis can be detected by the quantitative increase of viral DNA over time post-infection.

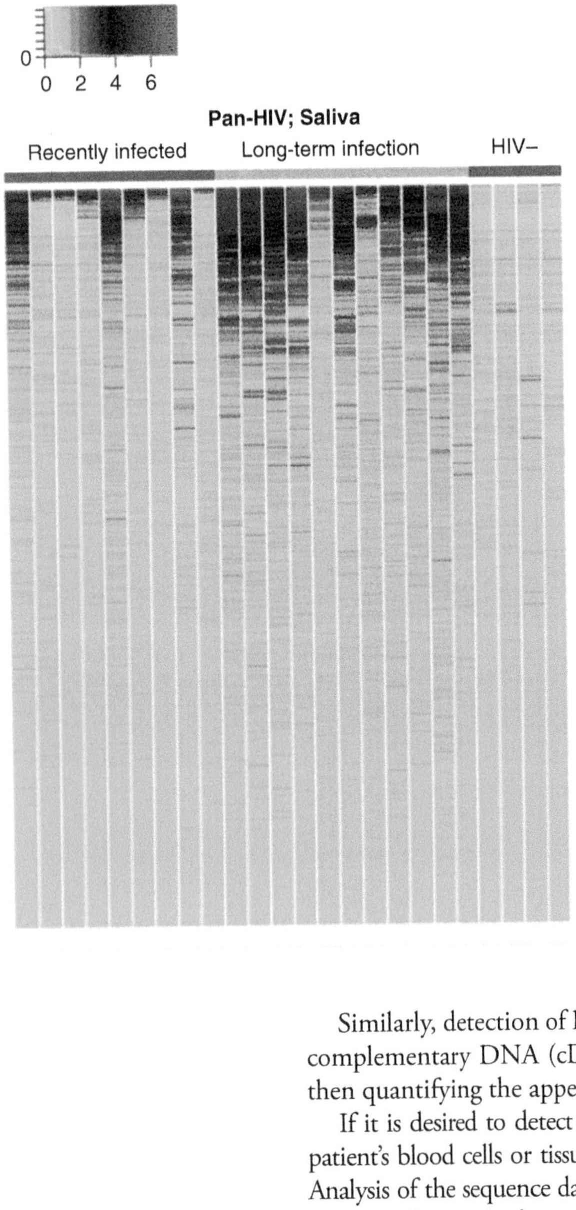

Figure 12.9 Saliva from recently HIV-infected people, individuals with long-term HIV infection, and those without HIV infection was incubated with a Pan-HIV protein microarray containing over 180 immunoreactive HIV proteins, protein fragments, or epitopes from 10 types or subtypes of HIV. (a) Heat map indicating strength of antibody reactivity by color, from green indicating lowest reactivity to red indicating highest reactivity. The 180 HIV proteins are ranked on the vertical axis from highest reactivity at the top to lowest at the bottom. Figure courtesy of Antigen Discovery Inc., Irvine, CA, USA.

Similarly, detection of RNA virus genome replication can be accomplished by first creating a complementary DNA (cDNA) copy of the viral RNA genome with reverse transcriptase and then quantifying the appearance of viral RNA genomes with quantitative PCR.

If it is desired to detect all the virus genomes replicating in a complex sample of cells such as a patient's blood cells or tissue biopsy, then high-throughput, "deep" DNA sequencing may be used. Analysis of the sequence data obtained by comparison to all know viral sequences will yield a list of viruses replicating in the sample. Quantification can be inferred by the number of viral DNA sequences obtained per thousand or per million cellular DNA sequences obtained. The diploid human genome is approximately six billion base pairs, while viral genomes vary from a few thousand bases or base pairs to a few hundred thousand base pairs, and are therefore 1 million- to 10 000-fold smaller.

Characterization of viral mRNA expressed during infection

Viral mRNA expressed during infection also can be analyzed and characterized using gel electrophoresis for size fractionation followed by nucleic acid hybridization. Without hybridization, detection of viral mRNA against the background of cellular RNA is difficult because individual mRNA molecules are not present in high abundance.

Hybridization requires a DNA or RNA probe that is complementary to the mRNA sequences to be detected. Such probes can be prepared readily by use of molecular cloning of viral DNA fragments in bacteria and one of a number of methods for making radioactive probes. This use of recombinant DNA technology provides a convenient and inexpensive source of pure material in large quantities. Some basic methods for cloning viral DNA fragments are briefly outlined in Part V. One of the most important things to remember about nucleic acid hybridization or annealing is that *under the proper conditions, the presence of large (even overwhelming) amounts of RNA or DNA with sequences different from that of the test DNA or RNA has no effect on the rate or amount of hybrid formed.* This makes nucleic acid hybridization an exquisitely sensitive method for detecting RNA and DNA, and much understanding of the regulatory processes involved in viral and cellular gene expression has a basis in the ability to precisely measure such expression at any given time during the viral replication cycle.

An experiment showing the different mRNAs expressed from two regions of the HSV genome is described in Figure 12.10. In this experiment, mRNA was isolated from HSV-infected cells at six hours after infection. Aliquots then were fractionated by gel electrophoresis and blotted onto a membrane filter. Replicate blots were hybridized with radioactive total viral DNA probe, or with a probe made from cloned DNA fragments from specific regions of the viral genome (as shown).

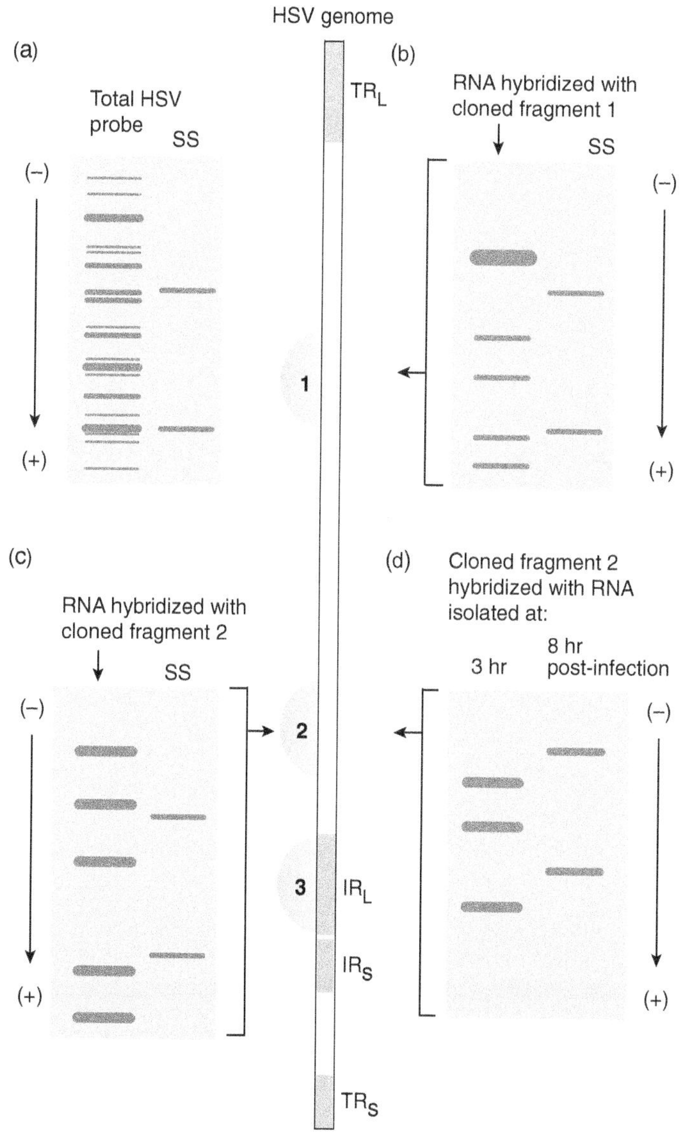

Figure 12.10 Different viral mRNA molecules are encoded by different regions of a viral genome. The diagram shows the 150 000-base pair HSV genome and the location of three cloned DNA fragments that can be used to hybridize to total infected cell RNA. More detailed information concerning the HSV genome and specific genes can be found in Chapter 17, Part IV. A number of fractionation gels are shown. (a) The total viral mRNA species expressed at five hours following infection. The RNA was isolated and fractionated, and a Northern blot made of the RNA. This was hybridized with radioactive viral DNA to locate the viral mRNA species. (b) The RNA species expressed in region 1 by hybridization with radioactive DNA from this region only. (c) The different RNAs seen with a probe for region 2. (d) The RNA expressed from region 2 changes in character between three and eight hours following infection (at the intermediate time shown in [c], all species are being expressed). The lanes marked "SS" contain radioactive ribosomal RNA included as a size standard.

In another experiment also shown in Figure 12.10, HSV mRNA was isolated at two different times (three and eight hours) following infection. At three hours, viral DNA replication has not yet begun. At six hours after infection, it is taking place at a high rate. The two RNA samples were fractionated by gel electrophoresis, subjected to Northern blotting, and then hybridized with a fragment of radioactive DNA from a specific region of the HSV genome. One can see that the amount of RNA present at the three-hour time point (i.e., early mRNA) is much reduced by six hours, and new RNA – late mRNA – is present at this later time.

Once the mRNA molecules produced by any virus have been characterized, they can also be detected by first creating cDNA and then using quantitative PCR or high-throughput sequencing as described in the "Detecting the Synthesis of Viral Genomes" section for viral genomes. These more modern methods are now widely used.

In situ hybridization

Hybridization of a cloned fragment of viral DNA to viral RNA (or DNA) in the infected cell can be achieved. The process is similar in broad outline to that for carrying out immunofluorescent analysis of antigens in a cell. In this type of hybridization, called *in situ hybridization*, the cells of interest are gently fixed and dehydrated on a microscope slide. Denatured probe DNA labeled with ^{3}H- or ^{35}S-labeled nucleosides is incubated with the cells on the slide. Following this, the slide is coated with liquid photographic emulsion that will detect radioactivity bound to the RNA or DNA of interest. The use of ^{3}H- and ^{35}S-labeled probes is favored because their decays are relatively low energy and the particle emitted is easily captured by x-ray emulsion near the site of its decay. Alternatively, a nonradioactive reagent can be incorporated into the probe DNA and detected with a secondary color or fluorescent reagent (FISH). When the micrograph is developed and observed, areas of RNA or DNA hybridizing to the specific probe are visible.

An example of *in situ* hybridization is shown in Figure 12.11. For this study, human brain tissue from an individual latently infected with HSV was taken at autopsy, preserved in formalin, embedded in paraffin, and sliced into thin sections. One set of sections was incubated with

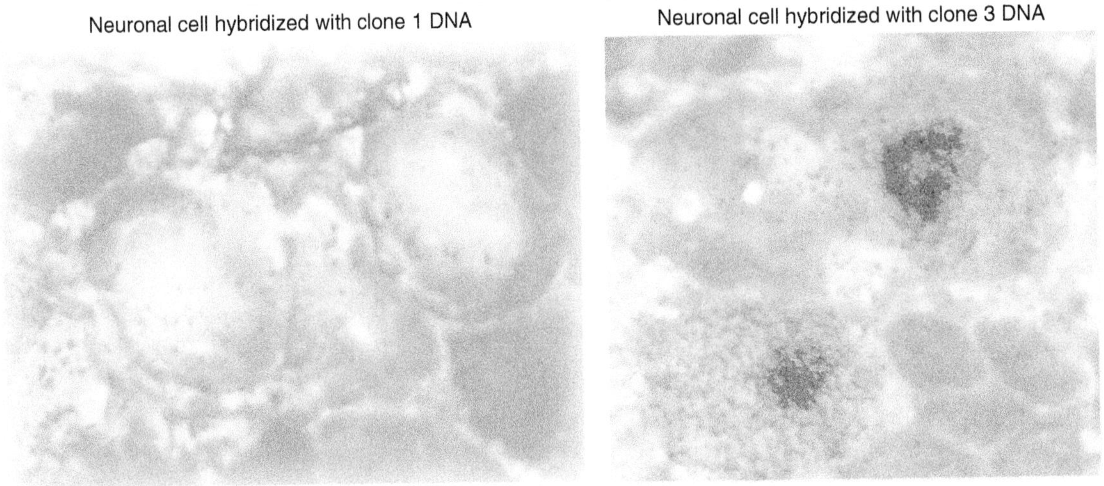

Neuronal cell hybridized with clone 1 DNA
Neuronal cell hybridized with clone 3 DNA

Figure 12.11 *In situ* hybridization of human neurons latently infected with HSV. The trigeminal nerve ganglion was taken at autopsy from a middle-aged man killed in an automobile accident. The tissue was sectioned, and individual slices were incubated with labeled probe DNA from either region 1 or region 3 of the HSV genome shown in Figure 12.10 under hybridization conditions. The left panel shows no hybridization; the dark spot in the neuronal nucleus is the nucleolus, which is the site of ribosomal RNA synthesis. The right panel shows positive hybridization due to the expression and nuclear localization of the HSV latency-associated transcript.

radioactive viral DNA probe from a region of the genome not expressed during latent infection, and another was incubated with a probe covering a region of the viral genome that is transcribed into **latency-associated transcripts** during latent infection. The nature of these latent-phase transcripts is described in more detail in Chapter 17, Part IV. But here, it is necessary to point out that the positive hybridization signal is only seen with probes complementary to it.

Like immuno-histochemical methods, *in situ* hybridization analysis can also be applied to larger scales. A histological section of a tissue or organ can be made, fixed, frozen, and/or embedded and then hybridized with an appropriate probe in order to locate areas where a specific viral transcript or viral genomes are being replicated. Indeed, the method can be applied to whole animals if they are small enough to allow sectioning. L. P. Villarreal and colleagues determined the effect of site of infection on the involvement of organs in which mouse polyomavirus will replicate in a suckling mouse using this method. An example is shown in Figure 12.12.

For this study, suckling mice were infected with polyomavirus by nasal or by intraperitoneal injection. After six days of replication, the mice were killed and carefully sectioned after freezing using a *microtome*, which is essentially a very sharp knife designed to cut thin sections of frozen or paraffin-embedded tissue. The slices were then placed on a membrane filter, stained, and hybridized with a radioactive polyoma-specific probe. The radioactivity was measured using a technique called *fluorography*, which is a way of visualizing low-energy radiation.

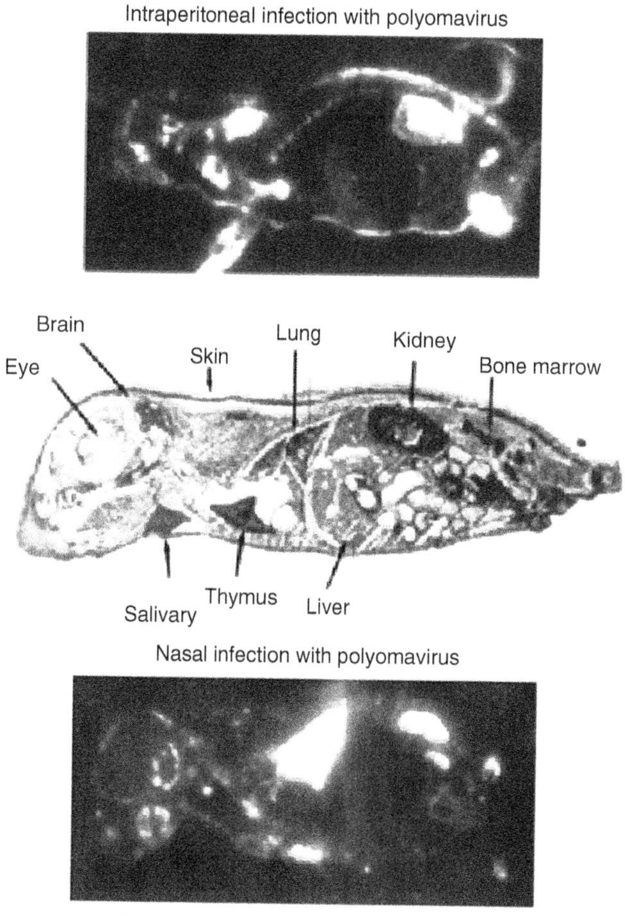

Figure 12.12 *In situ* hybridization of sections of suckling mice infected with polyomavirus. A stained section showing the location of major organs of the mouse is shown in the center. Fluororadiographs of sections showing tissues in which virus is replicating are shown above and below this section. Source: Photographs courtesy of L. P. Villarreal.

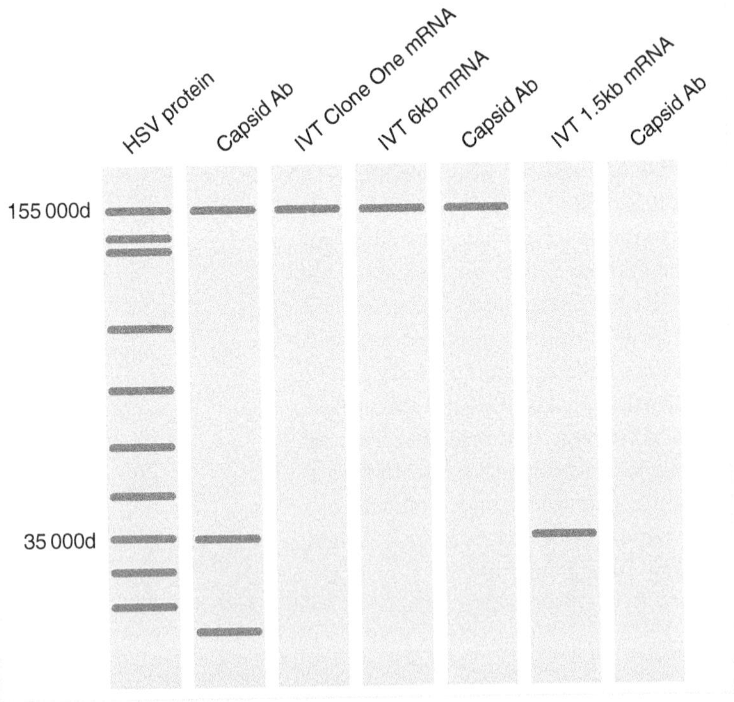

Figure 12.13 Characterization of isolated viral mRNA by *in vitro* translation (IVT). Total protein labeled in a one-hour pulse was isolated at six hours after infection from HSV-infected cells and fractionated on a denaturing gel. The capsid antibody (Ab) used in this experiment reacted specifically with only the 155 000-Da major capsid protein. The third lane shows the fractionation of protein synthesized *in vitro* using a rabbit reticulocyte system and mRNA hybridizing to DNA from region 1 of the HSV genome shown in Figure 12.9. Two proteins are seen: one migrating at 155 000 Da and the other at 35 000 Da. Demonstration that the large protein is, indeed, the major capsid protein is made by use of the antibody, as shown in the other lanes.

It is very clear from Figure 12.12 that virus inoculated in the nose replicates mainly in the lung, kidney, and thymus. In contrast, virus infected into the animal's peritoneum replicates efficiently in the kidney, brain, and bone marrow.

Further characterization of specific viral mRNA molecules

Different viral mRNA molecules encode different proteins. This is shown by the technique of *in vitro* translation (IVT). For such an experiment, either total infected cell mRNA or a purified fraction of such RNA is combined with radioactive amino acids and mixed with an extract isolated from rabbit reticulocytes, which contain ribosomes and all other requirements for protein synthesis. Any synthesized proteins can be fractionated by size on denaturing gels, or the protein products reactive with a specific antibody or antibodies can be isolated using one of the techniques outlined in Figure 12.7, and then fractionated.

An example shown in Figure 12.13 demonstrates that the 6-kb mRNA detected with cloned DNA probe of HSV (fragment 1, shown in Figure 12.10) encodes the 155 000-Da HSV capsid protein. In this experiment, the two mRNA species hybridizing to the specific region (6 and 1.5 kb) were subjected to IVT in the same sample. The translation products were then tested for reactivity with a polyclonal antibody monospecific for this capsid protein to yield the results shown. It can be concluded that the large capsid protein must be encoded by the large mRNA because the smaller mRNA encoded in this region is not large enough.

USE OF MICROARRAY TECHNOLOGY FOR GETTING A COMPLETE PICTURE OF THE EVENTS OCCURRING IN THE INFECTED CELL

Early virologists called the time period between when infectious virus entered the host cell and when progeny virus was produced the **eclipse period** of infection because they could not readily determine what was going on using the techniques they had at hand. The experimental techniques outlined in this chapter have allowed modern virologists to visualize the eclipse period with the illumination of increasingly detailed knowledge. While the experimental analysis of virus infection takes time and money, state-of-the-art application of micro-robotic techniques, laser-guided detection of target macromolecules interacting with substrates, and computer-enhanced quantitative measurement of such interactions, collectively termed *microarray analysis*, now provide the means of obtaining real-time measures of the intracellular environment as infection proceeds.

The basic idea behind microarray analysis is quite simple, and one example is illustrated in Figure 12.14.

In the most common versions, a large number of very small samples of individual target molecules, either nucleotide sequences complementary to cellular and viral genes or peptides known or thought to interact with host and virus-modified proteins, are bound to an inert substrate such as glass or a nylon membrane – the smaller the dimensions of the spots, the more samples that can be spotted on the matrix. Currently, sizes as small as 80 μ can be spotted, which means that a microscope slide can accommodate 10 000 or more different samples that are in use. This matrix containing the test material, with each variant spotted in a known location, is known as a microarray or microchip.

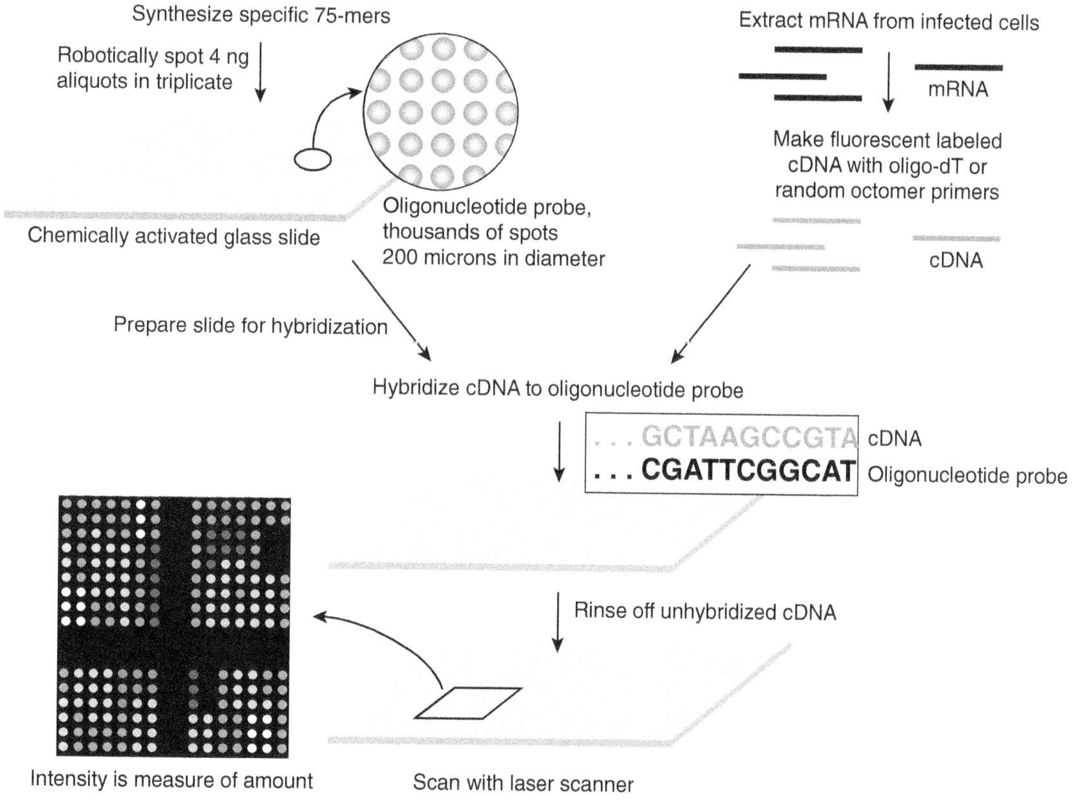

Figure 12.14 The application of microarrays or the study of viral products produced in an HSV-infected cell. (a) A DNA microarray for analysis of mRNA in HSV-infected cells. Oligonucleotides complementary to each viral transcript are bound to a glass slide along with oligonucleotides complementary to a number of diagnostic cellular transcripts. Samples of mRNA are isolated from cells under different conditions of infection, and cDNA copies are made using a dye-substituted deoxy-base; a different fluorescent dye is used for each condition. Then the cDNA is hybridized to the chip, unhybridized material washed away, and bound material is localized by scanning with a laser of a wavelength that only excites one or the other of the two dyes. The color and intensity of the signal in each spot can then be directly related to the amount of mRNA present in the original sample.

The microchip is then incubated with a small sample of a solution containing mixtures of macromolecules known or suspected to interact with the chip substrates. This could be mRNA or cDNA synthesized from mRNA if the chip contained fragments of DNA, or it could be a mixture of proteins from infected cells if the chip contained antibodies or peptides known to bind to a subset of the experimental mix. The volume of the experimental solution is kept very small by doing the incubation in a very small chamber. For example, if the probes were bound to a glass slide, layering a glass cover slip over the entire array could form the incubation chamber. Obviously, the tighter the patterning of spots in the array, the smaller the total volume needed. Ideally, a solution of materials from a few or even a single individual infected cell could be the source.

Following incubation and rinsing, interactions between chip-bound molecules and the experimental mixture added can be assayed by laser scanning. If a DNA chip was being used, fluorescent-tagged cDNA molecules made from the mRNA present in the infected cell mixture would only bind to complementary sequences on the chip after hybridization, and these could be detected by fluorescence upon laser illumination at the proper wavelength. If DNA–protein interactions were under investigation, laser power could be adjusted to partially atomize some of the protein bound to each spot, and its nature could be determined by mass spectroscopy.

QUESTIONS FOR CHAPTER 12

1 Antibodies against HSV-1 glycoproteins are tagged with a heavy metal in the Fc region. Virus is allowed to infect a cell, and immediately following this, the antibody is added. Then the cell is sectioned and an electron micrograph of this cell is taken. Where would you expect to see the heavy metal?

2 Which of the following methods can be directly applied to investigate the properties and characteristics of a viral protein?

(a) Electrophoresis in a sodium dodecyl sulfate polyacrylamide gel
(b) Western blot analysis with specific antibodies
(c) *In situ* hybridization with a specific antibody
(d) Immunohistochemistry with a cloned DNA fragment
(e) Determination of the sequence of the viral gene encoding it
(f) Nucleic acid hybridization

3 How is radiolabeling with amino acids used to examine the patterns of viral protein synthesis within infected cells? Give one specific example.

4 What are the ways in which a monoclonal antibody might be used in the analysis of a specific viral protein?

Viruses Use Cellular Processes to Express their Genetic Information

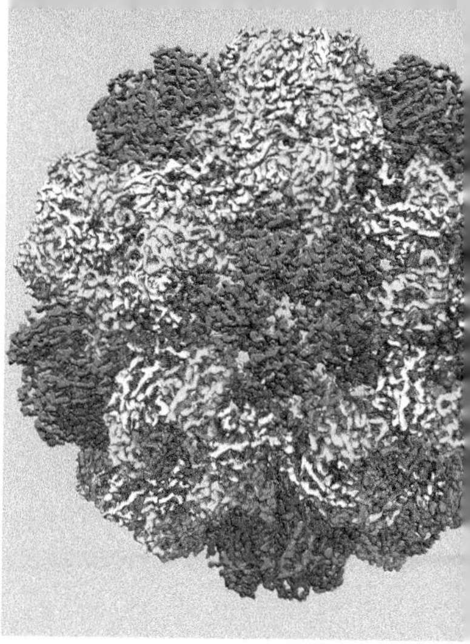

CHAPTER 13

- PROKARYOTIC DNA REPLICATION IS AN ACCURATE ENZYMATIC MODEL FOR THE PROCESS GENERALLY
- The replication of eukaryotic DNA
- The replication of viral DNA
- The effect of virus infection on host DNA replication
- EXPRESSION OF mRNA
- PROKARYOTIC TRANSCRIPTION
- Prokaryotic RNA polymerase
- The prokaryotic promoter and initiation of transcription
- Control of prokaryotic initiation of transcription
- Termination of prokaryotic transcription
- EUKARYOTIC TRANSCRIPTION
- The promoter and initiation of transcription
- Control of initiation of eukaryotic transcription
- Processing of precursor mRNA
- Location of splices in eukaryotic transcripts
- Posttranscriptional regulation of eukaryotic mRNA function
- Virus-induced changes in transcription and posttranscriptional processing
- THE MECHANISM OF PROTEIN SYNTHESIS
- Eukaryotic translation
- Prokaryotic translation
- Virus-induced changes in translation
- QUESTIONS FOR CHAPTER 13

Basic Virology, Fourth Edition. Martinez "Marty" Hewlett, David Camerini, and David C. Bloom.

Since viruses must use the cell for replication, it is necessary to understand what is going on in the cell and how a virus can utilize these processes. A virus must use cellular energy sources and protein synthetic machinery. Further, many viruses use all or part of the cell's machinery to extract information maintained in the viral genome and convert it to messenger RNA (mRNA) – the process of **transcription**. While cellular mechanisms for gene expression predominate, virus infection can lead to some important variations. RNA viruses face special problems that differ for different viruses. Also, many viruses *modify* or *inhibit* cellular processes in specific ways so that expression of virus-encoded proteins is favored.

To understand how viruses parasitize cellular processes, these processes should be understood. Indeed, the study of virus gene expression has served as a basis for the study and understanding of processes in the cell. All gene expression requires a mechanism for the exact replication of genetic material and the information contained within, as well as a mechanism for "decoding" this genetic information into the proteins that function to carry out the cell's metabolic processes.

Whether prokaryotic or eukaryotic, the cell's genetic information is of two fundamental types: ***cis*-acting genetic elements** or signals, and ***trans*-acting genetic elements** or genes. Genetic elements that act in *cis* work only in the context of the genome in which they are present. These include the following:

1 Information for the synthesis of new genetic material using the parental genome as template.
2 Signals for expression of information contained in this material as RNA.

Trans-acting elements are just that, information expressed to act, more or less freely, at numerous sites within the cell. Such information includes the genome sequences that are transcribed into mRNA and ultimately *translated* into proteins, as well as the sequences that are transcribed into RNA with specific function in the translational process: ribosomal RNA (rRNA) and transfer RNA (tRNA). Certain other regulatory RNA molecules such as microRNAs and siRNAs also can be included in this category.

While both prokaryotic and eukaryotic cells utilize DNA as their genetic material, a major difference between eukaryotic and prokaryotic cells is found in the way that the double-stranded DNA (dsDNA) genome is organized and maintained in the cell. Most bacterial chromosomes are circular, and whereas they have numerous proteins associated with them at specific sites, bacterial genomic DNA can be considered as "free" DNA (i.e., not associated with any chromosomal proteins).

In contrast, eukaryotic DNA is tightly wrapped in protein, mainly histones. Thus, the eukaryotic genome is the protein–nucleic acid complex, chromatin. The unique structure of this chromatin and its condensed form of chromosomes, and the ability of these to equally distribute into daughter cells during cell division, are manifestations of the chemical and physical properties of the **deoxyribonucleoprotein** complex.

There are also differences in the way that genetic information is stored in bacterial and eukaryotic chromosomes. In bacterial chromosomes, genes are densely packed, and only about 10–15% of the total genomic DNA is made up of sequences that do not encode proteins. Non-protein-encoding sequences include mainly short segments that direct the transcription of specific mRNAs, short segments involved in initiating rounds of DNA replication, and the information-encoding tRNA and rRNA molecules.

In some eukaryotic genomes including humans, on the other hand, 98% or more of the DNA does not encode any stable product at all! Some of this DNA has other functions (such as the DNA sequences at the center and ends of chromosomes), but some of these non-protein-coding DNA sequences have accumulated over evolutionary time, and their current function (if any) is the subject of continuing experimental investigation and vigorous debate. While the function of such

DNA sequences may not be clear, the origin of much of it is much better understood. Detailed sequence analysis of the human genome has demonstrated that about 8% is composed of simple sequence repeats and segment duplications, another 8% comprises retroviral complementary DNA (cDNA) elements that no longer can replicate, and 30% or so are transposable elements clearly related to retroviruses. Thus, about 20 times more DNA in the human chromosome is related to retroviruses than is involved in protein-encoding information! This is a striking demonstration of the importance of viruses in the biosphere and of their coevolution with their hosts.

PROKARYOTIC DNA REPLICATION IS AN ACCURATE ENZYMATIC MODEL FOR THE PROCESS GENERALLY

Despite differences in the nature of bacterial and eukaryotic genomes, the basic process of genomic DNA replication is quite similar. The two strands of cellular DNA are *complementary* in that the sequence of nucleotide bases in one determines the sequence of bases in the other. This follows from the Watson–Crick base-pairing rules. In the process of DNA replication, the following rules are useful to keep in mind:

1 A pairs with T (or U in RNA); G pairs with C.
2 The newly synthesized strand is antiparallel to its template.
3 New strands of DNA "grow" from the 5′ to 3′ direction.

The replication of a DNA molecule using the basic Watson–Crick rules is outlined in Figure 13.1, where the proteins involved in prokaryotic as well as eukaryotic replication are noted. The parental DNA duplex "unwinds" at the growing point (the **replication fork**) and two daughter DNA molecules are formed. Each new daughter contains one parental strand and one new strand, shown in different colors in Figure 13.1. Each base in the new strand and its **polarity** are determined by the three rules just presented.

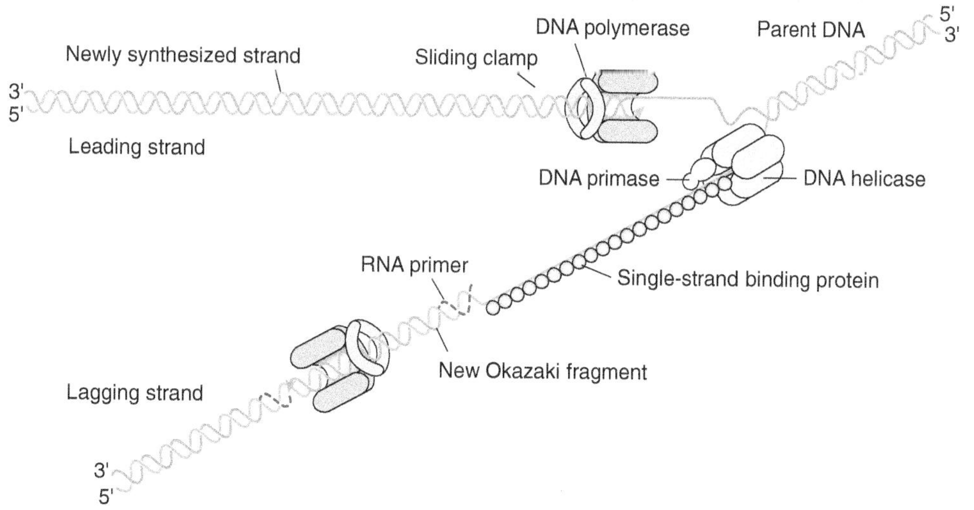

Figure 13.1 The enzymes and other proteins associated with DNA around a growing replication fork. The process is described in the text. Each new DNA chain must initiate with an RNA primer that forms in the vicinity of the unwinding DNA duplex. The unwinding is mediated by enzymes termed *helicases* that are complexed with primases. One DNA strand grows continuously; this is the "leading strand." Replication on the other ("lagging strand") is discontinuous due to the requirement for DNA synthesis to proceed from 5′ to 3′ antiparallel to the template strand. These discontinuous fragments are also called **Okazaki fragments** after the man who first characterized them. The primers are then removed, the gaps filled in, and the DNA fragments are ligated.

The replication of DNA can be divided into two phases: *initiation* of a round of DNA replication, which leads to generation of a complete daughter strand, and the *elongation* of this strand following initiation. Each round of DNA replication is initiated with a short piece of RNA that functions as a primer to begin the DNA chain. The first priming reaction takes place at a specific region (or site) on the DNA called the **origin of replication (ori)**. This origin of replication comprises a specific sequence of bases where an **origin binding protein** interacts and causes local denaturation to allow the replication process to begin.

Following denaturation of the DNA duplex at the origin of replication mediated by the ori binding protein, a **primosome** made up of helicase and primase enzymes synthesizes a short RNA primer in the 5′ to 3′ direction antiparallel to the DNA template, and a DNA melting enzyme (**helicase**) associates with the growing replication fork, leading to unwinding of the template duplex. The process differs on the two template strands following the initiation of replication. One strand (the **leading strand**) is antiparallel to the growing DNA chain, and **DNA polymerase** and associated sliding clamp proteins remain associated with this template, leading to the synthesis of a continuous strand of new DNA. The other strand, however, is in the wrong direction to serve as a template and remains single stranded by virtue of being stabilized by single-stranded DNA (ssDNA) binding proteins until a sufficient stretch of DNA forms. At this juncture, a primosome associates with the ssDNA near the replication fork and a new primer is synthesized, leading to association of DNA polymerase and associated clamp proteins and the synthesis of the second progeny strand of DNA (the **lagging strand**). As replication proceeds, enzymes and proteins required for unwinding the DNA duplex and maintaining it as single-stranded material interact with the denatured "bubble" to keep open the growing fork. Thus, at the DNA growing point, DNA synthesis is continuous in one direction, but discontinuous in the other direction. Lagging strand synthesis is required because in order to maintain proper polarity, new primer must be placed upstream of the growing point. In other words, priming must "jump" ahead on the template to continue synthesis of the new DNA strand, because DNA polymerase can only generate newly synthesized product in the 5′ to 3′ direction (reading the template strand 3′ to 5′). As lagging strand synthesis proceeds, primer RNA must be removed, gaps repaired, and discontinuous fragments ligated together to make the full strand. These final steps require the action of an **exonuclease** for removing RNA primer and **DNA ligase** for linking together the fragments of growing DNA on the lagging strands.

The replication of eukaryotic DNA

While the basic enzymatic processes outlined in the "Prokaryotic DNA Replication Is an Accurate Enzymatic Model for the Process Generally" section are the same in both prokaryotic and eukaryotic DNA replication, certain details differ significantly. There are several eukaryotic DNA polymerase enzymes, each encoded by genes that were generated by gene duplication during the evolution of eukaryotic cells (see Figure 1.2 for an example). One of these (Pol alpha) makes up part of the eukaryotic primosome, and another (Pol delta) is involved in chain elongation. The timing of eukaryotic DNA replication is tightly controlled and confined to the S-phase of the cell cycle, during which the clamp proteins, termed **PCNA (proliferating cell nuclear antigen),** are expressed. As was outlined in Part II, Chapter 10, and in more detail in Part V, the cell has a number of controls to ensure that chromosome and cellular replication only occur at appropriate times during the life of the organism, and unscheduled appearance of PCNA is an important marker indicating mistimed cell proliferation.

The nature of the eukaryotic genome also places strictures on the process of replication not encountered in prokaryotic cells. Chromatin – a deoxyribonucleoprotein made up of chromosomal DNA tightly associated with both histone and nonhistone proteins – must be disassociated into genomic DNA and associated proteins prior to replication, and reassociated following the passage of the growing fork. This often requires reversible chemical modifications of histones such as acetylation.

Also, as noted in Part II, Chapter 10, the telomeric ends of eukaryotic chromosomes are linear with the telomeres themselves being comprised of GC-rich sequences. Linear ends of DNA cannot be fully replicated by DNA polymerase because at the end there is no way for a primer to anneal with DNA upstream to initiate the process. Telomeres are replicated by the action of the telomerase ribonucleoprotein complex, which contains an integral RNA primer. As noted in Part I, Chapter 1, the sequence of the telomerase protein is related to that of retroviral reverse transcriptase.

The replication of viral DNA

The replication of viral DNA generally follows the same basic rules as for cellular DNA, and the replication of small DNA-containing viruses – such as the parvoviruses, polyomaviruses, and papilloma viruses – uses cellular DNA replication enzymes, although in the case of the latter two, the virus encodes a specific ori binding protein. Herpesviruses such as herpes simplex virus (HSV) encode a number of the proteins required in DNA replication, with the process being virtually identical to the cellular patterns outlined here, as shown in Figure 13.2. The same is true with the poxviruses. These DNA replication proteins have clear genetic relationships with cellular enzymes having the same function. Adenoviruses also encode their own DNA polymerase but, as outlined in Part IV, Chapter 16, use a protein primer for initiation of replication and carry out only leading-strand synthesis.

Those viruses that have circular DNA genomes do not face the problem of replicating linear ends of DNA molecules, but adenoviruses, herpesviruses, poxviruses, many bacteriophages, and other large DNA viruses have linear genomes. Each of these solves this **DNA end problem** with a unique mechanism described in the appropriate chapters of this book.

The effect of virus infection on host DNA replication

Virus infection can lead to degradation of cellular DNA with attendant recycling of nucleotides into the general viral biosynthetic pool. This is readily seen in productive infections by many bacteriophages. In infections with many eukaryotic viruses, degradation of host DNA is not complete – notably in infections leading to apoptosis. Cellular DNA replication is tightly regulated in differentiated eukaryotic cells, and infections with many tumor viruses lead to blockage of the regulatory circuits and unscheduled DNA replication and cell proliferation. In the case of infections with retroviruses, the viral genome becomes integrated in the host chromosome following its conversion from RNA into DNA by reverse transcriptase. The integrated viral DNA (the **provirus**) is then expressed as a cellular gene – often under strict regulation. This is also the case in **lysogenic** infections with bacteriophages such as phage λ. Such infections are described in the appropriate chapters of Part IV.

EXPRESSION OF mRNA

The expression of mRNA from DNA involves transcription of one strand of DNA (the mRNA coding strand that is the complementary sense of mRNA). Following initiation of transcription, RNA is polymerized with a DNA-dependent RNA polymerase using Watson–Crick base-pairing rules (except that in RNA, U appears in place of T). Although similar in broad outline, many details of the process differ between prokaryotes and eukaryotes. One major difference is that the bacterial enzyme can associate directly with bacterial DNA, and the enzyme itself can form a pre-initiation complex and initiate transcription. In eukaryotes, a large number of auxiliary proteins assembling near the transcription start site are required for initiation of transcription, and RNA polymerase can only associate with the template after these proteins associate. The process of transcription termination also differs significantly between the two types of organism.

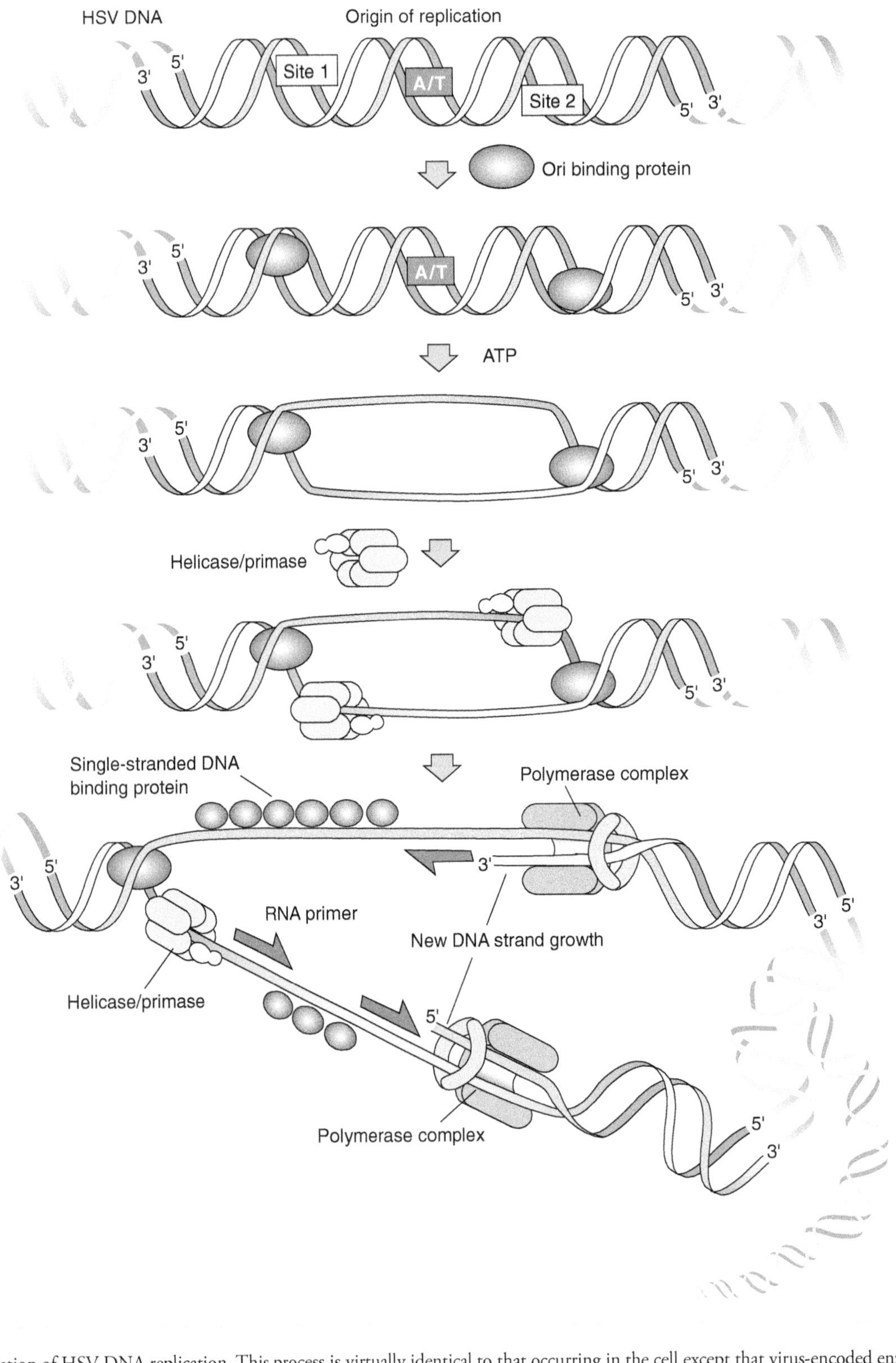

Figure 13.2 Initiation of HSV DNA replication. This process is virtually identical to that occurring in the cell except that virus-encoded enzymes and proteins are involved. The initial step is denaturation of the DNA at the replication origin with origin binding protein. Following this, the helicase–primase complex and ssDNA-binding proteins associate to allow DNA polymerase to begin DNA synthesis. Ori: Origin of replication; A/T: AT-rich sequence.

PROKARYOTIC TRANSCRIPTION

Regions of prokaryotic DNA to be expressed as mRNA are often organized such that a message is transcribed from which two or more proteins can be translated. The ability of bacterial RNA polymerase to transcribe such mRNA is often controlled by the presence or absence of a DNA-binding protein, called a repressor. The DNA sequence to which the repressor can bind is called the operator, and the genes expressed as a single regulated transcript are called operons. This is schematically shown in Figure 13.3. The operon model for bacterial transcription was first proposed in the early 1960s by Jacob, Monod, and Wollman from their genetic analyses of mutants of *Escherichia coli* unable to grow on disaccharide lactose. Since then, this operon model has been shown to be valid for a large number of prokaryotic transcriptional units.

In addition to organization into operons, prokaryotic gene expression differs from that of eukaryotes as a result of a fundamental structural difference between the cells: the lack of a defined nucleus in prokaryotes. In prokaryotic cells, transcription takes place in the same location and at the same time as translation. This coupling of the two events suggests that the most efficient regulation of gene expression in these cells will be at the level of initiation of transcription. The operon model also takes this into account.

Prokaryotic RNA polymerase

The DNA-dependent RNA polymerase of prokaryotic cells, especially that of *E. coli*, is well studied. The enzyme shown in Figure 13.4 contains five subunit polypeptides: two copies of **α**, one of **β**, one of **β**′, one of **σ**, and one of **ω**. The functions of all the subunits except **ω** are known quite precisely. The core enzyme, which can carry out nonspecific transcription *in vitro*, consists of the **β**′ subunit for DNA binding and the two **α** subunits and the **β** subunit for initiation of transcription and for interaction with regulatory proteins. The addition of the **σ** subunit creates the *holoenzyme* that transcribes DNA with great specificity, since this subunit is responsible for correct promoter recognition. It is the holoenzyme that is active *in vivo* for initiation of transcription.

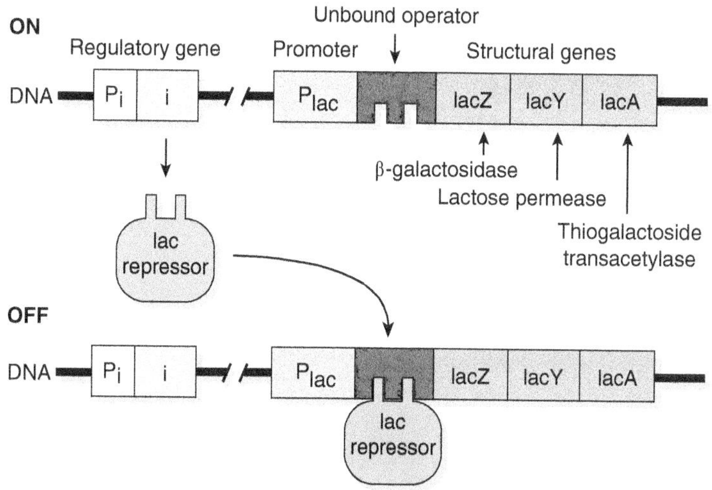

Figure 13.3 The *E. coli* lac operon. The promoter is always "on," but normally the lac repressor (lac I gene product) is bound to the operator that blocks transcription. The repressor can be inactivated by addition of lactose. The operator is also sensitive to cAMP levels as explained in the text. All the genes controlled by this operon are expressed as a single mRNA that can be translated into three separate proteins due to internal ribosome initiation.

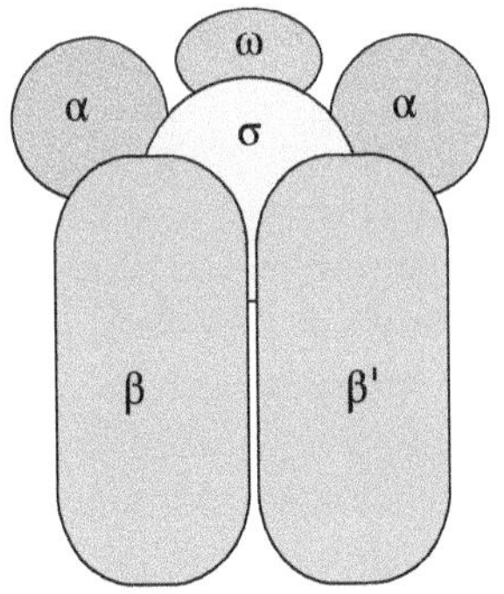

Subunit	Size	#/molecule	Function
α	36.5 kd	2	Chain initiation and interaction with regulatory proteins
β	151 kd	1	Chain initiation and elongation
β'	155 kd	1	DNA binding
σ	70 kd	1	Promoter recognition
ω	11 kd	1	Helps assemble and stabilize the complex

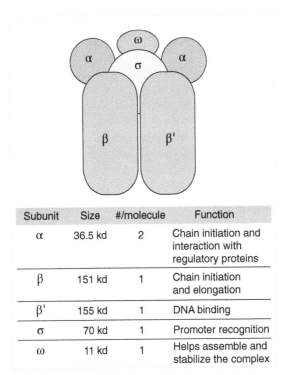

Figure 13.4 The bacterial RNA polymerase molecule. The enzyme is made up of six subunits with different functions. The complete enzyme is called the *holoenzyme*.

The prokaryotic promoter and initiation of transcription

The DNA to which the RNA polymerase holoenzyme binds to begin transcription looks very much like its eukaryotic counterpart. Consensus sequences are present at specific locations upstream from the start site of transcription. A sequence with the consensus TATAAT is found at the −10 position, and a sequence TTGACA at −35. The former sequence is often called the Pribnow box after its discoverer and is similar in function to the TATA box of eukaryotes. The RNA polymerase holoenzyme binds to the promoter, causing a transcription bubble to form in the DNA. Just as in eukaryotes, transcription begins with a purine triphosphate, and chain elongation proceeds in the 5′ to 3′ direction, reading the DNA template from the antisense strand in the 3′ to 5′ direction. The polymerase catalyzes incorporation of about 10 nucleotides into the growing mRNA before the σ subunit dissociates from the complex. Thus, σ is required only for correct initiation and transcription of the RNA chain's first portion.

Control of prokaryotic initiation of transcription

As mentioned, the bacterial RNA polymerase holoenzyme will form a transcription complex and begin to copy DNA, given the presence of a correct promoter sequence. Since the strategy of prokaryotic regulation dictates that gene expression be regulated at the level of this initiation, many inducible genes (genes whose expression goes up or down with given cellular conditions) have the general structure of the operon diagrammed in Figure 13.3. Binding of the repressor protein to the operator sequence of DNA, positioned at or immediately downstream of the initiation site, effectively provides a physical block to progress of the RNA polymerase. The repressor–operator combination acts, in effect, like an "on-off" switch for gene expression, although it should be understood that this binding is not irreversible and that there is some finite chance of transcription taking place even in the "off" state.

Presence of the appropriate inducing molecule, such as the metabolite of lactose responsible for inducing the *lac* operon, will cause a structural change in the repressor such that it can no longer bind to the operator. In cases such as the tryptophan operon, the repressor protein assumes the correct binding conformation only in the presence of the co-repressor (e.g., tryptophan). The overall situation is that regulated prokaryotic gene expression takes place unless the binding of a protein that blocks movement of RNA polymerase prevents it.

Enhancement of prokaryotic transcription is also seen. Using the example of operons for the genes required to utilize unusual sugars such as lactose, upregulation of gene expression can be observed. In this case, the response involves a system that can "sense" the amount of glucose presented to the cell and thus the overall nutritional state of that cell. Since the enzymes that metabolize glucose (the glycolytic pathway) are expressed constitutively (unregulated) in most cells, the availability of this sugar is a good signal for the cell to use in regulating the expression of enzymes for the metabolism of other sugars. The level of glucose available to the cell is inversely proportional to the amount of 3′,5′-cyclic adenosine monophosphate (**cAMP**) within the cell. This nucleotide can interact with a protein called the cyclic AMP receptor protein (CRP). A complex of cAMP–CRP binds to a region of DNA just upstream of the promoter but only in genes that are sensitive to this effect. When the complex binds, the DNA is changed in such a way that the rate of transcription is raised many fold. If the repressor protein is the "on-off" switch of this gene, then the cAMP–CRP complex is the "volume control" fine-tuning transcription as metabolic need arises. This regulation of the rate of transcription by the level of glucose is called **catabolite repression**.

Termination of prokaryotic transcription

Bacterial RNA polymerase terminates transcription by one of two means: in a ρ-dependent or ρ-independent fashion. The difference between these two involves the response of the system to the **termination factor (ρ factor)** and structural features near the 3′ terminus of the RNA.

In the case of ρ-dependent termination, the mRNA being transcribed contains, near the intended 3′ end, a sequence to which the ρ factor binds. The protein ρ is functional as a hexamer and acts as an adenosine triphosphate (ATP)-dependent helicase to unwind the product RNA from its template and terminate polymerization.

For ρ-independent termination, the sequence near the intended 3′ terminus of the transcript contains two types of sequence motifs. First, the RNA transcript contains a GC-rich region that can form a base-paired stem loop structure. Immediately downstream from this feature is a U-rich region. The presence of the GC-rich sequence slows progress of the polymerase. The stem loop that forms interacts with the polymerase subunits to further halt their progress. Finally, the AU-rich sequences melt and allow the transcript and template to come apart, terminating transcription.

EUKARYOTIC TRANSCRIPTION

The promoter and initiation of transcription

In eukaryotes, all transcription occurs in the nucleus except for that taking place in organelles. RNA polymerase II (pol II) is "recruited" into the pre-initiation complex formed by association of accessory transcription-associated factors assembling at the site where the transcript is to begin; the process is outlined in Figure 13.5. Transcription initiates in a "typical" eukaryotic promoter at a sequence of 6–10 bases made up of A and T residues (the TATA box), which occurs about 25 bases upstream (5′) of where the mRNA starts (cap site). The proteins making up this pre-initiation complex make a complex just large enough to reach from this region to the cap site, as shown in Figure 13.5. Formation of the pre-initiation complex around the TATA box can be modulated and facilitated by association of one of many transcription factors that bind to specific sequences usually upstream of – but within close proximity to – the cap.

These upstream transcription elements can interact with and stabilize the pre-initiation complex because dsDNA is flexible and can "bend" to allow transcription factors to come near the TATA-binding protein complex; this bending is diagrammed in Figure 13.6. The whole promoter region (containing the cap site, TATA box, and proximal transcription factor–binding sites) generally occupies the 60–120 base pairs immediately upstream (5′) of the transcription start site.

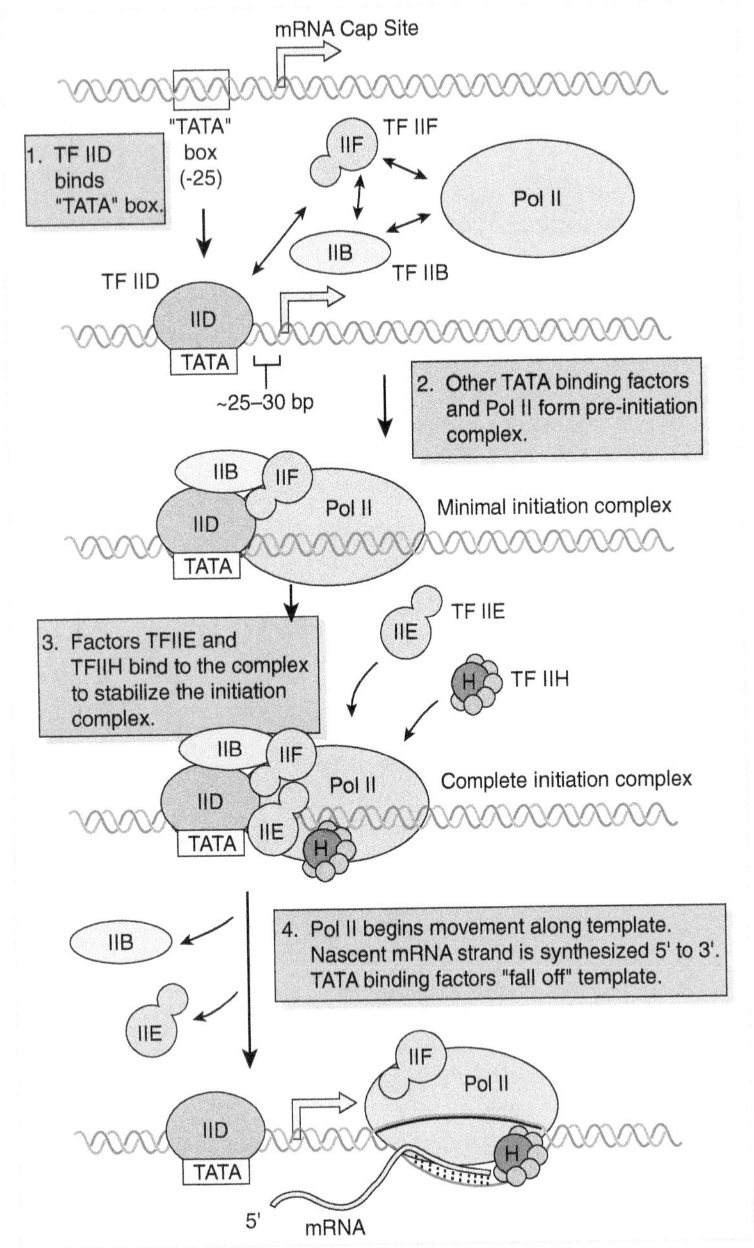

Figure 13.5 The multistep process of transcription initiation at a eukaryotic promoter. With most promoters, the process begins as shown at the top with assembly of the initiation complex at the TATA box. Upon its full assembly, the DNA template is denatured, and RNA synthesis antiparallel to the template strand (antisense strand) is initiated. The relative sizes of the proteins involved show how location of the pre-initiation complex at the TATA box is spaced to allow RNA polymerase to begin RNA synthesis about 25–30 bases downstream of it. TF: Transcription factor; Pol II: RNA polymerase II.

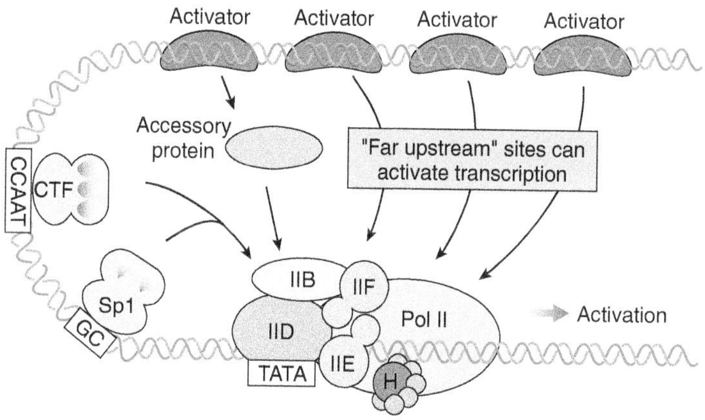

Figure 13.6 The flexibility of DNA allows transcription factors to bind at sites upstream of the TATA box to stabilize formation of the pre-initiation complex. Enhancer elements even further upstream (or, in some cases, downstream) can also bind activating proteins that can further facilitate and modulate the process.

Other control regions or **enhancers** can occur significant distances away from the promoter region. Such enhancers also interact with specific proteins and ultimately act to allow transcription factors to associate with the DNA relatively near the promoter; the process is also shown in Figure 13.6. Enhancers appear to help displace histones from the transcription template and therefore facilitate the rate of transcription initiation from a given promoter. Unlike the core promoter element itself, however, enhancers serve only to regulate and augment transcription, and the promoter that they act on can mediate measurable transcription in their absence. Enhancers themselves may be subject to modulation of activity by factors stimulating the cell to metabolic activity, such as cytokines and steroid hormones.

Control of initiation of eukaryotic transcription

Like with prokaryotes, controlling the access of RNA polymerase and associated enzymes with the gene to be transcribed to a large measure controls eukaryotic transcription. The process is complicated by the nature of the transcription template (chromatin), the number of regulatory proteins whose access must be controlled, and the need to transmit signals from the cell exterior through the cytoplasm to the nucleus where transcription occurs. The broadest level of control of transcription, one that mediates transcriptional regulation during development, is at the level of chromatin structure. Expression of large portions of cellular chromosomes can be essentially permanently repressed by methylation of DNA followed by its association with specific DNA-binding proteins and nonacetylated histones into condensed heterochromatin (Figure 13.7a). Although such regions of the chromosome can be derepressed under certain conditions, such as the abnormal release from replication control associated with tumor growth, normally this is an irreversible process.

Dispersed, euchromatin contains those genes that are normally transcribed in the cell. As described in this chapter, the binding of transcription factors, RNA polymerase, and the acetylation of the associated histones (allowing for an open, accessible structure) is regulated by the presence of regulatory proteins binding to DNA and controlling the expression of the gene in question.

Many regulatory proteins are sequestered in the cytoplasm in an inactive form until an extracellular signal interacts with its cellular receptor, leading to a **signal transduction cascade** involving covalent modification of target proteins followed by their migration to the nucleus and activation of the target genes. While there are a huge number of extracellular signals, a given cell can only recognize those for which it has specific receptors. These receptors and the cascades that they induce can be divided into a relatively small number of pathways leading to the reversible modification of target proteins (often by phosphorylation). The activation of the type I interferon response described in Part II, Chapter 7, is shown in Figure 13.7b. As shown, the presence of interferon leads to the formation of a heterodimer on the cell surface. Each of the subunits has a kinase protein associated with it (Jak1 or Tyk2), and the dimerization allows these proteins to phosphorylate each other – the phosphorylated kinase proteins then phosphorylate receptors on the C-terminal regions of the α interferon receptors themselves. These in turn serve as sites for the association of STAT proteins with the complex, and the kinases phosphorylate them leading to their disassociation from the complex, dimerization, and migration to the nucleus where they bind to regulatory sequences of α interferon–inducible transcripts such as those encoding 2′,5-oligoA synthetase, PKR, and so on. Thus, a single signal outside the cell can lead to multiple responses (a cascade) directed at that signal.

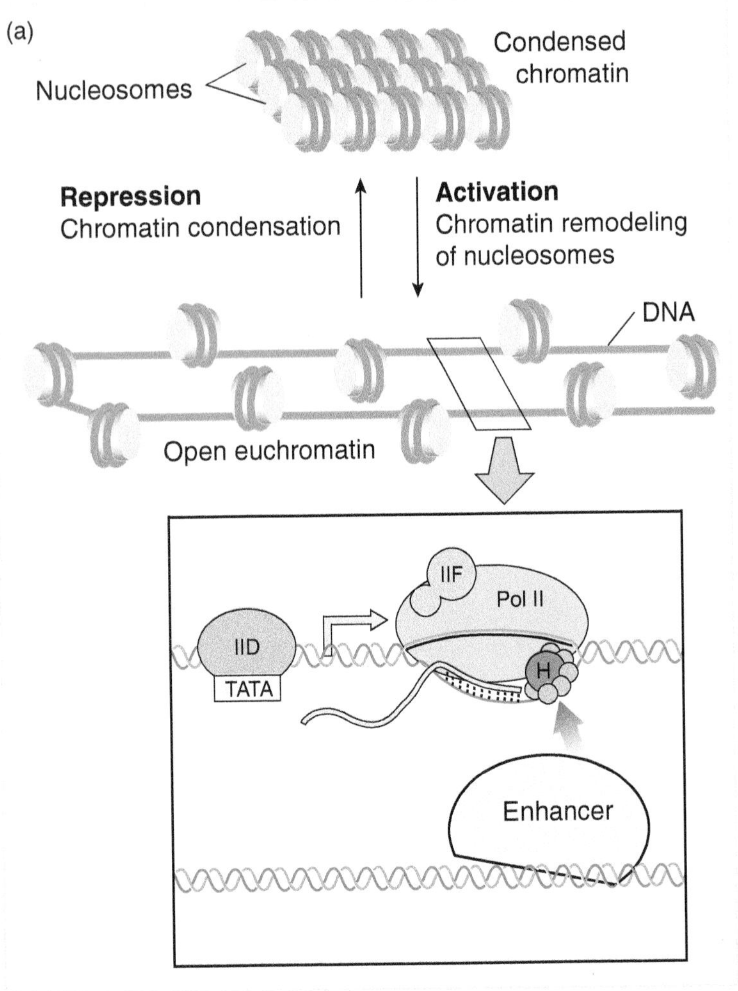

Figure 13.7 Control of eukaryotic transcription. (a) The availability of the transcription template is controlled by chromatin structure. This is generally a developmental process, and condensation of chromatin into heterochromatin is often essentially irreversible in a differentiated cell. Such condensed chromatin is transcriptionally silent and contains unacetylated histones and methylated DNA. Euchromatin is more loosely organized with acetylated histones; in such regions, genes are essentially off, but some transcription can occur at irregular intervals or during chromosome replication. Activation of genes in euchromatin, as a result of the binding of transcriptional activators to regulatory sequences such as enhancers, leads to high levels of transcription. (b) The Jak–Stat signal transduction cascade induced by alpha (α) interferon resulting in transcriptional activation of interferon-induced genes. As discussed in the text, the presence of α interferon leads to dimerization of specific receptors at the cell surface, leading first to the phosphorylation of Jak and Tyk bound to the C-terminal cytoplasmic domains of these receptors, followed by the phosphorylation of the receptor peptides themselves. This results in the phosphorylation of cytoplasmic STAT proteins, which dimerize and migrate to the nucleus where they activate appropriate transcription.

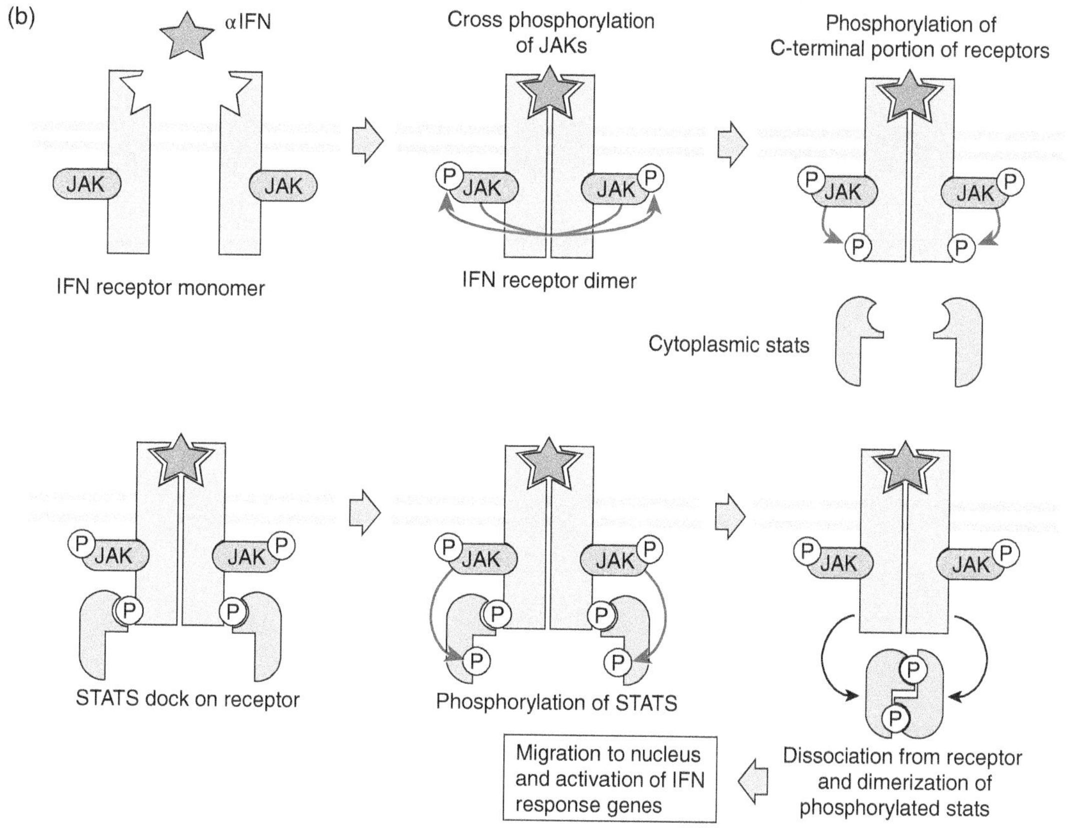

Figure 13.7 *Continued*

Processing of precursor mRNA

Following initiation of transcription, transcript elongation proceeds. RNA is also modified following initiation by addition of a **cap** at the 5′ end. Capping takes place by addition of a 7-methyl guanine nucleotide in a 5′-5′ phosphodiester bond to the first base of the transcript. This cap has an important role in initiation of protein synthesis. Transcription proceeds until the pol II–nascent transcript complex encounters a region of DNA containing sequences providing **transcription-termination/polyadenylation signals** that occur over 25–100 base pairs. A major feature of this region is the presence of one or more **polyadenylation signals**, AATAAA in the mRNA sense strand.

Other short *cis*-acting signals also are present in the polyadenylation region. A specific enzyme (terminal transferase) adds a large number of adenine nucleotides at the 3′ end of the RNA just downstream (3′) of the polyadenylation signal as it is cleaved and released from the DNA template. Interestingly, the polymerase itself can continue down the template for a short or long distance before it finally disassociates and falls off.

In addition to capping and polyadenylation, most eukaryotic mRNAs are spliced. In **splicing**, internal sequences (**introns**) are removed and the remaining portions of the mRNA (**exons**) are ligated to form the mature mRNA. Splicing takes place via the action of small nuclear RNA (**snRNA**) in complexes of RNA and protein (ribonucleoprotein) called **spliceosomes**. The process is complex, but the result is that most mature eukaryotic mRNAs are somewhat or very much smaller than the pre-mRNA precursor or primary transcript.

The generation of mature mRNA in the nucleus is shown diagrammatically in Figure 13.8. Although splicing is shown to occur after cleavage/polyadenylation in this schematic, the actual process may occur as the nascent RNA chain grows. The maturation of RNA and splicing are shown in a somewhat higher-resolution view in Figure 13.9. All modifications occur on the RNA itself: first capping, then cleavage/polyadenylation of the growing RNA chain, then splicing (if any). Thus, almost all eukaryotic mRNAs are capped, polyadenylated, and spliced. Because splicing can occur within or between sequences of mRNA encoding peptides, it can result in the generation of complex "families" of mRNA encoding related or totally unrelated proteins. Some general patterns of splicing known to be important in virus replication are shown in Figure 13.10a.

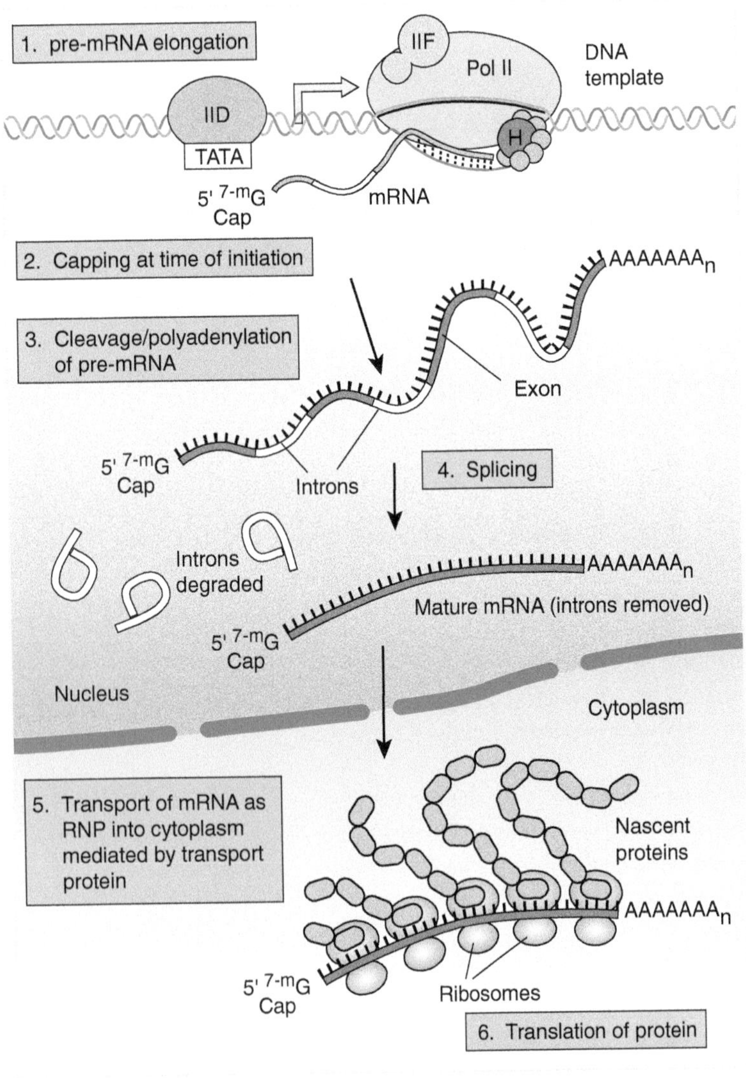

Figure 13.8 Steps involved in transcription and posttranscriptional modification and maturation of eukaryotic mRNA. The sequence of events is indicated by the numbers 1 through 6. RNP: Ribonucleoprotein; $^{7\text{-m}}$G: 7-methyl guanine.

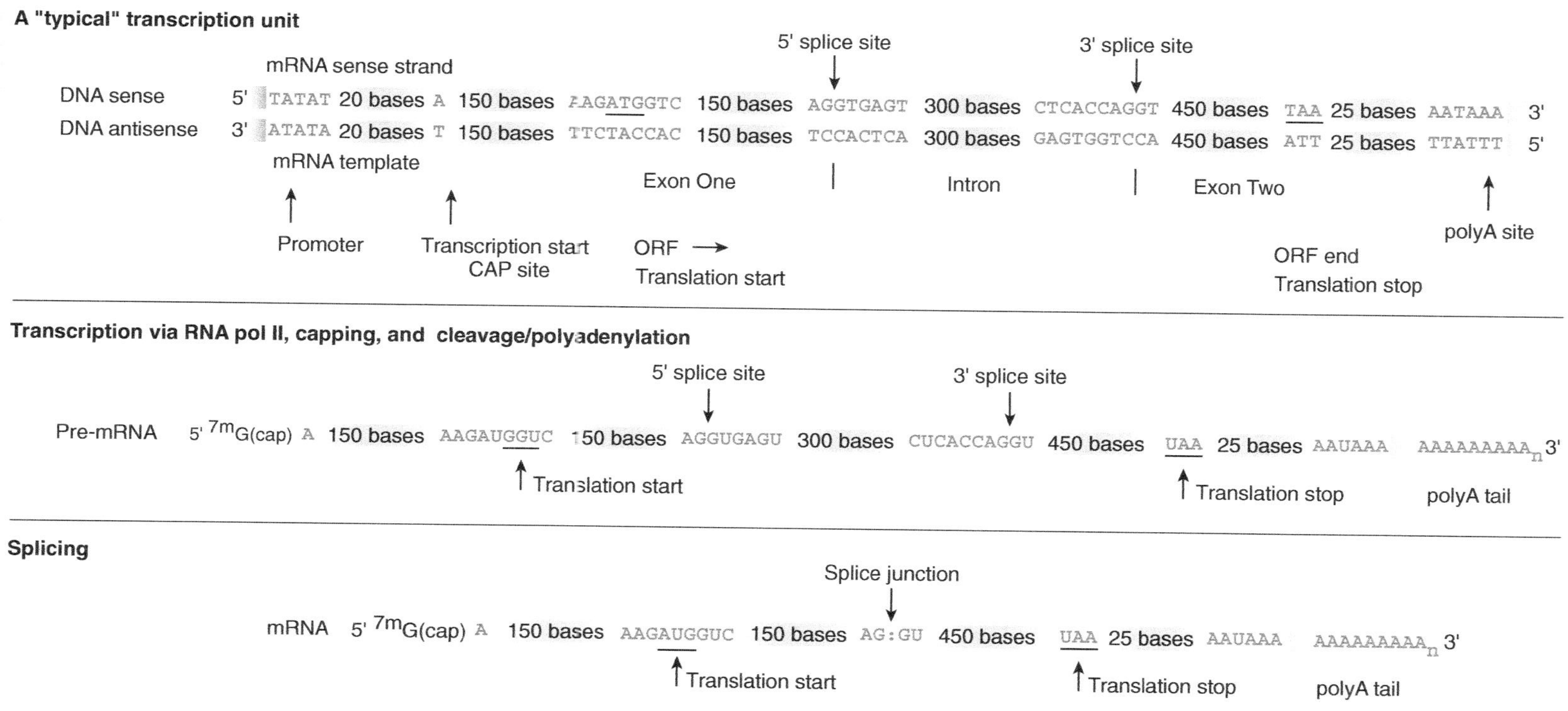

Figure 13.9 A "high-resolution" example of mRNA processing. The sequence of a hypothetical pre-mRNA transcript is shown. The transcript is capped and polyadenylated, and splicing removes a specific sequence of bases (the intron). This results in the formation of a translational reading frame as shown.

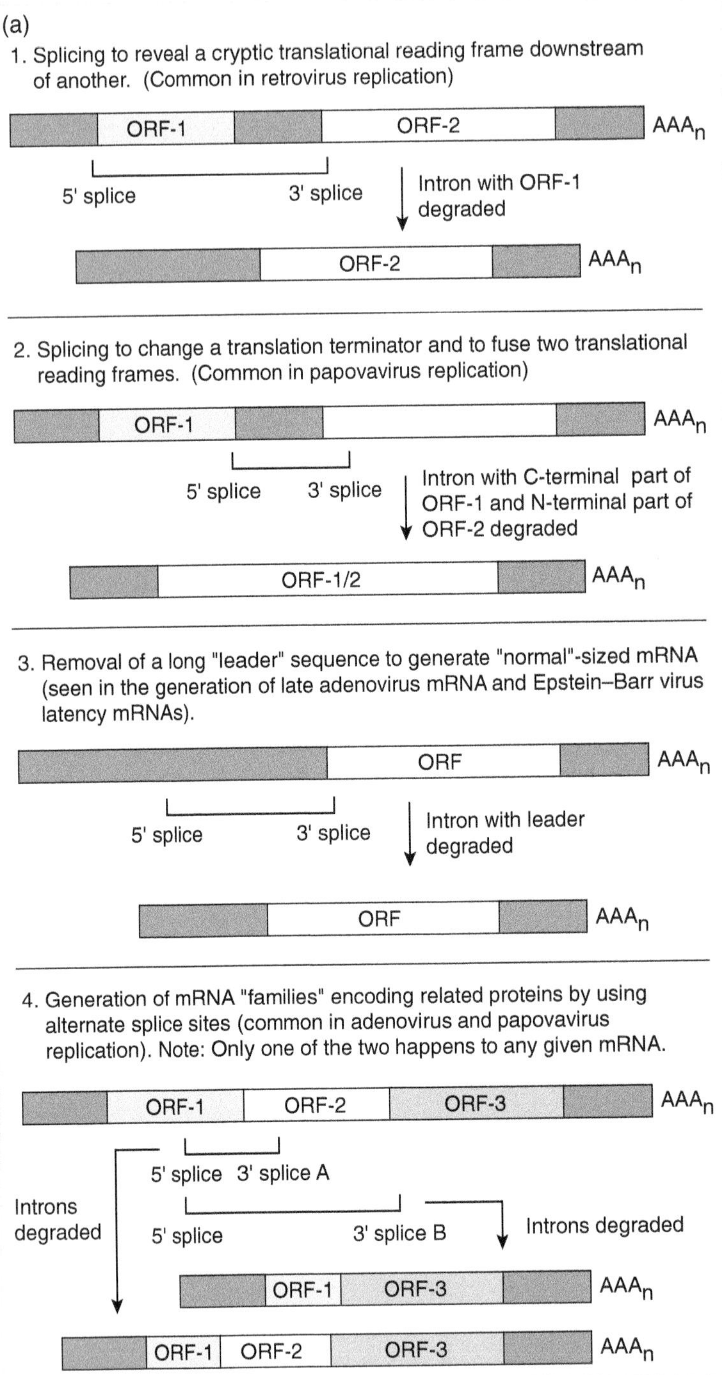

Figure 13.10 Some splicing patterns seen in the generation of eukaryotic viral mRNAs. (a) Schematic representation of different splice patterns that have been characterized. (b) Generation of a polymerase chain reaction (PCR) product from HSV latency-associated (LAT) RNA by using primers annealing to regions 5′ and 3′ of an intron. The gene encoding the HSV LAT transcript is about 9 kbp long, and there is a 2-kb intron that is located about 600 base pairs 3′ of the transcription start site (see Chapter 17 and Figure 17.2). PCR amplification of HSV DNA using the first primer set [P1:P(−1)] shown produces a product about 150 nucleotides (nt) long. Amplification using the second primer pair [P2 and P(−2)] will produce a fragment that is longer than 2000 nucleotides and cannot be seen. Next, LAT RNA from latently infected cells is used as a template for the synthesis of cDNA complementary to it (see Chapter 19). When the first primer pair is used for PCR amplification of LAT cDNA, a product the same size as that formed using genomic DNA as a template is formed. In contrast, however, when primer set 2 is used, the product of the cDNA is only about 160 base pairs long since the 2000-base intron has been spliced out. If the product of PCR primer set 2 were subjected to sequence analysis and compared to the sequence of viral DNA, a discontinuity at the splice sites would be revealed.

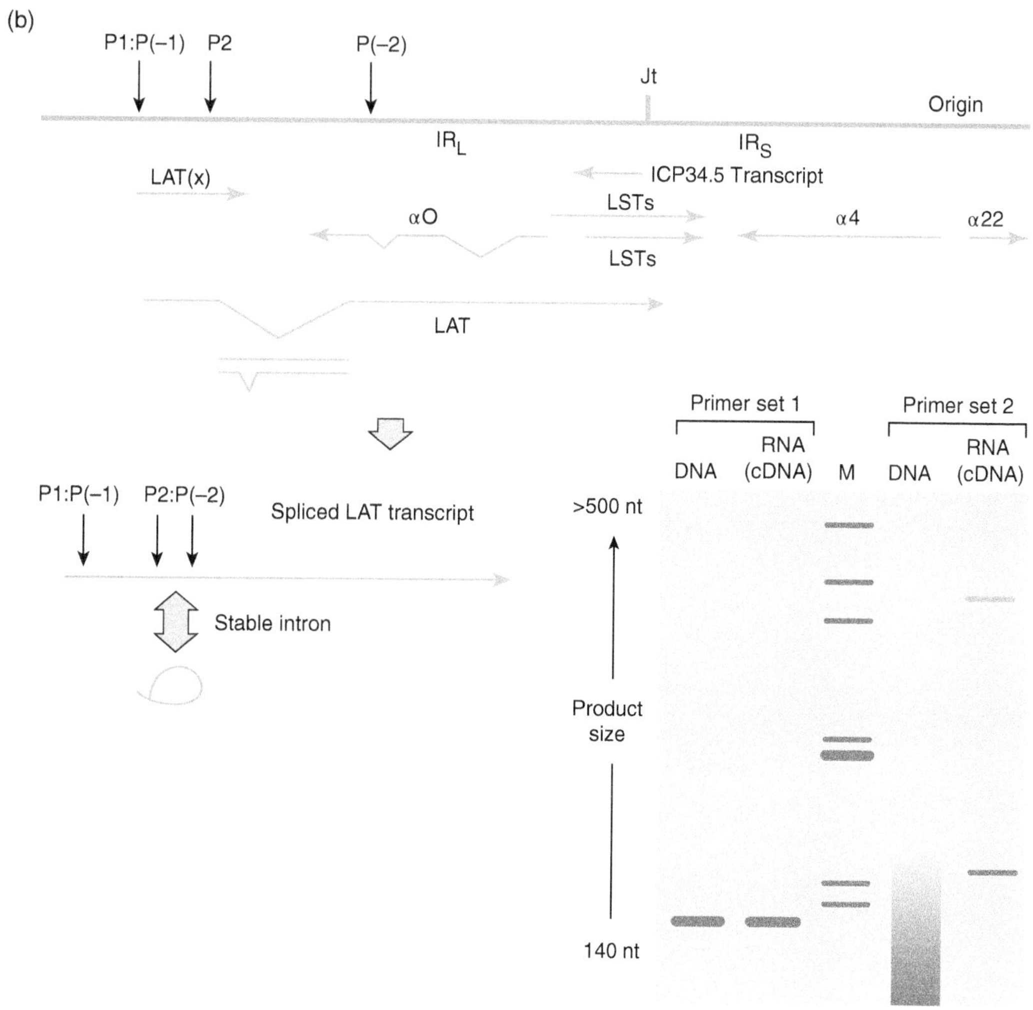

Figure 13.10 *Continued*

Location of splices in eukaryotic transcripts

Characterization of a spliced transcript, and its relationship to the DNA encoding it, can be done by comparative sequence analysis of the gene and the cDNA generated from the transcript. This cDNA can be detected by polymerase chain reaction (PCR) amplification of even extremely rare transcripts; therefore, a detailed picture of the splicing patterns of very-low-abundance mRNAs is technically quite feasible. An example of the generation of a PCR-amplified cDNA from a low-abundance latency-associated transcript of HSV is shown in Figure 13.10b. Latent-phase transcription by herpesviruses is discussed in more detail in Chapter 17, Part IV.

Posttranscriptional regulation of eukaryotic mRNA function

The many steps in the biogenesis of eukaryotic mRNA are outlined in Figure 13.11; each is subject to control. Thus, chromatin structure controls the availability of the transcription template, while the interaction between regulatory sequences within the overall transcription unit and regulatory proteins activates transcription. The processing and transport of mRNA in the nucleus are also subject to control, as is the ability of mRNA to associate with ribosomes in the cytoplasm. Finally, as noted briefly in Part II, Chapter 8, small cellular RNAs (microRNAs) with double-stranded regions and double-stranded RNA expressed by aberrant transcription from exogenous sources such as viruses are cleaved by a cellular enzyme complex termed a **dicer** into short 20–30-base sequences, which can interact with complementary sequences of mRNA, leading to blockage of protein synthesis and mRNA degradation.

Virus-induced changes in transcription and posttranscriptional processing

Many RNA-containing viruses completely shut down host transcription. Specific details are described in Chapters 14 and 15. The ability of DNA viruses to transcribe predominantly viral transcripts is usually a multistep process with the earliest transcripts encoding genes that serve

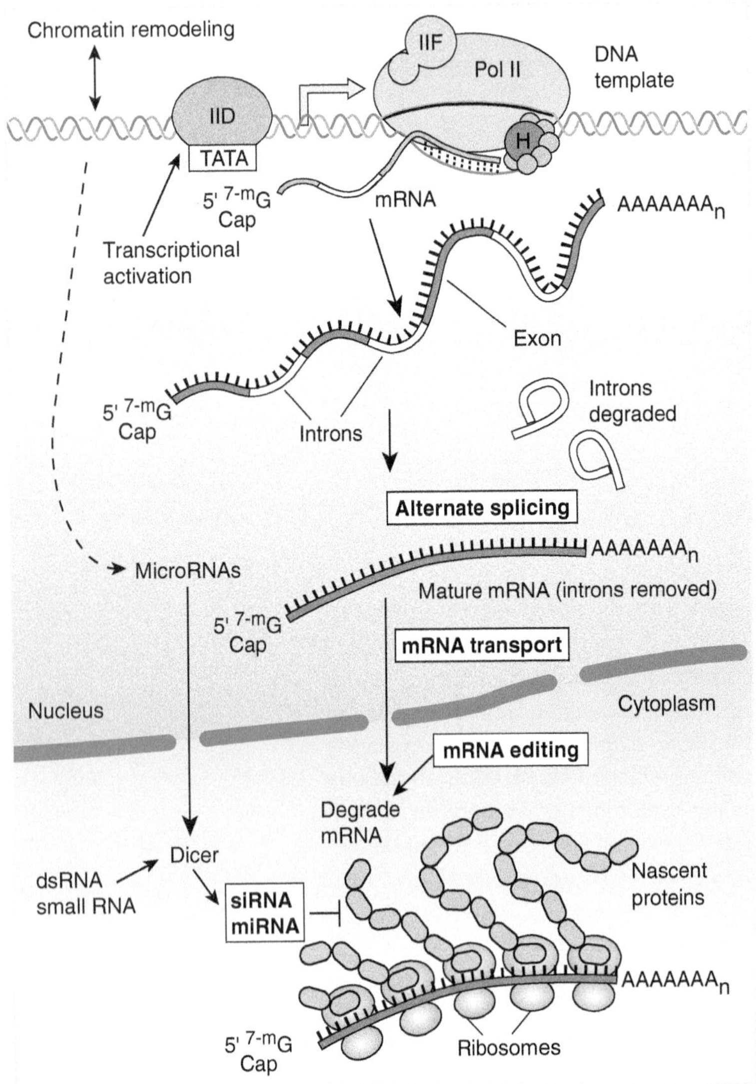

Figure 13.11 Posttranscriptional regulation of eukaryotic mRNA. Once transcribed, eukaryotic mRNA must be processed by splicing, and transported to the nucleus. Both of these processes can be regulated. In addition to modulation of the precise nature of mRNA sequence being expressed from a transcription unit, and the rate at which it is released to the nucleus, the small double-stranded RNAs either expressed in the cell or introduced by another process such as viral replication can be processed into very small RNA sequences called microRNA (miRNA) that bind to complementary mRNA sequences, leading to inhibition of protein synthesis and degradation of the mRNA. Transcribed mRNA can also be edited in the cytoplasm, leading to alterations in sequence. These processes are marked by asterisks in the figure.

regulatory functions, causing expression of viral genes to be favored. The mechanism of transcriptional shutoff varies among viruses. With nuclear-replicating DNA viruses, the process often involves the earliest transcripts being expressed from a viral promoter that has a powerful enhancer, allowing active transcription without extensive modification of the cell. This is followed by changes in the structure of cellular chromatin and increases in viral genomes so that viral transcription templates begin to predominate relatively rapidly. A major factor in usurpation of the cell's transcriptional capacity by these viruses is the fact that, in general, the uninfected cell has much more transcriptional capacity than it is using at the time of infection. Consequently, increases in the availability of viral templates, along with alterations of the host chromatin structure, result in virus-specific transcription predominating.

Some nuclear-replicating viruses also encode regulatory proteins that affect posttranscriptional splicing and transport of transcripts from the nucleus to the cytoplasm. Such alterations in splicing do not affect the basic mechanism of splicing, but can specifically inhibit the generation and transport of spliced mRNA at specific times following infection. This inhibition involves the ability of viral proteins to recognize and modify the activity of spliceosomes. The alteration of splicing and transport of mRNA has especially important roles in aspects of the control of herpesvirus gene expression and in the regulation of gene expression and viral genome production in lentivirus (retrovirus) infections. Specifics are described in Chapters 17 and 20.

While no nuclear-replicating DNA viruses of vertebrates that have been characterized encode virus-specific RNA polymerase, at least one – baculovirus, which replicates in insects – does. Further, this is a very common feature in DNA-containing bacteriophages. Indeed, as outlined in Chapter 18, changes in the infected bacteria's polymerase population is the major mechanism for ensuring virus-specific RNA synthesis and the change in types of viral mRNA expressed at different times after infection. This is also seen in the replication of the eukaryotic poxviruses, which replicate in the cytoplasm of the infected host cell, and thus do not have access to cellular transcription machinery (see Chapter 18).

One other posttranscriptional modification, **RNA editing**, has been observed in the replication of some viruses. RNA editing is an enzymatic process that is commonly seen in the biogenesis of mitochondrial mRNAs. One form of editing is found in the replication of hepatitis delta virus (see Chapter 15, Part IV). This editing reaction results in the deamination of an adenosine base in the viral mRNA and its conversion to a guanosine, which leads to alteration of a translation signal and expression of a larger protein than is expressed from the unmodified transcript. A second form of RNA editing that occurs as the RNA is expressed is the addition of extra bases to regions of the RNA. This is seen in the replication of some paramyxoviruses and in Ebola virus.

THE MECHANISM OF PROTEIN SYNTHESIS

Like transcription, the process of protein synthesis is similar in broad outline in prokaryotes and eukaryotes; however, there are significant differences in detail. Some of these differences have important implications in the strategies that viruses must use to regulate gene expression. Viruses use the machinery of the cell for the translation of proteins, and to date, no virus has been characterized that encodes ribosomal proteins or rRNA. However, some viruses do modify ribosome-associated translation factors to ensure expression of their own proteins. A notable example of such a modification is found in the replication cycle of poliovirus described in Chapter 14.

Eukaryotic translation

In a nucleated cell, processed mRNA must be transported from the nucleus. The mRNA does not exist as a free RNA molecule, but is loosely or closely associated with one or several

RNA-binding proteins that carry out the transport process and may facilitate initial association with the eukaryotic ribosome. This provides yet another point in the flow of information from gene to protein that is subject to modulation or control, and thus is potentially available for viral-encoded mediation.

The features of translational initiation in eukaryotic cells reflect the nature of eukaryotic mRNAs, namely, that they have 5′-methylated caps, that they are translated as monocistronic (single-gene) species, and that ribosomes usually do not bind to the messages at internal sites. Initiation involves assembly of the large (60S) and small (40S) subunits of the ribosome along with the initiator tRNA (met-tRNA in most cases) at the correct AUG codon. These steps require the action of several protein factors along with energy provided by ATP and guanosine triphosphate (GTP) hydrolysis. The process is shown in Figure 13.12.

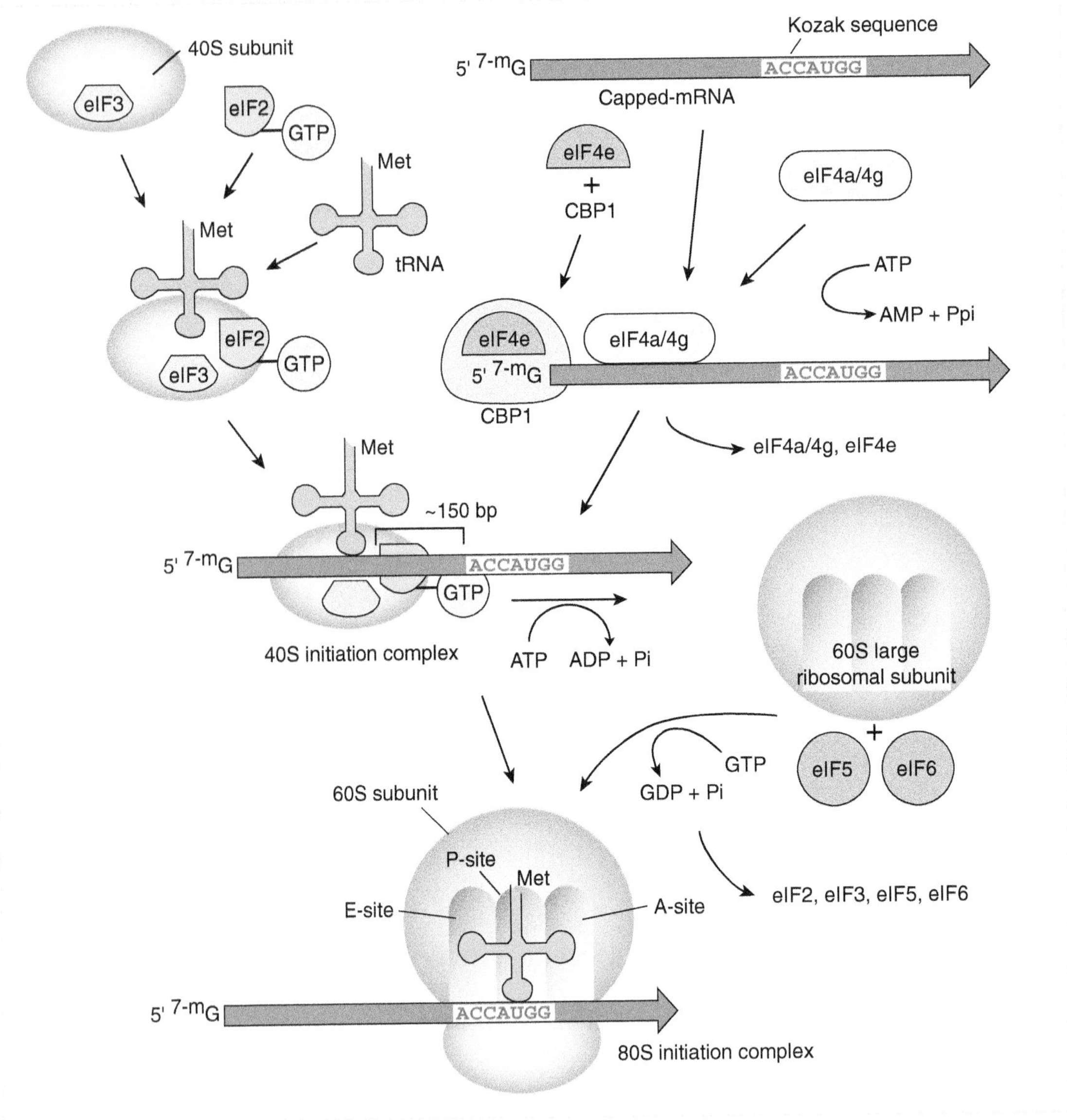

Figure 13.12 Initiation of eukaryotic translation. Note the initiation complex contains the 40S ribosomal subunit and must interact with the 5′ end of the mRNA molecule via the cap structure or an equivalent. The 60S subunit only becomes associated with the complex at the Kozak (or equivalent) sequence. The ribosome dissociates back into the two subunits at the termination of translation. This means that internal initiation, especially if an upstream open reading frame has been translated, is impossible or at least extremely rare. Pi: Inorganic phosphate; PPi: pyrophosphate.

The first phase of this process involves association of the 40S subunit with met-tRNA and is carried out by three **eukaryotic translation initiation factors** (eIF-2, eIF-3, and eIF-4C) along with GTP. This complex then binds to the 5′-methylated cap of the mRNA through the action of eIF-4A, eIF-4B, eIF-4F, and **CBP1 (cap-binding protein)**, requiring the energy of ATP hydrolysis.

The 40S–tRNA complex then moves in the 5′ to 3′ direction along the mRNA, scanning the sequence for the appropriate AUG that is found within a consensus sequence context (the **Kozak sequence**). Movement of the complex requires energy in the form of ATP. Finally, the 60S subunit joins the assembly through the activity of eIF-5 and eIF-6, GTP is hydrolyzed, all of the initiation factors are released, and the ribosome–mRNA complex (now called the 80S initiation complex) is ready for elongation.

The new peptide "grows" from N-terminal to C-terminal and reads the mRNA 5′ to 3′. Translation proceeds to the C-terminal amino acid of the nascent peptide chain, the codon of which is followed by a translation termination codon (UAA, UGA, or UAG). The sequence of bases, starting with the initiation codon, containing all the amino acid codons, and finishing with the three-base termination codon, defines a **translational open reading frame (ORF)**. In mature mRNA, any ORF will have a number of bases evenly divisible by three, but an ORF may be interrupted by introns in the gene encoding the mRNA.

Several ORFs can occur or overlap in the same region of mRNA, especially in viral genomes. Overlapping ORFs can be generated by splicing or by AUG initiation codons being separated by a number of bases not divisible by three. An example might be as follows (where lowercase bases represent those not forming codons):

5'–...AUGAAAUGGCCAUUUUAACGA...–3'

Translated in "frame 1," the sequence would be read as:

AUG AAA UGG CCA UUU UAA

but in "frame 3," it would be read as:

augaa AUG GCC AUU UUA A

In such an mRNA, ribosomes might start at one or the other translational reading frame, *but a given ribosome can only initiate translation at a single ORF*. Thus, if the ribosome initiates translation of, say, the second ORF, this is because it has missed the start of the first one; and if it has started at the first one, it cannot read any others on the mRNA. In other words, when a eukaryotic ribosome initiates translation at an ORF, it continues until a termination signal is encountered. Translation termination results in the ribosome falling off the mRNA strand, and any other potential translational reading frames downstream of the one terminated are essentially unreadable by the ribosome.

This simply means that a eukaryotic mRNA molecule containing multiple translational reading frames in sequence will not be able to express any beyond the first one translated (or, possibly two, if the ribosome has a "choice") from the 5′ end of the transcript. Any ORFs downstream of these are considered hidden or **cryptic ORFs**. This property of eukaryotic translation has important implications both in the effect of splicing on revealing "cryptic" or hidden translational reading frames, and in the generation of some eukaryotic viral mRNAs.

Prokaryotic translation

Prokaryotic messages have three structural features that differ from eukaryotic versions. First, mRNA is not capped and methylated at the 5′ end. Second, mRNA may be translated into more than one protein from different coding sequences and is, thus, **polycistronic (multigene) mRNA**. Finally, ribosome attachment to mRNA in prokaryotes occurs at internal sites rather than at the 5′ end. In addition, prokaryotic mRNAs are transcribed and translated at the same time and in the same place in the cell (**coupled transcription/translation**).

Features of prokaryotic translation reflect these structural and functional differences. Initiation, shown in Figure 13.13, begins with the association of **initiator tRNA** (*N*-formylmethionine-tRNA, or F-met-tRNA) with the small (30S) ribosomal subunit, together with mRNA through the action of three factors (initiation factor 1 [IF-1], IF-2, and IF-3), along with GTP. The complex that forms involves direct binding of the 30S subunit with its F-met-tRNA to the AUG that initiates translation of the ORF.

This AUG is defined by the presence of a series of bases (called the **Shine–Dalgarno sequence**) in the mRNA upstream from the start codon that is complementary to the 3′ end of the 16S rRNA in the 30S ribosomal subunit. The large (50S) ribosomal subunit now binds, accompanied by GTP hydrolysis and release of factors, to form the 70S initiation complex. From this point, elongation and termination occur in much the same manner as seen in eukaryotic cells.

Virus-induced changes in translation

Many viruses specifically alter or inhibit host cell protein synthesis. The ways they accomplish this vary greatly, and are described in Part IV where the replication cycle of specific viruses is covered

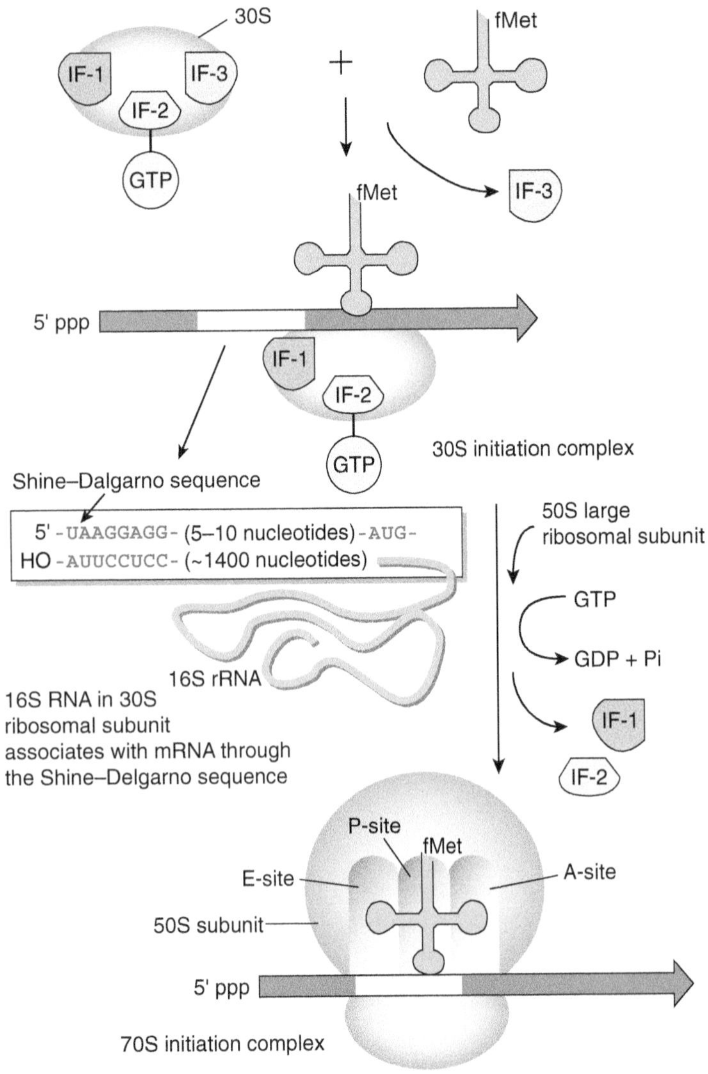

Figure 13.13 Initiation of translation of a prokaryotic mRNA. This can occur anywhere there is a Shine–Dalgarno sequence in the mRNA since the 30S ribosome associates with the mRNA at that site by virtue of the presence of a complementary sequence in the 3′ end of the ribosomal RNA. fMet: Formylmethionine.

in detail. Some viruses, notably retroviruses and some RNA viruses, can also *suppress* the termination of translation at a specific stop codon. The mechanism for such suppression may involve the ribosome actually skipping or jumping a base at the termination signal. When this happens, the translational reading frame being translated is shifted by a base or two. Other modes of suppression are not so well characterized, but all involve the mRNA at the site of suppression having a unique structure that facilitates it. This suppression is not absolute but occurs with either high or low frequency, resulting in a single mRNA being able to encode multiple, related proteins.

QUESTIONS FOR CHAPTER 13

1 A given mRNA molecule has the following structure. What is the maximum number of amino acids that the final protein product could contain?

Cap - 300 bases - AUG - 2097 bases - UAA - 20 bases - AAAA

2 Assume the following sequence of bases occurs in an open reading frame (ORF) whose reading frame is indicated by grouping the capitalized bases three at a time:

...AUG...(300 bases)...CGC AAU ACA
UGC CCU ACC AUG AAU AAU ACC UAA
gguaaaug...

What effect might deletion of the fourth A in the above strand of mRNA have on the size of a protein encoded by this ORF?

3 Both prokaryotic and eukaryotic cells transcribe mRNA from DNA and translate these mRNAs into proteins. However, there are differences between the two kinds of cells in the manner in which mRNAs are produced and utilized to program translation. In the table below, indicate which of the features applies to which kind of mRNA. Write "Yes" if the feature is true for that kind of mRNA or "No" if it is not true.

Feature	Eukaryotic mRNA	Prokaryotic mRNA
The small ribosomal subunit is correctly oriented to begin translation by association with the Shine–Dalgarno sequence.		
Open reading frames generally begin with an AUG codon.		
The 5′ end of the mRNA has a methylated cap structure covalently attached after transcription.		
During protein synthesis, an open reading frame can be translated by more than one ribosome, forming a polyribosome.		
Termination of transcription may occur at a site characterized by the formation of a GC-rich stem loop structure just upstream from a U-rich sequence.		

4 Which of the following statements is/are true regarding the primer for most DNA replication?
(a) It is degraded by an exonuclease.
(b) It is made up of ribonucleic acid.
(c) It is synthesized by a primase.

5 All of the following are characteristics of *eukaryotic* mRNA, except:
(a) A 5′-methylated guanine cap.
(b) Polycistronic translation.
(c) A polyadenylated 3′ tail.
(d) Nuclear splicing of most mRNAs.
(e) The use of AUGs instead of ATGs.

6 In what cellular location would one find viral glycoproteins being translated?

7 What is the *minimum size* of a viral mRNA encoding a structural protein of 1100 amino acids?

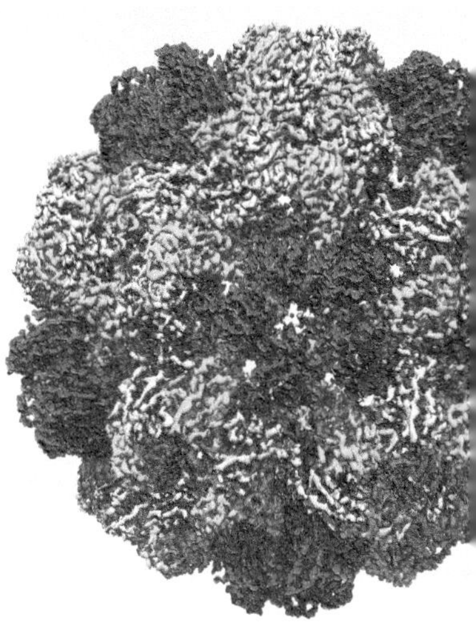

Problems

PART III

1 The drug acycloguanosine, sold as acyclovir, has been one of the most successful antiviral compounds produced. Acyclovir is used in the treatment of herpes simplex virus infections. These viruses replicate their double-stranded DNA genomes using a virus-specific DNA polymerase. The structure of acyclovir is shown here:

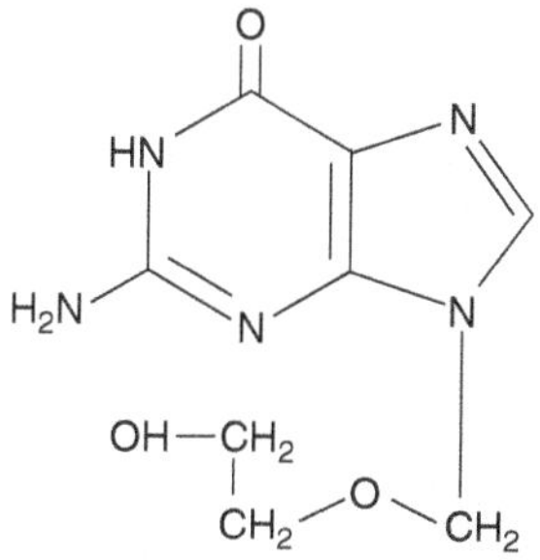

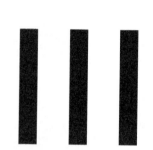

(a) Given the structure of this drug, what is the specific effect of this nucleoside analogue on herpesvirus DNA replication? Your answer should refer to a particular structural feature of the drug.
(b) Acyclovir is administered to patients in the form shown. What must happen to this drug inside the cell before it can inhibit viral DNA replication? Again, your answer must refer to a particular structural feature of the drug.

2 You have prepared two highly purified suspensions of poliovirus, each grown on a different host cell. The first (stock A) was grown on HeLa cells (a human cell line). The second (stock B) was grown on AGMK cells (a monkey kidney cell line). Each suspension has a total volume of 10.0 ml. The virus stocks were titered by diluting the stocks and performing plaque assays. Dilutions were made by taking 1 part of the virus stock and mixing it with 9999 parts of buffer (Dilution 1). A further 1 to 10^4 dilution was made, using this same procedure. The resulting suspension (Dilution 2) was plaque assayed in duplicate on a lawn of susceptible cells. In addition, you measured the optical density of the original virus stocks at a wavelength of 260 nm. You know that an optical density (OD_{260}) of 1.0 equals 10^{13} poliovirus particles/ml. The data you obtained are shown in the following table:

Virus Stock	Host Cell	Plaques in 1 ml of Dilution 2 (Replicate Plates)	OD_{260} of Virus Stock
A	HeLa	190 and 210	0.1
B	AGMK	48 and 52	0.5

(a) What are the plaque-forming unit (PFU)-to-particle ratios for the two viral stocks?

Stock A: ________

Stock B: ________

(b) Which host cell line produced the most *total virus particles?*

(c) Which host cell line produced the most *total infectious virus?*

3 The Svedberg equation, which describes the motion of a molecule through a solution under the influence of a centrifugal field, is:

$$S = \frac{v}{\omega^2 r} = \frac{M\left(1-\bar{n}r_{sol}\right)}{N_{av} f}$$

where S, the Svedberg coefficient, is a function of the molecular weight (M) and the frictional coefficient (f). The constants in the equation are Avogadro's number (N_{av}), the partial specific volume of the molecule (ν), and the density of the solution (ρ_{sol}).

This table gives some relevant data for several DNA molecules:

DNA	Molecular Weight	Configuration[a]
pBR322 DNA	2.84×10^6	DS, circular[b]
pBR322 DNA digested with *Eco*RI[c]	2.84×10^6	DS, linear
Phage T4 DNA	1.12×10^8	DS, linear
Phage T7 DNA	2.5×10^7	DS, linear
Phage ΦX174 RF DNA	3.76×10^6	DS, circular

DS: Double-stranded; RF: replicative form.

[a]Assume that $f_{ds,linear} >> f_{ds,circular}$.

[b]Assume the same degree of supercoiling for all of the circular molecules.

[c]pBR322 has only one recognition site of EcoRI.

Predict the sedimentation behavior (sedimentation rate) of the following pairs of molecules. In each case, state whether the indicated molecule of the pair will move "faster" or "slower" relative to the other member of the pair. (Note: You do not need to calculate a sedimentation rate. You need to determine the relative behavior of the pair of molecules in each case.)

Molecules		Relative Sedimentation Rate	
1	2	1	2
Phage T4 DNA	Phage T7 DNA		
pBR322 DNA	Phage ΦX174 RF DNA		
pBR322 DNA	pBR322 DNA digested with *Eco*RI		
Phage T7 DNA	pBR322 DNA digested with *Eco*RI		

RF: Replicative form.

Additional Reading for Part III

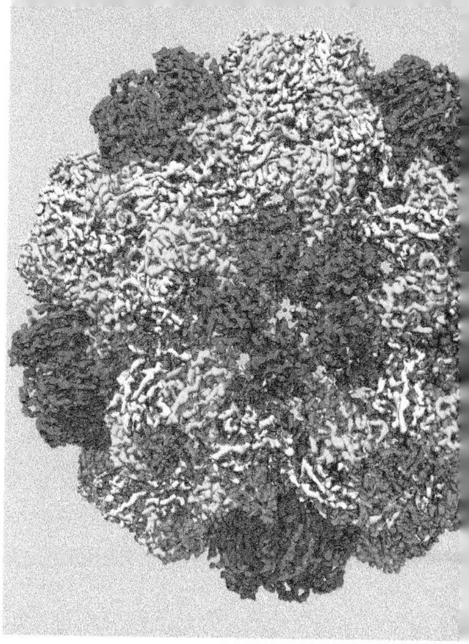

Note: See Resource Center for relevant websites.

Ausubel, F.M., Brent, R., and Kingston, R.E. (eds.) (1994–1999). *Current Protocols in Molecular Biology*. New York: Wiley.

Celis, A. and Celis, J.E. (1998). General procedures for tissue culture. In: *Cell Biology: A Laboratory Handbook*, 2e, vol. 1 (ed. J. Celis), 5–15. San Diego, CA: Academic Press.

Celis, A., Dejgaard, K., and Celis, J.E. (1998). Production of mouse monoclonal antibodies. In: *Cell Biology: A Laboratory Handbook*, 2e, vol. 2 (ed. J. Celis), 392–397. San Diego, CA: Academic Press.

Celis, J.E. and Olsen, E. (1998). One-dimensional sodium dodecyl sulfate–polyacrylamide gel electrophoresis. In: *Cell Biology: A Laboratory Handbook*, 2e, vol. 4 (ed. J. Celis), 361–370. San Diego, CA: Academic Press.

Cristofalo, V.J., Charpentier, R., and Philips, P.D. (1998). Serial propagation of human fibroblasts for the study of aging at the cellular level. In: *Cell Biology: A Laboratory Handbook*, 2e, vol. 1 (ed. J. Celis), 321–326. San Diego, CA: Academic Press.

Janeway, C., Travers, P., Hunt, S., and Walport, M. (1997). *Immunobiology*, 3e, chap. 2. New York: Garland.

Kroes, A. and Kox, L. (1999). Detection of viral antigens, nucleic acids and specific antibodies. In: *Encyclopedia of Virology*, 2e (eds. R.G. Webster and A. Granoff), 388–395. New York: Academic Press.

Landry, M.L. and Hsiung, G.-D. (1999). Isolation and identification by culture and microscopy. In: *Encyclopedia of Virology*, 2e (eds. R.G. Webster and A. Granoff), 395–403. New York: Academic Press.

Leland, D.S. (1996). Concepts of immunoserological and molecular techniques. In: *Clinical Virology*, 21–50. Philadelphia: W.B. Saunders.

Leland, D.S. (1996). Virus isolation in traditional cell cultures. In: *Clinical Virology*, 51–78. Philadelphia: W.B. Saunders.

Maunsbach, A.B. (1998). Fixation of cells and tissues for transmission electron microscopy. In: *Cell Biology: A Laboratory Handbook*, 2e, vol. 2 (ed. J. Celis), 249–259. San Diego, CA: Academic Press.

Osborn, M. (1998). Immunofluorescence microscopy of cultured cells. In: *Cell Biology: A Laboratory Handbook*, 2e, vol. 2 (ed. J. Celis), 462–468. San Diego, CA: Academic Press.

Pawley, J.B. and Centonze, V.E. (1998). Practical laser-scanning confocal light microscopy: obtaining optimal performance from your instrument. In: *Cell Biology: A Laboratory Handbook*, 2e, vol. 3 (ed. J. Celis), 149–169. San Diego, CA: Academic Press.

Whitaker-Dowling, P. and Youngner, J.S. (1994). Virus-host cell interactions. In: *Encyclopedia of Virology* (eds. R.G. Webster and A. Granoff), 1587–1591. New York: Academic Press.

VOLUME 1

Section 1. *Escherichia coli*, plasmids, and bacteriophages: Part II. Vectors derived from plasmids.

Section 2. Preparation and analysis of DNA: Part IV. Analysis of DNA sequences by blotting and hybridization.

Section 3. Enzymatic manipulation of DNA and RNA:

Part I. Restriction endonucleases.

Part II. Enzymatic manipulation of DNA and RNA: restriction mapping.

Section 4. Preparation and analysis of RNA: Part 4. Analysis of RNA structure and synthesis.

Section 7. DNA sequencing: Part I. DNA sequencing strategies.

Section 9. Introduction of DNA into mammalian cells: Part I. Transfection of DNA into eukaryotic cells.

VOLUME 2

Section 10. Analysis of proteins: Part III. Detection of proteins:

Subsection 10.7. Detection of proteins on blot transfer membranes.

Subsection 10.8. Immunoblotting and immunodetection.

Section 11. Immunology: Part I. Immunoassays:

Subsection 11.2. Enzyme-linked immunosorbent assay (ELISA).

Section 14. *In situ* hybridization and immunohistochemistry:

Subsection 14.3. *In situ* hybridization to cellular RNA.

Section 15. The polymerase chain reaction:

Subsection 15.1. Enzymatic amplification of DNA by the polymerase chain reaction: standard procedures and optimization.

Alberts, B., Johnson, A., and Lewis, J. (2002). *Molecular Biology of the Cell*, 4e. New York: Garland. The following chapters and pages: Chapter 5: 238–266. Basic genetic mechanisms: DNA replication. Chapter 6: 300–372. Basic genetic mechanisms: RNA and protein synthesis. Chapter 8: 491–513. Recombinant DNA technology: the fragmentation, separation, and sequencing of DNA molecules. Chapter 8: 495–500. Recombinant DNA technology: nucleic acid hybridization.

Lodish, H., Baltimore, D., Berk, A. et al. (1995). Transcription termination, RNA processing, and post-transcriptional control: mRNA processing in higher eucaryotes. In: *Molecular Cell Biology*, 3e, chap. 12. New York: Scientific American.

Lewin, B. (1997). *Genes VI*. New York: Oxford Press. The following chapters and pages: Chapter 6: 117–134. Isolating the gene: a restriction map is constructed by cleaving DNA into specific fragments. Restriction sites can be used as genetic markers. Obtaining the sequence of DNA. Chapter 7: 153–178. Messenger RNA. Chapter 8: 179–212. Protein synthesis.

Hartl, D.L. and Jones, E.W. (1999). Mutation, DNA repair, and recombination. In: *Essential Genetics*, 2e, 234–263. Boston: Jones and Bartlett.

Davis, R.H. and Weller, S.G. (1997). The mutational process. In: *The Gist of Genetics*, 133–142. Boston: Jones and Bartlett.

Coen, D.M. and Ramig, R.F. (1995). Viral genetics. In: *Virology*, 3e (eds. B.N. Fields and D.M. Knipe), chap. 5. New York: Raven Press.

Replication Patterns of Specific Viruses

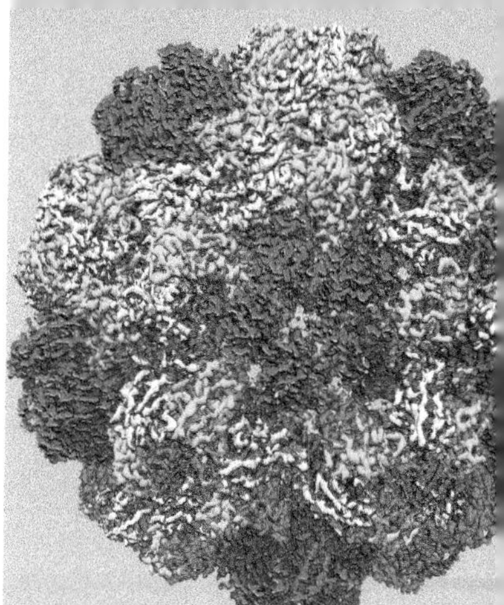

PART IV

Replication of Positive-Sense RNA Viruses

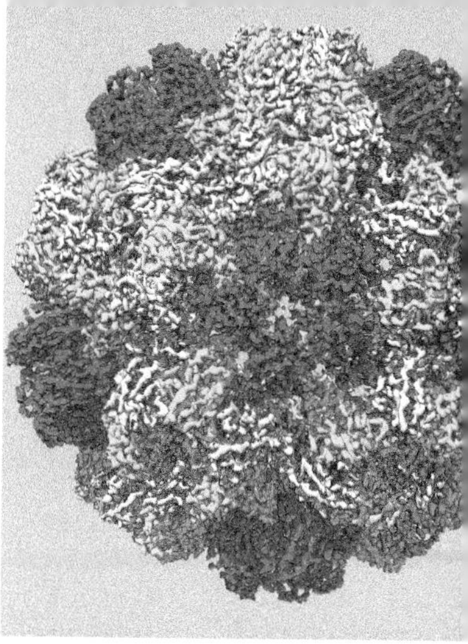

CHAPTER

14

Basic Virology, Fourth Edition. Martinez "Marty" Hewlett, David Camerini, and David C. Bloom.

RNA VIRUSES – GENERAL CONSIDERATIONS

By definition, RNA viruses use RNA as genetic material, and thus must use some relatively subtle strategies to replicate in a cell since the cell uses DNA. Ultimately, to express its genetic information, any virus must be able to present genetic information to the cell as translatable messenger RNA (mRNA), but the way this happens with RNA viruses will depend on the type of virus and the nature of the encapsidated RNA.

According to Watson–Crick base-pairing rules, once the sequence of one strand of either RNA or DNA is known, the sequence of its complementary strand can be inferred. The complementary strand serves as a template for synthesis of the strand of RNA or DNA in question. While the sequence of a strand of RNA is in a sense equivalent to its complement, the actual "sense" of the information encoded in the virion RNA is important for understanding how the virus replicates. As noted in Chapter 1, viral mRNA is the obligate first step in the generation of viral protein; therefore, an RNA virus must be able to generate something that looks to the cell like mRNA before its genome can be replicated.

The ways that viruses, especially RNA viruses, express their genomes as mRNA are limited by necessity and form an important basis of classification. The use of this criteria in the Baltimore classification of viruses was outlined in Chapter 5, Part II. The fundamental basis of this classification for RNA viruses is whether the viral genome can be directly utilized as mRNA or whether it must first be transcribed into mRNA. This classification breaks RNA viruses that do not utilize a DNA intermediate (an important exception) into two basic groups: the viruses containing mRNA as their genomes and those that do not. This second group, which comprises the viruses encapsidating an RNA genome that is complementary (antisense) to mRNA and the viruses that encapsidate a double-stranded RNA (dsRNA) genome, requires the action of a specific viral-encoded transcriptase. Such viral transcriptases are contained in the virion as a structural protein, and utilize the virion genomic RNA as a template for transcription.

The basic strategy for the initiation of infection by these two groups of viruses, members of which are described in some detail in this chapter and Chapter 15, is outlined in Figure 14.1a. This classification ignores a very significant complication: It makes no accommodation for the fact that a very important group of viruses with genomes that can serve as mRNA use DNA as the intermediate in their replication. These are the retroviruses. These viruses and their relatives use a very complex pattern of viral-encoded and cellular functions in their replication, and are described in Chapter 19 only after a full survey of the "simpler" RNA and DNA viruses is presented.

A general picture of RNA-directed RNA replication

With the exception of retroviruses and some unusual viruses related to viroids, single-stranded RNA (ssRNA) virus genome replication requires two stages; these are shown in Figure 14.1b. First, the input strand must be transcribed (using Watson–Crick base-pairing rules) into a strand of complementary sequence and **opposite polarity**. Replication occurs as a "fuzzy," multi-branched structure. This complex, dynamic structure contains molecules of viral transcriptase (replicase), a number of partially synthesized product RNA strands ("nascent" strands), and the genomic-sense template strand. The whole **ribonucleoprotein** (RNP) complex is termed the type 1 **replicative intermediate** or **RI-1**. The single-stranded products generated from RI-1 are antisense to the genomic RNA.

This complementary strand RNA serves as a template for the formation of more genomic-sense RNA strands. This second replicative intermediate (**RI-2**) is essentially the same in structure as RI-1 except that the template strand is of opposite sense to genomic RNA and the nascent product RNA molecules are of genome sense.

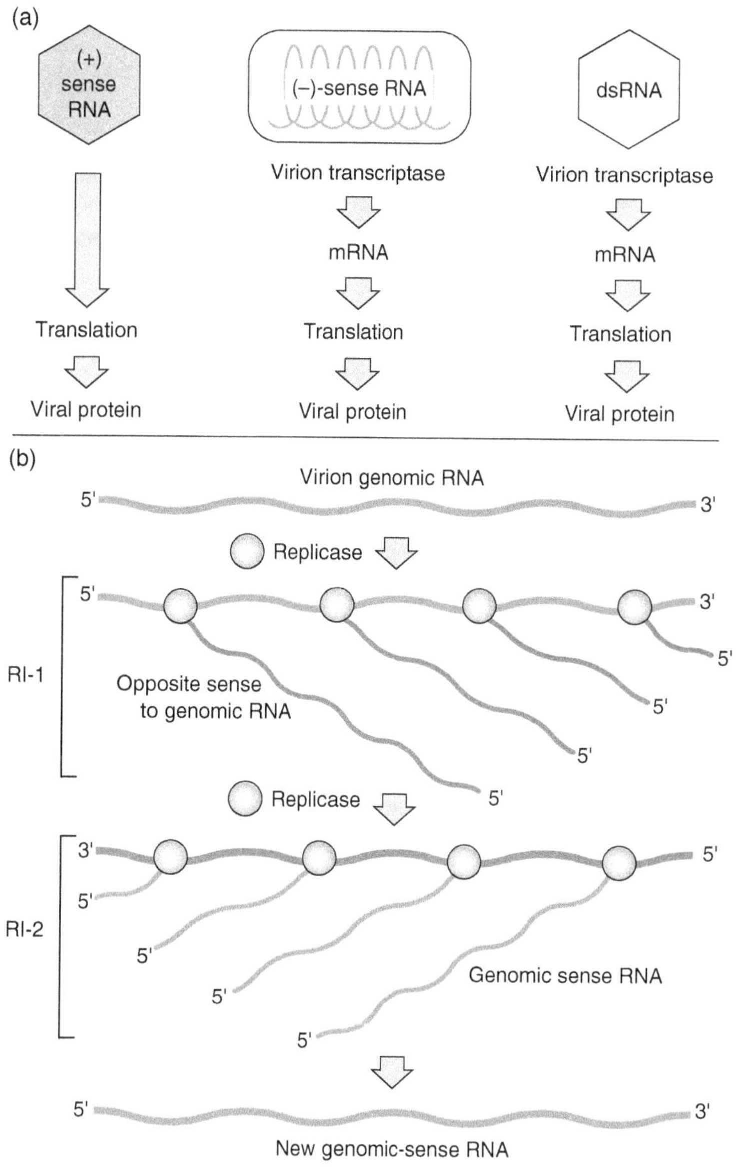

Figure 14.1 Some general features of viruses containing RNA genomes that use RNA-directed RNA transcription in their replication. (a) The general relationship between viruses containing a genome that can be translated as the first step in the expression of viral genes versus those viruses that first must carry out transcription of their genome into mRNA utilizing a virion-associated transcriptase. (b) The basic rules for RNA-directed RNA replication. As with DNA-directed RNA and DNA synthesis, the new (nascent) strand is synthesized 5′ to 3′ antiparallel to the template, and the Watson–Crick base-pairing rules are the same, with U substituting for T. However, the very high thermal stability of double-stranded RNA (dsRNA) leads to complications. The major complication is that newly synthesized RNA must be denatured and removed from the template strand to avoid its "collapsing" into a double-stranded form. Formation of such dsRNA is an effective inducer of interferon (see Chapter 8, Part II), and it appears to be refractory to serving as a template when free in the cytoplasm. A second complication is that in order to generate a single-stranded RNA (ssRNA) molecule of the same coding sense as the virion genome, *two* replicative intermediates (RIs) must be generated. These intermediates are dynamic structures of ribonucleoprotein containing a full length template strand, and a number of newly synthesized product RNA molecules growing from virion-encoded replicase that is traversing the template strand. RI-1 generates RNA complementary to the virion genomic RNA. This serves as a template for new virion genome RNA in RI-2.

Remember:

Virion RNA is the template in RI-1.
RI-1 produces template RNA of opposite sense to virion RNA.
RNA that is complementary to virion RNA is the template in RI-2.
RI-2 is the intermediate for expression of RNA of the same sense as the virion.

One further general feature of the replication of RNA viruses is worth noting. The **error frequency** (i.e., the frequency of incorporating an incorrect base) of RNA-directed RNA replication is quite high compared to that for dsDNA replication. Thus, typically DNA-directed DNA replication leads to incorporation of one mismatched base per 10^7–10^9 base pairs, while RNA-directed RNA synthesis typically results in one error per 10^5 bases. Indeed, the error rate in the replication of some RNA genomes can be as high as one error per 10^4 nucleotides.

Part of the reason for this error rate for RNA is that there is no truly double-stranded intermediate; therefore, there is no template for error correction or "proofreading" of the newly synthesized strand as there is in DNA replication. A second reason is that RNA polymerases using RNA templates seem to have an inherently higher error frequency than those utilizing DNA as a template.

For these reasons, infection of cells with many RNA viruses is characterized by the generation of a large number of progeny virions bearing a few or a large number of genetic differences from their parents. This high rate of mutation can have a significant role in viral pathogenesis and evolution; further, it provides the mechanistic basis for the generation of defective virus particles described in Chapter 21. Indeed, many RNA viruses are so genetically plastic that the term **quasi-species swarm** is applied to virus stocks generated from a single infectious event, as any particular isolate will be, potentially at least, genetically significantly different from the parental virus. The concept of a quasi-species, as applied to virus populations, has been important for the application of evolutionary models to such populations. As a result, the analysis of mutational changes over time can employ the models that are used in population genetics.

REPLICATION OF POSITIVE-SENSE RNA VIRUSES WHOSE GENOMES ARE TRANSLATED AS THE FIRST STEP IN GENE EXPRESSION

The first step in the infectious cycle of this group of positive-sense RNA viruses (also called **positive [+] strand viruses**) leading to expression of viral proteins is *translation of viral protein*. If the virion (genomic) RNA is incubated with ribosomes, transfer RNA (tRNA), amino acids, adenosine triphosphate (ATP), guanosine triphosphate (GTP), and the other components of an *in vitro* protein synthesis system, protein will be synthesized.

Further, if virion RNA is transfected into the cell in the absence of any other viral protein, infection will proceed and new virus will be produced. This can occur in the laboratory provided there are proper precautions to protect the viral RNA, which is chemically labile.

Positive-sense RNA viruses (other than retroviruses) do not require a transcription step prior to expression of viral protein. This means that the nucleus of a eukaryotic cell is either somewhat or completely superfluous to the infection process. All the replication steps can take place more or less efficiently in a cell from which the nucleus is removed.

For instance, removal of the nucleus can be accomplished in poliovirus infections by use of a drug, **cytochalasin B**, which breaks down the actin-fiber cytoskeleton that anchors the nucleus inside the cell. Cells treated with this drug can be subjected to mild centrifugal force, causing the nucleus to "pop" out of the cell. Such enucleated cells can be infected with poliovirus, and new virus can be synthesized at levels equivalent to those produced in normal nucleated cells.

A very large number of positive-sense RNA viruses can infect bacteria, animals, and especially plants, and the patterns of their replication bear strong similarities. The replication patterns of the positive-sense RNA important to human health can be outlined by consideration of just a few, if the replication of retroviruses is considered separately.

A basic distinction between groups of positive-sense RNA viruses involves whether the viral genome contains a single open translational reading frame (**ORF**) as defined in Chapter 13, Part III, or multiple ones. This difference correlates with the complexity of mRNA species expressed during infection.

POSITIVE-SENSE RNA VIRUSES ENCODING A SINGLE LARGE OPEN READING FRAME

Picornavirus replication

Picornaviruses are genetically simple and have been the subject of extensive experimental investigation owing to the number of diseases they cause. Their name is based on a pseudoclassical use of Latin mixed with modern terminology: *pico* ("small")-RNA-virus.

The replication of poliovirus (the best-characterized picornavirus, and perhaps the best-characterized animal virus) provides a basic model for RNA virus replication. Studies on poliovirus were initiated because of the drive to develop a useful vaccine against paralytic poliomyelitis. These studies successfully culminated in the late 1950s and early 1960s. Protocols developed for replicating the virus in cultured cells formed the basis for successful vaccine development and production. At the same time, the relative ease of maintaining the virus and replicating it in culture led to its early exploitation for molecular biological studies. It is still a favored model.

Other closely related picornaviruses include rhinoviruses and hepatitis A virus. These replicate in a generally similar manner, as do a number of positive-sense RNA–containing bacterial and plant viruses. Indeed, close genetic relationships among many of these viruses are well established.

The poliovirus genetic map and expression of poliovirus proteins

A schematic of the icosahedral poliovirus virion is shown in Figure 14.2. In accordance with its classification as a positive-sense RNA virus, the poliovirus genomic RNA isolated from purified virions is mRNA sense and acts as a viral mRNA upon infection. Full characterization and sequence analysis established that the genome is 7741 bases long with a very long (743-base) leader sequence between the 5′ end of the mRNA and the (ninth!) AUG, which initiates the beginning of an ORF extending to a translation termination signal near the 3′ end. There is a short untranslated trailer following the 7000-base ORF, and this is followed by a polyA tract. The polyA tail of the poliovirus mRNA is actually part of the viral genome; therefore, it is not added posttranscriptionally, as with cellular mRNA (see Chapter 13). A simple genetic map of the viral genome is shown in Figure 14.2a.

The entire genome of poliovirus was assembled from oligodeoxynucleotides as a double-stranded complementary DNA (cDNA) molecule and subsequently transcribed by RNA polymerase into infectious RNA. The experiment, reported in 2002, raised some issues of security with respect to possible bioterrorist implications of this work. In fact, the synthesis of infectious influenza virus, using preserved tissue material from the extremely virulent 1918 strain, led to an even stronger reaction. These issues highlight the increasingly sensitive nature of some aspects of modern virological research.

While poliovirus RNA *is* mRNA and can be translated into protein in an *in vitro* translation system, it has two properties quite different from cellular mRNA. First, poliovirus virion RNA has a protein VPg at its 5′ end instead of the methylated cap structure found in cellular mRNA. The VPg protein is encoded by the virus. The viral mRNA also has a very long leader that can assume a complex structure by virtue of intramolecular base pairing in solution. The structure of this leader sequence, especially near the beginning of the translational reading frame (the **internal ribosome entry site** [IRES]), mediates association of the viral genome with ribosomes. The IRES structure and its role in translation initiation comprise an alternate way in which eukaryotic ribosomes can initiate protein synthesis without binding at the 5′ end and transiting to an AUG codon. Subsequent to the characterization of its role in picornavirus replication, it also has been found to function in the translation of several cellular transcripts. With poliovirus RNA, the normal Kozak rules for the selection of the AUG codon to initiate translation in an mRNA (see Chapter 13, Part III) do not apply. Indeed, the AUG triplet that begins the large

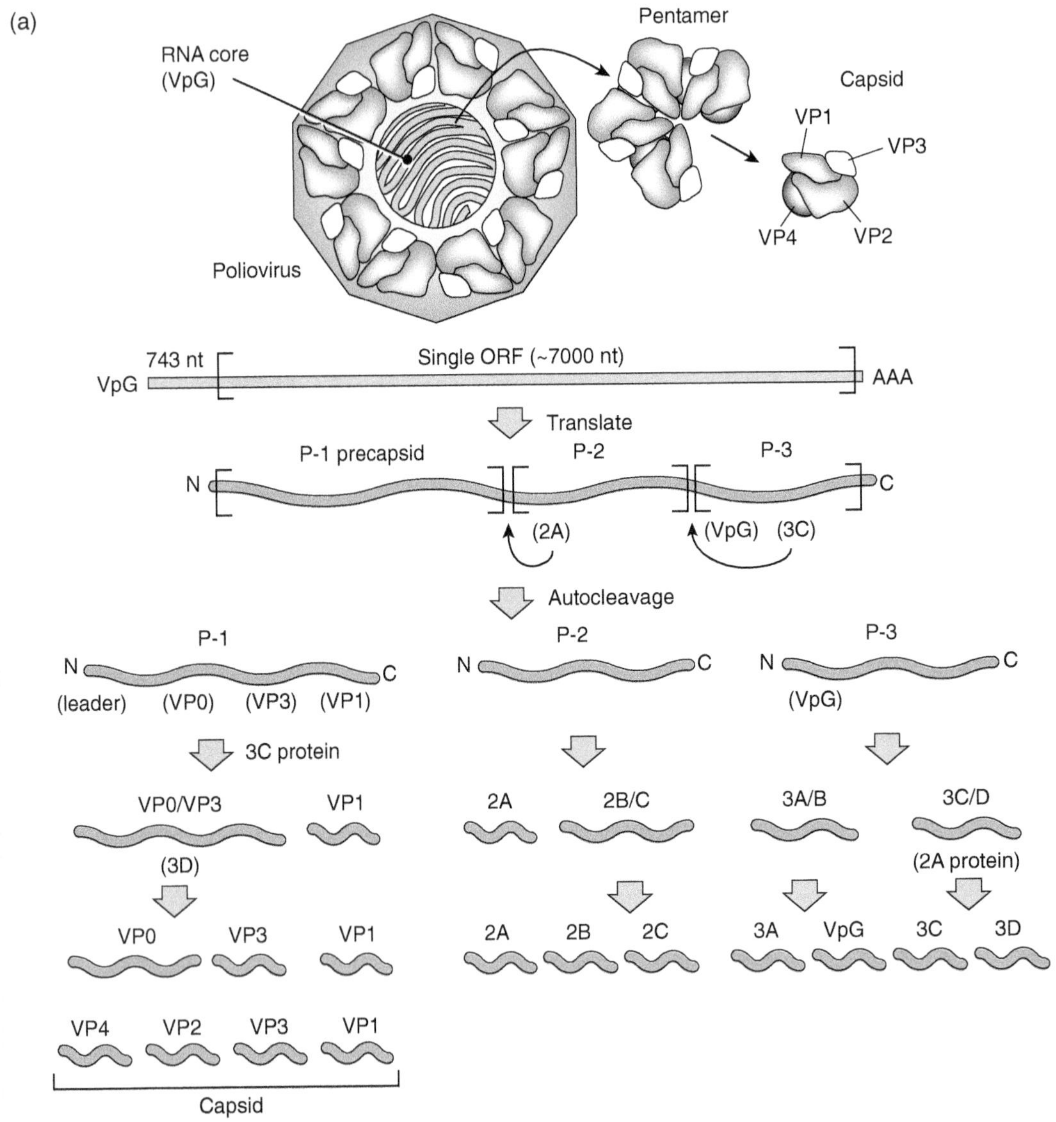

Figure 14.2 (a) Poliovirus, a typical picornavirus. The 30-nm-diameter icosahedral capsid comprises 60 identical subunits – each a pentamer of subunits (often called protomers) containing a single copy of VP1, VP2, VP3, and VP4. The map of the approximately 7700-nucleotide (nt) single-stranded RNA genome that serves as mRNA in the initial stages of replication is also shown. Unlike cellular mRNA, poliovirus genomic RNA has a viral protein (VPg) at its 5′ end instead of a methylated nucleotide cap structure. The RNA has a circa 740-nt sequence at the 5′ end that encodes no protein, but assumes a complex secondary structure to aid ribosome entry and initiation of the single translational reading frame. The single precursor protein synthesized from the virion RNA is cleaved by internal proteases (2A and 3C) initially into three precursor proteins, P1, P2, and P3. Protein P1 is then proteolytically cleaved in a number of steps into the proteins that assemble into the precapsid, VP0, VP1, and VP3. Proteins P2 and P3 are processed into replicase, VPg, and a number of proteins that modify the host cell, ultimately leading to cell lysis. With three exceptions, all proteolytic steps are accomplished by protease 3C, either by itself or in association with protein 3D. Protease 2A carries out the first cleavage of the precursor protein into P1 and P2 as an intramolecular event. It also mediates cleavage of the protease 3CD precursor into protease 3C and protein 3D. It is not known how the third cleavage that does not utilize protease 3C occurs. This is the maturation of the capsomers by the cleavage of VP0 into VP2 and VP4. The VP4 protein is modified by the addition of a myristyl residue at the amino terminus (myr = myristyl). (b) The structure of the poliovirus internal ribosome entry site (IRES). The diagram is a schematic of the predicted secondary structure in the 5′ proximal region of the poliovirus genome. The shaded secondary-structure features make up the IRES. Note that one of the mutations associated with attenuation of the Sabin vaccine strains is located in this region. The site of the AUG at which initiation of the large polyprotein occurs is also indicated.

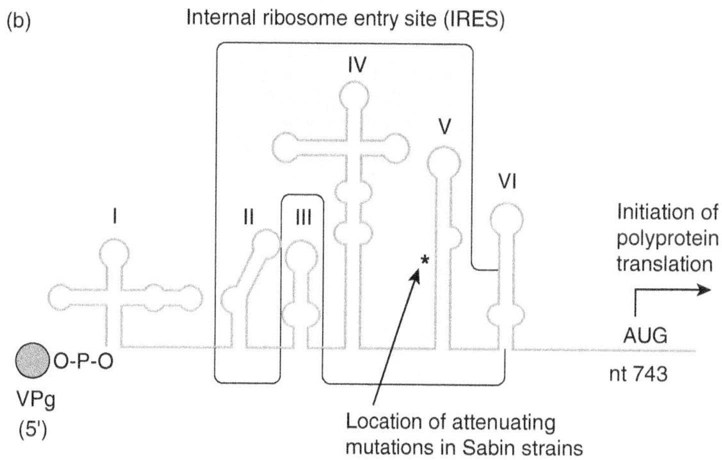

Figure 14.2 *Continued*

poliovirus ORF is preceded by eight other AUG triplets within the leader that are not utilized to initiate translation. The structure of this region of the poliovirus RNA genome is diagramed in Figure 14.2b. The IRES structure is now used routinely in the construction of plasmids where an internal ribosome initiation is needed.

Upon successful initiation of infection, viral genomic mRNA is translated into a single large protein that is the precursor to all viral proteins. This precursor protein is also shown in Figure 14.2a; it contains all the poliovirus proteins that are expressed during infection. Thus, all the viral proteins such as those shown in Figure 12.1 are derived from it.

The smaller proteins are cleaved from the precursor polyprotein by means of two proteases (2A and 3C) that comprise part of this large viral protein. As briefly outlined in Chapter 6, Part II, many viruses utilize proteolytic cleavage of large precursor proteins via virus-encoded proteases during the replication process, and such proteases are important potential targets for antiviral chemotherapy (see Chapter 8). Indeed, the development of protease inhibitors has had a very encouraging effect on attempts to treat AIDS.

The steps in processing are complex, and have yet to be fully worked out in complete detail. Both viral proteases utilize a cysteine residue as part of their active sites; thus, they are termed **C-proteases**. They exhibit a very high specificity, and although both cleave the precursor peptide at sites between specific amino acids (Tyr-Gly for protease 2A and Gln-Gly for protease 3C), neither cleaves all available sites and protease 2A does not cleave nonviral peptides with any efficiency at all. Clearly, secondary structure and other features of the substrate protein are important in determining cleavage sites.

The first two cleavages take place intramolecularly, that is, within the protein in which the proteases are covalently linked. These cleavages result in the formation of three large precursor proteins, P1, P2, and P3. Protein P1 contains the capsid proteins VP1, VP3, and VP0, as well as a short leader protein (L). While not established for poliovirus, the L protein of other picornaviruses has been associated with both virus assembly as well as cellular trafficking pathways. In addition, the P1 protein is **myristoylated** at the N terminal end, involving the covalent addition of the 14-carbon fatty acid myristic acid. As a result, the N-terminus of VP0 will have this modification, which is known to enable such modified proteins to associate efficiently with membrane structures. The P2 and P3 proteins are precursors for a number of nonstructural proteins, including the viral replicase enzyme and proteins and enzymes that alter structure of the infected cell. Protein P3 also contains the VPg protein. The general steps in derivation of mature viral proteins from the precursor protein are shown in the genetic map of Figure 14.2a.

The later stages in processing of the precursor proteins involve mainly protease 3C, although protease 2A cleaves the 3CD precursor of protease and replicase into variants then termed 3C′ and 3D′. It remains unknown whether these variants have any role in replication, given that they are not seen in infections with all strains of the virus. While protein 3D is not a protease (it is the replicase protein), it aids in cleavage of the VP0–VP3 precursor into VP0 and VP3. The 3CD precursor itself, however, can also act as a protease and may have a specific role in some of the early cleavage events.

Since the poliovirus ORF is translated as a single, very large protein, poliovirus technically has only one "gene." This is not strictly true, however, since different portions of the ORF contain information for different types of protein or enzyme activities. Further, different steps in processing of the precursor proteins are favored at different times in the replication cycle; therefore, the pattern of poliovirus proteins seen varies with time following infection, as shown earlier in Figure 12.1.

The demonstration of precursor–product relationships between viral proteins can be tricky and experimentally difficult, but the procedure's theory is simple and based on analysis of proteins encoded by the virus, consideration of the virus's genetic capacity to encode proteins, and a general understanding of the translation process itself. The separation and enumeration of viral proteins based on their migration rates in denaturing gels, which is a function of protein size, are outlined in Chapter 12, Part III.

For poliovirus, many years of analysis can be summarized as follows: The total molecular size of the proteins encoded by the virus cannot exceed approximately 2300 amino acids (7000/3). Despite this, the total size of viral proteins estimated by adding radioactive amino acids to an infected cell and then performing size fractionation on the resulting radiolabeled material is significantly greater. Further, it is known that poliovirus efficiently inhibits cellular protein synthesis, so most proteins detected by the addition of radioactive precursor amino acids to infected cells (also termed a *pulse* of radioactive material) are, indeed, viral.

This conundrum can be resolved by using a technique called a **pulse–chase experiment**, and by using *amino acid analogues,* which inhibit protease processing of the precursor proteins. In pulse–chase experiments, radioactive amino acids are added for a short time. This is the "pulse." Then a large excess of nonradioactive amino acids is added to dilute the label. This is the "chase."

Only the largest viral proteins isolated from a poliovirus-infected cell exposed only to the radioactive pulse for short periods (followed by isolation of the infected cell) had radioactivity. This finding suggests that these proteins are the first viral products synthesized. If the pulse period is followed by chase periods of various lengths, radioactivity is eventually seen in the smaller viral proteins. Such a result is fully consistent with a kinetic precursor–product relationship between large (precursor) proteins and smaller mature (product) viral proteins.

The relationship between precursor and product was confirmed by adding translation inhibitors at specific times following a pulse of radioactive amino acids. This step resulted in the loss of label incorporated into large proteins, but did not affect the appearance of label in the smaller proteins derived from the precursor proteins already labeled during the pulse. Finally, the addition of amino acid analogues that inhibited proteolysis of the precursor protein contributed a further confirmation of the process.

The poliovirus replication cycle

As shown in Figure 14.3, everything tends to "happen at once" during the poliovirus replication cycle. Viral entry involves attachment of the virions by association with the cellular receptor. For poliovirus the receptor, Pvr, is a specific CAM-like molecule (CAM = cellular adhesion molecule) called CD155. The binding of poliovirus virions to the receptor has been examined by x-ray crystallography and appears to involve insertion of a part of the receptor into "canyon" cavities on the surface of the virus particle.

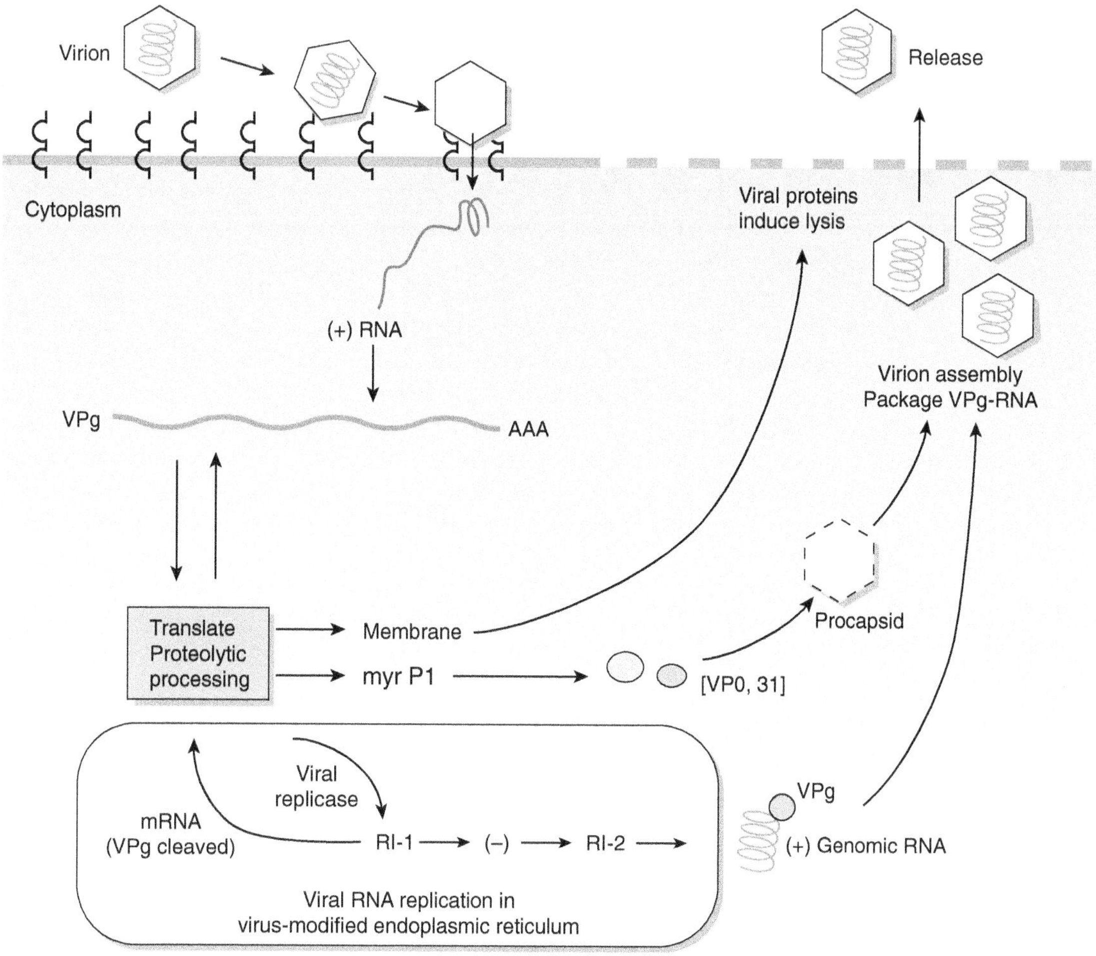

Figure 14.3 The poliovirus replication cycle. The schematic representation is broken into discrete steps. Viral entry is by receptor-mediated rearrangement of the virion proteins to form a transmembrane pore, releasing the positive-sense RNA genome into the cytoplasm of the cell. Viral entry involves attachment to a cell surface receptor, followed by rearrangement of the capsid with insertion of the myristoylated-VP4 terminus into the cell membrane, releasing the positive-sense RNA genome into the cytoplasm of the cell. This RNA is translated into a large polyprotein. Viral replicase released from the precursor protein then mediates generation of RI-1 and RI-2 to generate more mRNA that, unlike the original genomic RNA, has the VPg protein cleaved off. As infection proceeds, the replication complexes become associated with cellular membrane structures into replication compartments. Newly synthesized positive-sense RNA is also translated, and the process repeats many times until sufficient capsid protein precursors are formed to allow assembly of the procapsid. Procapsids associate with newly synthesized positive-sense RNA still containing VPg at its 5′ end, and entry of viral genomes results in capsid maturation. As the process continues, virions accumulate in the cytoplasm until viral proteins induce cell lysis and virus release occurs. The entire process can take place in the absence of a nucleus.

Since poliovirus is able to efficiently infect cells that are mutated in the protein dynamin, required for the function of clathrin-coated pits, it is now thought that poliovirus does not enter host cells by way of receptor-mediated endocytosis, as diagrammed in Figure 6.2, even though some other picornaviruses may depend upon this pathway for entry. The current model for poliovirus attachment and entry into the cell is as follows (Figure 14.3):

1 Virus particles attach to Pvr, the poliovirus receptor on the surface of the cell.
2 Receptor binding induces a rearrangement of the virus particle that results in the insertion of helical regions of VP1 into the cell membrane, along with the myristoylated amino terminal end of VP4, thus creating a channel into the cytoplasm.

3 Viral RNA is released into the cell cytoplasm after further particle rearrangements, perhaps triggered by ionic changes.

Viral RNA is translated into protein, portions of which are involved in replication of the viral genome by generation of the replication structures, RI-1 and RI-2. The protein VPg is a primer for this replication by having a uracil residue added to it, a process called uridylation. The initiation of replication requires an RNA secondary structure feature called the *cis*-acting replication element (CRE), located within the coding region in the genome for the 2C protein. Poliovirus replicase, protein $3D^{pol}$, catalyzes the generation of both negative- and positive-sense products. It has recently been demonstrated that *cis*-acting sequence elements that control replication are present in the poliovirus genome. Secondary structure features at the 5′ end as well as within coding regions appear to be required for efficient RNA replication. Other poliovirus proteins are also involved, as well as one or more host proteins, since much of the viral genome's replication takes place in membrane-associated compartments generated by these proteins within the infected cell's cytoplasm. Generation of new mRNA-sense (positive) strands of poliovirus RNA leads to further translation, further replication, and finally capsid assembly and cell lysis.

Details of the poliovirus capsid's morphogenesis were worked out several decades ago. While there is still some controversy concerning the timing of certain steps in the assembly process (especially the timing of the association of virion RNA with the procapsids), poliovirus assembly serves as a model for such processes in all icosahedral RNA viruses (see Chapter 6, Part II). Proteolytic cleavage of precursor proteins plays an important role in the final steps of maturation of the capsid. This cleavage does not involve the action of either protease 2A or 3C. Rather, it appears to be an intramolecular event mediated by the capsid proteins themselves as they assemble and assume their mature conformation. The molecular sizes of the poliovirus capsid proteins are given in Table 11.1.

The most generally accepted scheme is shown in Figure 14.4. In viral morphogenesis, myristoylated-P1 protein is cleaved from the precursor protein by the protease 2A segment. Five copies of this protein aggregate, and the protein is further cleaved by protease 3C into myristoylated-VP0, VP1, and VP3, which forms one of the 60 capsid *protomers*. Five of these protomers assemble to form the 14s pentamer. Finally, 12 of these 14s pentamers assemble to form an empty capsid (**procapsid**).

This procapsid is less dense than the mature virion, so its proteins can be separated readily by centrifugation. Analysis of the procapsid proteins demonstrates equimolar quantities of myristoylated-VP0, VP1, and VP3. Following formation of the procapsid, viral RNA associates with the particle, and a final cleavage of VP0 into VP2 and myristoylated-VP4 occurs to generate the mature virion. After virions are assembled, the cell lyses and virus is released.

Picornavirus cytopathology and disease

The most obvious cytopathology of poliovirus replication is cell lysis. But prior to this, the virus specifically inhibits host cell protein synthesis. Inhibition of host cell protein synthesis involves proteolytic digestion of the translation initiation factor eIF-4G so that ribosomes can no longer recognize capped mRNA (see Chapter 13, Part III). Such modification leads to the translation of only uncapped poliovirus mRNA because its IRES allows it to assemble the translation complex with the virus-modified ribosomes. Note that this rather elegant method of shutoff will not work with most types of viruses because they express and utilize capped mRNA!

There are three related types, or serotypes, of poliovirus. They differ in the particular antigenic properties of viral structural proteins. Most poliovirus infections in unprotected human populations result in no or only mild symptoms, but one serotype (type 3) is strongly associated with the disease's paralytic form. Infection with this serotype does not invariably lead to a

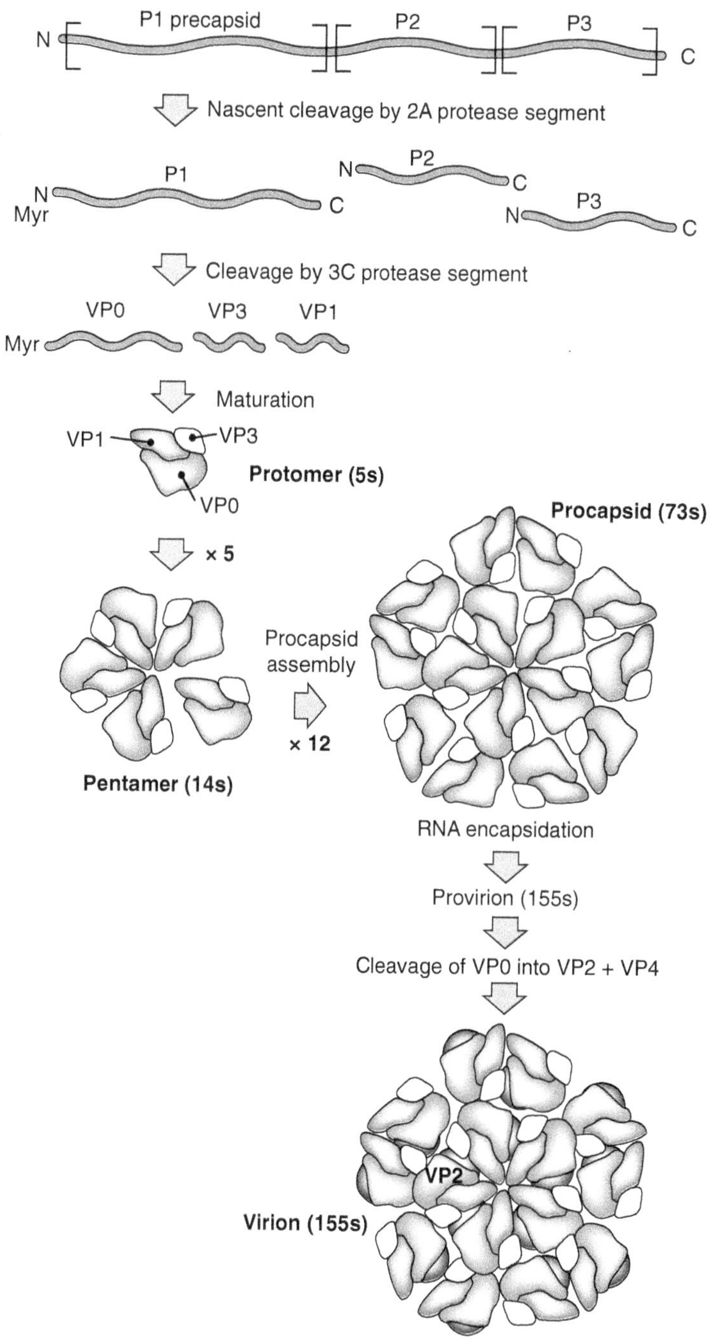

Figure 14.4 The steps in the assembly of the poliovirus virion. Precursor proteins associate to form 5s protomers, which then assemble to form pentamers. Twelve of these assemble to form the procapsid into which virion RNA is incorporated. Final cleavage of VP0 into VP2 and VP4 takes place to form the mature capsid that has a diameter of 28–30 nm.

paralytic episode, but the probability of such an episode is much higher than with the others. All serotypes are distributed throughout the regions where poliovirus is endemic in a population, although some predominate in some locations.

Poliovirus is spread by fecal contamination of food or water supplies. Receptors for the virus are found in the intestine's epithelium, and infection results in local destruction of some tissue in the intestine, which can result in diarrhea. Unfortunately, motor neurons also have receptors for poliovirus, and if the virus gets into the bloodstream, it can replicate in and destroy such neurons, leading to paralysis. This result is of no value to the virus since the virus initiating neuronal infection cannot be spread to other individuals and is eventually cleared; thus, the paralytic phase of the disease is a "dead end" for the virus. The virus stimulates an immune response, and the individual recovers and is resistant or immune to later infection.

Vaccination against poliovirus infections is accomplished effectively with both inactivated and attenuated live-virus vaccines, as described in Chapter 8, Part II. Since the only reservoir of poliovirus is humans, immunity through vaccination against the virus is an effective way of preventing disease. Currently, a major effort is underway to completely eradicate the disease from the environment (see Chapter 25).

A number of other picornaviruses cause disease; many are spread by fecal contamination and include hepatitis A virus, echoviruses, and coxsackievirus. Like poliovirus, these viruses occasionally invade nervous tissue. Coxsackievirus generally causes asymptomatic infections or mild lesions in oral and intestinal mucosa, but can cause encephalitis. Echoviruses are associated with enteric infections also, but certain echovirus serotypes cause infant nonbacterial meningitis, and some epidemic outbreaks with high mortality rates in infants have been reported.

Another widespread group of picornaviruses are the rhinoviruses, one of the two major groups of viruses causing common head colds. Unlike the other picornaviruses detailed here, rhinoviruses are transmitted as aerosols. Because of the large number (~100) of distinct serotypes of rhinovirus, it is improbable that an infection will generate immunity that prevents subsequent colds. There are no known neurological complications arising from rhinovirus infections.

Flavivirus replication

The success and widespread distribution of picornaviruses and their relatives demonstrate that the replication strategy found in translation of a single large ORF is a very effective one. If more evidence were needed on this score, the plethora of mosquito-borne flaviviruses should settle the matter completely!

Flaviviruses are enveloped, icosahedral, positive-sense RNA viruses. They appear to be related to picornaviruses, but clearly have distinct features, notably an envelope. Because mosquitoes and most other arthropods are sensitive to weather extremes, it is not surprising that arboviral diseases occur throughout the year in the tropics and subtropics, but occur only sporadically, and in the summer, in temperate zones.

Many flaviviruses demonstrate tropism for neural tissue, and flaviviruses are the causative agents of yellow fever, dengue fever, and many types of encephalitis. In the United States, the mosquito-borne St. Louis encephalitis virus leads to periodic epidemics in the summer, especially during summers marked by heavy rains and flooding, such as the summer of 1997 in northeastern states.

West Nile virus was first isolated in the Middle East, as suggested by its name. However, it has now invaded the Western Hemisphere and is firmly established throughout the United States. The scenario began in the late summer of 1999, when at least 1900 people in Queens, New York City, were infected with West Nile. Analysis of the virus suggested that it originated from a strain present in Israel. No one knows how this virus arrived in New York. However, it soon spread into the wild bird population and began its march across the country. At the end of 2005, the virus was present in all of the contiguous states, with most reporting both human and animal cases. We can now say that West Nile virus has established itself as endemic in North America.

Of recent interest is Zika virus, a member of the flavivirus family that has made its way from Central Africa, through Southeast Asia and into South America. The virus is transmitted by *Aedes aegypti* mosquitos. Normally a tropical and subtropical species, these mosquitos have moved north into Florida due to the effect of climate change on potential habitats. Zika virus fever, while normally a mild, self-limiting febrile disease, has been associated with cases of microcephaly as a result of infection of pregnant women. In addition, cases of the neurological illness Guillain–Barré syndrome have been linked to these infections. We will have more to say about this in later chapters when we discuss emerging viral diseases.

An abbreviated outline of the flavivirus replication cycle can be inferred from the genetic and structural map shown in Figure 14.5, and taken from work with yellow fever virus. The flavivirus genome is over 10 000 bases long, and unlike poliovirus, it is (i) capped at the 5′ end and (ii) not polyadenylated at the 3′ end. Like poliovirus, the large ORF is translated into a single precursor protein that is cleaved by integral proteases into individual proteins. Some of these cleavage steps are shown in Figure 14.5. The structural protein precursor includes an integral membrane protein (M) and an envelope glycoprotein. These membrane-associated proteins are translated by membrane-bound polyribosomes, and the process of insertion into the cell's membrane follows the basic outline described for togaviruses later in this chapter. The M protein contains a "signal" sequence at its N-terminal that facilitates the insertion of the nascent peptide chain into the endoplasmic reticulum. This signal is cleaved from the PreM protein within the lumen of the endoplasmic reticulum – probably by the action of cellular enzymes. The NS (non-structural) proteins encode the replicase enzymes and do not form part of the virion. Despite this, it is interesting that antibodies directed against the precursor, NS1, protect animals against infection.

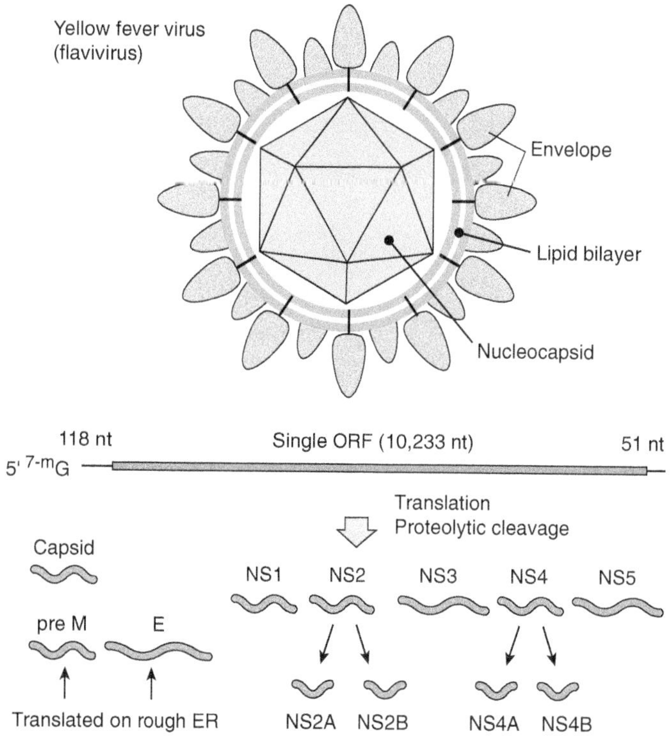

Figure 14.5 The yellow fever virus (a flavivirus) and its genome. This flavivirus has a replication cycle very similar in broad outline to that detailed for poliovirus. Unlike poliovirus, flaviviruses encode a single envelope glycoprotein, and the yellow fever virus's approximately 10 000-nucleotide (nt) genome is capped, although not polyadenylated. Also in contrast to poliovirus, the yellow fever virus precursor polyprotein is cleaved into a large number of products as it is being translated, so the very large precursor proteins of poliovirus replication are not seen. The enveloped capsid is larger than that of poliovirus, with a diameter of 40–50 nm. ER: Endoplasmic reticulum.

POSITIVE-SENSE RNA VIRUSES ENCODING MORE THAN ONE TRANSLATIONAL READING FRAME

A positive-sense RNA virus that must regulate gene expression while infecting a eukaryotic host faces a fundamental problem: The eukaryotic ribosome cannot initiate translation of an ORF following translation of one upstream of it. While a positive-sense RNA virus genome could (and some do) contain more than one ORF, these ORFs cannot be independently translated at different rates during infection without some means to overcome this fundamental mechanistic limitation.

One way to circumvent the problem is for a virus to encapsidate more than one mRNA (in other words, for the virus to contain a segmented genome). This approach is utilized by a number of positive-sense RNA viruses infecting plants, but has not been described for animal viruses. This finding is somewhat surprising since there are numerous negative-sense RNA viruses with segmented genomes that are successful animal and human pathogens. The list contains influenza viruses, hantaviruses, and arenaviruses.

Despite the disinclination of positive-sense RNA viruses that infect animal cells to encapsidate segmented genomes, another strategy for regulating mRNA expression is utilized successfully. This strategy involves the encoding of a cryptic (hidden) ORF in the genomic RNA, which can be translated from a viral mRNA generated by a transcription step during the replication cycle. With this strategy, viral gene expression from the full-length positive-sense mRNA contained in the virion results in translation of a 5′ ORF, and this protein (an enzyme) is involved in generation of a second, smaller mRNA by transcription.

The second mRNA (which is not found in the virion), in turn, is translated into a distinct viral protein. Such a scheme allows the nonstructural proteins encoded by the virus – the enzymes required for replication – to be expressed in lesser amounts or at different times in the infection cycle than the proteins ending up in the mature virion. Clearly, this approach is effective as witnessed by the number of important pathogens that utilize it.

Two viral mRNAs are produced in different amounts during togavirus infections

Togaviruses are enveloped RNA viruses that display a complex pattern of gene expression during replication. Sindbis virus is a well-studied example. This arthropod-borne virus causes only very mild diseases in (rarely) humans, but its size and relative ease of manipulation make it a useful laboratory model for the group as a whole.

Sindbis virus has a capsid structure similar to picornaviruses and flaviviruses, and like flaviviruses, the capsid is enveloped. The viral genome contains two translational ORFs. Initially, only the first frame is translated into viral replication enzymes. These enzymes both replicate the virion RNA *and* generate a second mRNA that encodes viral structural proteins.

The viral genome

Sindbis virus and its 11 700-base genome are shown in Figure 14.6. The virion genomic RNA (termed 49s RNA for its sedimentation rate in rate zonal centrifugation – see Chapter 11, Part III) has a capped 5′ end and a polyadenylated 3′ end. Both capping and polyadenylation appear to be carried out by viral replication enzymes, possibly in a manner somewhat analogous to that seen for the negative-sense vesicular stomatitis virus (VSV), which is discussed in Chapter 16.

The Sindbis virus genome contains two ORFs. The 5′ ORF encodes a replication protein precursor that is processed by proteases to generate four different replicase polypeptides. The 3′ ORF encodes capsid protein and envelope glycoproteins.

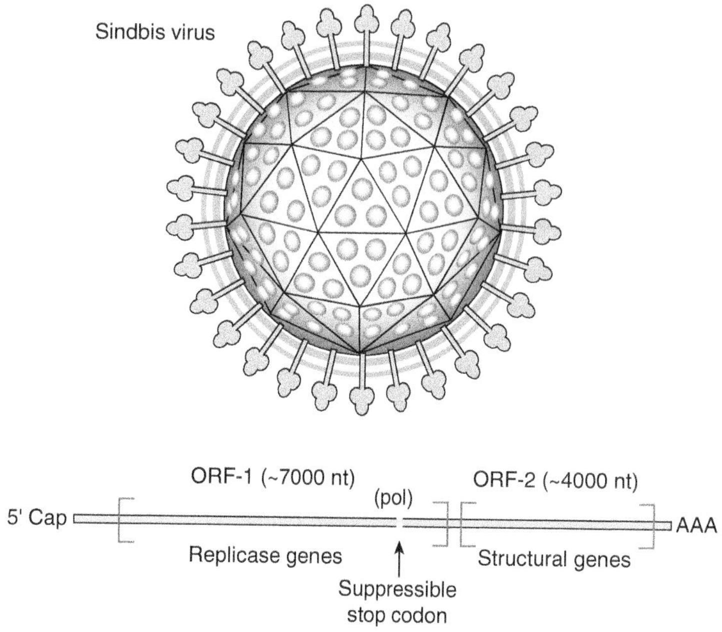

Figure 14.6 Sindbis virus – a typical togavirus. The virion (60–70 nm in diameter) and genetic map are shown. The Sindbis genome contains two translational reading frames; only the upstream (5′) one can be translated from the approximately 11 000-nucleotide (nt) capped and polyadenylated 49s (positive) virion-associated genomic RNA. This upstream translational frame encodes nonstructural proteins via expression of two precursor proteins. The larger, which contains the polymerase precursor, is translated by suppression of an internal stop codon in the reading frame.

The virus replication cycle

Virus Entry Viral entry is via receptor-mediated endocytosis as shown in Figure 14.7a. The entire virion, including envelope, is taken up in the endocytotic vesicle. Acidification of this vesicle leads to modification of the viral membrane glycoprotein. This allows the viral membrane to fuse with the vesicle, and causes the capsid to disrupt so that viral genomic mRNA is released into the cytoplasm.

Early gene expression As shown in Figure 14.7b, only the 5′ ORF can be translated from intact viral mRNA, because the eukaryotic ribosome falls off the viral mRNA when it encounters the first translation stop signal (either UAA, UAG, or UGA – see Chapter 13). With Sindbis virus, this situation is complicated by the fact that this first ORF in the genomic RNA contains a stop signal about three-fourths of the way downstream of the initiation codon. This termination codon can be recognized to generate a shorter precursor to the nonstructural proteins, but it can also be *suppressed*. (In genetics, the term **suppression** refers to the cell periodically ignoring a translation stop signal because of either an altered tRNA or a ribosomal response to secondary structure in the mRNA encoding it.) With Sindbis virus infection, the suppression is ribosomal, and results in about 25% of the nonstructural precursor protein containing the remaining information shown in ORF-1 in the genetic map. As discussed in Chapter 19, suppression of an internal stop codon also has a role in the generation of retrovirus protein.

In Sindbis virus infection, translation of infectious viral RNA generates replication enzymes that are derived by autoproteolytic cleavage (i.e., self-cleavage) of the replicase precursor protein. This can be considered an "early" phase of gene expression; however, things happen fast in the infected cell, and this may only last for a few minutes.

Viral genome replication and generation of 26s mRNA The replication enzymes expressed from genomic 49s positive-sense mRNA associated with genomic RNA to generate 49s negative-sense

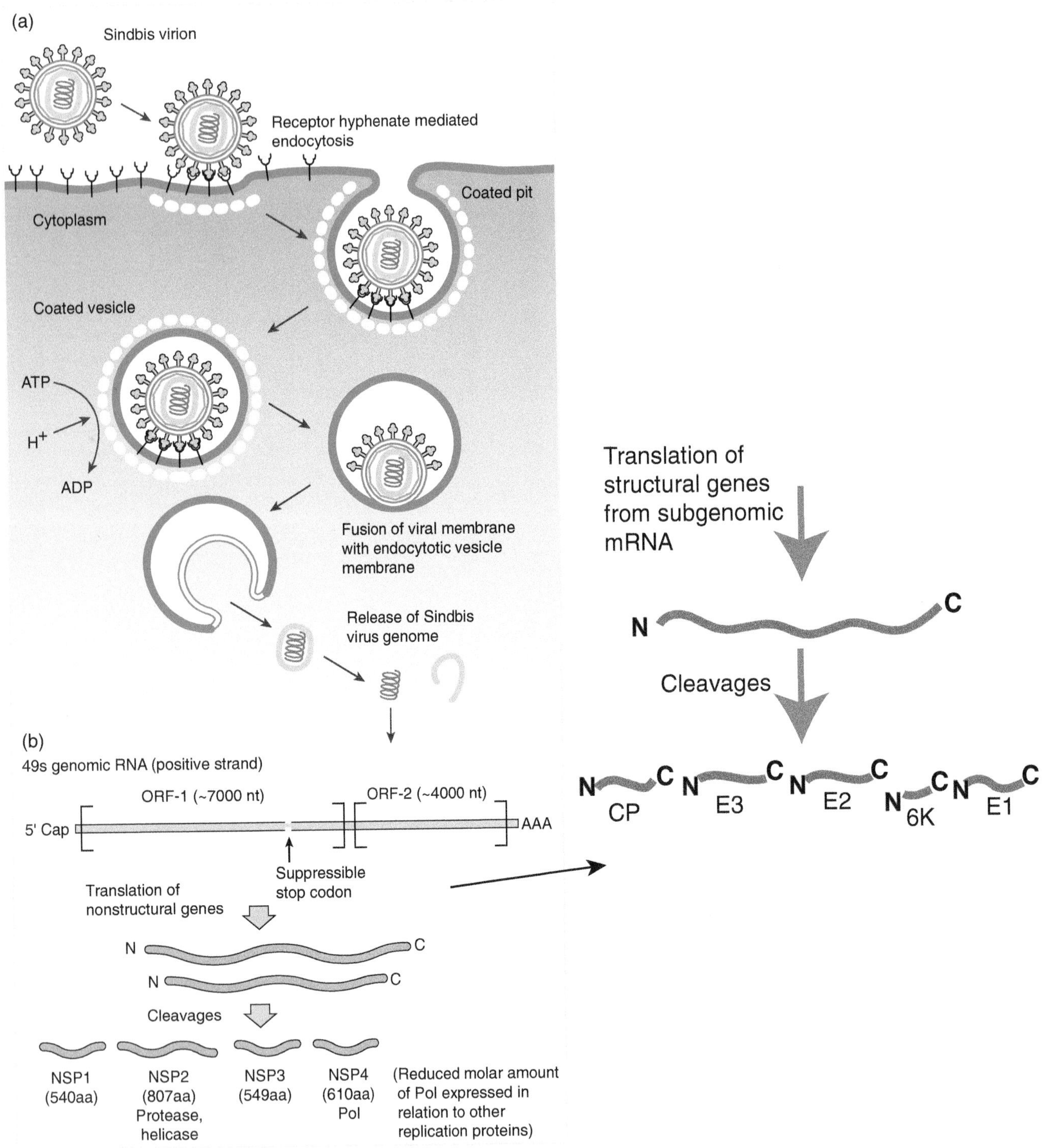

Figure 14.7 The early stages of Sindbis virus infection. (a) The first step is receptor-mediated endocytosis, leading to fusion of the viral membrane with that of the endocytotic vesicle, which leads to release of the Sindbis virus genome (mRNA) into the infected cell's cytoplasm. As outlined in Chapter 6, Part II, internalization of the enveloped virion within an endocytotic vesicle is followed by acidification and covalent changes in membrane proteins. This results in fusion of the viral membrane with that of the endocytotic vesicle and release of the viral genome. (b) Translation of the virion RNA results in expression of the precursors to the nonstructural replicase and other viral proteins encoded in the 5′ translational reading frame. These proteins mediate replicase, capping, and protease functions. Viral structural proteins are translated from ORF-2, using the 26s subgenomic mRNA (see Figure 14.8).

RNA through RI-1 are shown in Figure 14.8a. The next step in the process is critical to regulated expression of the two virus-encoded precursor proteins. With Sindbis, the negative-sense RNA complementary to genomic positive-sense RNA is the template for *two* different positive-sense mRNAs. Both are capped and polyadenylated. The first is more 49s positive-sense virion RNA. The second is 26s positive-sense RNA. The shorter 26s mRNA is generated by replicase beginning transcription of negative-sense RNA in the middle and generating a "truncated" or **subgenomic mRNA**. The region on the negative-sense strand where the transcriptase binds is roughly analogous to a promoter, but its sequence does not exhibit the features of promoters found in DNA genomes.

Generation of structural proteins

The short 26s mRNA contains only the second ORF contained in the full-length genomic RNA. This ORF was hidden or inaccessible to translation of the full-length virion mRNA. With the 26s mRNA, however, cellular ribosomes can translate the ORF into precursors of capsid and envelope proteins. Expression of structural proteins, thus, requires at least partial genome replication and is generally termed *late* gene expression, although it occurs very soon after infection. Translation of the 5′ region of late 26s mRNA generates capsid protein that is cleaved from the growing peptide chain by proteolytic cleavage. This cleavage generates a new N-terminal region of the peptide. The new N-terminal region of the peptide contains a stretch of aliphatic amino acids, and the hydrophobic nature of this "signal" sequence results in the growing peptide chain inserting itself into the endoplasmic reticulum in a manner analogous to synthesis of any cellular membrane protein. This process is shown in Figure 14.8b.

Following initial insertion of the membrane proteins' precursor, the various mature proteins are formed by cleavage of the growing chain within the lumen of the endoplasmic reticulum. This maturational cleavage is carried out by cellular proteins.

Posttranslational processing, such as glycosylation of membrane-associated components of the late structural protein, takes place in the Golgi apparatus, and viral envelope protein migrates to the cell surface. Meanwhile, capsid formation takes place in the cytoplasm, genomes are added, and the virion is formed by budding through the cell surface, as described in Chapter 6, Part II.

Togavirus cytopathology and disease

The replication process of togaviruses is a step more complex than that seen with picornaviruses, and the cell needs to maintain its structure to allow continual budding of new virus. Accordingly, there is less profound shutoff of host cell function until a long time after infection.

A major cytopathic change is alteration of the cell surface. This can lead to fusion with neighboring cells so that virus can spread without ever leaving the first infected cells. This alteration to the cell surface also involves antigenic alteration of the cell. Such types of cytopathology are found with many enveloped RNA viruses, whether they are positive or negative sense.

Based upon the number of viruses identified as belonging to the group, the togaviruses are an extremely successful group of viruses, and like the flaviviruses, many are transmitted by arthropods. As noted in Chapter 5, Part II, it is for this reason that these two groups of positive-sense RNA viruses are termed arboviruses (arthropod-borne viruses). While this terminology is convenient for some purposes, it does not recognize significant differences in the replication strategies of these two groups of viruses. Further, numerous other types of viruses that are spread by arthropod vectors, and some togaviruses and flaviviruses, are *not* transmitted by such vectors. A striking example is rubella (German measles) virus.

Many togaviruses cause sporadic outbreaks of mosquito-borne encephalitis because they have a propensity for replication in cells making up the brain's protective lining. Although such

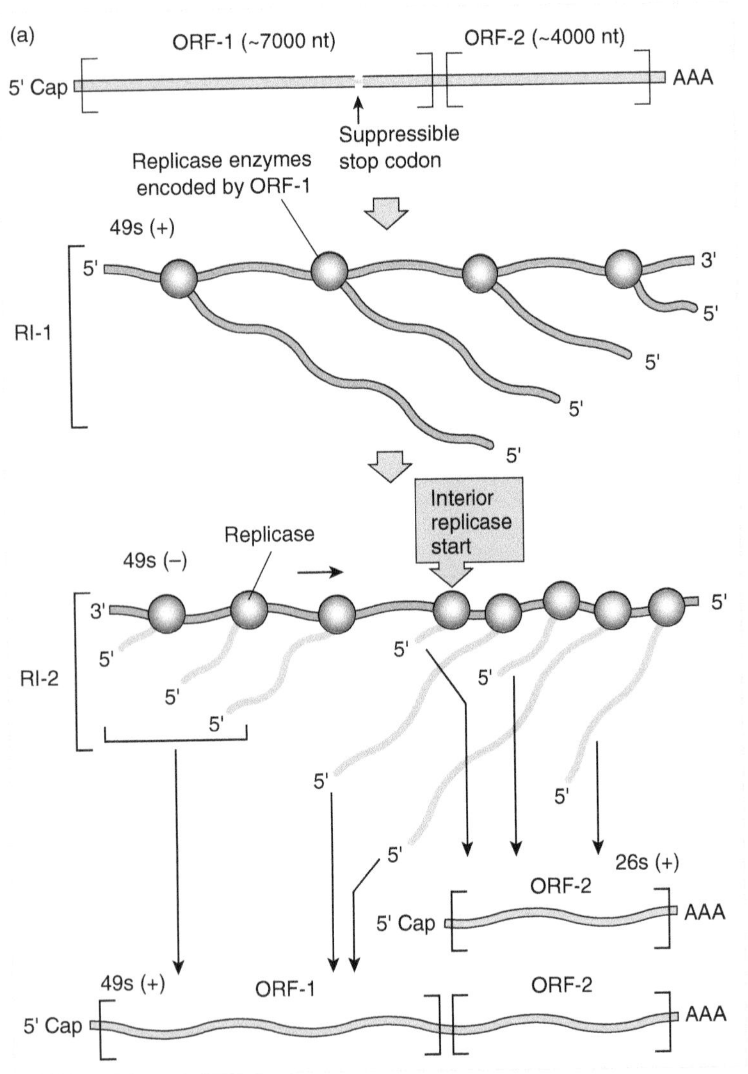

Figure 14.8 (a) The replication of Sindbis virus genome, and generation of the subgenomic 26s mRNA. This mRNA is expressed by an internal start site for viral replicase, and is translated into structural proteins since it encodes only the open reading frame (ORF) that was cryptic in the 49s positive-sense virion RNA. (b) The synthesis of Sindbis virus structural proteins. Structural proteins are translated as a single precursor. When the N-terminal capsid protein is cleaved from the precursor, a signal sequence consisting of a stretch of aliphatic amino acids associates with the endoplasmic reticulum. This association allows the membrane protein portion of the precursor to insert into the lumen of the endoplasmic reticulum. As the protein continues to be inserted into the lumen, it is cleaved into smaller product proteins by cellular enzymes. Cellular enzymes also carry out glycosylation.

disease can be severe, many forms have a favorable prognosis with proper medical care, as neurons are not the primary targets of infection.

The only known host for rubella virus is humans. The virus causes generally mild and often asymptomatic diseases in children and adults, although a mild rash may be evident. Despite the generally benign course of infection, it is remarkable that rubella is associated with a diverse group of clinical diseases, including rubella arthritis and neurological complications.

Periodic local epidemics are characteristic of rubella virus infections, and although the virus induces an effective immune response, the endemic nature of the virus ensures that once a large-enough pool of susceptible individuals arises, sporadic regional epidemics occur. The major problem with these periodic occurrences is the very fact that the disease is often so mild as to be asymptomatic in adults of childbearing age. While the symptoms are very mild for adults and children, this is not the case for fetal infections. Infection of the mother in the first trimester of pregnancy often leads to miscarriage, and a fetus who survives is almost inevitably severely developmentally impaired. Infection of the mother later in pregnancy has a more benign outcome.

The tragedy of rubella infections is that although there are effective vaccines, the disease is often so mild that an individual can be infected and can spread the virus without knowing it.

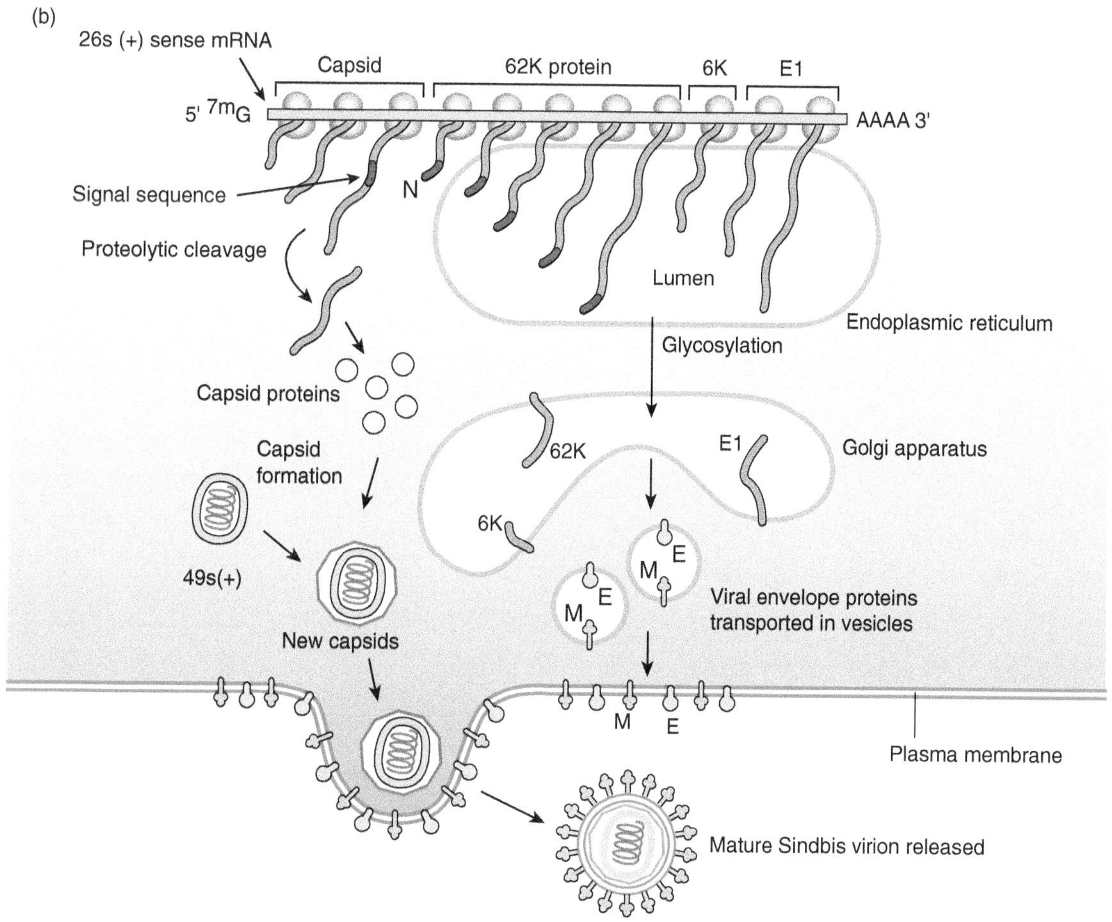

Figure 14.8 *Continued*

For this reason, women of childbearing age who are in contact with young children or other adults at risk of infection should be vaccinated.

A somewhat more complex scenario of multiple translational reading frames and subgenomic mRNA expression: coronavirus replication

Even more complex scenarios exist for expression and regulation of gene function in infections by positive-sense RNA viruses. The replication strategy of the coronaviruses is a good example of such complexity. Coronaviruses and toroviruses are members of the Coronaviridae family and, together with the Arteriviridae and Roniviridae families, make up the larger grouping called the order Nidovirales (nido = nested). The structure of coronaviruses is shown in Figure 15.9 – the helical nucleocapsid is unusual for a positive-sense RNA virus.

The nucleocapsid is helical within a roughly spherical membrane envelope, and the glycoproteins project as distinct "spikes" from this envelope. These glycoprotein spikes from the lipid bilayer appear as a distinctive crown-like structure in the electron microscope, hence the name *corona* (crown)-viruses.

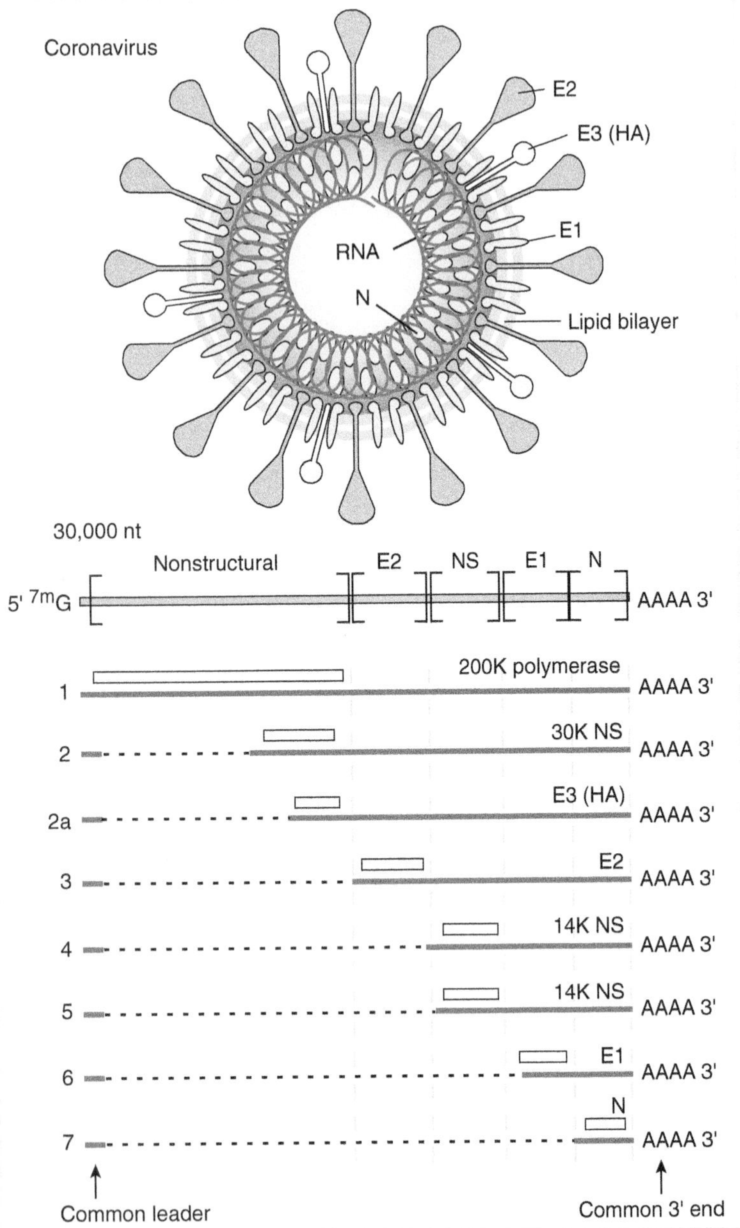

Figure 14.9 A schematic representation of the coronavirus virion. This is the only known group of positive-sense RNA viruses with a helical nucleocapsid. The name of the virus is derived from the appearance of the glycoproteins projecting from the envelope, which gives the virus a crown-like shape. The diameter of the spherical enveloped virion ranges between 80 and 120 nm depending on experimental conditions in visualization. The 30 000-nucleotide (nt) capped and polyadenylated positive-sense genome encodes eight translational reading frames that are expressed through translation of the genomic RNA and seven subgenomic positive-sense mRNAs. These capped and polyadenylated subgenomic mRNAs each have the same short 5′ leader and share nested 3′ sequences. Although two models exist for the production of this nested set, the most likely at this time appears to be that they are derived by transcription of subgenomic negative-sense templates, produced by discontinuous copying of the viral genomic RNA.

The 30-kb coronavirus genome encodes at eight separate translational reading frames, and is the template for the synthesis of at least seven subgenomic mRNAs. Each subgenomic mRNA contains a short, identical leader segment at the 5′ end that is encoded within the 5′ end of the genomic RNA. All subgenomic mRNAs have the same 3′ end, and thus are a nested set of transcripts, giving the name to the order Nidovirales. Only the 5′ translational reading frame is recognized in each, and the others are cryptic. These features are also shown in Figure 14.9.

Coronavirus replication

Coronavirus replication involves the generation and translation of genomic and subgenomic viral mRNAs as shown in Figure 14.10. Virus entry is by receptor-mediated fusion of the virion with the plasma membrane, followed by release of genomic RNA. A good deal of work

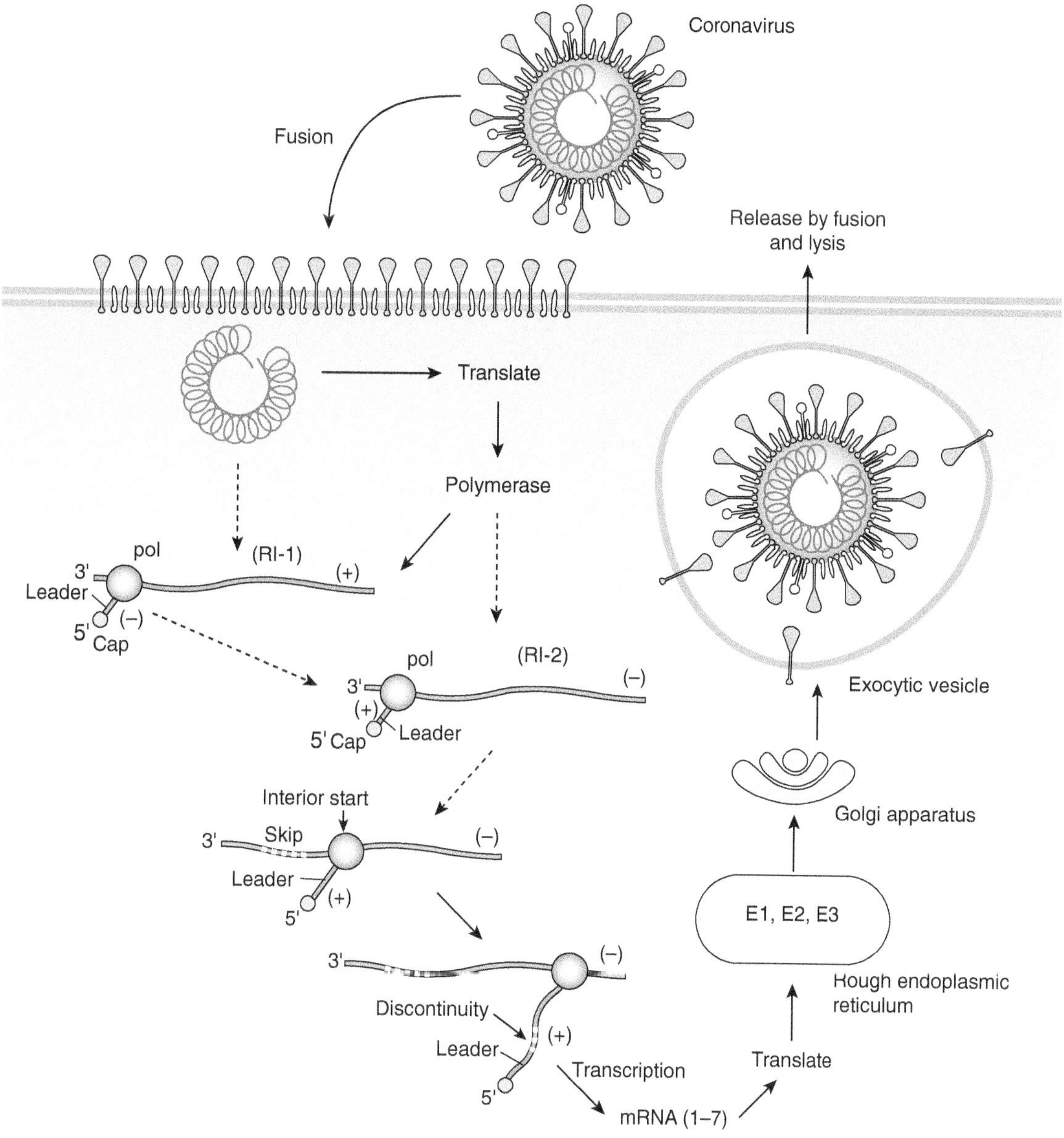

Figure 14.10 The replication cycle of a coronavirus. Replication is entirely cytoplasmic. Infection is initiated by receptor-mediated membrane fusion to release the genomic mRNA. This RNA is translated into the very large (>200 kd) polymerase/capping enzyme. The interaction between full-length virion positive-sense RNA and replicase generates the templates for the mRNAs. Two models are proposed for the synthesis of subgenomic mRNA: leader-primed synthesis and discontinuous negative-strand synthesis. The second of these two models is shown in the figure. The result of both models is the synthesis of a nested set of mRNAs that contain the same 5′ leader sequence and overlapping 3′ ends. Translation of the various subgenomic mRNAs leads to synthesis of the various structural and nonstructural proteins encoded by interior translational reading frames. The mature virions assemble and become enveloped by budding into intracytoplasmic vesicles; these exocytotic vesicles then migrate to the cell surface where virus is released. At later times, cell lysis occurs.

concerning viral replication has been stimulated by the identification of a coronavirus as the agent of the emerging diseases severe acute respiratory syndrome (SARS) and Middle East respiratory syndrome (MERS) (discussed in more detail here and in Chapter 25), as well as the more recent coronavirus disease 2019 (COVID-19).

Virus entry is by receptor-mediated fusion of the virion with the plasma membrane, followed by release of genomic RNA. The receptor for the SARS coronavirus is angiotensin-converting enzyme 2 (ACE2). The importance of this virus–receptor interaction will be discussed in this chapter with respect to the pathogenesis of SARS coronavirus (SARS-CoV) and MERS coronavirus (MERS-CoV).

This RNA (one of the largest mRNAs characterized) is translated into a replication protein that, interestingly, is encoded in an ORF encompassing 70% of the virus's coding capacity. The reason why coronavirus replication proteins are encoded by such a large gene is not yet known.

The mature replication proteins derived from the first translation product are used to produce all subsequent mRNA species. There are two competing models that have been presented for coronavirus transcription (Figure 14.10): leader-primed transcription and discontinuous transcription during negative-strand synthesis.

Leader-primed transcription proposes that the replication proteins first produce a full-length negative-strand copy of the genome, using a standard RI-1 structure. From this template is then transcribed multiple copies of the extreme 3′ end, called the leader region. These leader transcripts then function to prime synthesis of subgenomic mRNAs, initiated at homologous regions in between each of the genes (intergenic sequences).

Discontinuous transcription during negative-strand synthesis proposes that the replication proteins transcribe negative-strand copies of the genome, using RI-1 structures. Some of these products are subgenomic. These subgenomic species are produced when the replicase complex in the RI-1 pauses at the intergenic regions and then jumps to the end of the genome, copying the leader sequence. The result of this step is a subgenomic negative-strand RNA that is the complement of the mRNA. Subsequent transcription of this template produces the mRNA itself, using RI-2 structures that are also subgenomic.

Evidence can be obtained in support of both of these models, and both result in mRNAs that have common 5′ sequences (the leader) and common 3′ regions. This nested set of mRNAs is observed during coronavirus infection. Both full-length and subgenomic replicative intermediates can be found in cells at various times after infection. Much of the evidence obtained with the SARS coronavirus and with other related viruses tends to support the second of these models, that is, discontinuous transcription during negative-strand synthesis.

The specific mechanism of the transcriptase jumping in each model is proposed to involve transcriptional regulating sequences that contain core elements recognized in protein–RNA interactions. The net result, however, is that each mRNA has the same 5′ leader sequence and therefore has only one sequence of RNA needing to be capped. The addition of the polyA tracts onto the individual mRNAs also only requires the recognition of one sequence on the positive-sense template by viral replicase since all mRNAs have the same 3′ end. An alternative possibility is that the polyA is template-derived, coming from transcription of a common polyU sequence present at the 5′ end of the subgenomic negative strands.

Cytopathology and disease caused by coronaviruses

Certain coronaviruses, along with the rhinoviruses, can cause mild and localized respiratory tract infections (head colds). The mildness of colds results from a number of both viral and cellular factors. First, the viruses causing the common cold have a very defined tissue tropism for nasopharynx epithelium. Spread of the virus is limited by ill-defined localized immune factors of the host. The ability of a cold virus infection to remain localized at the site of initial infection is a great advantage to the virus. Local irritation leads to sneezing, coughing, and runny nose – all important for viral spread. Mildness and localization of the infection tend to limit the immune response, which is another distinct advantage. A mild infection results in short-lived immunity, and this, along with the fact that a large number of serotypes exist as a result of the high error frequency of the genome replication process, mean that colds are a common and constant affliction.

In the late winter and spring of 2003, a new illness called SARS broke out, focused in China and Singapore. SARS proved to be more than the common cold, having a case fatality rate of 10–20%. The etiologic agent of SARS is SARS-CoV. Although the original transmission to humans was apparently from the civet cat, a recent report suggests that the natural reservoir host for SARS-CoV is one of several species of bats.

SARS-CoV has been shown to utilize the cellular protein ACE2 as a receptor to initiate viral entry into the cytoplasm of the infected cell. One model that has been proposed for the high mortality induced by this virus involves the role of angiotensin in acute lung injury. ACE2 converts angiotensin from a form that induces tissue damage and lung edema into a form of the protein that is more benign. Infection with SARS-CoV appears to cause downregulation of this enzyme, an event that is proposed to be significant in the pathogenesis of this virus.

Coronaviruses were in the news again in 2012 when an outbreak of SARS occurred in the Middle East. In this case, the virus in question was different and, after isolation, has been named the Middle East respiratory syndrome coronavirus (MERS-CoV). This virus, transmitted to humans from camels, is quite similar to SARS-CoV and has the same molecular features for replication and transcription.

On December 31, 2019, researchers in Wuhan, China, reported to the World Health Organization (WHO) concerning a cluster of respiratory illnesses that were determined to be due to a novel coronavirus. It was clear from the outset that this virus was more like SARS and MERS in its pathogenicity than like the other four human coronaviruses, all agents of the common cold.

Over the next month, the virus would shut down Wuhan and much of the surrounding region, eventually making its way to other parts of the world to become a declared pandemic. WHO named this virus SARS-CoV-2, recognizing its close relationship to what is now known as SARS-CoV-1, the agent of the 2002–2004 outbreak. The disease syndrome was designated coronavirus disease 2019 (COVID-19), a name that soon became synonymous with an international public health crisis. On January 30, 2020, WHO declared COVID-19 a public health emergency of international concern. The outbreak was declared a pandemic by WHO on March 11, 2020.

The speed at which research on SARS-CoV-2 moved was a tribute to modern molecular science and rapid research communication. By the end of January 2020, the sequence of the viral genome had been determined and detailed structural analyses of the particles were carried out using high-resolution cryo-electron microscopy. In April 2020 the first Phase I clinical trial of a vaccine candidate was initiated, and by July 2020 over 100 vaccines had entered clinical trials worldwide, including two entering Phase III trials. This represented unprecedented speed by the scientific community, regulators, governments, and industry to mobilize against this new pathogen.

Evidence from comparative genomic sequencing indicated that this virus originated in bats and, after likely passing through an intermediate host, became infectious for humans. This pathway of emergence has also been shown for SARS-CoV-1 (bat to civet cat to human) and MERS (bat to camel to human).

Like SARS-CoV-1, this new coronavirus uses ACE2 as a cell surface feature to which the viral spike protein binds to gain entry. Unlike the earlier virus, however, the receptor-binding domain of the SARS-CoV-2 spike is buried until proteolytic activation by host cell enzymes such as furin or the serine protease TMPRSS2. The virus then enters the cell by fusion with the host membrane.

Similar to SARS and MERS, the COVID-19 syndrome is characterized by a high fever and cough during the early stages of infection. While all three of these viruses can progress to an acute respiratory disease (ARD) in the lower lungs, which can be fatal, COVID-19 does not have the mortality rate of SARS or MERS, with SARS and MERS killing roughly 25% of those that were hospitalized, compared to 1–2% for COVID-19. Nonetheless, because SARS-CoV-2 may be more efficient at human-to-human spread, it resulted in many more people worldwide dying of COVID-19 than died of SARS or MERS.

Like all human coronaviruses, robust humoral and cell-mediated immune responses are generated in response to the natural infection, but protective immunity may be relatively short-lived. In fact, neutralizing antibody levels in the serum drop to undetectable levels within several months after infection in a large number of people. As a result, reinfections can occur, though there is evidence that reinfections are less severe, and less likely to be life-threatening.

REPLICATION OF PLANT VIRUSES WITH RNA GENOMES

A large number of plant viruses contain RNA genomes, and many of the early discoveries in virology were accomplished with plant viruses. The discovery of viruses as specific infectious particles at the end of the nineteenth century focused on work to elucidate the cause of tobacco mosaic disease, culminating in the first description of the tobacco mosaic virus (TMV). This virus took center stage for a number of important early events in biochemical virology, including the first crystallization of a virus particle by W. M. Stanley at University of California, Berkeley; demonstration of the infectious nature of a positive-sense RNA genome by Gierer and Schramm; and *in vitro* assembly from isolated protein and RNA of an infectious particle by H. Fraenkel-Conrat.

The majority of plant RNA viruses are nonenveloped and have single-stranded genomes. The exceptions are two groups of plant viruses with negative-sense genomes (the plant rhabdoviruses and the *Tospovirus* genus of the bunyavirus family) and one group with dsRNA genomes (e.g., the wound tumor virus).

All of the positive-sense plant RNA viruses have genomes that can be translated entirely or in part immediately after infection. Structure of the genome RNA is varied (Table 14.1). The 5′ end may be capped or may have a covalently linked genome protein similar to picornavirus VPg. The 3′ end may be polyadenylated or not, or may be folded into a tRNA-like structure that can actually be charged with a specific amino acid. There appears to be no role in virus

Table 14.1 Genomic structure of some positive-sense RNA viruses infecting eukaryotes.

Virus	Number of Genome Segments	5′ End	3′ End
Poliovirus	1	VPg	PolyA (genome encoded)
Yellow fever virus	1	Methylated cap	NonpolyA
Sindbis virus	1 (expresses subgenomic mRNA)	Methylated cap	PolyA (A)
Coronavirus	1 (expresses nested subgenomic mRNA)	Common leader with methylated cap	PolyA (A)
Tobacco mosaic virus	1	Methylated cap	tRNAhis
Potato virus Y	1	VPg	PolyA
Tomato bushy stunt virus	1	Methylated cap	NonpolyA
Barley yellow dwarf virus	1	VPg	NonpolyA
Tobacco rattle virus	2	Methylated cap	NonpolyA
Cowpea mosaic virus	2	VPg	PolyA
Brome mosaic virus	3	Methylated cap	tRNAtyr

translation for this tRNA, but the fact that the cytoplasm of eukaryotic cells has an enzyme that functions to regenerate the CCA at the 3′ end of tRNA molecules suggests that the tRNA structure may provide the viral genome with a means of avoiding exonucleolytic degradation from the 3′ end.

While expression of the positive-sense RNA genomes of plant viruses follows the same general rules outlined for replication of corresponding animal viruses, there is an added complication. A number of plant virus RNA genomes are segmented. This segmentation means that individual mRNA-sized genomic fragments can be (theoretically, at least) independently replicated and translated. Independent replication and translation allow the virus to maintain a replication cycle in which individual viral genes can be expressed at significantly different levels.

Use of this strategy in virus replication adds the complication that the packaging process is potentially very inefficient. This is certainly true for the packaging of influenza virus described in Chapter 15. Alternatively, the packaging process might be controlled in some way to ensure that each viral particle gets its requisite number of genomic fragments. Despite this complication, segmented genomes are a viable strategy for RNA virus replication, and it is not clear why it is not used in the replication of any known positive-sense animal viruses.

With viruses of vascular plants, the limitations in the size of objects that can pass through the cell wall led to another adaptation. The plant viruses with segmented positive-sense RNA genomes package each segment *separately*. Although this separate packaging means that each cell must be infected with multiple virions, plant viruses seem to thrive using this approach, probably for the following reason: Plant viruses are often transmitted mechanically and then spread from cell to cell via the plant's circulation without involvement of a specific immune defense; therefore, high concentrations of virus at the surface of the cell can be maintained.

Viruses with one genome segment

TMV has a helical capsid that encloses a single RNA genome segment of 6.4 kb. Primary translation of the genome produces the replicase complex consisting of the 126-kd and 183-kd replication proteins. Two subgenomic mRNAs are transcribed from negative-sense RNA generated from RI-1. The translation of these two species yields the 17.5-kd coat protein and a 30-kd protein involved in movement of the virus within the infected plant.

Tomato bushy stunt virus has a single RNA genome of 4.8 kb packaged into an icosahedral capsid. Translation of the capped genome results in production of the 125-kd viral replicase. Two subgenomic mRNAs are transcribed from the full-length negative-sense strand generated from RI-1. Translation of these two species leads to synthesis of the 41-kd coat protein and two other proteins thought to be required for cell-to-cell movement of the virus.

Viruses with two genome segments

The genome of cowpea mosaic virus consists of two separate strands of RNA packaged into *separate* icosahedral particles. Since both strands are required for infection, a cell must be infected together by each of the two particles. The larger of the two RNAs (5.9 kb) is translated into a polyprotein that is cleaved into a 24-kd protease, the 4-kd VPg, a 110-kd replicase, and a 32-kd processing protein. The smaller (3.5-kb) RNA encodes a polyprotein that is cleaved into the 42-kd and 24-kd coat proteins and a set of proteins required for cell-to-cell movement of the virus.

Viruses with three genome segments

Brome grass mosaic virus has three separate RNA genome strands (3.2 kb, 2.8 kb, and 2.1 kb) contained in *three separate* icosahedral particles. Again, since all three genome segments are required for infection, cells must receive one of each of the particles. Each of the capped genome segments is translated into a protein. These products include the 94-kd viral replicase, a 109-kd capping enzyme, and a 32-kd cell-to-cell-movement protein. In addition, one of the RNAs is transcribed into a subgenomic mRNA that encodes the 20-kd viral coat protein.

REPLICATION OF BACTERIOPHAGES WITH RNA GENOMES

The great majority of well-characterized RNA bacteriophages have linear, single-stranded, positive-sense genomes enclosed within small, icosahedral capsids. These phages (grouped together as the **Leviviridae**) include the male bacteria-specific phage Qβ, MS2, and R17, which attach to the bacteria's F pili.

In broad outline, the replication process of these RNA-containing bacteriophages follows that described for eukaryotic viruses. Infection begins with a translation step, and replication of the viral genome occurs through production of the RI-1 and RI-2 intermediates described earlier in this chapter.

Regulated translation of bacteriophage mRNA

There is a major difference in the way protein synthesis occurs on bacterial ribosomes as compared to eukaryotic ribosomes, and this leads to a significant difference in the way that expression of viral-encoded protein is controlled. As discussed in Chapter 13, bacterial ribosomes can initiate translation at start sites in the interior of bacterial mRNA. This means that a bacterial mRNA molecule with several ORFs can be translated independently into one or all of the proteins. In an RNA bacteriophage infection, protein synthesis programmed by the incoming genome is characterized by synthesis of viral RNA replicase only. Later in infection, after genome replication begins, transition to synthesis of capsid and other proteins begins.

This temporal regulation is governed by the secondary structure of the genome, and initiation of protein synthesis encoded by interior ORFs by ribosomal mechanisms. This can be seen in the phage Qβ, which is diagrammed in Figure 14.11. This virus encodes three distinct

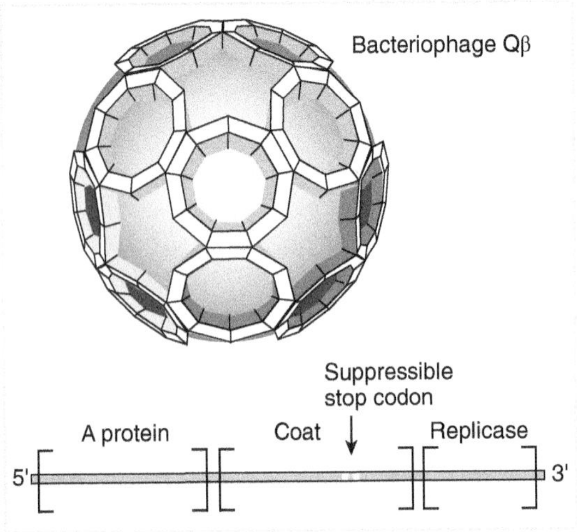

Figure 14.11 The approximately 25-nm-diameter icosahedral capsid of positive-sense RNA bacteriophage Qβ. The positive-sense RNA genome contains three separate open reading frames (ORFs). These ORFs can be independently translated from the full-length virion RNA because, unlike the situation in eukaryotic viruses, bacterial ribosomes can initiate translation at interior start signals provided that the ribosome can interact with them. With this bacteriophage, ribosome attachment and translation require active transcription to allow the nascent positive-sense RNA to be unfolded so that the translation start is accessible.

translational reading frames encoding genes for the A (maturational) protein, the coat protein, and replicase. The coat protein translational reading frame has a translation terminator that is misread (suppressed) as a tryptophan residue about 1% of the time, and when this happens, a larger capsid protein with additional amino acids is generated. Suppression of the termination is absolutely required for phage replication.

A portion of the replication cycle of Qβ is shown in Figure 14.12. Ribosomes can associate with the genomic RNA, but this positive-sense genome is folded in such a way that the only start codon available for interaction with a ribosome is the one that begins translation of phage RNA replicase. All other start codons are involved in base-pairing interactions as a part of the secondary structure. For this reason, replicase is the only phage protein expressed at the start of infection.

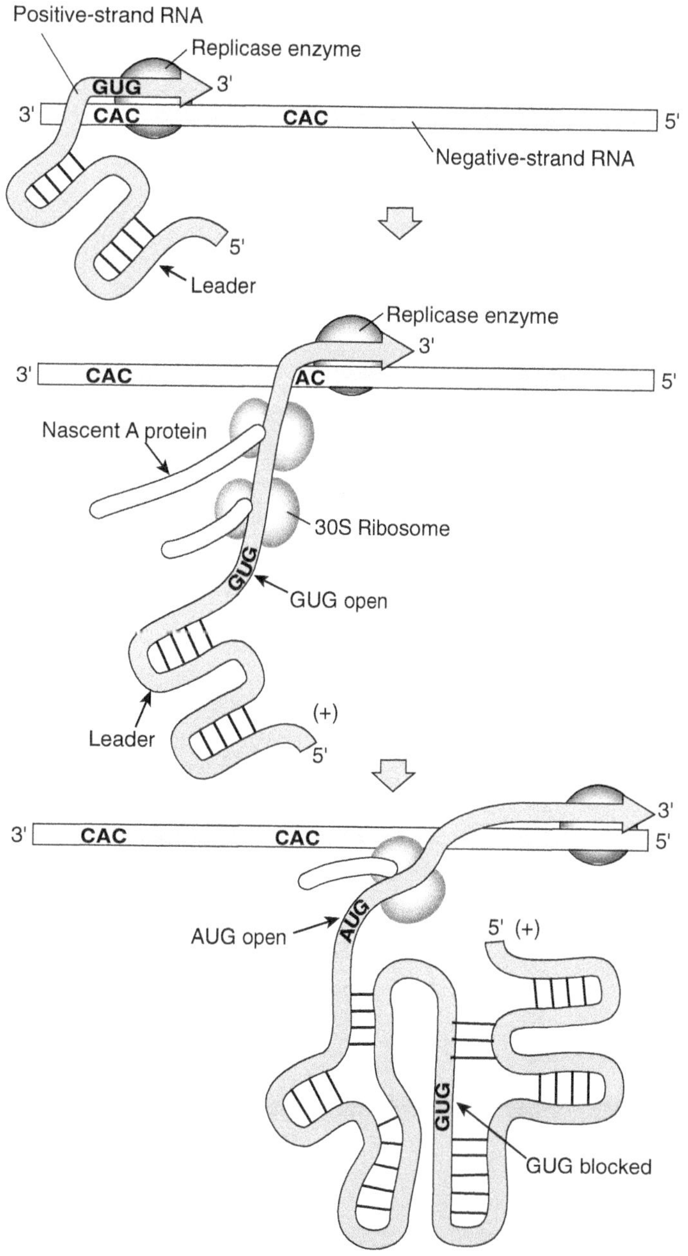

Figure 14.12 Coupled transcription–translation of bacteriophage Qβ RNA results in opening the blocked translational start site for the A (maturational) and coat proteins. As the replicase enzyme passes the region containing the translation start site on the negative-sense template (which is a GUG for the A protein), the nascent positive-sense mRNA can interact with a ribosome before it has a chance to fold into a structure in which this initiator codon is sterically blocked. Multiple-ribosome entry results in translation of a large number of copies of the maturational and coat proteins being synthesized. High levels of coat protein specifically inhibit translation of replicase from full-length genomic RNA so that replicase is only synthesized at early times in the replication cycle. For this reason, it is often termed an "early" protein or gene product.

Synthesis of new positive-sense genomes takes place through formation of RI-1 and RI-2. As new positive-sense genomic RNA disassociates from the negative-sense template near the replicase, secondary structure has not yet formed. This results in the start codon for the A and coat proteins being available to begin translation. The A protein uses a GUG instead of an AUG initiation codon. Similarly, newly replicated positive-sense strands immediately interact with ribosomes to yield the capsid proteins necessary for the formation of new virus particles.

This simple mechanism ensures that the earliest protein expressed will be replicase. Further, since a relatively large amount of RI-2 will need to be present, synthesis of A and capsid proteins will only occur when there are a large number of genomes waiting to be encapsidated. Multiple entry of ribosomes onto the nascent viral mRNA ensures that a large amount of structural protein will be available when necessary.

Finally, the phage controls the amount of replicase synthesized in infection so that progeny positive-sense strand does not end up recycling too long. Such control is accomplished by the capsid protein actually inhibiting synthesis of replicase from mature positive-sense RNA. Therefore, after about 20 minutes, increasing levels of capsid proteins shut off replicase synthesis.

Case 1: Enteroviruses

Clinical presentation/case history: Patient is an 18-year-old female who presented with a two- to three-week history of upper respiratory symptoms and myalgias (sore or aching muscles) that spontaneously resolved, followed by the development of severe headache, nausea, and vomiting two days prior to admission. She presented to the ER, where blood tests revealed a high white blood cell count of 38 000/µl (normal is 4000–12 000/µl). Further case history revealed that her entire family had also experienced similar symptoms. Her father had also developed a severe headache accompanied by delirium, which had spontaneously resolved. Her twin brothers had also developed similar symptoms accompanied by a rash.

Diagnosis: The patient was admitted for presumptive diagnosis of meningitis, supportive care, and intravenous (IV) antibiotics. Magnetic resonance imaging (MRI) of the brain showed diffuse, symmetric parenchymal edema of the cortical gray matter and brain stem consistent with meningoencephalitis. In order to differentiate between bacterial meningitis and viral meningitis, a spinal tap was performed. Analysis of the cerebrospinal fluid (CSF) revealed no evidence of bacterial antigens, normal levels of glucose, and the presence of neutrophils. This was consistent with a viral meningoencephalitis. Viral analysis (polymerase chain reaction [PCR] and reverse transcription PCR [RT-PCR]) of CSF detected coxsackievirus.

Treatment: There is no treatment for coxsackievirus infections, and only supportive care can be offered. Proper disinfection and handwashing practices are important to prevent transmission of enteroviruses to susceptible individuals.

Disease notes: Enteroviruses are transmitted by the oral-fecal route and are highly infectious. They often cause subclinical or clinically benign cold-like or mild gastrointestinal symptoms. However, a number of members of the *Enterovirus* genus are associated with a variety of more severe symptoms, including infections of the brain (meningitis and encephalitis), infections of the heart (myocarditis and pericarditis), muscle pains that can resemble a heart attack, and hand-foot-and-mouth disease, which is a vesicular rash associated with a fever that is common among young children, particularly in daycare settings.

Case 2: Zika-Induced Guillain–Barré Syndrome

Clinical presentation/case history: In December 2015, a 69-year-old Colombian male with no past medical history developed fever, rash, headache, joint and muscle pain, conjunctivitis, and fatigue for one week. On his seventh day of illness, he developed bilateral lower-extremity flaccid paralysis that ascended the following day to his upper extremities, trunk, and neck. At this time, the patient was hospitalized in the intensive care unit and found to have absence of neurologic reflexes, loss of sensation, and respiratory failure requiring intubation and mechanical ventilation for 10 days. Electromyography, which measures the electric activity of muscles, showed reduced activity consistent with sensory-motor nerve demyelination. The patient received intravenous immunoglobulin during this time and began to improve. One month after his admission to the hospital and after physical therapy, the patient made a complete recovery and was discharged.

Diagnosis: This patient was diagnosed with Guillain–Barré syndrome after viral infection during the Zika epidemic in Colombia. The diagnosis of Guillain–Barré syndrome was based on clinical symptoms of bilateral flaccid paralysis of extremities, loss of deep tendon reflexes, the time course of development of neurologic symptoms, and nerve conduction studies consistent with Guillain–Barré syndrome.

Guillain–Barré syndrome may be seen following infection with multiple viruses that present with similar symptoms, including dengue virus, chikungunya virus, influenza virus, and Zika virus (ZIKV). The clinical symptoms for these viral infections are virtually indistinguishable, highlighting the importance of virologic and serologic diagnostics to differentiate the infections. ZIKV RNA can be detected in the blood by reverse transcription followed by polymerase chain reaction during the first 7–10 days of infection, and in the urine for approximately 10–20 days. Due to cross reactivity between antibodies against dengue virus and ZIKV, there is currently no specific antibody-based diagnostic test for ZIKV. ZIKV infection is usually diagnosed clinically based on symptoms and on evidence of virologically confirmed cases in the same area during the same time period.

Treatment: Treatment of ZIKV infection involves supportive care, with treatment of fever and pain with acetaminophen. Treatment of standard Guillain–Barré syndrome is based on removal or dilution of cross-reactive antibodies that cause nerve demyelination by, respectively, plasmapheresis or administration of intravenous immunoglobulin. However, it is currently unknown if ZIKV-induced Guillain–Barré syndrome works via antibody-mediated mechanisms, and it is unclear if these treatments are beneficial in ZIKV-induced Guillain–Barré syndrome.

Disease notes: In 2015 and 2016, there was an epidemic of Zika fever in South and Central America that started in Brazil and spread to over 40 countries. This outbreak included over 1.3 million suspected cases in Brazil and over 50 000 suspected cases in Colombia. The initial symptoms of ZIKV infection seen in this patient were typical of Zika fever, which is similar to the symptoms of dengue fever. The development of Guillain–Barré syndrome, however, is much less common; it occurs in about 1 in 1000 cases of Zika disease. ZIKV infection is particularly detrimental to the developing fetus since it can infect and kill developing neurons, leading to neurological defects including microcephaly. ZIKV is spread primarily by *Aedes aegypti* mosquitoes but may also be spread by *Aedes albopictus*. ZIKV can also be transmitted sexually and by other exchange of body fluids.

QUESTIONS FOR CHAPTER 14

1 What are the steps in the attachment and entry of poliovirus in a susceptible host cell?

2 The Picornaviridae (e.g., poliovirus) have, as their genome, one molecule of single-stranded RNA. This genomic RNA functions in the cell as a monocistronic mRNA. However, picornavirus-infected cells contain 10 or more viral proteins.

(a) What mechanism have these viruses evolved such that this monocistronic mRNA produces this large number of translation products?

(b) The poliovirus mRNA does not have a 5′ methylated cap that is present on host cell mRNA. How do host cell ribosomes begin translation of this message?

3 Foot-and-mouth disease virus (FMDV) is a member of the family Picornaviridae. Based on your knowledge of the properties of members of this family, complete the following table with respect to FMDV and each of the characteristics listed. State whether the characteristic is present or absent.

Characteristic	Present or Absent for FMDV
5′ methylated cap	
Subgenomic RNAs	
3′ polyadenylation	
Single-stranded, positive-sense genome	
Expression of genome as a polyprotein	

4 The poliovirus genome is a single-stranded RNA of about 7500 nucleotides, with a covalently linked terminal protein, VPg, at the 5′ end and a polyA sequence at the 3′ end. The polyA tail is not added after replication but is derived from the template during replication. VPg is important for replication of this viral RNA, along with poliovirus polymerase and certain host enzymes.

There are two models for the action of VPg:

Model 1. VPg may act as a primer for RNA synthesis, being used as VPg-pU_{OH}.

Model 2. VPg may act as an endonuclease, attaching itself to the 5′ end of a new RNA chain. In this model, RNA synthesis is primed after addition of U residues to the 3′ A at the end of the genome by a host enzyme, followed by a loop-back and self-priming mechanism.

Given these two models, imagine that you have an *in vitro* system to test the properties of poliovirus genome replication. Your system contains viral genomic RNA as a template and all of the necessary proteins, except as indicated here:

(a) Assume that model 1 is true. What would you expect to see as the product of the reaction if VPg was left out of the mixture?

(b) Assume that model 2 is true. What would you expect to see as a product of the reaction if endonucleolytic activity of VPg was inhibited?

5 Draw the structures of the poliovirus RI-1 and RI-2. What are the similarities and differences for these two structures?

6 Which of the following statements is (are) true in regard to the poliovirus genome?

(a) It lacks posttranscriptional addition of repeating adenines.

(b) It is approximately 1400 bases long.

(c) It contains a VPg protein that is cleaved prior to packaging.

(d) It has a single precursor protein that is cleaved by cellular cytoplasmic nucleases.

7 How are the structural proteins of Sindbis virus generated during the infectious cycle?

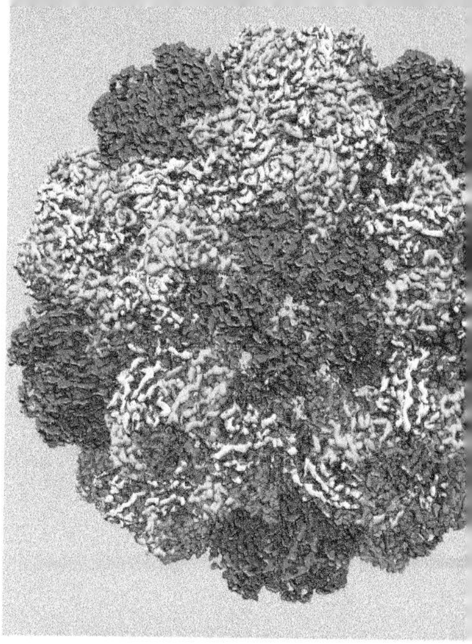

CHAPTER 15

Replication Strategies of RNA Viruses Requiring RNA-directed mRNA Transcription as the First Step in Viral Gene Expression

Basic Virology, Fourth Edition. Martinez "Marty" Hewlett, David Camerini, and David C. Bloom.

A significant number of single-stranded RNA (ssRNA) viruses contain a genome that has a sense *opposite* to messenger RNA (mRNA) (i.e., the viral genome is *negative-sense RNA*). To date, no such viruses have been found to infect bacteria, and only one type infects plants. But many of the most important and most feared human pathogens, including the causative agents for flu, mumps, rabies, and a number of hemorrhagic fevers, are negative-sense RNA viruses.

The negative-sense RNA viruses generally can be classified according to the number of segments that their genomes contain. Viruses with **monopartite** genomes contain a single piece of virion negative-sense RNA, a situation equivalent to that described for the positive-sense RNA viruses in Chapter 14. A number of groups of negative-sense RNA viruses have **multipartite** (i.e., *segmented*) genomes. Viral genes are encoded in separate RNA fragments, ranging from two for the arenaviruses to eight for the orthomyxoviruses (influenza viruses). As long as all RNA fragments enter the cell in the same virion, there are no special problems for replication, although the packaging process during which individual segments must all fit into a single infectious virion can be inefficient.

It is well to remember that there is a fundamental difference in the replication strategy of a negative-sense RNA virus as compared to a positive-sense RNA virus. Since the virus must have the infected cell translate its genetic information into proteins, it must be able to express mRNA in the infected cell. With a negative-sense RNA virus, this will require a *transcription* step: Genetic information of the viral genome must be transcribed into mRNA. This presents a major obstacle because the cell has no mechanism for transcription of mRNA from an RNA template.

The negative-sense RNA viruses have overcome this problem by evolving means of carrying a special virus-encoded enzyme – an **RNA-dependent transcriptase** – in the virion. Thus, viral structural proteins include a few molecules of an enzyme along with the proteins important for structural integrity of the virion and for mediation of its entrance into a suitable host cell. Clearly, the isolated genome of negative-sense RNA viruses cannot initiate an infection, in contrast to the positive-sense RNA viruses discussed in Chapter 14. Other groups of viruses (notably retroviruses, discussed in Chapter 19) include enzymes important to mRNA expression in their virion structures, but focusing on negative-sense RNA viruses' replication strategies provides useful general considerations.

One of the more interesting general questions concerning these viruses is, how did they originate? Sequence analyses of replicating enzymes encoded by different viruses often demonstrate similarities to cellular enzymes, implying a common function and suggesting a common origin. While the cellular origin of most viral enzymes can be established by sophisticated sequence analysis, this has yet to be accomplished with RNA-directed RNA transcriptases. Initially it was concluded that the RNA-to-RNA pathway was limited to the world of viruses. However, a number of examples of cellular enzymes that carry out the same or similar reactions have come to light. Among these are the complex of enzymes involved in RNA interference. To date, a good candidate for a common progenitor enzyme has yet to be identified. When this is accomplished, more definitive statements can be made concerning origins of these viruses.

The fact that no bacterial viruses with this replication strategy have been identified is at least consistent with the possibility that negative-sense RNA viruses are of recent origin. A recent origin would imply that all the negative-sense RNA viruses are fairly closely related to each other, and there is some evidence that this is the case.

REPLICATION OF NEGATIVE-SENSE RNA VIRUSES WITH A MONOPARTITE GENOME

There are eight "families" of negative-sense RNA viruses that package their genomes as a single piece of RNA:

Bornaviridae, Mymonaviridae, Filoviridae, Nyamiviridae, Paramyxoviridae, Pneumoviridae, Rhabdoviridae, and Sunviridae

They all share some similarities of gene order and appear to belong to a common "superfamily" or order: Mononegavirales.

Interestingly, despite genetic relatedness of these viruses, they do not share a common shape, although all are enveloped. Also, the rhabdovirus family contains several members that infect plants. Is this a "recent" radiation to a new set of hosts? Whatever the answer to this question, there is no doubt that the Mononegavirales viruses are a successful group with significant pathologic implications for humans and other vertebrates.

Human diseases caused by the viruses of this order include relatively mild flu-like respiratory disease (parainfluenza) caused by a paramyxovirus. More severe diseases include mumps, measles, hemorrhagic fevers with high mortality rates caused by Marburg and Ebola virus (filoviruses), and neurological diseases ranging from relatively mild ones caused by bornavirus to the invariably fatal encephalitis caused by rabies virus (a rhabdovirus). The diseases characterized by high mortality rates are not maintained in human reservoirs but rather are zoonoses – diseases of other vertebrates transmissible to humans (see Chapter 3, Part I).

The replication of vesicular stomatitis virus – a model for mononegavirales

Infection of humans with naturally occurring strains of rabies virus leads to fatal diseases. This and other factors make this virus difficult and dangerous to work with – indeed, much of the work on it is carried out in a few very isolated laboratories in the United States, including Plum Island in Long Island Sound. In contrast, the closely related rhabdovirus vesicular stomatitis virus (VSV) is one of the most carefully studied extant viruses. Its replication strategy forms a valid model for the replication of all Mononegavirales viruses and provides important insights for the study of replication of other viruses with negative-sense RNA genomes. Remember that negative-sense viruses must have some way to turn the viral genome (virion RNA) into mRNA before infection can proceed.

The vesicular stomatitis virus virion and genome

The VSV virion and genetic map are shown in Figure 15.1. Like most rhabdoviruses, it has a distinctive bullet-shaped structure. The VSV genome encodes five proteins, all present in the virion in different amounts. The viral genome is about 11 000 bases long. Since individual

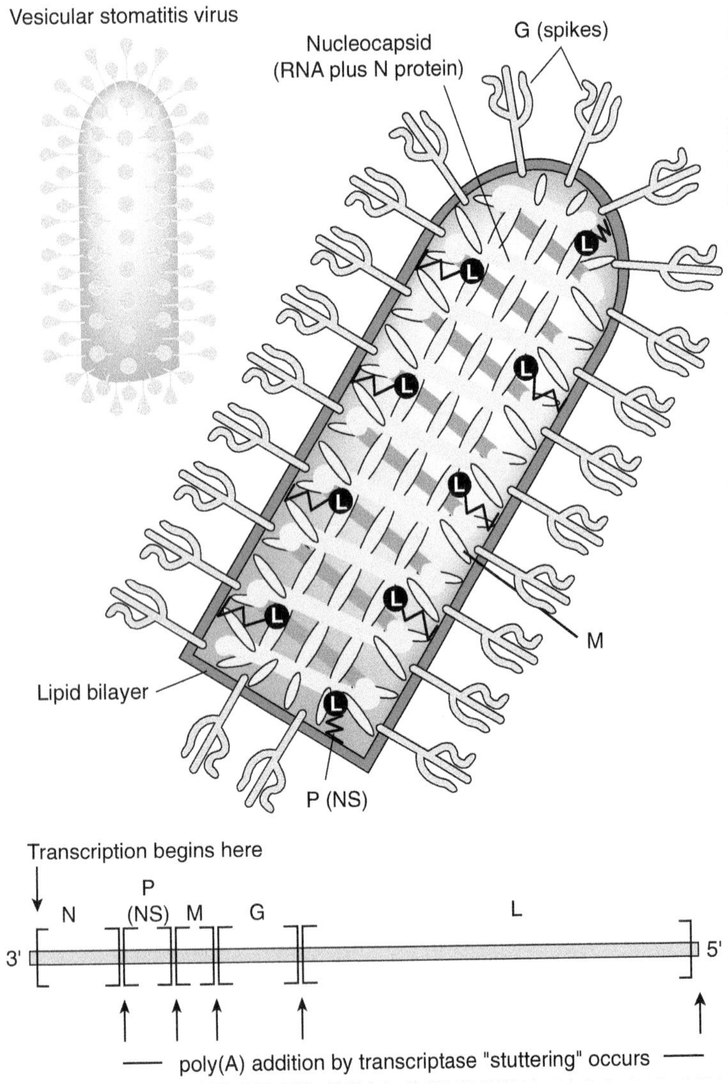

Figure 15.1 The vesicular stomatitis virus (VSV) virion. All rhabdoviruses have this characteristic bullet shape that appears to be due to the P (formally called NS) and L proteins interacting with the envelope in a specific way. The 70 × 180 nm VSV virion contains enzymes for RNA transcription that can be activated by mild detergent treatment and incubation with nucleoside triphosphates *in vitro*. The genetic map of VSV is also shown. The 11 000-nucleotide (nt) virion negative-sense strand RNA encodes five individual mRNAs; each is capped and polyadenylated by virion enzymes. *Note*: Because the genomic RNA serves as a template for mRNA synthesis, it is shown in 3′ to 5′ orientation instead of the conventional 5′ to 3′ orientation. N, Nucleocapsid; M, matrix; G, envelope glycoprotein; L, part of the replication enzyme; P (or NS), also part of the replication enzyme.

mRNAs are generated from the virion negative-sense RNA, viral genes in the genome have an order opposite to the order in which mRNAs appear in the cell. Locations of the viral genes are shown in the genetic map of Figure 15.1. The L and P proteins function together to cap mRNA, generate mRNA, polyadenylate positive-sense viral mRNA, *and* replicate the viral genome. (Remember: The virus must bring its own replication enzymes into the cell because the cell cannot deal with ssRNA that is not like mRNA.)

Generation, capping, and polyadenylation of mRNA

The first part of the VSV replication cycle is outlined in Figure 15.2. Virus attachment and internalization occur by receptor-mediated endocytosis. The virion does not fully disassemble in the infected cell. The intact ribonucleoprotein (RNP) nucleocapsid contains the genomic negative-sense template and transcription/replication enzymes. The virion-associated transcription/RNA replication enzyme initiates and caps each of the five discrete positive-sense mRNAs within this transcription complex. Like the positive-sense RNA viruses expressing capped mRNA, this capping takes place in the cytoplasm.

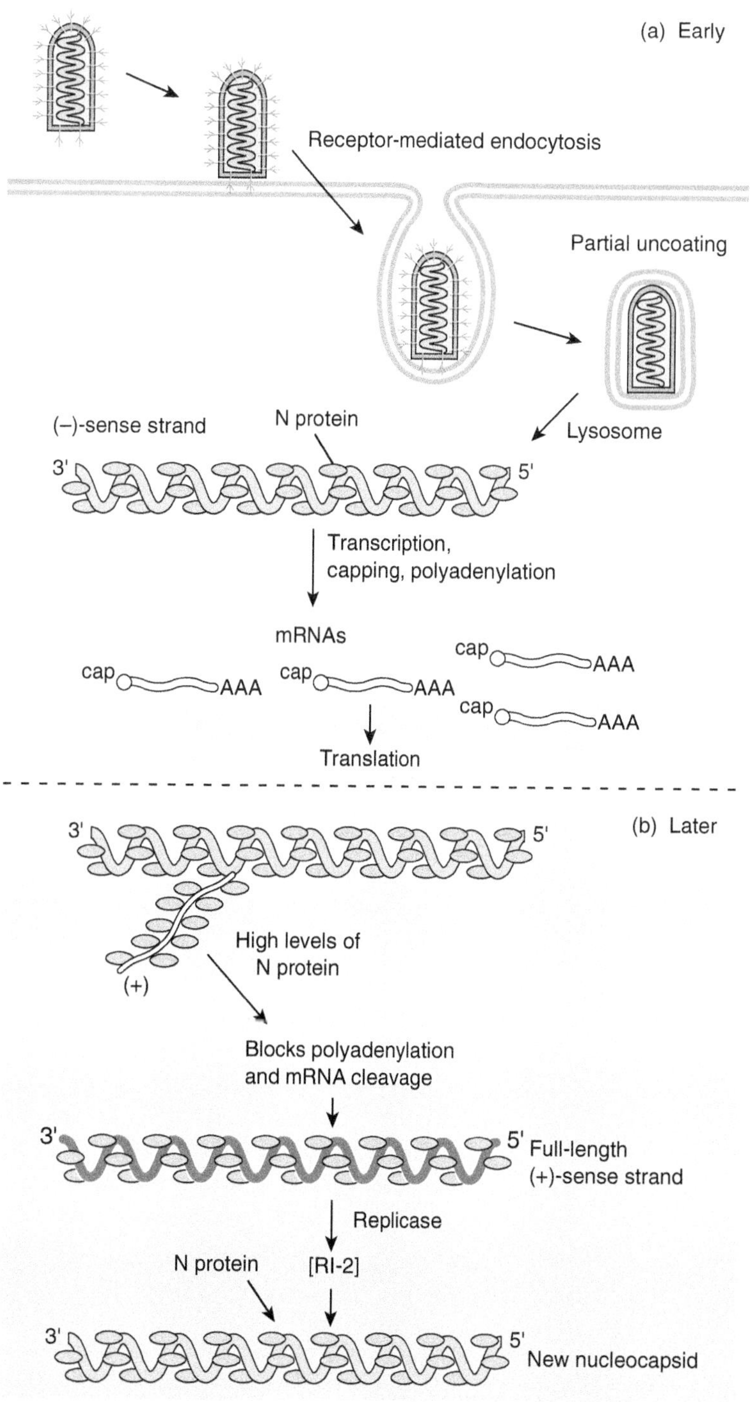

Figure 15.2 The VSV replication cycle. (a) Early events in infection begin with virus attachment to the receptor followed by receptor-mediated endocytosis and partial uncoating to virion ribonucleoprotein (RNP). This is transcribed into mRNAs that are translated in the cytoplasm. (b) Later, as protein synthesis proceeds, levels of the N (nucleocapsid) protein increase, and some nascent positive-sense strand from RI-1 associates with it. This association with N protein blocks the polyadenylation and cleavage of individual mRNAs, and the growing positive-sense strand becomes a full-length positive-sense strand complement to the viral genome that serves as a template for negative-sense RNA synthesis via RI-2. (c) At still later times in the replication cycle, viral proteins associate with the nucleocapsids made up of newly synthesized negative-sense genomic RNA and N protein. These migrate to the surface of the infected cell membrane, which has been modified by the insertion of viral G protein translated on membrane-bound polyribosomes. M protein aids the association of the nucleocapsid with the surface envelope, and virions form by budding from the infected cell surface.

Interestingly, while the cap structure is identical to that found on cellular mRNA, the specific phosphodiester bond cleaved in the cap nucleoside triphosphate is different from that cleaved in the nucleus by cellular enzymes. This difference suggests that this enzymatic activity was probably not simply "borrowed" by the virus from an existing cellular capping enzyme, but was derived from some other enzymatic activity of the cell.

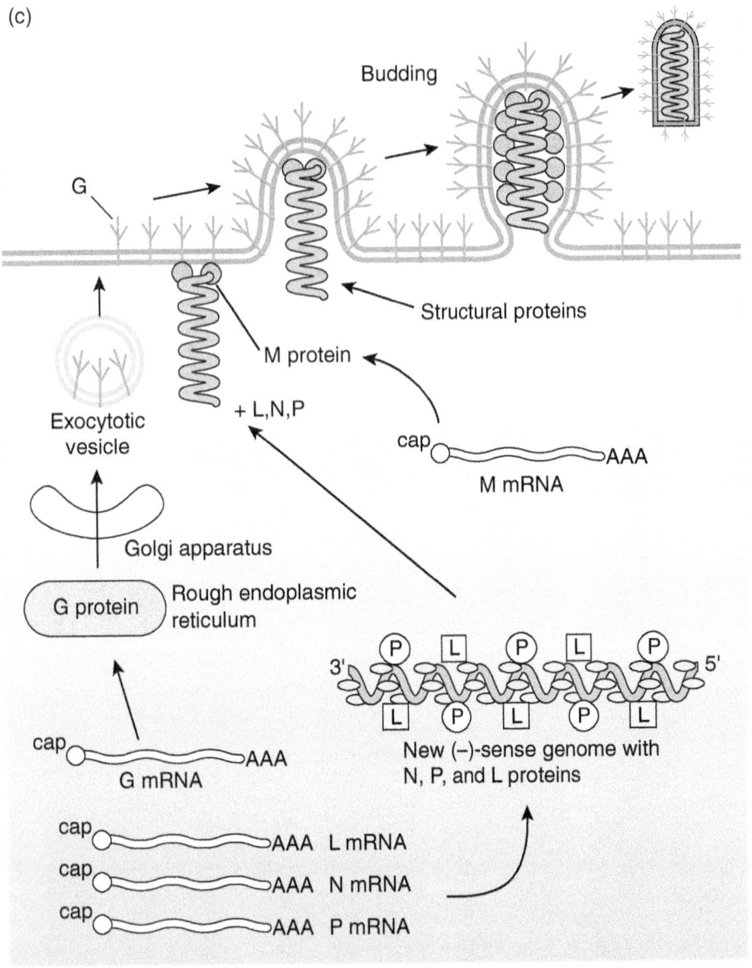

Figure 15.2 *Continued*

The polyA tails of the mRNA species are generated at specific sites on the negative-sense genome by a "rocking" mechanism in which the enzyme complex "stutters" and generates a long polyA tail and releases mRNA. The enzyme can then release, start over, or continue on. The process is outlined in Figure 15.3; this biochemical "decision" has resulted in a polarity of abundance of viral mRNA and the proteins encoded: nucleocapsid (N) protein > P (NS) protein > matrix (M) protein > envelope glycoprotein (G) > L protein.

The generation of new negative-sense virion RNA

Negative-sense virion RNA can only be generated from a full-length positive-sense template in an RI-2 complex. But the partially disrupted virion generates mRNA-sized pieces of positive-sense strand. As shown in Figure 15.3, full-length negative-sense strand is only generated when levels of N protein become high enough in the cell so that newly synthesized positive-sense RNA can associate with it. This association prevents the rocking–polyadenylation–cleavage–re-initiation process used in the generation of mRNA and allows the formation of full-length template. This genome-length positive-sense template serves as the template for new virion negative-sense strand, which also associates with N, and the other structural proteins encoded by the virus.

While the process and biochemical "choice" between production of mRNA and full-length positive-sense template RNA are best characterized in the replication of VSV, it appears that very

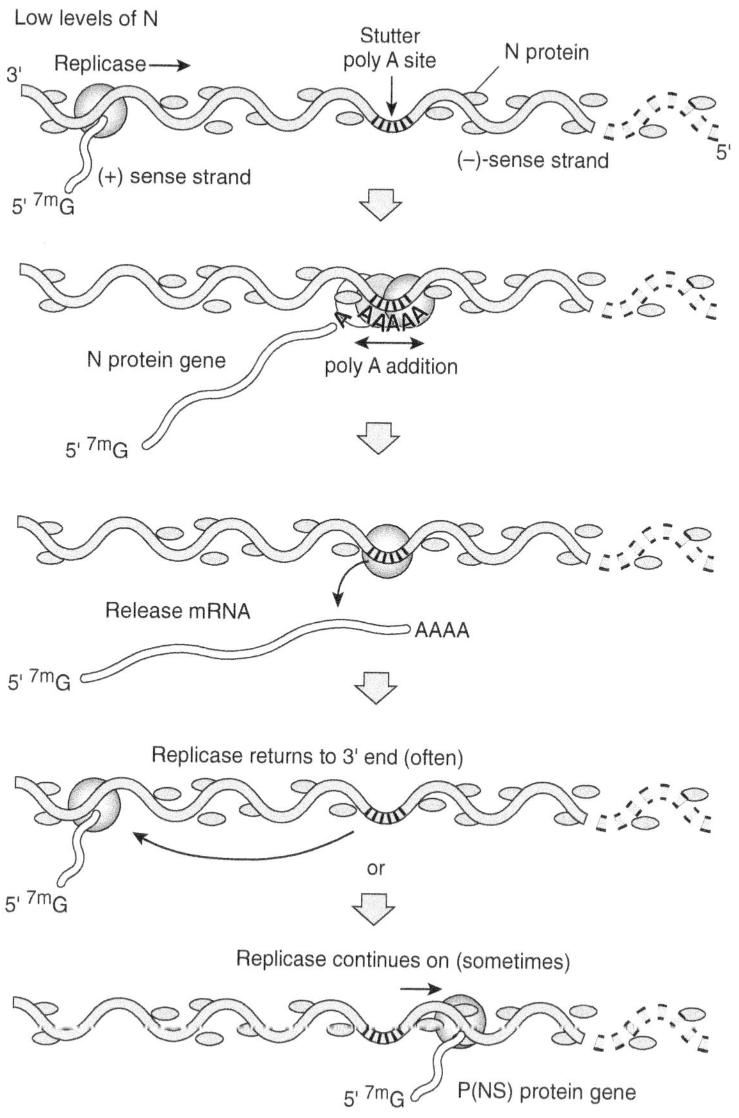

Figure 15.3 A higher-resolution schematic of the generation of positive-sense strand mRNA from genomic negative-sense strand RNA template in the absence of N protein. Polymerase associates with the template at the extreme 3′ end, and "tunnels" or "burrows" under the N protein. Transcription begins with capping of the nascent mRNA, and proceeds through the first gene on the template (the N protein) gene. At the end of this gene, the transcriptase encounters an intergenic "pause" or stutter site. The enzyme pauses here and adds a number of A residues to the nascent mRNA, which is then released. The transcriptase then either dissociates from the template and begins the process over again at the extreme 3′ end, or continues on to synthesize a transcript encoding the next gene on the genomic template. At the end of this gene, the same process occurs. Since the transcriptase has a higher probability of returning to the extreme 3′ end of the template, the mRNAs are synthesized in decreasing amounts, with those encoding N protein > P (NS) protein > M protein > G protein > L protein.

similar mechanisms exist for other viruses in the Mononegavirales order. Further, other negative-sense viruses that have multipartite genomes probably utilize equivalent mechanisms since (where characterized) their positive-sense genome templates are larger than the positive-sense mRNA expressed during infection.

The details of VSV infection and morphogenesis are generally similar to those discussed for positive-sense enveloped viruses, which are described in some detail in Chapter 14. The process of mRNA synthesis, template synthesis, and new negative-sense genome formation continues for an extended period until sufficient levels of the viral structural proteins are attained to form the virion RNPs. Virion RNP then buds through the plasma membrane and is released. These late events are outlined in Figure 15.3.

The mechanism of host shutoff by vesicular stomatitis virus

As noted, many virus infections are characterized by virus-mediated inhibition of host mRNA and protein synthesis. The mechanism of this shutoff varies with the virus in question. For example, poliovirus, which does not utilize a capped mRNA, actually inhibits the ability of capped mRNA to be translated by modification of a translation initiation factor following infection. Obviously, this mechanism cannot work with viruses that express capped mRNA.

Since VSV does not utilize the cell's nucleus during its replication, it essentially "enucleates" the host cell following infection by blocking host transcription. This enucleation is another function carried out by the viral M protein. In this role, the protein specifically interferes with the export of mRNA from the nucleus by inhibiting the nuclear transport proteins of the cell (see Chapter 13). Since some negative-sense RNA viruses, such as bornaviruses and flu viruses, utilize the nucleus for replication, this mechanism cannot be universal for negative-sense RNA viruses.

The cytopathology and diseases caused by rhabdoviruses

The disease caused by VSV involves formation of characteristic lesions in the mouth of many vertebrates (hence the name "vesicular stomatitis"). Although humans can be infected by VSV, this virus is primarily a disease of cattle, horses, and pigs. Such a wide host range seems to be a common feature of rhabdovirus infection. VSV-induced disease can be severe in animals because they cannot eat during the acute phase of infection. The course is generally self-limiting and mortality rates are not significant, provided proper care is given the affected animal. Such is obviously not possible with free-ranging cattle, and VSV outbreaks can have severe economic consequences if not properly managed.

The disease caused by the related rabies virus demonstrates a completely different strategy for virus pathogenesis and spread. The essentially 100% mortality rate for rabies is in distinct contrast to mortality rates for most viral diseases. The pathogenesis of rabies is briefly described in Chapter 4, Part I. It should be remembered that since rabies is spread by animal bites, the behavioral changes induced by the virus are important for its spread. Except under the stress of mating or in territorial disputes, vertebrates (especially carnivores, the general host for rabies) do not randomly attack and bite other members of their own species. The high replication of rabies virus in the salivary glands of the rabid host, along with excitability and other induced behavior changes, makes the infected animal a walking "time bomb." This is an excellent example of how a virus of submicroscopic proportions and encoding only a few genes can direct the billions of cells of its host animal to a single purpose: propagation of the virus.

Paramyxoviruses

Paramyxoviruses have large genomes (approximately 15 000 bases) and their replication cycle is reminiscent of that described for rhabdoviruses. One notable exception is that several (including mumps) generate mRNAs that have been edited by the addition of extra G nucleotides as the mRNAs for specific genes (notably, the mRNA for the phosphoprotein P) are expressed. The addition of these nucleotides is apparently accomplished by a stuttering step similar to that involved in the addition of polyA residues at the end of transcripts. This editing results in several variant mRNAs being expressed from a single viral gene.

Paramyxoviruses can be subdivided further into paramyxovirus proper, parainfluenza virus, mumps virus (*Rubulavirus*), measles virus (*Morbillivirus*), and pneumoviruses such as respiratory syncytial virus. The structure of Sendai virus, a typical paramyxovirus that causes respiratory disease in mice, and its genetic map are shown in Figure 15.4.

The pathogenesis of paramyxoviruses

Mumps, measles, canine distemper, and rinderpest are all caused by paramyxoviruses. Mumps is classified as a relatively benign "childhood" disease; the infection usually occurs in children just when they begin to socialize in preschool or daycare facilities. The virus spreads rapidly,

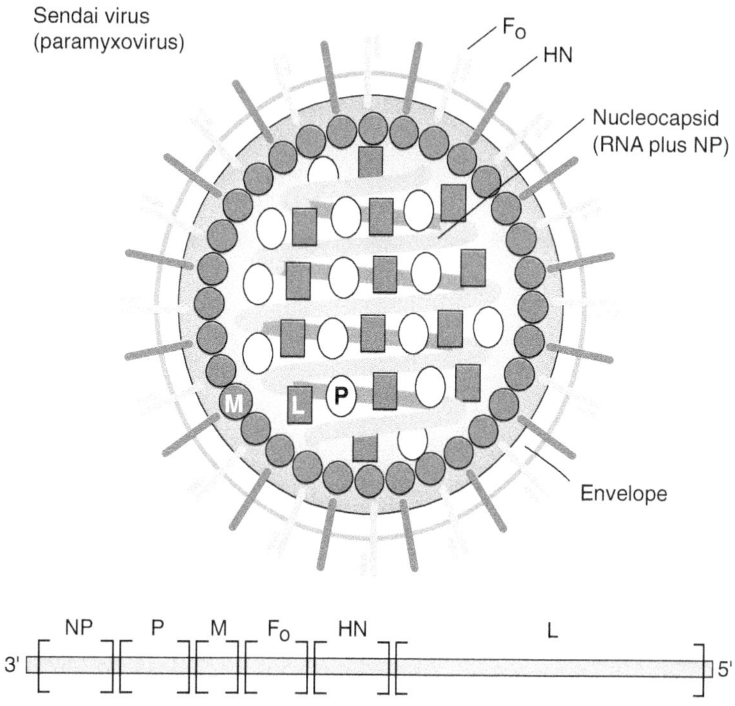

Figure 15.4 The genetic map and virion structure of Sendai virus, a typical paramyxovirus. The Sendai virion is a flexible, helical nucleocapsid that contains the 15 000-nucleotide (nt) genome and is about 18 nm in diameter and 1000 nm in length. The roughly spherical enveloped virion is about 150–200 nm in diameter. The gene marker "HN" is a membrane protein that contains both neuraminidase and hemagglutination activity. The replication strategy is similar to that outlined for VSV. Also like VSV, the negative-sense strand genomic RNA is shown 3′ to 5′ instead of in the conventional 5′ to 3′ orientation.

generally causes a mild inflammation of glandular tissue in the head and neck, and leads to lifelong immunity. Since the symptoms are generally forgotten and do not lead to any notable physiological consequences, the disease is considered mild.

Infection of postpubescent children or adults, however, can be a significantly different story. Here, the virus can infect gonadal tissue and lead to major discomfort, and occasionally to permanent reproductive damage.

The pathology of respiratory syncytial virus also is quite different for infants and adults. This virus establishes a mild, cold-like infection in an adult's nasopharynx. Following recovery, the virus can persist in the throat as a relatively normal member of the microbe population that coexists in this moist, warm environment. Since it is not invasive, the persistent infection is usually asymptomatic unless there is a complicating environmental factor. Such a factor can be very dry air in heated buildings during winter in temperate zones throughout the world. This dry air can lead to chronic respiratory irritation and mild infections by respiratory syncytial virus, as well as other pathogens. Unfortunately, the virus can spread from adults to children and infants. In hospitals, an active infection in nursery health care workers can lead to fatal epidemics in newborns whose undeveloped immune systems cannot cope with the infection.

Measles, as described in Chapter 4, Part I, although often termed a childhood disease, can cause major neurological damage to infected children, and its introduction into unprotected populations has resulted in high mortality. Measles and the closely related distemper and rinderpest viruses cause serious and often fatal diseases at all ages. Distemper infections cause high mortalities in domestic and wild animals, and the broad host range and easy transmission of canine distemper have resulted in its being a major infectious agent in marine mammals. Another related virus, rinderpest, is a serious disease of domestic cattle that has spread to wild ungulates in sub-Saharan Africa. Indeed, it was considered a greater threat than human habitat encroachment to the survival of much African wildlife, both because of its pathology and because of human efforts to stop the natural and necessary seasonal migration of wild ungulates that harbor the virus to prevent reinfection of domestic cattle. Thankfully, international efforts to eradicate this virus were mounted in the later part of the twentieth century, led by the United Nation's Food and Agricultural Organization (FAO). This culminated in the announcement by the FAO on June 28, 2011, that rinderpest virus had been eliminated globally, joining smallpox virus as the only two for which this has been accomplished. Nevertheless, the rinderpest effect on wild populations remains as a prime example of human habitat disruption leading to ecological distress. Such disruption can be a major factor in evolution of viral disease, as discussed in Chapters 1 and 22.

Filoviruses and their pathogenesis

In 1967 some medical researchers working with Ugandan African green monkeys (an important experimental animal and source of cultured cells) in Marburg, Germany, and in Yugoslavia contracted a severe hemorrhagic fever that was highly infectious to clinical staff via blood contamination. A total of 7 of 25 of these workers subsequently died of the infection. Since its first appearance, the infectious agent, termed Marburg virus, has caused several outbreaks of hemorrhagic fevers with similar mortality rates in sub-Saharan Africa, notably in Zimbabwe, South Africa, and Kenya.

In 1976 an outbreak of a similar disease with a significantly higher mortality rate (50–90%) occurred in Zaire and Sudan. Eventually, over 500 individuals were infected. A virus related to Marburg virus, named Ebola virus, was proved to be the infectious agent by identification of specific antibodies in the blood of victims and survivors. Several sporadic outbreaks of this disease have been reported in Africa since then, with the two most recent being the 2014–2016 epidemic in the West African countries of Guinea, Liberia, and Sierra Leone, as well as the 2019 epidemic in the Democratic Republic of Congo.

The high mortality rate of Ebola virus infection and its proclivity for spread to hospital workers via contaminated blood, respiratory aerosols, and body fluid contamination have made it a favorite subject for doomsayers and sensationalists in the media. Hollywood entered the scene with the 1995 movie *Outbreak*, which was generally inaccurate and misleading. Still, the properties of the disease and its ease of spread have served as a warning to public health workers and epidemiologists that acute infectious disease is a continuing threat to human society. This threat is generally discussed in Chapter 1. A major source of concern in assessing the risk posed by filoviruses is that the natural reservoir for these viruses has yet to be identified; surveys of antibody titers in a number of wild monkey populations argue against these monkeys being a reservoir. In addition, bats have been investigated and, while they can be infected with the virus in laboratory settings, no virus has been recovered from bats in the endemic areas. As of this writing, ecological studies are underway in the endemic areas of Africa in an attempt to identify the natural reservoir. Results at this writing are not conclusive, but evidence begins to indeed point toward bats as the likely reservoir hosts.

Marburg virus and the five strains of Ebola virus (including Reston virus, which only causes disease in nonhuman primates) are members of a group of nonsegmented, negative-sense RNA viruses called filoviruses. These viruses are characterized by a very flexible virion that assumes characteristic comma and semicircular shapes in the electron microscope. The viral genome is about 19 kb long and encodes a polymerase (Pol), a glycoprotein (G), a nucleoprotein (NP), and four other structural proteins (VP40, VP35, VP30, and VP24) in the following order: 5′-Pol-VP24-VP30-G-VP40-VP35-NP-3′.

This gene order and the general structure of the genome are quite reminiscent of those seen with other viruses of the Mononegavirales superfamily, and while there is little known about the details of the replication cycle, it can be assumed to be similar to the cycles described for rhabdoviruses and paramyxoviruses. Indeed, workers in Germany recently showed that the mRNA for a variant of the viral glycoprotein is modified by an editing reaction similar to that described for mumps virus. The Ebola virus glycoprotein has been implicated as a necessary but not sufficient requirement for the full expression of virulence during productive infection.

Bornaviruses

The bornaviruses are a fourth member of the Mononegavirales superfamily. They have only recently been subjected to careful molecular biological study, but the following facts are known. They cause a variety of neurological symptoms in all warm-blooded vertebrates infected by them. Infection can also lead to behavioral modifications ranging from minor to severe, although the aggressive frenzy seen in the late stages of rabies is not seen.

The bornavirus genome is approximately 9 kb long and encodes six genes, including envelope proteins, other structural proteins, and a viral polymerase. Bornavirus mRNAs are capped and polyadenylated, and are the only nonsegmented negative-sense RNA viruses that use the nucleus of the infected cell as a site of replication. The best-characterized group of negative-sense RNA viruses that do this are the orthomyxoviruses described later in this chapter – these have segmented genomes. Like mRNA expression by these viruses, some bornavirus positive-sense RNAs generated from genomic negative-sense strand are spliced in the nucleus, but in contrast, bornavirus mRNAs are capped by the viral-encoded polymerase instead of utilizing cellular caps.

Interest in further characterization of these viruses has been heightened by the finding that they can infect humans. Since horses, sheep, and cattle are frequent reservoirs, this puts agricultural workers at risk. Between 2011 and 2013 three fatal cases of human encephalitis were reported in Germany. Molecular analysis revealed the infectious agent to be a previously unknown bornavirus transmitted from variegated squirrels.

Other mononegavirales families

Members of three of the other four families (Nymaviridae, Pneumoviridae, and Sunviridae) are all enveloped, spherical viruses with very similar genome organization, differing in their host organisms. The best known of these is human respiratory syncytial virus, which is the causative agent of lower respiratory tract infections during infancy and early childhood. The fourth family (Mymonaviridae) is a filamentous virus that infects fungi.

NEGATIVE-SENSE RNA VIRUSES WITH A MULTIPARTITE GENOME

The negative-sense RNA viruses with monopartite genomes share enough similarities to allow their grouping into a superfamily, the Mononegavirales. In contrast, there are two orders of negative-sense RNA viruses with segmented genomes: Articulovirales, which includes the Orthomyxoviridae (the influenza viruses), and Bunyavirales, which includes 10 different virus families. Despite this, the negative-sense RNA viruses with multipartite genomes do share some features in replication strategies and genomic sequence.

Due to periodic and frequent spread through the human population, influenza (flu) virus infections are almost as familiar to the human population as are colds. Influenza virus is the prototype of the orthomyxovirus group. There are three distinct types of influenza virus: types A, B, and C. Type A is usually responsible for the periodic flu epidemics that spread through the world, although type B can also be an agent. Influenza types A and B have eight genomic segments, and type C has seven.

The suffix *-myxovirus* ("myxo" is from the Greek word for mucus) was originally coined to group these viruses with the paramyxoviruses, since both were associated with respiratory infections, and both are enveloped and therefore readily inactivated with lipid solvents. While these two groups share some general features of structural organization and proteins of related sequence, they are not at all closely related. Similarities and differences between these two groups of viruses are shown in Table 15.1.

The influenza A (flu virus) virion, which is shown along with the genes encoded in its eight genomic negative-sense RNA segments in Figure 15.5, looks somewhat like a small version of a paramyxovirus virion. As noted in Table 15.1, several of the membrane envelope proteins in these two virus groups clearly are related. Despite this, the replication details are quite different. Flu virus mRNA is generated from transcription of separate and individual flu RNPs in the infected cell's nucleus.

Table 15.1 Similarities and differences of orthomyxovirus and paramyxovirus.

Similarities	Differences
• RNA genome is single stranded, negative sense. • They both have a helical nucleocapsid. • They both have virion-associated transcriptase. • Virion buds from the cell surface. • They both have two related glycoproteins: neuraminidase and hemagglutinin.	• Orthomyxovirus mRNA can be spliced. • Orthomyxoviruses have a segmented genome. • Orthomyxoviruses require a nucleus for replication. • Orthomyxovirus mRNA requires cellular 5′ caps (cap snatching).

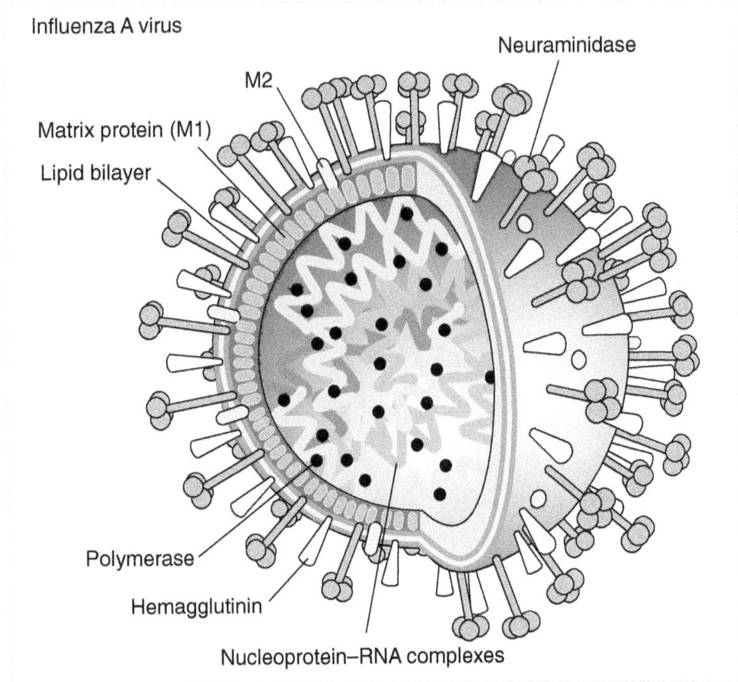

Figure 15.5 The structure of influenza virus A. The virion is about 120 nm in diameter, and the genome is made up of eight helical nucleocapsid segments that total about 13 600 nucleotides of negative-sense strand RNA. The virus requires the nucleus for replication. Although these virions also exhibit neuraminidase and hemagglutinin, the glycoproteins responsible are separate.

Involvement of the nucleus in flu virus replication

Despite some general similarities with VSV in transcription of the genomic negative-sense strands of influenza virus to generate mRNA, there are important differences in the overall replication process. A major difference is that influenza virus mRNA synthesis and genome replication require the cell's nucleus. There are two readily apparent reasons for this. First, flu replicase cannot cap mRNA; therefore, each flu virus mRNA generated has to use a cellular mRNA cap as a "primer." Synthesis of each flu virus mRNA begins with a short stretch of cellular mRNA with its 5′ methylated cap. This **cap snatching** is a form of intermolecular splicing, and is accomplished by the flu virus replication–transcription complex as it associates with actively transcribed cellular mRNA. Thus, the virus inhibits cellular mRNA transport and protein synthesis, but not initiation of transcription.

Second, influenza A virus utilizes the intramolecular splicing machinery of the host cell's nucleus. Two of the RNPs of the flu virus express mRNA precursors that are spliced in the nucleus. Each of these gene segments, then, can encode two related proteins. This splicing takes place via cellular spliceosomes in a manner identical to that described in Chapter 13. The result of the splices is that two segments of the viral genome actually generate four distinct mRNAs. Thus, with influenza A, the eight flu virus negative-sense genomic segments encode 10 specific mRNAs that are translated into distinct viral proteins.

Generation of new flu nucleocapsids and maturation of the virus

An abbreviated schematic of the influenza A virus replication cycle in a susceptible cell is shown in Figure 15.6. Infection is initiated by virus attachment to cellular receptors followed by **receptor-mediated endocytosis**. The separate RNPs with their negative-sense genome segments are transported to the nucleus where viral mRNA synthesis begins.

Viral mRNA synthesis requires the activity of at least two influenza virus polymerase subunits: PB1 and PB2. PB1 has active sites that bind the conserved 3′ and 5′ sequences of viral RNA, as well as the endonuclease activity necessary to cleave the host cap sequence. In addition, PB1 has the polymerizing activity of the complex. PB2 has cap-binding activity, and it is to this subunit that the host pre-mRNA binds. Cleaving of the small (1–13 nucleotide) cap structure from the host begins the process of mRNA synthesis, during which a capped, subgenomic copy of the viral RNA is produced. Synthesis stops about 15–22 nucleotides short of the 5′ end of the viral RNA, where a small (4–7 nucleotide) U region serves to cause stuttering or reiterative synthesis, producing a poly[A] tail, a mechanism similar to the one we saw earlier for the rhabdoviruses.

At some point, viral RNA synthesis must switch from making mRNA to making full-length template RNA and then new viral RNAs. This switch requires the presence of multiple copies of the viral protein NP, as well as the polymerase subunit PA. A complete model of this change to full-length synthesis has not yet been completely worked out. However, the synthesis would require the formation of RI-1 and RI-2 intermediates. Recent evidence suggests the involvement of a second copy of the flu polymerase in the replicative events.

Since all viral RNA synthesis takes place in the nucleus, it is necessary that newly replicated genomes be transported to the cytoplasm for maturation of new virus particles. This transport takes place when new viral RNA molecules complex with two viral proteins: M1 and NS2. The NS2 protein contains a nuclear export signal that interacts with a cellular nuclear export protein (an exportin) and likely also overrides the nuclear localization signals present on the NP and polymerase proteins. Flu nucleocapsids that have been assembled in the nucleus and transported into the cytoplasm migrate to the cell's surface where virions bud off. While the particle-to-PFU ratio of influenza A is always high compared to, say, poliovirus, or even to influenza B, packaging is not random and involves specific packaging signals.

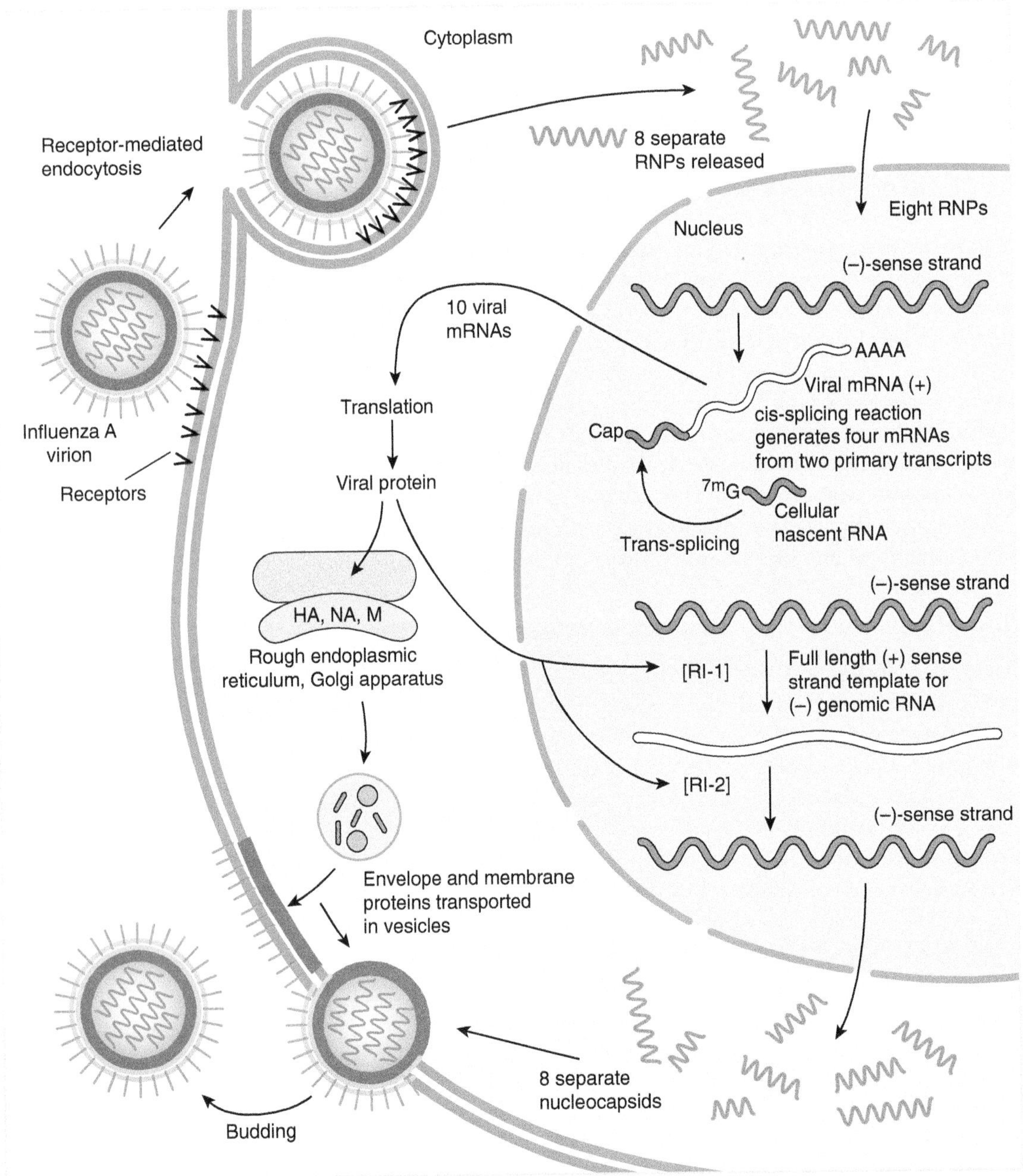

Figure 15.6 An outline of the replication cycle of influenza. Following virus attachment to its cellular receptor(s) and endocytosis, the envelope fuses with vesicular membrane. The released ribonucleoprotein (RNP) capsid segments, each containing a specific negative-sense genomic segment, migrate to the nucleus where transcription of positive-sense RNA takes place using virion-associated transcriptase. The transcription and formation of mRNA require the "snatching" or "stealing" of caps of nascent cellular mRNA by a *trans*-splicing mechanism. Two of the pre-mRNAs generated in this way are further subjected to one of two alternative *cis*-splicing reactions using cellular machinery, so that each generates two separate mRNAs. Translation of viral proteins leads to proteins that modify the cell and its plasma membrane. The viral proteins associated with the nucleocapsid RNPs migrate to the nucleus, where they mediate the synthesis of full-length positive-sense template and synthesis of negative-sense strand genomic RNA. Viral membrane-associated proteins are translated on the rough endoplasmic reticulum and processed in the Golgi apparatus. New virions form by the association of the nucleocapsids with virus-modified membrane and budding. Influenza A virus does not control this aspect of packaging; therefore, phenotypic mixing is frequent following mixed infection. NA, Neuraminidase; HA, hemagglutinin; M, matrix protein.

Influenza A epidemics

Flu is generally considered to be a mild disease, but influenza can be a major killer of the aged and the immune compromised. Even though the body mounts a strong and effective immune reaction to influenza infections, and the individual is immune from reinfection upon recovery, the virus is able to mount periodic epidemics in which prior immunity to related strains is no protection. The solution to this apparent enigma is found in the broad host range of influenza A and the unique ability of influenza A (but not B or C) genomic segments to be independently packaged into individual virion particles during infection. Such a situation leads to a very inefficient packaging process, but allows for rapid dissemination of a favorable mutation. If there is a mixed infection of two different influenza A virus strains in the same cell, significant genetic changes can arise and will provide a significant evolutionary advantage to the progeny.

Since most immune protection against a viral infection is directed against surface components of any virus (the membrane glycoproteins in the case of influenza virus), one can predict that the antigenic properties of these surface proteins will change or "drift" over time. This drift is due to the random accumulation of amino acid changes (mutations), along with the slight selective advantage of a virus that has a surface protein not as efficiently recognized by the immune system as those of the virus that induced immunity in the first place. Such drift is found in many viruses and other pathogens.

With influenza A, however, independent packaging of the individual RNPs in the infected cell provides a more rapid means of antigenic variation. There is always the possibility that an individual can be infected at the same time with two different influenza A viruses. This will not happen very often, but if it does, one result of the mixed infection will be the generation of a new hybrid virus that might have, say, a hemagglutinin membrane glycoprotein from one parent and all the other components from the human virus. To add to this, swine influenza virus strains recognize some of the human cell receptors utilized by their human influenza A virus counterparts. This means that in a farm where pigs are intensively cultivated, a multiple infection could involve a swine virus as well as a human virus. This abrupt change in antigenic nature of the membrane protein is termed **antigenic shift**.

The problem with antigenic shift is complicated by the fact that pigs (but not humans) have efficient receptors for avian influenza viruses. Therefore, a multiple infection in pigs with different avian strains or avian and porcine strains can lead to a very significant reassortment of different markers. This can happen with some frequency in areas in which there is very intensive farming and animal husbandry in relatively limited spaces, which is typical of many small farms in East Asia where pigs, ducks, chickens, and other animals are all tended together.

Upon antigenic shift, the resulting successful virus is essentially a "new" virus, and is relatively unaffected by the immune defenses mounted against earlier forms of virus. Thus, the new virus can spread throughout the population despite the high level of immunity to prior forms of influenza A. The timing of the occurrence of such new viruses cannot be predicted, but can be readily quantified by measuring the antigenic reactivity of viral components to various standard immune reagents generated against earlier forms of the virus. When such a new virus is seen, an epidemic can be predicted.

The immunological variation of various flu virus proteins from virus isolated over a considerable period of time is shown in Figure 15.7. Flu strains are designated by indication of the hemagglutinin (H) and the neuraminidase (N) types. At least 15 H and 9 N subtypes have been identified. However, at this point only three H (H1, H2, and H3) and two N (N1 and N2) subtypes have been circulating in the human population. When the hemagglutinin and neuraminidase components change together (as in generation of the influenza A2 virus in 1957), a major worldwide epidemic (pandemic) can occur. In fact, this happened in April 2009 when a new version of influenza A, called H1N1/09, was identified in an outbreak in Mexico. By June

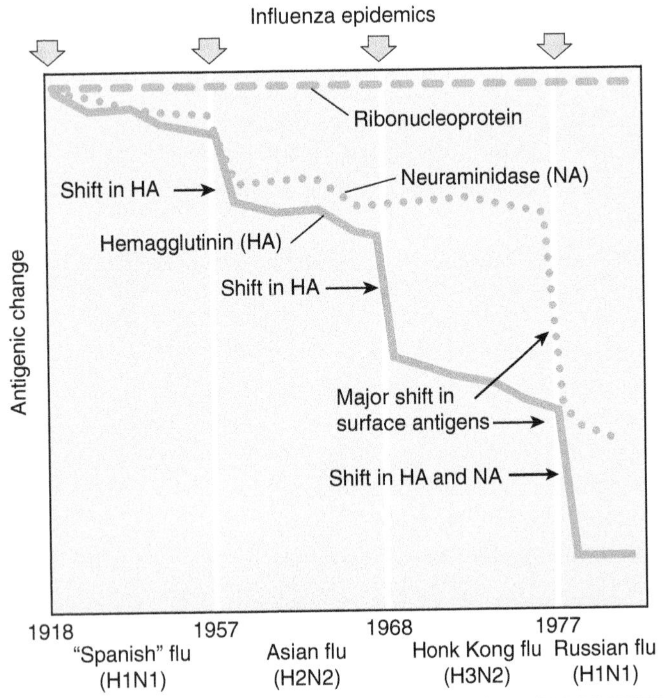

Figure 15.7 Antigenic changes in the surface glycoproteins of influenza A virus between 1933 and 1968. Abrupt changes in these antigens (antigenic shifts) are the result of mixed infections and random assortment of nucleocapsids to generate novel genotypes. Such shifts, which occur with random frequency, lead to epidemics worldwide. Strain designations at the bottom of the figure indicate hemagglutinin (H) and neuraminidase (N) genotypes.

of that year, the World Health Organization (WHO) had declared a pandemic. The virus in this case was found to be a re-assortant, containing RNA segments from human (PB1), avian (PB2 and PA), and swine (HA, NA, NP, M, and NS) influenza A viruses. The estimated global mortality from this pandemic, using probabilistic models, was 284 500 people, with 80% of these younger than 64 years of age.

Note that the interior RNP is antigenically stable. One reason for this stability is that there is little humeral immune reaction to these components of the virus because they are not efficiently presented at the infected cell surface by major histocompatibility complex (MHC) type I; therefore, there is little or no pressure to change. Indeed, it is the antigenic stability of the RNP that defines the major influenza types. Another factor contributing to the stability of the sequence of the capsid proteins forming the RNP is that most changes to these interior proteins would interfere with their function, and thus lead to a virus with impaired ability to replicate.

In 1997, a strain of avian influenza was diagnosed in humans in Hong Kong. The virus, designated H5N1, was transmitted directly from birds to humans, a rare and inefficient event. In fact, this strain of the virus did not transmit from human to human. In this first incident with this strain, 18 people were infected and 6 of them died. Notwithstanding the inefficient passage between humans, the high concentration of people in Hong Kong, along with their proclivity for purchasing live poultry for home butchering, led to a worrisome outbreak of the disease. The draconian measures of wholesale slaughtering of all live poultry within the confines of the former British colony were initially thought to have been effective in stopping this outbreak. However, this was not to be the case. By 2003, H5N1 had reappeared in Hong Kong, both in birds and in human infections, and was soon after reported in several other Southeast Asian countries. By 2004, the virus was identified in birds in China, Korea, Japan, and Mongolia. By 2005, the virus was reported in birds in Turkey and Romania. Given that this avian virus is carried by migratory birds, there is little doubt that it will make its way to virtually every area of the planet.

Human cases have been reported in Thailand, Indonesia, Vietnam, China, and Cambodia. As of May 2006, 216 human cases had been diagnosed, with 122 deaths, for a case-fatality rate of greater than 56%. Since this virus does not seem to be infecting pigs thus far, it has been suggested that, in order for the development of a strain that can be transmitted from human to human, mixing and reassortment during a case of human infection with both H5N1 and a circulating human virus would have to take place. At this writing, no human-to-human transmission has been proved.

A more recent example of an avian strain is H7N9, first reported in 2013 in China. This virus has produced a series of six epidemics, the most recent of which occurred in early 2019. As with H5N1, transmission seems to take place from birds to humans by direct contact during rearing or marketing of infected birds. Once transmitted to humans, the virus has a high case-fatality rate. The six epidemics resulted in (by February 2019) a total of 1565 human infections from contact with infected birds, with a 39% fatality rate. This is no evidence in these data of human-to-human transmission.

OTHER NEGATIVE-SENSE RNA VIRUSES WITH MULTIPARTITE GENOMES

Bunyavirales

The most recent taxonomic classification groups all of the other segmented-genome negative-sense RNA viruses into an order, the Bunyavirales. This includes viruses formerly called the bunyaviruses, as well as those formerly called arenaviruses. In terms of the number of members, this order is one of the largest known, with 330 serologically distinct viruses. The order itself consists of five separate families. Many members of this diverse order are arboviruses, being transmitted by mosquitoes, ticks, sandflies, leafhoppers, or thrips. Other members are vectored by rodents. Examples of some of these vectored agents are listed in Table 15.2.

Virus structure and replication

The order Bunyavirales is all segmented, ssRNA, negative-sense viruses. Of these, members of the Peribunyaviridae family have been studied most extensively. La Crosse encephalitis orthobunyavirus is an example. Orthobunyaviruses all have tripartite, negative-sense RNA genomes. As outlined in Figure 15.8, the enveloped virions are about 90–110 nm in diameter. The membrane contains two viral glycoproteins: G1 and G2. Within the particle are three size classes of circular nucleocapsids, each consisting of one of the genomic RNAs in a helically symmetric complex with the nucleocapsid (N) protein and the viral polymerase (L).

Table 15.2 Some members of the Bunyavirales and their vectors.

Family	Vector	Examples
Arenaviridae	Rodents	Lassa fever virus, Junin virus, Machupo virus
Fimoviridae	Leaf mite	Fig mosaic emaravirus, raspberry leaf blotch emaravirus
Hantaviridae	Rodents	Hantaan virus, Sin Nombre virus
Nairoviridae	Ticks	Nairobi sheep disease virus, Crimean-Congo hemorrhagic fever virus
Peribunyaviridae	Mosquitoes, ticks	Bunyamwera virus, La Crosse encephalitis virus
Phenuiviridae	Ticks, sand flies, leafhoppers	Rift Valley fever phlebovirus, Uukuniemi phlebovirus, rice stripe virus

Since these are negative-sense viruses, the first event after infection is transcription. For La Crosse virus, a typical orthobunyavirus, viral mRNAs are produced from each genome segment, as also shown in Figure 15.8. Viral messages have 5′ capped termini and 3′ ends with no polyA. The cap structures are derived from cytoplasmic host mRNA by endonucleolytic cleavage. This cap-snatching reaction, although similar to that described for influenza virus, takes place *outside* the nucleus.

The viral mRNAs are subgenomic, as with influenza. Replication of the orthobunyavirus genomic (and antigenomic) RNA occurs in the cytoplasm. These RNAs have 3′ and 5′ inverted complementary sequences of about 10–14 nucleotides that may play a role in the replication event. The nucleocapsids themselves have a circular form that may reflect base pairing of these sequences.

The three genome segments demonstrate a variety of expression strategies; some of these are also shown in Figure 15.8. The largest segment expresses a single protein, the viral polymerase (L). The middle-sized segment encodes two or three proteins, depending on the specific virus in question. Expressed proteins are the two glycoproteins G1 and G2, along with – where present – a nonstructural protein NS_M. These proteins are translated as a precursor polyprotein that is posttranslationally cleaved.

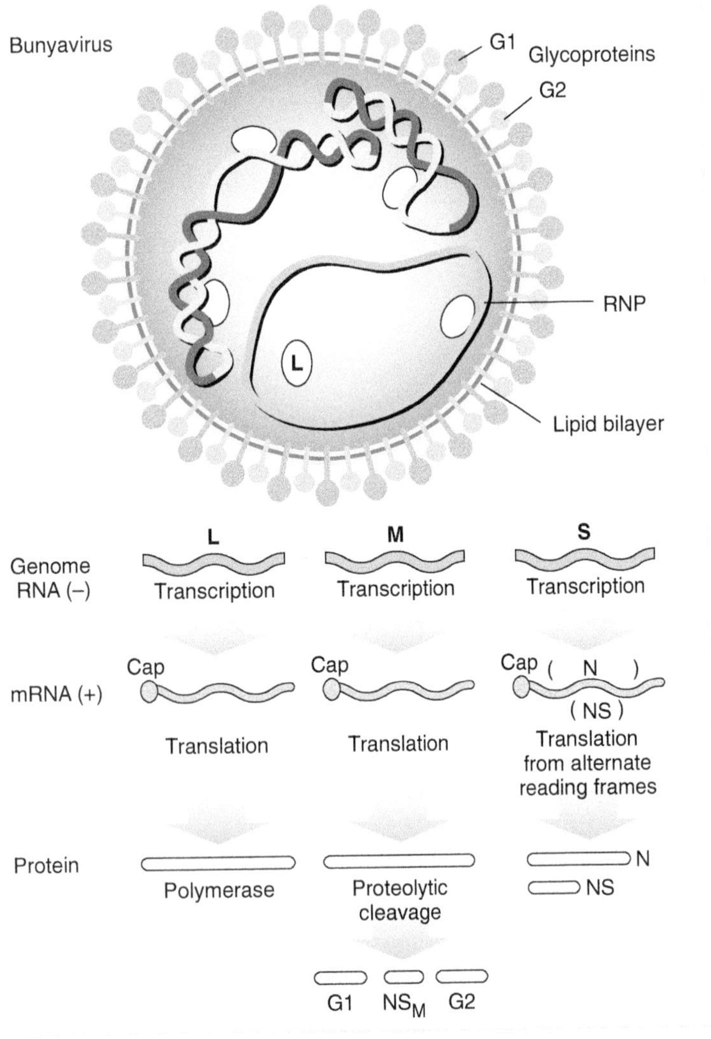

Figure 15.8 The bunyavirus virion. The three ribonucleoprotein (RNP) segments, each associated with both L and N protein, are contained within a well-defined envelope made up of two glycoproteins. The virion diameter ranges from 80 to 120 nm. The size of the RNPs is determined by their sedimentation rates (see Chapter 11). The general scheme of gene expression and genome replication of La Crosse virus is also shown. Expression and replication take place in the cytoplasm, but have many similarities to the process outlined for influenza virus. The positive-sense strand mRNA expressed from the S genomic segment contains two partially overlapping translational reading frames that are out of phase with each other. Alternative recognition of one or the other translation initiation codons by the cellular ribosomes leads to the expression of two proteins with a completely different amino acid sequence.

The smallest RNA genome segment encodes one or two viral proteins. For the nairoviruses and hantaviruses, this segment expresses mRNA for the N protein. In the *Orthobunyavirus* genus, the subgenomic RNA from this segment can be translated into the N protein or, using a separate, alternate reading frame, into another nonstructural protein, NS. Apparently the "decision" as to which reading frame is utilized in this small mRNA is entirely random. Sometimes the ribosome starts at one AUG and sometimes at the other.

The small genomic segments of the Phleboviruses and the Tospoviruses (members of the Phenuiviridae family) are **ambisense genomes**: They contain both positive- and negative-sense genes. The term "ambisense" refers to the fact that the open reading frames defining the two proteins are oriented in opposite directions in the genome RNA, and their expression requires a strategy that is vaguely reminiscent of that utilized in the expression of Sindbis virus subgenomic RNA. This is shown in Figure 15.9. The small (S) virion-genomic RNP is transcribed into a positive-sense mRNA that is translated as the N protein encoded within the negative-sense portion of the ambisense virion genomic segment. The genomic ambisense RNA also serves as the template for the transcription of a separate ambisense antigenomic RNA that acts as a template for the transcription of capped mRNA encoding the NS_S (nonstructural S) protein. This RNA is the same sense as the virion RNA; thus, even though Phlebo- and Tospoviruses are negative-strand RNA viruses, a portion of their genome is mRNA (i.e., positive) sense.

Pathogenesis

Members of the Bunyavirales order infecting vertebrates cause four kinds of disease in humans and other animals: encephalitis, hemorrhagic fever, hemorrhagic fever with renal involvement, and hemorrhagic fever with pulmonary involvement. La Crosse encephalitis orthobunyavirus is transmitted by mosquitoes and is one of the main causes of viral encephalitis during spring and summer in the upper Midwest. Rift Valley fever phlebovirus, transmitted by the sandfly, causes recurring zoonoses and epidemics of hemorrhagic fever in sub-Saharan Africa. Hantaan virus, transmitted by rats, is the prototype of the *Hantavirus* genus and causes Korean hemorrhagic fever, a disease complicated by renal failure.

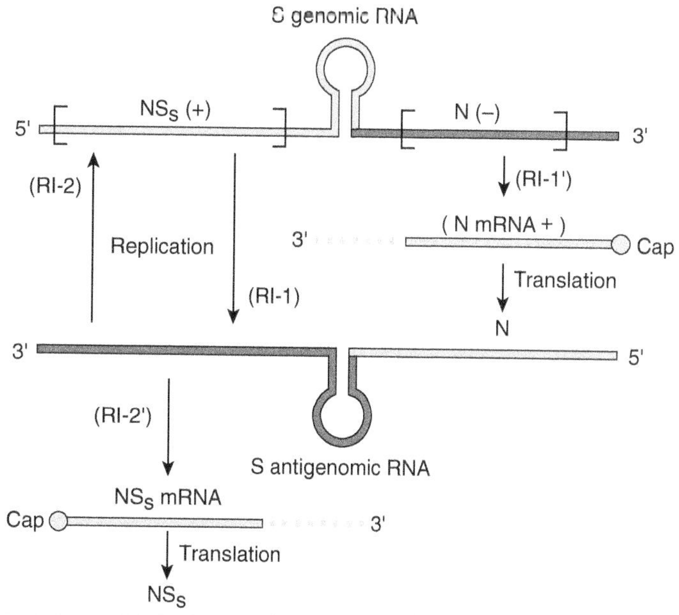

Figure 15.9 The ambisense strategy of gene expression exhibited by some bunyaviruses and by arenaviruses. The expression of the small genomic segment of a Tospovirus and a Phlebovirus is shown. With these viruses, full gene expression requires the generation of subgenomic mRNA of the same sense as the genomic RNA. Thus, even though the genomic RNA is nominally negative sense, it has regions of positive-sense information in it! This strategy is referred to as *ambisense* since both senses are present in the genome.

Sin Nombre virus, another member of the *Hantavirus* genus, was identified as the causative agent of outbreaks of a relatively fatal hemorrhagic fever with pulmonary involvement, originally termed *hantavirus adult respiratory distress syndrome (HARDS)* in the early 1990s and now referred to as hantavirus pulmonary syndrome (HPS). This and related viruses, transmitted by aerosols from fecal pellets of small rodents such as the deer mouse, are found distributed throughout the United States, although localized epidemics of HPS have occurred in areas such as the Southwest. Epidemiological investigations of these outbreaks suggest that increases in the rodent vector population (aided by sporadic mild wet winters that increase forage for the rodents) result in increasing likelihood of transmission to humans. As of early 2017, the CDC has reported that there have been 728 confirmed cases of HPS since the identification of the syndrome, with a case-fatality rate of 36%.

Arenaviruses

Members of the Arenaviridae family of the order Bunyavirales have bipartite, single-strand, negative-sense RNA genomes contained as helical nucleocapsids within an enveloped particle 90–100 nm in diameter. The virions also contain a number of host cell ribosomes accidentally packaged with the finished particles. These ribosomes play no role in the virus infectious cycle. The presence of these ribosomes gives the virus particles a "sandy" appearance in electron micrographs, leading to the name of the family (*arena* is the Latin word for "sand").

Virus gene expression

The largest genome segment (7.2 kb) encodes two proteins: the viral polymerase, L, and a smaller regulatory protein, Z. The small genome segment encodes the glycoprotein precursor, ultimately cleaved into the two membrane proteins, GP1 and GP2, as well as the nucleocapsid protein, NP. In each case, the two open reading frames contained within the genome segment are arranged in an ambisense fashion. In each case, there is a stretch of RNA between the two genes that consists of a hairpin loop structure that may play a role in regulating the termination of mRNA's transcription.

Primary transcription of the genome produces subgenomic mRNAs for the L and NP proteins. This is followed by transcription from the antigenome RNAs to yield the subgenomic mRNAs for Z protein and the glycoprotein precursor. The virus's mRNAs have methylated 5′ caps that may be derived from host messages. The 3′ ends of viral mRNAs are not polyadenylated. Replication of viral genomes may involve inverted terminal complementary sequences, as described for the bunyaviruses.

Pathogenesis

Lymphocytic choriomeningitis virus (LCMV) causes a mild, influenza-like disease in mice and humans, although rare and severe encephalomyelitis has been observed. At the other end of the spectrum are severe and often fatal diseases caused by agents such as Lassa fever virus in West Africa and agents of the South American hemorrhagic fevers: Junin virus (Argentina), Machupo virus (Bolivia), and Guanarito virus (Venezuela).

A very interesting aspect of these viruses' pathogenesis (as outlined in Chapter 7, Part II) is that infection of infant animals (whose immune system is still developing) generally leads to persistent infections. If, however, the virus infects an adult animal with a fully functioning immune system, rapid death follows. Wild populations harboring the virus can secrete large amounts of virus that can be lethal to humans or other animals interacting with them. This is one of the reasons why habitat destruction in Africa with its accompanying disruption of native rodent populations that are chronic carriers of the virus has led to periodic outbreaks of arenavirus-induced fatal disease.

VIRUSES WITH DOUBLE-STRANDED RNA GENOMES

The family Reoviridae contains 15 distinct genera with infectious agents specific for vertebrates (reoviruses and rotaviruses), invertebrates (cytoplasmic polyhedrosis virus), and plants (wound tumor virus). Members of this family have genomes consisting of 10, 11, or 12 segments of double-stranded RNA (dsRNA). There is a group of bacterial viruses, many infecting *Bacillus subtilis*, that also contains segmented, double-stranded genomes. The replication strategy employed by these viruses must take into account that the genome is dsRNA, which is extremely stable, and consequently difficult to dissociate into a form exposing a single-stranded template for RNA-directed mRNA transcription.

Orthoreovirus structure

Orthoreoviruses contains 10 dsRNA segments. A schematic of the virion and the protein-coding strategy of the genomic segments is shown in Figure 15.10. These genome segments of the reoviruses are packaged into an icosahedral capsid that consists of two – or, in some members, three – concentric shells, each having icosahedral symmetry. The capsid is made up of three major structural proteins, as well as a number of low-abundance structural proteins, including virion-associated transcriptase, as the virions contain all of the enzymatic machinery necessary for the production of viral mRNA, including activities involved in capping and methylation. Genome segments range in size from about 4 kb to about 1 kb. The genomic RNAs have 5′ methylated caps on the positive-sense strand of the duplex and a 5′ triphosphate on the negative-sense strand. Neither strand is polyadenylated.

The orthoreovirus replication cycle

Some features of the replication of orthoreovirus in the infected cell are shown in Figure 15.11. After attachment and entry into the host cell cytoplasm via receptor-mediated endocytosis, reovirus particles are partially uncoated, leaving behind an inner-core subviral particle. This subviral particle contains the 10 genome segments and transcriptional enzymes. Production of mRNAs occurs by the copying of one strand of each duplex genome into a full-length strand. The mRNAs are capped and methylated by viral enzymes but do not have polyadenylated 3′ termini. These transcriptional events require six viral enzymes: a polymerase, a helicase, an RNA triphosphatase, a guanyltransferase, and two distinct methyltransferases. The latter three enzymes are all involved in the capping reaction.

Each of the genome segments encodes a single transcript that is translated into a single protein, except for one of the smaller segments (S1) of the *Orthoreovirus* genus. This segment encodes two proteins encoded in two nonoverlapping translational reading frames. Both proteins are encoded by the same mRNA by virtue of random recognition of either of the two translation initiation codons by cellular ribosomes. Most of the gene products are structural, either forming one of the multiple capsids or comprising the transcriptional complex of enzymes found within the core.

Replication of the double-stranded genomes and final assembly of progeny virions are not completely understood. It is thought that 10 unique mRNAs associate to form a core progeny virion, associating with the appropriate capsid proteins. These positive-sense RNAs then serve as templates for the synthesis of negative-sense strand, leading to the production of progeny double-stranded genomes within the nascent particle.

This rather convoluted means of generating the double-stranded genome is a consequence of the fact that dsRNA will not readily serve as a template for its own synthesis because of its very

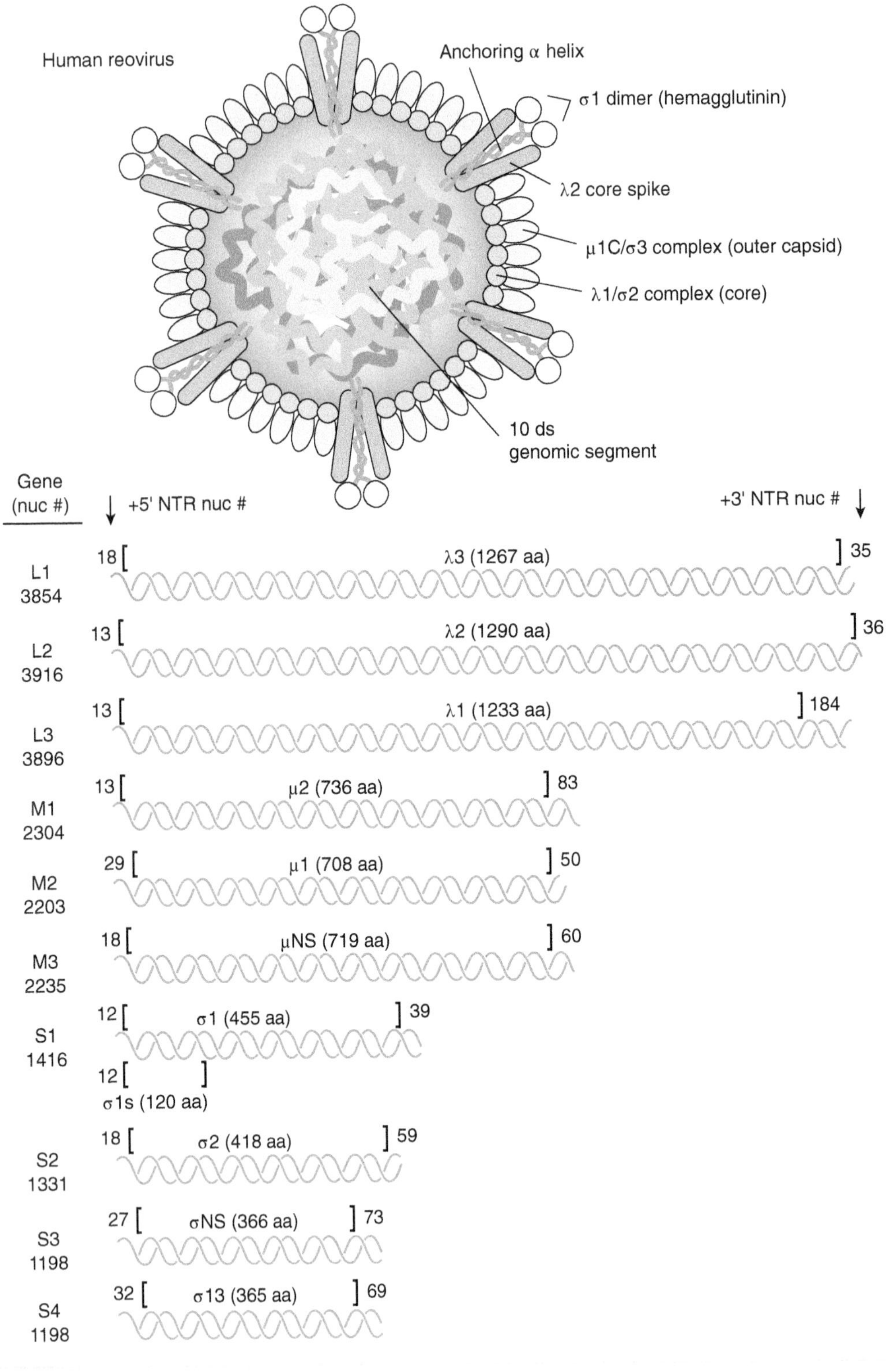

Figure 15.10 The 60-nm-diameter human reovirus with its double shell. The 10 segments of the reovirus genome and the proteins encoded are shown. Note that the S1 segment encodes two overlapping translation frames. Like the situation with the La Crosse virus mRNA encoded by the S genomic fragment, these proteins are expressed by alternate initiation sites for translation. Thus, the virus encodes 11 proteins. The total size of the genome is 23 549 base pairs.

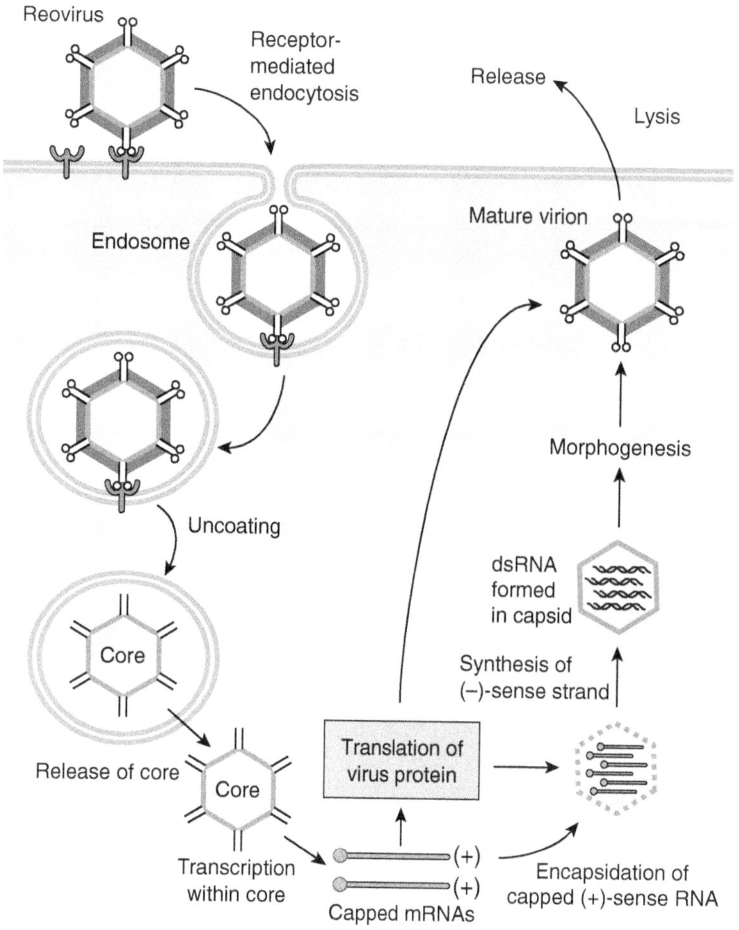

Figure 15.11 The reovirus replication cycle. Virus attachment is followed by receptor-mediated endocytosis. Virion "core" particles are formed by the degradation of the outer shell in the endosome, and this core particle expresses capped mRNA using a virion transcriptase. Various viral proteins are translated, and structural proteins assemble around newly synthesized viral mRNA. This process is apparently random, since random assortment of genetic markers following mixed infection is readily observed (see Chapter 3, Part I). The complementary strand of the double-stranded genomic RNAs is synthesized in the immature capsid while morphogenesis proceeds. Virus release is by cell lysis.

great stability. The environment inside the capsid is apparently relatively nonaqueous, and in this nonpolar space, the dsRNA is more readily denatured due to charge repulsion between the phosphate backbones of the two RNA strands. Thus, the double-stranded genome is able to partially denature to serve as a template to generate large quantities of positive-sense mRNA that is extruded from the inner core.

Replication of orthoreovirus RNA, then, does not involve RI-1 or RI-2 intermediates. Further, ideally, no free dsRNA is formed inside the cytoplasm of the infected cell, precluding the induction of interferon. In practice, however, this situation is not realized, and many cells infected with reovirus produce significant interferon. While the yield of virus is quite sensitive to the interferon-mediated antiviral state in cells, apparently the major induction occurs rather late in the replication cycle when cellular organization is deteriorating. Thus, the virus is able to keep ahead of the response for a period of time sufficient for efficient replication in the host.

Pathogenesis

The prototype viruses of this family (now grouped in the genus *Orthoreovirus*), although originally isolated from human sources, are not known to cause clinical disease in humans. The name *reovirus* stands for "respiratory enteric orphan virus," an orphan virus being one for which no disease is known.

In contrast, members of the *Rotavirus* genus are perhaps the most common cause of gastroenteritis with accompanying diarrhea in infants, and they remain among the leading causes of early childhood death worldwide. Other significant pathogens of humans and domestic animals found in this family include Colorado tick fever virus (*Coltivirus* genus) and bluetongue virus of sheep.

SUBVIRAL PATHOGENS

As touched on in Chapter 1, viruses, as efficient and compact as they may be, are not in fact the simplest infectious agents. A number of other entities that are smaller than viruses can cause disease in animals and plants. These agents can be collectively considered to be subviral pathogens. They may contain genetic information for the expression of a protein, or they may express no gene products at all. A number of them may not even be contained within a capsid, and one group, the prions, while able to replicate themselves, does not appear to contain nucleic acid.

Subviral pathogens are parasitic on cellular processes, but if viruses parasitize the ability of a cell to express protein from information contained in nucleic acids, subviral pathogens can be considered to be parasitic on other macromolecular processes in the cell, including transcription and protein assembly and folding.

A large number of subviral pathogens lacking capsids are parasitic on plants, and many can cause plant pathology without expressing protein. These agents can be differentiated by a detailed characterization of their modes of replication, but only the viroids are considered in this text because of this group's relationship to the human pathogen hepatitis delta virus (HDV).

Viroids

Plant viroids are infectious agents that have no capsid and have an RNA genome that encodes no gene product; they do not require a helper for infectivity. Viroids are classified into two families, Pospiviroidae and Avsunviroidae, and are included as part of the taxonomy of subviral agents. Potato spindle tuber viroid is the prototype of this class of agents. The viroids are covalently closed, circular ssRNAs, 246–375 nucleotides long, whose sequence is such that base pairing occurs across the circle, as shown in Figure 15.12. As a result, these agents have the form of a dsRNA rod with regions of unpaired loops. Their replication is carried out by plant RNA polymerase, and likely proceeds through an antigenome. Large multimeric structures can be

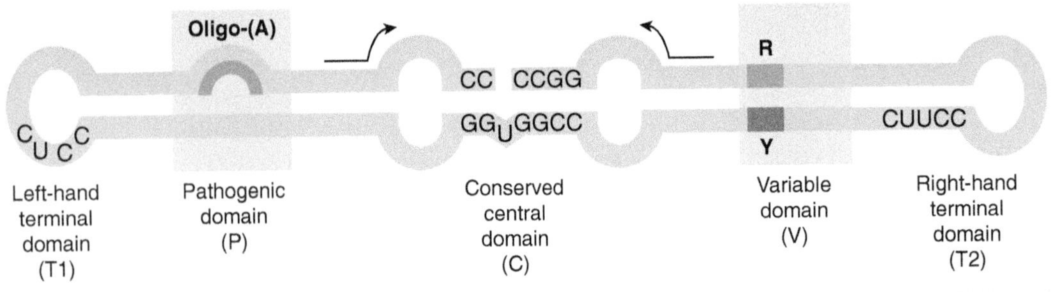

Figure 15.12 The potato spindle tuber viroid genome. Various pathogenic strains range from 250 to 360 nucleotides in length. This circular RNA does not encode a protein, but the sequences indicated as pathogenic are required to cause the disease. Modification of these sequences leads to a viroid that is nonpathogenic and can protect the plant from pathogenesis by the original viroid. Viroid RNA is replicated with cellular RNA polymerase, forming large multimeric structures of both positive and negative sense. Individual viroid RNA is released by RNA self-cleavage.

observed in infected plant nuclei, and self-cleavage of such multimers into unit-length RNA molecules is involved in "maturation" of the infectious form.

Viroids spread from plant to plant through mechanical damage caused by insects or by cultivation. They are also spread by propagation of cuttings from infected plants. Viroids may also be present in seeds. Very often, viroids are transmitted during the manipulation of crop plants for harvest, as is the case with the coconut cadang-cadang viroid, transmitted from tree to tree on the metal spikes that harvesters wear on their shoes to climb the trunk.

More than 30 viroids have been described infecting a wide variety of plant species. Many of these have great agricultural significance and are known to destroy fields of economically important crops. The actual mechanism of their pathogenesis is obscure but it clearly involves specific sequences within the viroid RNA, as there are examples where a viroid RNA with sequence very similar to a pathogenic one is not pathogenic and can provide some protection to the host plant. More recent evidence has suggested that viroid pathogenesis is due to the synthesis of RNA interference (RNAi), leading to gene silencing in the host plant cell.

Prions

As noted earlier, HDV utilizes an envelope borrowed from a helper virus, and itself encodes only one gene product. Pathogenic and nonpathogenic plant viroids are able to propagate their genomes without encoding capsid or any other protein. Prions form a logical limit to how simple a pathogen's structure can be. Prions are infectious agents that do not appear to have nucleic acid genomes! They are also included in the taxonomy of subviral agents.

Unfortunately, this simplicity does not mean that investigation of the problem of prion pathogenicity is itself simple. Prion-based diseases have a very long incubation time, and the biological assay is slow and expensive. Further, the fact that prion-induced disease is mediated by protein means that the infectious agent is extremely difficult to inactivate. Most methods for sterilization of infectious agents are ineffective for prions.

The name *prion* was coined by Stanley Prusiner (who won the 1997 Nobel Prize in medicine for his studies) as an acronym for *proteinaceous infectious particle*. Prions are the causative agents of a series of spongiform encephalopathies, including scrapie disease of sheep, Kuru and Creutzfeld–Jakob disease (CJD) of humans, and bovine spongiform encephalopathy (BSE), popularly termed "mad cow disease."

It is fair to argue that these infectious agents are not viruses in any real sense of the word. Still, the fact remains that many techniques for the study of their structure, propagation, and pathogenesis are based on the study of viruses, and prions, perhaps arbitrarily, are included in most compendiums describing virus replication and virus-induced disease.

Prions are most consistently characterized simply as copies of a single host protein that can assume more than one structure (or **isoform**) upon folding after translation. Thus, the DNA sequence that originally encodes the prion is a part of the host genome itself. One isoform is benign, while the other induces cytopathology.

Scrapie, the prion-based disease of sheep, has been investigated most thoroughly, but it is assumed that the agents of all the other diseases are similar if not identical. The protein in question, called PrP, is a normal gene product found in the brain where it is synthesized and degraded in a manner similar to many other proteins characterized by dynamic turnover in the cell. When PrP is changed to the infectious form, called PrP_{Sc} (in the case of scrapie) or PrP_{CJD} (in the case of Creutzfeld–Jakob disease), the protein is converted into the pathogenic isoform (Figure 15.13).

Whereas PrP is normally stable in its benign configuration, certain alterations in a single amino acid caused by a heritable mutation can lead to an unstable protein. This unstable protein can spontaneously convert to the pathogenic form with some low frequency. The properties of

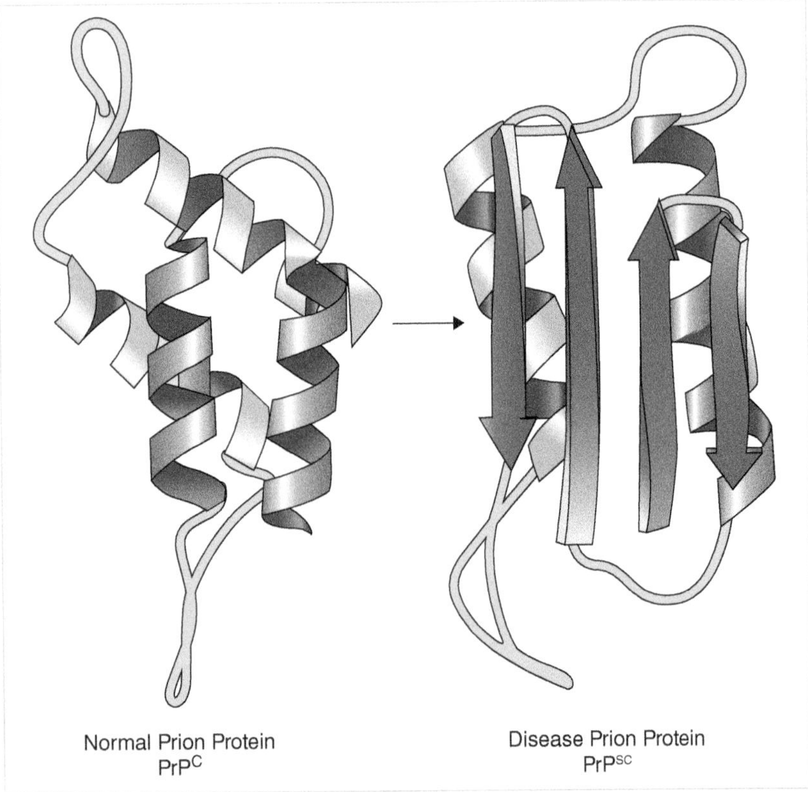

Figure 15.13 Prion-specific protein.

this converted protein differ in many ways from those of the normal form (for instance, in solubility and protease resistance). It is thought that accumulation of this abnormal form in the brain leads to cell death and the characteristic neurological symptoms of prion-based disease.

What is most important to spread of the disease is that the abnormal PrP_{Sc} protein is able to catalyze the conversion of normal PrP to the disease isoform. While this conversion is most efficient in the original animal, the protein can also induce the conversion when introduced into another animal, especially if it, too, contains the critical amino acid.

Although the exact mechanism of this conversion is not clearly understood, models to explain the phenomenon suggest that interaction between the normal and disease forms of the proteins can result in replication of the abnormal form through an intermediate that may normally be part of this protein's degradation pathway.

Transmission of these infectious agents has been clearly demonstrated. For instance, on mink farms, animals given feed that contains waste material from sheep slaughter may contract a prion disease called *transmissible encephalopathy*. Likewise, Creutzfeld–Jakob disease is transmittable from patient to patient by an iatrogenic route, due to contaminated instruments.

As predicted from this model, susceptibility to prion-based diseases in humans and animals is a genetic trait. Still, given a high enough inoculum, conversion of benign PrP to the pathogenic form can take place even when the original protein substrate does not contain the critical amino acid. Transmission via contamination of neurological probes that have been sterilized normally has been well documented, and occurs with enough frequency to excite real concern.

Beginning in 1986, Great Britain experienced an outbreak of BSE (mad cow disease) resulting from feeding dairy and beef cows with dietary supplements synthesized from the offal and carcasses of scrapie-infected sheep. The practice of using slaughterhouse renderings as a feed supplement has been widespread in animal husbandry, and since scrapie is a relatively common disease in some herds of sheep in Great Britain, the use of contaminated carcasses was well established. The problem arose because of the way this material was rendered. In the past, the offal was rendered by extensive heat treatment, which apparently was sufficient to destroy PrP_{Sc}. In the 1980s, however, the high cost of fossil fuel led English suppliers to use a chemical method of rendering the carcasses that ineffectively inactivated the prion material. The very long incubation period of prion-induced BSE resulted in a long delay before symptoms appeared in British herds.

As damaging as this has been to the English cattle industry, there is an even more serious possibility. There is good documentation that the disease can be transmitted to domestic and zoo cats, and a number of young people in Britain have developed Creutzfeld–Jakob disease. This was never reported to occur in young adults in England previously, and it has been suggested that the cattle disease is transmissible to humans. This possibility has been difficult to substantiate because while the normal incidence of spontaneous Creutzfeld–Jakob disease is very low, the number of new cases does not represent a statistically significant increase. Disturbingly, however, the disease was formerly confined to the elderly, and the occurrence of the disease in young people is worrisome. This concern is enhanced by the fact that the form of prion isolated from young patients has a glycosylation pattern similar to the PrP_{BSE} found in cattle and is significantly different from the glycosylation pattern of PrP_{CJD} isolated from older victims of the disease. This form is now termed *variant Creutzfeld–Jakob disease* (vCJD).

For this reason, the British beef-processing industry has been sorely tested. New national policies concerning the feeding of cows were implemented, and it is currently illegal to purchase certain cuts of beef in England that are considered to be potential carriers of the disease, including cuts with large amounts of bone marrow and nerve tissue. Other countries have banned the importation of British beef.

The rate of occurrence of youth-associated vCJD has not increased since public health officials have become aware of the problem. As of 2015 there have been 229 cases worldwide. An analysis of the genetics of these patients indicates that those most susceptible to vCJD have a specific polymorphism of the normal human protein that allows conversion to the pathogenic form.

Another prion disease of importance that is present in the United States is chronic wasting disease (CWD), affecting deer, elk, and moose. CWD was first identified in 1960 in deer in captivity, and in 1981 in free-ranging deer or elk. Sometimes called the "zombie disease" in the popular press, CWD produces symptoms much like scrapie in sheep, such as weight loss, stumbling gait, and drooling. The disease is fatal. As of January 2019, the CDC reports the presence of CWD in deer, elk, or moose in 24 of the continental United States and in 2 Canadian provinces. While there is no evidence of transmission to humans and resulting disease, the CWD prion has been successfully transmitted to squirrel monkeys with subsequent disease presentation. As a result, CDC advises caution to hunters when handling or eating meat from affected animals.

Case 3: Respiratory syncytial virus (RSV)

Clinical presentation/case history: A three-month-old infant was brought to the ER with severe respiratory symptoms, labored breathing, nostril flaring, and weakness. A history revealed that several days before, she had developed cold-like symptoms that gradually worsened. She had a fever of 102 °F, rapid heart rate, normal blood pressure, and cyanosis (bluish coloration of the skin due to insufficient oxygen in the blood). Examination of her chest with a stethoscope revealed rattles and crackles upon breathing, an indication of mucus in the deep airways. A computed tomography (CT) scan of the lungs was ordered to determine the extent of the pathology, and nasal swabs and blood samples were sent for screening in an attempt to identify the pathogen.

Continued

Diagnosis: Because of infants' immature immune systems, advanced respiratory illness such as this is a serious concern. If the infection cannot be controlled quickly and the patient's condition stabilized, death can occur. The CT scan revealed large opaque (dense) areas in the lungs, indicating inflammation. The viral and bacterial analyses are an important tool to differentiate between bacterial etiologies such as *Bordetella pertussis* (causative agent of whooping cough) and viral etiologies such as influenza or respiratory syncytial virus (RSV). In this case a rapid enzyme-linked immunosorbent assay (ELISA)-based test quickly identified RSV antigens, providing a definitive diagnosis.

Treatment: The patient was immediately started on ribavirin, a pro-drug that is converted to a 5′ triphosphate nucleotide that acts as an antiviral by interfering with RNA metabolism. For this reason it is effective against a number of RNA viruses. For RSV infections of infants, the drug is given in aerosol form so that high doses can be delivered to the respiratory tract without causing systemic toxicity. The infant responded to the ribavirin as well as to supportive care, which included oxygen and IV fluids.

Disease notes: RSV, a paramyxovirus, is the leading cause of bronchiolitis and pneumonia in infants. Almost all children seroconvert by age three and have lifelong protection. The virus is highly contagious by aerosol transmission, and it is common for the infections to spread rapidly among siblings and in daycare settings. Most RSV infections resemble a bad cold, but occasionally they progress rapidly to severe respiratory disease. Immunocompetent children between two and six months of age and immunosuppressed children under three years are at the highest risk for severe disease, and infections of the immunosuppressed have a high incidence of mortality.

QUESTIONS FOR CHAPTER 15

1 What features of the viral replication cycle are shared by measles virus, vesicular stomatitis virus, and influenza virus?

2 When the *genomes* of negative-sense RNA viruses are *purified* and introduced into cells that are permissive to the original intact virus, what will occur?

3 The Rhabdoviridae are typical negative-sense RNA viruses and must carry out two types of RNA synthesis during infection: transcription and replication. *Briefly* describe each of these modes of viral RNA synthesis.

4 Sin Nombre virus is the causative agent of the outbreak of hantavirus-associated disease that was first identified in a cluster of cases originating in the Four Corners area of the southwestern United States.

(a) To which virus family does this virus belong?

(b) Which animal is the vector for transmission of this virus to humans?

(c) What feature of the disease caused by this virus makes it different from other members of its genus?

5 Bunyavirus gene expression includes three different solutions to the problem of presenting the host cell with a "monocistronic" mRNA. For each of the genome segments (L, M, and S), describe in a simple drawing or in one sentence how this problem is solved.

6 Your laboratory has now become the world leader in research on the spring fever virus (SpFV), especially the debilitating variant SpFV-4 that causes senioritis. Your team has determined that these viruses are members of the family Orthomyxoviridae, but an international commission on virus nomenclature has suggested that they be assigned to a subgenus of the influenza viruses. While you agree with the family designation, you are convinced that they belong to a new genus that you have tentatively called the *Procrastinoviruses*.
The following table lists properties of SpFV strains that your laboratory has investigated.

Viral function	Results for SpFV	
A	Virion membrane glycoproteins	Two major proteins, one with hemagglutinin activity and the other with neuraminidase activity
B	Matrix proteins in virion	One matrix protein
C	Genome segments	Eight single-stranded RNA molecules
D	Viral mRNA synthesis	Nuclear location, with cap scavenging from host mRNA precursors and RNA splicing to produce some species of viral mRNA
E	Nonstructural (NS) proteins in infected cells	Three NS proteins, two encoded by RNA segment 8 and one encoded by RNA segment 6
F	Site of infection	Generalized neuromuscular locations, ultimately targeting higher neural functions associated with memory and motivation

(a) Which of these features justify inclusion of SpFV in the Orthomyxoviridae family?
(b) Which of these features justify your proposal that SpFV should be considered a new genus of this family?

You have just received an isolate of SpFV-4 obtained from a severe outbreak of senioritis at a large East Coast university. The epidemic began among a group of students who had just returned from a semester abroad in Paris.

(c) As an expert virologist, which viral proteins do you predict are most likely to distinguish this isolate of SpFV-4 from those you have investigated in your laboratory?
(d) What phenomenon could account for these differences?

7 What are two differences between the members of the *Hantavirus* genus and members of the other genera of the family Bunyaviridae?

8 Influenza virus will *not* grow in a cell from which the nucleus has been removed. Although influenza virus does not have a DNA intermediate in its life cycle, there is still a requirement for nuclear functions.
(a) List two molecular events during the influenza virus life cycle that require something provided by the host cell nucleus.
(b) For which of these events is the *physical presence* of the nucleus in the cell absolutely required? Why?

9 The data shown in the figure below were obtained for three different isolates of influenza type A virus. The three viruses (designated 1, 2, and 3) were grown in cell culture in the presence of radioactive RNA precursors. The radiolabeled RNA genome segments were then separated by electrophoresis through a polyacrylamide gel. The drawing below shows the relative migration in this gel of each of the genome segments. In addition, the segment number and the viral gene product or products produced by that segment are shown.
(a) In isolates of influenza virus, H and N numbers refer to the genotypes of the hemagglutinin and neuraminidase, respectively. Suppose that virus 1 is found to be H1N1 and virus 2 is found to be H2N3. What would be the designation for virus 3?
(b) The antiviral drug amantidine is used to stop or slow down an influenza virus infection. Virus 1 is sensitive to amantidine, while virus 2 is resistant to this antiviral agent. Your mentor predicts that you will find virus 3 to be sensitive to amantidine. What evidence in this electropherogram leads your mentor to suggest this?
(c) By what genetic mechanism (typical for the Orthomyxoviridae) did virus 3 arise?

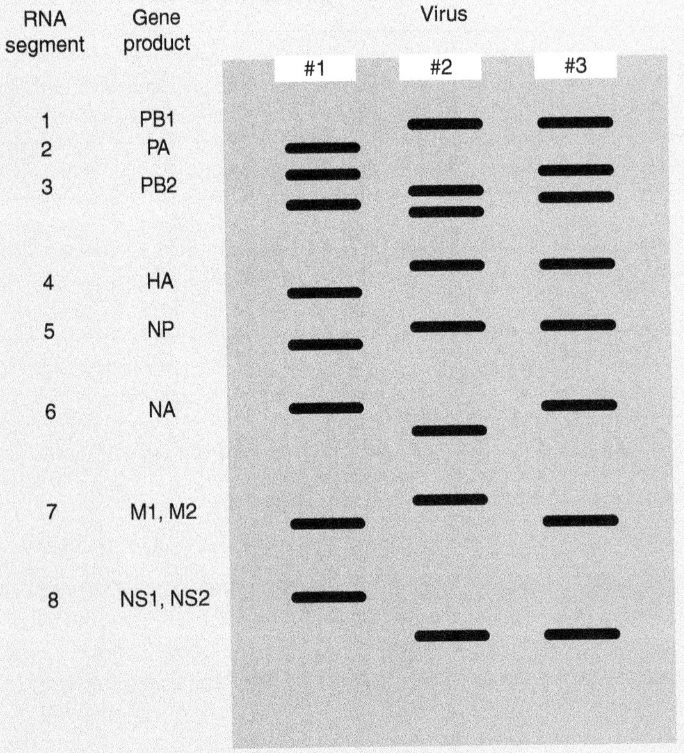

10 Reovirus is the prototype member of the family Reoviridae. Describe the features of this virus that make it different from other RNA genome viruses.

11 Viroids are infectious agents of plants and are circular, single-stranded RNA molecules. Describe the features of infection of a plant with this kind of agent.

12 In what sense can a prion be described as a "self-replicating entity"?

Replication Strategies of Small and Medium-sized DNA Viruses

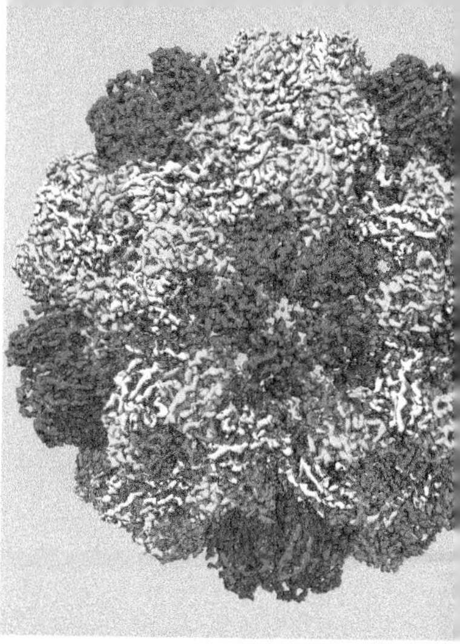

CHAPTER

16

- ✱ DNA VIRUSES EXPRESS GENETIC INFORMATION AND REPLICATE THEIR GENOMES IN SIMILAR, YET DISTINCT, WAYS
- ✱ PAPOVAVIRUS REPLICATION
- ✱ Replication of SV40 virus – the model polyomavirus
  - The SV40 genome and genetic map
  - Productive infection by SV40
  - Abortive infection of cells nonpermissive for SV40 replication
- ✱ The replication of papillomaviruses
  - The HPV-16 genome
  - Virus replication and cytopathology
- ✱ THE REPLICATION OF ADENOVIRUSES
- ✱ Physical properties of adenovirus
  - Capsid structure
  - The adenovirus genome
- ✱ The adenovirus replication cycle
  - Early events
  - Adenovirus DNA replication
  - Late gene expression
  - VA transcription and cytopathology
  - Transformation of nonpermissive cells by adenovirus
- ✱ REPLICATION OF SOME SINGLE-STRANDED DNA VIRUSES
- ✱ Replication of parvoviruses
  - Dependovirus DNA integrates in a specific site in the host cell genome
  - Parvoviruses have potentially exploitable therapeutic applications
- ✱ DNA viruses infecting vascular plants
- ✱ Geminiviruses
- ✱ The single-stranded DNA bacteriophage ΦX174 packages its genes very compactly
- ✱ QUESTIONS FOR CHAPTER 16

Basic Virology, Fourth Edition. Martinez "Marty" Hewlett, David Camerini, and David C. Bloom.

DNA VIRUSES EXPRESS GENETIC INFORMATION AND REPLICATE THEIR GENOMES IN SIMILAR, YET DISTINCT, WAYS

Given that DNA is the universal genetic material of cells, it is not particularly surprising that viruses utilizing DNA as their genome comprise a significant proportion of the total number of known viruses. It also is not particularly surprising that such viruses will often use a significant proportion of cellular machinery involved in decoding and replicating genetic information encoded in double-stranded DNA (dsDNA), which is, after all, the stuff of the cellular genome.

While it might be expected that all viruses with DNA genomes would follow a generally similar pattern of replication, this is not the case. Indeed, viruses with DNA genomes utilize as many variations on a general replication strategy as do RNA viruses. There are both naked and enveloped DNA-containing viruses, and a number of DNA viruses encapsidate only a single strand of DNA. One group of animal viruses utilizing DNA as genetic material replicates in the cytoplasm of eukaryotic cells, and some DNA viruses infecting plants contain multipartite genomes. A major and extremely important group converts RNA into DNA, while a related group converts RNA packaged in the virion into DNA as the virus matures!

While one can make useful generalizations concerning the replication of DNA viruses (indeed, one must if the material is to be readily mastered), it is wise to treat such generalities as only basic guides. Thus, viruses of eukaryotic cells that replicate using the nucleus express their RNA using cellular transcription machinery, but bacterial DNA viruses as well as at least one group of insect DNA viruses (the baculoviruses) encode one or a number of novel RNA polymerases or specificity factors to ensure that only viral messenger RNA (mRNA) is expressed following infection. Similarly, the cytoplasmic-replicating DNA genome–containing poxviruses of eukaryotes encode many enzymes involved in transcription and mRNA modification.

Many DNA viruses use DNA replication enzymes and mechanisms that are generally related to the processes seen in the uninfected cell, but there is one major complexity when DNA replication of viruses is considered. This is the fact that while all viral DNA replication requires a primer, some groups do not utilize RNA primers! Thus, one of the basic tenets of the process outlined in Chapter 13 of Part III is violated.

Viruses with linear genomes face a major problem that also affects the replication of cellular chromosomal DNA. This "end problem" derives from the fact that the primer for DNA replication must be able to recognize short stretches of the viral genome – either through base pairing or through specific DNA–protein interactions. Consider the problem for discontinuous strand DNA synthesis as shown in Figure 13.1 in Chapter 13, Part III. When the primer anneals to the very 3′ end sequences of the template, DNA replication can proceed 5′ to 3′ down to the next fragment. But how is the primer to be removed and replaced with DNA? There is no place for a new primer to anneal upstream of this last gap to be filled. This situation means that the viral genome would have to become shorter every time it replicated and would rapidly disappear!

Eukaryotes have solved the end problem in the replication of their linear chromosomal DNA by using the enzyme telomerase to replace lost sequences. DNA viruses with linear genomes have evolved different means to overcome this end problem. Herpesviruses and many bacterial DNA viruses have genomes with repeated sequences at their terminals so that the viral genome can become circular via a recombination event following infection. Thus, even though the virion DNA is linear, replicating viral DNA in the cell is either circular or joined end to end in long **concatamers**. These structures are then resolved to linear ones when viral DNA is encapsidated.

Adenovirus, on the other hand, has solved the problem by using a primer that is covalently bound to a viral protein that binds to the viral DNA's end. Further, adenovirus DNA replication proceeds only continuously; there is no discontinuous strand synthesis.

Small single-stranded DNA (ssDNA) viruses, like parvoviruses, have solved the problem by encoding a complementary repeat sequence at the end that allows the genome to form a "hairpin loop" at the end; thus, the end of the molecule is not free. A similar solution is seen in the genome structure of poxvirus. Like chromosomal DNA, this linear DNA genome is covalently closed at its ends. Thus, in effect, replication just proceeds "around the corner" onto the complementary strand.

Another important general strategy found in the replication of nuclear-replicating eukaryotic viruses and many bacteriophages is the establishment of infections where the viral genome remains in lifelong association with its host. Such a process has tremendous evolutionary advantages to any pathogen, but again, the specifics of the process in terms of mechanism differ greatly between the groups.

Given these variations, it is important to describe the basic processes of DNA virus replication in a logical way, and this is perhaps best done by consideration of how much cellular function and cellular transcriptional machinery are needed for productive replication. This is roughly correlated with overall size of the viral genome. The usefulness of such a grouping is that the viruses in each group share certain similarities in their replication strategies. Equally important, they share similarities in the way they can alter cells during the replication process. Such alterations can have profound and far-reaching effects on the host's health.

The discussion of three unrelated families of viruses infecting eukaryotic cells in this chapter follows this (admittedly flawed but convenient) organizational strategy. The unifying features of these viruses are that they replicate in the nucleus of the host cell, and each strictly relies on one or another related function found in actively replicating animal cells for their successful propagation. Two other families of viruses, one infecting plants and one bacteria, are included to demonstrate some of the strategies viruses can utilize to ensure that their DNA genomes are as physically compact as possible.

PAPOVAVIRUS REPLICATION

The term *papovavirus* stands for "*pa*pilloma, *po*lyoma, *va*cuolating" viruses. Actually, members of the group fall into two distinct families: the papillomaviruses and polyomaviruses. These two groups are similar regarding icosahedral capsids, circular genomes, and the ability to remain associated with the host for long periods, as well as their requirement to specifically alter cell growth in the host cell's response to neighboring cells for virus replication. They differ in genome size and in many details of host cell specificity.

One unusual structural feature of the polyomavirus capsid is that although it is an icosahedron, the capsid subunits do not form hexon and penton arrays as is normal for such a structure (see Chapter 5, Part II). Rather, all 60 pentameric subunits are equivalent and can assemble in an asymmetrical fashion to form the capsid. This is shown in Figure 16.1a. Any functional or genetic strictures on polyomaviruses that might require this unusual structure are not clear.

Replication of SV40 virus – the model polyomavirus

The polyomaviruses have genomes of approximately 5000 base pairs. Capsids are made up of three proteins, usually called VP1, VP2, and VP3. Polyomaviruses can cause tumors in animals and can transform the growth properties of primary cells in culture, especially the cells from animals different from the virus's natural host (see Chapter 10, Part III). Polyomaviruses also stay persistently associated with the host, often with little evidence of extensive pathology or disease. Although these viruses kill the cells in which they replicate, this process is slow. In

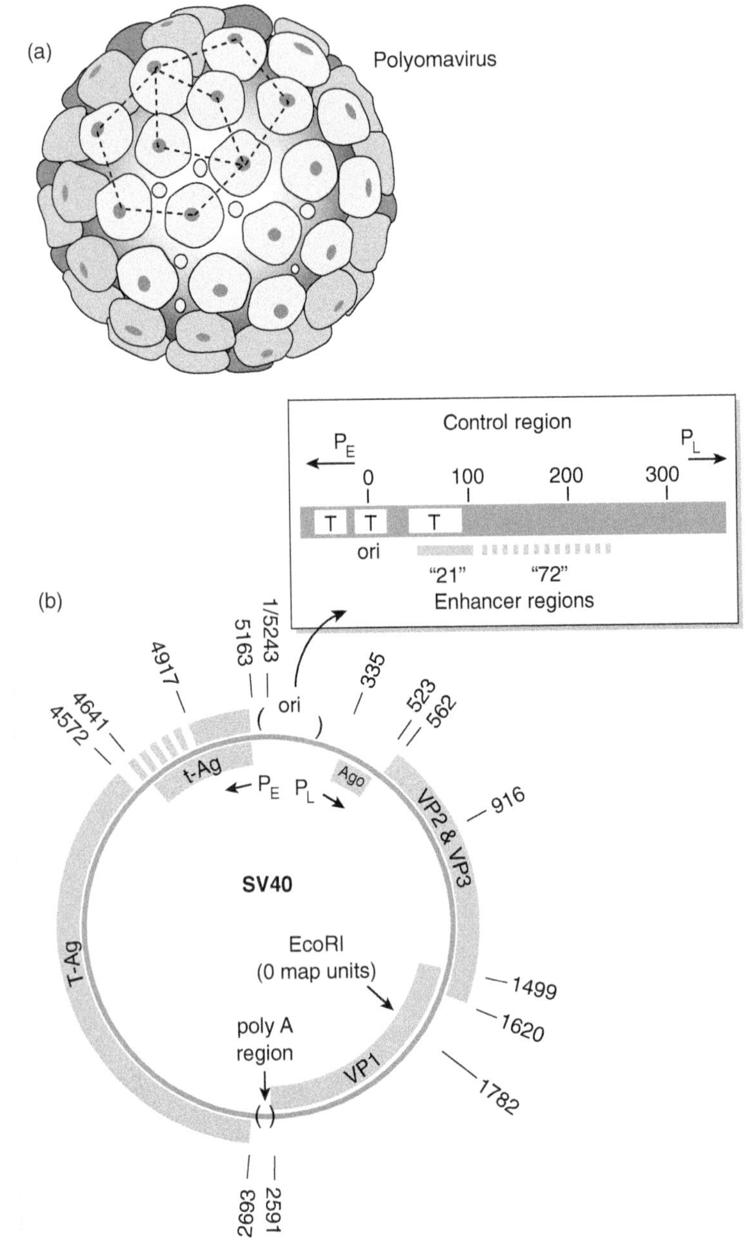

Figure 16.1 Polyomavirus and the genetic and transcript map of SV40 virus. (a) The 60 pentameric subunits of the capsid proteins are arranged in an unusual fashion so that the packaging of individual capsomers is not equivalent in all directions. The drawing is based on computer-enhanced analysis using electron microscope and X-ray diffraction methods (see Chapter 5). The 5243-base pair dsDNA genome is condensed with host cell histones and packaged into the 45-nm-diameter icosahedral capsid. *Source:* Based onSalunke, D.M., Caspar, D.L., and Garcea, R.L. (1986). Self-assembly of purified polyomavirus capsid protein VP1. *Cell* 46: 895–904. (b) The early and late promoters, origin of replication, and bidirectional cleavage/polyadenylation signals are shown along with the introns and exons of the early and late transcripts. A high-resolution schematic of the approximately 500-base-pair control region with the early and late promoters is also provided. Two early promoter enhancers, one containing the 21-base-pair repeats and the other containing the 72-base-pair repeats, are shown. The origin of replication (ori) is situated between the enhancers and the early promoter, and the three binding sites for large T antigen (T) are indicated. (c) A higher-resolution schematic of the processing of early viral mRNAs. Splice sites, translational reading frames, and other features are indicated by sequence number. Details are described in the text. Note that the 3′ end of the pre-mRNA occurs just beyond the early polyadenylation site (2590) that is situated in the 3′ transcribed region of the late pre-mRNA. (d). A higher-resolution schematic of the processing of late viral mRNAs. Splice sites, translational reading frames, and other features are indicated by sequence number. Details are described in the text. Note that the 3′ end of the pre-mRNA occurs just beyond the late polyadenylation site (2650) and is situated in the 3′ transcribed region of the early pre-mRNA. T-Ag, Large T antigen; t-Ag, small t antigen.

keeping with the requirement for extensive cellular function during replication, there is no global virus-induced shutoff of host function.

One widely studied polyomavirus is murine polyomavirus (MPyV), originally isolated from wild mice and named for its ability to cause many types of small tumors in some strains of newborn mice. Another widely studied polyomavirus is SV40 virus, which was originally named *simian vacuolating agent 40*. There is no evidence that this virus causes tumors in its natural host, but it can readily cause growth transformation of cells in culture. SV40 virus originally was found as a contaminant of African green monkey kidney (AGMK) cells in which poliovirus was being grown for vaccine purposes. Early recipients of the Salk polio vaccine got a good dose of the virus, but no pathology has been ascribed to this, at least to the present time.

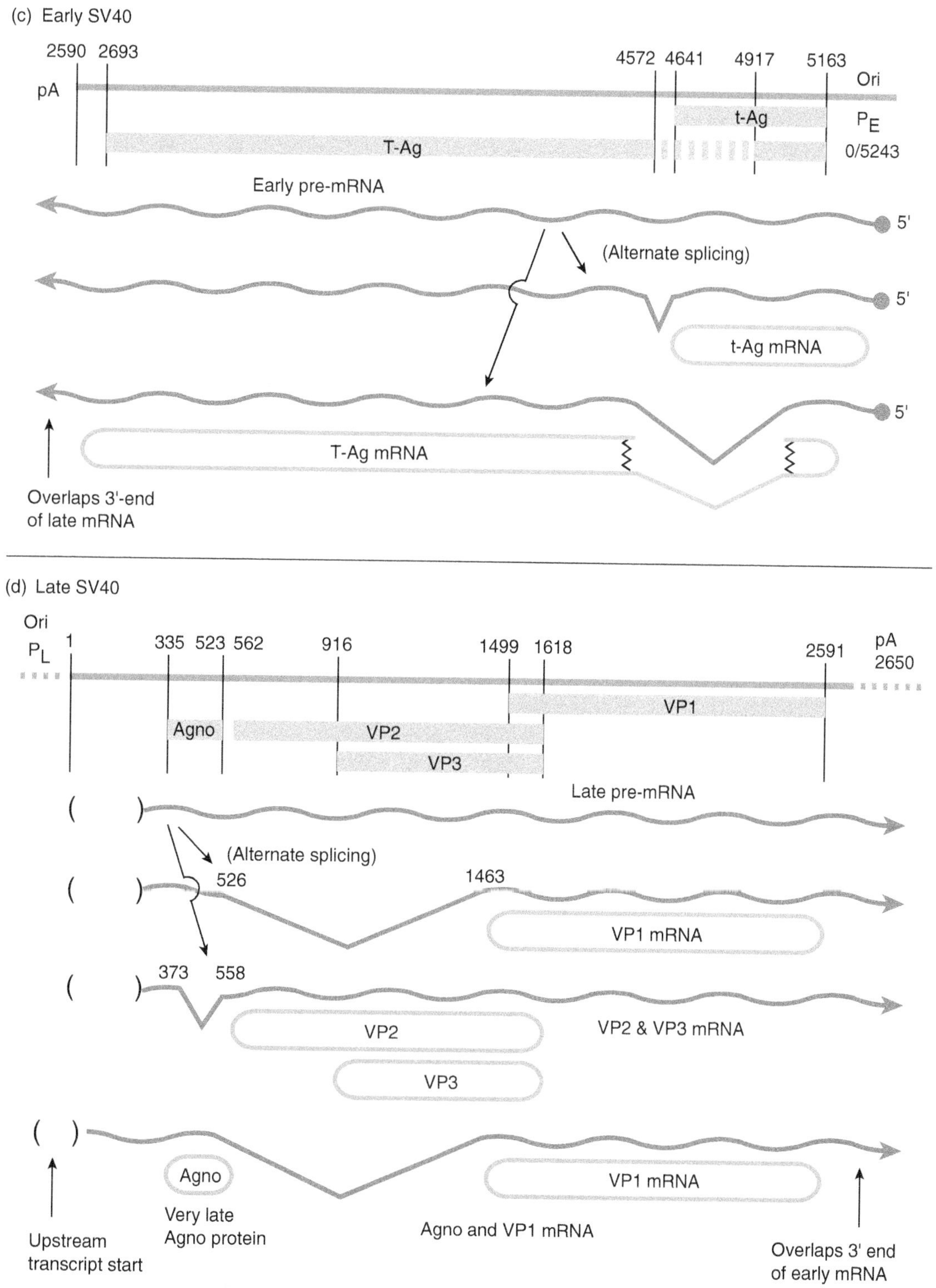

Figure 16.1 *Continued*

Whereas Rous sarcoma virus (a retrovirus) had been known to cause tumors in chickens since the early part of the twentieth century, the fact that its genome is RNA made understanding of its mechanism of oncogenesis out of the reach of molecular biologists working in the 1950s and 1960s. Indeed, major progress awaited the discovery of reverse transcriptase by Howard Temin and David Baltimore in 1970. In contrast, the fact that the DNA-containing SV40 and mouse polyomaviruses cause growth transformation and tumors in the laboratory provided a model for the study of the process that could be exploited with the techniques available at the time. The study of these viruses essentially launched the molecular biological study of carcinogenesis and eventually led to the discovery of tumor suppressor genes and their important role in regulating cell growth and division.

The SV40 virus's importance in fundamental research in oncogenesis, ease of manipulation in the laboratory, and convenient genome size have contributed to its status as, arguably, the most extensively studied of all DNA viruses. While MPyV and SV40 replication differ in some important features, the overall strategy is the same. Thirteen human polyomaviruses are currently known. The BKPyV and JCPyV viruses were the first two of these viruses associated with human disease and are closely related to SV40; they are spread by body fluids such as blood and urine, or saliva. Primary infection occurs in children with little obvious pathology. In the United States, most children are infected with BKPyV virus by the age of five to six years, and the only signs of infection may be a mild respiratory illness. Infection with JCPyV virus occurs somewhat later, with most children being infected between the ages of 10 and 14 years.

Resolution of infection is complete in children with normally functioning cell-mediated immunity. Despite resolution, the virus persists for the life of the individual – one primary site of persistence being the kidney from which BKPyV virus can be periodically shed. In addition, JCPyV virus can be recovered from brain biopsy specimens. While this persistence has no known clinical manifestations in the healthy individual and is thought to be the result of viral genomes persisting in an inactive state in nondividing, terminally differentiated cells, immunosuppression by HIV infection or prior to organ transplantation can lead to severe consequences. In immune-compromised individuals, JCPyV virus is associated with a rare progressive destruction of neural tissue in the **central nervous system (CNS)** called progressive multifocal leukoencephalopathy (PML). This neuropathology is the result of the fact that transcription of the JCPyV virus RNAs can take place in oligodendrocytes (but not other cells) in the adult brain, but just what aspect of immune suppression (such as that engendered by HIV infection, or certain monoclonal antibody therapies; see **Case Study: Polyomaviruses**) reactivates this dormant virus is unknown. BKPyV virus infections can lead to kidney pathologies in immune-compromised individuals, especially organ transplant recipients.

The exact sequence of human JCPyV virus isolated from individuals in various parts of the world varies enough to allow its use as a genetic population marker. Extensive studies on natural isolates show that individual variants are strongly associated with individual ethnic and racial population groups, and their movements throughout the world can be traced by the occurrence of specific virus variants. This means that the virus has been associated with the human population for an extremely long time, and that variants have arisen as populations have diverged.

The pattern of infection of young animals followed by virus persistence and shedding is quite characteristic of the infection of laboratory strains of mice with MPyV. Infections of suckling mice can lead to the formation of tumors, hence, the name polyomavirus. Initially the ability of MPyV to cause tumors was thought to be a unique property of polyomaviruses of mice, but it has been discovered recently that Merkel cell polyomavirus (MCPyV) is the cause of Merkel cell cancer (a highly aggressive type of skin cancer) in humans. Given that 11 of the 13 human polyomaviruses have only been discovered in the past 10 years, it remains to be seen whether other of these viruses are linked as a co-factor in other human cancers.

The SV40 genome and genetic map

The SV40 virus genome contains 5243 base pairs, and its map showing essential features is displayed in Figure 16.1b. The genome is organized into four functional regions, each of which is discussed separately in this section.

The control region This region covers about 500 bases and consists of the origin of replication, the early promoter/enhancer, and the late promoter. The sequence elements in this region overlap to a considerable extent, but the bases specifically involved with each function can be located precisely on the genome. This has been done by making defined mutations in the sequence and analyzing their effects on viral genome replication and on expression of early and late genes.

The early promoter region contains a TATA box and enhancer regions (noted by 72-base and 21-base repeats). Surprisingly, the late promoter does not have a TATA box, and late mRNA initiates at a number of places within a 60- to 80-base region. The multiple start sites for late mRNA expressed from this "TATA-less" promoter provided one of the early clues that the TATA box functions to assemble transcription complexes at a specific location in relation to mRNA initiation. It is not clear exactly what substitutes for the TATA box in the late promoter, but it is thought that transcription complexes can form relatively readily throughout the region.

The origin of replication (ori) is about 150 base pairs in extent and contains several elements with a sequence critically linked by "spacers" whose length but not specific sequence is important in function of the origin. The ori elements have some dyad symmetry; that is, the sequence of the far left region is repeated in the inverse sense in the far right region. This symmetry is thought to have a role in allowing the DNA helix to "melt" at the origin, facilitating the entry of replication enzymes to begin rounds of DNA replication. The general process was described in Chapter 13, Part III.

The early transcription unit The SV40 genome's early region is shown in high resolution in Figure 16.1c. It is transcribed into a single mRNA precursor that extends about halfway around the genome, and contains two open translational reading frames (ORFs). The single early pre-mRNA transcript can be spliced at one of two specific sites (i.e., the pre-mRNA is subject to alternative splicing – see Chapter 13, especially Figure 13.7). If a short intron is removed, an mRNA is generated that encodes a relatively small (approximately 20 000 Da) protein (**small t antigen**), which has a role in allowing the virus to replicate in certain cells.

A slightly smaller (and more abundant) mRNA is generated by the splicing of a larger intron in the pre-mRNA. This removes a translation terminator that terminates the small-t-antigen ORF. The smaller (!) mRNA encodes the **large T antigen** (approximately 80 000 Da). The large T antigen has a number of functions, including the following:

1 Activation of cellular DNA and RNA synthesis by binding to the cellular growth control gene products named Rb and p53. This binding stops these control proteins from keeping the cell contact inhibited. This function causes the infected cell to begin a round of DNA replication.
2 Blockage of apoptosis that is normally induced in cells where p53 is inactivated at inappropriate times in the cell cycle.
3 Binding to the SV40 ori to initiate viral DNA replication.
4 Shutting off early viral transcription by binding to regions in and near the early promoter.
5 Activating late transcription.
6 Playing a role in virion assembly.

The late transcription unit Late mRNA is expressed from a region extending around the other half of the genome from the late promoter; this is shown in Figure 16.1d. The late region contains two large ORFs that encode the *three* capsid proteins. Part of the expression of late proteins, then, requires alternate splicing patterns, just as is seen with the generation of early mRNA. Splicing of a large intron from the primary late pre-mRNA transcript generates an mRNA that encodes the 36 000-Da major capsid protein (VP1). A small amount of mRNA is generated by splicing a small intron near the 5′ end of the mRNA, allowing the first ORF to be translated into the 35 000-Da VP2 protein.

The third capsid protein, VP3, is also expressed from the same mRNA encoding VP2 by utilization of an alternative translation initiation site. Ribosomes sometimes "miss" the first AUG of the 5′ ORF in the mRNA expressing VP2. When this happens, the ribosome initiates translation at an AUG in phase with the first one but downstream, producing the 23 000-Da VP3 protein. Thus, one mRNA encodes both VP2 and VP3, depending on where the ribosome starts translation. This "skipping" does not violate the general rule that a eukaryotic ribosome can only initiate a protein at the 5' ORF, as the first AUG is not seen and thus is in the operational leader sequence of the mRNA.

There is a fourth late protein expressed from the late region, but this is only seen very late in infection. This basic protein, the "agnoprotein," is encoded in a short ORF upstream of that encoding VP2. Very late in infection, some mRNAs are produced by initiation of transcription farther upstream than at earlier times, and these can be translated into this protein. The role of this product is not fully understood, but it may be involved in allowing the virus to replicate in certain cells that are normally nonpermissive for viral replication.

The polyadenylation region About 180° around the circular SV40 genome from the ori/promoter region lies a second *cis*-acting control region. It contains polyadenylation signals on both DNA strands so that transcripts transcribed from both the early and late regions terminate in this region. It is notable that the polyadenylation signals for the mRNAs are situated such that the early and late transcripts have a region of 3′ overlap. This can lead to the generation of dsRNA during the replication cycle, with attendant induction of interferon in infected cells (see Chapter 7).

Productive infection by SV40

Productive infection by SV40 in its normal host can be easily studied in cell culture using monkey kidney cells. The replication cycle is quite long, often taking 72 hours or more before cell lysis and release of new virus occur. One reason for this "leisurely" pace is that the virus is quite dependent on continued cellular function during most of its replication. The virus replicates efficiently in cultured cells that are actively dividing either because they have not yet reached confluence or because the cells are growth transformed and not subject to contact inhibition of growth. (The basic growth properties of cultured cells are discussed in Chapter 10, Part III.)

While the virus replicates efficiently in replicating cells, it also is able to replicate well in cells that are under growth arrest. This is by virtue of T-antigen expression early in infection. Manifestations of this ability provide many useful insights into the nature of the cell's ability to control and regulate its own DNA replication, and led to the discovery of the tumor suppressor genes p53 and Rb discussed later in this chapter.

Virus attachment and entry The replication cycle of SV40 is outlined in Figure 16.2. Virions interact with a specific cellular receptor. This leads to receptor-mediated endocytosis, and the partially uncoated virion is transported in the endocytotic vesicle to the nucleus where viral DNA is released.

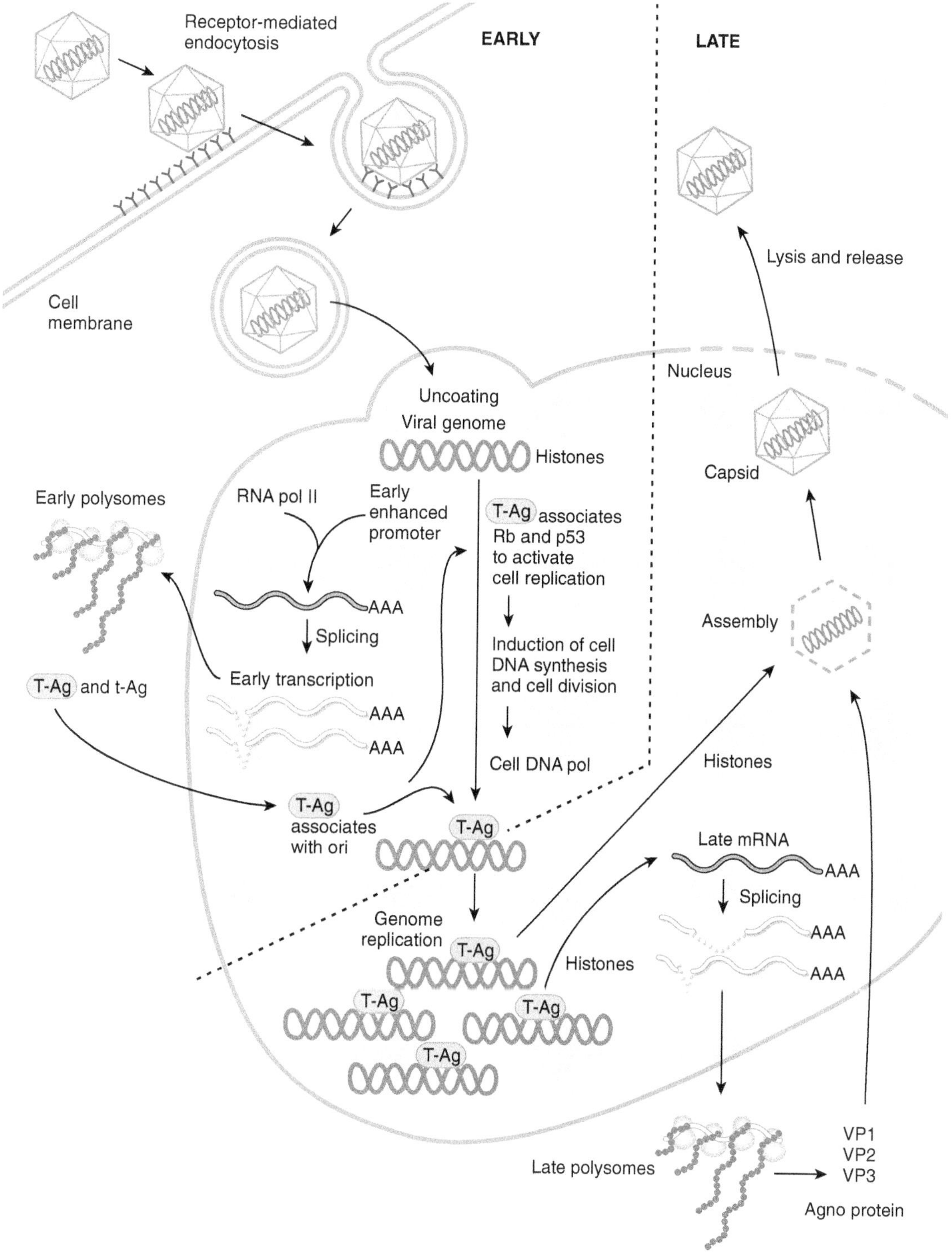

Figure 16.2 The replication cycle of SV40 virus in a permissive cell. The replication is divided into two phases, early and late. During the early stages of infection, virus attaches and viral genomes with accompanying cellular histones are transported to the nucleus via receptor-mediated endocytosis. RNA polymerase II (pol II) recognizes the enhanced early promoter, leading to transcription of early pre-mRNA, which is processed into mRNAs encoding small t antigen (t-Ag) and large T antigen (T-Ag). These mRNAs are translated into their corresponding proteins. Large T antigen migrates to the nucleus where it carries out a number of functions, including inactivation of the cellular growth control proteins p53 and Rb, and binding of the SV40 origin of DNA replication (ori). Viral DNA replication takes place by the action of cellular DNA replication enzymes, and each round of DNA replication requires large T antigen to bind to the ori. As genomes are replicated, the late stage of infection begins. High levels of large T antigen suppress the expression of early pre-mRNA and stimulate expression of late pre-mRNA. This is processed into two late mRNAs; the smaller encodes both VP2 and VP3, while the larger encodes VP1. At very late times, some transcripts are expressed and can be translated into the small agnoprotein. Viral capsid proteins migrate to the nucleus where they assemble into capsids with newly synthesized viral DNA. Finally, progeny virus is released by cell lysis.

The association of viral genomic DNA with cellular chromosomal proteins is a common feature in the replication of nuclear replicating viruses discussed in this chapter and Chapter 17. In the case of SV40 and other papovaviruses, the viral DNA is associated with histones and other chromosomal proteins when it is packaged into the virion. It remains associated with chromosomal proteins upon its entry into the nucleus. This means, in effect, that viral DNA is actually presented to the cell as a small or "mini"-chromosome. Essentially, then, the cell's transcriptional machinery recognizes the viral chromosome and promoters therein merely as cellular genes waiting for transcription.

Early gene expression Early gene expression results in formation of large quantities of large-T-antigen mRNA, and less amounts of small-t-antigen mRNA. The amounts of protein synthesized are roughly proportional to the amount of mRNA present. The small t antigen contains the same N-terminal amino acids as does large T antigen because of the way early pre-mRNA is spliced into the two early mRNAs, as shown in Figure 16.1c. The splice-generating mRNA that encodes the T antigen removes a translation stop signal. In contrast, the splice in the t-antigen mRNA is beyond the ORF, and thus does not affect protein termination. Generation of two proteins with major or minor differences in function but with a shared portion of amino acid sequence is quite common with many viruses. It is very important in the expression of adenovirus proteins.

The role of T antigen in viral DNA replication and the early/late transcription switch As outlined in this chapter, T antigen alters the host cell to allow it to replicate viral DNA. The T antigen also binds to the SV40 ori to allow DNA replication to begin, *and* to shut off synthesis of early mRNA. Each round of DNA replication requires T antigen to bind to the origin of DNA replication and initiate a round of DNA synthesis. DNA replication then proceeds via leading and lagging strand synthesis using cellular enzymes and proteins as described in Chapter 13. Since the SV40 genome is circular, there is no end problem, and the two daughter circles are separated by DNA cleavage and ligation at the end of each round of replication. This resolution of the interlinked supercoiled DNA molecules into individual genomes is mediated by cellular enzymes, notably topoisomerases and resolvases. The process is illustrated in Figure 16.3. It is important to note that association of the daughter DNA genomes with cellular histones is not shown in Figure 16.3, but this association is necessary for the virus to be efficiently encapsidated.

While DNA replication proceeds, the relative rate of early mRNA synthesis declines owing to accumulation of increasing amounts of large T antigen in the cell, which represses synthesis of its own mRNA by binding at the ori and early promoter. While the relative amount of early mRNA declines in the cell at late times, its production never entirely ceases because there is always some template that has not yet bound large T antigen available for early mRNA expression.

At the same time that this versatile protein is modulating and suppressing its own synthesis, it activates transcription of late pre-mRNA from replicating DNA templates. Late transcripts have heterogeneous 5′ ends, and as noted previously, very late in infection, the start of late mRNA transcription shifts to a point upstream of that previously used, and the **agnogene** protein (the agnoprotein) can be encoded and translated from a novel subset of late mRNAs.

Abortive infection of cells nonpermissive for SV40 replication

Relatively early in the study of polyomavirus replication, infection of cells derived from a species other than the natural host of SV40 was observed to lead to an abortive infection where no virus was produced. Despite this, virus infection was shown to stimulate cellular DNA replication and

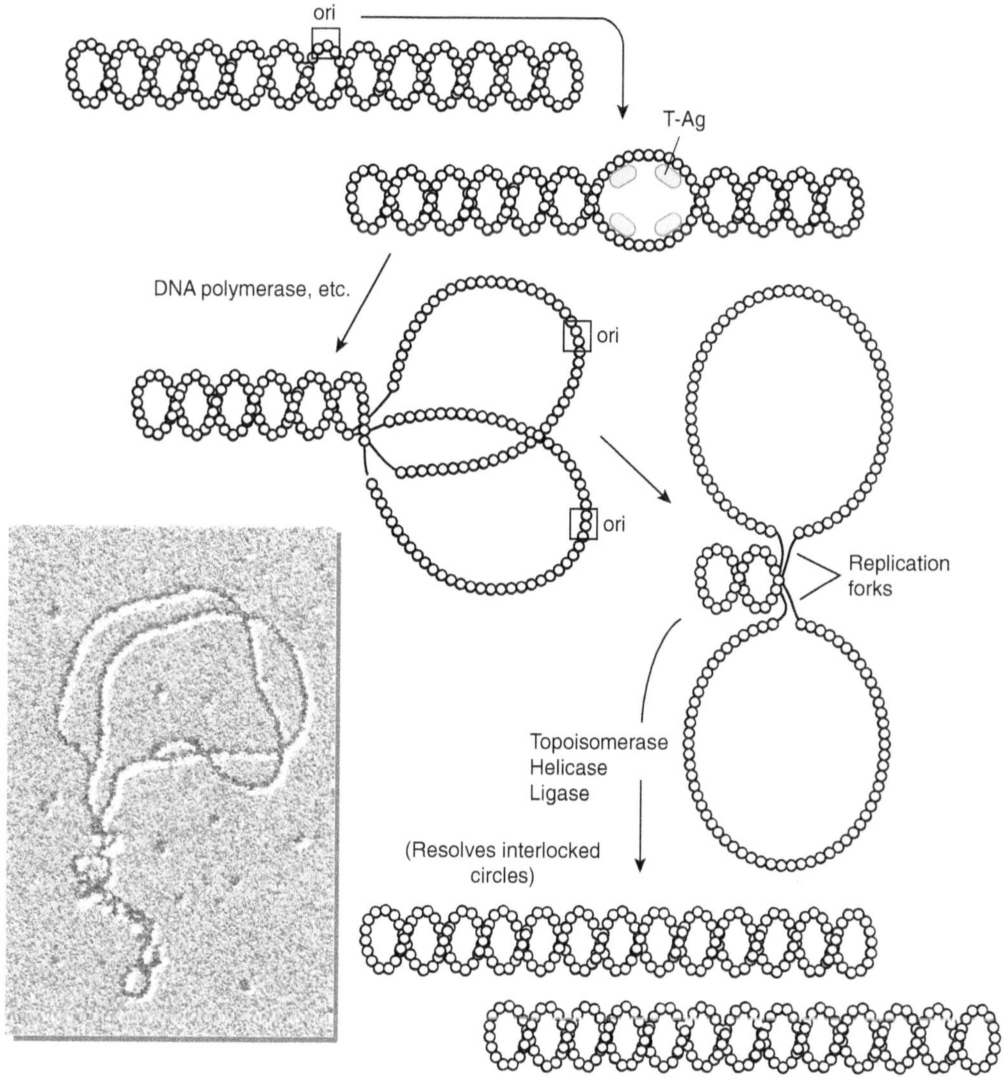

Figure 16.3 The replication of SV40 DNA. The closed circular DNA has no end problem, unlike the replication of linear DNAs. Structures of the replication fork and growing points are essentially identical to those in replicating cellular DNA, and use cellular DNA replication enzymes and accessory proteins. Replication results in the formation of two covalently closed and interlinked daughter genomes that are nicked and re-ligated into individual viral genomes by the action of cellular topoisomerase and other helix-modifying enzymes. T-Ag, Large T antigen; ori, origin of replication.

cell division, and study of this phenomenon provided early important models for the study of carcinogenesis. While such abortive infections may be purely a laboratory phenomenon, the information derived from them provided an important foundation for understanding the pathogenesis of papovaviruses in their natural hosts and viral oncogenesis.

In rodent (and some other nonprimate) cells, SV40 virus can infect and stimulate cellular RNA and DNA synthesis by expressing the large T antigen. As noted, this viral protein inactivates at least two cellular tumor suppressor or growth control genes (p53 and Rb). The role of such **oncogenes** in controlling cell growth is briefly touched on in Chapter 10, Part III, and is discussed in more detail in Chapter 19.

The two proteins in question (p53 and Rb) have two basic functions. First, they mediate an active repression of cell division by binding to and thus inactivating cellular proteins required to initiate such division. Second, levels of the free proteins above a critical level induce apoptosis (programmed cell death; see Chapter 10) in the cells that escape repression and begin to divide. As in the early phase of productive infection, in the first stages of infection of the nonpermissive cells, large T antigen displaces active replication–initiation proteins bound to p53 by binding this protein with higher avidity. The proteins thus liberated are free to initiate cellular DNA replication, but since there is no free p53, there is no induction of apoptosis.

These are the same steps that occur in the early stages of productive infection; however, viral DNA cannot be replicated in the nonpermissive cells. This failure is due to the inability of T antigen to interact effectively with one or more of its other cellular targets important in the early phases of infection. In this abortive infection, the cells in which T antigen is expressed do not die, but they replicate even while in contact with neighboring cells; this process is shown in Figure 16.4. The continued stimulation of cellular DNA replication by expression of viral T antigen can lead to continual cell replication (i.e., transformation). Stable transformation will require the viral genome to become stably associated with cellular DNA by *integration* of viral DNA into the cellular genome. Such viral DNA replicates every time the cell replicates, and thus keeps the cell transformed.

The integration of viral DNA into a host cell chromosome is not a function of T antigen or any other viral product. Indeed, most abortively infected cells will divide for a round or so until the viral DNA is lost, and then they will revert to their normal growth characteristics. This is sometimes termed **transitory (transient or abortive) transformation**. The integration of viral DNA into the host cell is the result of an entirely random recombination event and occurs at sites where a few bases of the circular viral DNA can anneal to a few bases of chromosomal DNA. This must be followed by breakage and re-ligation of the chromosome with the incorporated viral DNA. Obviously, this does not occur very frequently, but if a large number of cells are abortively infected with the polyomavirus in question and one or more integrate the viral chromosome and continue to express T antigen, those cells will form a focus of transformation. Such a focus is a clump of transformed cells growing on the surface of a culture dish of contact-inhibited cells. These foci can be counted and are subject to similar statistical analyses as are plaques formed by productive infection. Some typical foci of transformation are shown in Figure 10.5.

The replication of papillomaviruses

Cell transformation by SV40 appears to be a laboratory phenomenon, and many of the tumors caused by polyomaviruses can be thought of as dead-end artifacts of virus infection. In such infections, persistence appears to be due to the stability of histone-associated viral genomes in nonreplicating cells marked by occasional episodes of low-level viral replication as a result of immune crisis or other events that lead to changes in the transcriptional environment of the host cell.

In contrast, a related group of viruses, papillomaviruses, follow a natural replication scheme in their host that requires the formation of tumors, albeit usually benign ones, in their replication cycle. In this strategy of virus replication, persistence is a consequence of the continued replication of cells bearing viral genomes!

Papillomavirus replication combines some aspects of both the abortive and productive schemes just discussed. These viruses cause warts or papillomas, and there are many different types, with most showing no antigenic cross-reactivity with each other. Infections with most papillomavirus types are completely benign (although irritating or occasionally painful), but some can be spread by sexual intercourse, leading to persistent genital infections, especially in females. Statistical analyses comparing the incidence of cervical carcinoma and the patterns of persistent infection by some of these papillomaviruses (including human papillomavirus

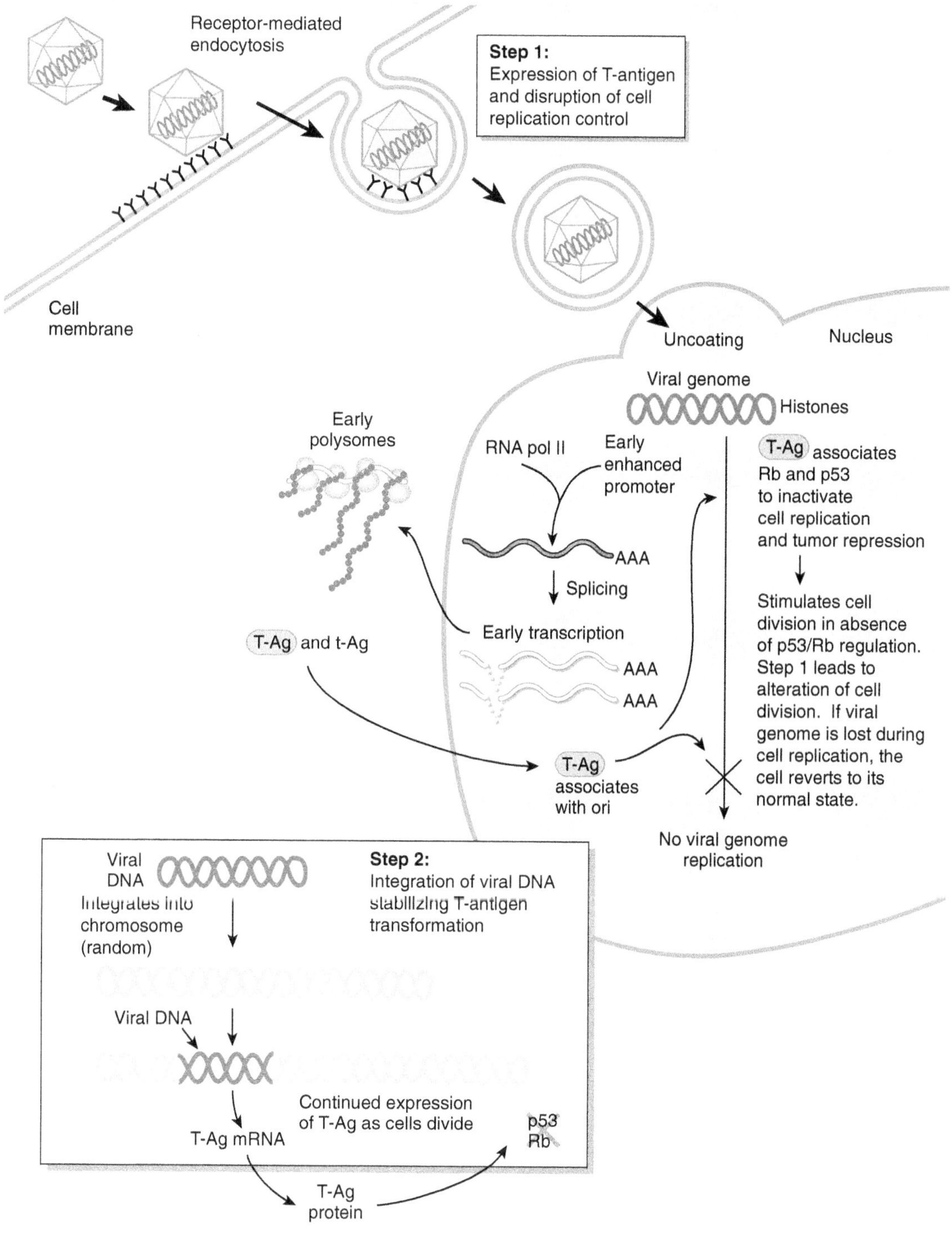

Figure 16.4 Representation of the two steps in transformation of a nonpermissive cell by SV40. The infection begins as described in Figure 16.2, and early mRNA is expressed into early proteins. The infection is abortive in that DNA replication and late gene expression cannot occur in the nonpermissive cell. Still, the large T antigen (T-Ag) is able to interfere with cellular growth control (tumor suppressor) proteins, leading to cell replication. Stable transformation requires a second step, the integration of the viral DNA. This is a random (stochastic) occurrence with SV40, and integration is random throughout the genome. A similar path is followed in the transformation of nonpermissive cells by other polyomaviruses. t-Ag, Small t antigen.

16 [HPV-16], HPV-18, HPV-31, and HPV-35]) demonstrate a highly significant correlation despite the fact that only a small number of infected individuals actually get the disease. In addition, HPV is also associated with a significant number of cancers of the vulva, vagina, penis, anus, mouth, and throat. Thus, these viruses are clearly human cancer viruses.

The HPV-16 genome

The circular genome of HPV-16 is shown in Figure 16.5. It is about 7900 base pairs long and is vaguely reminiscent of that of SV40, except there are many more early ORFs. Note that the region marked "LCR" corresponds to the promoter/origin region of SV40. Since the replication of papillomaviruses is difficult to study in cultured cells, a full characterization of the splicing patterns and transcripts expressed during infection has been and continues to be a very laborious effort. It requires analysis of DNA copies made of viral RNA using retrovirus reverse transcriptase, followed by cloning of the complementary DNA (cDNA) copies. Polymerase chain reaction (PCR) amplification of cDNA for direct sequence analysis also has been used. General methods for such analysis are covered in Chapters 11 and 12. Use of high-throughput sequencing (HTS) technologies as described in Chapter 11 is likely to greatly speed up these analyses in the near future.

Sequence analysis of the bovine papillomavirus genome and the transcripts expressed indicates that early and late transcripts are expressed from a single or limited number of early and late promoters as pre-mRNAs. While the extensive splicing of pre-mRNAs is reminiscent of infections with polyomaviruses, papillomaviruses differ in that early and late promoters are found in several regions within the genome.

Virus replication and cytopathology

Formation of a wart by infection with papillomavirus is outlined in Figure 16.6. It involves virus entering the basal cells of the epithelium (the skin in the case of warts). The virus expresses early genes that induce cells to replicate their DNA rather more frequently than would an uninfected

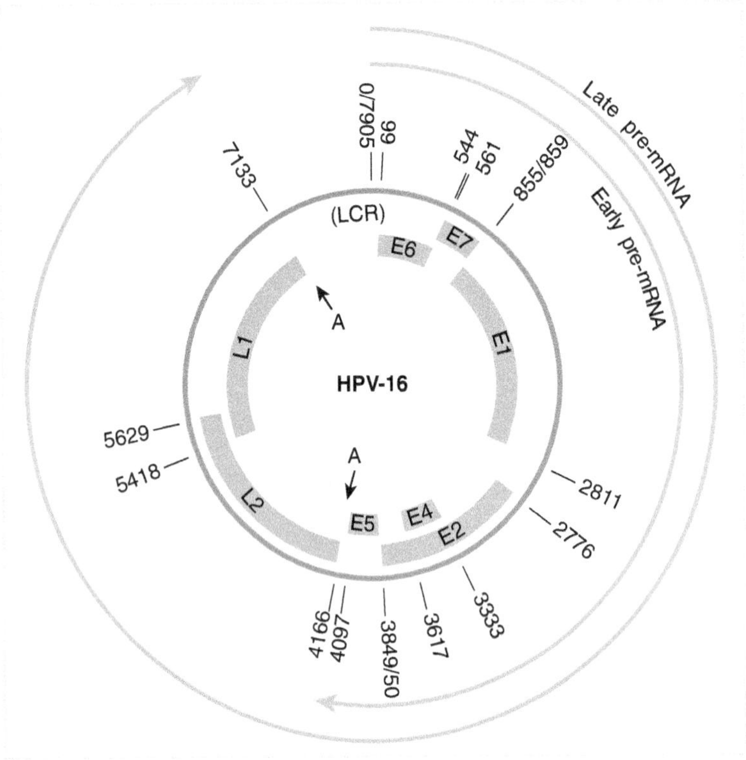

Figure 16.5 The human papillomavirus 16 (HPV-16) genome. The 7-kb circular genome contains a number of translational reading frames that are expressed from spliced mRNAs. Unlike the related polyomaviruses, papillomaviruses encode all proteins on the same DNA strand. The actual details of mRNA expression also appear to differ among different papillomaviruses. For example, HPV-16 has only one known promoter, which appears to control expression of both early and late transcripts. The locations of cleavage/polyadenylation signals for early and late transcripts are shown. All mRNAs appear to be derived by splicing of one or two pre-mRNAs. The characterization of transcripts has required heroic efforts of isolating small amounts of RNA from infected tissue, generating cDNA clones by use of reverse transcriptase and polymerase chain reaction, and then sequence analysis. This is necessary because many are present in very small amounts in tissue, and the virus does not replicate in cultured cells. The transcripts shown are three of nine that have been fully characterized, and it can be expected that others are also expressed. The region marked "LCR" encodes both the constitutive (plasmid) origin of replication and an enhancer. Location of the vegetative origin of replication is not known. Specific details of papillomavirus replication are described in the text.

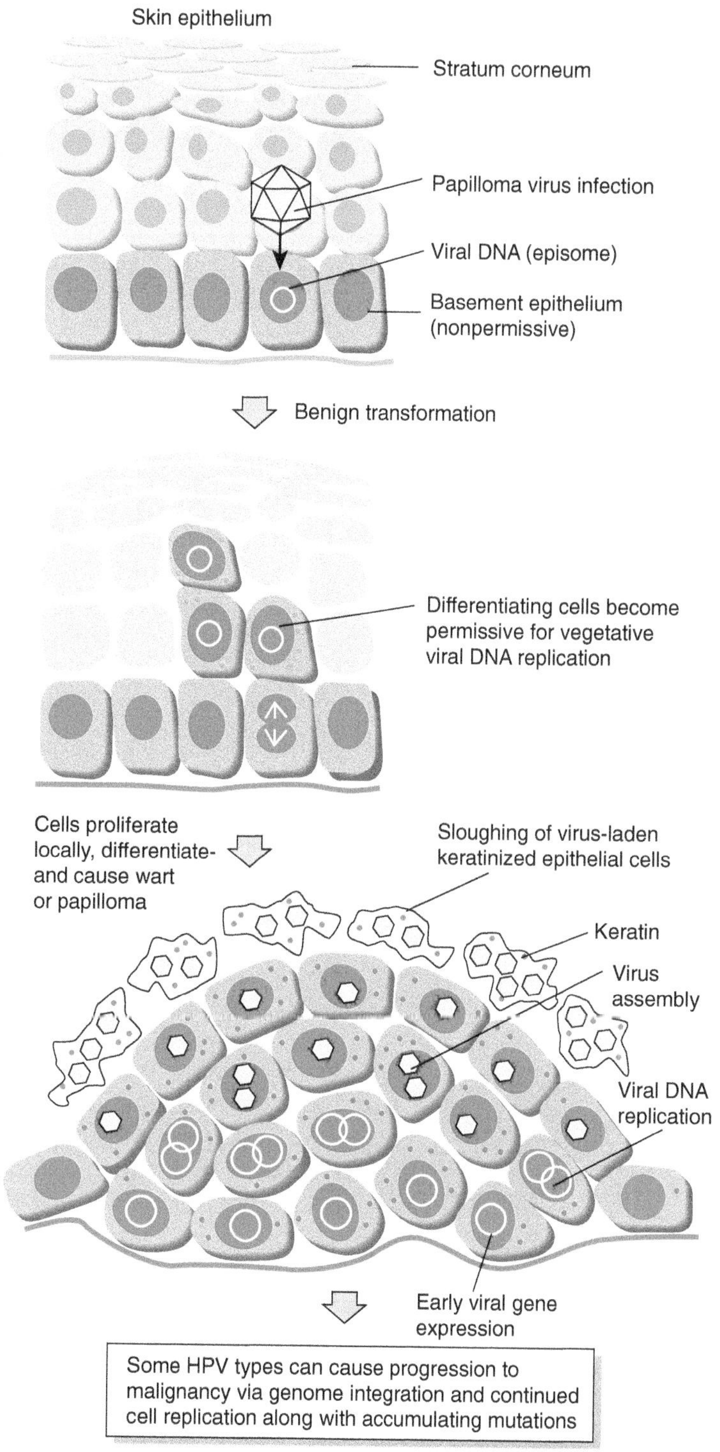

Figure 16.6 The formation of a wart by cell proliferation caused by infection of basement epithelial cells with human papillomavirus (HPV). Early gene expression leads to stimulation of cell division and terminal differentiation. This results in late gene expression and virus replication in a terminally differentiated, dying cell, which produces large quantities of keratin.

epithelial cell. Thus, one set of early functions is analogous to those of SV40 T antigen. But in marked contrast to SV40 replication in permissive cells where infection leads to vegetative viral genome replication and cell death, papillomavirus DNA remains in the infected cell nucleus as an **episome** or "mini"-chromosome where it can replicate when cell DNA replicates, but it does not replicate to the high numbers seen in viral DNA replication of a productive infection.

Such cell-linked replication is often termed **plasmid-like replication**. It involves the interaction of cellular DNA replication proteins with the viral origin of replication, which during persistent infection acts like an origin of cellular DNA replication and is subject to similar control. As the cells are stimulated to divide, they differentiate, and as they differentiate, they change their function and begin to produce proteins typical of terminal epithelial differentiation. For example, synthesis of K5 and K14 keratins characteristic of basal cells is terminated and keratin K1 and K10 characteristic of suprabasal skin cells are expressed. At some point in this terminal differentiation, some of these cells become fully permissive for high levels of viral DNA replication and late gene expression to generate capsid proteins. Such cells produce new virus while they die. Since this phenomenon is highly localized, and the virus infection normally just speeds up normal terminal differentiation of the epithelial cells, a benign wart is formed.

For HPV-16 and HPV-18 (and other "high-risk" HPV types that are associated with cancer), this growth enhancement is known to be a function of the actions of proteins encoded by the E5, E6, and E7 gene products that associate with and inactivate normal functions of the p53 and Rb proteins in a manner analogous to large-T-antigen activity in SV40. Presumably, chronic infection of cervical epithelium with either of these viruses can (rarely) generate a true cancer cell by further mutations of other control circuits in the cell. This oncogenic transformation is coincident with integration of papillomavirus DNA into cellular DNA, and it is speculated that oncogenesis involves a process similar to the transformation stabilization seen in abortive SV40 infection of the appropriate nonpermissive cell.

In such a transformed cell, no virus is produced, so formation of the cancer can be looked at as a dead-end accident induced by the continued stimulation of cell division caused by the virus's persistent infection. As these transformed cells continue to divide, they accumulate mutations that eventually allow them to spread to and invade other tissues, and form disseminated tumors (**metastasis**). In the case of benign warts in the skin and elsewhere, either inactivation of the p53 and Rb proteins is not so profound, or the stimulated cells are so close to death in their terminally differentiated state that they cannot become cancerous.

We have already mentioned that HPV-16 and HPV-18 can be sexually transmitted and have a high correlation with cervical cancer in women who are persistently infected with these strains. Other strains, notably HPV-6 and HPV-11, cause painful but generally non-cancerous genital warts. As a result, a quadrivalent vaccine was developed to protect against these four strains. The recombinant vaccine, trade name Gardasil, has been produced by Merck. The vaccine uses the major capsid protein L from each virus, produced by recombinant DNA techniques, to form self-assembled virus-like particles (VLPs). Gardasil was approved by the US Food and Drug Administration (FDA) for general use in June 2006, and a similar product, trade named Cervarix, was approved in 2009. The overwhelming success of these vaccines has led to a second generation (Gardasil 9) that protects against HPV types 6, 11, 16, 18, 31, 33, 45, 52, and 58, and has the potential to prevent over 90% of cervical, vulvar, vaginal, and anal cancers, as well as a significant number of oral and throat cancers.

The potential impact of these vaccines is quite large. If used prophylactically and administered to girls and boys before they are sexually active, the vaccine could reduce the worldwide incidence of HPV-associated cancers dramatically. This disease currently has a yearly prevalence of 16 per 1000 women, with an annual death rate of 9 per 1000, making it the third leading cause of death for women, behind breast and lung cancers. In the United States, the prevalence and mortality rates are lower, probably due to the widespread use of Pap smears for early detection. The use of this vaccine does not suggest that Pap smears can be abandoned, however, since at least 10% of cervical cancers are not linked to infection by these viruses.

THE REPLICATION OF ADENOVIRUSES

The adenoviruses comprise a large group of complex icosahedral, nonenveloped viruses of humans and other mammals. In humans, they generally are associated with cold or mild flu-like respiratory diseases, but some serotypes also are associated with gastrointestinal upsets. While adenoviruses are not at all closely related to the papovaviruses, they share with them a long replication cycle due to the need to stimulate and utilize many cellular functions to carry out virus replication. They also share the ability to transform cells in the laboratory via abortive infection. Also like the papovaviruses, adenovirus replication involves extensive splicing of a limited number of pre-mRNAs. The usage of alternative splicing sites leads to the expression of a nearly bewildering number of partially overlapping mRNAs encoding related proteins.

Despite these similarities to papovaviruses, there are striking differences in the details of replication and in the organization and replication of the viral genome. The relatively mild course of adenovirus infection, and some convenient properties in manipulation of the virus, make it an attractive candidate for use as a therapeutic agent (see Chapter 22).

Physical properties of adenovirus

Capsid structure

Adenoviruses have complex icosahedral capsids whose proteins are not present in equimolar amounts (see Figure 11.5), with projecting spikes or *fibers* at the 12 vertices (pentons). The viral genome is encapsidated with core protein that acts a bit like histone to provide a chromatin-like structure that is condensed in the interior of the nucleocapsid.

The adenovirus genome

The genome of adenoviruses is linear with specific viral protein (*terminal protein*) at the 5′ ends. The genome is about 30 000 base pairs, and the sequence at the genome's end (100–150 base pairs, depending on virus serotype) is inversely repeated at the other end. This is the ori for viral DNA.

The genome map with location of the many transcripts expressed during infection is shown in Figure 16.7. The genome is divided into 100 map units; therefore, each map unit is 300 base pairs. Transcript location is complicated by complex splicing patterns and the presence of a number of promoters. There are four early transcription "units" termed E1 through E4; each of these contains at least one promoter and polyadenylation signal. A single late promoter produces five "families" of late mRNAs, and there is also an unusual RNA called "VA" that is transcribed by the action of host cell RNA polymerase III (pol III).

The adenovirus replication cycle

Early events

Adenovirus enters the cell via receptor-mediated endocytosis in a manner analogous to that of papillomaviruses. Cellular receptors interact with the virion fiber proteins to initiate infection. Adenovirus DNA with a specific terminal protein bound to each 5′ end is released into the nucleus where it associates with cellular histones. In order to initiate gene expression, adenoviruses must stimulate the infected cell to transcribe and replicate its genes. This is accomplished by expression of the spliced mRNAs encoding the immediate-early (or "pre-early") gene E1A and E1B protein "families." The promoters for these are enhanced and can act in the cell in the

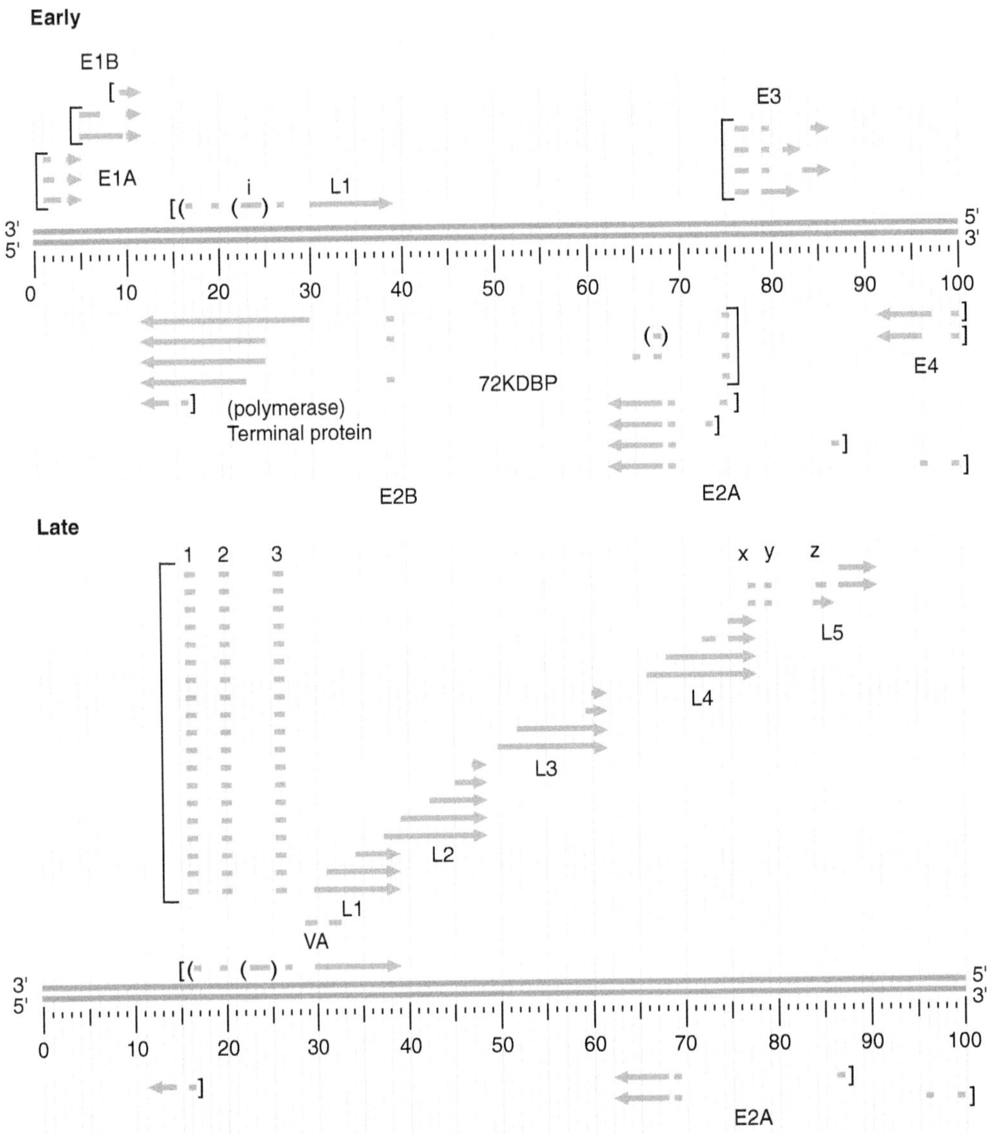

Figure 16.7 The genetic and transcription map of the 30-kb adenovirus genome. There are three kinetic classes of transcripts. The E1 transcripts are controlled by enhanced promoters and require no modification of the host cell because some functions of their expression are similar to those of T antigen in SV40 virus replication. These functions include stimulating cellular transcriptional activity and cell replication. Early in infection, only early transcripts are expressed. These include mRNAs encoding viral DNA polymerase and terminal proteins. There are a number of early promoters and transcription units. The E2 transcription unit also has a 72-kd DNA-binding protein (72KDBP) that shuts off early transcription. Two primary transcripts, E2A and E2B, are expressed from the same E2 promoter. The mRNA for the DBP continues to be expressed late because there is a second promoter upstream of the E2 promoter that is not shut off by the DBP. The major late promoter at map position 15 is always "on," but polyadenylation and splicing patterns change markedly as infection proceeds. Late in infection, the late transcription unit extends to one of five polyadenylation signals, and differential splicing results in generation of a myriad of late mRNAs encoding structural proteins as well as proteins involved in host cell modification and virus maturation.

absence of any viral modification (like the SV40 early promoter). The E1A gene products block the ability of the p53 and Rb growth suppressor genes to suppress cell division, while one or several E1B proteins inhibit apoptosis in the stimulated cell. Thus, these two proteins work in concert in a manner similar to that of polyomavirus large T antigen.

Stimulation of the infected cell's transcriptional machinery leads to expression of the four early pre-mRNAs that are spliced in various ways to produce early proteins, including a DNA polymerase protein (140-kd pol), a terminal protein, and a 72-kd DNA-binding protein (DBP). The latter shuts off most early promoters, but the E2 region is not shut off because a second promoter becomes active at times when 72-kd DBP is at high levels. Interestingly, the major late promoter is "on" early in infection, but only the L1 region is expressed as mRNA because all transcripts are terminated at the polyadenylation signal at 40 map units. This termination is due to the inhibition of splicing downstream of the L1 region through binding of cellular splice factors. Further, late transcripts downstream of L1 are not transported from the nucleus.

Adenovirus DNA replication

Adenovirus genome replication takes place via an unusual mechanism that involves formation of ssDNA as intermediates; the process is shown in Figure 16.8. Adenovirus DNA replication begins at either or both ends of the DNA and uses as a primer an 80 000-Da precursor of the 50 000-Da viral genome-bound terminal protein. The large priming terminal protein is proteolytically cleaved to the smaller terminal protein found in capsid-associated genomes during packaging. This is the only known instance where DNA replication initiates without a short RNA primer. However, the terminal protein does contain a covalently bound cytosine residue from which DNA replication proceeds. Note that replication utilizes the adenovirus-encoded DNA polymerase and is continuous – there are no short Okazaki fragments seen. The process can liberate the other strand as ssDNA, which can become circular by association of inverted repeat sequences at the end, and replication proceeds. Thus, adenovirus DNA replication can proceed via two routes shown in Figure 16.8. If DNA synthesis initiates at both ends of the genome at about the same time, type I replication occurs. If only one end of the genome is used to initiate a round of DNA synthesis, then type II replication occurs.

Late gene expression

With the increase in levels of early 72-kd DBP, much early gene expression shuts off. At the same time, E4 protein interferes with the inhibition of splicing downstream of the L1 region; effectively, this results in *polyadenylation site usage* changes so that transcription from the major late promoter generates transcripts covering as much as 24 000 bases. Differential polyadenylation and splicing generate the five families of late mRNAs that are translated into the structural proteins that will make up the capsids. Other late proteins alter aspects of cellular structure and metabolism to ensure efficient virus assembly and release. In addition to altering splicing patterns, some species of E4 protein actively mediate the transport of late mRNA from the nucleus to the cytoplasm.

VA transcription and cytopathology

The complex interaction between human adenovirus infection and the host cell requires that the cell remain functional for a long period following infection. This precludes extensive virus-induced shutoff of host cell function; hence, virus-induced cytopathology is slow, and cell death takes a long time. During this period, the cell can mount defenses against viral gene expression such as the induction of interferons, cellular gene products that can render neighboring cells resistant to virus infection (see Chapter 7, Part II). The human virus gets around this problem by synthesis of **VA RNA** (viral-associated RNA), which is a short, highly structured RNA molecule that interferes with the cell's ability to produce interferon and, most likely, other defense mechanisms. Indeed, this RNA molecule has many features of small interfering RNAs (siRNAs), which were discussed briefly in Chapter 8. VA RNA is expressed via cellular RNA pol III, which is the same polymerase used to transcribe cellular amino acid transfer RNAs (tRNAs).

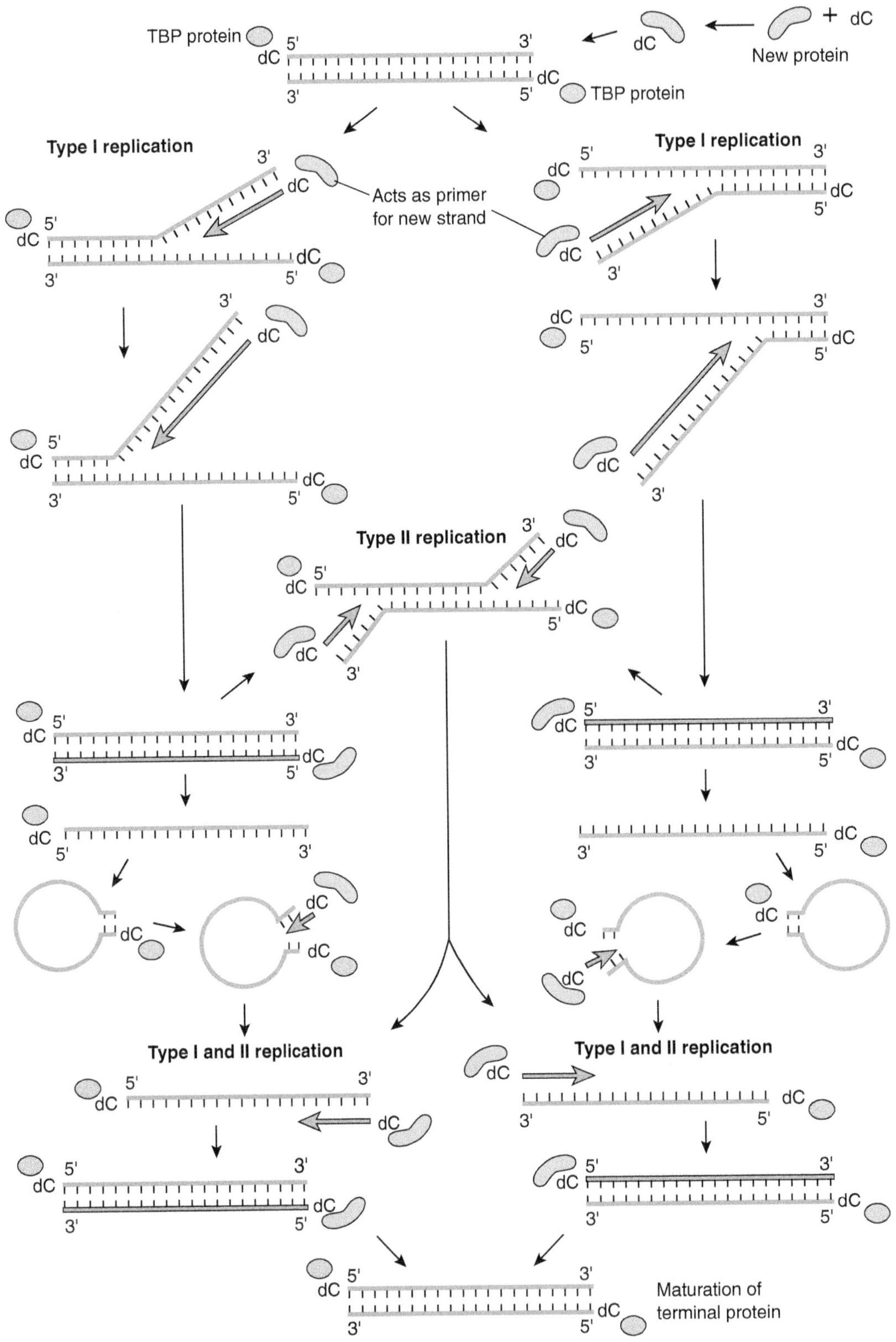

Figure 16.8 Adenovirus DNA replication. The 5′ ends of the viral genome have 50 000-Da terminal proteins bound to them. Adenovirus does not have discontinuous strand synthesis, and it exhibits other features that are at variance with the general scheme for viral DNA replication outlined in Chapter 14. Of major interest is the fact that there is no discontinuous strand synthesis. The process is marked by the accumulation of a large amount of single-stranded DNA (unusual in eukaryotic DNA replication). Further, the initial priming event requires the first nucleotide of the new DNA strand to be covalently bound to the 80 000-Da precursor of the 50 000-Da terminal protein. Following complete second strand synthesis, the precursor end proteins are proteolytically cleaved to form the mature terminal proteins. TBP, Precursor to terminal binding protein.

Interestingly, while the human Epstein–Barr herpesvirus expresses an analogous transcript (see Chapter 17), suggesting that this is an important feature in virus-mediated immune evasion, a number of adenoviruses of domesticated animals do not express a homologue to VA RNA.

A second aspect of the interaction between adenovirus and the host is reminiscent of papillomavirus replication. Adenovirus remains associated with the host for long periods of time as a persistent infection, especially in the epithelium of the adenoidal tissue and the lungs. The virus infects basement cells, but initiates DNA replication and viral assembly only in terminally differentiated cells. The virus actually induces an acceleration of apoptosis of these differentiated cells. One apparent advantage of eliminating dying infected cells is that more room is made available for the differentiation and growth of basement cells. This provides a ready and continuing source of cells in which the virus can initiate new rounds of replication. This stimulation of apoptosis presumably occurs because the relative levels of E1A and E1B are different in critical cells as compared to cells in which apoptosis is blocked by the latter viral protein.

Transformation of nonpermissive cells by adenovirus

As with SV40, infection of nonpermissive cells by at least some adenovirus types can lead to cell transformation and tumor formation. While there is currently no evidence for any involvement of adenovirus infection in human cancers, transformation seems to be accomplished by mechanisms very similar to those outlined for papovaviruses. Indeed, under some conditions, adenovirus gene products can substitute for early papovavirus gene products in mixed infections.

REPLICATION OF SOME SINGLE-STRANDED DNA VIRUSES

With many plant viruses, and some animal and bacterial ones, a relatively small capsid size provides some advantages. With plant viruses, this advantage is tied to the limitations of virus capsid size that can "fit" in pores of the plant's cell wall. The advantages for animal and bacterial viruses are less clear, but must exist.

Replication of parvoviruses

The parvoviruses are very small, nonenveloped, icosahedral viruses. Two of the three known groups infect warm-blooded animals, while the third group has members that infect insects. The parvovirus capsid diameter is 26–30 nm, significantly smaller than the polyomaviruses even though the viral genome is approximately 5 kb long. The virus is able to package the genome into such a small virion because the virus encodes only a single DNA strand. Interestingly, many parvoviruses can package the DNA strand of sense either opposite to mRNA or equivalent to mRNA in equal or nearly equal numbers. This means that the packaging signals utilized by the virus to encapsidate the genome must occur on both strands – this is probably through the interaction of the unique end structures of both strands with capsid proteins.

The genome of adeno-associated virus (AAV), a typical parvovirus, is shown in Figure 16.9. It encodes two protein translational reading frames that are expressed by a variety of transcripts. The first reading frame encodes nonstructural protein involved in replication, and the second encodes the capsid protein. The genome ends contain 120–300 bases of inverse repeated sequences so that they can form hairpin loops in solution and in the infected cell's nucleus. These terminal hairpins serve as primers for initiation of DNA replication, and since they are repeated at the ends of both positive (+) and negative (–) sense DNA strands, both can serve as templates for DNA replication.

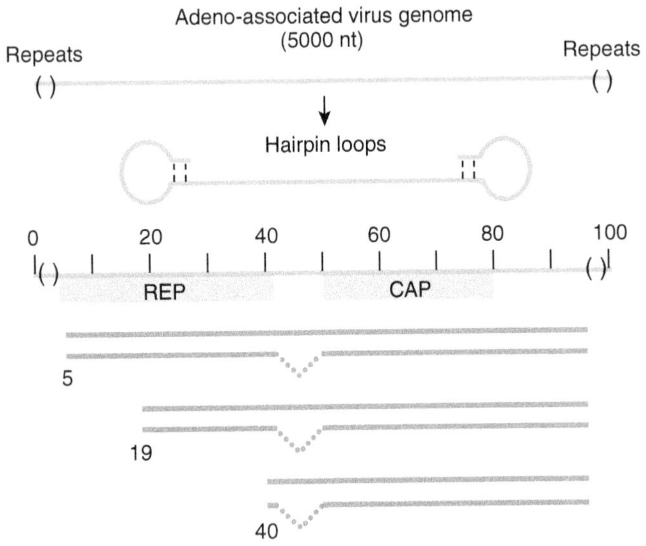

Figure 16.9 The 5000-nucleotide (nt) linear genome of adeno-associated virus (AAV). This ssDNA has repeated sequences on both ends that allow it to form a "hairpin" structure. This serves as the template for conversion into dsDNA by cellular enzymes. Cellular enzymes also mediate replication of the viral genome. Three families of coterminal mRNAs are expressed from the three AAV promoters; the genome encodes replication proteins and a capsid protein but depends on cell replication for its ability to replicate its genome. This cellular replication is induced by a helper virus such as adenovirus in the animal, but the virus can replicate in cultures of some actively replicating cells. Other groups of parvoviruses, such as minute virus of mice (MVM), are able to replicate in some actively replicating cells of their natural host.

Parvovirus replication is absolutely dependent on the host cell undergoing DNA replication. Thus, the virus can only replicate in actively replicating cells. Despite this, and unlike papovaviruses and adenoviruses, parvovirus has no ability to stimulate cell division via the action of a viral-encoded protein. This inability results in a very tight restriction of virus replication in the host's dividing cells, especially cells of the immune system. This can be devastating to young animals, and parvovirus infection of dogs is a major problem in kennels. Parvovirus infection can also be very destructive to actively growing cells in adult animals. For example, **feline panleukopenia**, a parvovirus disease characterized by destruction of the immune system, is a significant pathogen of domestic cats.

Upon infection, the ssDNA is converted into full dsDNA by cellular DNA repair enzymes following its entry into the nucleus. The double-stranded viral DNA template is transcribed into a number of 3′-coterminal transcripts from one of three viral promoters just 5′ of the transcript starts. Some of these transcripts are spliced, so each translational reading frame is translated into several proteins of related sequence. As noted, viral genome replication can only take place in cells in which there is active cellular DNA replication (i.e., in the S phase of cell division). The viral replication enzyme is involved in cleavage of the covalently closed replicating viral DNA into single-stranded genomic DNA and has no polymerase activity.

Dependovirus DNA integrates in a specific site in the host cell genome

AAV is representative of one major group of parvoviruses, the **dependoviruses**. It is usually found associated with active infections of adenoviruses, and occasionally with herpesviruses. The human parvovirus, AAV, is a well-characterized example. While the dependoviruses can be grown in culture in fetal cells or following proper chemical stimulation of some adult host cells, they depend on the adenovirus or herpesvirus helpers to stimulate the cell in such a way that they can divide. Thus, like viroids, these viruses are parasitic on other viruses.

The dependence on a helper virus might be expected to be a great impediment to virus replication for AAV, but this is overcome in part by its ability to integrate into chromosome 19 of the host when it infects a cell in the absence of the helper. The integrated viral DNA allows AAV to remain latent in host tissue for long periods of time, but to "reactivate" if and when that cell is infected with a virus that can act as a helper.

Integration takes place at short stretches of homologous sequences within a region of several hundred bases in the host chromosome. While it allows the viral genome to remain associated with the host for long periods, integrated viral DNA serves as a biological "time bomb" – ready to replicate and kill the cell when it is infected with the appropriate helper. Since the replication of AAV interferes with the efficiency of replication of the helper virus, it may be that this process has the ultimate effect of limiting infection of the helper, thus providing some benefit to the host!

Parvoviruses have potentially exploitable therapeutic applications

The strict requirement for actively replicating cells, and the competition between AAV and adenovirus and herpesvirus infections, suggest that such viruses might be exploitable as antiviral or anticancer agents. Laboratory studies showed this to be feasible. For example, breeds of laboratory mice have high occurrences of certain tumors. Infection of young mice with minute virus of mice (MVM), a murine parvovirus, results in a significant increase in the animal's lifespan and fewer occurrences of tumors at young ages! It should be clear, however, that an effective application of such a result to human cancers is not a straightforward undertaking.

Another potential use for AAV stems from the discovery that if the two genes that AAV encodes (rep and cap) are removed, then the virus loses its ability to integrate into chromosomes, and instead allows the virus to be maintained in the cell as circular episomes or plasmids for long periods of time. This property has prompted the use of AAV as a safe and efficient gene therapy vector for delivering genes as a therapy for disease. One of the first viral-vector-based gene therapy treatments to be licensed for use in humans is alipogene tiparvovec (trade name Glybera), which is being used to treat a lipoprotein lipase deficiency.

DNA viruses infecting vascular plants

While DNA viruses infecting vascular (i.e., "higher") plants might be expected to display genetic variability equivalent to that seen within animal and bacterial viruses, they do not appear to. The reason for this is that plant viruses must traverse a relatively thick and dense cell wall to approach and breach the plant cell's plasma membrane. Although at least one algal virus can insert its genome like bacterial viruses inject genomes, apparently the dimensions of the vascular plant's cell wall preclude this accommodation. This results in the viruses of higher plants having a strict limitation on the size of their genomes, and although such viruses are not fully characterized, they may require a significant number of cellular functions for replication.

Geminiviruses

One group of viruses that infects plants have single-stranded, covalently closed circular DNA genomes and are packaged into unusual twinned capsids. These "twin" capsid structures give the group its genus name, *Geminivirus* (from the Latin word *geminae*, for "twins"). The number of genes encoded and their arrangement on the genome distinguish the three major groups of these viruses. Two of the groups encapsidate the same genome in both of the twinned capsids; thus, they have a monopartite genome. In contrast, the third group contains a bipartite genome, and the two different genomic segments are packaged separately in each of the capsid halves. Rather astonishingly, one geminivirus isolated from bananas contains capsids bearing eight distinct genomic segments. How the virus accomplishes the rather remarkable feat of packaging different genomic segments into different subcapsids is an open question.

Representatives of geminiviruses include maize streak virus (a monopartite genome) and tomato golden mosaic virus (a bipartite genome). The genome (2.7–3.0 kb) organization of the geminiviruses has ORFs oriented in both directions around the circle, much like the papovaviruses. Since geminiviruses are single stranded, the input genome strand must be converted into dsDNA following infection, in order to obtain the appropriate template for transcription of mRNA whose translational reading frames are antisense to the virion DNA.

The geminiviruses are transmitted from plant to plant by leafhoppers or white flies. The virus can remain in the insect for long periods, but unlike the RNA-containing arboviruses, geminiviruses do not replicate in their insect vectors. Replication and transcription of these viruses take place in the nuclei of infected plants, using a rolling circle scheme. The exact function of the gene products predicted from sequence analysis has not been determined. Therefore, it is not yet possible to say which of the viral proteins might be specifically involved in this DNA replication.

The single-stranded DNA bacteriophage ΦX174 packages its genes very compactly

The gene packaging of bacteriophage ΦX174 suggests that genomic size compression also offers distinct advantages in the prokaryotic world. This icosahedral virus has a structure very similar to that of adenovirus, but with shorter fibers. It contains a circular ssDNA genome approximately 3.4 kb long. Upon infection of a bacterial cell, the ssDNA genome is converted into dsDNA. This has been termed the *replicative intermediate* or *replicative form* (*RF*), but is quite unlike the complex ribonucleoprotein complex with this name seen in the replication of ssRNA viruses.

Viral-encoded mRNA expression, protein synthesis, and genome replication occur following patterns that are generally simple examples of the more complex replication programs of DNA-containing bacteriophages described in Chapter 18. A striking demonstration of the extent this virus has gone to compress its genome comes from examination of its genetic map, shown in Figure 16.10. The virus encodes nine distinct genes, but where one might expect about 200–300 bases of the DNA sequence to contain nonprotein information, only 36 bases (<1%) of the genome are free of translational reading frames. This arrangement means that all transcriptional control sequences are contained within translational reading frames.

The start and stop signals for translation of individual neighboring ORFs often overlap. Further, two genes are *completely* contained within the translational reading frames of other, larger ones. This configuration is accomplished by having the translational frames in different phases (outlined in Chapter 22, Part V). While such overlapping genes are found in many viruses, including even the largest ones, such as herpesviruses and poxviruses, ΦX174 has taken this tendency to an extreme.

Such compactness provides some useful advantage to this bacteriophage, but as with all dynamic systems, there is a price. In a viral genome with such overlaps, one base change in a region of overlapping genes can affect *two* rather than one gene function. For this reason, more mutations would be expected to be lethal than is generally seen in viral genomes. This is indeed the case with ΦX174, whose sequence is more strongly conserved during replication than is the case with other DNA viruses, and generation of mutations in this virus for genetic analysis is a laborious task.

Overlapping genes probably result in the virus being less adaptable to host and other changes in its natural environment. This conservatism could have a negative survival value in the prokaryotic world, but the survival of the virus is clear evidence that deleterious effects are compensated by the efficiency of gene packaging.

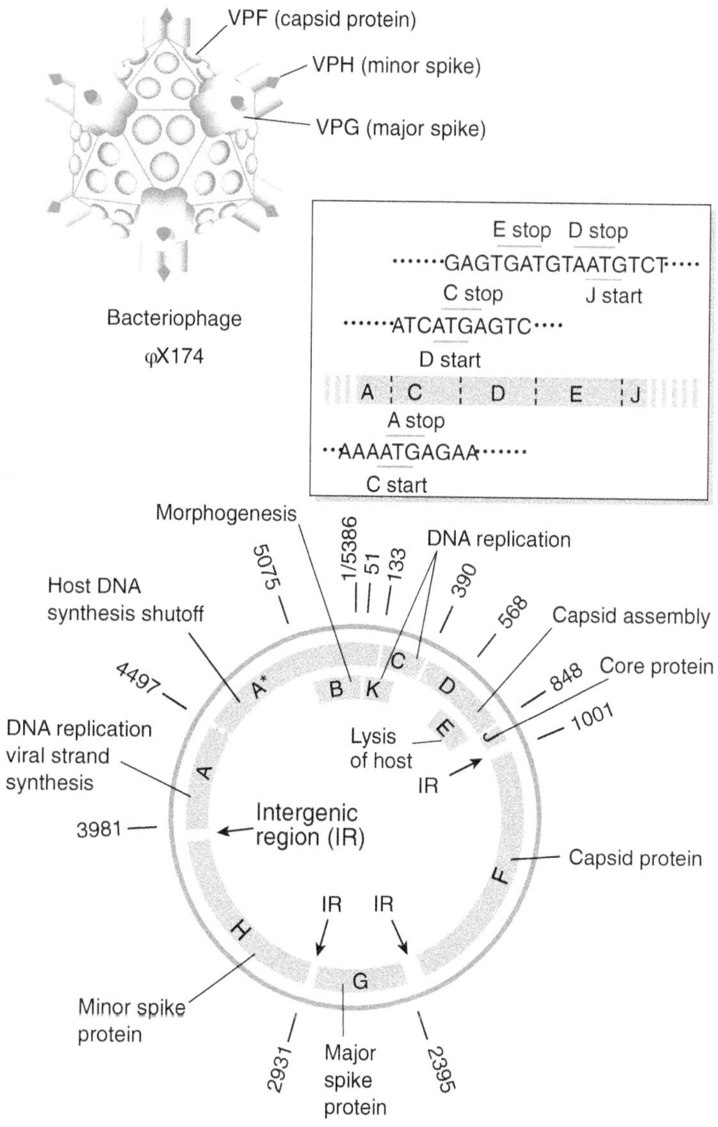

Figure 16.10 The capsid structure and compressed genome of bacteriophage ΦX174. The capsid is made up of three proteins: major capsid, major spike, and minor spike. In all, a total of 10 genes are compressed into 3.4 kb of ssDNA! This is accomplished by very short intergenic regions, and two completely overlapping genes. The functions of the proteins encoded by these genes are listed.

Case 4: JC virus (chapter 16)

Clinical presentation/case history: A 42-year-old female with multiple sclerosis (MS) is taken to the emergency room by her sister due to concerns over dramatic personality and mood changes. Upon admission, she was incoherent and confused. A history provided by her sister indicated that her mental status changes started about one week prior. She had complained that her face hurt, and she demonstrated right arm and leg weakness and slurring of her speech. The patient had no headache, neck stiffness, or fever. The attending physician ordered blood work, as well as a computed tomography (CT) scan (x-ray).

Diagnosis: The attending physician was concerned about the evidence of central nervous system (CNS) deficits and suspected an encephalopathy of the brain. The CT scan of the head

Continued

had been ordered to look for evidence of lesions in the brain. The CT showed a left temporoparietal cerebral edema (area of inflammation). This indicated a zone of pathology. Serology did not provide useful information for diagnosis. Because a number of viruses (and fungi) can cause lesions in the brains, a spinal tap was ordered, and polymerase chain reaction (PCR) analysis of the cerebrospinal fluid identified JC virus DNA. This suggested the likely etiology was progressive multifocal leukoencephalopathy (PML), which is caused by a reactivation of latent JC virus that is present in some B lymphocytes that can cross the blood–brain barrier. There they can form a focus of active infection, and result in sometimes severe cognitive deficits.

Treatment: There is no specific treatment for JC virus infection or PML; however, the underlying cause of this infection was determined to be due to the drug natalizumab that the patient was taking to control her MS. Natalizumab is a monoclonal antibody that targets the cell adhesion molecule alpha-integrin. The use of this medication reduces a specific aspect of the immune response that decreases the inflammation that exacerbates MS (and Crohn's disease). Unfortunately, taking this drug causes an increased risk of the "reactivation" of JC virus and the relatively rare brain lesions of PML in some patients. This complication of treatment with natalizumab has shed some light on the specificity and intricacies of the immune response that are critical for controlling different viruses!

Disease notes: JC virus is a polyomavirus that is usually acquired as a subclinical infection in childhood. The virus initially infects the tonsils and then spreads through the blood, where it infects the kidneys and becomes latent in the renal epithelium. JC virus has also been shown to be latent in the CNS, bone marrow, and peripheral blood cells. Greater than 80% of adults have antibodies to JC virus, and this infection normally causes no clinical problems except in the immunocompromised (and especially patients with AIDS) and people taking some monoclonal antibody immunosuppressants as described here.

QUESTIONS FOR CHAPTER 16

1

(a) The drawings in the following table represent possible structures for replicating DNA molecules. Indicate which ones might be found if you examined replicating adenovirus DNA isolated from an infected host cell.

(b) Adenovirus DNA replication proceeds in two stages. Suppose that you have an *in vitro* system that allows you to examine features of this synthesis. The reaction mixture has all the required viral and host proteins. Predict the effect of the following modifications on the process of the two stages. Use a "+" sign if the stage will occur normally, and a "−" sign if the stage will be blocked by the treatment.

Structure	Possible for Adenovirus?
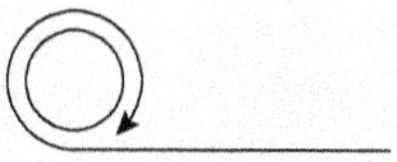	
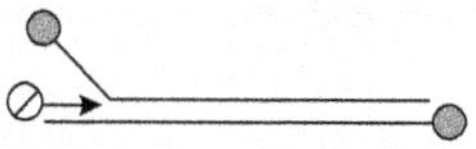	
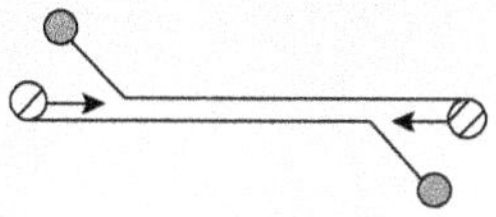	
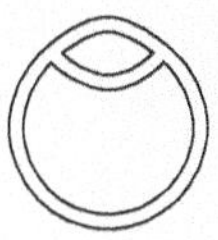	
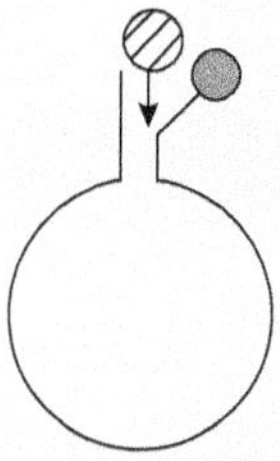	

Modification	First Stage	Second Stage
Control (no treatment)	+	+
Removal of the terminal protein from both 5′ ends of DNA genome		
Removal of the terminal complementary sequences from one end of the DNA genome		
Prevention of maturation of terminal protein from 80-kd to 55-kd form		

2 Cells that have been infected with adenovirus 2 (Ad2) are treated with the chemicals shown in the accompanying table. In each case, treatment inhibits the production of progeny Ad2 virus in the cell. Briefly give a reason why the Ad2 life cycle is blocked in each case.

Chemical	Effect on Cell	Reason for Ad2 Inhibition
NH_4Cl	Blocks acidification of secondary lysosomes and endosomes	
Vinblastine	Disrupts the microtubular cell cytoskeleton	
Emetine	Inhibits protein synthesis	

3 A papilloma (wart) virus enters a cell and does not produce progeny virus; however, episomal DNA is maintained within the cell, and some gene expression occurs. Of which kind of infection is this an example?

4 What are the functions of T antigen during the SV40 infectious cycle?

5 Which of the following about the life cycle of SV40 is false?

(a) It expresses three transcripts encoding three capsid proteins late.
(b) The genome contains a specific sequence of nucleotides that acts as a polyadenylation signal for transcripts using either strand of DNA as templates.
(c) It has specific promoters controlling expression of early and late transcripts.
(d) It replicates in the nucleus.
(e) It replicates using mostly cellular enzymes.

Replication of Some Nuclear-replicating Eukaryotic DNA Viruses with Large Genomes

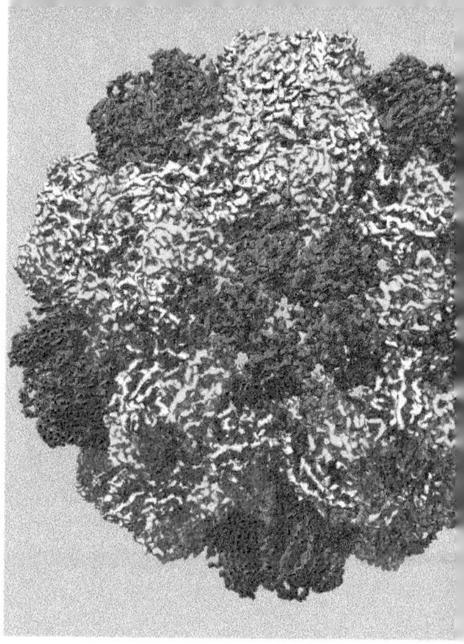

CHAPTER 17

- HERPESVIRUS REPLICATION AND LATENCY
- The herpesviruses as a group
  - Genetic complexity of herpesviruses
  - Common features of herpesvirus replication in the host
- The replication of the prototypical alphaherpesvirus – HSV
  - The HSV virion
  - The viral genome
  - HSV productive infection
- HSV latency and the LAT
  - HSV transcription during latency and reactivation
- How do the LAT and other specific HSV genes function to accommodate reactivation?
- EBV latent infection of lymphocytes: a different set of problems and answers
- Pathology of herpesvirus infections
  - Herpesviruses as infectious co-carcinogens
- BACULOVIRUS: AN INSECT VIRUS WITH IMPORTANT PRACTICAL USES IN MOLECULAR BIOLOGY
- Virion structure
- Viral gene expression and genome replication
  - Pathogenesis
- Importance of baculoviruses in biotechnology
- QUESTIONS FOR CHAPTER 17

Basic Virology, Fourth Edition. Martinez "Marty" Hewlett, David Camerini, and David C. Bloom.

The term *large*, when applied to DNA virus genomes, must be relative. The genomes of large DNA viruses encode anywhere from 50 to more than 1000 distinct genes, and on the upper end of size, the viral genomes can contain more genes than the simplest "free-living" organisms: the mycoplasmas.

Much of the genetic complexity of large, nuclear-replicating DNA viruses is due to viral genes devoted to providing the virus with the ability to replicate and to mature in differentiated cells, as well as viral defenses against or accommodations to host defense mechanisms. These genes are often not required for virus replication in one or another type of cultured cells, at least under certain conditions, and can be termed "dispensable for virus replication." While this designation is in relatively common use, it is misleading, because no virus gene maintained in a wild strain that replicates efficiently in the population at large is dispensable.

Stripped of "dispensable" genetic functions, a large-genome DNA virus must contain the same essential components as one with a small genome: genes devoted to subverting the cell into a virus-specific transcription factory, enzymes for viral genome replication, and the proteins and enzymes required to form the capsid and to assemble and release new infectious virions. Given these requirements, it is not too surprising that the replication basics of these large-genome, nuclear-replicating DNA viruses follow the same basic strategies as seen with smaller DNA viruses.

It is important to keep in mind, however, that there are many different ways a virus can modify a cell to result in a site favorable for its replication – "the devil is in the details"!

HERPESVIRUS REPLICATION AND LATENCY

The herpesviruses as a group

The herpesviruses are extremely successful enveloped DNA viruses. They have been identified in all vertebrate species studied, and extend into other classes of the animal kingdom (oysters, for example). Their replication strategy involves a close adaptation to the immune defenses of the host, and it is possible that their evolutionary origins as herpesviruses lie in the origins of immune memory. Eight discrete human herpesviruses are known at this time; each causes a characteristic disease.

Many herpesviruses are neurotropic (i.e., they actively infect nervous tissue); all such viruses are collectively termed *alphaherpesviruses*. Three human herpesviruses belong to this group: the closely related herpes simplex viruses 1 and 2 (HSV-1 and HSV-2), which are the primary agents of recurrent facial and genital herpetic lesions, respectively; and varicella zoster virus (VZV), which is the causative agent of chickenpox and shingles. VZV is more distantly related to HSV. Pseudorabies virus (PrV), an important animal pathogen that has many similarities with HSV, is also an alphaherpesvirus.

Five human herpesviruses are lymphotropic, meaning that they replicate and establish latency in tissues associated with the lymphatic system. These herpesviruses have been subdivided into beta- and gammaherpesvirus groups based on the specifics of their genome structure and replication. Viruses in these two groups share features that suggest they are more closely related to each other than they are to the three neurotropic herpesviruses.

Infections with human cytomegalovirus (HCMV) (the prototype of betaherpesviruses) are linked both to a form of infectious mononucleosis and to congenital infections of the nervous system. This virus can be devastating in individuals with impaired immune function, such as those suffering from HIV/AIDS or being clinically immune suppressed for organ transplantation or by cancer chemotherapy. The two other lymphotropic herpesviruses – the closely related human herpesviruses 6 and 7 (HHV-6 and HHV-7) – cause roseola, a ubiquitous and generally mild early-childhood rash.

Infections with human gammaherpesviruses, Epstein–Barr virus (EBV) and Kaposi's sarcoma herpesvirus or human herpesvirus 8 (KSHV or HHV-8), are convincingly linked to human cancers. Despite the high frequency of EBV infection in the general population, carcinogenesis is linked to additional environmental and possibly genetic factors, and the infection in most humans either is asymptomatic or results in a form of mononucleosis that is very similar in course to that caused by HCMV.

Genetic complexity of herpesviruses

Typically, a herpesvirus genome contains between 60 and 200 genes. Unlike adenoviruses, all of which share a basic genomic structure as well as general architecture, a comparative survey of the various herpesviruses' genomic structures displays a staggering array of individual variations on a general theme. Still, within this variation, gene order is generally maintained within large blocks of the genome and varying degrees of genetic homology are clearly evident. The most striking areas of homology are seen among those genes that provide basic replication functions.

One general feature of the complex herpesvirus genome arrangement is that herpes genomes contain significant regions of inverted repeat sequences. The size of herpesvirus genomes varies from 80 to 240 kb. Given that all the viruses share basic features of productive infection, this range in size means that different herpesviruses differ greatly in the number of "dispensable" genes they encode that are devoted to specific aspects of the pathogenesis and spread of the virus in question. Examples of such differences are described a bit further along in this chapter.

Common features of herpesvirus replication in the host

The replication strategies of all herpesviruses appear to share some basic features. The viruses establish a primary infection during which virus replicates to moderate or high titers, yet with generally mild symptoms that are fairly rapidly resolved. One outcome of this primary infection in the host is efficient and effective immunity against reinfection. Following initial infection, however, virus is not completely cleared from the host. Instead, one or another specific cells infected by the virus are able to maintain the viral genomes without a productive virus infection. This maintenance is at least partially a result of the virus being dependent upon specific cellular transcriptional machinery for high-efficiency replication. The presence of critical components of this machinery is highly dependent upon the state of differentiation and the intercellular environment of cells in those tissues in which the virus replicates and establishes latency. As with other DNA viruses that exhibit a similar pattern of persistence without apparent active infection, this is termed a *latent infection*. While definitions of *latency* vary with the virus in question, the strictest definition (which can be readily applied to herpesvirus latency) requires that no infectious virus be detectable in the majority of host cells during the latent phase.

With appropriate stress to those cells harboring viral genomes along with stress to the host's immune system, the activity of critical components of the cell's transcriptional machinery is activated, and virus can reactivate from latently infected tissue. Provided host immunity is sufficiently suppressed, a generally milder version of the primary infection ensues. This reactivation results in virus being available for infection of immunologically naive hosts, and establishes the infected individual as a reservoir of infection for life. Notably, most of these reactivation events result in the release of virus at the primary site of infection with little or no clinical symptoms! This attests to the stable balance between host and virus that has evolved.

Since the major groups of herpesviruses have evolved to utilize different terminally differentiated cell types as a reservoir in which virus replication must occur at some low level to initiate **recrudescence**, it follows that those viral genes devoted to the ability of the virus to replicate in the immune-competent host will show much divergence. At the same time, the basic similarity of the productive replication cycle, once it occurs, suggests that – as is the case – those viral genes involved in high titer replication will be recognizably similar.

The replication of the prototypical alphaherpesvirus – HSV

The HSV virion

All herpesviruses possess similar enveloped icosahedrons. The envelope of HSV contains 10 or more glycoproteins. The matrix (called the *tegument* for obscure reasons) lies between the envelope and the capsid and contains at least 15–20 proteins. The capsid itself is made up of six proteins; the major one, VP5, is the 150 000-Da major capsid protein. VP5 is also called U_L19 for the position of its gene on the viral genetic map. A computer-enhanced model of the HSV capsid structure is shown in Figure 9.3. A more conventional electron microscopic view is shown in Figure 17.1. The molar ratio of HSV capsid proteins is tabulated in Table 11.2 – various capsid proteins are present in widely differing amounts.

The viral genome

While each herpesvirus is different, a number of general features can be illustrated with the HSV-1 genome. The HSV-1 genome is linear, and is ~152 000 base pairs (bp) long. With HSV, the left end of the genome is set as 0 map unit and the right is 1.00 map unit; therefore, each 0.1 map unit is 15 200 bp. Although the virion DNA is linear, the genome becomes circular upon entering latency.

A high-resolution genetic and transcription map of the HSV genome is shown in Figure 17.2. Because the HSV genome spends the majority of its life in the latent state, the genome becomes circular, and therefore the map is shown as a circle, but note that the genome's ends are indicated at the top of the circle. Since the virus encodes nearly 100 transcripts and more than 70 open translational reading frames (ORFs), the map is complex. Still, the basic methods of interpreting it are the same as with the simpler simian virus 40 (SV40) map. Interpretation of the HSV genetic and transcription map is aided by the fact that few viral transcripts are spliced and most ORFs are expressed by a single transcript, each with a contiguous promoter.

The genetic map of HSV-1 is summarized in Table 17.1, where viral proteins and other genetic elements are listed. The number of viral proteins that are not required for replication of

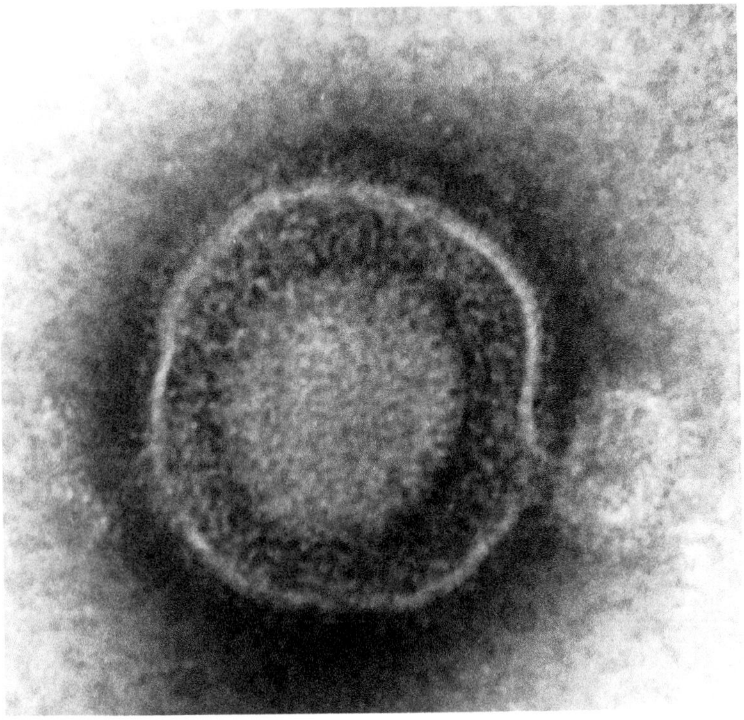

Figure 17.1 Electron micrograph of an enveloped HSV-1 virion revealing specific features, especially glycoprotein spikes projecting from the envelope. The capsid has a diameter of about 100 nm and encapsidates the 152 000-base pair viral genome. The interior of the capsid does not contain any cellular histones, in contrast to smaller DNA viruses. Rather, it contains relatively high levels of polyamines such as spermidine and putrescine, which serve as counterions to allow compact folding of the viral DNA needed in the packaging. *Source:* Courtesy of Jay Brown.

the virus in cultured cells is large. Many of these "dispensable" proteins have a role in aspects of the pathogenesis of the virus. The exact function of such proteins, in theory, can be established by studying the effect of the deletion of the genes encoding them on the way the virus replicates in its natural host. Because the natural host of HSV is humans, this analysis must be carried out in animal models instead. This study can be a difficult task, and the actual biological functions of many virus-encoded proteins and enzymes are still unknown.

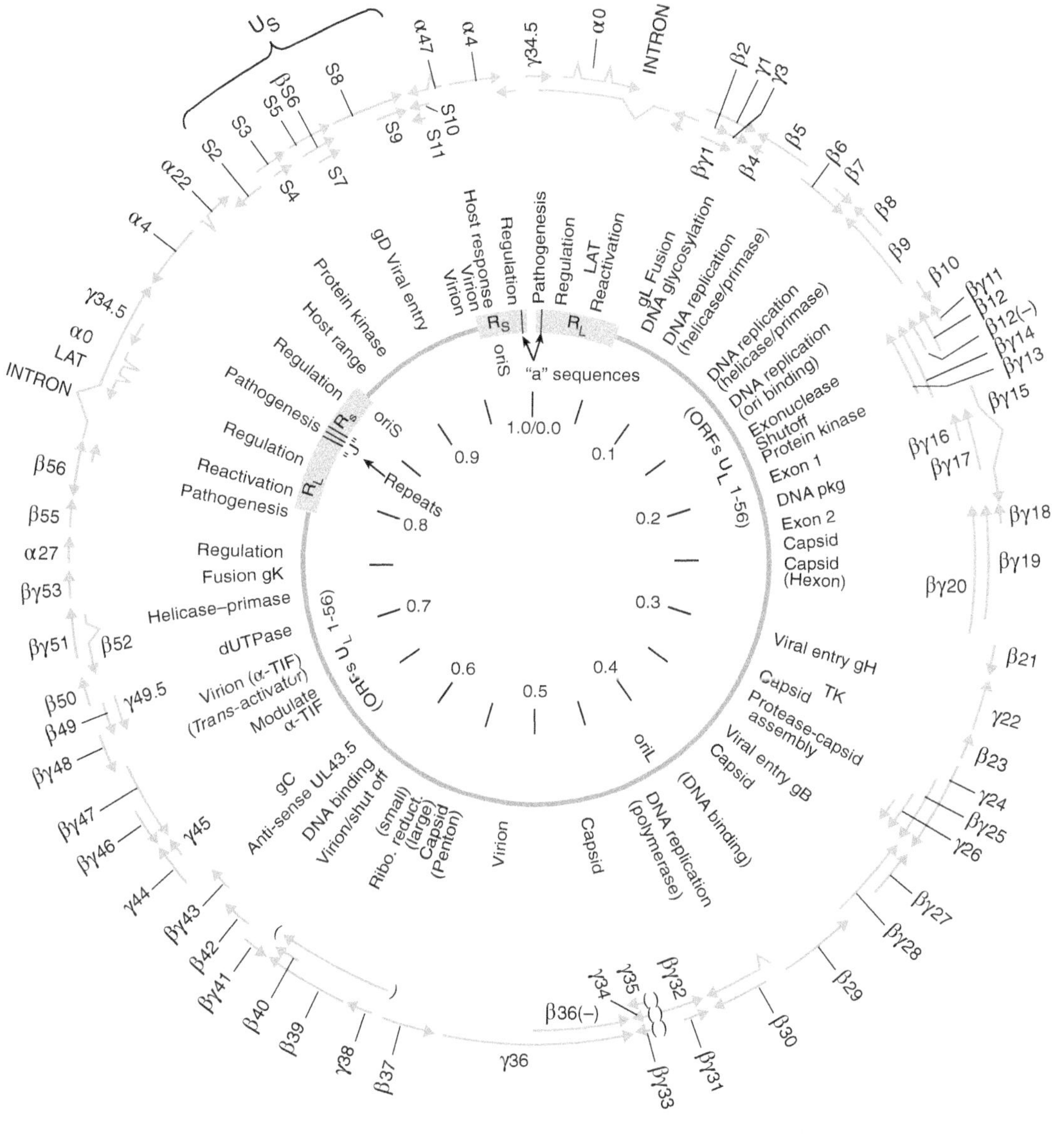

Figure 17.2 The HSV-1 genetic and transcription map. Specific features of the genome are discussed in the text, and tabulated in Table 17.1. Individual transcripts are controlled by their own specific promoters, and splicing is uncommon. Each transcript is headed by its own promoter, and most are terminated with individual cleavage/polyadenylation signals. The time of expression of the various transcripts is roughly divided into immediate-early (α), early (β), late (βγ), and strict-late (γ). This is, in turn, based on whether the transcripts are expressed in the absence of viral protein synthesis (α), before viral DNA replication and shutoff following this (β), before viral DNA replication but reaching maximum levels following this (βγ), or only following viral DNA replication (γ). The genome is about 152 000 base pairs and contains extensive regions of duplicated sequences.

Table 17.1 Some genetic functions encoded by herpes simplex virus type 1.

Location (map unit) (Figure 18.2)	Required for replication in culture?	Name of element or protein	Function
0.0	Yes	"a"	*Cis* genome cleavage, packaging signal
0.00–0.06	Yes	R_L	See below
0.05	No	ICP34.5	Neurovirulence
0.01 (R_L)	Yes	α0	Immediate-early transcription regulator (mRNA spliced) and interferon inhibitor
0.02 (R_L)	No	LAT	Approximately 600 bases in 5′ region facilitate establishment of latency, reactivation, and block apoptosis; no protein involved
0.04 (R_L)	No	LAT-intron	Stable accumulation in nucleus of some latently infected neurons, unknown function
0.06	Yes	gL	Viral entry, associates with gH
0.07	No	U_L2	Uracil DNA glycosylase, DNA repair
0.08	No	U_L3	Nonvirion membrane-associated protein
0.09	No	U_L4	Tegument protein, unknown function
0.1	Yes	Helicase–primase	DNA replication
0.1	Yes	U_L6	Capsid protein, capsid maturation, DNA packaging
0.11	No	U_L7	Unknown
0.12	Yes	Helicase–primase	DNA replication
0.13	Yes	Ori-binding protein	DNA replication
0.14	No	gM	Glycoprotein of unknown function
0.14	Yes	U_L11	Tegument protein, capsid egress and envelopment
0.16	Yes	Alkaline exonuclease	DNA packaging, capsid egress
0.15	No	$U_L12.5$	C-terminal two-thirds of U_L12, expressed by separate mRNA; specific function unknown
0.17	No	Protein kinase	Tegument associated
0.18	No	U_L14	Unknown
0.16/0.18	Yes	U_L15	DNA packaging, cleavage of replicating DNA (spliced mRNA)
0.17	No	U_L16	Unknown
0.2	Yes	U_L17	Cleavage and packaging of DNA
0.23	Yes	Capsid	Triplex
0.25	Yes	Capsid	Major capsid protein, hexon
0.27	Yes	U_L20	Membrane associated, virion egress
0.28	No	U_L21	Tegument
0.3	Yes	gH	Viral entry, functions with gL
0.32	No	U_L23	Thymidine kinase
0.33	No	U_L24	Unknown
0.33	Yes	U_L25	Tegument protein, capsid maturation, DNA packaging
0.34	Yes	U_L26	Maturational protease
0.34	Yes	$U_L26.5$	Scaffolding protein
0.36	Yes	gB	Glycoprotein required for virus entry
0.37	Yes	U_L28	Capsid maturation, DNA packaging
0.4	Yes	U_L29	ssDNA-binding protein, DNA replication
0.41	No	Ori_L	Origin of replication
0.42	Yes	DNA po1	DNA replication
0.45	No	U_L31	Nuclear phosphoprotein, nuclear budding
0.45	Yes	U_L32	Capsid maturation, DNA packaging
0.46	Yes	U_L33	Capsid maturation, DNA packaging
0.47	No	U_L34	Membrane phosphoprotein, nuclear budding

Table 17.1 *Continued*

Location (map unit) (Figure 18.2)	Required for replication in culture?	Name of element or protein	Function
0.47	Yes	U_L35	Capsid protein, capsomer tips
0.50	No	U_L36	ICP1/2, tegument protein
0.55	No	U_L37	Tegument phosphoprotein
0.57	Yes	U_L38	Capsid protein, triplex
0.58	Yes	U_L39	Large-subunit ribonucleotide reductase
0.59	Yes	U_L40	Small-subunit ribonucleotide reductase
0.6	No	U_L41	vhs (virion-associated host shutoff protein) destabilizes mRNA, envelopment
0.61	Yes	U_L42	Polymerase accessory protein, DNA replication
0.62	No	U_L43	Unknown
0.62	No	$U_L43.5$	Antisense to U_L43
0.63	No	gC	Initial stages of virus-cell association
0.64	No	U_L45	Membrane associated
0.65	No	U_L46	Tegument associated, modulates α-TIF
0.66	No	U_L47	Tegument associated, modulates α-TIF
0.67	Yes	α-TIF	Virion-associated transcriptional activator, enhances immediate-early envelopment transcription through cellular Oct-1 and CTF binding at TATGARAT sites
0.68	No	U_L49	Tegument protein
0.68	No	$U_L49.5$	Unknown
0.69	No	dUTPase	Nucleotide pool metabolism
0.7	No	U_L51	Unknown
0.71	Yes	Helicase-primase	DNA replication
0.73	No	gK	Virion egress
0.74	Yes	α27	Immediate-early regulatory protein, inhibits splicing
0.75	No	U_L55	Unknown
0.76	No	U_L56	Tegument protein, affects pathogenesis
0.76–0.82	Yes	R_L	See R_L above
0.82	Yes	R_L–R_S junction	Joint region, contains "a" sequences
0.82–0.86	Yes	R_S	See below
0.82–0.86 (R_S)	Yes	α4	Immediate-early transcriptional activator
0.86 (R_S)	Yes	Ori_S (*cis*-acting)	Origin of replication
0.86	No	α22	Immediate-early protein, affects virus's ability to replicate in certain cells
0.87	No	U_S2	Unknown
0.89	No	U_S3	Tegument-associated protein kinase, phosphorylates U_L34 and U_S9
0.9	No	gG	Glycoprotein of unknown function
0.9	No	gJ	Glycoprotein of unknown function
0.91	Yes	gD	Virus entry, binds HVEM
0.92	No	gI	Glycoprotein that acts with gE, binds IgG Fc, and influences cell-to-cell spread of virus
0.93	No	gE	Glycoprotein that acts gI, binds IgG-Fc, and influences cell-to-cell spread of infection
0.94	No	U_S9	Tegument-associated phosphoprotein
0.95	No	U_S10	Tegument-associated protein
0.95	No	U_S11	Tegument-associated protein phosphoprotein, RNA binding, posttranscriptional regulation
0.96	No	α47	Immediate-early protein that inhibits MHC class I antigen presentation in human and primate cells
0.96–1.00	Yes	R_S	See R_S above
1	Yes	"a"	*Cis* genome cleavage, packaging signal

α-TIF: Alpha-*trans*-inducing factor protein; HVEM: herpesvirus entry mediator; ICP: infected cell protein; IgG: immunoglobulin G; LAT: latency-associated transcript; MHC: major histocompatibility complex; mRNA: messenger RNA; Oct1: octamer-binding protein 1; R_L: long repeat; R_S: short repeat; ssDNA: single-stranded DNA; U_L: long unique; U_S: unique short; vhs: virion host shutoff.

The genome can be divided into six regions, each encoding a specific function as follows:

1 The ends of the linear molecules. The ends of the genome contain repetitive DNA sequences made up of various numbers of repeats of three basic patterns or groupings termed "a," "b," and "c." The "a" sequences also are found at the junction between the long and short segments of the genome (discussed later in this chapter). They also contain the signals used in the assembly of mature virions for packaging of the viral DNA.
2 The long repeat (R_L) region. The 9000-bp repeat (R_L) encodes both an important immediate-early regulatory protein (α0) and the promoter of most of the "gene" for the latency-associated transcript (LAT). This transcript functions in facilitating both efficient establishment and reactivation from latency by as-yet-unknown mechanisms.
3 The long unique (U_L) region. The long unique region (U_L), which is 108 000 bp long, encodes at least 56 distinct proteins (actually more because some ORFs are spliced and expressed in redundant ways). It contains genes for the DNA replication enzymes and the capsid proteins, as well as many other proteins.
4 The short repeat (R_S) regions. The 6600-bp short repeats (R_S) encode the very important α4 immediate-early protein. This is a very powerful transcriptional activator. It acts along with α0 and α27 (in the U_L region) to stimulate the infected cell for all viral gene expression that leads to viral DNA replication.
5 The origins of replication (ori's). HSV contains three short regions of DNA that serve as ori's. In the laboratory, any two can be deleted and virus replication will occur, but the three ori's are always found in clinical isolates. Ori_L is in the middle of the U_L region; ori_S is in the R_S, and thus, is present in two copies. All sets of ori's operate during infection to give a very complicated network of concatemeric DNA and free ends in the replication complex.
6 The unique short (U_S) region. The 13 000-bp unique short region (U_S) encodes 12 ORFs, a number of which are glycoproteins important in viral host range and response to host defense. This region also encodes two other proteins, α22 and α47, which are expressed immediately upon infection. The latter serves to block the infected cell's ability to present viral antigens at its surface.

HSV productive infection

HSV has a very complex genome, and the herpesviruses are the first ones described that have diploid copies of some of their genes. Still, the pattern of productive infection is roughly similar to that seen for smaller DNA viruses. In an HSV infection, the virus supplies most of the components it needs to replicate, and each HSV gene is encoded by an mRNA that has its own promoter and polyadenylation signal. Most (but not all) HSV transcripts are unspliced, and the relationship between gene structure and encoded polypeptide is relatively simple.

During the productive replication (**vegetative**) cycle, HSV gene expression is characterized by a progressive *cascade* of increasing complexity where the earliest genes expressed are important in "priming" the cell for further viral gene expression, in mobilizing cellular transcriptional machinery, and in blocking immune defenses at the cellular level. This phase is followed by the expression of a number of genes that are either directly or indirectly involved in viral genome replication. And, finally, upon genome replication, viral structural proteins are expressed in high abundance. The time of maximum expression of each viral gene is shown in Figure 17.2, and the cascade of increasingly complex transcription is shown schematically in Figure 17.3. The time required for completion of a replicative cycle of HSV and other alphaherpesviruses is fast compared with beta- and gammaherpesviruses as well as smaller nuclear-replicating DNA viruses such as adenoviruses and papovaviruses. HSV is able to replicate in a wide selection of animals, tissues, and cultured cells.

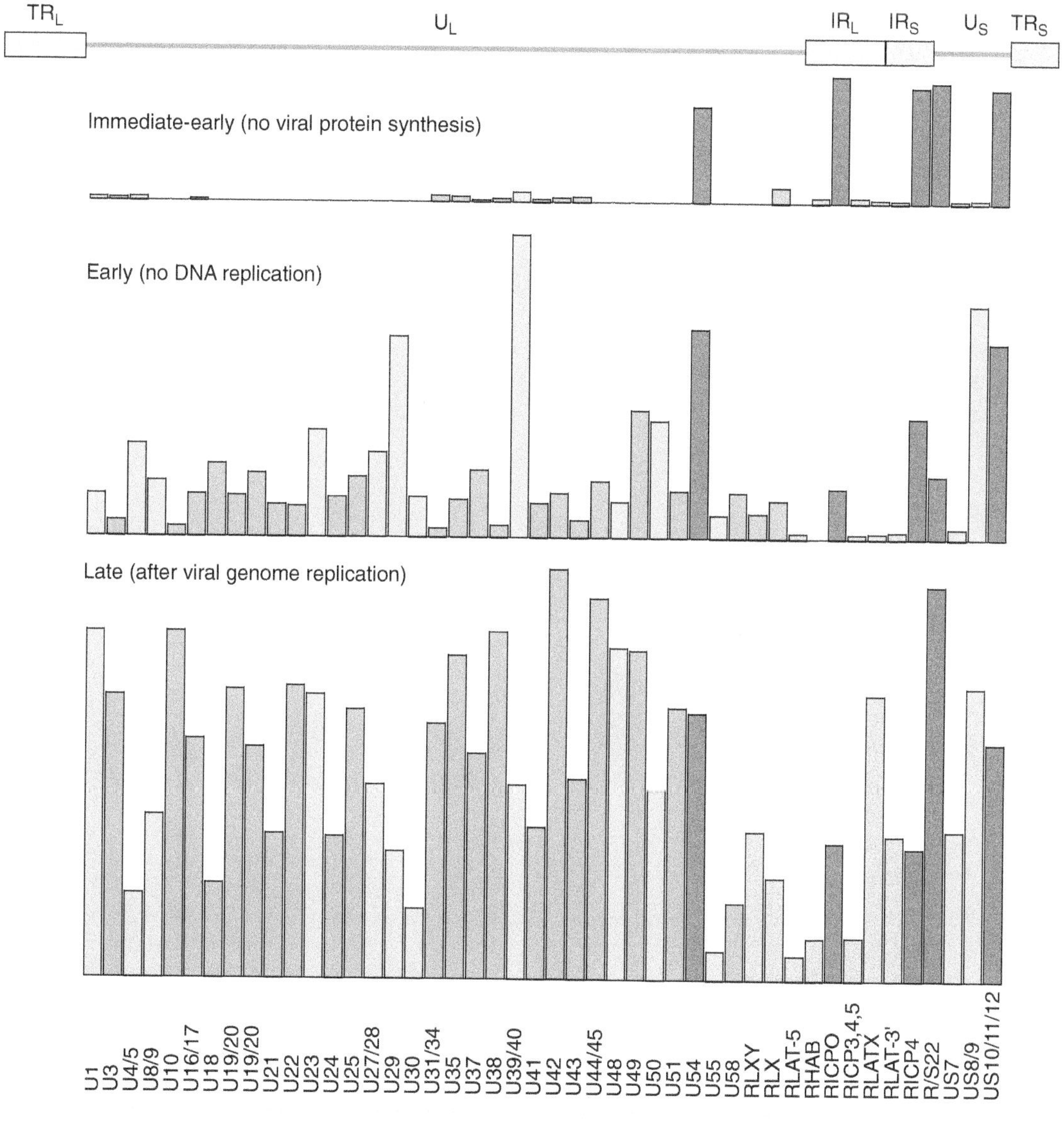

Figure 17.3 The programmed cascade of HSV transcription at different stages of the infection cycle. The details of this cascade are virtually identical for HSV-1 and HSV-2. The linear viral genome with its repeat and unique regions are shown at the top, and the levels of gene expression during the three basic phases of transcription as determined by DNA microarray analysis (see Chapter 12) are shown. Note that transcription from only one copy of each repeat region is shown for clarity. Immediately following infection, and prior to the expression of any viral proteins, five transcripts are expressed as immediate-early transcripts (red). The expression of these immediate-early transcripts is ensured by virtue of the interaction between their enhancers, the viral α-TIF, and cellular octamer-binding protein. These function during the early period and in the absence of any viral genome replication to allow relatively high levels of transcription of a number of early genes involved in replicating viral DNA as well as certain responses to host defense (green). The replication of the viral genome results in high levels of expression of the late transcripts (blue), which encode structural proteins and other responses to host defenses. The only clustering of genes into kinetic class is seen with the immediate-early transcripts, which are concentrated in the repeat regions. This may be important in reactivation from latency.

Initial steps in infection: virus entry The process of HSV infection and transport of viral DNA to the nucleus is shown in outline in Figure 17.4a. Virus attachment and entry require sequential interactions between specific viral membrane glycoproteins and cellular receptors. A group of related receptors are termed *herpesvirus entry mediators*, or HVEMs. Based on sequence analysis, these proteins, which occur widely but in varying proportions in different cell types, are related to cellular proteins that interact with the tumor necrosis factor (TNF) and the poliovirus receptor. Their function in the uninfected cell is unknown.

The virion membrane fuses with the host cell's membrane and capsid, and some tegument proteins are transported to the nucleus along cellular microtubules. The initial stages of infection and the fate of the viral envelope are shown in Figure 6.3. The virion-associated host shutoff protein (vhs, or U_L41) appears to remain in the cytoplasm where it causes the disaggregation of polyribosomes and degradation of cellular and viral RNA.

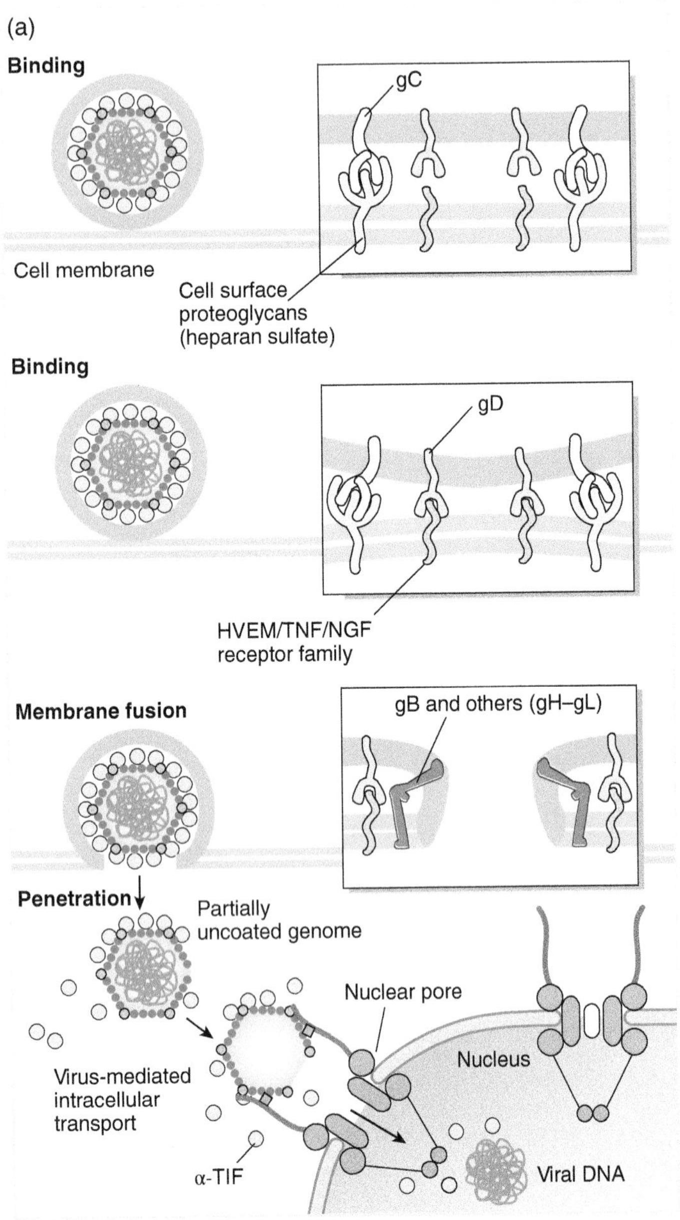

Figure 17.4 The entry of HSV-1 into a cell for the initiation of infection. (a) Outline of the process. The initial association is between proteoglycans of the surface and glycoprotein C (gC); this is followed by a specific interaction with one of several cellular receptors collectively termed herpesvirus entry mediators (HVEMs). These are related to receptors for nerve growth factors (NGFs) and tumor necrosis factor (TNF). The association requires the specific interaction with glycoprotein D (gD). Fusion with the cellular membrane follows; this requires the action of a number of viral glycoproteins, including gB, gH, gI, and gL. An electron micrographic study of herpesvirus fusion with the infected cell is shown in Figure 6.3. The viral capsid with some tegument proteins then migrate to nuclear pores utilizing cellular transport machinery. This "docking" is thought to result in the viral DNA being injected through the pore while the capsid remains in the cytoplasm. Some tegument proteins, such as α-TIF, also enter the nucleus with the viral genome. (b) Electron micrographic analysis of pseudorabies virus capsid "docking" and genome injection at the nuclear pore. A logical sequence is shown progressing from the (dark) full capsid to an empty one. The process is quite similar to injection of bacteriophage DNA into a bacterial cell (see Chapter 6, Part II). *Source:* Reprinted with the kind permission of the American Society for Microbiology from Granzow, H., Weiland, F., Jons, A. et al. (1997). Ultrastructural analysis of the replication cycle of pseudorabies virus in cell culture: a reassessment. *Journal of Virology* 71: 2072–2082.

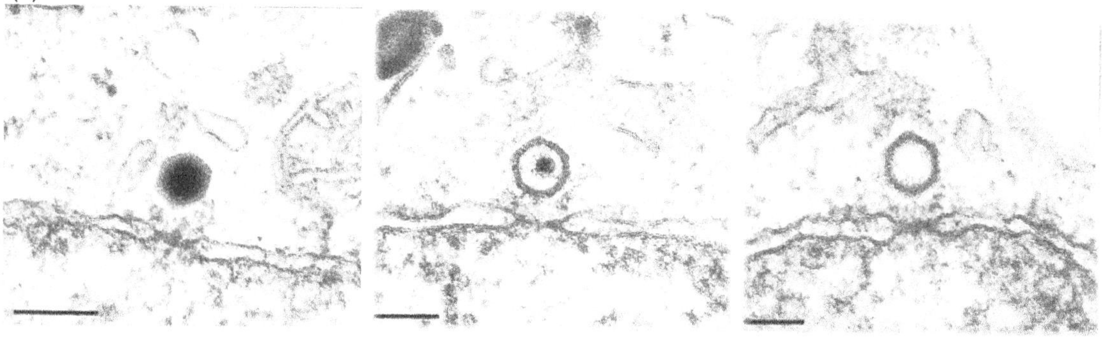

Figure 17.4 *Continued*

Unlike the genomes of smaller nuclear-replicating eukaryotic viruses, the HSV genome is not encapsidated with cellular chromosomal proteins, and, while it associates with such proteins upon infection, a highly regular packaging of histones on the viral genome during productive replication has not been demonstrated. Upon entry into the cytoplasm, the nucleocapsid is transported to the nuclear pores, where viral DNA is released into the nucleus. Elegant electron micrographs showing the docking and release of the genome of the closely related PrV into the nucleus are seen in Figure 17.4b. The viral genome is accompanied by the **α-TIF** protein (alpha-*trans*-inducing factor protein, also called VP16 or U_L48), which functions in enhancing immediate-early viral transcription. It does this by interacting with cellular proteins such as **Oct1** (octamer-binding protein 1) and **HCF-1** (host cell factor 1). Interestingly, it is the cellular protein that binds to the specific HSV-1 immediate-early gene promoter enhancer! The binding is to an eight-base stretch of nucleotides that has the nominal sequence TATGARAT, where R represents any purine. We will see in this chapter that this functional articulation between cellular DNA-binding proteins and the viral transcriptional activator is important in determining whether a specific cell will go on to productive infection and death or latent infection.

Immediate-early gene expression The process of viral gene expression during productive infection can be subdivided into a number of stages schematically shown in Figure 17.5. The process starts with immediate-early gene expression, also termed the "alpha" phase of gene expression, which is functionally similar to the immediate or pre-early stage of gene E1A and E1B expression in adenovirus infection. As seen in Figure 17.3, five HSV genes (α4 – ICP4; α0 – ICP0; α27 – ICP27/U_L54; α22 – ICP22/U_S1; and α47 – ICP47/U_S12) located in or near the repeat regions are expressed and function in this earliest stage of the productive infection cycle.

In HSV infection, immediate-early transcription is mediated by action of the virion tegument (matrix) protein α-TIF through its interaction with cellular DNA-binding proteins at specific enhancer elements associated with individual alpha-transcript promoters. The α-TIF protein (also known as VP16) is an extremely powerful transcriptional activator with very broad specificity. Its C-terminal region contains a very large number of acidic amino acids that activate transcription by mobilizing RNA polymerase bound to the pre-initiation complex at the promoter in the vicinity of the enhancer region. The activator is tethered to the DNA in this region by interaction between cellular DNA-binding proteins and elements within its N-terminal domain. This type of transcriptional activator is termed an **acid blob activator**, and α-TIF is the prototype of the group.

The result of this interaction between viral transcription factors and cellular DNA-binding proteins is that even in a cell that is not transcriptionally active, such as one that is not actively replicating, the virus can stimulate expression of its own genes. This is an alternative and highly regulated counterexample to the induction of cellular DNA synthesis and associated metabolic activation carried out by papovaviruses and adenoviruses through interaction of their early (or immediate-early) gene products with cellular growth regulators.

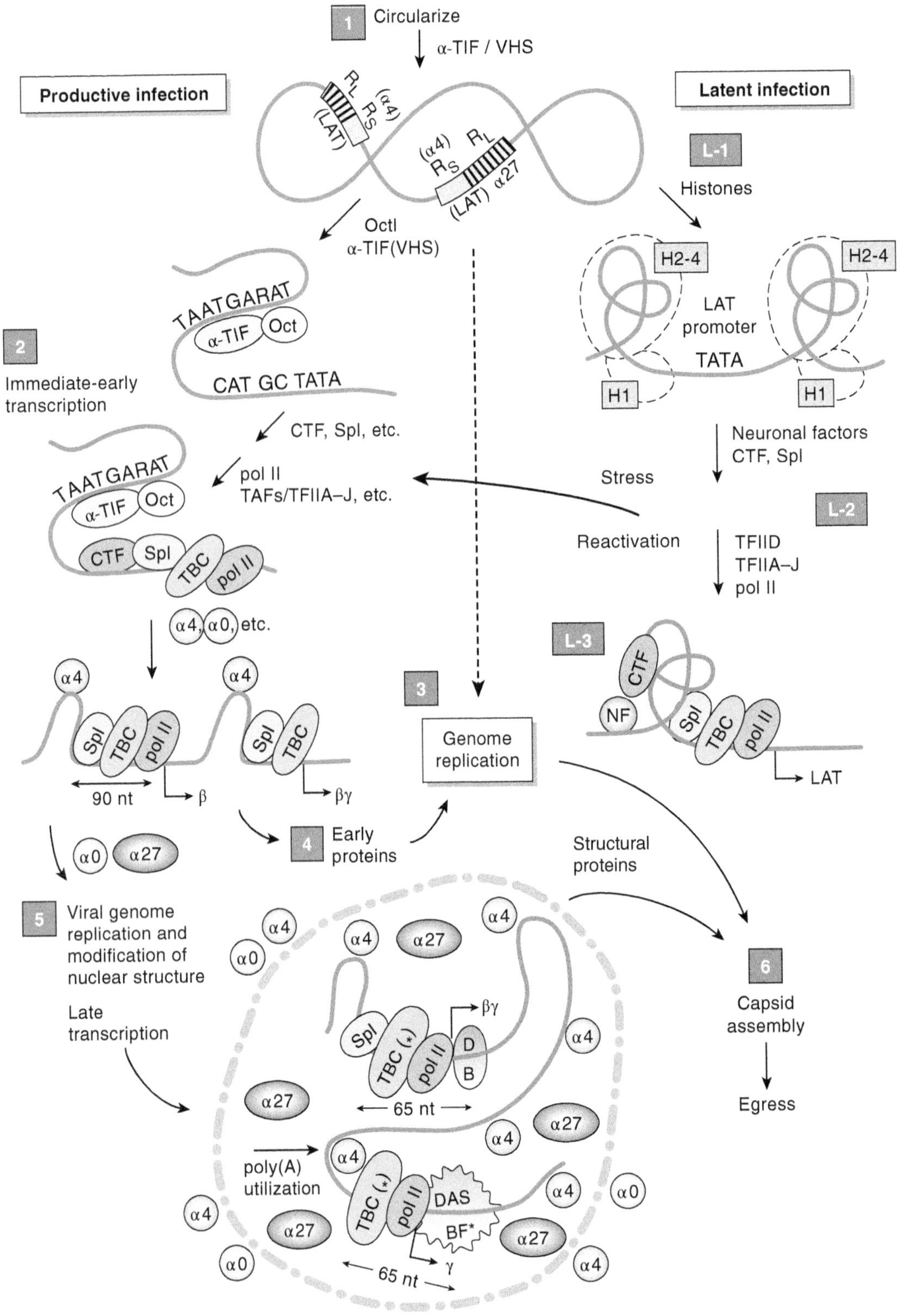

Figure 17.5 The HSV-1 productive and latent infection cycles. In productive infection, the viral genome becomes circular but does not associate with chromatin proteins (1). This is followed by immediate-early transcription that requires the association of cellular factors (Oct-1) binding to the TATGARAT sequence element within the immediate-early promoter enhancers and with α-TIF to enhance transcription of immediate-early transcripts (2). These are controlled with promoters with specific enhancers. This process results in transcriptional activation that leads to early transcription and, ultimately, to viral genome replication (3, 4). Viral genome replication is accompanied by rearrangement of nuclear structures and late transcription (5), and this is followed by capsid assembly (6). In latent infection, the earliest transcription does not occur and the viral genome becomes associated with histones to form a mini-chromosome (L-1). This essentially shuts down productive transcription but allows expression of the latency-associated transcript (LAT) (L-2). LAT facilitates the stress-induced reactivation of virus by an unknown mechanism (L-3). Reactivation reinitiates the productive cascade. vhs: Virion-associated host shutoff protein; α-TIF: alpha-*trans*-inducing factor protein; CTF, Spl, TAFs, TFIID, TFIIA-J, and TBC: all components of eukaryotic transcription machinery, as explained in Chapter 13.

Three of the proteins encoded by the immediate-early HSV genes – the α4, α0, and α27 proteins – are transcriptional regulators and activators of broad specificity. They function throughout the replication cycle. The mechanism of action of these transcriptional activators is complex. The α4 protein appears to interact with the basal transcription complexes forming at the TATA boxes of viral (and cellular) promoters and making the process of initiation of transcription more efficient. The α0 protein does not bind directly to DNA, and part of its function may be to mobilize cellular transcriptional machinery by induction of structural changes to the organization of the host cell nucleus. The α27 protein exhibits a number of functions, including mediating the transport of unspliced viral mRNA from the nucleus to the cytoplasm, inhibiting cellular splicing, influencing polyadenylation site usage, and activating transcription by an unknown mechanism.

The two other α proteins, α22 and α47, are dispensable for virus replication in many types of cultured cells. But α22, which has a role in the posttranscriptional processing of some transcripts, is required for HSV replication in some cell types and may have a role in maintaining the virus's ability to replicate in a broad range of cells in the host. Perhaps this is achieved by providing some types of cells with the capacity to express a group of late transcripts. The α47 protein appears to have a role in modulating host response to infection by specifically interfering with the presentation of viral antigens on the surface of infected cells by the major histocompatibility complex (MHC) class I complex (see Chapter 7, Part II).

Early gene expression Activation of the host cell's transcriptional machinery by the action of alpha gene products results in expression of the early or beta genes. Seven of these are necessary and sufficient for viral genome replication under all conditions: DNA polymerase (U_L30), DNA-binding proteins (U_L42 and U_L29), ori-binding protein (U_L9), and the helicase–primase complex (U_L5, U_L8, and U_L52). When sufficient levels of these proteins accumulate within the infected cell, viral DNA replication ensues.

Other early proteins are involved in increasing deoxyribonucleotide pools of the infected cells, while still others appear to function as repair enzymes for the newly synthesized viral genomes. These accessory proteins are "nonessential" for virus replication in that cellular products can substitute for their function in one or another cell type or upon replication of previously quiescent cells. However, disruptions of such genes often have profound effects on viral pathogenesis or viral ability to replicate in specific cells.

Genome replication and late gene expression Viral DNA replication at high levels under the control of virus-encoded enzymes is termed **vegetative DNA replication**. The vegetative replication of HSV DNA occurs in a number of stages that tend to occur simultaneously in the infected cell nucleus. First, HSV-encoded ori-binding and DNA-denaturing proteins bind to one or all of the ori's, and a replication fork carrying out DNA synthesis is generated. This process is shown in Figure 17.6.

During the replication process, this circular replication structure is "nicked" at a replication fork, and a "rolling circle" intermediate is formed. As shown in Figure 17.6, such a rolling circle (in theory) generates a continuous concatemeric strand of newly synthesized viral DNA that is available for encapsidation. In actuality, as DNA is being replicated, new synthesis begins at any one of a number of ori's, and highly concatenated, linked networks of DNA are formed in the infected cell. Although these networks are difficult to visualize, and appear as a "tangled mess" in the electron microscope, the packaging process for viral DNA allows individual, genome-sized pieces of viral DNA to be encapsidated. It should be noted that there is growing evidence that the majority of HSV-1 replication ensues from these linear and branched structures, and that initial rolling-circle replication might be limited to the earliest rounds of replication and/or during reactivation.

The encapsidation process involves viral maturation and encapsidation proteins associating with the "a" sequences of the newly synthesized genomes, simultaneously cleaving them from the growing replication complex and packaging them into mature capsids. This process also is shown in Figure 17.6.

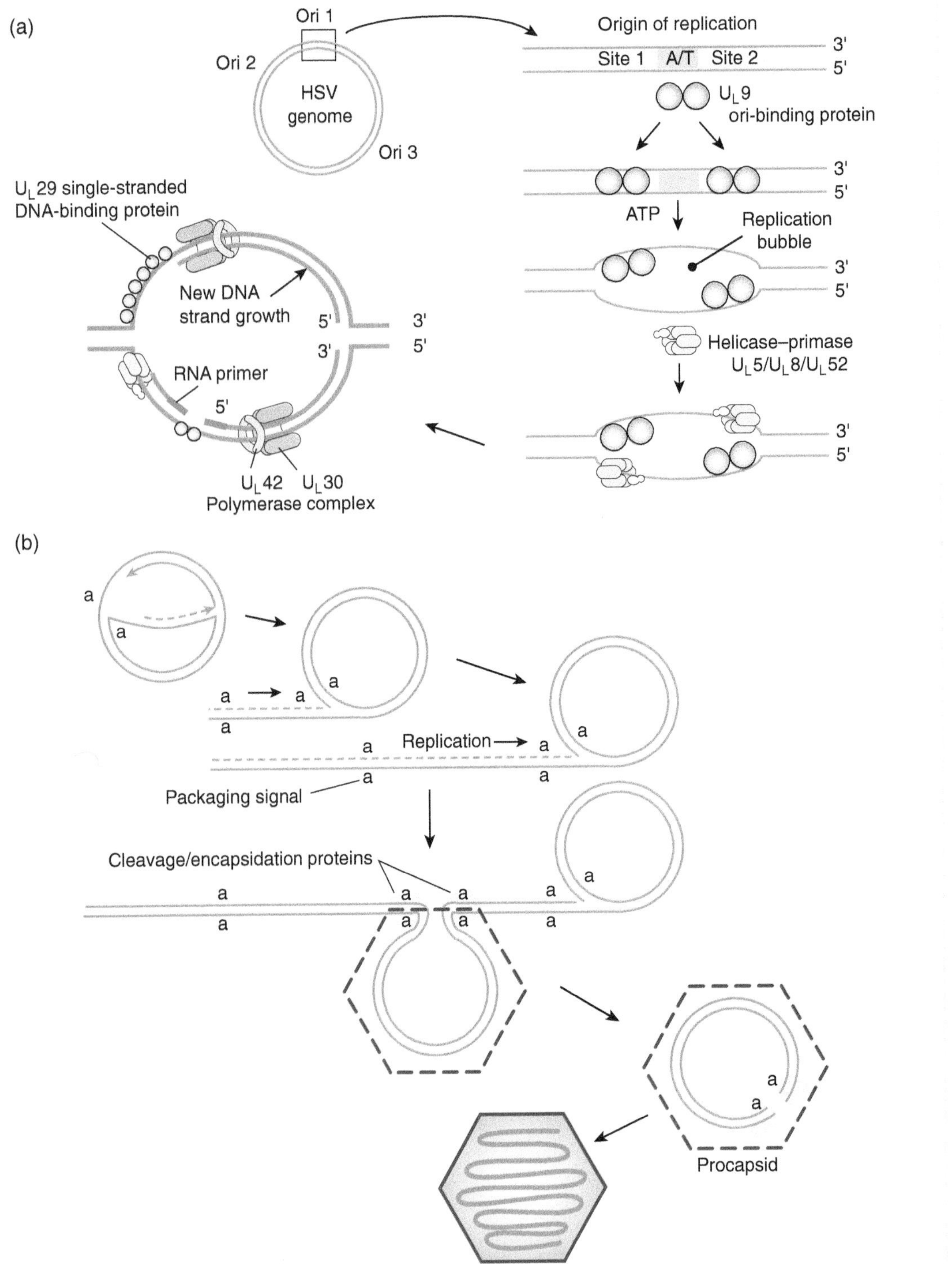

Figure 17.6 Replication and encapsidation of viral genomes. (a) HSV DNA initiates rounds of DNA replication at one of three origins of replication (ori's). (b) The genome is circular in the cell during latency and possibly the initial rounds of lytic replication, which leads to a structure that is nicked to form a rolling circle. Long concatemeric strands of progeny DNA are encapsidated by the interaction of cleavage/packaging proteins with the specific packaging signals at the end of the viral genomes (the "a" sequences).

Viral DNA replication represents a critical and central event in the viral replication cycle. High levels of DNA replication irreversibly commit a cell to producing virus, which eventually results in cell destruction. DNA replication also has a major influence on viral gene expression. Early expression is significantly reduced or shut off following the start of DNA replication, while late genes begin to be expressed at high levels.

Immunofluorescence studies using antibodies against specific viral proteins involved in DNA replication and transcription, such as shown in Figure 17.7, demonstrate that DNA replication and late transcription occur at discrete sites, or "replication compartments," in the nucleus. Prior to DNA replication, the α4 protein and the single-stranded (ss) DNA-binding protein ICP8 (U_L29) are distributed diffusely throughout the nucleus. Concomitant with viral DNA replication, distribution of these proteins changes to a punctate (point-like) pattern. In the case of α4, this change involves interaction with α0 and α27.

Virus assembly and release More than 30 HSV-1 gene products are structural components of the virion, and all are expressed with late kinetics. As outlined in Chapter 6, Part II, HSV capsids assemble around viral scaffolding proteins in the nucleus, and then other viral proteins interact with replicated viral DNA to allow DNA encapsidation. The encapsidated DNA is not associated with histones, but highly basic polyamines (perhaps synthesized with viral enzymes) appear to facilitate the encapsidation process.

Mature capsids bud through the inner nuclear membrane that contains the viral glycoproteins (Figure 17.8). In the early maturation process in the nucleus, capsids appear to be surrounded by the "primary" tegument protein, U_L31, and this directs the budding through the inner nuclear membrane into which the U_L34 phosphorylated membrane protein has been inserted. These "primarily enveloped" capsids then bud through the outer nuclear membrane where the primary envelope is lost. The cytoplasmic capsids then associate with the numerous tegument proteins of the mature virion, including α-TIF and vhs, which appear to functionally interact to help final envelopment. Final envelopment takes place as the mature capsids and associated tegument proteins bud into exocytotic vesicles, the membranes of which contain all the glycoproteins associated with the mature virions. Infectious virions can either remain cell-associated within these vesicles, and spread to uninfected cells via virus-induced fusion, or can be released from the cell in exocytotic vesicles for reinfection such as shown in Figure 6.9b. Obviously, in the latter case, the virion itself is subject to immune surveillance and host-mediated immune clearance.

HSV latency and LAT

All herpesviruses can establish latent infections in their natural host. Such an infection is characterized by periods of highly restricted (or no) viral gene expression in the cells harboring the latent genomes, interspersed with periods of virus replication and infectivity (reactivation or *recrudescence*). The different types of herpesviruses carry out latency and reactivation processes in generally similar ways, but details depend on the virus in question.

Neurotropic herpesviruses like HSV establish a latent infection by entering a sensory nerve axon near the infection site. The virus particle then can migrate to the neuron's nucleus in the nerve ganglion. HSV-1 tends to favor the lip and facial areas for initial infection; hence, the sensory nerve invaded is the trigeminal ganglion. For infections of genital mucosa, HSV-2 invades the sciatic nerve (sacral) ganglia. If virus gene expression occurs normally, viral DNA replication and cell death will occur; however, in most infected neurons, the earliest stages of gene expression appear to be blocked. Thus, these neurons are, in a sense, nonpermissive for replication. This likely results from the fact that one or several cellular transcription factors required to express the immediate-early genes are absent or present in a modified form that is different from that found in epithelial and other cells fully permissive for HSV replication.

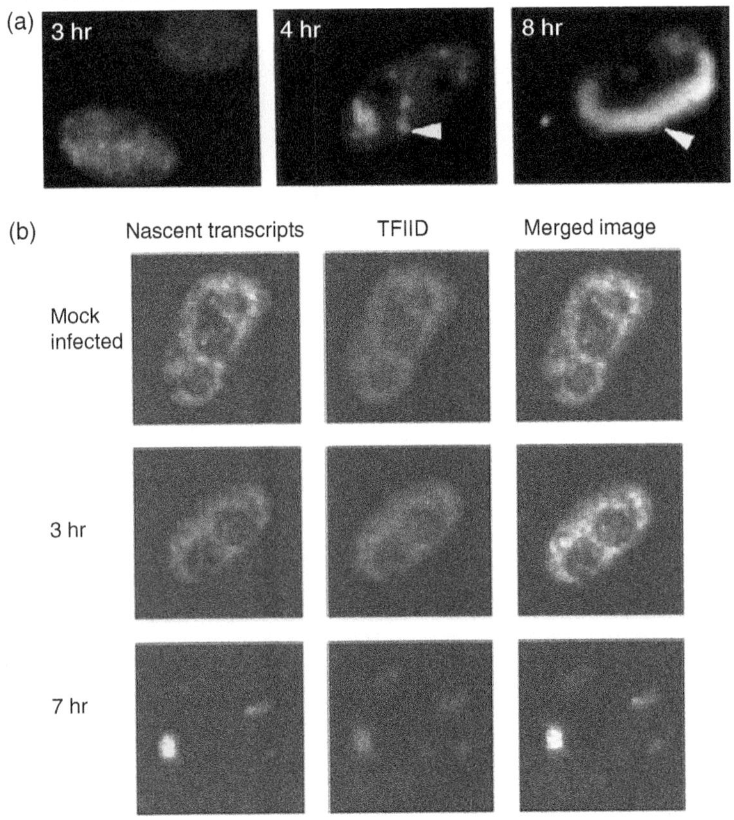

Figure 17.7 Immune fluorescence analysis of the rearrangement of nuclear structures following HSV-1 infection. (a) The localization of an antibody to the early single-stranded DNA–binding protein. The antibody is found diffusely distributed in the nucleus of infected cells at three hours after infection, but it (and the associated replicating viral DNA molecules) rapidly becomes concentrated in "replication factories" following this. *Source:* Courtesy of R. M. Sandri-Goldin. (b) Confocal microscopy of infected cell nuclei. In this series of views, an uninfected (mock infected) cell, and cells at three hours and seven hours following infection, were incubated with two antibodies bound to different chromophores. The first (green) is specific for newly synthesized RNA. This trick is accomplished by incubating the infected cells in medium containing a modified nucleotide that is incorporated into RNA. The base (5-Br-uridine) is antigenic, and there is a good commercial antibody available against it. The second (red) is an antibody specific to a component of the pre-initiation complex TATA-associated factor TFIID, and will react with this complex as it forms at the transcription start site. The merged colors (yellow) show that RNA and the transcription complex are confined to localized areas in the infected cell nucleus at the late time. Confocal microscopy is described in Chapter 12.

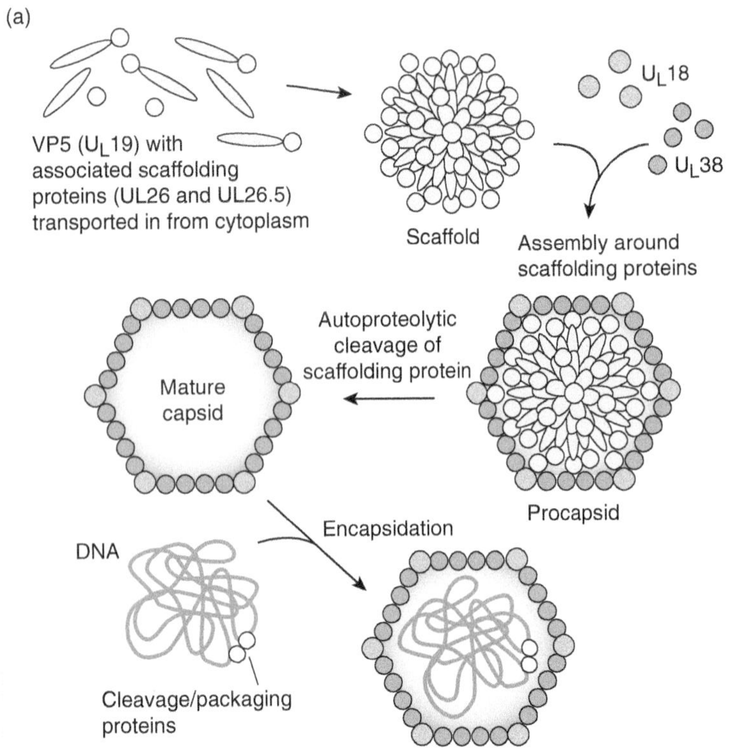

Figure 17.8 Maturation of the HSV capsid and its envelopment by tegument and virus-modified nuclear membrane. The procapsid assembles around scaffolding proteins that are then digested away, and the empty capsid incorporates DNA by means of the action of cleavage/packaging proteins. The filled capsid then migrates through the double nuclear membrane by first budding into the intercisternal space between the inner and outer membranes, and then fusing this initial membrane with the outer nuclear envelope, releasing the capsids into the cytoplasm. Final enveloping is by budding through the walls of the exocytotic vesicle, and final release of the enveloped virion is by fusion of this vesicle with the cytoplasmic membrane, as shown in Figure 6.9.

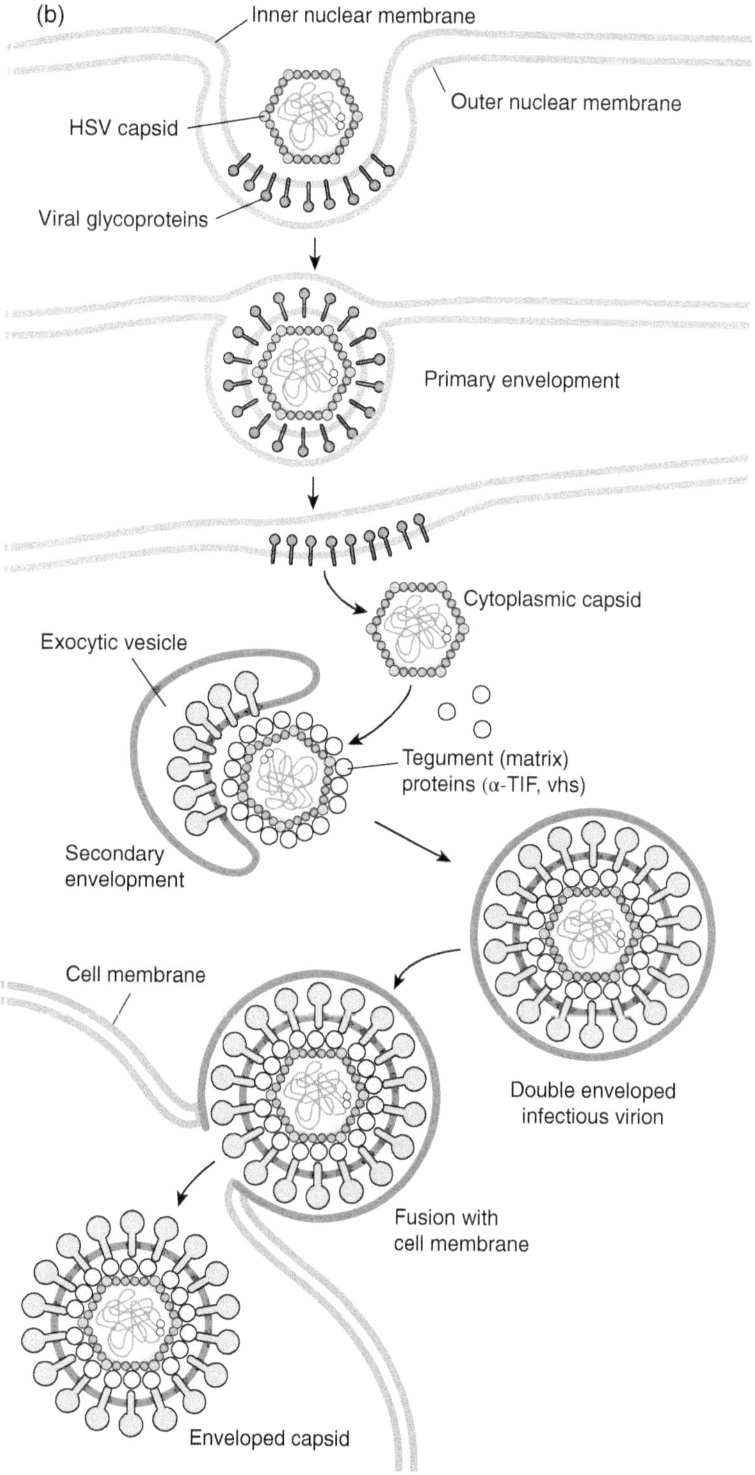

Figure 17.8 *Continued*

In the latently infected neuronal nucleus, HSV DNA is present as a histone-wrapped mini-chromosome or episome, which is in distinct contrast to the viral genome's physical state during productive infection. Since viral DNA is in the neuronal nucleus and fully differentiated neurons do not replicate, the DNA can remain for long periods of time – probably for the host's life. Experimental studies of viral gene expression in neurons during reactivation suggest that when the host is stressed with certain agents such as HMBA (hexamethylene bisacetamide), epinephrine (adrenaline), or sodium butyrate, the cellular transcriptional environment changes, and immediate-early as well as other viral transcripts are expressed at detectable levels. This is sufficient for the replication of virus in at least a few neurons, which results in the transitory appearance of a small amount of virus that travels down the axon and re-infects the area where the virus first infected.

Reactivation can happen many times with no apparent damage to the trigeminal nerve ganglion. It is not known whether one or a few neurons die with each reactivation, but there is such a very large number of neurons that the loss of a few during reactivation does not have a significant effect on enervation of the lip. The process of establishment of and reactivation from latent infection is outlined schematically in Figure 17.9, and the transcriptional switches involved are indicated in Figure 17.4.

HSV transcription during latency and reactivation

The viral genome of HSV and many (but not all) other alphaherpesviruses is not fully shut off during latent infection. One transcript family, collectively termed the LATs, is expressed from a single latent-phase active promoter located in both copies of the R_L (see Figure 17.2). The HSV LAT is a large transcript that is weakly expressed during productive infection, but unlike all other known productive-cycle transcripts, it does not "shut down" in latently infected cells. This transcript is spliced to generate an unusual 2-kb intron, which – unlike most introns – is stable and accumulates in the infected cell. Indeed, it is this intron that is responsible for the nuclear *in situ* hybridization signal seen in about one-third of latently infected neurons. An example of such a signal is shown in Figure 12.10. The transcriptional complexity of the LAT region does not stop here! There are eight microRNAs (miRNAs; see Chapter 7) encoded within the LAT region, a number of which have been shown to regulate the expression of critical viral gene products such as ICP0, ICP4, and ICP34.5, as well as cellular genes.

Epinephrine induction of rabbits latently infected with HSV via the ocular route leads to relatively efficient shedding of virus from the eye (see the "Rabbit Models" section of Chapter 3, Part I). This induction also results in a transitory expression of productive-cycle viral transcripts. This transitory expression can be readily detected using polymerase chain reaction (PCR) amplification of complementary DNA (cDNA) generated from polyadenylated RNA isolated from rabbit ganglia. An example of such an experiment is shown in Figure 17.10.

In this experiment, rabbit ganglia were isolated either before induction or at eight hours after induction. RNA was isolated, and cDNA was synthesized with retrovirus reverse transcriptase from the 3′ end of the RNA using a primer made up of oligodeoxythymidine, which will anneal to the mRNA polyA tail. Then, specific viral cDNA sequences corresponding to those found in ICP4, ICP27, DNA polymerase, and LAT mRNA were amplified with specific primers. The results clearly show the continued presence of LAT mRNA in latent and reactivating ganglia, but the productive-phase transcripts are seen only during a short "window" during reactivation. The simplest interpretation of such an experiment is that the epinephrine changes the transcriptional "program" of a few latently infected neurons that can then produce a small amount of virus. This virus migrates down the nerve axon and establishes a low-level infection in the rabbit cornea, which results in the ability to isolate infectious virus.

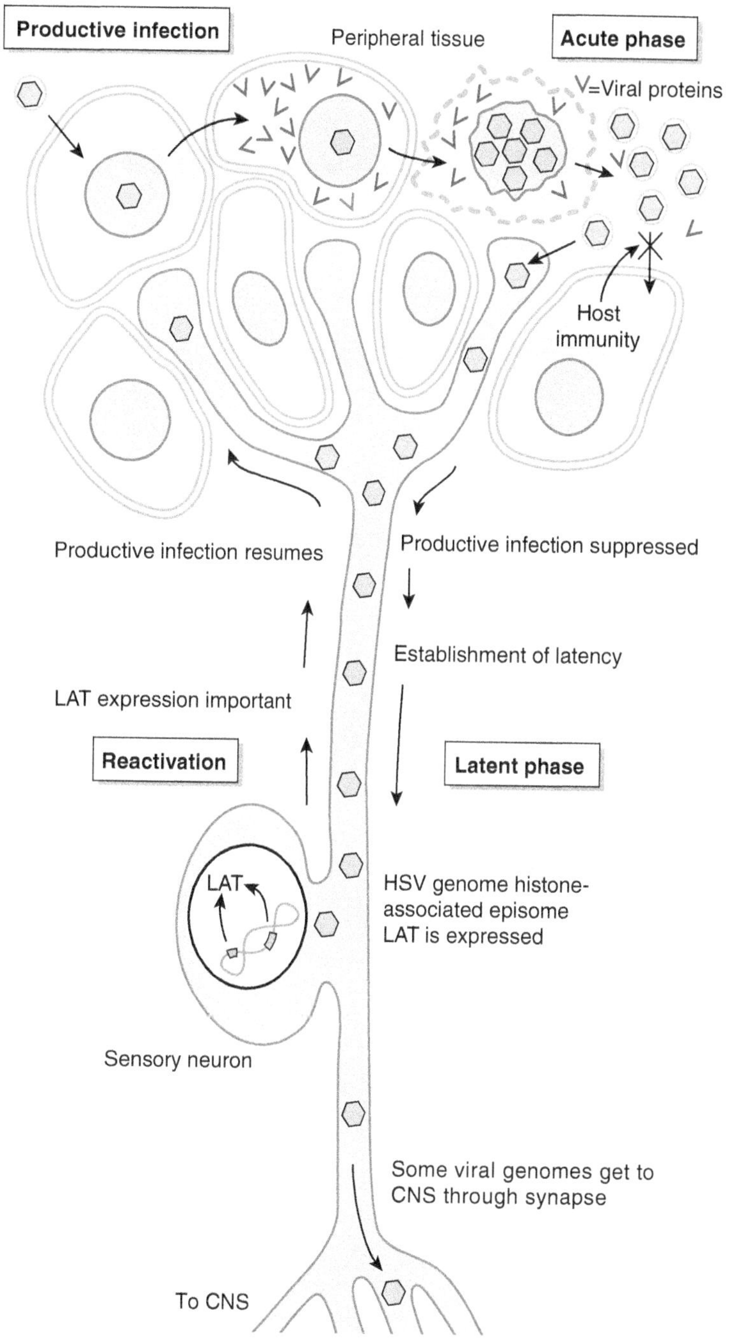

Figure 17.9 The "decision" made by HSV upon infection of epidermal tissue enervated with sensory neurons. Productive infection follows infection of peripheral tissue, but entry of the virus into neurons leads to latent infection in a significant proportion of them.

How do the LAT and other specific HSV genes function to accommodate reactivation?

Successful reactivation should not be viewed as merely involving the function of a limited number of genes expressed during the latent phase of infection by HSV. Rather, it is clear that the reactivation process involves a highly orchestrated interaction between a number of viral genes specifically directed toward ensuring efficient replication of small amounts of virus in an immune host. It is not known (yet) just how they all work, but they make the reactivation process more efficient so that more virus is produced when the animal is stressed.

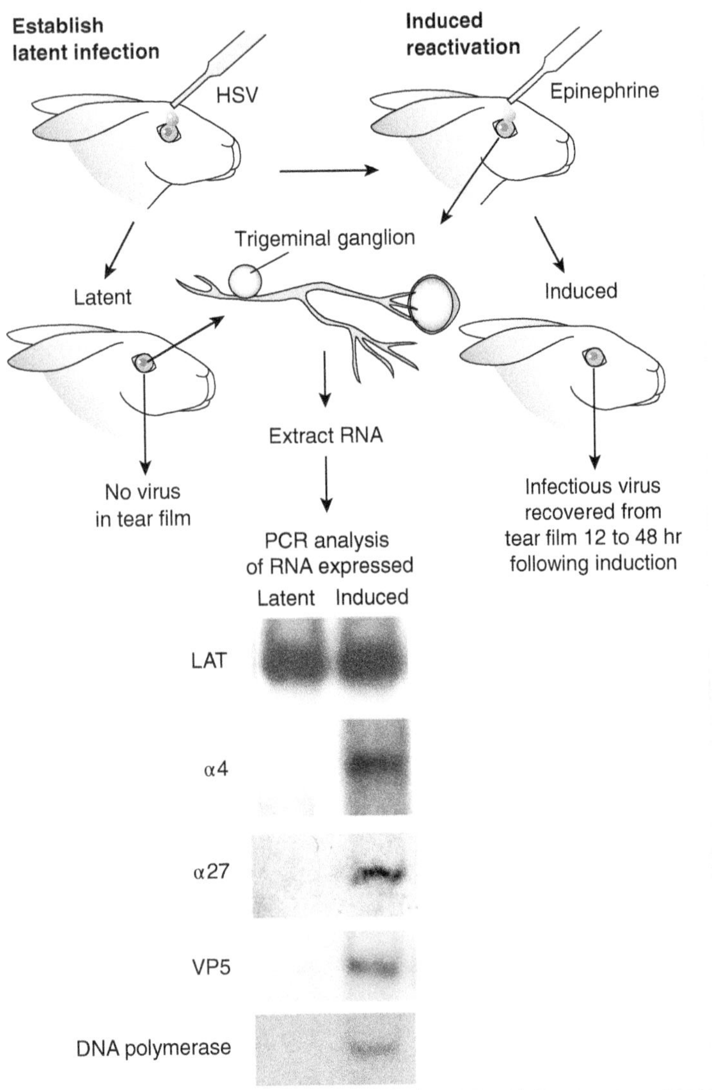

Figure 17.10 The expression of HSV transcripts during latent infection and reactivation in the rabbit. A standard approach toward experimental investigation, as was briefly described in the "Rabbit Models" section of Chapter 3, Part I, is outlined. If a latently infected rabbit is killed and RNA in the trigeminal ganglion extracted and subjected to reverse transcription–polymerase chain reaction (RT-PCR) using primers specific for the latent-phase transcript (the latency-associated transcript, or LAT) or several productive-cycle transcripts (α4, α27, DNA polymerase, or VP5), the only signal seen is with the LAT primer pair. This indicates that only the LAT region is being transcribed with any frequency. Following induction of the rabbit with epinephrine, however, all transcripts are expressed, suggesting that at least some viral genomes are induced to enter productive-cycle replication.

A large number of experimental studies have focused on the role of the LAT in the latency process, and they suggest that it plays a number of roles. The problem in understanding LAT function is complicated by two factors. First, deletion of the LAT promoter and significant portions of the LAT transcript does not have an absolute role in latency – LAT(–) viruses still establish latency reasonably effectively in animal models, and reactivation, while less efficient, is still readily measurable. Second, the regions of the HSV LAT that do affect latency and reactivation do not encode proteins! Thus, the transcript itself or portions of the LAT region DNA must mediate its action(s). To date, various laboratories have presented experimental evidence that one region or another of the LAT has a role in efficiency of establishment of latency – possibly by protection of the infected neuron via inhibition of apoptosis, repressing lytic gene transcription as well as affecting the efficiency of reactivation itself through modulation of apoptosis.

While the actual role of the LAT in reactivation is still unclear, it is now fairly well established that expression of the LAT during latency is the result of its unique position in the viral genome. Chromatin analyses have shown that the LAT locus is bounded by short DNA sequence elements, which form chromatin insulators and insulate the LAT region from the rest of the genome. This suggests that the LAT region serves as an "idling" enhancer, insulated and functioning to maintain a transcriptionally permissive chromatin configuration during latency. Stress, however, can lead to changing the patterns of histone modifications through chromatin remodeling, and possibly reversing the insulating activity, thus opening the rest of the viral genome to transcription. In addition, LAT expression is transiently turned off following stress, consistent with the model that one function of the LAT is to help maintain latency through lytic gene repression.

If the precise role(s) of the LAT in this switch from latency to productive infection remains to be established, some of the next events in reactivation of HSV from a latently infected neuron can be envisioned to be similar to those events that might be seen in the expression of viral genes in a cell transfected with infectious HSV DNA. Since transfected DNA does not contain αTIF, enhancement of immediate-early genes will not take place, but limited expression of early viral genes leads to some viral DNA replication and production of a few infectious virions that can infect neighboring cells. While the process is inefficient, there is no barrier to it occurring in cultured cells, and so plaques will form and virus infection will spread.

However, a similar process in a reactivating host will quickly encounter some profound problems. First, inefficient replication of virus in peripheral cells can induce interferon, which will block further virus replication. Second, the host immune memory will quickly marshal all the immune defenses available to suppress and clear the active replication of virus.

HSV encodes a number of genes to counter these host defenses. First, the inhibition of MHC class I–mediated antigen presentation at the surface of the infected cell by the α47 protein is thought to be effective in slowing the host's ability to detect the earliest stages of productive replication. Second, HSV encodes a protein, ICP34.5, that specifically blocks interferon's inhibition of translation in the infected cell by inhibiting phosphorylation of the translational initiation factor eIF-2. This has the result of ensuring efficient translation of the small amounts of viral transcripts expressed in this limited infection. Third, the virus encodes a protein that inhibits the infected cell's tendency to undergo apoptosis, thus ensuring a higher yield of infectious virus.

Other viral proteins also have been identified as having potential roles in interfering with cellular and host defenses, which, if allowed to function, would be very effective in inhibiting replication of the small amount of virus produced by the reactivation step itself.

EBV latent infection of lymphocytes: a different set of problems and answers

EBV is a lymphotropic herpesvirus infecting primarily B lymphocytes, although it also infects epithelial cells of the nasopharynx. Although difficult to study in cell culture, it is an important human pathogen that has a well-established role in causing several important types of human cancers. Like the human alphaherpesviruses, a high percentage of adult humans have antibodies to EBV indicating prior infection, and the virus is able to persist in a latent state for the life of the host. Since this latency is maintained in a cell that is relatively short-lived (the B lymphocyte), the virus must express a number of specific functions to ensure that it can maintain itself efficiently.

As with all herpesviruses, the reservoir of the virus is a previously infected individual who has been induced to shed infectious virus that can invade an immunologically naive individual. Symptoms of primary infection in children and infants are mild and often inapparent, but infection of adolescents and adults can lead to a sometimes severe mononucleosis, which is the result of antigen-independent proliferation of infected B cells that tend to crowd out red blood cells and cause anemia. EBV-induced mononucleosis is temporary, but a number of EBV-infected B lymphocytes survive as replicating, essentially immortal, reservoirs of latent viral genomes.

It should be remembered that normal circulating B lymphocytes are short-lived. After a period of time, apoptosis is induced, causing the cell to die. As briefly discussed in earlier chapters, the process of apoptosis or programmed cell death is a natural stage in the life cycle of many types of differentiated cells. One benefit of such a process is that cells generated for a specific purpose (such as B lymphocytes) can be eliminated as the need for them decreases. Another benefit is that cells that have undergone a number of replication cycles, and may have accumulated deleterious mutations, can be eliminated. Both primary and latent infections with EBV are characterized by the expression of a viral gene product interfering with the induction of apoptosis in B lymphocytes – this serves to stabilize the pool of cells bearing viral genomes.

The EBV genome, shown in Figure 17.11, is 172 000 bp, which is divided into five unique DNA regions bounded by internal repeat regions. The unique regions encode genes similar to those expressed during productive infection by alphaherpesviruses, and their structure and the regulation of their expression are generally similar to a productive HSV infection, although the efficiency of the process is lower and the time required for replication is longer. Primary infection of B lymphocytes often leads to the establishment of a latent infection in which the viral genome is circularized and histone-associated and from which only a limited number of transcripts are expressed. The latent genome and the transcripts expressed from it are shown in Figure 17.11.

The initial stages of latency in B lymphocytes are characterized by the expression of a number of transcripts. One very long primary transcript is expressed from one of two juxtaposed promoters; this primary transcript, which encompasses about half the viral genome, generates six transcripts encoding six separate **Epstein–Barr nuclear antigens (EBNAs):** EBNA-1, 2, 3A, 3B, 3C, and LP by alternative splicing. Two other transcripts encoding latent membrane protein 1 (LMP-1) and LMP-2 proteins are also expressed, as are two small RNAs (**Epstein–Barr-encoded RNAs**, or **EBERs**) expressed from RNA polymerase III promoters.

When all latent proteins are expressed in B lymphocytes, the cells are stimulated to divide and apoptosis is blocked; with time, some of the immortalized B cells shut down latent EBV transcription and become dormant, similar to memory B cells. Such "EBV memory cells" may serve as a secondary reservoir of viral genomes. Other latently infected lymphocytes continue to divide, and in these cells EBNA-1 binds to the viral oriP, which functions analogously to the oriP of papillomaviruses allowing viral genomes to replicate in synchrony with the host cell chromosomes. EBNA-1 also binds to other sites in the viral chromosome, allowing it to bind to cellular proteins leading to efficient segregation of newly replicated viral DNA into each daughter cell.

The mechanism by which EBV immortalizes B cells appears to involve inactivation of p53 and Rb tumor suppressor genes in a manner roughly similar to that seen in infection with papovaviruses and adenoviruses. Other specific virus-induced processes are also involved, however. One important result of immortalization by EBV is that the chromosomal telomeres are stabilized, perhaps by alteration of the activity of telomerase enzyme in B lymphocytes. This enzyme has an important role in cell mortality (see Chapter 10), and functions to allow the progressive loss of DNA at the ends of chromosomes until a critical point is reached, after which cell death ensues.

Reactivation of EBV from latent infection requires specific stimulation of the latently infected lymphocyte. This reactivation event faces a number of the same obstacles outlined for HSV reactivation. In response, like HSV, EBV inhibits cell-based interferon defenses against virus infection. In contrast to the mechanism utilized by HSV and analogous to that utilized by human adenoviruses through the expression of viral-associated (VA) RNA (see Chapter 16), however, EBV accomplishes this by the expression of a set of small virus-encoded RNAs, the EBERs. These function to inhibit the interferon response and block other cellular defense responses by mechanisms thought to be similar to those mediated by small interfering RNAs (siRNAs) (Chapter 8).

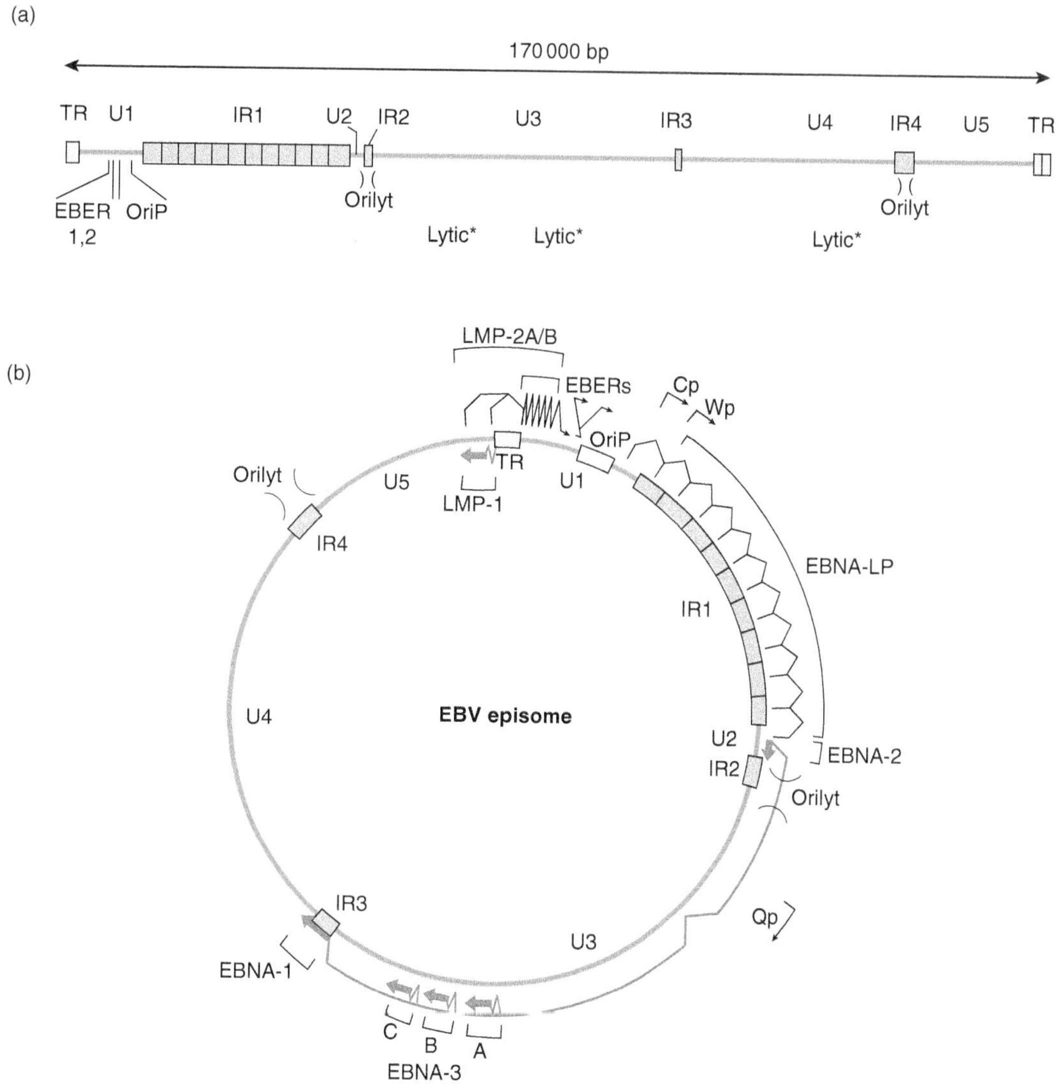

Figure 17.11 The Epstein–Barr virus (EBV) genome and the latency transcripts. (a) The 170 000-bp EBV genome contains five regions of unique sequence separated by regions of DNA containing numerous repeated sequence elements. The IR1 region contains a number of copies of a basic repeat element. The virus has two origins of replication, the OriP, which mediate latent genome replication and segregation in concert with cellular chromosome replication, and two copies of the OriLyt, which functions during productive infection for vegetative DNA replication. Three regions of the genome contain genes homologous to those expressed during the productive replication of the alphaherpesviruses. The structure of these genes is similar in the two classes of herpesviruses, with each gene being expressed under the control of its own cognate promoter. (b) During the latent phase of infection, the genome circularizes and becomes a histone-associated mini-chromosome. A number of latent transcripts are expressed, and one family, the Epstein–Barr nuclear antigens (EBNAs), is derived by alternative splicing of a very large precursor transcript that extends from one of two promoters (Cp and Wp) rightward to the end of the IR3 region. The latent membrane protein (LMP) transcripts are expressed from their own cognate promoters, and LMP-2A and LMP-2B are also derived by alternate splicing of a small precursor RNA. The EBERs are expressed from RNA pol III promoters.

Pathology of herpesvirus infections

HSV, EBV, and other herpesvirus latent infections are relatively benign conditions, and latency does not appear to cause many serious problems. However, if the host's immune system does not function properly owing to disease or clinically induced immunosuppression for organ transplantation, there can be significant medical problems, including disseminated herpesvirus infections in the brain or other tissues. For example, HSV keratitis, which can result in blindness, is a major cause of complications in corneal transplants. Reactivation of the betaherpesvirus, HCMV, is a major factor in mortality of organ transplants, as is reactivating VZV.

Primary herpesvirus infection in a newborn (neonate) can be devastating because the infant's immune system has not yet developed fully. It has also been shown that the inflammation of chronic HSV-2 reactivation has a role in increasing susceptibility to HIV infection, resulting in the development of AIDS.

Herpesviruses as infectious co-carcinogens

In areas of the world where there are many cases of malaria, EBV coinfection with a malaria infection can lead to a type of cancer called *Burkitt's lymphoma*. This cancer is found in portions of tropical Africa. In Japan and China, eating some very complex hydrocarbons found in foods pickled by fungal fermentation can interact with EBV-infected cells to lead to *nasopharyngeal carcinoma*. Both types of cancers are, then, associated with EBV infection, but require a **co-carcinogen** for development. This is probably true for papillomavirus-caused cancers, and is also true for hepatitis B virus–associated liver carcinomas. The co-carcinogen functions in some way to induce mutations in cells' growth-transformed latent EBV gene products. These mutations eventually lead, as is the case with papillomavirus carcinogenesis, to metastasizing cancer cells.

While many transformed cell lines contain integrated EBV DNA, like the situation with papillomavirus, the viral genome can become integrated into the cellular genome during the process of cell transformation. However, unlike papillomavirus integration, EBV DNA integration tends to be at a specific chromosomal site. There are, however, transformed cell lines in which the EBV is maintained as an episome, and most primary tumors do not contain integrated DNA; therefore, integration of the genome is not a strict requirement for transformation. Following transformation, no further expression of viral genes is necessary for maintenance of the viral genome.

A second human herpesvirus, HHV-8, is also associated with a human cancer called *Kaposi's sarcoma* (KS). This cancer was first described in the last part of the nineteenth century as a rare disease of very old men. It is marked by a slow progression of the formation of sarcomas made up of highly pigmented epithelial cells, is not particularly invasive, but eventually leads to death. Its occurrence in the general population is known to be associated with extensive loss of cellular immune capacity and specific geographical and genetic factors.

A high incidence of KS was observed beginning in the early 1980s in homosexual men in a number of gay communities, notably San Francisco, New York, and Los Angeles. Victims were found to have advanced immunodeficiency, and the study of this growing epidemic led to the discovery of HIV as a cause of AIDS.

Despite the high frequency of KS in some gay communities in which AIDS was common, a number of factors demonstrate that the decline in immunity in the late stages of HIV-induced disease is not the sole causal factor in this cancer. The most obvious one is that many groups of HIV-positive individuals go on to exhibit the symptoms of AIDS without any development of KS. Epidemiological analysis strongly suggests the action of a co-carcinogen or possibly genetic factors working along with HIV in development of the disease.

HHV-8 is found to be present in 70–80% of individuals at risk for development of KS, while it is present in less than 2% of the general population. Further, HHV-8 DNA is readily found in endothelial cells of AIDS patients, while it is difficult, if not impossible, to find it in other tissues of the same individual.

This finding, along with the known ability of EBV to act as a co-carcinogen in the formation of Burkitt's lymphoma and nasopharyngeal carcinomas, suggests the strong possibility that HHV-8 infection, along with HIV-induced loss of immune capacity, is a contributing factor in the development of KS. Consistent with this possibility is the fact that HHV-8 encodes a number of cell-derived genes known to function in the oncogenic transformation of specific cells. These include the *bcl-2* gene, which inhibits apoptosis; a G protein–coupled receptor, which is active and can cause transformation of cultured cells; and a gene related to cellular K-cyclin, which can induce contact-inhibited cells to enter the S phase and begin the process of division.

BACULOVIRUS: AN INSECT VIRUS WITH IMPORTANT PRACTICAL USES IN MOLECULAR BIOLOGY

Arthropods are infected by a wide variety of viruses. Some of them, the classic arboviruses such as yellow fever virus, dengue virus, and La Crosse encephalitis virus, replicate within the cells of their arthropod hosts but cause no cytopathology and no disease. Other viruses infect their arthropod hosts and cause significant pathology and disease. Among these are the baculoviruses, grouped into the nuclear polyhedrosis viruses (NPVs) and the granulosis viruses (GVs). Some baculoviruses infect insects such as the alfalfa looper (*Autographa californica*), while others infect arthropods such as the penaeid shrimp. The virus infecting the alfalfa looper (*Autographa californica* NPV, or AcNPV) has been extensively studied because of its value in biotechnology (discussed in the "Pathogenesis" and "Importance of Baculoviruses in Biotechnology" sections).

Virion structure

Baculoviruses are large and complex in structure. The genome is large (80–230 kb), circular, double-stranded (ds) DNA contained in a nucleoprotein core within a capsid. The rod-shaped capsid is composed of ring-shaped subunits that are 30–60 nm in diameter and stacked longitudinally to give an overall length of 250–300 nm.

Virions can be found in two forms. Budded viruses (BVs) have acquired a virus-modified envelope to surround the capsid as they exit the infected cell through the plasma membrane. This form of the virus is involved in secondary infection from the initial site of entry into the insect.

Occluded viruses (OVs) have an envelope that appears to be derived from virus-modified nuclear membrane. This form of the virus is found embedded within a matrix consisting of a crystalline lattice of a single protein, polyhedrin for NPVs and granulin for GVs. NPVs often exhibit several nucleocapsids embedded in this matrix; however, only a single capsid is embedded in the OV matrix, resulting from GV infections.

The OV form is transmitted horizontally during feeding by the insects. The very stable matrix serves to protect the virions from the environment, but readily dissolves in the insect's midgut, with its high pH, just prior to infection of cells at that site.

Viral gene expression and genome replication

Viral gene expression and genome replication take place in the cell's nucleus. The genome size of these viruses predicts a large coding capacity, and indeed, there are well over 100 viral proteins expressed.

The infectious cycle is divided into early and late phases, followed by the occlusion phase. During the early phase, viral mRNAs are transcribed by the host's RNA polymerase II. Normal posttranscriptional modifications take place to viral mRNAs, although only one transcript is RNA spliced. As would be expected, some of the proteins encoded by these early genes are required for replication of the viral DNA.

Late gene expression occurs following the onset of viral DNA replication. This switch in transcription is mediated by a virus-encoded RNA polymerase that recognizes a unique set of promoter sequences. The encoding of a unique viral RNA polymerase is unusual for eukaryotic DNA viruses, but is seen in the replication of cytoplasmic DNA viruses and in many DNA-containing bacteriophages. These are discussed in Chapter 19.

Polyhedrin protein is synthesized very late in the replication cycle, and it is during this stage that the occluded form of the virus begins to accumulate. The production of polyhedrin is not required for viral replication, and the protein gene can be deleted without affecting production of progeny virus. This property has been exploited as a useful tool for applications in biotechnology, as discussed in the "Importance of Baculoviruses in Biotechnology" section.

Pathogenesis

Some baculoviruses infect insect pests, but some infect ecologically and economically important insects, including silkworms. Infection of a susceptible insect by a virulent virus such as AcNPV leads to a distinctive cytopathology and ultimately to death. Larvae of the leafhopper infected with AcNPV eventually have viral replication carried out in virtually every tissue. This results in the larvae disintegrating into a liquid that consists of mostly virus particles within their polyhedron matrices. This phenomenon is called *melting* and is essentially macroscopic lysis.

Importance of baculoviruses in biotechnology

Because of their high infectivity for insects, and because they display narrow host ranges, baculoviruses have become important tools in the battle against pests that feed on important plant species. Many researchers hope that baculoviruses can replace chemical pesticides as a biological control agent, for instance, for control of the apple maggot in Europe, the coconut beetle in the South Pacific, and soybean pests in Brazil. There is hope that baculoviruses may also be effective in controlling the tussock moth larvae in the forests of the Pacific Northwest.

A second role for these agents is in the laboratory. Since polyhedrin protein is not essential for viral replication, the gene encoding it can be deleted from viral DNA with no effect on virus replication. Also, the promoter controlling expression of the polyhedrin gene is quite active. Accordingly, deletion of the coding region of the polyhedrin gene allows the insertion of foreign genes of reasonably large size whose expression is under control of the baculovirus promoter. Thus, infecting insect cells in culture allows expression of the cloned gene in an invertebrate cell that processes the protein in much the same way as a vertebrate cell. This system is useful for producing large amounts of correctly modified and folded eukaryotic gene products.

Case 5: Cytomegalovirus (CMV) (for Part IV, Chapter 17)

Clinical presentation/case history: A 55-year-old male with a past medical history of ulcerative colitis (an immune-mediated inflammatory condition of the bowel) was admitted with a two-week history of bloody diarrhea and crampy abdominal pain, and associated decreased appetite. The bowel movements were described as liquid in consistency with a bright red color, but no mucus. The patient had not traveled recently, had no animal contacts, and had not recently used antibiotics. The patient had been recently treated with 6-mercaptopurine (an immunosuppressive that reduces lymphocytes) and high-dose anti-inflammatory steroids to treat a flare-up of the ulcerative colitis. A blood count upon admission revealed a white blood cell count of 1000/μl (4000–10 000/μl is normal). The immunosuppressive therapy was stopped, and lab tests were ordered on the stool samples to determine the cause of the diarrhea.

Diagnosis: The fact that the patient was on immunosuppressive therapy made the differential diagnosis of this prolonged and severe diarrhea challenging due to the fact that a number of normally self-resolving infections of the intestine also needed to be considered. Tests for bacterial pathogens (including *Shigella*, *Salmonella*, and *Camphylobacter*) were negative; however, a polymerase chain reaction (PCR) test for viral pathogens revealed a large number of copies (10 500/ml) of cytomegalovirus (CMV).

Treatment: The patient was immediately started on intravenous (IV) ganciclovir, which reduced the CMV load, and the diarrhea eventually cleared as the inflammation of the intestine subsided.

Disease notes: CMV is a persistent infection of lymphocytes that is commonly acquired early in life. Over 50% of the US population is seropositive for CMV by adulthood. Normally, CMV is a clinically benign infection and is usually subclinical when acquired as a child. If primary infection occurs as an adolescent or adult, it can present as mononucleosis with clinical symptoms of sore throat, lethargy, and lymphadenopathy that closely resembles Epstein-Barr virus (EBV) mononucleosis. Once infected, the virus remains latent in a subpopulation of lymphocytes and is persistently shed by others. Infectious virus can be spread by contact with blood or saliva, and it is in mother's milk. CMV can cause clinically significant infections in the immunocompromised, where it can cause gastrointestinal infections, hepatitis, and severe infections of the retina. In most cases these infections can be controlled by ganciclovir or foscarnet, drugs that target the activity of the viral polymerase. CMV infections are a particular concern for organ transplant patients, who must undergo immunosuppression or risk transplant failure or other severe complications due to a CMV flare-up. These patients are usually treated with antivirals before and after the transplant to prevent this complication.

QUESTIONS FOR CHAPTER 17

1 What feature in herpes simplex virus type 1 (HSV-1) allows the virus to evade the immune system and establish a latent infection?

2 HSV does not alter tumor suppressor gene products. Considering this, how does HSV get around the fact that the host cell is not transcriptionally active?

3 You have isolated the viral DNA of two separate cell cultures infected with virus. You know that one viral infection is due to SV40, while the other is due to HSV. In each case, the DNA was isolated at a time exhibiting a great deal of cytopathology. Further, the DNA you have isolated from one of the cultures contains almost no single-stranded material. In order to determine which culture was infected with which virus, you can do one or more of the following:

(a) Check the other culture. It should have a lot of single-stranded DNA and would be the one infected with HSV.

(b) Measure the density of DNA isolated from the cultures. If one has a significant amount with a density indicative of a high G+C content, that is the one infected with HSV.

(c) Take total DNA from the infected cultures and gently sediment it. Try to isolate relatively small circular DNA molecules, which would be indicative of SV40 infection.

(d) Isolate the DNA from the infected cells, digest with *Eco*RI restriction enzyme, and do a Southern blot on the digest. The culture infected with SV40 should yield a fragment of about 5400 base pairs, which will hybridize to SV40 DNA probe.

Which of these methods would not give you the information you need?

4 HSV is a member of the Herpesviridae family. The virus enters the cell by membrane fusion, and the dsDNA genome is transported into the nucleus of the cell where viral gene expression begins.

(a) Delivered to the nucleus, along with the DNA genome, is the viral DNA-binding protein α-TIF. What is the function of this protein?

(b) The protein α-TIF is actually made late during viral replication (it is one of the gamma class of genes). What will be the effect on herpesvirus gene expression if a cell is infected with a temperature-sensitive α-TIF mutant and the cells are placed at 39.5 °C (nonpermissive temperature) early during infection (within the first hour)? Late during infection (after 16 hours)?

5 Briefly explain why the herpesvirus α-TIF protein is the product of a late (gamma) gene whose action is required early during infection.

6 The drug acyclovir is a guanosine analogue that is a specific antiviral agent for certain members of the family Herpesviridae. When HSV-1 is grown in cell culture in the presence of this drug, acyclovir-resistant mutants of HSV-1 can be selected. Name two HSV-1 genes that can be mutated to make the virus acyclovir resistant, and give a brief reason for the resistance in each case.

7 HSV-1 can infect epithelial cells and then go on to establish a latent infection in the sensory ganglia. In the table below, indicate (with a "Yes" or a "No") which viral feature will be found in each kind of cell. Assume that the sensory ganglia cells are in the latent state and have not yet been reactivated.

HSV-1 Feature	Epithelial Cells	Sensory Ganglia Cells
Viral DNA in the nucleus		
Expression of the alpha class of viral transcripts		
Expression of LAT RNA		
Production of viral capsid proteins		

Replication of Cytoplasmic DNA Viruses and "Large" Bacteriophages

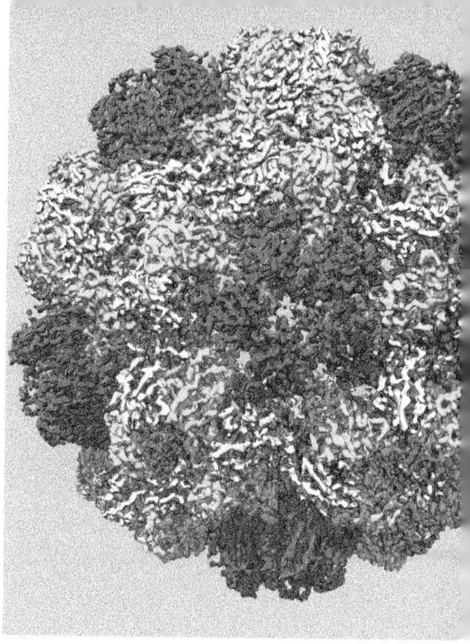

CHAPTER 18

Basic Virology, Fourth Edition. Martinez "Marty" Hewlett, David Camerini, and David C. Bloom.

If the classification of DNA viruses into large and small is arbitrary, and in the terms of this text, it certainly is, then the grouping of large DNA viruses that replicate in the cytoplasm of eukaryotic cells with large DNA-containing bacteriophages is even more so. Indeed, this grouping is only defensible in that bacteria have no nuclei, so *any* virus infecting them by necessity will replicate in the cytoplasm. Indeed, there is evidence based on the details of capsid assembly and some very limited genetic sequence homologies that the bacteriophages discussed in this chapter have a distant relationship to the herpesviruses discussed in Chapter 17! Still, organizational criteria can be satisfied best by inclusion of these viruses in this chapter. This inclusion can be operationally defended because these bacteriophages – like poxviruses, and unlike herpesviruses and many other DNA viruses utilizing the nucleus as a site for genome replication – encode many transcriptional enzymes required for replication.

POXVIRUSES – DNA VIRUSES THAT REPLICATE IN THE CYTOPLASM OF EUKARYOTIC CELLS

The poxviruses are a very successful group of double-stranded DNA (dsDNA)-containing viruses that have evolved a highly specialized mode of replication and pathogenesis in animal hosts: the ability to replicate in the cytoplasm of infected cells. Study of smallpox disease and attempts to control it over the past three centuries are responsible for generating much of our basic understanding of virus-induced immunity, virus epidemiology, and viral pathogenesis.

The pox virion is complex and contains virus-coded transcription enzymes

Poxviruses are physically complex, ovoid or brick-shaped objects with an ill-characterized but apparently complex subvirion structure, which contains an inner core surrounded by a double membrane derived from the host. A schematic diagram of the structure is shown in Figure 18.1. They comprise the largest known animal viruses and contain blocks of genes homologous with very large insect and protist viruses such as iridoviruses, ascoviruses, and mimiviruses (see Chapter 1). Poxvirus virions have dimensions of 250–300 nm by 250 by 200 nm. This size is just large enough to be resolved with ultraviolet light, and if great care is taken in sample preparation, poxvirus virions can be observed as refractile points with a high-quality optical ultraviolet microscope.

One consequence of the cytoplasmic site of poxvirus replication is that the virus has no access to cellular transcription and DNA replication machinery – at least during the earliest times after infection. It must, then, supply all or most of the nuclear functions that other nuclear-replicating DNA viruses appropriate from the cell; therefore, it is hardly surprising that poxviruses have large genomes. The genomes of orthopox viruses, which include smallpox and vaccinia (the Jennerian vaccine against smallpox – see Chapter 8), contain the replication genes clustered within the center 50% or so of the genome. This is flanked on either side by genes specific to the actual type and strain of poxvirus. The core replication sequences are highly conserved, but the flanking sequences diverge widely.

The complex virions contain all the enzymes necessary for transcription, polyadenylation, and capping of a specific class of viral messenger RNAs (mRNAs) – those encoding the enzymes required to begin the replication process in the cell. In this, the replication strategy is vaguely reminiscent of some negative-sense RNA viruses. A partial list of enzymes found in the virion is as follows:

- RNA polymerase
- Early transcription factors
- mRNA capping enzymes

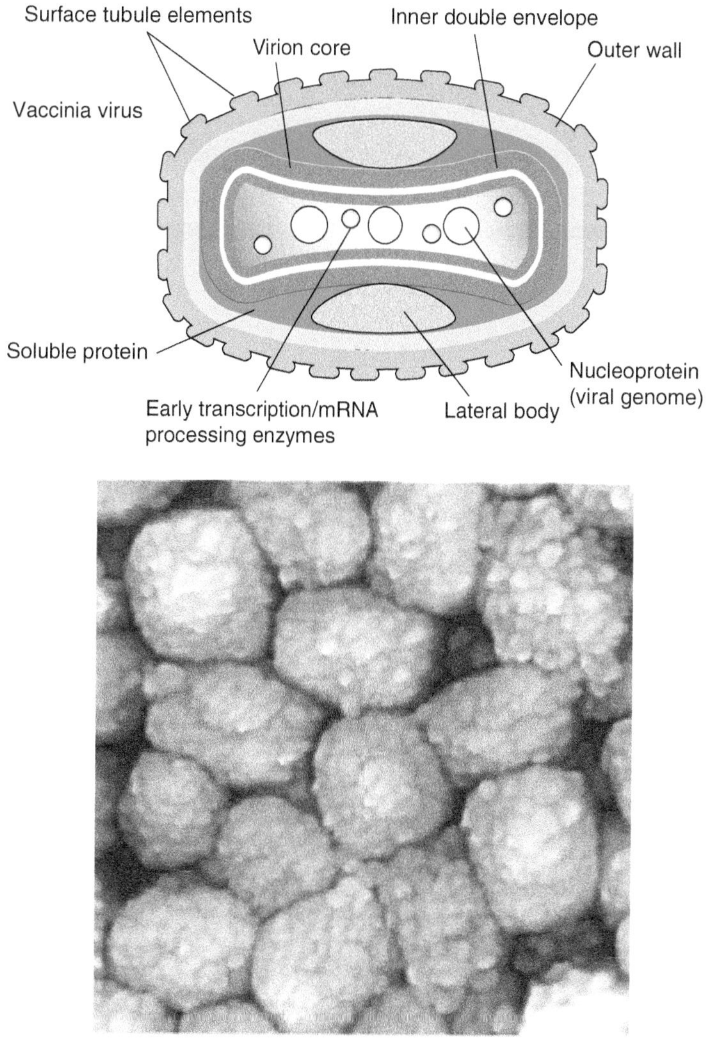

Figure 18.1 The vaccinia virus virion. The structure of poxviruses is the most complex known among the animal viruses, and rivals that of some bacterial ones. The particles are on the order of 400 nm in their longest dimension. The virion contains numerous enzymes involved in RNA expression from the viral genome concentrated as a nucleoprotein complex in the core. The lateral bodies have no known function. The inset shows an atomic force micrograph of mature vaccinia virions in which the lateral bodies are observable as "bulges" on the surface of the virion. *Source:* Courtesy of A. McPherson.

- mRNA poly(A) polymerase
- RNA helicase
- DNA helicase, ligase, and topoisomerase
- Protein kinase(s)

The poxvirus replication cycle

The replication cycle of a typical poxvirus is shown in Figure 18.2. Most details of this replication cycle were established by studying vaccinia. Interestingly, neither the ultimate origin of vaccinia or cowpox virus itself is known. The reservoir for cowpox virus is thought to be rodents, and the virus is unusual in its extremely broad host range. Vaccinia for vaccine production has been traditionally grown on the skin of horses, and may have originated as a horse virus. The high degree of conservation of core replication genes for all poxviruses means that the molecular details of their replication are very similar; this is not necessarily the case for the ways that the different types of orthopox viruses evade complex host defenses.

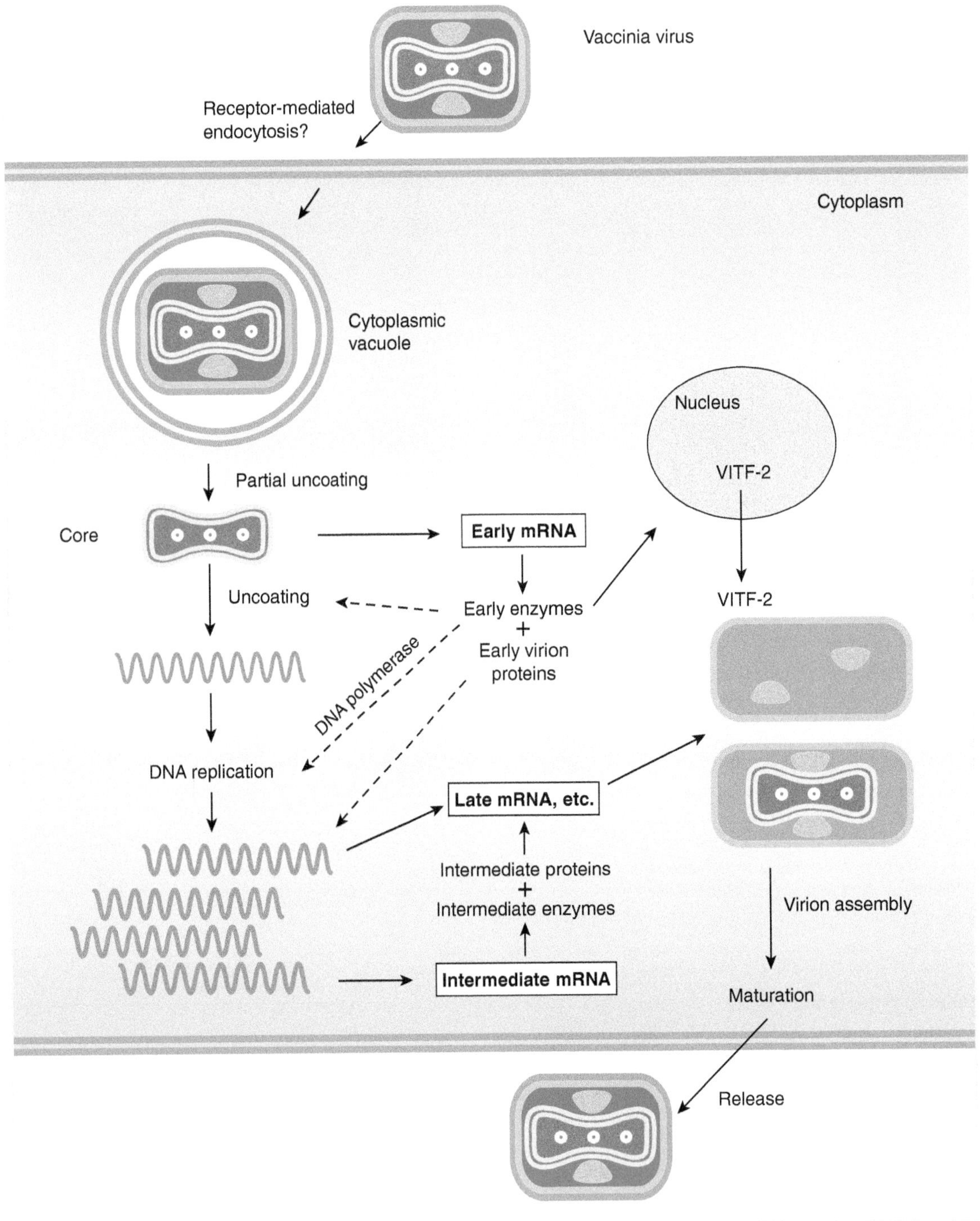

Figure 18.2 The replication cycle of vaccinia virus. Following viral attachment to cellular receptors, internalization is thought to be by receptor-mediated endocytosis. Virions are partially uncoated in the vesicles, and core particles are then released into the cytoplasm, where early mRNA synthesis and expression of early viral proteins occur. These proteins function to continue the uncoating of the core and to replicate viral DNA. Late mRNA expression from replicated genomes leads to expression of structural and other proteins involved with virus maturation. Viral gene expression and genome replication cease by approximately 6 hours after infection, but morphogenesis of the complex virion requires a further 14–16 hours. VITF = virus intermediate transcription factor

Viral replication can be separated into three specific temporal phases or stages in the infected cells, as was the productive replication cycle of other DNA viruses described in Chapters 16 and 17.

Early gene expression

Productive infection involves the virus interacting with specific receptors on the surface of susceptible cells, followed by virus entry into the cytoplasm, which is concomitant with *partial uncoating*. This partial uncoating results in the virion-associated transcription enzymes beginning expression of early viral mRNA, which is translated by the host's cellular translation machinery.

Partial uncoating can be accomplished *in vitro* by treating infectious virions in the laboratory with nonionic detergents such as NP-40 in the presence of a reducing agent such as mercaptoethanol (EtSH). The detergents solubilize lipid-associated membranes, and the treatment of pox virions with such reagents results in **core particles** that will transcribe viral DNA in the core into early mRNA as long as ribonucleoside triphosphates and metal ions are supplied. While not completely established experimentally, it is thought that the early enzymes expressed from the core accomplish complete uncoating of the infected cell's core particles. While the virus replicates in the cytoplasm, one early protein facilitates the transport of a cellular nuclear protein, VITF-2 (viral intermediate transcription factor), into the cytoplasm where it has a major role in expression of the next set of viral transcripts – the delayed-early or intermediate class.

The virus also expresses a number of early genes located toward the genome termini, which are clear homologs to cellular genes. These interfere with host response by inhibiting chemokine and/or cytokine responses, interferon, complement-mediated cell lysis, and apoptosis. Further, various poxviruses encode a gene, closely related to a cellular growth factor gene, which causes localized cellular proliferation – one of the characteristics of the distinct pock or pustule containing infectious virus that erupts on the skin, and (perhaps) providing further cells for continued virus replication.

Genome replication

Complete uncoating of the core particles and initiation of viral genome replication signal the end of the early period of infection, which (depending on the exact poxvirus) occurs during the first three to eight hours following infection. Poxvirus genome ranges in size from 134 kb to 260 kb, depending on the virus under study, but all viral genomes share some basic features. The dsDNA molecules have covalently closed ends, and with relatively long stretches of inverted terminal repeat sequences at the ends. As a result, the ends of double-stranded molecule form loops. Viral DNA replication (Figure 18.3) initiates at nicks near these closed ends.

After helicase unwinding, the 3′ end of the exposed molecule acts as a primer for extension, resulting in self-priming and recreation of two copies of the inverted sequences that allow loop formation. DNA synthesis then proceeds along the length of one strand, around the distal loop, and then along the other strand, creating a concatamer that is two genome lengths in size. Resolution of this structure by endonuclease and ligase produces two molecules of the poxvirus genome. It is clear, however, that poxvirus DNA replication does not require association of a specific origin-binding protein with viral DNA, and *any* circular DNA molecule present in the cytoplasm at the time of infection is replicated to high copy numbers. This is of great practical value, since recombination also takes place in the cytoplasm during DNA replication, and the system can be exploited to generate defined recombinant viruses that have a number of potential medical and research uses. Some promising applications of such viruses are described in Part V.

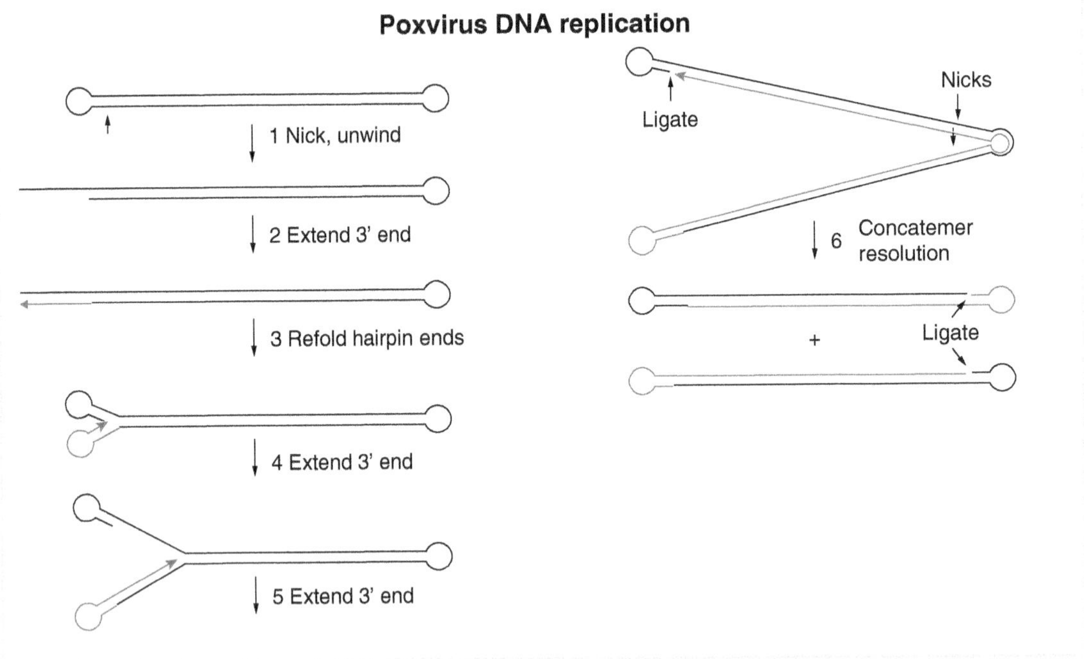

Figure 18.3 Replication of poxvirus DNA. The covalently closed, circular genome is "nicked" by an endonucleolytic cleavage. The exposed 3′ end of the DNA is extended by the viral DNA polymerase, using the opposite strand as a template. When the terminus is reached, each of the two strands can then fold into the hairpin structures, as shown. DNA polymerase then continues the extension of the 3′ end of the molecule until a complete copy has been produced. The two daughter strands are then released by endonucleolytic cleavage and finished by ligation.

Intermediate and late stages of replication

As with other DNA viruses we have examined, viral genome replication serves as an important dividing line in the kinetic cascade of viral gene expression. While early genes are expressed from core particles and parental genomes, intermediate and late genes are expressed from newly replicated DNA. During the intermediate stages of replication, a series of intermediate mRNAs are expressed by virtue of early proteins interacting with viral polymerase, VITF-2, and other factors. Some enzymes synthesized during the intermediate stage facilitate the transcription of late genes, which are translated into late proteins. These late proteins include a large number of virion proteins as well as the proteins required for the morphogenesis of complex pox virion. Virion-associated transcription enzymes required to initiate the next round of viral replication are also made at this time. Morphogenesis is a multistep process: First the inner core particles assemble and are enveloped with a double membrane derived from the Golgi apparatus. These mature into cytoplasmic virions, which can spread to neighboring cells through cellular junctions, and which can be released upon cell lysis.

Pathogenesis and history of poxvirus infections

Poxvirus infections are relatively unusual among the viruses presented in this text in that they cause a high rate of mortality during the natural course of spread in the host population. This is well illustrated by a consideration of smallpox in humans. This disease was a scourge for thousands of years, and occurred in two forms: variola major, with a mortality rate greater than 20% in immunologically naive populations, and variola minor, with lower mortality rates (2–5%). The high mortality rate is associated with the pattern of virus spread and pathogenesis in the host, as described in Chapter 4, Part I, but remember that in humans, the virus is spread by inhalation. While primary infection is in the lungs, virus travels through the host via viremia, leading to secondary epidermal infections resulting in the characteristic skin eruptions making pox virus available for further spread. Aerosols from lung lesions may also be an important factor in transmission. The formation of open sores on the skin leads to a high incidence of superinfection with opportunistic pathogens, as well as relatively nonspecific physiological responses to a disseminated infection of the skin.

The mode of spread of the virus in humans is facilitated by the fact that unlike almost all other animal viruses, pox virions are very resistant to inactivation by desiccation. Infectious virus can be recovered in contaminated clothing, bedding, housewares, and soils for significant periods following the resolution of infection in a particular individual. As noted, the pustules characteristic of pox virus infections are the result of expression of an early virus-encoded protein related to epidermal growth factor. Also, the virus is able to modulate host immunity by virtue of expressed viral proteins that sequester cytokines important in mediating the antiviral immune response, including tumor necrosis factor (TNF) and interferon-γ.

While the pathogenesis of the poxviruses results in their being very successful pathogens in a number of animal hosts, they carry an "Achilles heel." Like some RNA viruses, the poxviruses do not remain associated with the host after primary infection; whether or not the host is killed, the virus is cleared. Hosts that survive have permanent immunity to reinfection. Further, many animal poxviruses are immunologically closely related to human smallpox virus, and the viral proteins expressed by such viruses can induce immunity to smallpox in humans. Indeed, Jenner's original regularization and characterization of vaccination techniques described in Chapter 8, Part II, were based on common knowledge that dairy workers infected with the relatively mild cowpox virus were refractory (immune) to infection with human smallpox.

Is smallpox virus a potential biological terror weapon?

The characteristics of its pathogenesis and spread make smallpox uniquely subject to public health prevention measures, and a careful worldwide program of vaccination, disease reporting, and isolation of infected individuals led to eradication of smallpox from the human population at large in the 1970s. Currently, the only known smallpox viruses exist in public health laboratories in Russia and the United States.

While this success story is heartening, political instability in the Middle East as well as the degradation of Russian security measures with the collapse of the Soviet Union have made it very evident that declaration of full victory in the war against human suffering caused by poxviruses is premature. With the discontinuation of active vaccination campaigns, susceptibility to infection is now widespread throughout the world. Convincing scenarios have been discussed in which organized and trained terrorists could, for whatever motivation, penetrate within a large open population such as those in the developed world and instigate widespread outbreaks of infection. These would not need to lead to a mass epidemic to cause severe political and economic dislocations, since modern health care facilities are ill equipped to deal with the active disease and its containment. Further, mass vaccination would require large stocks of vaccine, and stocks have waned since victory against smallpox was declared.

The US government has stockpiled supplies of both vaccines and an antiviral agent designed to combat an outbreak of smallpox. The vaccine, ACAM2000, is derived from a plaque-purified vaccinia stock and has been tested extensively to show its efficacy in comparison to the classic vaccinia preparations. Tecovirimat is a maturation inhibitor that has been shown in preliminary studies to have both prophylactic and therapeutic potential. These two developments mitigate to some degree the concerns for the use of this virus as a weapon of terrorism.

Such considerations point, again, to the fact that no scientific or medical program of prevention is anywhere near as effective against a biological threat of this magnitude as is the absence of motivation to develop the threat in the first place. If the political efforts and international organization needed for the eradication of the stocks of smallpox in the world had been effective and complete, the problem would be greatly lessened, but still would not be solved, because there exists plenty of expertise available for the construction of virulent pathogens such as smallpox from its component genes. Again, the only real defense against potential biological terror weapons is for societies to ensure that there are no strong motivations for an organized effort by an actual or self-considered disenfranchised minority to resort to them. This would not, of course, guard against the threat of an occasional psychopath, but an isolated outbreak of a virulent disease in the face of an informed and empowered populace is no real threat.

REPLICATION OF "LARGE" DNA-CONTAINING BACTERIOPHAGES

As briefly outlined in Chapter 1, the study of three groups of DNA-containing bacterial viruses infecting *Escherichia coli* (the T-even, T-odd, and λ phages) has occupied a central and seminal place in the development of molecular biology and the functional understanding of gene expression and gene manipulation. Aspects of the replication and structure of these viruses are still studied for their own sake, and the λ phages are in general use as agents for molecular cloning of large DNA segments.

Components of large DNA-containing phage virions

Bacteriophages display a variety of structures, as briefly outlined in Chapter 5, Part II. While the size and complexity of specific structural features vary with the size of the viral genome and the nature of the virus, the phages discussed here all share similar structural features. The structure of bacteriophage T7, shown in Figure 18.4, illustrates some of these features. Notable structural elements include a complex icosahedral head containing the viral genome, a noncontractile tube or sheath for transmission of the viral genome into host bacteria, and a **base plate** and fiber structure for attachment of the phage to the host.

Replication of phage T7

The genome

Phage T7 has a linear, dsDNA genome 39 936 base pairs long. Its genetic map is shown in Figure 18.4. An interesting feature not seen in the other DNA viruses discussed herein is that the viral genes are "clustered" into specific regions that are expressed at specific times. As with the T-even phages discussed in this chapter, phage T7's DNA genome is terminally redundant – the 160 base pairs occurring at the beginning of the genome are directly repeated at the other end. This redundancy serves to allow the viral genome to circularize during replication so that no sequences are lost during the replication process (see Chapter 13).

Phage-controlled transcription

As with other DNA viruses, replication of T7 can be divided into temporal phases. The bacteriophage T7 transcription program is augmented by a phage-encoded RNA polymerase with a specificity different from that of the host cell. Infection is initiated by insertion of the genome into a host bacterial cell. The first molecular event occurring after the beginning of the insertion of the phage genome is transcription of a set of phage genes called the early genes. These are

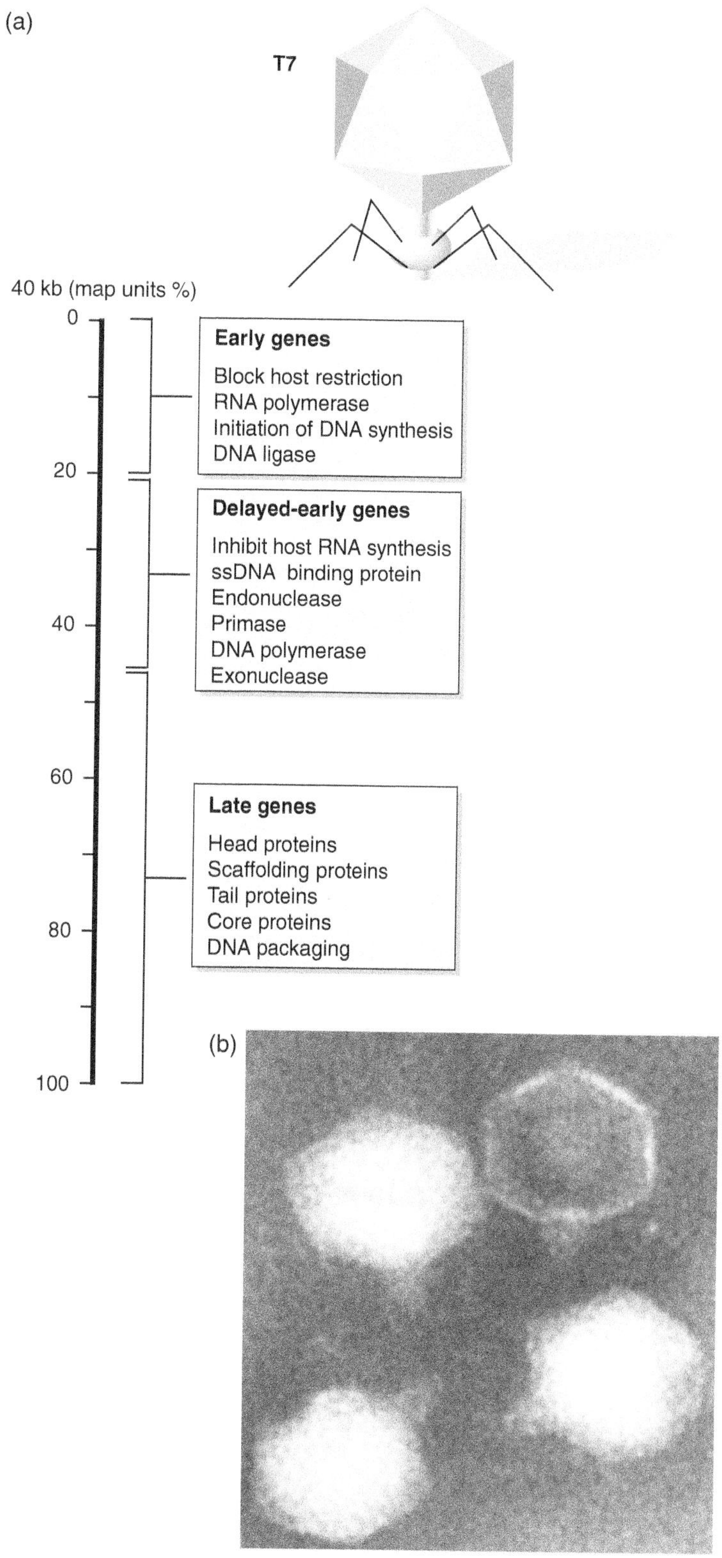

Figure 18.4 The structure and genetic map of T7 bacteriophage. (a) The 40-kb gene map shows that genes are clustered according to function, with those involved in the earliest stages of infection shown at the top. Transcription begins as the DNA is injected into the host, so the top portion of the genome must be injected first. The early genes include an RNA polymerase that transcribes later genes from the viral genome. (b) Negative-stained electron micrograph of purified T7 phage. *Source:* Courtesy of Ian Molineux.

generally equivalent to the immediate-early or pre-early genes expressed by adenoviruses and herpesviruses. Early transcription is carried out by host RNA polymerase as soon as the first 20% of the genome has entered the cell, and results in production of five mRNAs encoding phage proteins. Included in this set of phage genes is the T7-specific RNA polymerase, the enzyme that will carry out the balance of gene expression (delayed early and late) for the virus. T7 RNA polymerase differs from the host enzyme in the nucleotide sequence of the promoter regions that it recognizes. The phage RNA polymerase only recognizes promoter sequences found upstream of the delayed-early and late classes of T7 genes. Another immediate-early gene, encoding a protein kinase, catalyzes phosphorylation and inactivation of host RNA polymerase, effectively stopping transcription of host mRNAs. Thus, within a few minutes after infection, phage T7 takes over the host cell and converts it into a factory for production of new virus particles.

Further phage genome insertion occurs with 45% of the DNA now in the cell and the onset of the delayed-early phase, in which genes essentially equivalent in function to the early genes described for numerous DNA-containing animal viruses are expressed. These include genes encoding a T7 DNA polymerase that carries out replication of the viral genome. Late genes include genes encoding structural proteins for the phage capsid, as well as a gene required for cell lysis. The infectious cycle continues with the replication of phage DNA and assembly of progeny virus. This assembly process is generally described in Chapter 6, Part II.

The Practical Value of T7

Phage T7 provides another look at how a virus can completely take over a host cell and convert it into a factory for making progeny particles. In this case, rather than altering an existing host enzyme for transcription, the virus encodes a completely new RNA polymerase that recognizes an entirely different set of promoter sequences. So effective is this system that virtually no host mRNA is transcribed. Because of this specificity, the T7 polymerase and its promoter sequences have become the basis for several expression vectors used in recombinant DNA technology.

T4 bacteriophage: the basic model for all DNA viruses

The study of bacteriophage T4, along with other related T-even phages, occupies a unique place in the annals of molecular biology and molecular genetics. The speed and ease of manipulation and the facility of doing genetics with mixed-phage infections, plus the convenience of using *E. coli* as a host, made these viruses a major subject of experimental investigation from the 1930s through the mid-1960s. Many "firsts" were recorded in the study of T-even phages, and even now, their study continues to offer new insights and concepts, especially with regard to understanding large-scale macromolecular structures. Further, gene products encoded by phage T4 provide a valuable resource for biotechnology and genetic manipulation.

The T4 genome

The genome of T4 has some unique structural features. A complete genetic map and model of phage structure are shown in Figure 18.5. The map is really no more complex than that of herpes simplex virus (HSV), whose genome is about 90% the size of T4 (150 versus 168 kb). One obvious difference is that nearly 50% of T4's genetic complexity is devoted to encoding proteins of the complex capsid; this value is nearer 30% for HSV. Also, for obvious reasons, T4 does not need to encode a large number of genes important for dealing with host immune defenses. Although the viral genome is a linear, dsDNA molecule, the genetic map is circular. This results from the DNA being terminally redundant – a situation similar to that seen for phage T7. As with other linear DNA virus genomes, terminal redundancy allows the genome to become circular by recombination.

The T4 genome is circularly permuted. Circular permutation means that the starting point in the linear genome (one end of the molecule) differs for various members of a particular virus

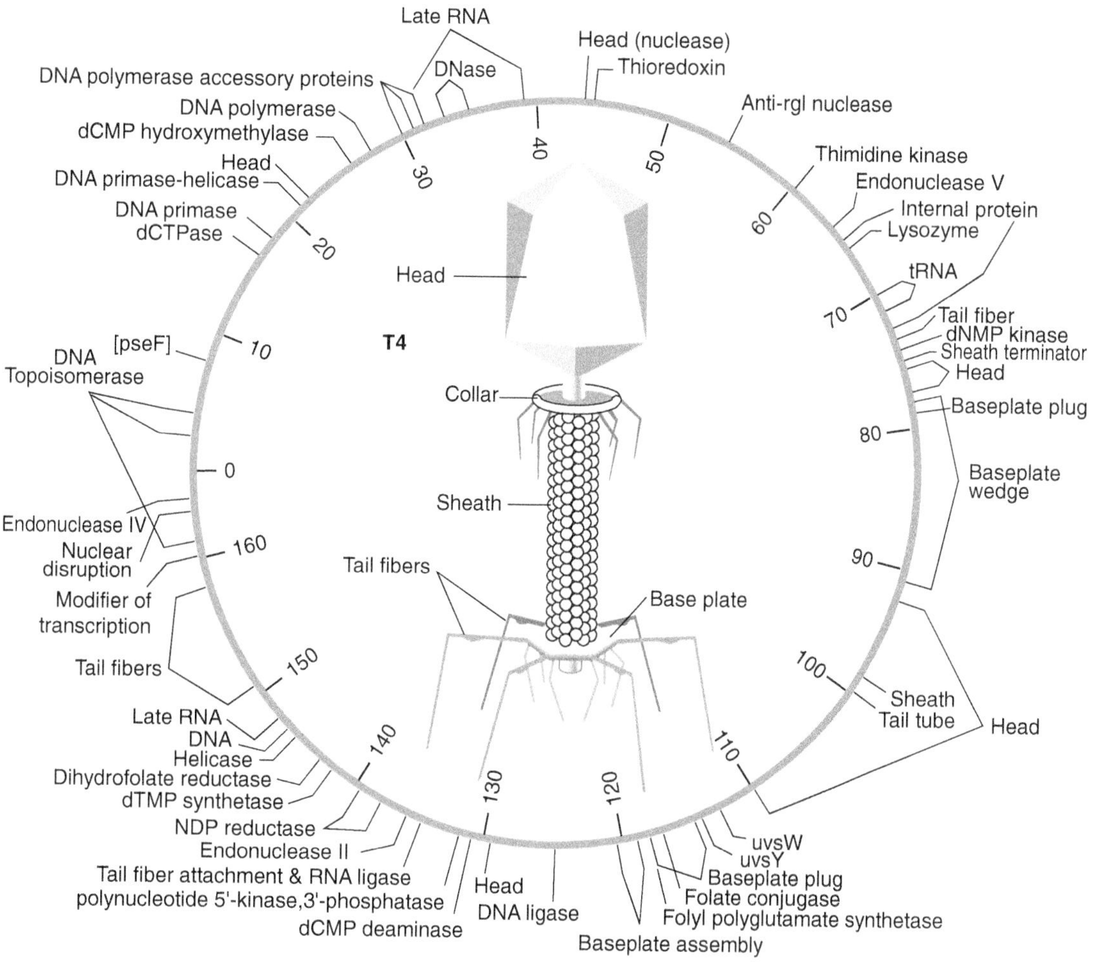

Figure 18.5 The genetic map and structure of bacteriophage T4. By convention, this map is divided into kilobases instead of map units. Since the viral DNA ends are redundant, the starting point shown is entirely arbitrary. Considerably larger than T7, the phage particles have similar head shapes, but T4 has a contractile sheath and a base plate important in attachment (see Chapter 6, Part II). The viral genetic map is as complex as, but not more so than, that of HSV shown in Figure 17.2. Note that unlike HSV but similar to T7, many genetic elements are functionally clustered.

population. Essentially, all possible starting points in the linear sequence are represented, as shown diagrammatically in Figure 18.6. Circular permutation is a consequence of the viral genome being replicated by a complex rolling circle mechanism (much like that of HSV discussed in Chapter 18). Each phage head encapsidates one full genome length of DNA *plus a bit more.* The generation of such circularly permuted genomes also means that there is no unique packaging signal for T4 DNA. A possible mechanism for this packaging of a "genome-plus" piece of DNA is shown in Figure 18.6.

In addition to the distinctive genome structure of these viruses, the base composition of the DNA differs from that of the host cell. T-even phage DNA contains the unusual base 5-hydroxymethyl cytosine (5′-OHMeC) in place of cytosine. The precursor triphosphate, 5′-OHMeCTP, is synthesized in the cell by phage-specific enzymes. In fact, T4 hydroxymethylase, identified by Seymour Cohen in 1957, was the first viral-encoded enzyme ever to be described. T-even phage DNA is further modified in that a portion of the 5′-OHMeC base has one or two glucose residues covalently linked to the hydroxymethyl residue. This glycosylation has a role in the virus abrogation of host restriction defenses.

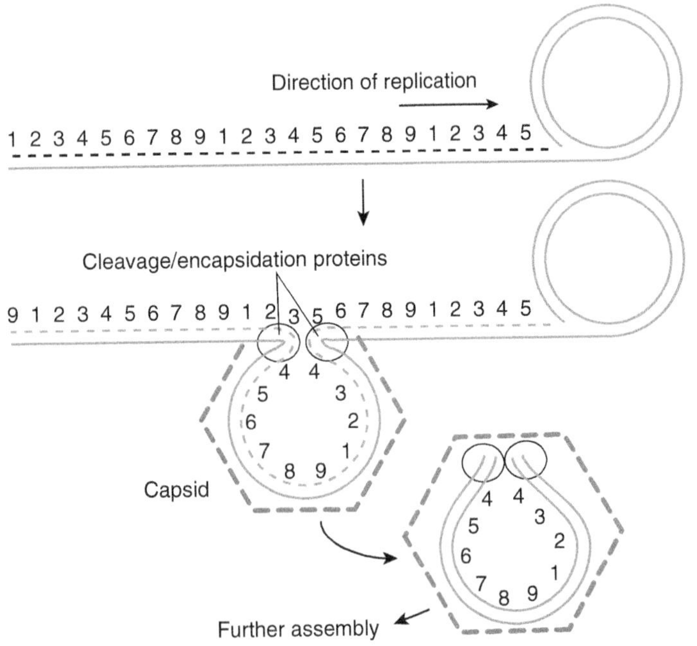

Figure 18.6 Rolling circle replication and packaging of phage T4 DNA. The process has similarities to that of HSV (shown in Figure 18.5), but there is no specific packaging signal in the viral genome. Packaging begins at random sites, and once the phage head is filled with an amount of DNA that is equivalent to 110% or so of the full genome, the ends are cleaved and packaging is completed. This results in the encapsidated DNA ends being redundant. This redundancy leads to this linear DNA molecule producing a circular genetic map, as shown in Figure 18.4.

Regulated gene expression during T4 replication

As is usual in DNA virus replication, the transcription of T-even phage genes is temporally controlled during infection; thus, expression of T4 genes varies with time after infection and can be divided into four stages: immediate-early, delayed-early, quasi-late, and late (or strict-late). This temporally regulated pattern is shown in Figure 18.7. The transcriptional switching in T4 infection involves use of the bacterial host's DNA-dependent RNA polymerase; this mechanism is distinct from the encoding of a novel RNA transcription enzyme like T7. During T4 infection, the specificity of RNA polymerase for particular promoters is altered by expression of phage-specific σ factors, and by modification of the core enzyme by phage-encoded enzymes. The role of such factors in bacterial RNA polymerase specificity is outlined in Chapter 13.

T4 gene expression begins with transcription of the immediate-early genes. This takes place utilizing the host's RNA polymerase and sigma (σ) factors. The transition to delayed-early gene expression involves recognition by the host enzyme of certain phage promoter sequences and modification of host enzyme, possibly by the phage-catalyzed covalent addition of adenosine diphosphate (ADP)-ribosyl groups to each of the RNA polymerase alpha (α) subunits. This change occurs in two steps: The first is catalyzed by a phage gene (*gp alt*) that enters the cell along with viral DNA, and the second is catalyzed by a phage gene (*gp mod*) that is itself the product of immediate-early expression. ADP-ribosylation may not be required for the transitions, since double mutants in both *gp alt* and *gp mod* are able to carry out delayed-early gene expression in a normal manner.

The two other phases of viral gene expression – quasi-late and strict-late – require replicating DNA structures and involve sequential replacement of host σ factors by phage proteins.

Capsid maturation and release

The patterns of T4 maturation are known in exquisite detail. A broad outline is shown in Figure 18.8. In essence, the process of head assembly and DNA encapsidation is very similar to that of the other large DNA viruses described herein. After filling of the head, the other components of the complex virion, which have preassembled to form subassemblies, come together to form the

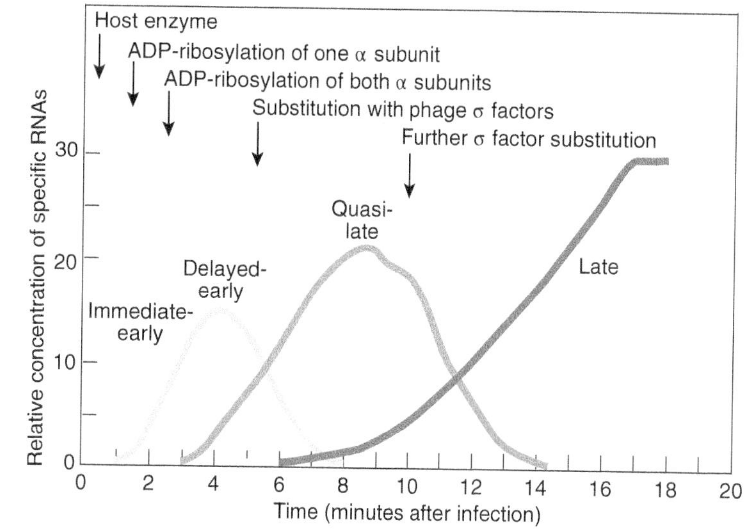

Figure 18.7 Time of appearance of various functions encoded by T4 bacteriophage. Each class of transcripts is transcribed by modified *E. coli* RNA polymerase. The modifications are sequential: First one and then the second alpha (α) subunit of the RNA polymerase are modified by covalent linkage of an ADP molecule; then various phage-encoded proteins displace first the host sigma (σ) factor and then one another to generate enzymes of altered specificity.

complete particle. As complex as it is, all these steps are simple biochemical reactions driven by mass action, and can be mimicked in a test tube. Mature phage is released from the infected cell by expression of a late lysozyme that disrupts the bacterial cell wall, releasing virus.

Replication of phage λ: a "Simple" model for latency and reactivation

Many eukaryotic viruses, especially those with DNA genomes, have evolved complex mechanisms for remaining associated with the individual that they have infected long after the disease caused by the initial infection is resolved. A good example of this is found with the latent phase of infection by herpesviruses; another is the limited transformation of cells induced by papillomavirus infections of epithelial cells.

The mechanism for continued association between virus and host varies with the virus and host in question, but nowhere is it more fully described than in the replication cycle of bacteriophage λ. In phage λ, one encounters a virus with two very different outcomes of infection that each have very different consequences for the host bacterial cell – either lysis or lysogeny.

Productive or lytic infection entails expression of the phage genes required for replication of phage DNA, synthesis of phage structural proteins, assembly of viral particles, and lysis of the host cell. Although different in some details from the process outlined for T7 and T4 phages, the process is essentially the same.

Lysogeny, on the other hand, is characterized by the virus suppressing its own vegetative DNA replication. Instead, the viral genome becomes integrated into the bacterial chromosome, where the virus exists in a lysogenic or latent state until a set of metabolic stimuli reinitiates productive infection. When integrated, the phage λ genome is called a **prophage**. It is clear that there are phenomenological parallels with herpesvirus and papillomavirus latency, as well as with retroviruses discussed in Chapter 19, but the details are unique to bacteriophage λ.

Infection of a bacterial cell by phage λ leads to a competition between the mutually exclusive processes leading to lytic or lysogenic phases of infection, and the outcome is just a matter of which process happens first. Both outcomes result from the action of proteins encoded by a small subset of phage genes expressed immediately upon infection. Their interaction is complex; indeed, the biochemical "decision" between lysogeny and lytic replication is among the most complex biochemical control pathways known. The original phenomenon of latency was discovered in the 1920s, and only after 60 years or so of development of ever-more sophisticated biochemical, physical,

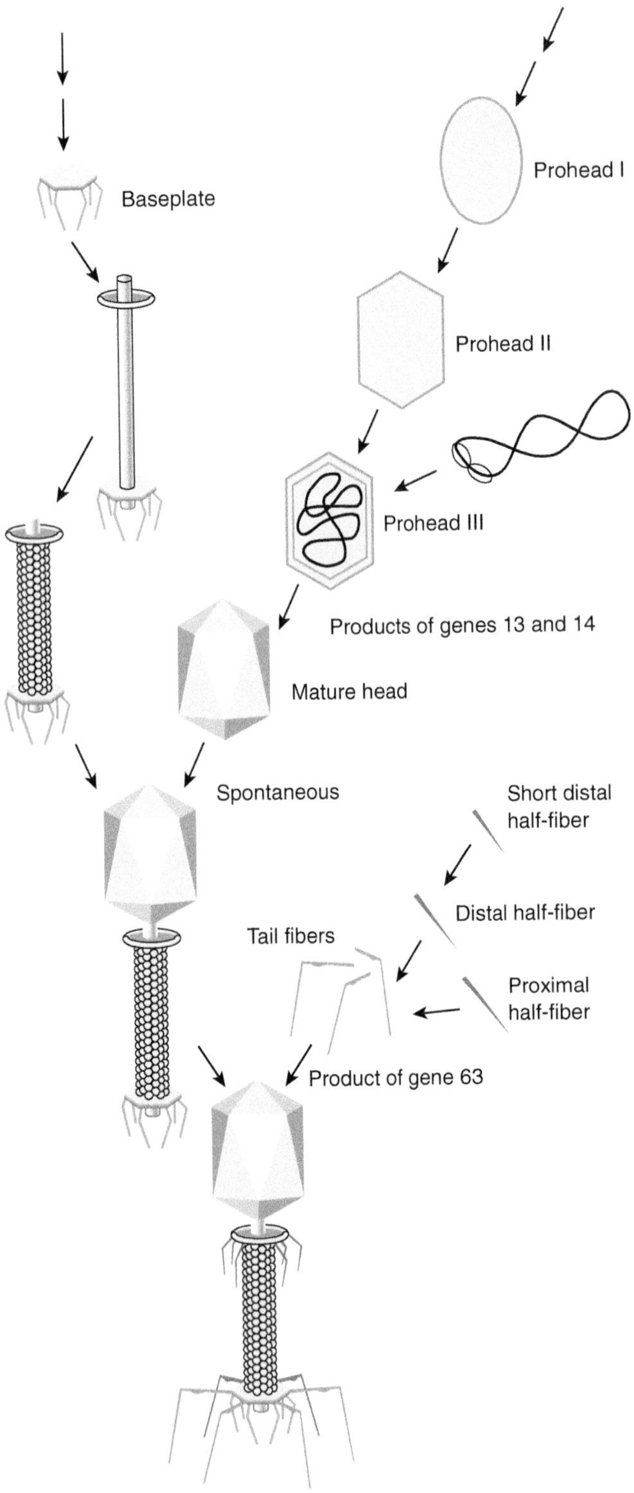

Figure 18.8 The assembly of T4 bacteriophage. Note that assembly of the phage head is similar to the process seen with HSV. Other components of the virion are assembled as "subassemblies" brought together sequentially to form the full phage. *Source:* Modified from Wood, W. B. 1979. Bacteriophage T4 assembly and the morphogenesis of subcellular structure. *Harvey Lectures*, Series 73: 203–223.

and genetic analyses have the details been fully worked out. It is not too much to say that the successful melding of biochemistry, molecular biology, classic genetics, and molecular genetics was needed to fully decipher the complex and elegant pathways involved in the biochemical decision made by phage λ each time it initiates an infection in *E. coli*. This melding stands both as a triumph of modern biology and as a model for understanding all biological processes.

The phage λ genome

The phage λ genome, whose map is shown in Figure 18.9, is a linear dsDNA molecule 48.5 kb in size. Unlike either T7 or the T-even phages, there is no terminal redundancy. Despite this, the genetic map is represented as a circle because λ DNA can become circular following infection. The genome has a complementary stretch of single-stranded bases at each end – the *cos* (cohesive ends) sites. These sites can anneal in the phage head to form a noncovalently bonded circle, which is converted to a covalently closed molecule by ligation shortly after infection. They also serve as packaging signals for replicated viral DNA.

Phage λ gene expression immediately after infection

Upon injection of viral DNA, unmodified *E. coli* RNA polymerase can recognize two λ promoters and can transcribe two mRNAs from the viral DNA, as shown in Figure 18.9. This is the immediate-early phase of gene expression. The rightward promoter, termed P_R, transcribes mRNA encoding the cro protein, which is so named because of its control function (the acronym stands for "control of repressor and other things"). The leftward promoter, called P_L1, controls expression of mRNA for the N protein, which acts to modulate the utilization of transcription termination signals (see Chapter 13, Part III). The promoters in this case are not specifically named in Figure 18.9.

The earliest stage of transcription from these two promoters terminates at two sites, both of which have ρ-dependent termination signals. Much of the early regulation events involve the action of competing phage genes on these termination signals to suppress their activity. If one set is suppressed, transcription proceeds through the signal and a further set of transcripts is expressed; if the other set is effectively suppressed, then a *second* set of extended transcripts encoding competing functions is expressed.

The action of cro: lytic growth The λ cro protein, under appropriate metabolic conditions of growth, can repress transcription of the mRNA encoding another phage regulatory protein called the λ CI repressor by binding to control sites just upstream of these two promoters. The

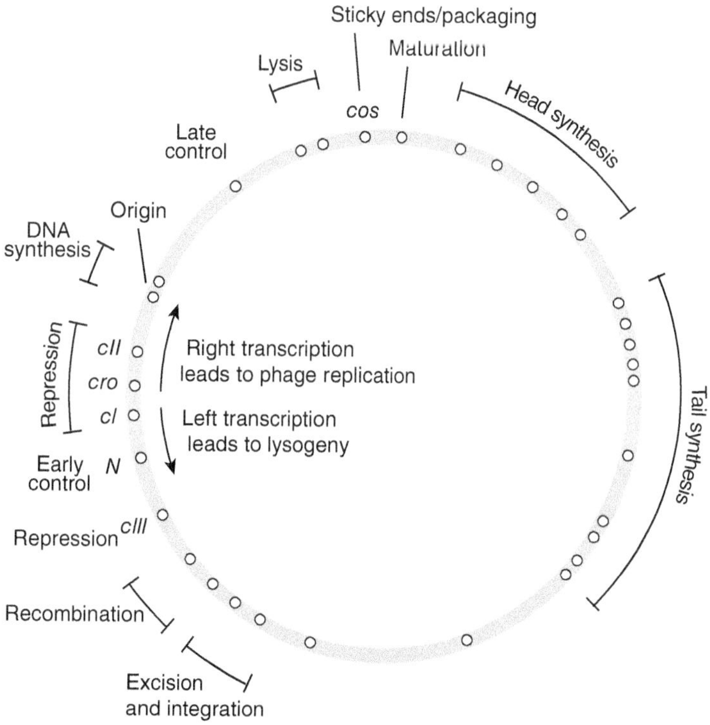

Figure 18.9 The bacteriophage λ genetic map. Specific clustered functions are indicated. The primary decision of whether to replicate or integrate involves the single question of whether leftward or rightward transcription occurs first. This process is entirely stochastic (random). If transcription takes place to the left, cI repressor is expressed and blocks lytic replication. At the same time, integrase and recombination functions lead to the phage DNA being integrated into the host bacteria's genome.

C1 repressor is involved in establishing lysogeny, and its repression serves as a "green light" for lytic growth, especially when sufficient N protein has been made. The lytic genes include DNA replication and capsid proteins. The result of this rather Byzantine process is production of progeny phage and lysis of the cell.

Modulating the activity of the N protein: priming the cell for lysogeny Expression of high levels of the λ phage N protein leads to expression of a set of delayed-early genes via the interaction with a host protein called N-utilizing substance A (nusA). This interaction allows RNA polymerase to read through the two immediate-early termination sites, producing longer transcripts in both directions. The transcripts expressed from rightward transcription encode two further phage proteins, cII and cIII, which together enhance the expression of cI, the λ repressor.

Action of cI, cII, and cIII: establishment of lysogeny Extension of leftward transcription as well as additional rightward transcription will eventually lead to onset of lytic-phase gene expression. Essentially simultaneous with the earliest events on this path, however, phage λ cII and cIII proteins can act in concert to stimulate transcription of mRNA for the cI repressor. Both cII and cIII are encoded in the mRNA that begins at P_R and continues through T_R1. This "read-through" expression requires N protein action, and the cII protein stimulates transcription of cI mRNA beginning at P_{RE} (promoter for repressor establishment), while the cIII protein functions to stabilize cII against degradation by inhibiting host cell protease.

The cI repressor serves two roles: (i) It represses transcription of cro mRNA from P_R and of N mRNA from P_L1, and (ii) it stimulates transcription of its own mRNA. Thus, it blocks expression of genes required to initiate the lytic cycle. Notice that by blocking cro and N synthesis, cI also blocks synthesis of cII and cIII. Therefore, *all subsequent transcription of cI mRNA during the lysogenic* phase takes place from P_{RM} (promoter for repressor maintenance). Recall that promoter switching in response to repression by viral products is seen in the replication of adenovirus, discussed in Chapter 16.

Integration of λ DNA: generation of the prophage Lysogeny is established with λ integrase action, which catalyzes phage genome recombination into a specific site in the *E. coli* chromosome. The phage genome is then found as a linear sequence within host DNA, and the host cell is now called a λ *lysogen*. The name comes from the fact that while phage replication is generally repressed, occasionally lytic growth can by triggered; thus, the bacteria harboring the prophage can give rise to lysis. The only phage gene expressed in a λ lysogen is cI, and in most lysogenic cells, phage DNA remains stably integrated, replicating along with the cellular chromosome.

Biochemistry of the decision between lytic and lysogenic infection in E. coli

Competition for binding by cro and cI at the operator OR Both cro and cI bind to the operator OR as dimers. The three binding sites in OR are called OR1, OR2, and OR3, and are shown in Figure 18.10. The affinity of these sites differs for each protein dimer; cro repressor binds in the order OR1, OR2, and OR3. Binding to OR3 effectively blocks transcription of cI from PRM. Conversely, cI binds in the reverse order of OR3, OR2, and OR1. The cI protein's binding to OR1 blocks transcription of cro mRNA from PR. The additional binding of cI to OR2 acts to stimulate transcription of cI mRNA from PRM. Thus, as noted, cI has the dual ability to repress cro transcription as well as to stimulate transcription of its own mRNA.

This competitive binding and competing repression of the divergent promoters P_R and P_{RM} result in a control system that is quite sensitive to relative concentrations of cro and cI. During the early stages of infection, the rates of synthesis and degradation of these two proteins will dictate which pathway – cro dominated (lytic) or cI dominated (lysogenic) – will be followed. The variety of metabolic factors controlling these rates are described further here.

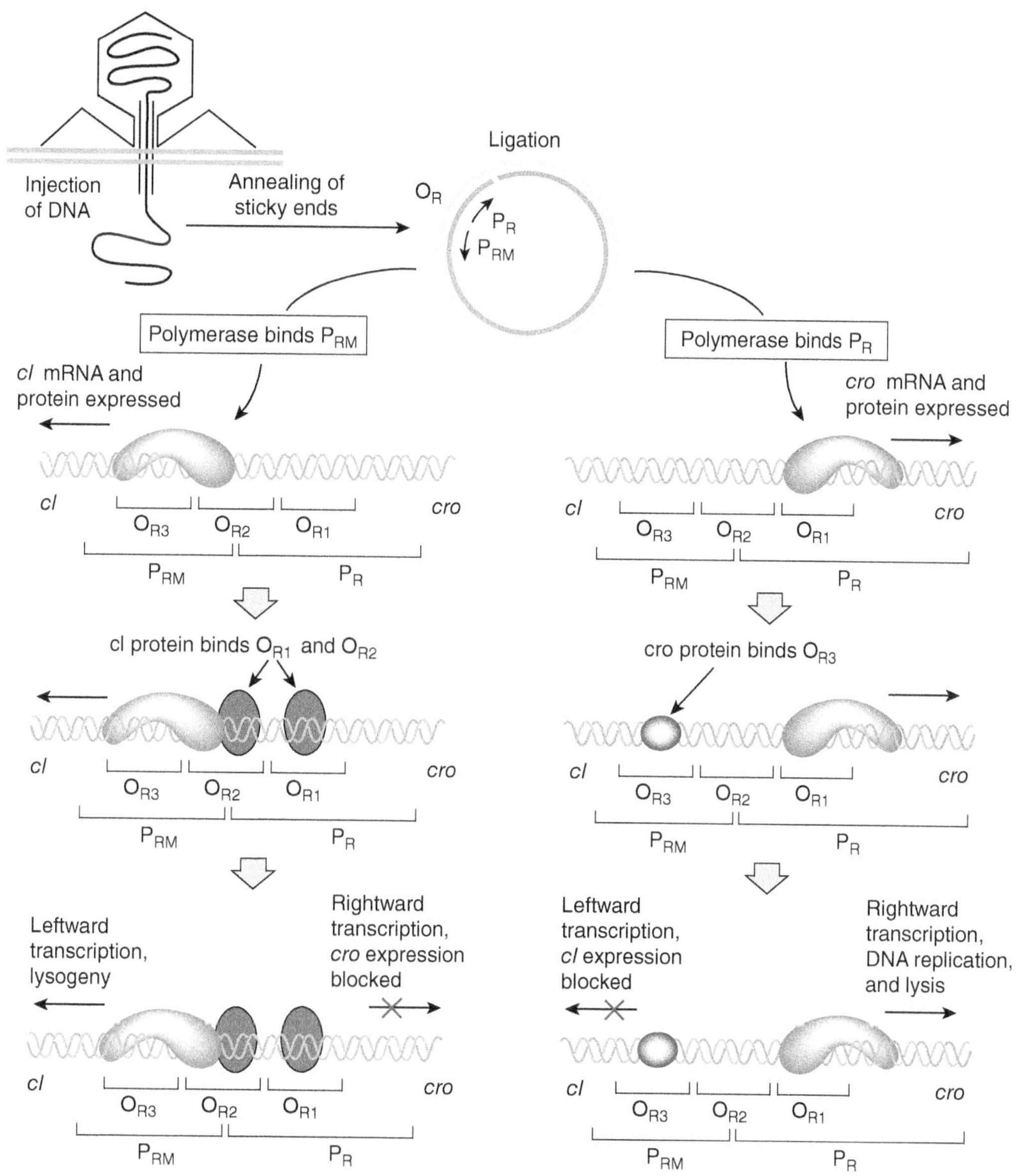

Figure 18.10 The earliest events in the infection of a bacteria by phage λ. Details of the biochemical decision concerning lysogenic or lytic replication are outlined here and described in the text. *Source:* Based on Ptashne, M. (1986). *A Genetic Switch: Gene Control and Phage λ.* Palo Alto, CA: Blackwell Science and Cell Press, especially Chapters 2 and 3.

Factors affecting the lytic/lysogenic "decision" The relative transcription rate of mRNAs for the two critical phage proteins cro and cI is altered by a variety of metabolic conditions that a lysogenic cell may encounter. The strategy of viral gene expression is such that lysogeny will occur in a healthy, rapidly growing cell. When, however, the cell encounters changes that threaten survival of a particular cell such as DNA damage or starvation, λ phage "jumps ship" by inducing lytic replication – the resulting phage particles may survive until a more fruitful time for bacterial growth and lysogeny.

Stability of the cI protein itself can determine the balance between the two competing repressors. The cI protein can be proteolytically cleaved and inactivated by the host's recA protease, which is induced by DNA damage. In the so-called SOS repair system, the recA protease destroys cellular repressor proteins, resulting in expression of a series of bacterial enzymes involved in repair of DNA damage. This cellular response also results in the destruction of cI and induction of the lytic phage pathway in a λ lysogen. Indeed, exposure to ultraviolet light has long been a favored method for the induction of lytic infection in lysogenic strains of *E. coli*.

Action of the cellular protease HflA is sensitive to the cell's nutritional state as moderated by the catabolite repression system. At low glucose levels in the cell (i.e., when metabolic activity is slow), the cell generates high levels of 3′,5′-cyclic adenosine monophosphate (cAMP) as an intracellular signal of metabolic stress. Indeed, similar signals are used in eukaryotic cells in response to a number of extracellular stimuli. At high cAMP concentrations, activity of the HflA protease increases. The result is an increased inactivation of cII and a concomitant decrease in cI synthesis during the early phase of infection.

A GROUP OF ALGAL VIRUSES SHARES FEATURES OF ITS GENOME STRUCTURE WITH POXVIRUSES AND BACTERIOPHAGES

As mentioned, there is a strong resemblance between aspects of the capsid structure and assembly of herpesviruses and T-even bacteriophages. Other viruses of the animal, bacterial, and plant kingdoms also share certain details of structural and replication proteins. This suggests either that they form a large superfamily (see Chapter 1), or that the genes in question are all derived from a common source or both (Figure 18.11).

Viruses of vascular plants are subject to a strict limitation of genome size because of their problems of penetrating the cellulose cell wall of such plants (see Chapter 17). This limitation does not apply to viruses of "lower" plants such as algae, however. Recently *Paramecium bursaria chlorella-1 virus* (PBCV-1), a virus that infects a type of algae that often lives symbiotically with a species of paramecium, was extensively characterized and shown to have some suggestive similarities to large DNA-containing phages, poxviruses, and other large DNA viruses.

PBCV-1 has the appearance of an iridovirus (see Chapter 5, Part II). It has a genome of 330 kb, which makes it among the largest viruses known. The genome is linear DNA with closed ends like the genomes of poxviruses. The virus does not infect its algal host like an animal virus; rather, the virion attaches to the cell's surface and the DNA is injected into the cytoplasm (a process analogous to a number of bacteriophages, including those described in this chapter). Also, similar to bacterial viruses, the PBCV-1 genome is extensively methylated. This methylation is in response to the presence of restriction enzymes of limited specificity that are encoded by its host. The ability of algae to encode such enzymes is a rarity among eukaryotes and is indicative of the degree of coevolution between virus and host.

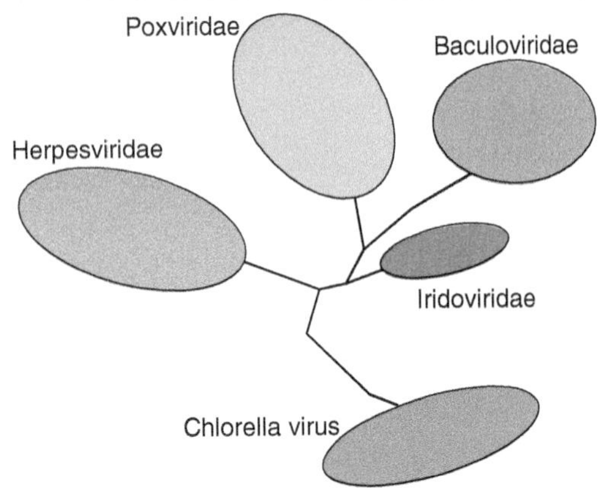

Figure 18.11 A phylogenetic tree of selected large-DNA-containing virus families based upon sequence divergence in conserved regions of DNA polymerase genes. Compare with Figure 1.2. *Source:* Based on figure based upon analyses described in the Phycodnaviridae section of *Virus Taxonomy, Eighth Report* (Fauquet et al., eds.). Amsterdam: Elsevier, 2005.

This virus is found at concentrations as high as 4×10^4 PFU/ml in freshwater throughout the United States, China, and probably elsewhere in the world. The virus can only infect free host cells, and since it cannot infect its algal host when the latter is symbiotically living with the paramecium, it is a good guess that existence of the virus is a strong selective pressure toward the symbiotic relationship. This virus represents a good example of the extensive coevolutionary interaction between viruses and their hosts throughout many ecosystems, and the importance of viruses in forming those ecosystems.

QUESTIONS FOR CHAPTER 18

1 *E. coli* K12 is a λ lysogen, meaning that the λ bacteriophage genome has been stably incorporated into the genome of the host cell.

(a) In this cell, which viral protein, if any, will you find being expressed?

(b) When this cell is irradiated with ultraviolet light, the resulting damage to the cellular DNA induces the SOS response. This system results in the proteolytic destruction of several repressor proteins, including λ cI. What is the effect of this treatment on expression of the λ genome?

2 You plan to carry out an experiment in which you infect *E. coli* cells with bacteriophage T4. You have a 10-ml culture of cells containing 3×10^8 cells/ml. You have a stock of phage T4 containing 10^{10} PFU/ml. To start the infection, you add 0.3 ml of this virus stock to the 10-ml culture of cells.

(a) What is the multiplicity of infection?

(b) If phage T4 normally produces about 200 virus particles per infected cell, what will be the total yield of virus from this infection *at the end of one cycle of virus growth?*

(c) You repeat the experiment with four identical *E. coli* cultures. To three cultures you add nalidixic acid, an inhibitor of DNA synthesis, at the times indicated in the table below. The fourth culture receives no inhibitor and is a control. Predict the results of this experiment by completing the following table. (The entire life cycle of bacteriophage T4 takes 20 minutes.) Use a "+" to indicate normal function or activity and a "−" to indicate inhibition.

Time of Addition of Nalidixic Acid	Phage DNA Synthesis	Immediate-Early Gene Expression	Yield of Progeny per Cell
Control (no inhibitor)			
0 minutes			
5 minutes			
18 minutes			

3 Which of the following treatments will inhibit the *complete* expression of immediate-early genes of the bacteriophage T4 during infection of an *E. coli* host cell? Use "+" to indicate expression and "−" to indicate inhibition of expression.

Treatment	Immediate-Early mRNA Expression
Rifampicin (an inhibitor of host RNA polymerase)	
Nalidixic acid (an inhibitor of DNA replication)	
A mutation that inactivates the phage DNA polymerase	
A mutation that inactivates the phage lytic enzyme	
Chloramphenicol (an inhibitor of protein synthesis)	

Continued

4 The following table lists a series of mutants of bacteriophage T4. Predict the properties of infection of a host cell by each of these mutants with respect to expression of each class of genes shown in the table. Indicate "+" if the class *is* expressed and "–" if the class is *not* expressed for each mutant.

Phage Mutant	Immediate-Early	Delayed-Early	Quasi-Late	Late
Control (wild type)	+	+	+	+
A phage mutant that cannot catalyze ADP-ribosylation				
A nonfunctional mutant of the phage DNA polymerase				
A nonfunctional mutant of the phage protein gp55				
A nonfunctional mutant of the phage receptor-binding proteins on the tail fibers (infected under permissive conditions and then changed to nonpermissive conditions)				

5 In an experiment with bacteriophage T7, you have isolated mutants that are defective in specific phage genes. In each case, a nonfunctional version of the phage protein is produced. Indicate by "+" or "–" which of the classes of T7 genes are transcribed in the case of cells infected with each of the indicated phage mutants.

T7 Phage	Early (Class I)	Delayed-Early (Class II)	Late (Class III)
Control (wild type)	+	+	+
T7 RNA polymerase mutant			
T7 DNA polymerase mutant			

6 Bacteriophage T7 has a linear, dsDNA genome. Gene expression of this phage after infection of its *E. coli* host cell occurs in three different phases. Genes that are expressed early are class I, genes that are expressed in a delayed-early fashion are class II, and genes that are expressed late are class III.

(a) For each of the T7 functions listed in the following table, identify into which class of genes they are most likely to be classified: class I, class II, or class III.

(b) Suppose that a cell has been infected with bacteriophage T7 and is in the middle of delayed-early (class II) gene expression. At this point you add the drug rifampicin, which is an inhibitor of the host RNA polymerase. What will be the effect of this treatment on the expression of bacteriophage T7 genes in this cell?

T7 Function	Class
Head capsid protein	
RNA polymerase	
DNA polymerase	

7 Bacteriophage λ infects *E. coli* and can follow one of two pathways: lytic or lysogenic. Several viral genetic locations are involved in these pathways. Assume that you have the phage mutants shown in the table below. Predict which of the pathways, if any, each mutant can follow. Indicate your prediction with a "Yes" or a "No." In each case, assume that the phage enters the cell under conditions where the mutant phenotype will be expressed.

Mutant	Lytic?	Lysogenic?
Wild type	Yes	
N-minus		
CII-minus		
Deletion of the P_{RM} region		
CIII-minus		
Mutated O_R3 that fails to bind cro		
Deletion of the cro gene		

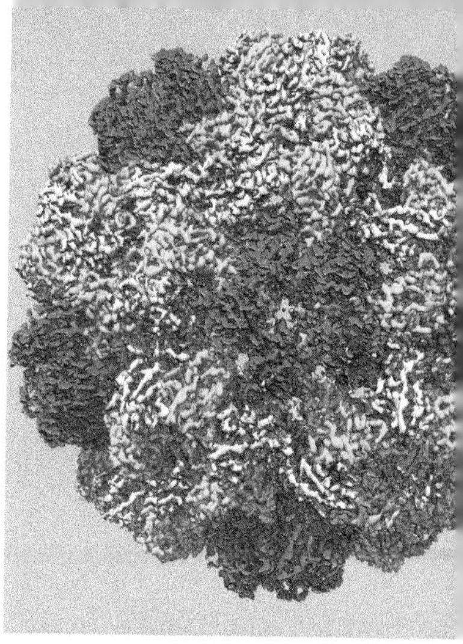

Retroviruses: Converting RNA to DNA

CHAPTER 19

A survey of the nuclear-replicating DNA viruses demonstrates many diverse evolutionary adaptations to the host's immune response and host defenses against virus infection. No matter what the complexity or details of the actual productive replication cycle are, viruses that can establish and maintain stable associations with their hosts following a primary infection have a great survival advantage. The lysogenic phase of infection by phage λ discussed in Chapter 18 is an excellent example of the genetic complexities a virus can utilize to avoid the chancy cycle of

Basic Virology, Fourth Edition. Martinez "Marty" Hewlett, David Camerini, and David C. Bloom.

maturation followed by the random process of establishing infection in a novel host. If able to stably associate with its host cell, a virus will need only occasional breaching of the environmental barriers between individual host cells or host organisms. Further, once infection is established, the host serves as a continuing source of infectious virus.

To stay associated with the host cell, retroviruses use the strategy of λ phage – they integrate their genomes into that of the host. Thus, they become, for all intents and purposes, cellular genes. Retroviruses accomplish their strategy of attack from within with only a few genes. Their survival strategy requires great specialization, but the ability of retroviruses to replicate and stay associated with the host has been a profoundly successful adaptation. This event occurred very early in the history of life. There is evidence of retroviruses or genetic elements related to them in all eukaryotic organisms and in some bacteria.

The retroviruses (and some close relatives) owe their uniqueness to one important ability: *the conversion of genomic viral RNA into cellular DNA*. The key to this ability lies in the activity of one enzyme, **reverse transcriptase (RT or Pol)**. The RT encoded in all retroviruses and retrovirus-like genetic elements that utilize it shares much similarity in structure and function. As noted in Chapter 1, RT is related in structure and mechanism to telomerase, a critical eukaryotic cellular enzymatic activity. Telomerase is absolutely required for accurate replication of chromosomal DNA in eukaryotic cells, and therefore has been in existence as long as such cells. The occurrence of retroviruses in all eukaryotes suggests that RT has been present for as long.

RETROVIRUS FAMILIES AND THEIR STRATEGIES OF REPLICATION

There are a large number of different retroviruses. Each has its own special features, but all share similarities of structure and replication cycle. Various groupings have been made as the details of genetic capacity and replication strategies have become elucidated. The simplest grouping is into simple and complex retroviruses based on the number of genes encoded: The simple retroviruses encode only those genes essential for replication, while the complex ones encode various numbers of other genes that regulate the interaction between the virus and the host cell.

Retroviruses also have been grouped into three broad groups based on the general details of their pathogenesis in the host: oncornaviruses, lentiviruses (immunodeficiency viruses, notably HIV), and spumaviruses. The spumaviruses (also called foamy viruses since *spuma* means "foam" in Latin and because of how infected cells appear in culture) apparently cause completely benign infections and are not nearly as well characterized as are members of the other two groups. Infections with many oncornaviruses are also completely benign, although a significant number cause serious disease, including cancer. The course of lentivirus (*lenti* is the Latin word for "slow") infections is characterized by a relatively long incubation period followed by severe and usually fatal disease.

The most accurate current classification of retroviruses is more complex, however. Currently, seven groups can be distinguished based on their genetic relatedness as measured by genome sequence similarity. Five of these have oncogenic potential and fit into the oncornavirus subgrouping. In reality, however, some of these groups are more closely related to lentiviruses or spumaviruses than they are to each other.

The productive infection cycles of all the viruses studied to date have general similarities in the processes of entry, gene expression, assembly of new virus, and release. The cycle often, but not always, involves cell death. Retroviruses demonstrate some exceptions to this general pattern. First, viral genomes are expressed as cellular messenger RNA (mRNA); therefore, replication does not always involve a period of exponential increase in viral genomes within the infected cell. Further, the replication of many types does not *directly* lead to cell death. Infection may generate a cell that sheds viruses for many weeks, months, or years.

This prolonged interaction between virus and cell is because the retrovirus genome, which is originally RNA, is converted to DNA and *integrated* into the host cell's chromosomal DNA. This viral DNA (the **provirus**) serves as a cellular gene whose sole function is to replicate virus, and this replication occurs by the simple process of transcription. There is no need for the virus to induce major metabolic and organizational changes to the cell. Because of the way retroviruses use unmodified cellular processes, there is very little latitude for cells to evolve means of specifically countering the expression of viral genes.

Many oncornaviruses have evolved another very successful strategy to interact with their hosts. They have evolved methods to stimulate the replication of cells into which their genomes are integrated. This ensures a continuing reservoir of virus-producing cells. Although it may eventually lead to cancer and death, the process is a long one. During the extended period while a tumor-causing retrovirus is continually expressed and available for spread, it is important that the host's immune defenses do not eliminate the infected cells. Retroviruses have evolved to induce subtle changes to the cell surface that do not induce cytolytic immune responses that would eliminate infected cells. Thus, many, if not most, retrovirus infections are inapparent, at least during early stages.

The lentiviruses as exemplified by HIV, which causes AIDS, use a different strategy to evade immunity. HIV targets and kills cells of the immune system, and actively avoids the immune response while it defeats it. Some of the unique aspects of the pathogenesis of this important virus are discussed in Chapters 20 and 23.

The molecular biology of retroviruses

Retrovirus structural proteins

The structure of two "typical" retroviruses is shown in Figure 19.1. The virion contains a membrane envelope with two viral proteins: the transmembrane protein (TM) and the surface protein (SU). Both are products of the envelope gene and are collectively referred to as Env protein; they are translated as one precursor protein. The SU protein is important in receptor recognition, and the TM protein is required for entry into cells. All retrovirus-infected cells express some Env protein on the cell's surface. While the host can generate antibody and T-cell responses to this Env protein, such immune responses, while cytotoxic, do not effectively clear infectious virus from the host.

The virus capsid is derived from a second virus precursor protein: the Gag protein, which is cleaved by the viral protease to create the matrix (MA), capsid (CA), and nucleocapsid (NC) proteins. This name Gag is derived from early work showing that various groups of retroviruses could be distinguished by capsid proteins that induced cross-reactive antibodies. Thus, the capsid protein was termed *group-specific antigen* (Gag). The capsid is often shown as an icosahedron, but the actual shape of the capsid differs somewhat for different types of retroviruses. Some mature retroviruses are distinguished by capsids that are "collapsed" like a partially deflated soccer ball, while others form a cone shape with rounded ends.

The interior of the capsid contains a few copies of three extremely important viral enzymes: protease (PR), reverse transcriptase (RT), and integrase (IN). The RT and IN enzymes are required for the early stages of retrovirus infection, while the PR enzyme is required for later steps in viral replication. All are derived by a pattern of proteolytic cleavage from precursor proteins. This maturational cleavage and that of the Gag precursor protein occur only late in viral assembly, sometimes following encapsidation of the viral genome and release of the virion from the infected cell. This strategy neatly limits reinfection of the producer cell. Reinfection of some retroviruses is further limited by their inability to interact with a cell bearing the envelope protein that they encode due to suppression of the appearance of the cellular receptor on the surface of the cell.

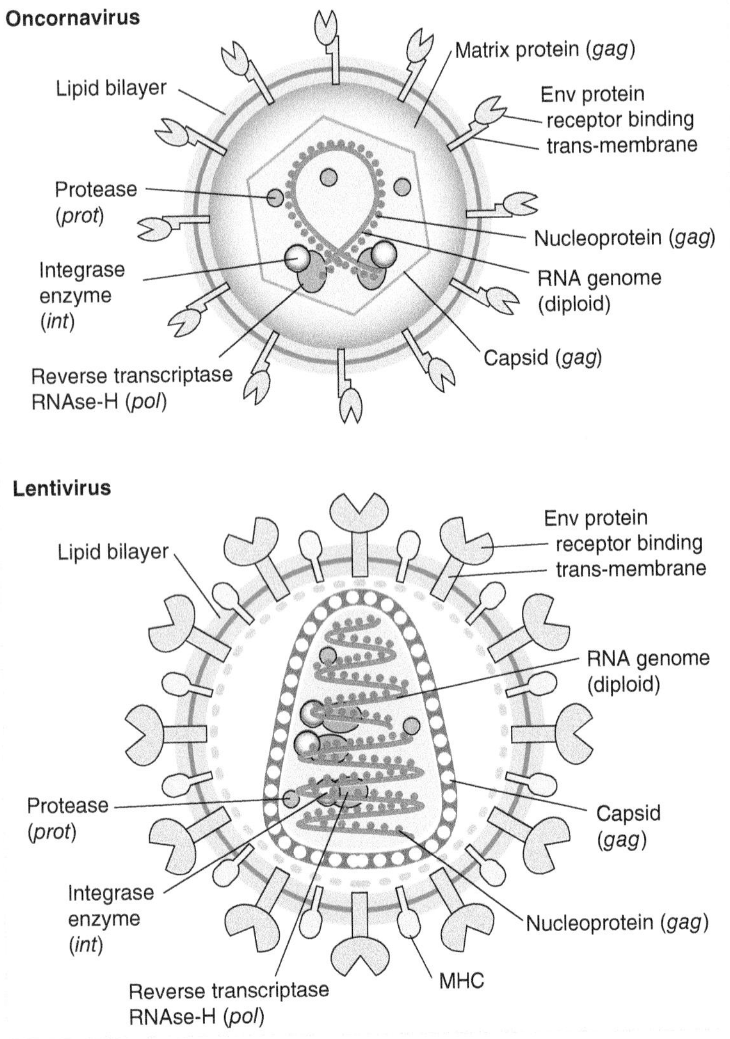

Figure 19.1 The structures of an oncornavirus and a mature lentivirus. Virion diameter varies between 80 and 120 nm for different oncornaviruses, and lentiviruses are slightly larger with a diameter of 100 to 120 nm. Virion proteins are all derived by proteolytic processing of the Gag precursor protein, while envelope glycoproteins are derived by processing of the Env protein. The approximate molar amounts of viral structural proteins are indicated by the numbers of copies shown. The "cone" shape of the mature lentivirus capsid is a result of a structural change as proteolytic processing occurs in the immature capsid budded from the infected cell.

The precursor for the three enzymes is a protein called the Gag-Pol fusion protein. It is generated by one of two different mechanisms, depending on the exact virus in question. Some retrovirus genomes encode a suppressible stop codon between the ***gag*** and *pol* genes. The expression of the fusion protein follows a mechanism similar to that discussed for the expression of the Pol protein by Sindbis virus in Chapter 14. Other retroviruses encode the *gag* and *pol* genes in two different reading frames, and the generation of the Gag-Pol precursor requires an unusual ribosomal-skipping mechanism that is outlined in the "The *gag, pol,* and *env* Genes" section.

The retrovirus genome

The positive-sense strand RNA genome is between 7000 and 13 000 bases long, depending on the particular retrovirus in question. This genomic RNA is capped and polyadenylated, as expected for an RNA molecule that is expressed by transcription of a cellular gene by cellular transcription machinery. In the case of a retrovirus, however, the cellular DNA is the provirus generated by reverse transcription of the viral genome, followed by integration into the cellular genome.

The retrovirion contains two copies of the RNA genome (i.e., the virus has a diploid genome). This feature is found in all retroviruses, and the diploid genome's actual function is not clear. It is not strictly required for full reverse transcription of the viral genome, and virus in which the two copies are genetically different have been generated in the laboratory. Since reverse transcriptase is very prone to error in its conversion of RNA into DNA, speculation is that the diploid genome provides a biological buffer against too-rapid mutational change of the viral genome during initial stages of infection.

As noted, a fundamental classification of retroviruses is based on the fact that different groups have significantly different genetic complexities. Despite this, all retrovirus genomes contain three essential genes from which structural proteins of the virus are derived. The viral RNA also contains required untranslated sequences at the 5′ and 3′ ends. These sequences are important both in generating the provirus and, when in the provirus, in mediating expression of new viral genomes and mRNA. The order of genes and genetic elements in the virion-associated RNA genome of all retroviruses is as follows:

$$5'\text{cap} : \text{R} : \text{U5} : (\text{PBS}) : (\text{leader}) : gag : pol : env : (\text{PPT}) : \text{U3} : \text{R} : \text{polyA}_n : 3'$$

Here the cap and polyA sequences are added by cellular enzymes.

The R:U5:(PBS):leader region The R sequence is so named because it is repeated at both ends, and is between 19 and 250 bases, depending on the virus in question. These sequences contain important transcriptional signals that are only utilized in the proviral DNA. Following the repeat region at the 5′ end of genomic RNA, there is a sequence called U5, for unique at the 5′ end, ranging from 75 to 190 bases, depending on the virus, that does not encode protein but has important *cis*-acting regulatory signals. The U5 sequences have transcription signals utilized in the proviral DNA. This is followed by the primer binding site (PBS), which is where a specific cellular transfer RNA (tRNA) binds and serves as a primer for the initiation of reverse transcription. The leader sequence (50–400 bases in different viruses) follows and contains the genome packaging signals (Ψ site) important in the virus's maturation. This sequence also contains splice donor signals (5′ splice signals), which are important in generating spliced retrovirus mRNAs.

The gag, pol, and env genes The *gag* gene encodes coat proteins and is always terminated with a translation termination signal. This signal is followed by the *pol* (or *pro* and *pol*) genes, which occur either in the same translational reading frame or in another one in different viruses. When *pol* genes are in the same reading frame as *gag*, they are expressed as a Gag-Pol precursor protein by virtue of a suppressible stop codon in a manner analogous to that described in the expression of the nonstructural protein precursors from the 42S genomic RNA of Sindbis virus (see Chapter 14).

Many retroviruses, however, contain the *pol* genes in another translational frame. Here, the fusion protein is expressed by a ribosomal-skipping, frame-shifting suppression mechanism. This occurs because the structure of the mRNA is such that ribosomes can occasionally miss this termination signal, skip, and go on to continue translation. Indeed, some retroviruses encode *gag*, *pro*, and *pol* gene products in three different reading frames, and two ribosome skips must occur.

By either process, the Gag-Pol fusion proteins are expressed in much lower amounts than Gag alone (5% or less), but they can be incorporated into capsids, allowing for maturation and generation of protease by self-cleavage. The protease can then digest the Gag and Gag-Pol precursor proteins to generate the mature internal structural proteins and viral enzymes.

The other retrovirus protein, Env, is always present as a hidden or cryptic translational reading frame downstream of *gag-pol* in the virion RNA. Again, like other instances of eukaryotic mRNAs containing multiple translational reading frames, only those nearest the 5′ cap can be initiated since ribosomal binding is at or near the cap site (see Chapter 13, Part III). With retroviruses, as with the late VP1 protein of SV40 virus (see Chapter 17), Env is only translated from spliced mRNA.

The 3′ end of the genome There is a variable-length sequence following the *env* translational reading frame. It contains a polypurine tract (PPT) important in generating DNA from virion RNA, an untranslated sequence unique to the 3′ end of the RNA (U3), and a second copy of the R sequence.

Genetic maps of representative retroviruses

Oncornaviruses Some representative genetic maps of retroviruses are shown in Figure 19.2. Many oncornaviruses have a gene map identical to the basic map discussed above: These are the simpler retroviruses. For example, avian leukosis virus (ALV) and murine leukemia virus (MLV), both able to cause tumors in animals, albeit slowly, have this gene arrangement.

Some oncornaviruses encode a further unique translational reading frame downstream of *env*: the viral oncogene, v-*onc*. This gene, which is related to one of the many cellular replication control genes (cellular oncogenes, c-*onc* genes), is expressed during virus mRNA production by virtue of an alternate splicing pattern of the unspliced pre-mRNA. The first retrovirus shown to have a v-*onc*, Rous sarcoma virus (RSV), contains this additional gene (*v-src*) within the unique sequences of the viral RNA, but mouse mammary tumor virus (MMTV) contains this additional open reading frame (*v-sag*) extending into the **long terminal repeat (LTR)**.

The presence of a v-*onc* gene in a retrovirus is often correlated with the ability of the virus to rapidly cause tumors in infected animals. RSV, the first virus definitely shown to cause cancer and the first characterized retrovirus, is the basic prototype for all *rapid-transforming retroviruses*. Analysis of its oncogene, *src*, was seminal in developing an understanding of the relationship between viral oncogenes and cellular growth control genes. Viral oncogenes were essentially "stolen" by ancestral retroviruses from the genes of an infected cell. Thus, all oncornaviruses bearing an oncogene are classified as complex-genome retroviruses, but were derived from the simpler type.

Human T-cell leukemia virus (HTLV) There are actually two distinct human T-cell leukemia viruses (HTLV-1 and -2), but for the purposes of this discussion they can be discussed together. The viruses encode a complex set of regulatory genes in addition to *gag*, *pol*, and *env*. These genes are in a set of overlapping translational reading frames, within and 3′ of *env*, and the mRNAs expressing them display relatively complex splicing patterns. The *tax* gene product acts something like activators, such as the SV40 T antigen, to stimulate cell division and metabolic activity. It is thought that this stimulation of T cells leads only indirectly to transformation. The *tax* and *rex* genes also regulate viral gene expression. This is described in a bit more detail in this chapter.

Replication of retroviruses: an outline of the replication process

Initiation of infection

While there are specific differences between the mechanisms of entry of the various lentiviruses and oncornaviruses, most of their replication patterns are generally similar to that diagrammed in Figure 19.3. Infection begins with entry of the virus after recognition of specific cell surface receptors, followed by fusion of the viral and cellular envelopes and capsid entry into the cytoplasm. This leads to partial uncoating of the viral capsid.

Generation of cDNA Virion mRNA conversion into complementary DNA (cDNA) by the action of RT initiates while the virion is in the cytoplasm and may be completed there or when the virus reaches the nucleus. Conversion takes place in the nucleoprotein environment of the partially opened capsid. During this complex process, which is outlined in some detail here, the R, U5, and U3 regions of virion RNA are fused and duplicated in the cDNA. This generates two copies of a

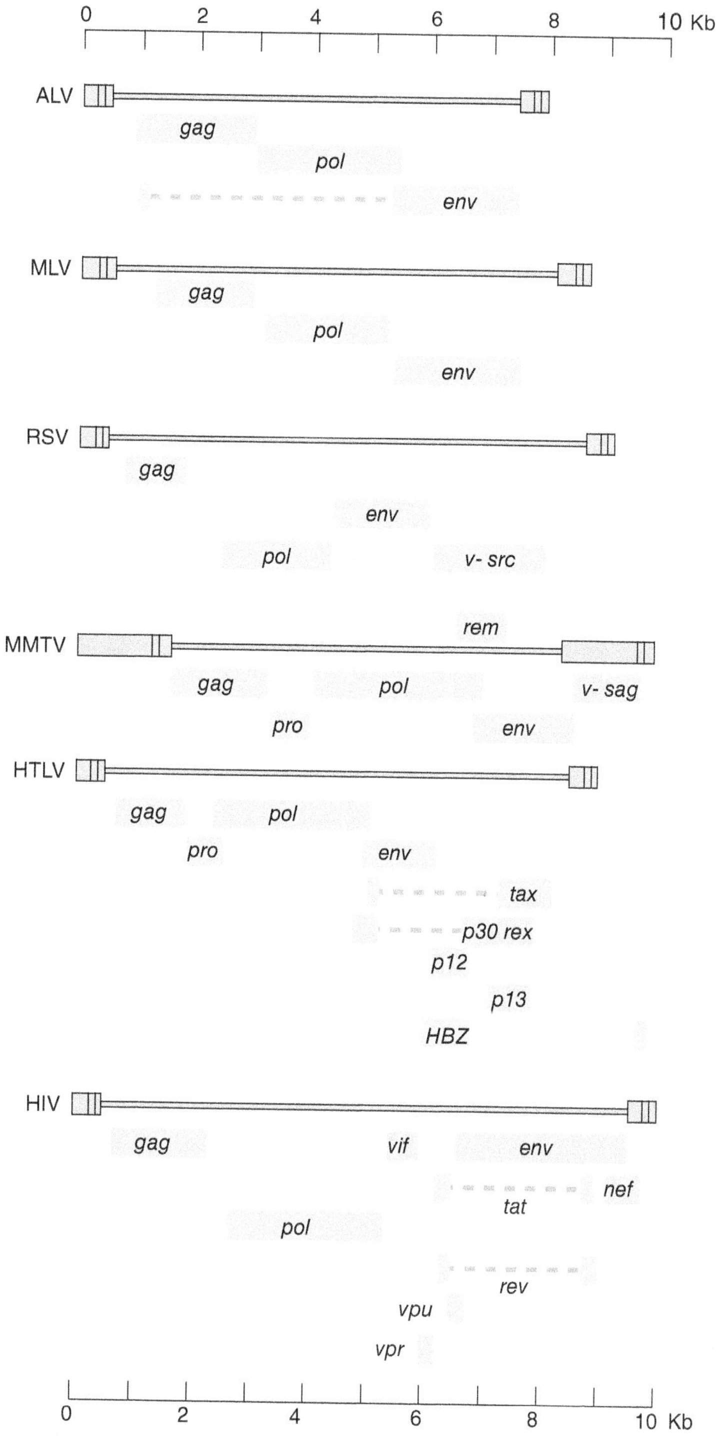

Figure 19.2 Genetic maps of various retroviruses. Specific examples are discussed in the text. Avian leukosis virus (ALV) and murine leukemia virus (MLV) are slow-transforming oncornaviruses. Note: The Env protein of ALV has a short region at its N-terminal that is the same as the N-terminal of Gag. Rous sarcoma virus (RSV) and mouse mammary tumor virus (MMTV) are rapid-transforming oncornaviruses; they have an additional v-*onc* gene. In the MMTV genome, the v-*sag* gene is encoded in the U_3 region. Human T-cell leukemia virus (HTLV) is an example of a slow-transforming oncornavirus that encodes extra regulatory proteins in addition to Gag, Pol, and Env.

sequence that only occurs in the DNA of the provirus, the LTR. The LTR contains a promoter/enhancer for transcription of viral mRNA, and polyadenylation/transcription termination signals.

Migration of the cDNA (with integrase) into the nucleus The details concerning the mechanism of migration of the cDNA into the nucleus are not the same in lentivirus and oncornavirus infections. In the latter, migration requires a dividing cell, and the breakdown of the nuclear

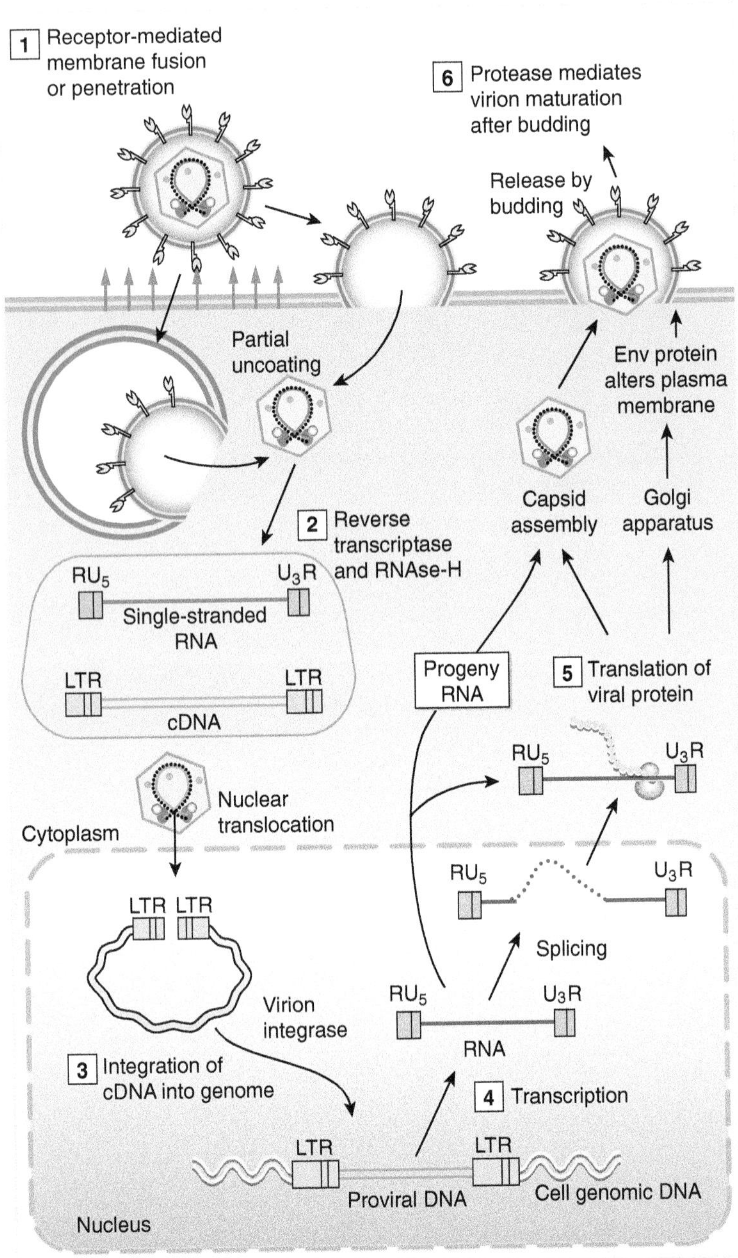

Figure 19.3 The replication cycle of a typical retrovirus. (1) Adsorption and penetration by receptor-mediated membrane fusion result in partial uncoating of the viral capsid. (2) The generation of cDNA takes place by action of virion reverse transcriptase and RNase-H in the reverse transcription complex (RTC), and it results in formation of two copies of the long terminal repeat (LTR) made up of the R, U3, and U5 regions. (3) The cDNA now joins with viral integrase and certain host proteins to form the pre-integration complex (PIC). This is followed by integration of the proviral cDNA into the genome by the action of virion integrase. (4) The integrated provirus is transcribed from the viral promoter contained in the LTR. Transcription terminates at the other LTR at the end of the provirus. Some viral transcripts are spliced, and all are exported from the nucleus to the cytoplasm. (5) Viral mRNA molecules are translated into structural proteins. Immature capsids are assembled, associate with viral glycoproteins at the cell surface, and bud from the cell membrane. (6) Following this, the final stages of capsid maturation occur in the virion by means of encapsidated protease after release from the infected cell.

membrane allows passage of the cDNA to close association with cellular chromatin. On the other hand, lentivirus migration as evidenced by studies on HIV involves the action of the matrix protein (MA), viral protein r (Vpr), and the integrase itself. Genetic studies have demonstrated that HIV integrase alone can mediate transport of cDNA through the nuclear pore, and this and other experiments suggest that there are two or three redundant mechanisms for such transport. The complete viral cDNA joins with the integrase and certain cellular proteins to form the pre-initiation complex (PIC), which migrates into the nucleus.

Integration of the retroviral cDNA into the host genome Integration proceeds in three stages. First the retroviral integrase protein (IN) cleaves two nucleotides from the 3′ ends of the retroviral cDNA genome. Second, IN cleaves the host DNA (usually at a site in open chromatin) and ligates it to the 3′ ends of the retroviral cDNA. Finally host DNA repair proteins remove the two-base overhangs and ligate the 5′ ends of the retroviral cDNA to the host DNA. The ability of lentiviruses to infect and integrate their proviral DNA into nondividing cells is a major factor in their pathogenesis as it allows them to infect and integrate their genomes into terminally differentiated macrophages. This will be discussed further in Chapter 20. The ability of integrase to carry this out has major applications in the use of retroviruses to deliver genes to cells (discussed in Chapter 25).

Expression of viral mRNA and RNA genomes Following integration, the LTR serves both as a promoter and as a polyadenylation/transcription stop signal. Transcription generates full-length virion RNA, which then either is transported to the cytoplasm for translation or encapsidation into virions, or can be spliced in the nucleus to generate *env* mRNA and mRNAs encoding v-*onc* or regulatory proteins.

Capsid assembly and maturation

The Env protein is incorporated into the cell's plasma membrane. Meanwhile, expression of the *gag* and *pol* genes leads to assembly of capsids in the cytoplasm. Immature virus particles bud through the plasma membrane, and final maturation of the virion enzymes and capsid proteins takes place in released virions.

Action of reverse transcriptase and RNase H in synthesis of cDNA

The generation of cDNA involves RNA-primed DNA synthesis from a primer that is a specific cellular tRNA bound to the virion genomic RNA and RT. Since the primer is bound near the 5′ end of the linear RNA molecule, cDNA synthesis must switch from the 5′ to the 3′ end of the genome to continue. During the synthesis of cDNA, the LTR is formed. The LTR of the provirus contains only information encoded by the virus; however, it has this information rearranged and duplicated compared to the RNA genome found in the virion. This duplication, in a sense, is a functional equivalent to circularization of a linear DNA virus, to ensure that no sequences are lost during the RNA priming step. Duplication also enables the virus to encode its own promoter/regulatory sequence.

The process of cDNA synthesis and LTR generation is shown schematically in Figure 19.4 and can be broken into five steps. Because cDNA is synthesized in the ribonucleoprotein environment of the partially uncoated reverse transcription complex (RTC), the RT enzyme can remain associated with the RNA and RNA–DNA hybrid templates during the times that it must be transferred from one site to another.

1 Priming cDNA synthesis. The first step is synthesis of the minus strand of cDNA from the positive-sense RNA genome, beginning at the tRNA primer and ending at the 5′ end of the genome. This generates a short segment of single-strand cDNA encoding R and U_5 still linked to the tRNA primer. At the same time, the RT enzyme exhibits a second activity: RNase-H. This specific RNase activity only destroys RNA from a DNA–RNA hybrid molecule as the cDNA is synthesized. RNase H activity degrades the 5′ end of the virion RNA as it is reverse transcribed forming minus strand cDNA.

2 The first template switch of RT. The RT complexed to R:U5-cDNA with the bound primer is then somehow transferred to the 3′ end of the RNA template, where the R region of the single-stranded cDNA anneals to the complementary 3′ R sequence of the RNA genome. It has been suggested that this event may occur following formation of a genomic RNA lariat, similar to an intron lariat, which covalently attaches the 5′ end of the genome to the 3′ nucleotide of the U3 region by a unique 2′–5′ phosphodiester bond.

3 Completion of the negative-sense cDNA strand. Reverse transcription then continues until there is a complete cDNA copy of the residual RNA template. RNase H activity removes all of the RNA template, except the PPT, which is resistant to degradation.

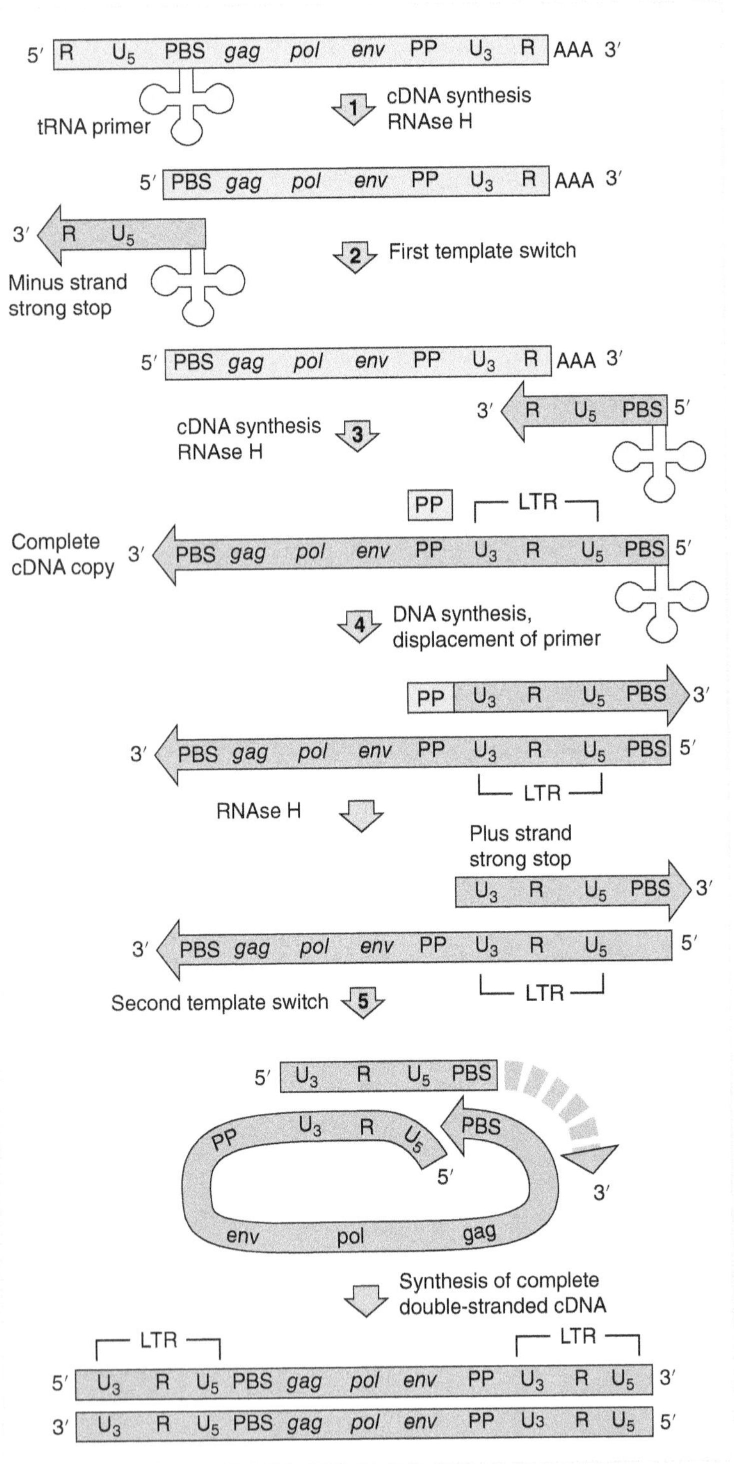

Figure 19.4 The detailed mechanism for formation of retroviral cDNA from viral RNA. The individual steps shown in the schematic are discussed in the text.

4 Start of the positive-sense cDNA strand. The polypurine region of the +-sense RNA serves as a primer for synthesis of the positive-sense cDNA strand, this time using the newly synthesized negative-sense cDNA as a template proceeding to the region of the primer, which is still the tRNA. Degradation of the last bit of RNA from the tRNA primer follows.

5 The second RT template switch. The partially double-stranded cDNA then anneals to its own tail, which contains a cDNA PBS sequence. Eventually, the process leads to formation of two free 3′ ends that can be used to complete synthesis of *both* cDNA strands. This synthesis results in a complete double-stranded cDNA molecule with LTRs of sequence U3:R:U5 at both ends.

Retrovirus gene expression, assembly, and maturation

Transcription and translation of viral mRNA

Following integration of cDNA, as described in the "Integration of the Retroviral cDNA into the Host Genome" section, viral mRNA is expressed from the 5′ LTR, which is acting as a promoter. Many LTRs also contain sequences that act as enhancers to ensure that transcription is efficient even in nondividing cells that have low overall transcriptional activity. Thus, the 5′ LTR serves a similar function to the early promoter/enhancer of SV40, or the E1A promoter of adenovirus, or the immediate-early promoters of herpesviruses.

Transcription ends near the poly-A signal in the 3′ LTR. Note that the LTR's enhancer/promoter sequences are actually encoded by the virion RNA's U3 region, while the polyadenylation signal is in the R region. This means that there is a polyadenylation signal near the full virion mRNA's cap site, but the proximity of the 5′ splice site and the structure of the 5′ end of the nascent mRNA insure that this site is not utilized efficiently.

Unspliced viral RNA can migrate to the cytoplasm for translation, or it can serve as a precursor to generate spliced *env* and other viral mRNAs. Typical splicing patterns are schematically shown in Figure 19.5. The translation of full-length viral mRNA displays some features that are either unique to the expression of retroviruses or rarely seen in expression of cellular mRNA into protein. In many retroviruses, the translation terminator at the *gag* translational reading frame's end is embedded in a region of the mRNA that has a highly specific secondary structure. This structure sometimes allows the ribosome to skip a base at or near the termination codon. Also, the *gag* translational reading frame of some retroviruses can start either at the normal AUG

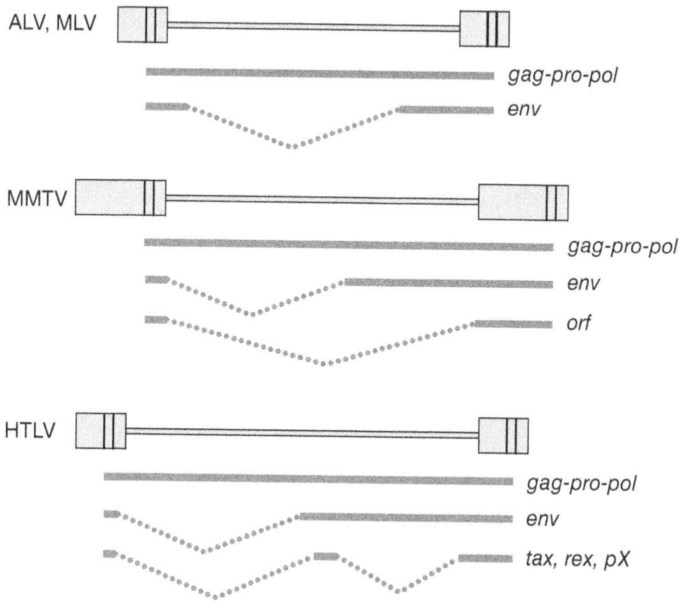

Figure 19.5 Splicing patterns of various retrovirus RNAs to generate subgenomic mRNAs. The genes that can be translated in each mRNA are shown. Note: In each case the unspliced genomic RNA serves as the mRNA to encode the Gag and Gag-Pol precursor proteins. ALV: Avian leukosis virus; MLV: murine leukemia virus; MMTV: mouse mammary tumor virus; HTLV: human T-cell leukemia virus.

codon, or, more rarely, at a CUG codon a short distance upstream. When this happens, a Gag protein variant is synthesized. This variant has a leader signal that allows it to interact with the virion envelope's membrane, and thus serves as the matrix protein.

Capsid assembly and morphogenesis

Maturation of capsids takes place as the virion buds from the infected cell. Only after the virion is released are final proteolytic cleavages made to generate active RT. In most cases, the final maturation process is completed only after release of the immature virion, when the Gag-Pol precursor is cleaved in the capsid into free protease (and one *gag* subunit in the capsid). The free protease releases free and active RT and integrase in the virion. The Gag and Gag-Pol precursor proteins are also cleaved to form the mature structural Gag proteins, matrix (MA), capsid (CA), and nucleocapsid (NC). As a result, little free RT is expressed in the infected cell, and regeneration of progeny cDNA in the infected cell for further integration into the host genome is avoided. Moreover, the immature Gag proteins can't form a pre-integration complex. In this way, the new virus cannot reinfect the cell producing it.

MECHANISMS OF RETROVIRUS TRANSFORMATION

Integration of the retrovirus genome into the cellular chromosome does not necessarily lead to an alteration in cell metabolism. But, as outlined in the introductory portions of this chapter, viruses that can stimulate their resident cell to proliferate have some replication advantage over those that cannot. Strategies of transformation differ, but can be roughly broken into three basic types described in this section.

Transformation through the action of a viral oncogene – a subverted cellular growth control gene

Rapid-transforming retroviruses and some other oncornaviruses encode a v-*onc* gene that is related to one of many c-*onc* genes that function at different points in the control of cell replication, often in response to an external signal. The v-*onc* genes, which were originally "stolen" from the infected cell, act enough like the cellular gene to short-circuit the cell's growth regulatory system, causing the cell to divide out of control. Examples of cellular growth control genes (proto-oncogenes) "pirated" by retroviruses are shown in Table 19.1.

Figure 19.6 presents a schematic diagram of some sites of action of a cell's growth regulators. All of these regulators work as switch points, often by being able to undergo a reversible alteration in structure by a chemical modification. A critical mutation in any of these proteins can alter this reversibility and result in a dominant change in which the switch is "locked on." These regulators fall into one of five classes:

1 Growth hormones
2 Receptors for extracellular growth signals
3 G proteins, which act as transducers of extracellular signals by interaction with receptors and binding of guanosine triphosphate (GTP)
4 Protein kinases that regulate the action of other proteins and enzymes by phosphorylation of serine/threonine or tyrosine residues
5 Specific transcription factors that either turn on or turn off critical genes

The first two types of growth control elements (cell growth hormones and their receptors) work as matched pairs. Platelet-derived growth factor (PDGF), for example, will only bind to and stimulate its own specific receptor.

Table 19.1 Selected examples of oncogenes acquired by retroviruses.

Retrovirus	Oncogene	Proto-oncogene	Class of proto-oncogene products (signal transduction pathway factors)
Simian sarcoma virus	*sis*	Platelet-derived growth factor (PDGF)	Growth factor
Avian erythroblastosis virus	*erb B*	Epidermal growth factor (EGF) receptor	Growth factor receptor (tyrosine kinase)
Murine sarcoma virus	*ras*	Unknown (growth signal)	G protein (receptor signal)
Avian myelocytoma virus	myc	Regulates gene expression	Transcription factor (nuclear)
Mouse myeloproliferative leukemia virus	*mpl*	Hematopoietin receptor	Tyrosine kinase
Avian erythroblastosis virus-ES4	*erb A*	Hormone receptor	Thyroid hormone receptor
Harvey murine sarcoma virus	*H-ras*	G protein	GTPase
Kirsten murine sarcoma virus	*K-ras*	G protein	GTPase
Rous sarcoma virus	*src*	Src family tyrosine kinases	Tyrosine kinase (receptor associated)
Abelson murine leukemia virus	*abl*	Tyrosine kinase	Signal transduction
Moloney murine sarcoma virus	*mos*	Serine-threonine kinase	Germ-cell maturation
3611 murine sarcoma virus	*raf*	Serine-threonine kinase	Signal transduction
Avian sarcoma virus	*jun*	AP-1	Transcription factor
Finkel-Biskis-Jenkins murine sarcoma virus	*fos*	AP-1	Transcription factor
MC29 avian myelocytoma virus	*myc*	Regulates gene expression	Transcription factor

Oncornavirus alteration of normal cellular transcriptional control of growth regulation

The slow-transforming retroviruses like MLV work in a different way. These viruses usually integrate in a region of the cell's genome from which viral genes can be expressed with little or no effect on the animal. In rare cases, however, the virus can integrate near a cellular oncogene that is transcriptionally silent. This integration may interrupt transcriptional shutoff in one of several possible ways. There can be a direct promoter capture where the viral LTR is close enough to the oncogene so that it can direct transcription of the gene. Alternatively, the retroviral LTR enhancer could activate the quiescent promoter that normally expresses the oncogene transcript. Another possibility is that the integration event disrupts expression of a repressor of oncogene transcription.

No matter what the exact mechanism is, the result is that many months or years following infection with a slow-transforming retrovirus, a cellular oncogene is activated. This activation results in abnormal replication of a cell that can accumulate further mutational damage until a malignant tumor forms.

Oncornavirus transformation by growth stimulation of neighboring cells

HTLV-1 causes cancer in a fundamentally different way. In this mode of oncogenesis, the DNA of the provirus is not integrated into the cancer cell itself. With HTLV-mediated carcinogenesis, the proviral DNA is integrated into a lymphoid cell that produces increased amounts of specific cytokines (growth factors) as a response to viral gene expression. These cytokines are a normal component of the immune response that induces and stimulates T-cell proliferation, but their continued expression over a period of years can lead to mutations in growth control in target T

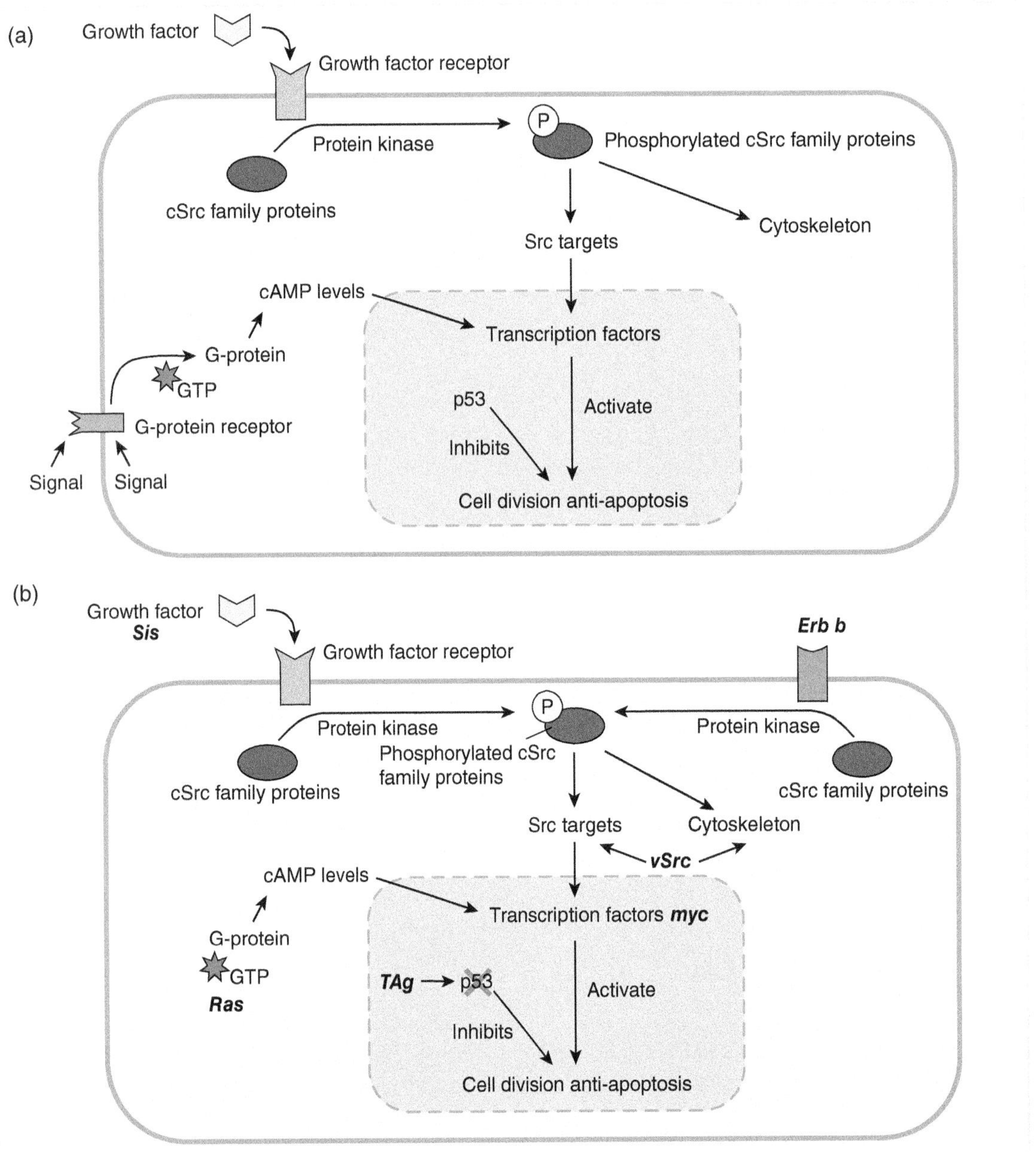

Figure 19.6 Cell division and oncogenes. (a) In the normal cell, division is controlled in a variety of ways, including signals from growth factors (such as platelet-derived growth factor [PDGF]) that trigger phosphorylation of cSrc family proteins or signals received by G-protein linked receptors that modulate cyclic adenosine monophosphate (cAMP) levels. Cell division is inhibited by proteins such as p53. (b) Retroviruses have "pirated" cellular genes to allow them to short-circuit control of cell division. *Sis* mimics PGDF, *erb B* mimics epidermal growth factor (EGF), *vSrc* bypasses the cSrc family of proteins, *Ras* acts as a continuously active G-protein, and T-antigen (TAg), a gene product of SV-40 virus, blocks the action of p53.

cells, which can lead to mutations that result in uncontrolled growth and cancer. It is thought that the mutations occurring are similar in nature to those occurring in cells that are constantly stimulated to divide by integration of a DNA tumor virus genome, such as seen in human papillomavirus type 16 (HPV-16)–induced cervical carcinomas (see Chapter 16).

CELLULAR GENETIC ELEMENTS RELATED TO RETROVIRUSES

Many genetic elements within all genomes appear to be either remnants of retroviruses or, at least, closely related to retroviruses. The discovery of these elements is tied to the relatively complex genetic analysis of certain genes that do not display strict Mendelian inheritance properties. For example, some genes or genetic elements can move around in the genome. The movement of a gene sequence from one location in the genome to another was first documented by Barbara McClintock, who studied the genetics of corn in the 1940s. It took considerable time before the importance of transposition was widely appreciated among biologists, but McClintock was finally awarded the Nobel Prize for this work. This ability to "transpose" genes has tremendous theoretical and practical importance because it can be used to insert genes of interest and to inactivate undesirable ones.

The molecular basis for transposition was first determined in bacteria, where certain drug resistance markers are located within sequences that have the property of being able to insert a copy of themselves at another location in the bacterial DNA. These bacterial **transposons**, as they came to be called, are known to fall into three main classes, depending on which set of transposition enzymes they express. Some properties of these classes are shown in Table 19.2.

Class I bacterial transposons are simple insertion sequences. They have inverted repeat sequences on either side of a **transposase** gene. This transposase gene encodes an enzyme that initiates the transposition event by cutting the target and transposon sequences. Class IB, the composite transposons, adds a drug resistance element, such as resistance to kanamycin (Tn5) or tetracycline (Tn10), to this structure.

Class II elements add another enzyme, a **resolvase**, which serves to complete the homologous recombination event begun during the transposition. (Resolution of recombination for the other transposons is carried out by the normal cellular enzymes.)

Class III consists of specialized bacteriophages such as Mu, which, during its infectious cycle, inserts copies of its genome in random positions throughout the bacterial chromosome.

Since the phenomenon was discovered in corn, it is clear that transposition occurs in eukaryotic cells. Some eukaryotic transposable elements are quite similar to class I elements of bacteria, but much eukaryotic transposition goes through an RNA intermediate, and the transposed sequence is a reverse-transcribed copy of a processed cytoplasmic RNA. The general name **retroelements** is used to describe four kinds of eukaryotic sequences that propagate using reverse transcriptase, including the retroviruses themselves. Table 19.3 lists these classes: retroviruses, *retrotransposons*, *retroposons*, and *retrointrons*.

Retrotransposons are closely related to retroviruses, and are discussed in a bit more detail in the "Retrotransposons" section. Retroposons, such as the long interspersed elements (LINEs), have no LTR but do encode reverse transcriptase (RT) and can insert copies of themselves in other places in the genome. Retrointrons were found in mitochondrial DNA in the gene for one

Table 19.2 Bacterial transposons.

Class	Inverted repeat?	Transposase?	Resolvase?	Drug resistance element?	Other enzymes?	Examples
IA	Yes	Yes	No	No	None	IS1, IS2
IB	Yes	Yes	No	Yes	None	Tn5, Tn10
II	Yes	Yes	Yes	Yes	None	Tn3, Tn7
III	No	Yes	No	No	Yes	Mu phage

Table 19.3 Some retroelements of eukaryotic cells.

Element present in?	*env* gene	Long terminal repeat (LTR)	Reverse transcriptase	Example
Retrovirus	Yes	Yes	Yes	Many
Retrotransposon	No	Yes	Yes	Ty1 (yeast), copia (drosophila)
Retroposon	No	No	Yes	Long interspersed elements (LINEs)
Retrointron	No	No	Yes	In mitochondrial DNA

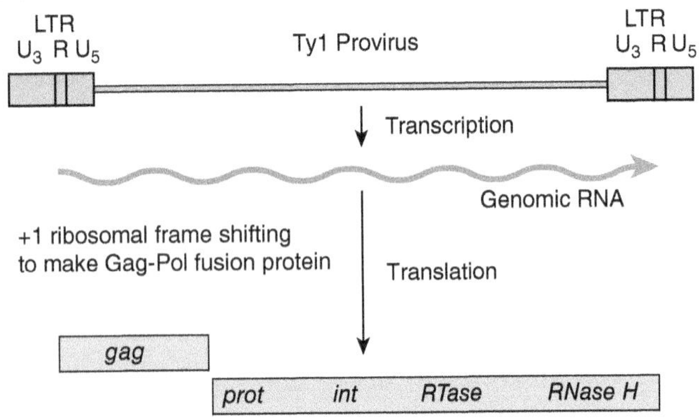

Figure 19.7 The genomic structure of yeast Ty1. The similarity to a retrovirus is evident.

of the subunits of cytochrome oxidase. These introns encode RT and are able to move copies of themselves into different locations.

Retrotransposons

The retrotransposons are most interesting in comparison to retroviruses. These transposable elements were first found in yeast cells, where sequences such as Ty1, Ty2, or Ty3 could be observed to transfer a cDNA-like copy of themselves to various sites in the chromosomes. The genomic structure of Ty1 is shown in Figure 19.7. These elements have a coding region that produces a Gag protein or a Gag-Pol fusion protein by ribosomal frame shifting. The coding region is flanked by LTRs containing U3, R, and U5 sequences with the expected transcriptional control regions.

The replication cycle of these elements involves the transcription of an mRNA from the inserted Ty element, using the cell's RNA polymerase II. The mRNA is exported to the cytoplasm where it is translated into Gag and Gag-Pol precursor proteins. Particles that resemble the cores of retroviruses are assembled. It is within these particles that reverse transcription takes place, using a cellular tRNA specific for methionine as a primer and producing a double-stranded cDNA copy of the Ty element, including the LTRs. The particles now reenter the nucleus where the new Ty DNA can be inserted into another location in the chromosome. Thus, virus-like particles can be isolated from cells and these particles have reverse transcriptase activity. Similar elements are found in *Drosophila*, where the copia and gypsy elements are retrotransposons.

The relationship between transposable elements and viruses

The close structural and biological relationship between retrotransposons and retroviruses is very clear. Unfortunately, this relationship does not necessarily establish a lineage between them. Strong arguments are made that all retrotransposons are derived from retroviruses, where some have lost more genetic material than others. They have survived by virtue of their ability to induce genetic changes that give them a survival advantage.

While this is certainly a defensible argument for which there is good support based on the spread of copia and gypsy, an opposite argument is just as defensible: that some retrotransposons are derived from the same cellular origins that gave rise to retroviruses, but never went the whole route toward independent existence. In this scenario, the retroviruses themselves are just the most complex manifestation of the action of a cellular reverse transcriptase.

These same arguments can be made for the other group of transposable elements: those that do not utilize RT. As there are relatively simple bacterial viruses that survive by moving around the bacterial genome utilizing transposase, it is possible that all bacterial elements are just different stages in the loss of genetic material from these viruses. However, the converse argument is just as compelling: The transposons have an independent existence in which bacterial viruses have captured the transposase and adopted a partial transposable element lifestyle.

No matter what the origin or origins of such elements are, they are an important factor in the evolutionary change of organisms. The fact that genes can move between organisms either as viruses or as transposons, or as both, means that once a gene is available for adaptation to a novel ecosystem, it potentially can be moved throughout the many organisms waiting to exploit such an environment by processes of infection and integration that can be understood in terms of the basic properties of viruses.

QUESTIONS FOR CHAPTER 19

1 Members of the family Retroviridae convert their RNA genomes into double-stranded cDNA prior to integration of the provirus into the host cell genome. This conversion into cDNA is carried out completely by the enzyme reverse transcriptase, encoded by each replication-competent member of this family.

(a) What are the three kinds of biochemical reactions catalyzed by reverse transcriptase? Please *be specific* in your answer.

(b) What host molecule serves as the primer for the synthesis of the initial strand of cDNA by reverse transcriptase?

2 At which steps in the life cycle of HIV might drug treatment hinder the progress of infection?

3 Rous sarcoma virus (RSV) is a nondefective member of Retroviridae and can cause tumors in birds. The following diagram shows the structure of the RSV genome.

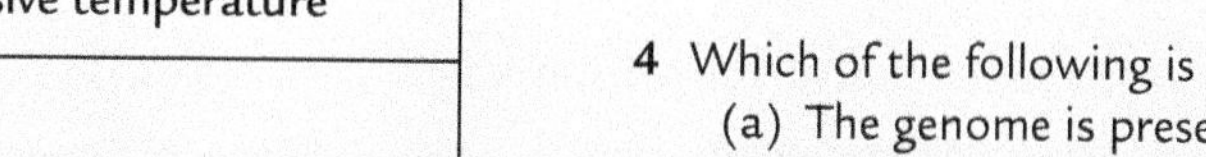

(a) Predict the effect that infecting cells with the following temperature-sensitive (ts) mutants of RSV at the *nonpermissive temperature* has on the virus replicative cycle. Be very specific about the effect of the mutation. An answer such as "has no effect" or "stops the virus" will not be acceptable.

Ts mutation in	Effect on RSV infection at nonpermissive temperature
Pol	
Gag	
Src	

(b) What is the function of the region of the genome labeled "PBS"?

4 Which of the following is true about retroviruses?

(a) The genome is present in two copies in the virion.

(b) There are repeated segments at both ends of the virion RNA.

(c) Reverse transcriptase is primed by cellular tRNA.

Continued

(d) Expression of the RT protein may require the ribosome to skip a translation termination signal.

5 What kinds of functions may be encoded by the v-*onc* of transforming retroviruses?

6 There is a great demand for pure reverse transcriptase (RT) for use in molecular biology laboratories around the country. To obtain this particular enzyme, you isolate virus by periodically harvesting the cell culture medium overlaying virus-infected cells. Virions are purified by centrifugation, and capsids are then disrupted. The enzyme is obtained by running the sample through a column with a bound antibody that recognizes an epitope of RT. Is this a good method for isolating RT? Why or why not?

7 How are retroviruses different from other viruses that have RNA genomes? How are transforming retroviruses such as Rous sarcoma virus different from lentiviruses such as HIV?

8 How is the proviral DNA of the retrovirus different from the RNA genome?

9 What part of the immune response is damaged and ultimately destroyed by HIV infection?

Human Immunodeficiency Virus Type 1 (HIV-1) and Related Lentiviruses

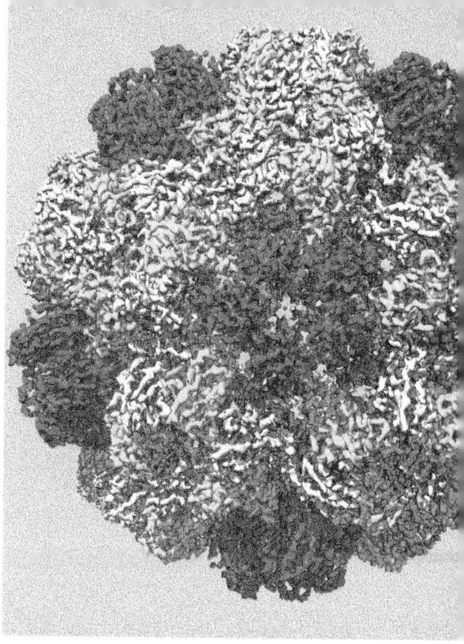

CHAPTER 20

- ✳ HIV-1 AND RELATED LENTIVIRUSES
- ✳ THE ORIGIN OF HIV-1 AND AIDS
- ✳ HIV-1 AND LENTIVIRAL REPLICATION
- ✳ DESTRUCTION OF THE IMMUNE SYSTEM BY HIV-1
- ✳ QUESTIONS FOR CHAPTER 20

HIV-1 AND RELATED LENTIVIRUSES

Lentiviruses are more complex than most other retroviruses such as oncornaviruses, in that lentiviruses have more genes, which in turn allow a more complex and efficient life cycle. Lentiviruses share most features of their replication with other retroviruses, but have additional mechanisms that facilitate their patterns of efficient replication and disease causation.

The lentivirus family includes two types of human immunodeficiency virus (HIV-1 and HIV-2), which infect humans and cause the acquired immune deficiency syndrome (AIDS); many types of simian immunodeficiency virus (SIV) that replicate in a variety of monkey and ape species; maedi/visna virus (MVV) of sheep; bovine immunodeficiency virus (BIV) of cows; feline immunodeficiency virus (FIV), which replicates in domestic and nondomestic cat species; equine infectious anemia virus (EIAV) of horses; and caprine arthritis-encephalitis virus (CAEV), which replicates in goats (Table 20.1). Unlike the oncornaviruses, lentiviruses can infect nondividing cells. In addition, most lentiviruses infect macrophages, including microglial cells in the brain, and cause neurological disease.

THE ORIGIN OF HIV-1 AND AIDS

HIV type 1 (HIV-1) is by far the most well-known and studied lentivirus because it is responsible for the worldwide AIDS pandemic. AIDS has affected people all over the world since it was first described in 1981, although it was likely present in Africa on a much smaller scale since the early twentieth century. AIDS is now one of the leading causes of death from an infectious agent

Basic Virology, Fourth Edition. Martinez "Marty" Hewlett, David Camerini, and David C. Bloom.

Table 20.1 Table of lentiviruses showing host species.

Virus	Host species
Visna/maedi virus	Sheep
Equine infectious anemia virus	Horse
Caprine arthritis encephalitis virus	Goat
Feline immunodeficiency virus	Cats
Bovine immunodeficiency virus	Cow
Simian immunodeficiency virus	Nonhuman primates
Human immunodeficiency virus type 1 (HIV-1)	Human
Human immunodeficiency virus type 2 (HIV-2)	Human

in the world, despite the fact that the disease is sexually transmitted and is therefore preventable. There are two types of HIV: HIV-1, which is responsible for the global AIDS pandemic, and HIV-2, which is less pathogenic and is primarily found in West Africa. Moreover, HIV-1 strains can be divided into several groups that were likely independently transmitted to humans from chimpanzees (*Pan troglodytes*) or gorillas (*Gorilla gorilla*). In contrast, HIV-2 is highly related to SIV found in sooty mangabeys (*Cercocebus atys*). Both viruses are much less pathogenic in their non-human primate hosts than in humans.

HIV-1 AND LENTIVIRAL REPLICATION

Since HIV-1 is the most studied lentivirus and is as complex as any lentivirus, we will discuss the additional genes of HIV-1 and features of its replication that are not common to all retroviruses and were therefore not discussed in Chapter 19. HIV-1 encodes six accessory genes, which flank or overlap the *env* gene: *vif, vpr, tat, rev, vpu,* and *nef* (Figure 20.1). These genes are expressed from a family of multiply spliced mRNAs. They function to regulate the replication of HIV-1, to inactivate host antiviral functions, and to make HIV-1 replication more efficient.

HIV-1 enters cells by a multistep process involving the viral surface and transmembrane glycoproteins (TMs); a cellular receptor, **CD4**; and a cellular co-receptor, usually **CCR5** or **CXCR4** (Figure 20.2). This dual-receptor entry mechanism helps HIV-1 avoid the immune response, by limiting exposure of crucial epitopes. The surface glycoprotein of HIV-1 (abbreviated SU) is approximately 120 kDa overall and is therefore also called gp120. The SU binds the primary receptor, CD4, with high affinity (with a binding constant of $\sim 10^{-9}$ M) via a site that is in a cleft too small for most antibodies to bind or block. Subsequent to binding CD4, gp120 undergoes a conformational change that exposes the co-receptor binding site. Thus, this site is only exposed for a short time while the virus is bound to the cell surface, and therefore the immune system has little chance to develop antibodies to the co-receptor binding site. HIV-1 binds its co-receptor, usually CCR5 or CXCR4, after the conformational change induced by CD4 binding. This second binding event induces a second conformational change in gp120 that triggers the TM, also known as gp41, to mediate fusion of the viral and cytoplasmic membranes causing the virion core to enter the cytoplasm (Figure 20.2).

After the HIV-1 particle enters the cell, uncoating begins (Figure 20.3). This process results in formation of the reverse transcription complex (RTC), consisting of two copies of the viral RNA genome, the matrix protein (MA) viral protein R (Vpr), and the enzymes **reverse transcriptase** (RT or Pol) and integrase (IN). Reverse transcription of the viral RNA genome into double-stranded DNA begins in the cytoplasm, continues during transport of

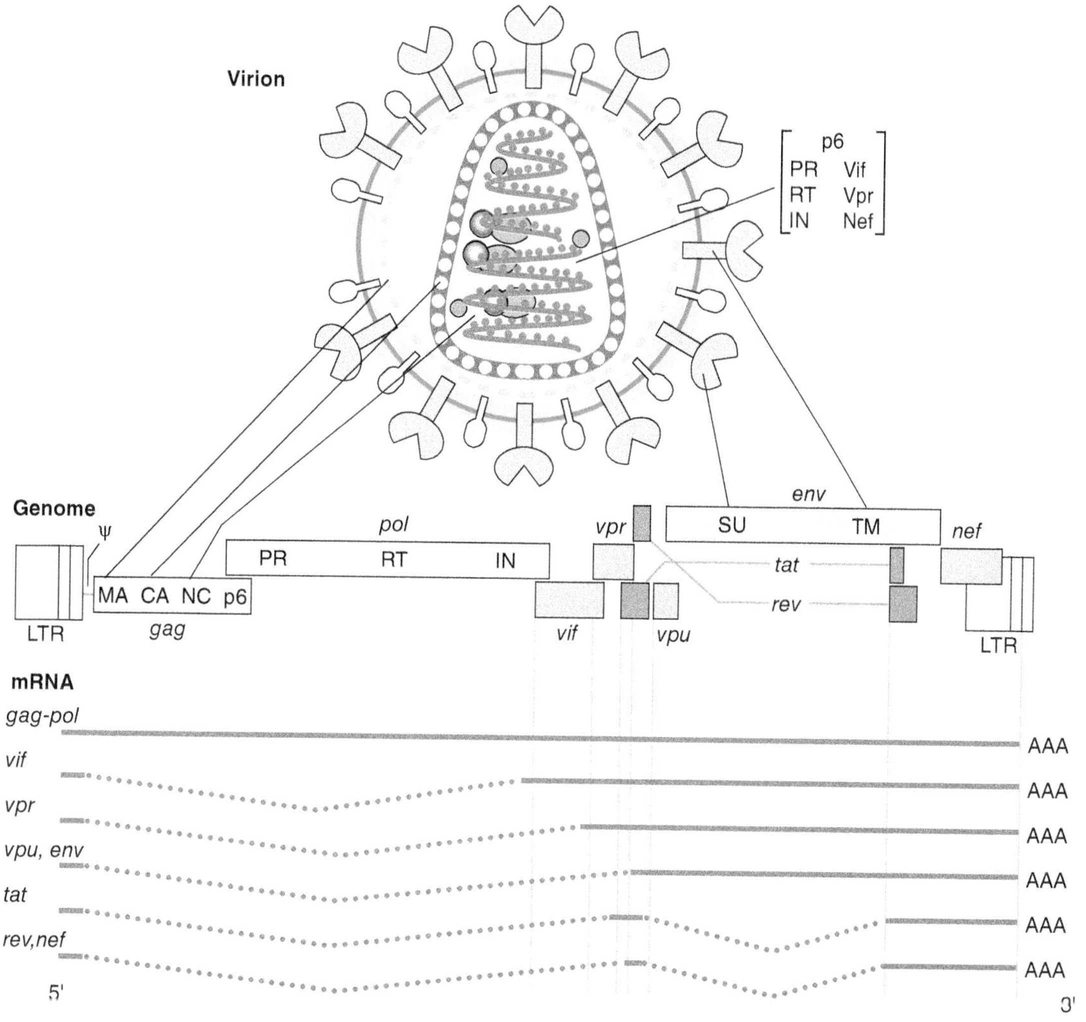

Figure 20.1 The HIV-1 genome, viral particle, and transcripts. The figure shows the HIV-1 genome, the location of virion proteins, and the mRNA used to express each gene. The virion is 100–120 nm in diameter. The "cone"-shaped mature capsid forms following proteolytic processing of the immature capsid after the virion buds from the infected cell. As in the case of other retroviruses, the final *gag*, *pol*, and *env* gene products are derived by proteolytic processing of precursor proteins. HIV-1 transcripts are processed into three size-groups of mature mRNA molecules as shown.

the RTC to the nucleus, and may be completed in the nucleus or at the nuclear membrane. Once reverse transcription is completed, the complex is called the pre-integration complex (PIC) since it is ready for nuclear import and integration of the viral complementary DNA (cDNA) genome into the cellular genome. Nuclear transport of the RTC and PIC is a property of lentiviruses that allows infection of nondividing cells such as macrophages and microglial cells of the brain. This is a property that is not shared with simpler retroviruses that require nuclear membrane dissolution during cell division for access to the host chromosomes, and whose infectious cycle therefore requires dividing cells. In contrast, lentiviral RTC/PIC nuclear transport allows access to the host cell chromosomes without cell division. This property also enables lentiviral gene transfer vectors to deliver genes to nondividing cells (see Chapter 22).

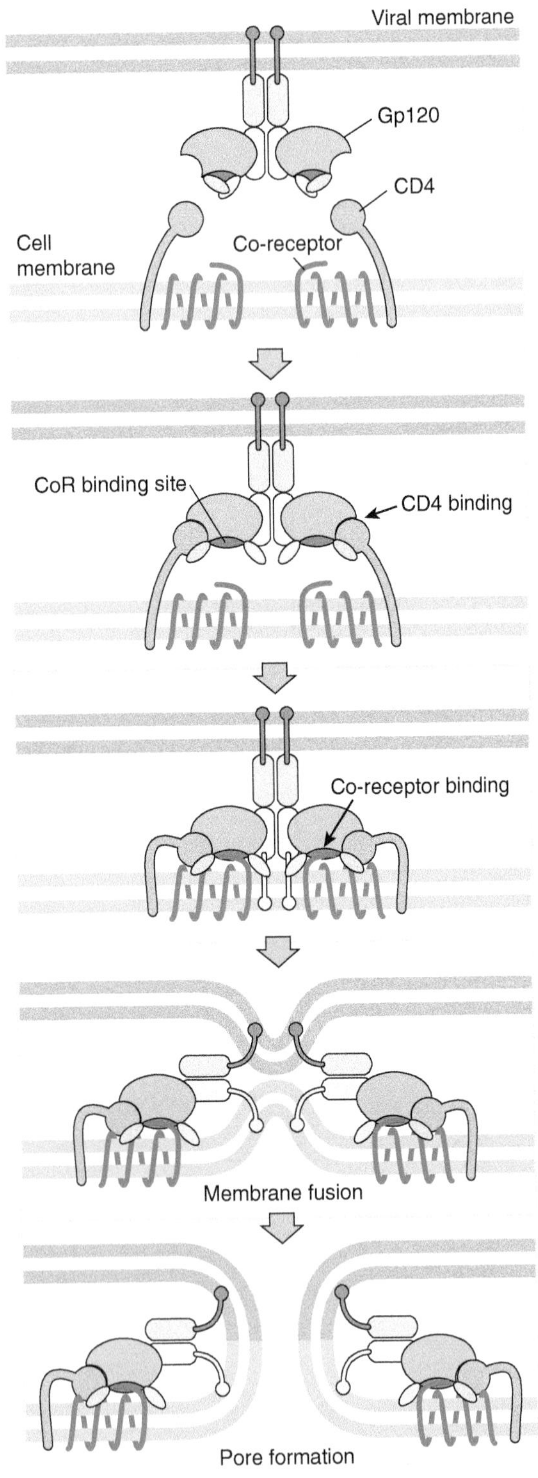

Figure 20.2 HIV-1 entry in detail. HIV-1 binds cell surface CD4 via its surface glycoprotein, gp120. Following CD4 binding, gp120 undergoes a conformational change that exposes the co-receptor binding site. HIV-1 then binds its co-receptor, CCR5 or CXCR4 or a related molecule, which triggers a second conformational change. This causes the virion transmembrane protein, gp41, to extend into the target cell membrane and then fold back on itself to initiate fusion of the viral and cytoplasmic membranes. Note that the HIV-1 gp41–gp120 heterodimer forms a trimeric complex on the virion surface, but only two heterodimers are shown here for clarity.

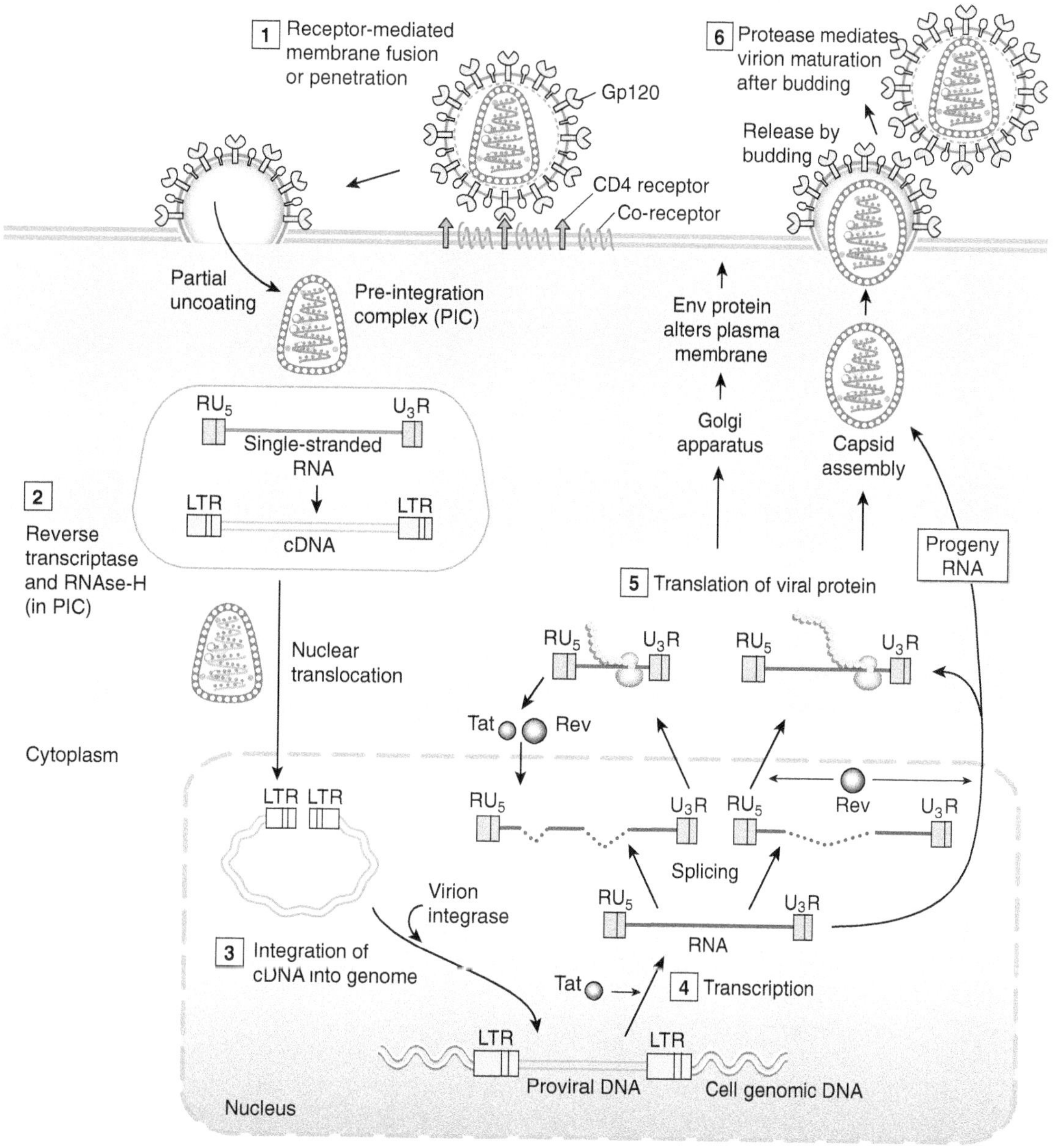

Figure 20.3 The HIV-1 life cycle. After receptor-mediated membrane fusion at the cell surface (1), partial uncoating of the viral capsid forms the pre-integration complex (PIC). The generation of cDNA takes place by action of virion reverse transcriptase and RNase-H (2). The generation of cDNA results in formation of two copies of the long terminal repeat (LTR) made up of the R, U3, and U5 regions. This is followed by integration of the proviral cDNA into the genome by the action of virion integrase (3). Cell division is not required for nuclear transport of HIV-1 cDNA where integrase has a major role in transit across the intact nuclear membrane. The integrated provirus acts as its own gene that is transcribed from the viral promoter contained in the 5′ LTR. Transcription terminates at the other LTR at the 3′ end of the provirus (4). Transcription of viral genes and splicing lead to expression of viral mRNAs, some of which are translated into structural proteins (5). The immature capsids are assembled and bud from the cell membrane. Following this, the final stages of capsid maturation (6) occur in the virion by means of encapsidated protease after release from the infected cell.

Mammalian cells have an intracellular defense against retrovirus infection mediated by the cytidine deaminase family of **APOBEC** proteins. These proteins are packaged into HIV-1 and deaminate cytidine residues in the negative-sense first strand of HIV-1 cDNA, resulting in the creation of uracil in the viral cDNA (Figure 20.4). This uracil-containing DNA is recognized by host enzymes that remove the uracil residues and may lead to the destruction or repair of the viral genome. Genomes that are repaired after second-strand cDNA synthesis, however, are hypermutated with G-to-A mutations on the sense strand, since A is paired with the newly created U residues. Both genome destruction and hypermutation contribute to loss of infectivity of new virions in the absence of Vif. The HIV-1 Vif protein (Vif stands for "viral infectivity factor") blocks APOBEC action by causing APOBEC to be ubiquitinated and degraded in infected cells, and thereby preventing APOBEC from being packaged into HIV-1 virions (Figure 20.4).

Integration of the HIV-1 cDNA, as for all retroviruses, is mediated by the viral integrase protein (IN) as described in Chapter 19. Once integrated into the host genome the HIV-1 provirus behaves in many ways like a cellular gene, as do the proviruses of all retroviruses. In particular, the HIV-1 long terminal repeat (LTR) has binding sites for nuclear factor kappa B (NF-kB), Sp1, TATA-binding protein (TBP), and other cellular transcription factors that activate viral transcription. In addition to cellular transcription and RNA transport factors, however, HIV-1 RNA synthesis and transport are regulated by the viral Tat and Rev proteins. HIV-1 Tat protein binds two cellular factors, cyclin-T1 and cyclin-dependent kinase-9 (CDK-9), and this complex binds to a stem-loop structure that forms at the 5′ end of HIV-1 transcripts called the trans-activating response element (TAR; Figure 20.5). In the absence of Tat, most HIV-1 transcripts are short and do not express HIV-1 genes. When Tat is present and bound to TAR with cyclin-T1 and CDK-9, it causes phosphorylation of the C-terminal domain of RNA polymerase II by CDK-9, thereby increasing the processivity of HIV-1 transcription and increasing the level of full-length HIV-1 transcripts 100-fold.

Like all retroviral transcripts, some HIV-1 transcripts are spliced and others are not, due to inefficient RNA splicing signals. Initial HIV-1 transcripts may remain unspliced, to express the *gag* and *pol* genes or become new virion genomes; they may be singly spliced to express *vif*, *vpu*, *vpr*, and *env*; or they may be doubly spliced to express *tat*, *rev*, and *nef*. Initially, however, only the doubly spliced mRNA molecules encoding Tat, Rev, and Nef are exported to the cytoplasm for translation, because the longer HIV-1 mRNAs contain nuclear retention signals (NRS) that block their export to the cytoplasm. The NRS sequences are spliced out of the shorter, doubly spliced messages so these messages are efficiently transported to the cytoplasm and translated to express the Tat, Rev, and Nef proteins. Rev is transported to the nucleus due to its **nuclear localization signal** (NLS). Rev binds unspliced and singly spliced HIV-1 RNAs since they contain a complex stem–loop structure called the Rev response element (RRE; Figure 20.6).

Rev returns to the cytoplasm escorting the longer HIV-1 mRNA molecules to the ribosome for translation, thereby increasing their expression more than 100-fold. Both Tat and Rev are absolutely required for HIV-1 replication since both proteins greatly increase the level of HIV-1 gene expression.

The HIV-1 viral protein U (Vpu) increases the efficiency of infectious virion production by causing degradation of tetherin and of intracellular CD4 when it is bound to the viral SU, gp120. Tetherin is a cellular protein that has membrane anchoring domains at both ends. It physically tethers HIV-1 and other enveloped viruses to the virus-producing cell, thereby restricting their release and maturation to the infective form (Figure 20.7). HIV-1 Vpu protein sequesters tetherin in the Golgi apparatus and targets it for degradation in cellular lysosomes. Similarly, the internal interaction of CD4 with the HIV-1 SU protein, gp120, prevents gp120 from reaching the cell surface so that it can be put on the surface of nascent virions. Vpu targets internal CD4 for degradation and releases this block to gp120 localization at the cytoplasmic membrane, where it is needed for viral assembly.

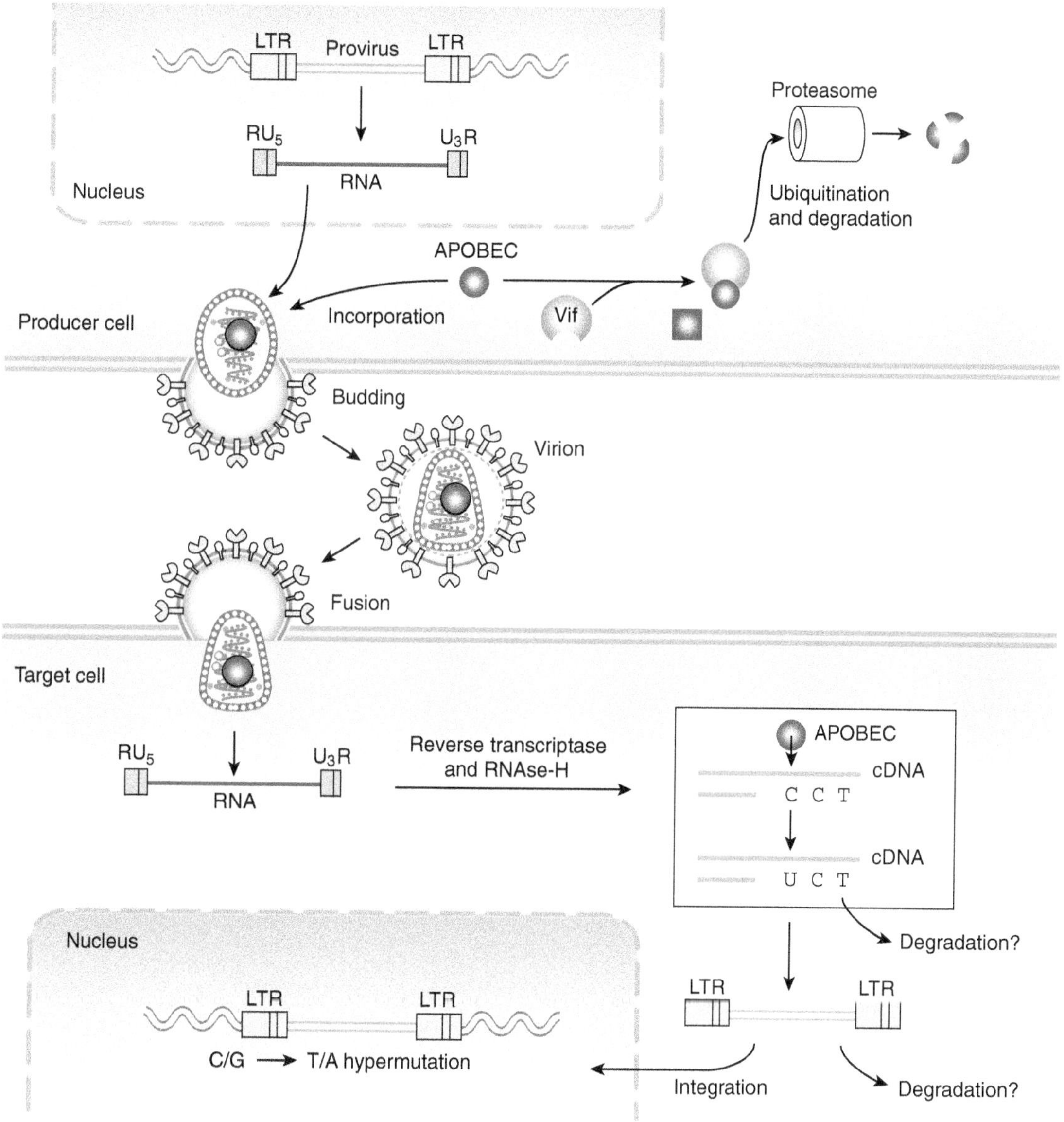

Figure 20.4 Vif and APOBEC action. In the absence of effective HIV-1 Vif protein, APOBEC (red) is incorporated into the virion in the virus-producing cell. Vif (green) blocks APOBEC incorporation into virions by targeting it for proteasomal degradation. If APOBEC enters the virion and subsequently reaches a target cell, it deaminates cytidine residues in the first strand of retroviral cDNA (blue). The resulting uracil residues function as a template for the incorporation of adenine, which, in turn, can result in strand-specific C/G to T/A transition mutations that affect virus viability. Uracil residues also trigger degradation of the retroviral DNA before it can integrate into the host cell's genome.

The Nef protein of HIV-1 downregulates the cell surface expression of CD4 to prevent interaction of CD4 with gp120 on the cell surface (Figure 20.8). Nef also downregulates cell surface major histocompability complex type I (MHC-I) molecules to aid in immune evasion by decreasing the ability of anti-HIV-1 cytotoxic T lymphocytes (CTLs) to recognize infected cells. In addition, Nef plays a role in stimulating cells to facilitate viral replication through interaction with the cellular Src family and p21-associated kinases (PAK). Nef is packaged in the HIV-1 virion and exerts its stimulatory effects at early stages of viral replication.

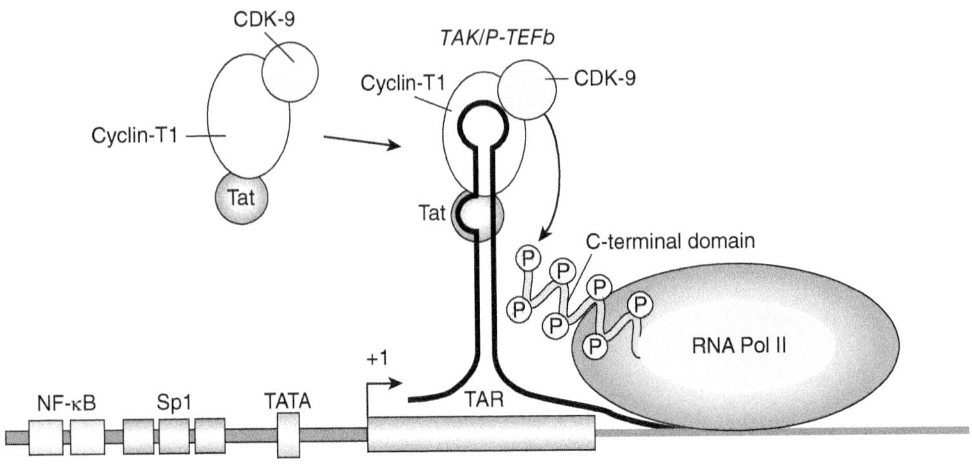

Figure 20.5 Tat activation of HIV-1 transcription. A tripartite complex consisting of Tat, cyclin-T1, and CDK-9 binds to the trans-activation response region (TAR), which is a stem loop that forms at the 5′ end of HIV-1 transcripts. Subsequently, CDK-9 phosphorylates the C-terminal domain of RNA polymerase II and stimulates elongation of HIV-1 transcripts by increasing the processivity of RNA polymerase II.

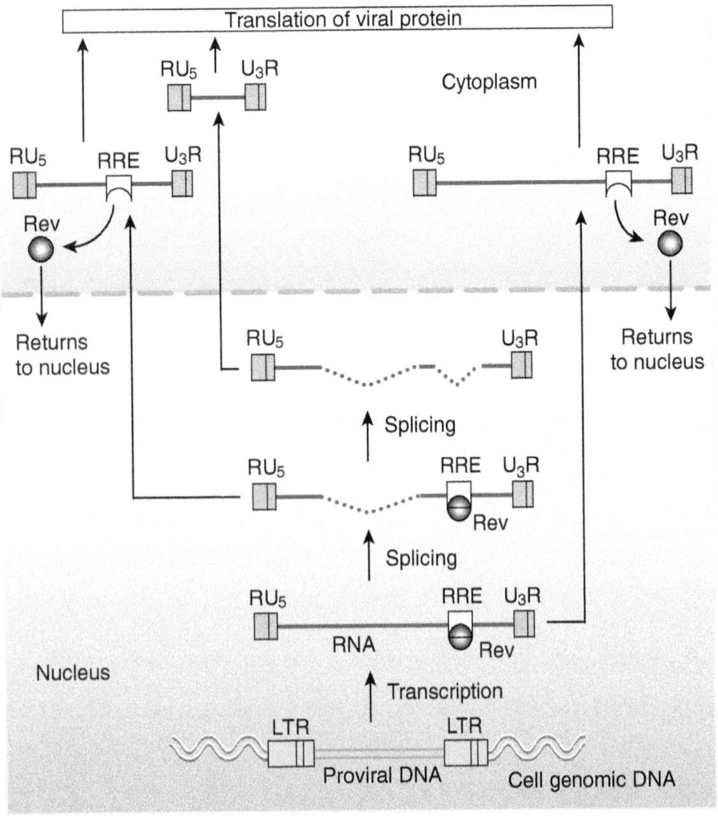

Figure 20.6 Rev binds to the Rev response element (RRE) to mediate nuclear export of unspliced and singly spliced HIV-1 RNA molecules. The HIV-1 Rev protein binds to a complex multi-stem-loop RRE that forms in unspliced and singly spliced HIV-1 RNA molecules. Rev mediates transport of these molecules from the nucleus to the cytoplasm, then discharges its cargo and returns to the nucleus. This is needed because these longer HIV-1 RNAs contain nuclear retention signals (NRS) that retain them in the nucleus in the absence of Rev. In contrast, doubly spliced HIV-1 mRNA molecules do not contain NRS or RRE and are therefore exported to the cytoplasm in the absence of Rev.

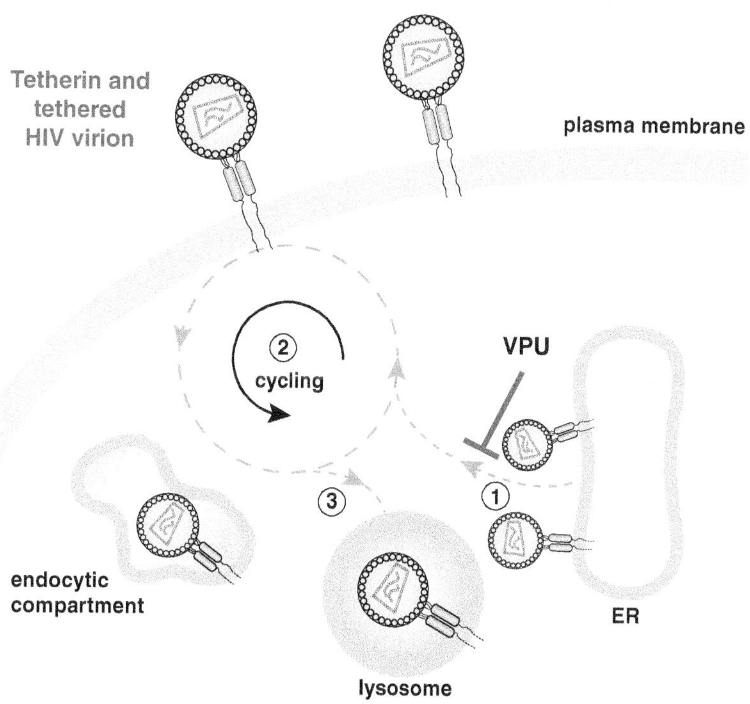

Figure 20.7 HIV-1 Vpu protein blocks the antiviral effect of tetherin. In the absence of effective Vpu protein, tetherin anchors HIV-1 particles at the cell surface, thereby blocking their release and maturation to the infectious stage. Vpu counteracts this by sequestering tetherin in the Golgi and targeting it for lysosomal degradation.

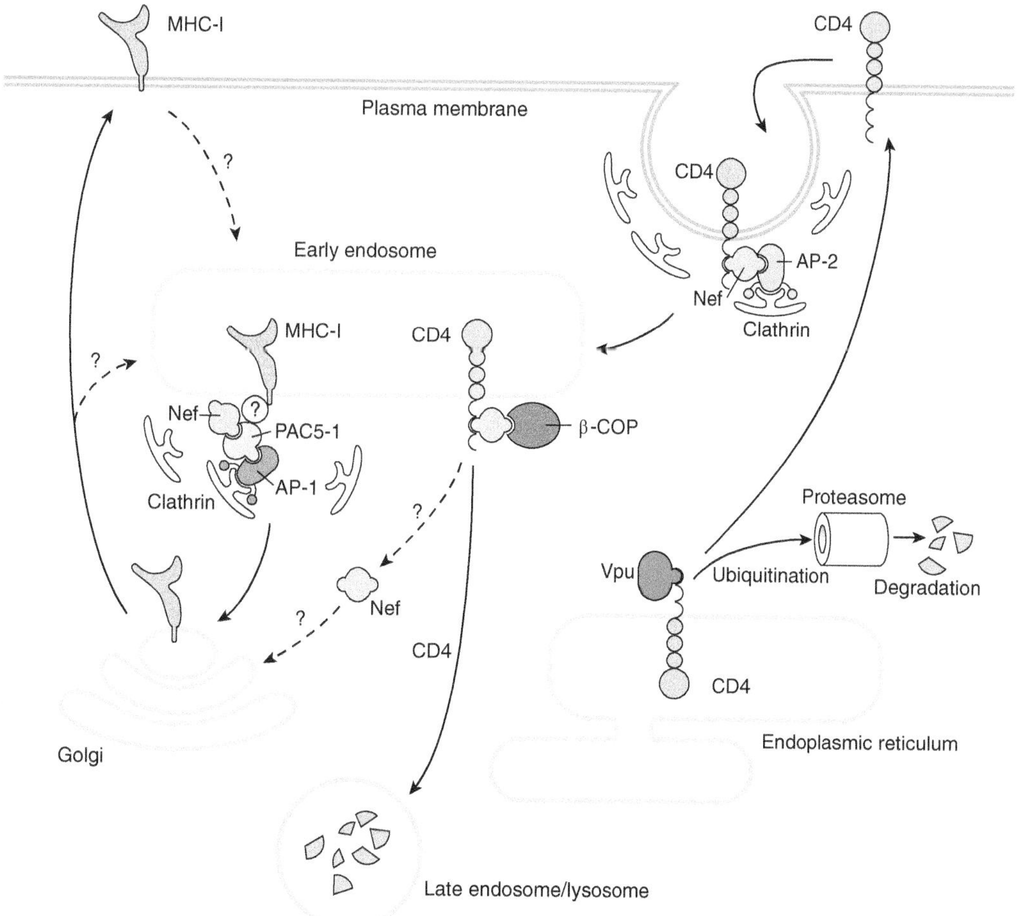

Figure 20.8 Nef and Vpu downregulate CD4 expression and Nef downregulates MHC-I expression in HIV-1-infected cells. HIV-1 Nef protein mediates internalization of CD4 by binding the cytoplasmic domain of CD4 and linking it to clathrin adaptor protein-2 (AP-2) and β-COP, leading to its internalization and degradation. In contrast, Vpu associates with CD4 in the endoplasmic reticulum (ER) during modification and transport. Vpu causes CD4 to be ubiquitinated, released from complexes with gp120, and degraded. Nef also mediates internalization of MHC-I molecules using clathrin adaptor protein-1 (AP-1) and phosphofurin acidic cluster sorting protein-1 (PACS-1).

HIV-1 Vpr is a component of the PIC, as noted, and is one of several proteins in the complex that contain NLSs, which allow infection of non-dividing cells. Vpr also mediates cell cycle arrest at the G2/M stage and induces apoptosis in infected cells. The G2/M arrest favors viral replication, and the induction of apoptosis after viral replication is complete may aid in immune evasion by eliminating immunogenic viral proteins and protein fragments.

DESTRUCTION OF THE IMMUNE SYSTEM BY HIV-1

HIV-1 is both a sexually transmitted and a blood-borne virus. An individual can be infected by transmission of the virus through intact genital or rectal mucosal membranes. In addition, the virus can also enter the body by direct injection into the bloodstream. This can be the result of mechanical abrasion of mucous membranes, sharing needles with HIV-1-positive injection drug users, or transfusion or injection of contaminated blood products. Replication of HIV-1 in cells throughout the body is a complex process, and it differs among individuals. The virus interacts with a specific receptor, CD4, which is present on T cells and macrophages as well as some other white blood cells (leukocytes). The virus must also interact with a co-receptor, which is one of a number of chemokine receptors found on specific lineages of leukocytes. As mentioned, two well-characterized and important co-receptors are CCR5 and CXCR4, which are both found on macrophages and T cells.

HIV-1 normally has higher affinity for CCR5 than for CXCR4, so it enters macrophages more efficiently via this co-receptor since they have 10-fold lower levels of CD4 on their surface than T cells. Both co-receptors can be readily used for entry into T cells, but CCR5 is preferentially used during acute infection and during the clinically latent period. The reason for this is not entirely known, but may in part be due to stimulation of resting T cells by HIV-1 when it binds CCR5 that does not occur by binding to CXCR4. Infection of mucosal T cells, macrophages, and/or dendritic cells may also be an important selection for CCR5 tropic HIV-1 (R5 HIV-1) during the early stages of HIV-1 infection *in vivo*. Both R5 HIV-1 and CXCR4 tropic HIV-1 (X4 HIV-1) cause AIDS. In about half of AIDS patients, X4 HIV-1 arises during the course of infection and leads to more rapid progression of disease. Nevertheless, many patients with AIDS are infected with R5 HIV-1 only. It is notable that individuals who have a homozygous 32-base-pair deletion in the CCR5 gene that results in the absence of CCR5 from the cell surface are highly (but not completely) resistant to HIV-1 infection. Individuals who are heterozygous for this mutation progress more slowly to AIDS than individuals with normal CCR5 expression.

The course of HIV-1 infection can be divided into three phases: acute infection, clinical latency, and AIDS. This is shown for one patient (a nine-year-old child) in Figure 20.9. During the acute phase of infection, HIV-1 replicates to high levels in the blood and lymphoid organs, often reaching 10^6 copies of HIV-1 RNA per milliliter of blood fluid (plasma). During this phase of infection, the majority of intestinal mucosal CD4+ T cells are destroyed since they are nearly all CCR5+, HIV-1 is cytolytic, and R5 HIV-1 predominates. The early loss of mucosal T helper cells is significant since they comprise around half of the total number of CD4+ T helper cells in the body! The acute phase of infection ends when anti-HIV-1 T helper cells, CTL, and antibody-producing B cells suppress viral replication to a level that is maintained for several to a dozen years depending on the patient. Moreover, this equilibrium level of replication, known as the viral "set-point," that is achieved at the beginning of the clinically latent stage of infection is predictive of the time required for progression to AIDS and death (in the absence of treatment) in each infected individual. Individuals with a viral load set-point of greater than 36 000 copies of HIV-1 RNA per milliliter of plasma have a median survival time of only 3.5 years in the absence of treatment, while those with a set-point level of less than 4500 copies of HIV-1 RNA per milliliter of plasma have a median survival time of 10 years if untreated.

During the clinically latent phase of HIV-1 infection, viral replication persists, especially in lymphoid tissues including the spleen, lymph nodes, and thymus. CD4+ T cells are killed by direct and

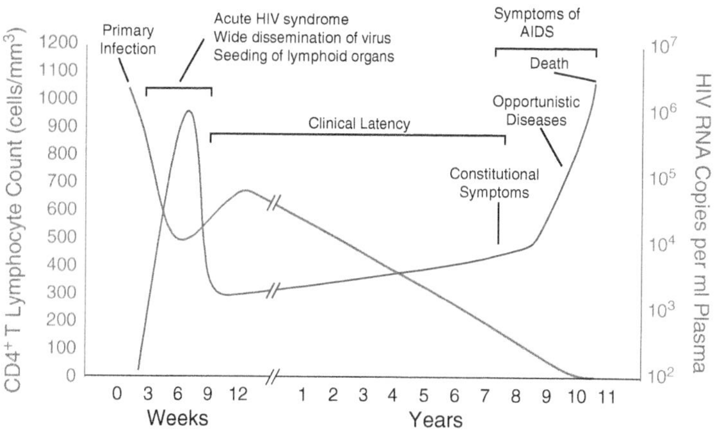

Figure 20.9 The pathogenesis of HIV infection leading to AIDS. Acute infection may be symptomatic, but is short-lived. No symptoms are present during clinical latency, but HIV is not eliminated. When the CD4 count is <200/µl of blood, the symptoms of AIDS appear and lead to death in the absence of treatment. *Source:* Adapted from Pantaleo, G., Graziosi, C., and Fauci, A.S. (1993). *NEJM* 328: 327–335.

indirect consequences of HIV-1 infection. CD8+ T cells also turn over more rapidly since HIV-1 stimulates a strong CTL response that contains viral replication but fails to eliminate it. The hematopoietic system responds to the loss of T cells by increasing their production, but gradually as more CD4+ T cells are lost, the B cell and CTL responses that contain viral replication become less effective due to the loss of T helper cell function. Moreover, HIV-1 readily evolves toward greater pathogenicity due to the high error frequency of HIV-1 RT, short viral life cycle, and relatively high viral load. Changes often occur in the *env* and *nef* genes, including greater CCR5 affinity or mutation to CXCR4 tropism, due to mutation in *env* and increased CD4 downmodulation due to mutation in *nef*. In the absence of treatment, these factors inexorably lead to AIDS.

AIDS is defined as a CD4+ T cell count of less than 200 per microliter of plasma or the presence of an opportunistic infection or malignancy in an HIV-1-infected individual. A normal CD4+ T cell count is around 1000 per microliter. HIV-1-infected individuals normally become sick when they have fewer than 200 CD4+ T cells per microliter of plasma, indicating that loss of CD4+ T cells is a fundamental cause of AIDS. Since 1995, when several effective anti-HIV-1 drugs (viral protease inhibitors and non-nucleoside analog RT inhibitors) were introduced, HIV-1 replication can be suppressed by medication, thereby blocking or at least delaying the onset of AIDS in HIV-1-infected individuals. Anti-HIV-1 drugs are usually used in combinations of two, three, or four drugs that have different patterns of resistance so that drug-resistant HIV-1 does not arise rapidly. Nevertheless, drug treatment is often ineffective over the long term, since drug-resistant HIV-1 may eventually be selected or the toxicity of anti-HIV-1 medications may necessitate cessation of treatment.

Following the initiation of effective treatment, HIV-1 viremia decays in a triphasic manner (Figure 20.10). The rapid first phase of decay ($t_{1/2}$ = 2 days) results from the death of infected T cells and lack of continued HIV-1 replication in T cells, the second phase of decay of viremia ($t_{1/2}$ = 30 days) results from loss of continued infection of macrophages and death of infected macrophages, and the third phase results from the lack of continued infection of resting memory T cells and their gradual loss. These latter cells turn over slowly, resulting in a very slow loss of residual HIV-1 ($t_{1/2}$ = 44 months). By extrapolation from these data, it may require over 60 years of treatment for complete elimination of HIV-1 from an infected individual!

Many challenges remain in preventing the spread of AIDS and caring for AIDS patients. Developing a vaccine against HIV-1 and better antiviral medicines with fewer side effects are both important goals. In the developed world, another important goal is to develop methods to eliminate the long-lived viral reservoir in resting memory T cells in infected individuals. Foremost, however, is the need for better education regarding the risk of AIDS, which is a preventable disease.

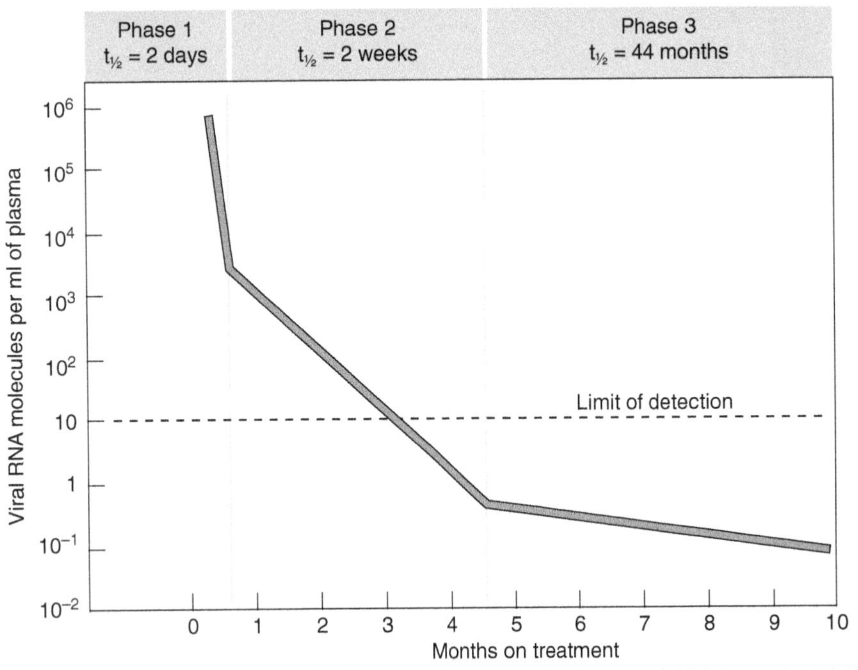

Figure 20.10 Triphasic decay of HIV-1 in plasma following initiation of effective antiviral therapy. The decay of HIV-1 in plasma following initiation of effective treatment occurs in three phases. The first phase has a half-life of approximately two days and results from the death of infected T lymphocytes. The second phase of viral decay has a half-life of about two weeks. During this phase, virus is released from infected macrophages and from resting, latently infected CD4 T cells stimulated to divide and develop productive infection. The third phase has a half-life of 44 months and results from the reactivation of integrated provirus in memory T cells and other long-lived reservoirs of infection. This reservoir of latently infected cells may require more than 60 years of treatment to eliminate.

QUESTIONS FOR CHAPTER 20

1 HIV-1, like all retroviruses, exports unspliced and incompletely spliced mRNA molecules from the nucleus to the cytoplasm. Mammalian cells, however, do not normally export incompletely spliced mRNA molecules. How does HIV-1 overcome this obstacle in its life cycle?

2 What feature of lentiviruses including HIV allows them to infect nondividing cells? What implications does this have for pathogenesis?

3 The discovery of reverse transcriptase by Howard Temin and David Baltimore was a milestone because it contradicted the central dogma of molecular biology. Explain this statement. What is the role of reverse transcriptase in the HIV-1 life cycle?

4 The mammalian APOBEC family of cytidine deaminases constitutes an intracellular defense against retrovirus infection. Explain how the action of human APOBEC protein leads to hypermutation of the HIV-1 genome. How does it cause degradation of the HIV-1 genome?

5 What benefit, in terms of immune evasion, does HIV-1 derive from requiring both a receptor and a co-receptor in order to bind to and enter a cell?

6 The HIV-1 Tat protein was the first transcription factor known to act by binding the nascent RNA transcript. How does Tat act to increase the level of full-length HIV-1 transcripts?

7 What is the function of the HIV-1 viral protein U (Vpu)? Why is this important to the HIV-1 life cycle?

8 During the course of HIV-1 infection, within each infected individual, the surface glycoprotein of HIV-1, gp120, evolves to have higher affinity for the co-receptor CCR5 or to use a second more abundant co-receptor, CXCR4. What are the selective forces that likely drive this evolution, and what does this imply about the importance of viral entry during natural infections?

9 The Nef protein of HIV-1 also evolves during infection in each patient from an early form that downregulates MHC-I proteins efficiently and downregulates CD4 less efficiently to a late-stage form that downregulates CD4 efficiently and downregulates MHC-I proteins less efficiently. Why might this evolution of Nef be favored?

Hepadnaviruses: Variations on the Retrovirus Theme

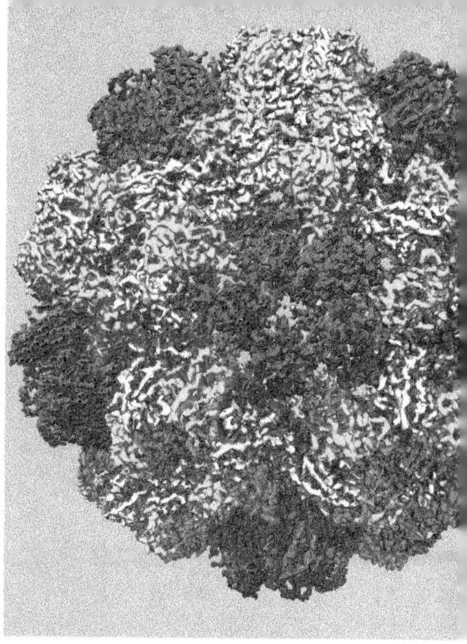

CHAPTER 21

- THE VIRION AND THE VIRAL GENOME
- THE VIRAL REPLICATION CYCLE
- THE PATHOGENESIS OF HEPATITIS B VIRUS
- PREVENTION AND TREATMENT OF HEPATITIS B VIRUS INFECTION
- HEPATITIS DELTA VIRUS
- A PLANT "HEPADNAVIRUS": CAULIFLOWER MOSAIC VIRUS
- Genome structure
- Viral gene expression and genome replication
- THE EVOLUTIONARY ORIGIN OF HEPADNAVIRUSES
- QUESTIONS FOR CHAPTER 21

It is appropriate to end the molecular biology of virus replication with a brief description of a virus group that combines a complex, even bizarre, replication strategy with a very small and compact genome. Add to this complexity the fact that the hepadnaviruses are clearly related to retroviruses, and you have quite a finale!

The hepadnaviruses are named after their propensity, illustrated by the human member–hepatitis B virus, to infect and damage the liver. The name also reflects the fact that they contain a DNA genome in mature capsids. Hepadnaviruses are related to retroviruses in that they encode and incorporate reverse transcriptase into the virion, and the genome is replicated by transcription of the viral genome followed by its conversion into DNA as the virion matures. It has been a difficult technical task to set up systems where hepadnaviruses replicate effectively in cultured cells, although they establish persistent (even lifelong) infections in their hosts. The lack of good cell culture models for their study makes them a difficult research subject, but the availability of duck and woodchuck hepatitis virus model systems alleviates this problem to some degree.

Basic Virology, Fourth Edition. Martinez "Marty" Hewlett, David Camerini, and David C. Bloom.

THE VIRION AND THE VIRAL GENOME

Hepatitis B virus (HBV) virions, also called *Dane particles* after the investigator who first described their characteristic appearance in the electron microscope, are small (35–45 nm) enveloped icosahedrons. The envelope contains three membrane-associated surface polypeptides (S), while the capsid (or core) is composed of a single core capsid protein, the C or "core" antigen, along with reverse transcriptase (termed P).

The viral genome is 3.2 kb long, making it one of the smallest replication-competent virus genomes known. The virion DNA is *partially* double stranded (ds).

As shown in Figure 21.1, virion DNA contains a full-length negative-sense strand that is complementary to the viral messenger RNAs (mRNAs), and a partially completed positive-sense strand;

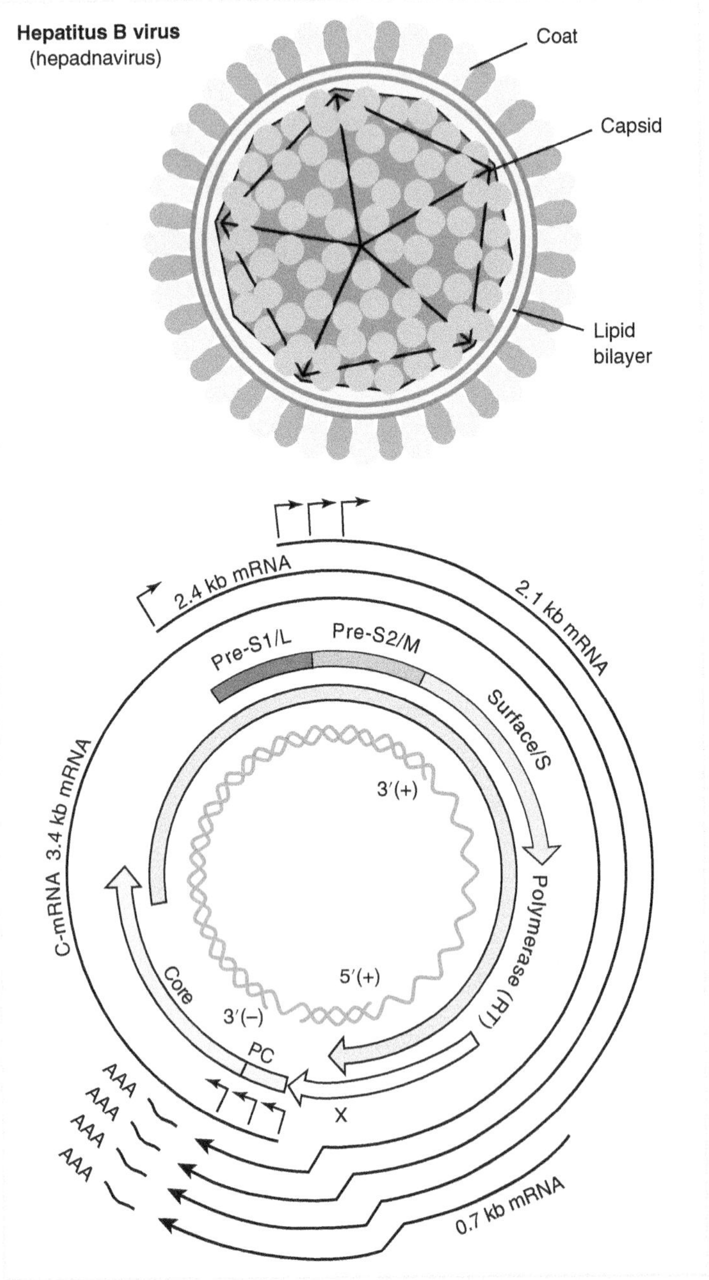

Figure 21.1 A diagram of the virion structure and a genomic and genetic map of human hepatitis B virus. The virion has a lipid envelope with a single exterior coat protein and a single interior capsid protein, which encloses the viral DNA genome. The virion DNA is partially double stranded and was derived by incomplete reverse transcription of full-length virion RNA transcribed during infection. The genetic map shows the compressed arrangement of the hepatitis virus genes and the transcripts expressed from the hepatitis B virus genome. The genome contains three specific promoters. All transcripts terminate at the same site in the interior of the core protein gene.

the single molecule of P protein in the core is covalently bound to the 3′ end of this partially completed strand. Remember that retroviruses contain between 50 and 100 copies of reverse transcriptase in their mature capsids (Chapter 19). The fact that P is actually bound to the partially formed second strand of virion DNA probably is a factor in this difference. The virion DNA is linear but arranged as a circle with a specific gap or nick in the full-length negative-sense strand.

The genetic map of the virus (also shown in Figure 21.1) is complex. A region at the gap in the negative-sense strand has two sets of repeated sequences, and this region encodes a polyadenylation site and three potential promoter sequences. Reading the sequence in a clockwise manner starting at the promoters (indicated by small arrows in Figure 21.1), one finds four translational reading frames that encode C (core protein), P (polymerase), S (envelope proteins), and the X protein. The core and polymerase open reading frames lie in the same orientation as *gag* and *pol* in a retrovirus genome, but the S translational reading frame overlaps the P reading frame. This occurs because the encoded S protein information is in a different reading frame from the P protein code.

THE VIRAL REPLICATION CYCLE

Following receptor-mediated adsorption, penetration by membrane fusion, and partial uncoating, the partially double-stranded virion DNA is completed by virion reverse transcriptase. The genome migrates to the infected cell's nucleus where the free ends are ligated (probably by cellular enzymes), and the small circular double-stranded DNA (dsDNA) molecule becomes associated with cellular histones to become an episome, or mini-chromosome.

Note that, unlike the replication process of retroviruses, viral genomes rarely become integrated into the cellular genome; and unlike the episomal DNA of some papovaviruses, the viral genome cannot be replicated by cellular DNA polymerase (see Chapter 16).

Cellular enzymes interacting with virion promoters transcribe four partially overlapping and unspliced mRNAs of 3.5, 2.4, 2.1, and 0.7 kb. These transcripts all have distinct 5′ ends, but terminate at the same polyadenylation site on the viral genome (see Figure 21.1).

The largest mRNA (also called C-mRNA), which is longer than the DNA template from which it is expressed by virtue of the location of the polyadenylation signal, has sequences repeated at both ends (like retroviral genomic RNA). It serves as a precursor to virion DNA, when it is encapsidated into immature cores or capsids. The 3.5 kb C-mRNA also encodes the core protein and the P protein from an internal translation initiator. Newly synthesized P can reverse transcribe the 3.5 kb mRNA in the cytoplasm, and some of this cDNA generates full-length double-strand cDNA, which migrates back to the nucleus of the infected cell where further transcription can take place. Unlike reverse transcription with retroviruses, that mediated by the P protein does not require a transfer RNA (tRNA) primer – the protein itself serves as a primer.

The S proteins, which are progenitors to the envelope proteins, are expressed by the 2.4 and 2.1 kb transcripts. These transcripts have no obvious counterpart in retroviruses. The 0.7 kb mRNA expresses the X protein.

Expression of viral proteins also leads to the encapsidation of C-mRNA. These immature cores then proceed to generate a complete negative-sense cDNA copy of the encapsidated RNA by the action of encapsidated reverse transcriptase (P). The RNase H activity of the P protein degrades encapsidated template RNA, and partial replication of positive-sense DNA occurs using the negative-sense cDNA as a template while the capsid matures to yield infectious virus.

THE PATHOGENESIS OF HEPATITIS B VIRUS

Hepatitis B virus is only one of a number of viruses targeting the liver. General aspects of the differential pathogeneses of these various hepatitis viruses were discussed in Part I, Chapter 4.

Although the hepatitis B virus genome rarely integrates into its host cell genome, the episomal viral transcription unit survives for a long time in infected liver cells. Continued production of new virus, perhaps modulated by the X protein, leads to a persistent infection. Immune-competent individuals who were infected as healthy adults usually can clear the virus after a long recovery period, although permanent liver damage can occur.

Such recovery does not always happen, however, and hepatitis B virus infections of adults can have serious and fatal sequelae. Individuals infected as infants or young children often do not clear the virus efficiently and become chronically infected. A further complication results from infection of immunocompromised adults, where high mortality rates from acute hepatic failure are not uncommon. Since the virus is spread by injection of contaminated blood, hepatitis B infections are a significant danger to medical personnel, especially those treating chronic intravenous drug users and AIDS patients.

It is notable that chronic hepatitis B virus infections acquired in early childhood rather convincingly are statistically correlated with the subsequent development of **hepatocellular carcinomas** (HCCs). Chronic hepatitis is endemic in areas of Southeast Asia, regions that also demonstrate a high occurrence of fatal liver cancer. The mechanism by which chronic hepatitis infections lead to hepatic carcinoma is not fully understood, but the fact that a very high proportion of infected woodchucks develop HCC when experimentally infected with the related woodchuck hepatitis virus adds strong support to the epidemiological evidence linking HBV and cancer.

Several models of oncogenesis are currently under investigation. It could be that the occasional integration of viral DNA into an infected cell genome leads to the interaction between a viral protein and cellular control circuits in a manner somewhat analogous to human papillomavirus (HPV)-induced carcinomas. The only virus gene product that currently is thought to be a potential candidate for having a direct role in induction of tumors is X protein, which has been shown to have some regulatory and transcriptional stimulatory activities in the laboratory. This model suffers, however, from the fact that a significant portion of cancer cells isolated from HCC patients do not contain any evidence of integrated HBV DNA, and many of those that have viral DNA integrated have extensive rearrangements and deletions in the X protein–encoding sequences.

Another plausible current model is that the continued destruction of liver tissue due to chronic infection leads to abnormal cell growth by a mechanism similar to that seen with chronic human T-cell leukemia virus (HTLV) infections of T lymphocytes (see Chapter 19). Chronic tissue damage leads to a proliferation of cytokines produced to encourage tissue regeneration, but this eventually leads to mutational damage to cells and, ultimately, cancer.

PREVENTION AND TREATMENT OF HEPATITIS B VIRUS INFECTION

A recombinant vaccine for HBV was developed in the 1980s and subsequently replaced the previous HBV vaccine, which was purified from blood of infected individuals. The new vaccine consists of the HBV surface antigen (S) produced in yeast cells. It was the first recombinant vaccine and is highly safe and effective. The recombinant HBV vaccine is now one of the World Health Organization (WHO)-recommended routine childhood vaccines used widely in over 180 countries throughout the world.

Despite the vaccine, HBV is still a major cause of hepatitis and liver cancer throughout the world. Most infections are cleared without treatment, but chronic or severe HBV infection can be treated with nucleoside or nucleotide analogs also used to treat retroviral infections, including HIV infection. The most commonly used medicines include entecavir and lamivudine. These drugs can't cure a patient of HBV, but they may lessen the symptoms and prevent the development of severe disease.

HEPATITIS DELTA VIRUS

As briefly outlined in Chapter 4, Part I, hepatitis delta virus (HDV) appears to be absolutely dependent on coinfection with HBV for spread. Despite this, there are a significant number of cases where it can be inferred that an individual was infected with HDV without any evidence of active or prior HBV infection.

The HDV genome, shown in Figure 21.2, has very significant similarities with plant viroid RNAs! It is difficult to come up with a convincing scenario that explains how a plant pathogen could become associated with human HBV, which has certain important similarities to retroviruses. The HDV particles are enveloped with a membrane containing the three envelope glycoproteins of HBV. Within the envelope is the HDV nucleocapsid containing a covalently closed, circular, single-stranded 1.7-kb RNA molecule of negative-sense orientation complexed with multiple copies of the major gene product of this RNA, the *delta antigen*.

The circular RNA can form base pairs within itself, forming a rod-like structure reminiscent of plant viroid agents. The delta antigen contains three major structural domains. There are two RNA-binding domains, a nuclear localization signal, and a multimerization domain characteristic of proteins in the **leucine zipper** family. Many of these proteins are known to have a role in regulating transcription.

After entry and uncoating, the genome and associated delta antigen are transported to the nucleus of the cell where the replicative cycle begins. The delta virus genome is transcribed and replicated by *host cell RNA* polymerases, including I, II, and III. This is truly unique in animal virus systems, and is a major exception to the rule that cells cannot copy RNA into RNA. Somehow this agent has evolved to co-opt all three host RNA polymerases for this job.

RNA is transcribed into an antigenome that is positive sense and also a covalently closed circle. Transcription also generates a subgenomic mRNA that is capped and polyadenylated and is translated into the delta antigen. The generation of the subgenomic mRNA may occur by transcription that does not continue to generate the full antigenomic template for transcription of further genomic RNA. Alternatively, it may be generated by the circular RNA acting as a **ribozyme** that autocatalytically cleaves itself into a linear form. This latter mechanism is known to be the way that unit-length genomic RNA is generated from circular intermediates generated

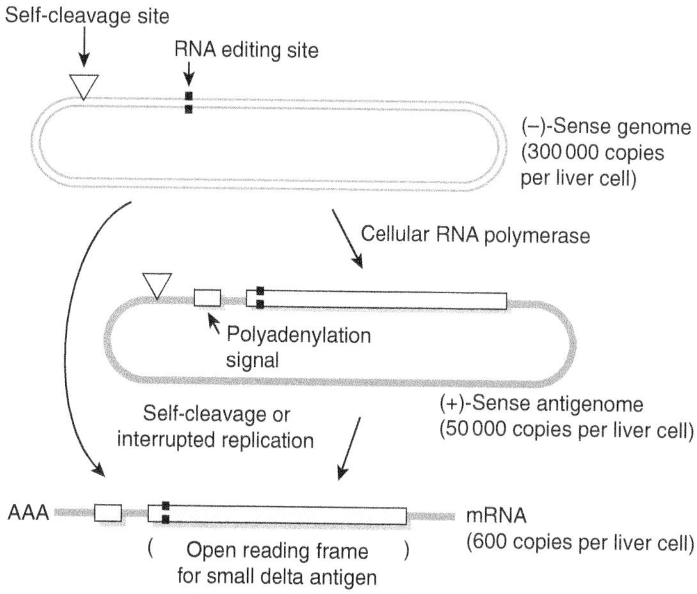

Figure 21.2 The three RNAs of hepatitis delta virus found in infected liver cells. The genomic negative-sense RNA, which is replicated by means of RNA polymerase II, encodes the antigenomic positive-sense RNA, which is the template for genomes, and a subgenomic positive-sense mRNA. This mRNA is cleaved from the antigenomic RNA by RNA self-cleavage. Further, the RNA can be edited by cellular enzymes so that the first translational terminator can be altered. With such edited RNA, a protein 19 amino acids larger than that expressed from unedited RNA is produced.

during the replication process. The term *ribozyme* was invented by Thomas Cech to explain the fact that in splicing of fungal pre-mRNAs, the RNA molecules can assume a structure so that they can hydrolyze an internal phosphodiester bond without the mediation of any protein at all. He was awarded the Nobel Prize for this discovery.

The delta antigen comes in two forms, a small version (195 amino acids) and a somewhat larger version (214 amino acids). The two forms differ by 19 amino acids, and translation of the larger form results from an RNA editing reaction that changes a UAG stop codon into a UGG. This editing suppresses the termination codon and allows continued translation. The short form of the delta antigen is required for genome replication, while the long form suppresses replication and promotes virus assembly.

HDV is spread by blood contamination and causes a pathology much like that of other hepatitis viruses, resulting in liver damage. The severity of this disease results from coinfection with HBV or superinfection of an HBV-positive patient with HDV. In this latter situation, fatality rates can be as high as 20% and virtually all survivors have chronic hepatitis.

While HDV pathology requires coinfection with HBV, this does not explain occurrence and spread of the virus. The virus is found in indigenous populations of South America and is prevalent in Europe, Africa, and the Middle East, but is relatively uncommon in Asia, where there is a high frequency of endemic HBV infections. There may be some way the virus can be maintained and spread without HBV, or it may be able to replicate asymptomatically in some hosts who are also asymptomatically infected with HBV.

A PLANT "HEPADNAVIRUS": CAULIFLOWER MOSAIC VIRUS

The known dsDNA viruses of plants include the caulimoviruses (cauliflower mosaic virus) and the badnaviruses (rice tungro bacilliform virus). Like the hepadnaviruses, each of these groups of plant viruses replicate their DNA genomes through a single-stranded RNA intermediate using a reverse transcriptase activity. Since cauliflower mosaic virus has been most widely studied, a few features of this agent are examined here, and its genome is shown in Figure 21.3.

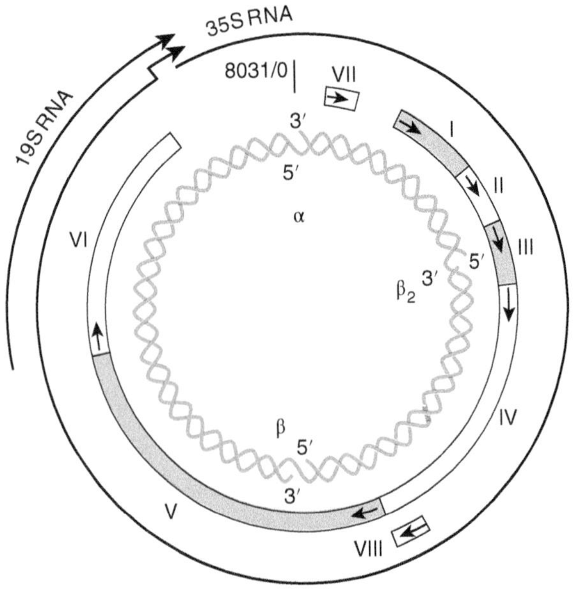

Figure 21.3 The genome of cauliflower mosaic virus. The three breaks in the genome are indicated by Greek letters α, β, and $β_2$. The translational reading frames are indicated by Roman numerals I through VIII. The two viral mRNAs are also indicated.

Genome structure

The genome of cauliflower mosaic virus is a circular dsDNA of about 8 kb that contains three single-strand breaks: one in the transcribed or negative strand, and two in the positive strand. The DNA is packaged into a 54-nm icosahedral particle assembled from a single coat protein. The genome itself has six major open reading frames transcribed in a single direction, as well as a large intergenic region.

Viral gene expression and genome replication

The virus's genome is uncoated and delivered to the cell's nucleus, where the strand breaks are repaired and a mini-chromosome is formed after association with cellular histones. Transcription takes place from this template. The genome is transcribed by host RNA polymerase into two species: a 35S RNA that is slightly longer than the genome and contains a short overlap, and a 19S RNA. These two transcripts are exported to the cytoplasm, where they are translated.

The 35S RNA is the message for five of the six proteins encoded. Exactly how these proteins are translated is not known, although some form of internal ribosome initiation is proposed. This set of proteins includes the coat protein as well as the viral reverse transcriptase.

The 19S RNA is translated into the sixth viral protein, a regulatory protein. Genome replication takes place by reverse transcription of the 35S RNA in the infected cell's cytoplasm. The primer for this replication is a host tRNA specific for methionine, and synthesis probably takes place within a precursor particle that will ultimately mature into a virion.

THE EVOLUTIONARY ORIGIN OF HEPADNAVIRUSES

The obvious relationship between the hepadnaviruses, caulimoviruses, badnaviruses, and retroviruses leads to a question that illustrates both the strengths and weaknesses of applying molecular genetics to questions of virus origins. Retroviruses (and their relatives, the retrotransposons) are widespread, if not ubiquitous, throughout the plant, animal, and bacterial kingdoms. The conservation of critical sequences of reverse transcriptase suggests that it is an enzyme whose appearance in the biological world was a unique event. It is not too much of a jump to suggest that its appearance occurred early in the evolutionary scene. Indeed, retroviruses may well have had a major role in some types of evolutionary processes, as they display the capacity to mediate horizontal transfer of genes or groups of genes among organisms.

The relationship between the replication strategy and gene order of the DNA viruses related to retroviruses is strong evidence that they derive from an ancestral retrovirus, but when this occurred is not known. It can be debated that the presence of hepadnaviruses in avian as well as mammalian species argues for a time of appearance before divergence of their hosts, but it is just as likely that a successful hepadnavirus progenitor arose in mammals and then spread to birds (or vice versa). The same argument holds for looking for predecessors of plant and animal retro-related viruses.

The point is that there is no real way to tell. Even if a clear picture of both bird and mammal or plant and animal ancestry vis-à-vis each other were established, it would not guarantee a resolution of the puzzle. Perhaps detection of the genetic remnants of an ancestral virus in the genomes of the various hosts could shed light on the question, but the fact that the viruses do not integrate as part of their productive life cycle makes the likelihood of finding such a remnant somewhat remote.

Case 6: Hepatitis B

Clinical presentation/case history: A 33-year-old male DJ develops symptoms of fever and malaise (tiredness) over a one-week period. He initially figures he had just "come down with a bug" but becomes concerned when he notices that his sclera (the white part of his eyes) are yellowish in color, and his urine has become dark brown in color. He goes to a doctor. Upon examination the doctor determines the patient has severe jaundice and draws blood to send to the lab for serology, chemistry, and cell count analysis. The doctor also learns, after talking to the patient, that he drinks 5–10 alcohol-containing drinks a week, and has a history of intravenous (IV) drug use and has shared needles "a few times."

Diagnosis: The symptoms of yellow pigmentation of the eyes and skin, along with the dark urine, are consistent with jaundice, a condition caused by damage to the liver, which releases bilirubin into the blood and is responsible for the yellow color. Jaundice can be caused by a number of pathologies, including alcohol- or drug-induced cirrhosis of the liver, hepatocellular carcinoma (tumor of the liver), and a number of infectious causes, including hepatitis viruses A, B, C, D, and E. Based on the case history of rapid onset of symptoms, fever, and the fact that the patient has shared needles, the doctor suspects an infectious cause of the hepatitis. The blood analysis, particularly the serology, will allow him to determine if this patient has evidence of an active hepatitis infection.

The patient's serology revealed the presence of immunoglobulin M (IgM) antibodies against hepatitis B virus surface antigen (HBsAg), as well as the presence of core antigen protein (HBcAg). The presence of the core antigen was a clear indication of an active hepatitis B virus (HBV) infection. However, since HBV can cause either an acute or chronic infection, this result alone could not distinguish between these two possibilities. However, the detection of the IgM class of antibodies against HBsAg indicated that the antibody response to HBsAg was fairly recent, since class switching had not yet occurred, and indicated to the doctor that the patient had a recently acquired acute infection with HBV.

Treatment: No treatment is required for acute HBV infection, which usually resolves with supportive care only. The doctor recommended plenty of rest and fluids, and to return for a follow-up visit in a month so he could determine if the infection resolves itself, or whether the patient develops a chronic HBV infection. For this determination, additional serology will be performed, and the doctor will look for the presence of viral proteins (HBsAg or HBcAg) as an indication that virus is still being produced in the liver. If these are not detected, then the infection has resolved and the patient will be immune to further infection by HBV. If the patient is found to have a chronic infection, he will be a carrier of HBV and will be at higher risk for cirrhosis of the liver and hepatocellular carcinoma. Chronic or severe HBV infection is treated with nucleoside or nucleotide analogs with or without immune modulators.

Disease notes: There were about 3000 new cases of HBV reported in the United States in 2014, a reduction of more than 99% since the recombinant HBV vaccine was introduced in 1991. Most of the fatalities are sequelae of chronic infections due to cirrhosis or hepatocellular carcinoma. Only 2–15% of HBV infections acquired as an adult result in a chronic infection; however, 30–90% of infections of children under the age of five result in chronic infection. This strikingly different outcome is likely due to the fact that T-cell immunity is required to clear an HBV infection, and children under five still have a relatively immature T-cell response. This allows the virus to establish a persistent, chronic infection in these children, and they will likely continue to shed the virus and be a source of new infections for their entire life.

In addition to treatment for chronic or severe HBV disease, there is a very effective and safe recombinant subunit vaccine that is now widely used in over 180 countries, including the United States. This vaccine, which consists of the HBV surface antigen produced in yeast, provides lifelong immunity.

QUESTIONS FOR CHAPTER 21

1 Compare translation of the hepadnavirus core and polymerase proteins with the translation of the retroviral *gag:pol* region (described in Chapter 19).

2 Describe the role of the viral reverse transcriptase in the replication of the DNA genome of hepadnavirus.

3 What is the best current model for how hepadnavirus causes hepatic carcinoma?

4 Justify the following statement: Hepadnaviruses evolved from an ancestral retrovirus.

5 Hepatitis delta virus (HDV) is classed as a subviral entity. What is a unique feature of the genome replication of this agent?

Problems

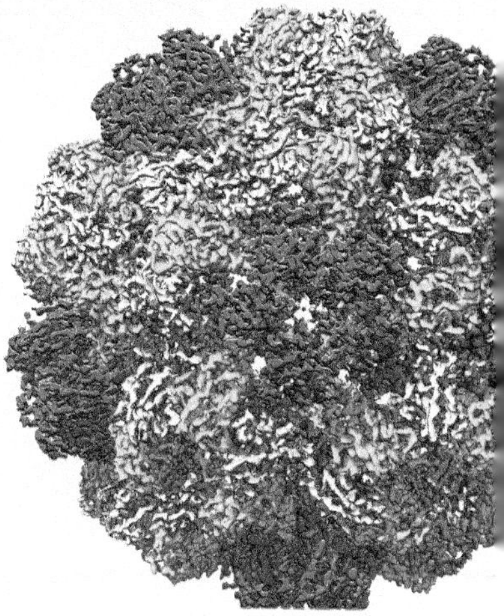

PART IV

1 Inhibition of protein synthesis in a cell can occur as a direct result of a virus infection or as a result of an infected cell being in an antiviral state due to interferon treatment. The following table describes data obtained for various viral infections (they are not all the same virus). Indicate whether the observed alteration of protein synthesis is a direct result of a virus or is the result of the antiviral state in the cell induced by interferon. Be careful! There may be effects that are true for both! Indicate your predicted results by placing a checkmark in the appropriate square.

Protein synthesis inhibited because ...	Directly a result of the virus?	Induced by interferon?
Cap structures are endonucleolytically removed from host mRNA in nucleus.		
RNase L is activated in the cytoplasm of the cell.		
Protein synthesis initiation factor eIF-2 is phosphorylated.		
Protein synthesis initiation factor eIF-4F is proteolytically degraded.		

2 For most DNA genome viruses, gene expression is classified as either "early" or "late."

(a) What event of the viral life cycle is used to distinguish early from late expression?

(b) For the following kinds of viral genes, indicate whether you would expect early or late expression.

Viral gene	Class (early or late)
Capsid protein	
DNA polymerase	
Inhibitor of host transcription	
Lytic enzyme	

3 Sin Nombre virus is the causative agent of the adult respiratory distress syndrome that is transmitted by deer mice in many regions of the western United States. From your knowledge of the family in which these viruses are classified, fill in the following table of properties.

Property	Data for Sin Nombre Virus
Sense of the RNA genome	
Number of genome segments	
Presence or absence of a virion-associated polymerase	
Site of replication in the cell	

4 The senioritis variant of spring fever virus (SpFV-4) is now at peak expression in a local epidemic at the University of Arizona. Even the most ambitious premedical student seems to be susceptible. However, while wandering through the nearly vacant science library, you come upon a study room with a group of seniors who are intently working behind an immense pile of books. As you watch, it becomes clear from observing their behavior that this group has not been infected. You convince them to donate some of their cells to your laboratory for further study. Remember that you have already determined SpFV-4 to be a member of the newly defined *Procastinovirus* genus of the family Orthomyxoviridae.

The following table of data compares the properties of the susceptible cells you have been using to grow the virus and the resistant cells you obtained from the uninfected study group:

	Results for SpFV-4 infection of	
Experiment to examine	**Susceptible cells**	**Resistant cells**
Virus attachment	Viral particles found attached to surface receptors in normal numbers	Viral particles found attached to surface receptors in normal numbers
Virus entry	Viral particles observed within endosomes in normal numbers	Viral particles observed within endosomes in normal numbers
Viral gene expression	Viral gene expression taking place in the cell nucleus; viral mRNAs in the cytoplasm	No viral gene expression in the nucleus; no viral nucleic acid present in the nucleus or cytoplasm

(a) Based on these data, what step in SpFV-4 infection do you think is blocked in the case of the resistant cells from the study group?

After interviewing the members of the study group, you find the following common characteristics of the members:

- They all take the bus to campus each morning.
- They all eat breakfast together each day at Louie's Lower Level, a student union restaurant.
- They are all biochemistry majors.

(b) Which of these behaviors suggests a possible origin for the resistance to SpFV-4?
(c) How would you begin to identify the source of the resistance?

5 For each of the following viruses, describe the most likely *normal route* of entry into the host cell, based on the described experimental observations.
(a) La Crosse encephalitis virus: At neutral pH (pH 7) the envelope of the virus particle does not fuse with a cell membrane, but at acidic pH (pH < 5) the envelope of the virus particle fuses with a cell membrane.
(b) Sendai virus: The envelope of the virus particle fuses with a cell membrane at either neutral or acid pH.

6 You have discovered that Mardi Gras virus (MGV) has a genome structure that most closely resembles members of the Bunyaviridae family (three single-stranded RNA segments). However, the symptoms of disease associated with MGV (behavioral abnormalities) are unlike any produced by the other members of the family, all of which cause either encephalitis or a hemorrhagic syndrome (except for the lone plant virus member). In spite of this, you wish to determine whether MGV has other molecular properties that would warrant inclusion into this family, possibly as the prototype member of a sixth genus.
(a) Complete the following checklist that you are preparing for members of your laboratory. In each case, if MGV is a member of the Bunyaviridae, predict the result you would expect when your laboratory investigates each property.

Molecular property	If MGV is a member of the Bunyaviridae
Number of membrane glycoproteins	
Mechanism of 5′-cap addition to viral mRNA	
Possible strategies for gene expression from the ssRNA	
Site of virus maturation in the host cell	

mRNA: Messenger RNA; ssRNA: single-stranded RNA.

(b) On a Tuesday afternoon, while you are away from the laboratory (teaching your virology class), one of your graduate students drops a vial containing MGV. The vial shatters on the desktop. Your technician, another graduate student, and an undergraduate student are present. The following Friday, none of them is in the laboratory. You discover that they are all at O'Malley's Tavern and therefore have been infected with MGV. What is the most likely route of infection in the laboratory accident?
(c) Given this accidental "experiment," is it likely that MGV is an arbovirus? Why or why not?

7 You have created two strains of *Escherichia coli*, genetically engineered to express λ bacteriophage proteins under certain conditions. In each case, the λ gene has been inserted into the bacterial DNA such that expression of the gene is under control of an inducible promoter (the promoter and operator sequence from the lac operon). In each case, assume that the uninduced cell behaves exactly as a normal *E. coli* cell would with respect to λ.

(a) The first strain is an *E. coli* cell with the gene for the λ cro protein, placed downstream from a copy of the lac promoter and operator. The cell is placed in the presence of IPTG, an inducer of the lac operon. Describe what will happen in such an induced cell to the λ bacteriophage in each of the following cases. Justify your answer with respect to control of the λ life cycle:

- The cell is infected with bacteriophage λ.
- The cell contains a λ provirus as a part of its DNA.

(b) The second strain is an *E. coli* cell with the gene for the λ cI protein, placed downstream from a copy of the lac promoter and operator. The cell is placed in the presence of IPTG, an inducer of the lac operon. Describe what will happen in such an induced cell to the λ bacteriophage in each of the following cases. Justify your answer with respect to control of the λ life cycle:

- The cell is infected with bacteriophage λ.
- The cell contains a λ provirus as a part of its DNA.

8 Messenger RNAs in eukaryotic cells are monocistronic. Viruses have evolved a variety of schemes for growth within this environment, given this constraint. Explain *briefly* (a short phrase will do) how each of the following viruses expresses all of the necessary viral proteins within its host cell.

(a) Poliovirus
(b) Vesicular stomatitis virus

9 In searching through the freezer in your advisor's laboratory, you discover a box containing three vials of material. In the bottom of the box, you find three labels that had fallen from the three now-unlabeled vials. The labels read as follows: "influenza virus type A," "La Crosse encephalitis virus," and "vesicular stomatitis virus." Your advisor gives you the task of identifying which of the vials contains which virus. You grow each virus in cell culture and prepare virions of each with the genome radiolabeled with phosphorus 32. After isolation of the genomic RNAs, you display them by electrophoresis in an agarose gel. The autoradiogram from this experiment is:

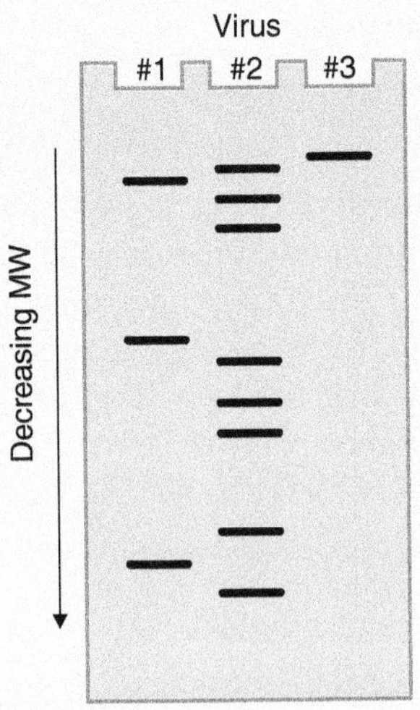

MW: Molecular weight.

Based on these data, indicate which of the vials originally had which of the labels found in the box. Use this table to indicate which label belongs on which vial:

Virus no.	Vial label
	Influenza virus type A
	La Crosse encephalitis virus
	Vesicular stomatitis virus

10 Methods are available for removing the nuclei from eukaryotic cells in culture. These enucleated cells can still carry out their cytoplasmic metabolic functions for a period of time. Predict which of the following viruses would be able to grow in such enucleated cells (assume that the enucleated cells can survive and metabolize long enough for a virus life cycle to be completed).

Virus (Family)	Growth in enucleated cells?
Poliovirus (Picornaviridae)	
Influenza Virus (Orthomyxoviridae)	
Human Immunodeficiency Virus (Retroviridae)	
Vesicular Stomatitis Virus (Rhabdoviridae)	

11 A physician is treating a patient who is having recurring outbreaks of herpes simplex virus type 2 (HSV-2) infection (herpes genitalis). She has decided to administer the drug acyclovir (acycloguanosine) to this patient during such outbreaks. The patient has a degree in molecular biology, but no training in virology. He asks the physician to explain the *mode of action and safety* of acyclovir to him.

(a) What should she tell him are the two reasons why acyclovir works specifically against HSV-2-infected cells?

(b) The physician has explained that these recurring outbreaks are due to an established latent infection in the basal ganglia. The patient wonders if the acyclovir treatment will cure him of this latent infection. Should the physician answer yes or no, *and* what reason should she give for her answer?

12 Human immunodeficiency virus (HIV) is a member of the *Lentivirus* genus of the family Retroviridae and is the causative agent of AIDS. In designing drugs that will inhibit this virus, it is important that the drug in question targets a specific viral function. For each of the drug types listed here, predict what stage of the HIV virus cycle will be blocked:

(a) An inhibitor of the viral protease, such as saquinavir
(b) An inhibitor of the viral integrase, such as raltegravir
(c) AZT (zidovudine) and related compounds
(d) An inhibitor of the viral protein Rev

13 You have been called in as a virological consultant in the case of the voles living at the site of the Chernobyl reactor. These rodents are apparently thriving in the midst of the radioactive waste and are mutating at a rate 10 times greater than normal. Your hypothesis is that viruses that might be vectored by these rodents could also mutate at a more rapid rate than normal. In a preliminary investigation, you have isolated viruses from this population of voles. You have discovered a virus from these rodents that you have tentatively named the Chernobyl vole virus (CVV). This table lists some of the features of this virus that you have determined:

	Virus feature	Result for CVV
A	Virion	Enveloped, with two membrane glycoproteins
B	Genome	ssRNA, negative sense, three segments
C	Virion-associated RNA polymerase	Present
D	mRNA synthesis	Nuclear cap scavenging to begin mRNA synthesis
E	Disease	Causes encephalitis in voles, with a case fatality rate of about 15%

mRNA: Messenger RNA; ssRNA: single-stranded RNA.

(a) Which of these features (A through D) would justify inclusion of CVV into the Bunyaviridae family? (Points will be deducted for features listed that *do not* justify this inclusion.)
(b) Which, if any, of these features make CVV different from other members of the family Bunyaviridae?
(c) Which, if any, of these features make CVV different from members of the *Hantavirus* genus of the Bunyaviridae?

14 Three of the strategies found in the translation of RNA genome viruses in eukaryotic cells are (i) monocistronic mRNAs translated into a polyprotein, (ii) monocistronic mRNA translated into a single protein, and (iii) the use of overlapping reading frames to produce two proteins from one mRNA. Indicate which of these strategies is employed for each of the following viruses. Indicate your choices with either “Yes” or “No.”

Virus: Family	Monocistronic to polyprotein	Monocistronic to single protein	Overlapping reading frames
Poliovirus: Picornaviridae			
Sindbis virus: Togaviridae			
Vesicular stomatitis virus: Rhabdoviridae			
La Crosse encephalitis virus: Bunyaviridae			

15 An experiment is designed to test the effect of infecting a cell with two different viruses. You wish to examine the effect of the coinfection on viral protein synthesis in the cell. Assume that in each case, the host cell is susceptible to each virus and could support the growth of either virus alone. In the following table, predict which of the two viruses will predominate or whether both will be normal with respect to viral protein synthesis, *and* give a *brief* reason in defense of your choice.

First virus	Second virus	Protein synthesis by which virus?	Why?
Adenovirus	Poliovirus		
Vesicular stomatitis virus	La Crosse encephalitis virus		
Influenza virus	Poliovirus		
Adenovirus	Herpesvirus		

16 This table has three viruses that have been considered in detail in this text: bacteriophage T4, poliovirus, and vesicular stomatitis virus. The table also includes events that may be steps in the assembly of one or more of these viruses. Write "Yes" or "No" in the table to indicate which of the events is associated with the assembly of which virus.

Event	Bacteriophage T4	Poliovirus	Vesicular stomatitis virus
Capsids or nucleocapsids are assembled before the insertion of the genome.			
Capsids or nucleocapsids have a helical symmetry.			
Final maturation of the particle requires proteolytic cleavage of one of the capsid or nucleocapsid proteins.			
New viral particles are released by budding from the surface of the cell.			

17 Both poliovirus (*Picornaviridae*) and Rous sarcoma virus (*Retroviridae*) have RNA genomes that are positive in sense. However, replication of the genome of these viruses differs drastically. In this table, indicate which feature applies to the replication of the RNA genomes of these viruses. Write "Yes" if the feature applies or "No" if the feature does not apply.

Replication feature	Poliovirus	Rous Sarcoma Virus
Replication requires the use of a host cell tRNA as a primer.		
Replication requires the use of a viral protein as a primer.		

Continued

Replication feature	Poliovirus	Rous Sarcoma Virus
Replication results in the conversion of ssRNA to dsDNA.		
The *final product* of replication is progeny single-stranded, positive-sense RNA.		

dsDNA: Double-stranded DNA; ssRNA: single-stranded RNA; tRNA: transfer RNA.

18 For each of the control points given here, identify at least one virus that can alter the cell such that it grows out of control and might result in a tumor. In each case, state how the virus changes the cell.

(a) Growth hormone/receptor
(b) G protein
(c) Tyrosine kinase
(d) Transcriptional regulator
(e) Tumor suppressor

Additional Reading for Part IV

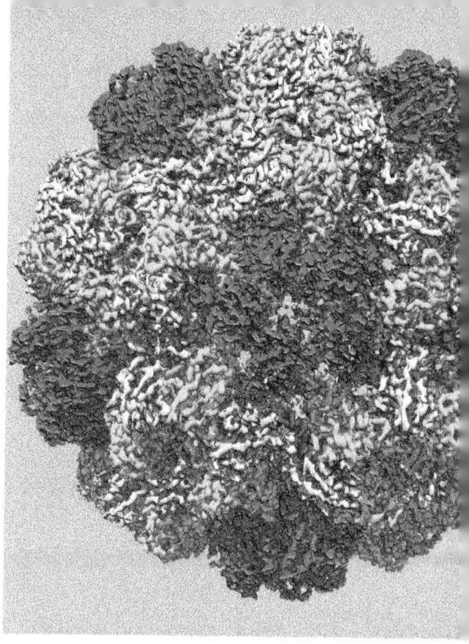

The best place to begin research and further reading on a specific virus or aspect of a specific virus is in the general reference:

- Mahy, B.W.J. and van Regenmortel, M. (eds.) (2008). *Encyclopedia of Virology*. New York: Elsevier.

More detailed information and specific recent citations of experimental articles and recent reviews can be found in the following:

- Flint, S.J., Racaniello, V.R., Rall, G.F., and Skalka, A.M. (2015). *Principles of Virology: Molecular Biology, Pathogenesis, and Control of Animal Viruses*, 4e. Washington, DC: ASM Press.
- Knipe, D.M. and Howley, P. (eds.) (2006). *Fields Virology*, 6e. New York: Raven Press.

Detailed aspects of pathogenesis of virus infections, again organized as a group of specific reviews by individual experts, is covered in the following:

- Nathanson, N. (ed.) (1997). *Viral Pathogenesis*. Philadelphia: Lippincott-Raven.

The Cold Spring Harbor Press publishes many books that contain detailed reviews of many aspects of the molecular biology of viruses and cells. Specific titles of interest to virologists include the following:

- DePamphilis, M.L. (ed.) (2006). *DNA Replication and Human Disease*. Cold Spring Harbor, NY: Cold Spring Harbor Press.
- Coffin, J.M., Hughes, H.H., and Varmus, H.E. (1997). *Retroviruses*. Cold Spring Harbor, NY: Cold Spring Harbor Press.

The ASM Press also publishes a large number of books of great use to molecular biologists, biomedical research workers, students, and the like. One recent title that contains chapters covering viruses discussed in this section is the following:

- McCance, D (ed.) (1998). *Human Tumor Viruses*. Washington, DC: ASM Press.

A very thorough (definitive, actually) discussion of the numerous genetic switches occurring during the replication of bacteriophage l and the biochemical basis of these switches can be found in the following:

- Ptashne, M. (1986). *A Genetic Switch: Gene Control and Phage l*. Palo Alto, CA: Blackwell Scientific Publications and Cell Press.

VIRUS RESOURCES ON THE INTERNET

As noted in the introduction to this text, the internet provides a very effective resource for the most recent information about specific viruses and recent publications concerning them. It is important, however, to keep in mind that many such internet sites are not formally reviewed or edited by any single group or body of experts, and addresses change; therefore, the reliability of any given site address or "factoid" within that site is subject to independent verification.

Some important sites for searching journals and publications include the following:

- The National Library of Medicine and government resources: http://www.ncbi.nlm.nih.gov
- American Society for Microbiology (ASM): http://www.asm.org
- Cold Spring Harbor Press: http://www.cshl.org
- *Nature*: http://www.nature.com
- *Science*: http://www.sciencemag.org
- *Scientific American*: http://www.sciam.com/index.html
- *Journal of General Virology*: http://vir.sgmjournals.org
- *Journal of Virology*: http://jvi.asm.org
- *Virology* (through Science Direct): www.sciencedirect.com

An increasingly large number of internet sites are devoted to individual virus topics related to virus replication, such as oncology and cell transformation. The following sites should be useful for beginning a study of any given virus.

General virus information and databases

http://www.tulane.edu/~dmsander/garryfavweb.html
http://www.virology.net/Big_Virology/BVHomePage.html
http://www.virology.net/ATVnews.html
http://www.epa.gov/microbes/index.html
http://www.wadsworth.org/databank/viruses.htm
http://www.diseaseworld.com
http://rhino.bocklabs.wisc.edu/virusworld

Specific virus sites

Adenoviruses
http://www.virology.net/Big_Virology/BVDNAadeno.html
http://www.cdc.gov/ncidod/dvrd/revb/respiratory/eadfeat.htm
http://www.ncbi.nlm.nih.gov/ICTVdb/ICTVdB/01000000.htm
http://www.stanford.edu/group/virus/adeno/adeno.html
http://www-micro.msb.le.ac.uk/3035/Adenoviruses.html

Arenaviruses
http://www.tulane.edu/~dmsander/WWW/335/Arboviruses.html
http://www.ncbi.nlm.nih.gov/ICTVdb/ICTVdB/00.003.htm
http://gsbs.utmb.edu/microbook/ch057.htm
http://www-micro.msb.le.ac.uk/3035/Arenaviruses.html

Baculoviruses
http://www.baculovirus.com
http://www.ncbi.nlm.nih.gov/ICTVdb/ICTVdB/06000000.htm
http://www-micro.msb.le.ac.uk/3035/kalmakoff/baculo/baculo.html

Bacteriophages
http://www.asm.org/division/m/M.html
http://www-micro.msb.le.ac.uk/3035/Phages.html

Bunyaviruses
http://www.ncbi.nlm.nih.gov/ICTVdb/ICTVdB/11000000.htm
http://www-micro.msb.le.ac.uk/3035/Bunyaviruses.html

Calicivirus
http://www.ncbi.nlm.nih.gov/ICTVdb/ICTVdB/12000000.htm
http://hgic.clemson.edu/factsheets/HGIC3720.htm
http://www.caliciviridae.com
http://www-micro.msb.le.ac.uk/3035/Caliciviruses.html

Coronaviruses
http://www.ncbi.nlm.nih.gov/ICTVdb/ICTVdB/index.htm
http://www-micro.msb.le.ac.uk/3035/Coronaviruses.html

Filoviruses
http://www.cdc.gov/ncidod/diseases/virlfvr/virlfvr.htm
http://www.ncbi.nlm.nih.gov/ICTVdb/Ictv/index.htm
http://www-micro.msb.le.ac.uk/3035/Filoviruses.html

Flaviviruses
http://www.tulane.edu/~dmsander/WWW/335/Arboviruses.html
http://www.cdc.gov/ncidod/dvbid/westnile/index.htm
http://www.cdc.gov/ncidod/diseases/hepatitis/c/index.htm
http://www.cdc.gov/ncidod/dvbid/yellowfever/index.htm
http://www-micro.msb.le.ac.uk/3035/Flaviviruses.html

Geminivirus
http://www.ncbi.nlm.nih.gov/ICTVdb/ICTVdB/index.htm

Hantavirus
http://www.cdc.gov/ncidod/diseases/hanta/hps/index.htm
http://www.ncbi.nlm.nih.gov/ICTVdb/ICTVdB/index.htm

Hepadnavirus
http://www.tulane.edu/~dmsander/Big_Virology/BVDNAhepadna.html
http://www.ncbi.nlm.nih.gov/ICTVdb/ICTVdB/index.htm

Hepatitis delta virus
http://www.hepnet.com/hepd/wormhdv.html

Herpesvirus
http://darwin.bio.uci.edu/~faculty/wagner
http://home.coqui.net/myrna/herpes.htm
http://www.ihmf.org
www.uct.ac.za/depts/mmi/stannard/herpes.html
http://www.cdc.gov/ncidod/diseases/ebv.htm
http://www-micro.msb.le.ac.uk/3035/Herpesviruses.html

Human immunodeficiency virus
http://www.bcm.tmc.edu/neurol/research/aids/aids1.html
http://www.ncbi.nlm.nih.gov/ICTVdb/ICTVdB/index.htm

Oncogenes
http://web.indstate.edu/thcme/mwking/oncogene.html

Oncornavirus
http://www.oncolink.upenn.edu

Orthomyxoviruses
http://www.cdc.gov/flu
www.uct.ac.za/depts/mmi/stannard/fluvirus.html
http://www-micro.msb.le.ac.uk/3035/Orthomyxoviruses.html

Papillomaviruses
http://www.cdc.gov/STD/HPV/STDFact-HPV.htm
http://www-micro.msb.le.ac.uk/3035/Papillomaviruses.html

Papovaviruses
http://www.tulane.edu/~dmsander/Big_Virology/BVDNApapova.html
http://www-micro.msb.le.ac.uk/3035/Polyomaviruses.html

Paramyxoviruses
http://www.cdc.gov/ncidod/dvrd/revb/index.htm
http://www-micro.msb.le.ac.uk/3035/Paramyxoviruses.html

Picornaviruses
http://www.cbs.dtu.dk/services/NetPicoRNA
http://www.iah.bbsrc.ac.uk/virus/picornaviridae
http://vm.cfsan.fda.gov/~mow/chap31.html
http://www-micro.msb.le.ac.uk/3035/Picornaviruses.html

Poxviruses
www.uct.ac.za/depts/mmi/jmoodie/pox2.html
http://www-micro.msb.le.ac.uk/3035/Poxviruses.html

Prions
http://www-micro.msb.le.ac.uk/3035/Prions.html

Reoviruses
http://www.iah.bbsrc.ac.uk/virus/Reoviridae
http://www.cdc.gov/ncidod/dvrd/revb
http://www-micro.msb.le.ac.uk/3035/Reoviruses.html

Retroviruses
http://www.retrovirus.info
http://library.med.utah.edu/WebPath/TUTORIAL/AIDS/AIDS.html
http://www-micro.msb.le.ac.uk/3035/Retroviruses.html

Rhabdoviruses
http://www-micro.msb.le.ac.uk/3035/Rhabdoviruses.html http://www.hhs.state.ne.us/epi/epirabie.htm

Rhinoviruses
http://www-micro.msb.le.ac.uk/3035/Picornaviruses.html#rhino

St. Louis encephalitis virus
http://www.vicioso.com/Health/disease/encephalitis/SLE.html

Togaviruses
http://www.ictvdb.iacr.ac.uk/ICTVdB/73020001.htm
http://www-micro.msb.le.ac.uk/3035/Togaviruses.html

Viral hepatitis
http://www.cdc.gov/ncidod/diseases/hepatitis/index.htm
http://www-micro.msb.le.ac.uk/3035/Hepatitis.html

Viroids
http://www-micro.msb.le.ac.uk/3035/Viroids.html

Molecular Genetics of Viruses

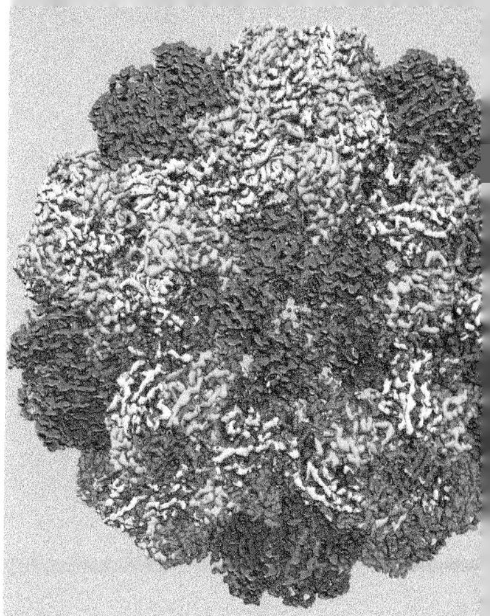

PART V

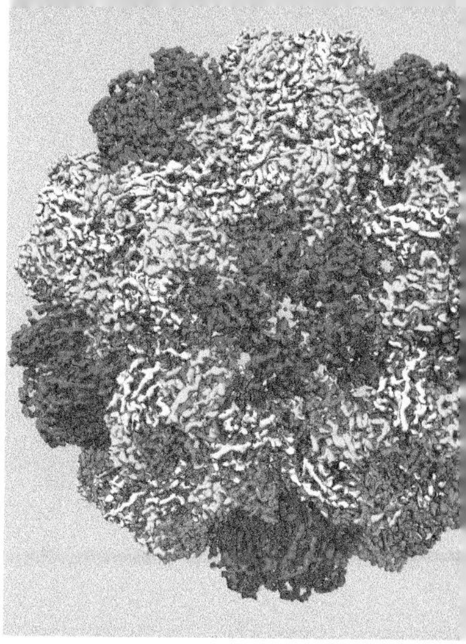

CHAPTER

22

The Molecular Genetics of Viruses

- MUTATIONS IN GENES AND RESULTING CHANGES TO PROTEINS
- ANALYSIS OF MUTATIONS
- Recombination
- ISOLATION OF MUTANTS
- Selection
- HSV thymidine kinase – a portable selectable marker
- Screening
- A TOOL KIT FOR MOLECULAR VIROLOGISTS
- Viral genomes
- LOCATING SITES OF RESTRICTION ENDONUCLEASE CLEAVAGE ON THE VIRAL GENOME – RESTRICTION MAPPING
- CLONING VECTORS
- Cloning of fragments of viral genomes using bacterial plasmids
- Cloning single-stranded DNA with bacteriophage M13
- DNA animal virus vectors
  - Baculovirus
  - Vaccinia
  - Adenovirus and adeno-associated virus provide vectors that can deliver genes to specific tissue
- RNA virus expression systems
  - Retrovirus vectors
  - A togavirus vector
- Defective virus particles
- DIRECTED MUTAGENESIS OF VIRAL GENES
- Site-directed mutagenesis
- GENERATION OF RECOMBINANT VIRUSES
- Homologous recombination
- Bacterial artificial chromosomes
- CRISPR-cas
- QUESTIONS FOR CHAPTER 22

Basic Virology, Fourth Edition. Martinez "Marty" Hewlett, David Camerini, and David C. Bloom.

The goal of molecular virology is the understanding of the mechanisms of viral replication and viral infection at the level of protein function. Ideally, and as outlined in Figure 22.1, this understanding requires a full molecular description of the viral and host genomes; the nature and function of the proteins they encode; as well as the mechanism of the interactions between viral and cellular proteins, especially those resulting in disease, recovery, and/or damage to the organism. This is a big order, but the ability to rapidly determine the sequence of any genome combined with the application of the techniques of molecular genetics have made this goal entirely feasible. The development of molecular genetics – the ability to not only describe the genes encoded by a biological entity, but also modify them and in so doing determine their functions – owes much to the study of virus replication cycles and virus–cell interactions. As pointed out in Chapter 1, the science of molecular biology has its roots in the science of virology; as we will see in this final section of this book, viruses and their study still play a leadership role in the development of biotechnology and the generation of useful information about the biological world.

The "tool kit" of molecular genetics and information technology allows a modern molecular virologist to identify, isolate, and determine the sequence of any gene and the protein encoded by it. Such sequence data can be compared to the sequence of other genes from related and, seemingly, unrelated organisms to determine basic features, which are not readily changed by evolutionary processes (as discussed in Chapter 24). Defined modifications can be made in any region of a protein of interest, and the effects of such modifications on virus replication, structure, and pathogenesis can be determined. The process is not easy, and detailed methodology is complex – the compendium *Current Protocols in Molecular Biology* edited by Ausubel and others comprises several thousand pages and takes up a good part of a library bookshelf, as well as being an extensive online resource. Still, the basic strategy behind such molecular manipulation is based on only a few basic principles of molecular biology – the characterization of viral genomes and gene products that has been described in Part III. Now, with the application of this information

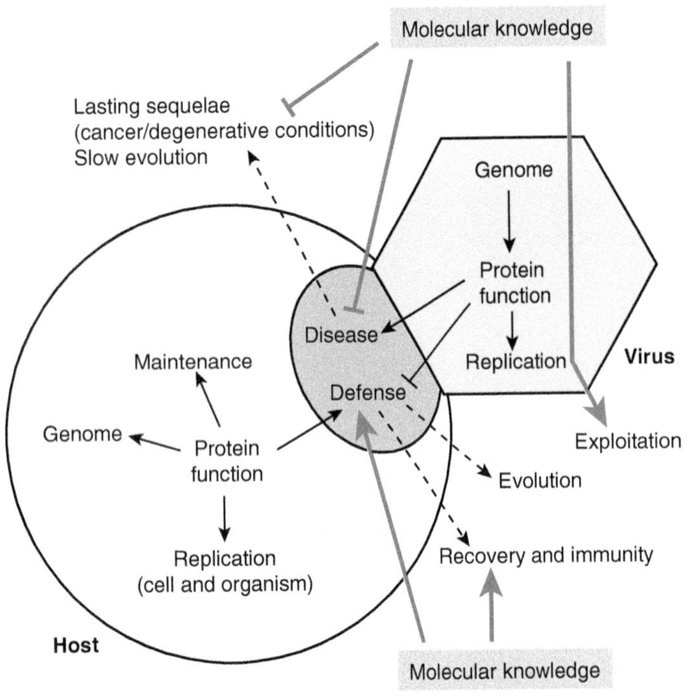

Figure 22.1 The impact of molecular understanding of viral and host genes on the interactions between virus and host. Different stages of the virus replication cycle and the interaction between virus and host are illustrated along with the outcomes of such interactions. Understanding the molecular basis of such interactions is important in controlling disease, aiding recovery and immunity, and preventing lasting sequelae. These interactions also can be exploited in biotechnology and medicine.

to the replication strategies of the major groups of viruses described in Part IV, we can illustrate both methods for manipulating viral genomes and for determining the effects of such manipulation of the parameters of viral infection.

MUTATIONS IN GENES AND RESULTING CHANGES TO PROTEINS

Sometimes, as nucleic acids replicate, a mistake occurs. This is a very rare event in organisms, but in viruses that replicate so rapidly and whose replication enzymes are often error-prone, such changes occur with appreciable frequency. With some viruses that have RNA genomes, such as HIV, the polymerase can generate one mistake for every 10 000 bases transcribed so that many changes are generated. Indeed, as outlined in Chapter 20, Part IV, some of these changes have a role in the virus' ability to avoid the body's immune defenses.

Changes in the genome usually result from insertion of the wrong base during genome replication. For example, a mismatched A:C pair could be formed during replication of DNA instead of A:T, or an A:T pair could be miscopied into G:T. More rarely, a piece of nucleic acid could be lost due to some slippage of polymerase.

Such changes in genomes are called **mutations** and can have effects on the function of the gene in which they occur. Despite this, mutations often lead to very minor changes in proteins. Over very many generations (usually over long time spans), these changes accumulate and lead to the formation of related but distinct organisms. In viruses, changes like these lead to the generation of small antigenic differences between strains or serotypes.

The effect of a mutation on the function of a protein can be profound if that protein is vital to the replication of a virus or has a very critical structure. Sometimes the change can be beneficial to the virus. For example, a viral DNA polymerase might be mutated so that the enzyme is no longer sensitive to a nucleoside analogue; hence, the replication of the mutant virus can no longer be inhibited with the antiviral drug.

Some examples of mutations are shown here where a hypothetical, very short, unspliced open reading frame is mutated. Only the messenger RNA (mRNA) sense DNA strand is shown, but remember that the DNA is double stranded:

Gene : ATG-GTT-GAT-AGT-CGT-TAT-TTA-CCT-CAA-TGG-CAG-TAA
Protein : Met-Val-Asp-Ser-Arg-Tyr-Leu-Pro-Gln-Trp-Gln

1 A mutation in the seventh codon (TTA) to TTC would lead to the substitution of phenylalanine for leucine. Such a change might affect the way the protein folded and change its function:

Met-Val-Asp-Ser-Arg-Tyr-*Phe*-Pro-Gln-Trp-Gln

2 A mutation in this same codon to GTA would lead to a protein with one aliphatic amino acid changed for another; this might have no effect at all on the protein's structure:

Met-Val-Asp-Ser-Arg-Tyr-*Val*-Pro-Gln-Trp-Gln

3 A mutation in this codon to TGA (a stop codon) would lead to a shorter protein; this would almost certainly destroy the protein's function, and if very close to the N-terminal amino acid, would result in the loss of all protein from the gene. Such translation termination mutations are sometimes termed *amber*, *ocher*, or *umber* (sometimes *opal*) mutations by geneticists for essentially historical reasons.

Met-Val-Asp-Ser-Arg-Tyr!

4 A mutation in the second codon (GTT) to GAT would lead to a change in protein charge, substituting asparagine for valine. This would almost certainly significantly alter the function of the protein:

Met-*Asp*-Asp-Ser-Arg-Tyr-Leu-Pro-Gln-Trp-Gln

All such mutations could be mutated further or mutated back by another base change. Such mutations are called *revertible*. If, however, a base were lost (deleted), the change is essentially permanent. Indeed, this type of mutation (addition or loss of genomic material) can be inferred if a mutation has essentially no ability to revert.

5 An example of a nonrevertible change can be seen if the first two bases of the second codon were lost. This **frame shift** mutation would lead to loss of the protein because the second codon is now a translation stop codon:

ATG-***TGA***-TAG-TCG-TTA-TTT-ACC-TCA-ATG-GCA-GTA-A

Other frame shifts can completely change a protein, and thus can completely change or lose the protein's function.

ANALYSIS OF MUTATIONS

Many mutations lead to significant differences in virus plaque morphology, growth rates, host range, cytopathic effects, interaction with the host immune system, and so on. Thus, one can distinguish a mutant virus from a normal or wild-type (*wt*) parent. Mutant viruses are useful because the lack or alteration of the function of a specific viral gene can lead to changes in virus replication, among other things, and this generates information about the function of mutated genes.

An example of the use of **complementation** to maintain a replication-deficient virus mutant is shown in Figure 22.2. If a virus, P^-, with a mutation in its polymerase gene is coinfected into a cell with a virus, C^-, containing a mutation in a capsid protein, for example, each virus can supply the function that the other virus is missing. Complementation allows enumeration of important genes in a virus and manipulation of mutant viruses. *In infections involving complementation, you get back the same virus that you put in.*

Recombination

Sometimes, two viruses that have different mutations can physically entangle each other during the infection process. This close association can lead to formation of a *recombinant* viral genome. The precise mechanism for the formation and resolution of recombinant genomes is still being investigated, but in many cases, the process happens most frequently as genomes are replicating. Also, the farther apart on the parental genomes the mutations lie, the more probability there is of forming a recombinant. Even so, closely separated mutations – even two different ones separated by only a few base pairs within a single gene – can recombine, albeit rarely.

A recombination event can lead to the formation of two novel viral genomes (one has both mutations, and one has neither and is thus *wt*). Such an infection therefore leads to generation of four types of virus: the two single-mutant input viruses, along with the *wt* and double-mutant virus generated from recombination. An example using herpes simplex virus (HSV) is shown in Figure 22.3. A virus with a temperature-sensitive (*ts*) mutation in DNA polymerase (U_L30 Pol^{ts})

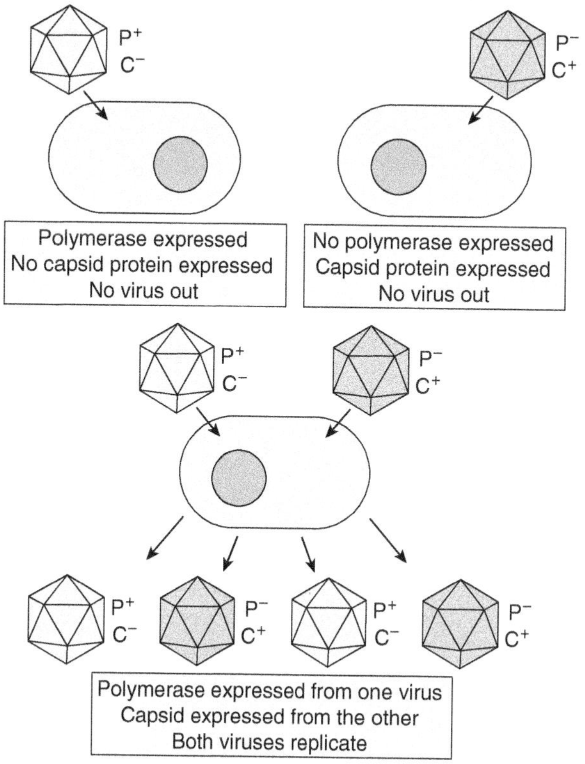

Figure 22.2 Complementation. Neither of two mutant viruses shown can replicate because each contains a lethal mutation in a required gene encoding an enzyme or structural protein. Still, if the two mutant viruses are infected into the same cell, each can supply functions missing in the other. This means that the infected cell will have all the necessary viral gene products for the replication of both mutant viruses – they can complement each other's growth in a mixed infection.

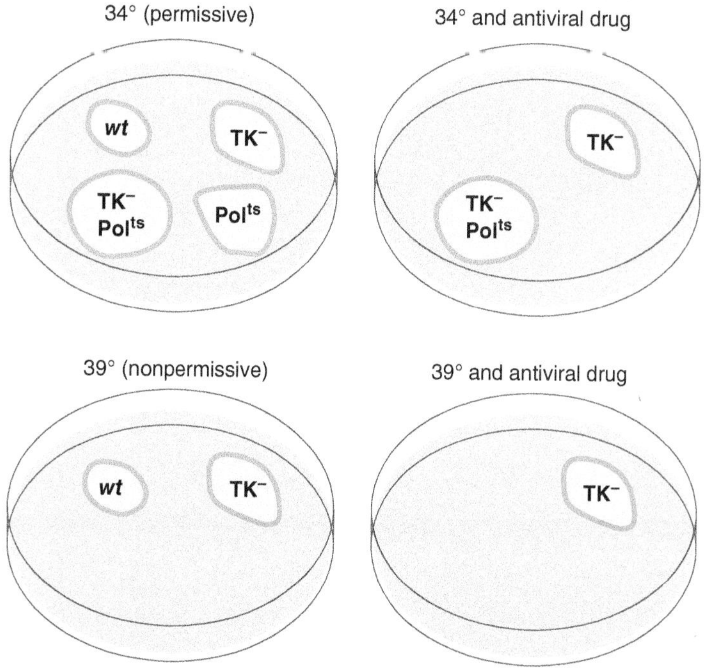

Figure 22.3 Replica plating of virus plaques to distinguish genotypes produced by recombination following a mixed infection. Mixed infection of two genotypes will (rarely) produce novel genotypes by recombination. The example shows replicas of virus plaques that were developed under different selective and screening conditions as described in the text.

is coinfected with a virus with a mutation in the thymidine kinase gene (U_L23 TK$^-$). Thus, one parent virus is not able to grow at an elevated (nonpermissive) temperature, and the other parent is resistant to inhibition of DNA synthesis with a nucleoside analogue.

Recombination results in four different viral genotypes being produced from the mixed infection: *wt*, double mutant (Pol^{ts}, TK$^-$), and the two single-mutant parents. Each genotype will produce plaques that can be distinguished from the others with proper **screening** or **selection** techniques. Replica plating of virus plaques and incubation under different conditions can distinguish between all four genotypes.

In the experiment, progeny viruses are plated and plaques are allowed to form. Then four identical replicas are made on other plates. These are incubated under various conditions that allow the genotypes of the viruses to be distinguished:

1 At high (nonpermissive) temperature, only *wt* and TK$^-$ will make plaques.

2 At normal (permissive) temperature, *wt*, Pol^{ts}, TK$^-$, and Pol^{ts}TK$^-$ all make plaques.

3 At permissive temperature with an antiviral nucleoside analogue drug, only virus lacking the TK gene (TK$^-$ and Pol^{ts}TK$^-$) will make plaques.

4 At nonpermissive temperature with an antiviral drug, only the mutant (TK$^-$) virus will form plaques.

ISOLATION OF MUTANTS

Selection

If a mutation in a virus leads to a significant alteration in growth properties, or resistance to a drug, or resistance to a neutralizing antibody, then the mutant can be selected by growth under conditions where the *wt* virus will not replicate or will replicate poorly. This is an ideal situation for isolation of viruses, but unfortunately, it is not easy to come up with appropriate selective conditions for many desirable mutations. In animal cells (as in bacteria), many mutations change aspects of the virus but do not markedly change growth properties.

HSV thymidine kinase – a portable selectable marker

The ability of the HSV TK$^-$ mutants to replicate in the presence of an antiviral drug is an example of a selective advantage for the virus under these conditions of growth. Further, a different selection scheme that was discussed in Chapter 12 allows for a selective advantage for growth of TK$^+$ organisms. The easy selection of cells that are TK$^+$ or TK$^-$ is an important tool because the HSV TK gene can be removed from the viral genome and used essentially as a portable selectable marker.

The HSV TK gene has been exploited for the last two decades to apply the power of selection to generation of recombinant viruses and cells bearing foreign genes. This is a result of the fact that the HSV TK gene is of convenient size (it is contained within a small piece of DNA – 2.3 kbp) that can be isolated and readily cloned from the HSV genome. This gene contains both the structural gene for the enzyme and the promoter controlling expression of the mRNA encoding this structural gene. Of equal significance, this promoter is expressed in uninfected cells and in viruses other than HSV.

To use the HSV TK gene for selection, a cell line that is constitutively TK$^-$ is constructed. This is accomplished by culturing cells in the presence of a toxic nucleoside analogue, as has been described in the selection process for hybridoma cells outlined in Chapter 12 and Figure 12.3. Since only TK$^-$ cells can replicate under these conditions, the cell lines can be readily established.

To construct a recombinant virus or cell line, one can incorporate the HSV TK gene into a plasmid bearing the gene of interest for transfection, or a TK plasmid can be added to the plasmid of interest. The process of transfection strongly favors cells taking up relatively large aggregates of DNA, so both plasmids will be incorporated.

After recovery from the transfection process, those cells or viruses that have undergone recombination with the added genome, and thus are TK^+, can be selected. For the selection of recombinant virus, the total virus recovered from the original transfection can be used to infect TK^- cells under selective conditions; only the viruses that express TK will be able to replicate because only those cells will survive the selection.

The TK selection technique can also be used in the generation of recombinant HSV. This follows from the fact that the TK gene is dispensable for HSV replication in most cells. Thus, a cloned plasmid of recombinant DNA that contains HSV sequences from either side of the TK gene recombined with the gene of interest, which replaces the viral TK gene itself, can be co-transfected with *wt* virus DNA. When the progeny virus of this transfection is used to infect TK^- cells being grown in the presence of the toxic nucleoside analogue, all cells infected with TK^+ parental virus will die. Thus, plaques formed will be greatly enriched for TK^- virus bearing the inserted gene of interest.

Screening

If the mutation leads to an observable change in the plaque morphology or some other aspect of infection, the virus mutant can be screened or picked. Here, one looks for the change and picks out the viruses with it. For example, a mutation in a single gene might alter the efficiency of packaging of virus and, thus, lower the burst size of virus from an infected cell. This mutant virus would make smaller plaques than would the *wt* parent, and one could easily pick out either large or small plaque-producing virus in a background of many of the others.

Of course, the HSV *ts* mutants described earlier in this chapter, which do not replicate at 39 °C, are also examples of mutants that can be isolated by phenotypic screening. Since they *do not* replicate under restrictive growth conditions, they cannot be selected for, but they can be picked or screened by going back to the replica plaque that does form at 34 °C.

A TOOL KIT FOR MOLECULAR VIROLOGISTS

Viral genomes

Since all the information required for a virus to replicate itself ultimately must be maintained as genetic information in the viral genome, it is important to represent this genetic information in a standardized way. Generally, the viral genome is represented as a schematic of general genomic organization (i.e., single- or double-stranded nucleic acid, linear or circular, etc.), with viral genes and *cis*-acting genetic elements represented. The schematic represents, then, both a physical and a genetic map of the virus (a shorthand summary of encoded genomic sequence information).

Viral genomes can also be expressed in terms of map units (mu). Map units vary in size depending on the genome in question; in other words, a map unit is a relative term. The term still has a great deal of use when describing a genome, especially when locating a specific genetic function within a genome. Map units can be on a scale from 0.0 to 1.0 or from 1 to 100. If one knows the size of the genome in bases or base pairs, one can determine the size of a map unit for any genome.

If the genome is linear, one end is arbitrarily defined as the start of the map (0.0 map unit), and the other is the end (1 or 100 map units). If the virus has a circular genome, a specific site within the genome must be designated by convention as the start and endpoint of the map. This may be in relation to a known genetic element, or in some cases, it is defined as the site where a highly specific restriction endonuclease cuts the genome.

A relatively straightforward example is the genetic and transcription map of SV40 virus shown in Figure 16.1. This 5243-base pair circular genome is cleaved once by the restriction enzyme *Eco*RI, which arbitrarily defines the start and end of the map. Since the map is divided into 100 map units, a single map unit is 52 base pairs. As another example, the HSV-1 genome (Figure 17.2) is 152 261 base pairs, and the map is expressed from 0.0 to 1.0 map unit. Thus, 0.01 map unit for HSV would be 1522 base pairs.

Most proteins found in a eukaryotic or prokaryotic cell – while more difficult to generalize – are larger than 100 but smaller than 1000 amino acids; therefore, one could estimate that a virus with a genome made up of 7000 bases could encode 10–20 "average"-size proteins. Such estimates are only that, and will not include protein information encoded in overlapping translational reading frames. However, these estimates are very useful when trying to determine whether or not all new or novel proteins seen expressed in a cell following virus infection are encoded by the virus. Further, it is possible to use such information to make inferences about whether certain viral proteins might be derived from precursors, or expressed unmodified.

LOCATING SITES OF RESTRICTION ENDONUCLEASE CLEAVAGE ON THE VIRAL GENOME – RESTRICTION MAPPING

As discussed in Chapter 8, Part II, restriction enzymes have a very high specificity for DNA sequence and, thus, cut DNA in only a limited number of locations. Locating sites on a viral genome where restriction endonucleases cleave relative to other genetic "landmarks" is of fundamental importance because the enzymes can be used to specifically break the genome into discrete pieces. Further, these pieces can be individually cloned into one or another bacterial or eukaryotic plasmid vectors so that they can be produced in large amounts. Cloned fragments of DNA can then be used for specific modifications, for probes for detecting viral genes and gene products as described in Chapters 11 and 12, or for expression of viral proteins for a variety of uses.

Mapping restriction fragments within a viral genome is like solving a jigsaw puzzle. The fragments are shuffled around until a consistent fit is obtained. A simple example for a linear viral genome is illustrated in Figure 22.4. Consider a viral genome with a size of 8 kbp, and two restriction enzymes (called "A" and "B" in the figure) that cleave this genome at a few sites. Cleavage of the viral DNA with enzyme A produces two fragments that are 5 and 3 kb and that can be readily separated by gel electrophoresis. This means that this enzyme cuts the DNA at only one site: either 3 or 5 kb in from one end. Cleavage of the same DNA with enzyme B generates three fragments that are 4.5, 2.5, and 1 kb. This means that this enzyme cuts the genome twice. A series of the eight possible arrangements of the sites in relationship to the ends of the viral genome and each other is shown in Figure 22.4a.

Further analysis can resolve the problem into one correct arrangement of the sites, as shown in Figure 22.4b. First, cutting with both enzymes together results in the generation of four fragments: 4, 2.5, 1, and 0.5 kb. This means that the viral genome can have the restriction sites located in one of two possible arrangements vis-à-vis each other and its unique ends. Finally, if one end can be specifically labeled, say by locating a specific gene there, then a single, unique arrangement can be deduced (Figure 22.4c).

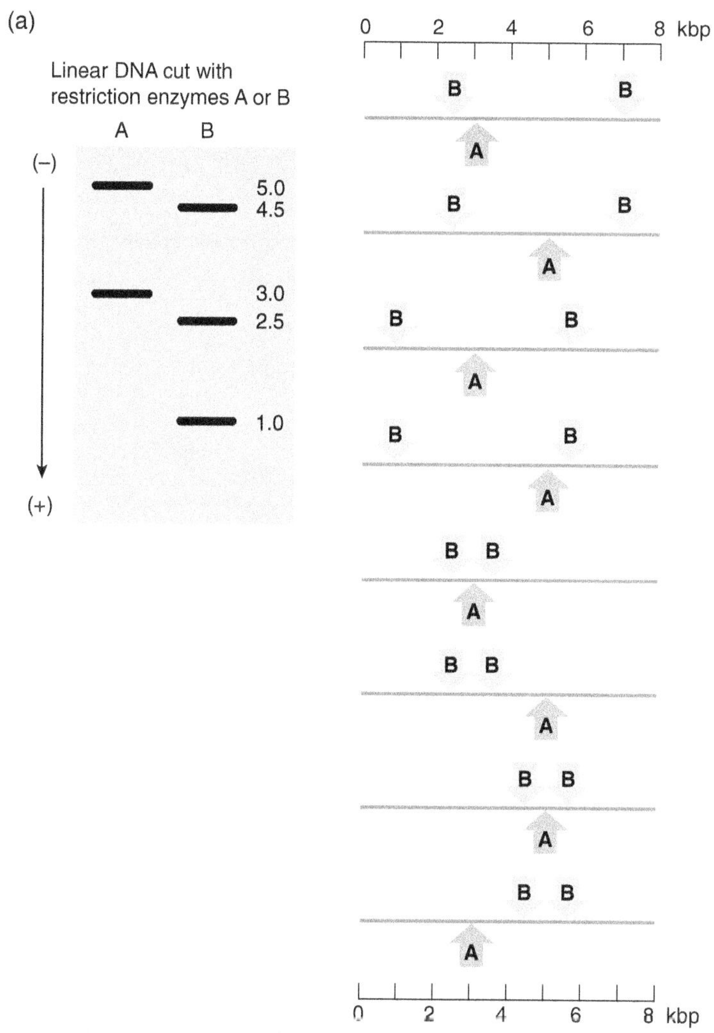

Figure 22.4 Mapping restriction endonuclease cleavage sites on a viral genome. The basic methods for mapping restriction sites resemble those used in putting a puzzle together. Essentially, all possible combinations are tried until one is found that satisfies the results of all combinations of cuts with the restriction enzymes being mapped. (a) The eight (8) possible arrangements of sites for enzyme A that cuts the genome once and enzyme B that cuts the genome twice. Since the two enzymes can cut the DNA into three pieces, there are 2^3, or 8, possible arrangements. (b) Cutting the DNA with both enzymes together followed by measuring the size of the resulting fragments eliminate all but two of the possible arrangements posited in (a). (c) Finally, a marker specific for one region of the DNA will allow choice of a single arrangement of sites.

Since cleavage of a circular DNA molecule will result in it becoming linear, the same principles can be applied to mapping circular genomes. Also, depending on the size of the genome in question and specificity of the restriction enzymes being mapped, a few large fragments or many small fragments can be generated by digestion with restriction enzymes. Separate digestion of large fragments can be used to provide information about the arrangement of sites of enzymes that cut the genome more frequently, so that continuing the process as described can be used to build a map of any genome and any specific enzyme.

While restriction enzymes do not generally cleave single-stranded DNA (ssDNA) or ssRNA, restriction maps can still be generated from genomes utilizing these types of nucleic acids. Single-stranded genomes can be enzymatically converted to double-stranded forms with DNA polymerase, and RNA genomes can be converted into DNA forms using reverse transcriptase isolated from retroviruses. The conversion of RNA into DNA also allows manipulation of any RNA virus genome using methods equivalent to those for DNA genomes – the general approach is termed **reverse genetics**.

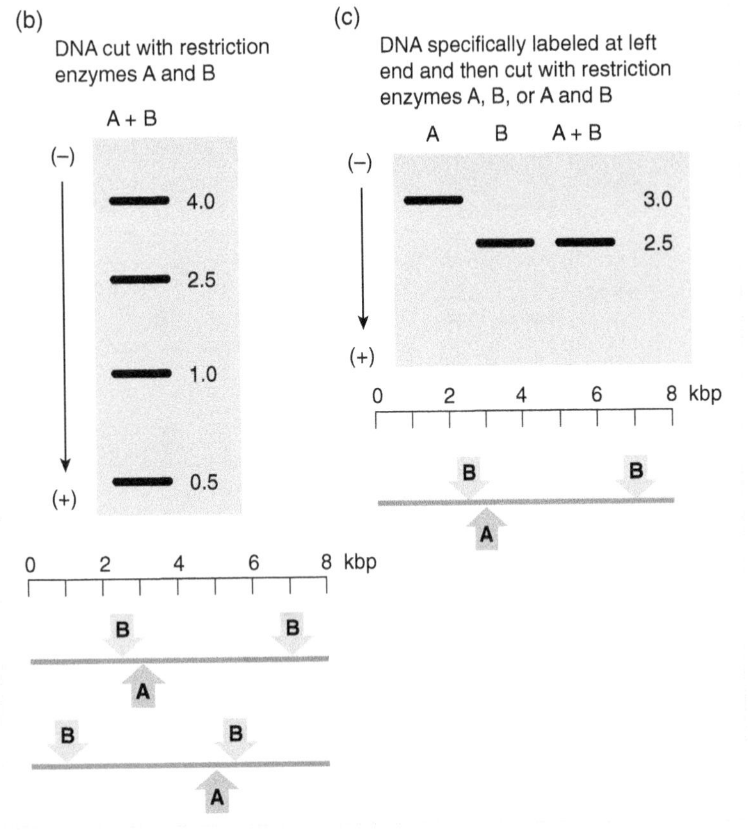

Figure 22.4 *Continued*

CLONING VECTORS

Generation of specific fragments of viral genomes by digestion with restriction enzymes and subsequent isolation allows them to be cloned and maintained separately from the genome itself. This process, which was originally developed using plasmids of *Escherichia coli*, requires a vehicle or vector for maintaining the fragment as DNA at convenient levels in a suitable organism, which now include bacterial and vertebrate cell lines. The size of the cloned fragment maintained was originally somewhat limited by the methods used to introduce the vector into the host cell and because high **copy number** replication of the vectors can lead to genetic modifications of the cloned sequences such as deletions and rearrangements. Current technology using replication vectors able to incorporate large pieces of DNA without genetic modification has, to a great extent, alleviated such problems. The application of one approach using **bacterial artificial chromosomes (BACs)** to clone the entire genome of large viruses is briefly outlined at the end of this chapter.

Cloning of a gene with a suitable promoter for its expression as mRNA in the host cell allows the synthesis of virtually unlimited amounts of a protein of interest. Although bacterial hosts limit the type of posttranslational modifications that can occur with such an *in vitro* expressed protein, the use of a number of vectors that can be replicated in vertebrate cells avoids such complications. Application of one of a number of methods for specifically mutating the expressed gene to generate a protein with defined modifications is an important method for determining the role of specific amino acid sequence elements in the functioning of the protein in question.

Cloning of fragments of viral genomes using bacterial plasmids

Generation of specific fragments of viral genomes with restriction enzymes allows them to be cloned and maintained separately from the genome itself. This process utilizes the fact that many bacteria maintain extrachromosomal plasmids bearing drug resistance markers (see Chapter 8, Part II). These plasmids contain a bacterial origin of replication and one or several genes encoding enzymes or other proteins that mediate the drug resistance. A number of plasmids can be grown to very high copy numbers inside bacteria, and many carefully engineered variants can be maintained in *E. coli* and are laboratory standards. The genetic maps of three widely used plasmids, pBR322, pUC19, and pBluescript (pBS), are shown in Figure 22.5. Plasmid pBR322 was one of the first widely used high-copy-number plasmids in *E. coli*, and the other two plasmids have been specifically constructed to incorporate a number of convenient features. The pBS plasmids, for example, contain promoters from T7 and T3 bacteriophages. These promoters are absolutely specific for virus-encoded RNA polymerases; and, since they point in opposite directions in the pBS plasmids, one can selectively transcribe genes cloned in either direction. In addition, this plasmid can be induced (again using sequences inserted from phage) to produce a single-stranded template (instead of its normal double-stranded structure). This is useful for sequencing or for mutagenesis.

These and most other plasmids have four important general features:

1 The *cis*-acting origin of replication, which allows plasmid replication in the carrier cell. The sequence of this origin of replication determines the actual number of plasmid genomes produced in each bacterium. Usually, one wants to maximize this number, but low-copy-number origins are useful for specialized purposes (e.g., for maintaining large segments of cloned DNA).
2 One or more drug resistance markers that can be used to allow the bacteria in which the plasmid is present to grow in the presence of a drug such as ampicillin or tetracycline that inhibits growth of *E. coli*, which does not contain the plasmid. This is a **selectable** genetic marker.
3 The presence of one or more restriction sites that can be used for cloning the viral (or other) DNA fragment of interest. Both the pUC and pBS plasmids have regions in which a large number of specific restriction enzyme cleavage sites are clustered. Such **multiple cloning sites (MCSs)** are very convenient for cloning a variety of fragments.
4 An enzymatic marker (*lacZ* gene encoding ß-galactosidase) that can be used as a genetic "tag" to differentiate the plasmids containing the cloned fragment from those that do not. This allows the rapid screening of host cells bearing the recombinant plasmid against a background of cells that do not contain the plasmid.

A favorite marker seen in the pUC and pBS plasmids is the bacterial β-galactosidase enzyme, which can turn a colorless substrate blue and thus cause bacterial colonies in which the enzyme is present to become blue under the proper growth conditions. When a fragment is cloned into the MCS, this enzyme is inactivated. This property allows one to rapidly screen a plate for the presence of either blue or white colonies in the background of the other. The pBR322 plasmid does not have β-galactosidase as a screenable marker. Despite this, there are a number of unique cloning sites within the drug's resistance markers, and the ability of a bacterial colony to grow in the presence of one drug but not in another can be used to select for plasmids containing inserts.

To clone a fragment of viral DNA, a suitable plasmid is purified and cleaved with a restriction enzyme that interrupts a screenable marker. Then this linearized fragment is mixed with a fragment of DNA that has been cleaved from the viral genome with the same (or occasionally another) restriction enzyme. The two fragments are then ligated using bacteriophage T4 DNA ligase to form a circular plasmid in which the desired fragment has been inserted. The resulting mixture of religated plasmid without fragment, plasmid with fragment, and other pieces of DNA that may be randomly joined is then mixed with bacterial cells under conditions in which cells will efficiently take up and internalize large circular fragments of DNA. This process, which is (unfortunately) often called transformation in bacteria, is just transfection – briefly described in Chapter 6, Part II.

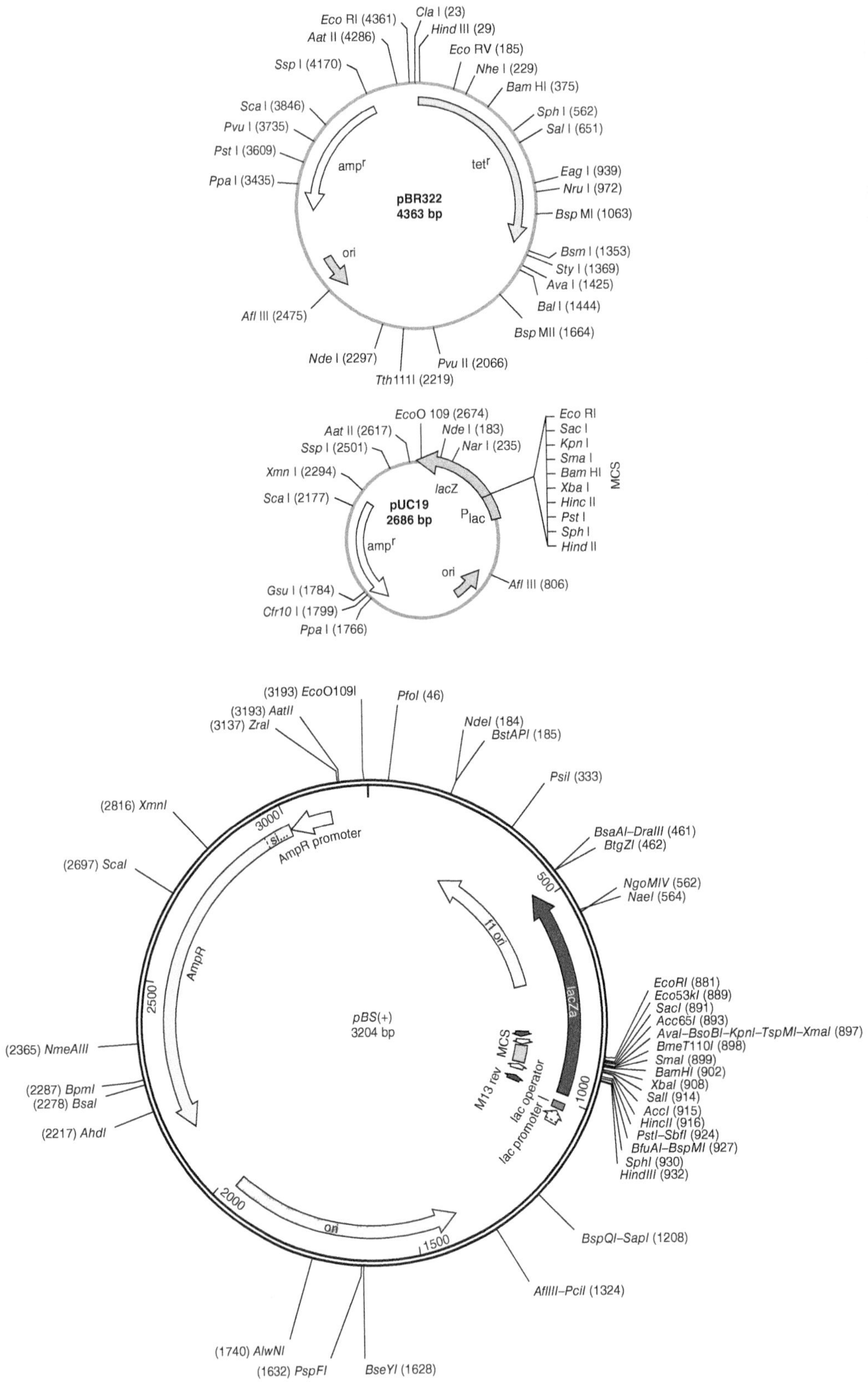

Figure 22.5 Three widely used cloning plasmids that replicate in *E. coli*. These plasmids contain drug resistance markers that can be used for screening or selection, high-copy-number origins of replication, and genetic markers for screening. pBR322 is one of the oldest plasmid vectors and is somewhat unique in that it contains two different antibiotic resistance markers. Cloning into unique restriction sites within either antibiotic allows inactivation of that antibiotic resistance gene, and allows one to select with the other one. Different variants of pUC-based plasmids have different sets of restriction sites in their multiple cloning sites. Unlike pBR322, when one clones a gene into a pUC vector, it inactivates the *lacZ* gene that enables one to screen for insertions using ß-galactosidase selection. The pPBS plasmids contain two bacteriophage promoter elements (T7 and T3, shown by the purple arrows on either side of the MCS) that can be used in conjunction with commercially available bacteriophage-encoded RNA polymerases to make specific transcripts from the cloned sequences. pBS also contains a filamentous phage origin of replication that allows phage (and single-stranded DNA) to be made from the plasmid. MCS: Multiple cloning site.

Individual bacterial cells are then plated on selective medium, and those able to grow in the presence of the drug will form colonies. It should be clear that such colonies must have come from bacteria that have incorporated either modified or unmodified plasmids, since all other bacteria will be drug sensitive. Finally, the colonies are screened either for their inability to produce blue colonies in the case of the β-galactosidase screening marker, or for their inability to grow on agar plates containing tetracycline, or ampicillin in the case of the pBR322 plasmid.

Ligation of the fragments into the plasmid is often aided by the fact that restriction enzymes cleave at palindromic DNA sequences to generate a break in the DNA with a short stretch of complementary ssDNA at each end. For example, in the following sequence, where N is any nucleotide and lowercase indicates the complementary base on the antiparallel strand, the restriction enzyme *Eco*RI cleaves the two strands at the stars, as indicated:

5′-NNNNNG*AATT CNNNNN-3′

3′-nnnnnc ttaa*gnnnnn-5′

to produce ends with 5′-single-stranded overlaps:

5′--NNNNNG--3′ 5′–*AATTCNNNNN--3′

3′--nnnnncttaa*--5′ 3′--gnnnnn--5′

The overlapping sequences allow any DNA fragments digested with this enzyme to be annealed together at the complementary overlaps and then religated to form a complete DNA strand. Note that this ligation will regenerate the restriction site at both ends of the insert:

Plasmid:′5′--NNNGAATTCNNN--3′

3′--nnncttaagnnn--5′

Insert:

5′-***NNGAATTCCACGGTCAGCGATTCNN***-3′

3′-***nncttaaggtgccagtcgctaagnn***-5′

Cleavage:

Plasmid:5′--NNNGAATTCNNN-3′

--nnncttaagnnn-

Insert:

5′***AATTCCACGGTCAGCG***-3′

ggtgccagtcgcttaa5′

Religation and generation of two *Eco*RI sites:

5'-NNNG***AATTCCACGGTCAGCG***AATTCNNN--3′

3′-nnncttaa***ggtgccagtcgcttaa***gnnn--5′

Some enzymes produce blunt-ended DNA fragments, making annealing impossible. However, mixing relatively high concentrations of such blunt-ended fragments together at relatively low temperature can optimize ligation. Even where two different restriction enzymes are used to generate the open plasmid and the DNA fragment, any overhanging ends can be filled in with the appropriate enzymes to produce blunt ends, and these can be ligated together. When such different enzymes are used, however, the restriction site will be lost when the ligation takes place.

An example of cloning a DNA fragment is outlined in Figure 22.6. In this experiment, HSV DNA was digested with the restriction enzyme *Sal*I, which cleaves at the sequence GTCGAC and generates more than 20 specific fragments, ranging in size from larger than 20 kbp to less than 1 kbp, from the 152-kbp HSV genome. A portion of the restricted DNA with a size range of approximately 4–9 kbp was then ligated with pBR322, which had also been cut at its single *Sal*I site, which is within the tetracycline resistance gene. The mixture was then ligated, and transfected into *E. coli*. Cells were plated on nutrient agar plates containing ampicillin, and nine colonies were chosen at random for screening. Each was stabbed with a sterile toothpick, and bacteria were then **replica plated** by streaking the toothpick onto a nutrient agar plate containing ampicillin and then onto a nutrient agar plate containing both ampicillin and tetracycline.

(a)

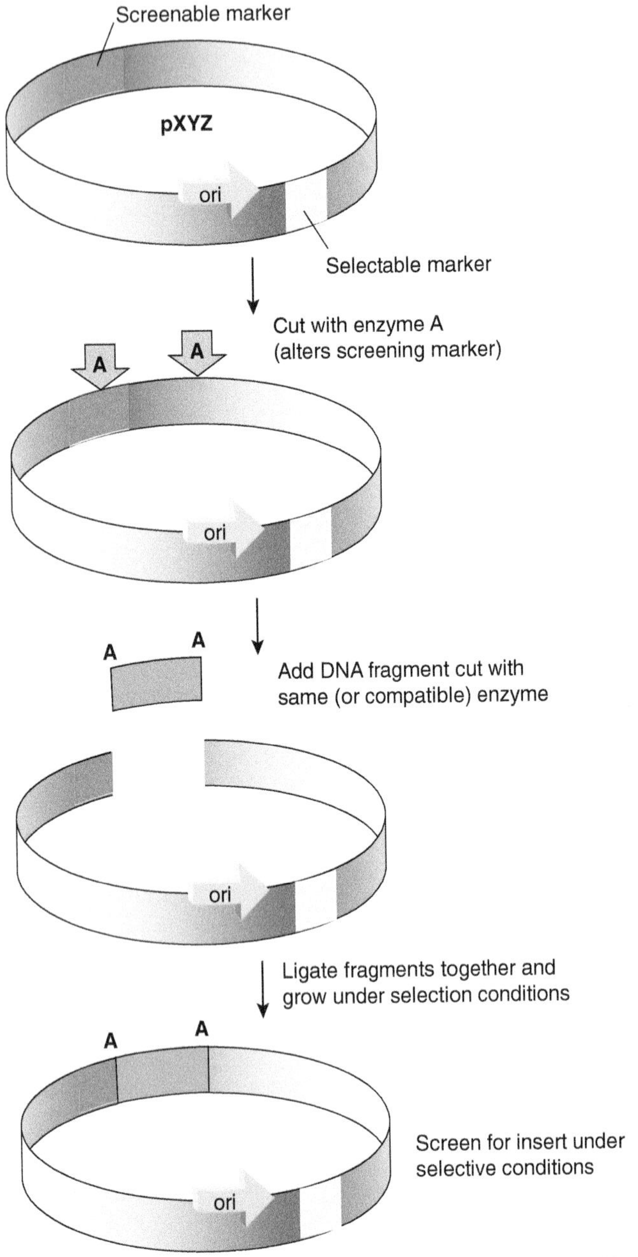

Figure 22.6 Isolation of a specific restriction fragment of viral DNA cloned into a bacterial plasmid. (a) Outline of the process of cloning a DNA fragment into a plasmid using selection and screening. (b) Cloning of a 6.3-kbp HSV DNA fragment. The electrophoretic separation of 20-μg aliquots of *Sal*I-digested HSV DNA, *Hin*dIII-digested bacteriophage λ DNA, and a pBR322 plasmid containing the cloned fragment are shown. The digestion of λ DNA with *Hin*dIII generates the six (6) fragments ranging from 23 to 2 kbp in size (lane "M"), and these serve as a convenient size marker. The screening of individual ampicillin-resistant colonies of bacteria that were formed by transformation (transfection) of a sensitive strain with pBR322, which had been ligated with a mixture of HSV DNA fragments, is shown. The cloned DNA fragment was isolated by *Sal*I digestion of plasmid DNA isolated from a bacterial colony that is ampicillin resistant but tetracycline sensitive.

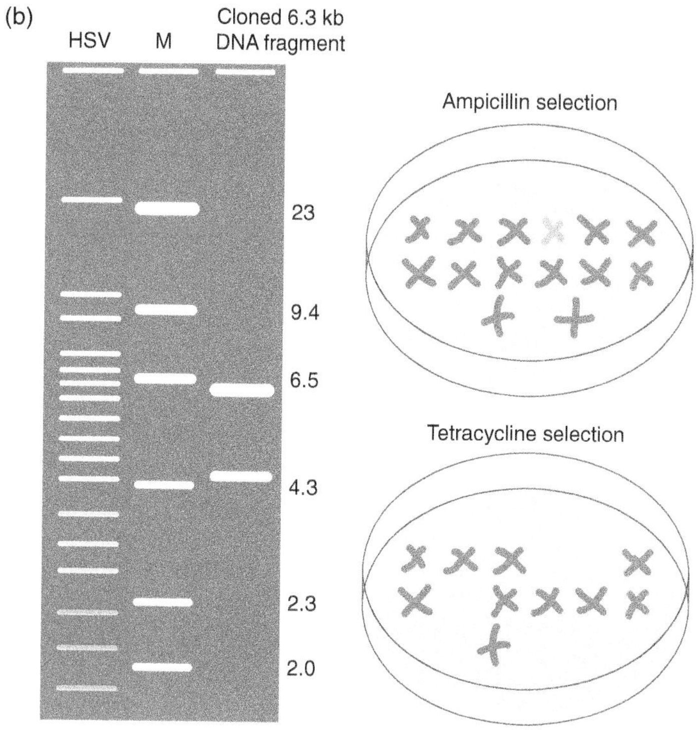

Figure 22.6 *Continued*

Bacterial colonies growing on the ampicillin-containing medium, but not the medium containing both ampicillin and tetracycline, were then chosen and further characterized. In Figure 22.6, plasmid from a tetracycline-sensitive colony was extracted, digested with *Sal*I, and fractionated on an agarose gel by electrophoresis next to a sample of the HSV DNA that had been digested. Two DNA species are present in the digested plasmid: the 4363-base-pair pBR322 molecule, and a 6.3-kbp DNA fragment corresponding to a specific fragment of HSV DNA. This DNA then can be characterized further by sequence analysis, and so on.

Cloning single-stranded DNA with bacteriophage M13

Many strains of bacteria can exchange DNA through conjugation mediated by the structures (pili) projecting from the cell wall. A group of filamentous viruses (*Inoviridae*) that contain ssDNA target such organelles for virus entry. The closely related phages f1, fd, and M13 infecting the F pili of *E. coli* are the best-studied examples. These viruses are able to initiate persistent infections in which 200–2000 new virus per bacterial generation are released from the modified pilis following replication and encapsidation.

The ability to isolate ssDNA encoding a gene or sequence of interest is often technically very useful, for example for generating a probe for a transcript expressed from a complex region of DNA, or (as described below) for generating defined mutations. The phage M13 has been adapted for cloning as a phagemid vector and is in wide use. A number of such a phagemids collectively termed M13mp/pUC vectors have been engineered to combine a modified M13 genome with the pUC or pBS plasmids (Figure 22.5) inserted in a non-essential region of the phage M13 genome. The pUC portion contains a plasmid origin of replication and a drug resistance gene and a polylinker retaining the translational reading frame of the β-galactosidase (*lacZ*) gene in which it is embedded. This phagemid also has the viral origin of replication.

When a fragment of DNA is cloned into the polylinker and then transfected into *E. coli* and plated on nutrient agar, bacterial colonies containing the plasmid will be colorless in the presence of a color reagent that turns blue when digested with the lacZ enzyme – in contrast, plasmids with no insert will be blue. Positive colonies can then be infected with *wt* phage, and infected bacteria will produce both *wt* phage and phage bearing the insert of interest. Depending upon the application, the DNA from the mixed phage can be utilized directly, or phage can be diluted and bacteria can be infected with individual phage – only those bacteria bearing the phagemid will be drug resistant and colorless.

DNA animal virus vectors

As noted earlier in this chapter, many proteins require specific posttranslational modifications for correct function, and such modifications are cell-specific. A number of animal virus vectors have been developed that will allow expression of the protein of interest in a cell that carries out such posttranslational modifications accurately. In general, such a vector will allow the insertion of the gene of interest into a region of the genome containing "non-essential" genes using recombination as described further in this chapter. The modified virus can be replication competent, or its generation may require specific functions provided by a **complementing cell line** or a helper virus. The actual isolation of the recombinant viruses can be quite a laborious process, but the use of BACs promises to simplify the process for several large DNA-containing viruses.

A number of viruses currently in use as cloning/expression vectors are listed in Table 22.1. Each is distinguished by specific features that are of value for particular applications.

Table 22.1 Some viral vector systems.

Virus Vector	Major Use	Insert Size (max)	Integrated?	Target Tissue	Issues
Baculovirus	Cloning/expression of animal genes	20 kb(?)	No	n/a	Requires special cells for maintenance
Poxviruses (including vaccinia and canarypox)	Cloning/expression of animal genes	>25 kb	No	Lung/liver/skin	Evokes strong immune response
Herpesvirus	Gene therapy	>50 kb	No	Neurons	Evokes strong immune response
Adenovirus	Gene therapy	8.2 kb	No	Lung/liver	Loss of long-term expression; toxicity
AAV	Gene therapy	4.7 kb	No	Vascular, muscle, liver	Small size of insert
Retrovirus	Gene therapy	7.5 kb	Yes	Variable	Loss of long-term expression; needs dividing cells for integration
Lentivirus	Gene therapy	7.5 kb	Yes(?)	Lymphocytes, neurons	Toxicity
Togavirus	Protein expression	4–5 kb (?)	No	n/a	No stable expression

Baculovirus

As noted, the polyhedrin gene of the insect baculovirus is expressed at high levels in infected cells and is not required for viral replication. Although an insect virus, insect cells carry out many posttranslational modifications of mammalian proteins accurately. The high level of expression from the polyhedrin gene promoter has made this a very popular cloning vector, and there are a large number of pUC-based plasmids containing the polyhedrin promoter, an intact lacZ gene, and some flanking viral sequences to facilitate recombination. A polylinker is present downstream of the polyhedrin promoter, and this can be used for insertion of the gene of interest. The plasmid (often called a **transfer vector**) is then grown to high copy numbers in bacteria using ampicillin resistance for selection. Recombinant virus is generated by co-transfection of *wt* viral DNA with the plasmid, and recombinant virus isolated by screening under conditions where lacZ expression generates a blue plaque. The resulting virus can then be used to infect cells for protein expression. Often the protein of interest is linked to some functional element that can aid in its isolation.

Vaccinia

Vaccinia virus contains a significant number of genes that can be replaced without affecting its ability to replicate in culture, and its properties are well known by virtue of its long medical history in humans. Further, its genome replicates in the cytoplasm of infected cells and has a propensity to recombine with any other homologous DNA present (Chapter 18). These features have made it a popular expression vector. Originally, recombinant virus was purified using drug selection and inactivation of the viral thymidine kinase gene, as described later in this chapter for the isolation of herpesvirus recombinants. Recently, the isolation of recombinants has been facilitated by the use of BACs. One of the major values of vaccinia-based expression of cloned recombinant proteins is the broad host range of the virus – it will replicate in almost any mammalian cell.

Adenovirus and adeno-associated virus provide vectors that can deliver genes to specific tissue

Up to 30 kbp of the adenovirus genome between the two ends, which contain the packaging signals and replication origins, can be replaced with foreign DNA. Adenovirus-based vectors can be grown in complementing cell lines, which provide all the required viral functions, and high titers (up to 10^{11} particles/ml) can be obtained. Expression of the cloned genes can continue for a limited but useful time in infected cells. Further, these adenovirus-like particles can efficiently infect a limited number of human cells, including those of the liver and lung. Initial studies suggested that adenovirus vectors could be used as **gene therapy** in delivering therapeutic genes to, for example, recipients with a genetic deficiency in immune responses, but toxic host responses have severely limited clinical trials at this juncture, and it is not clear just how useful the approach will be.

A major problem encountered in the use of viruses to deliver cloned genes has been that long-term expression of viral genes generally does not occur in recipient cells. The replication pattern of adeno-associated viruses and other dependoviruses (Chapter 16) provides a potentially useful solution to this problem, since infection of such viruses in the absence of a helper leads to genome integration and long-term viral gene expression. As much as 4.5 kb of foreign DNA can be incorporated into adeno-associated virus, replacing the viral replication and capsid genes, and if those are provided in a complementing cell along with requisite adenovirus gene products, adeno-associated-like viral particles bearing the cloned gene of interest can be produced. Such particles can then be used to infect vascular epithelial, muscle, and liver cells, and the cloned foreign gene contained in the vector DNA is integrated stably in random sites throughout the host genome, where prolonged expression has been reported.

RNA virus expression systems

Retrovirus vectors

Since the replication cycle of oncornaviruses requires integration of complementary DNA (cDNA) into the host genome and continued synthesis of viral RNA from the provirus, their use as expression vectors has been the subject of great experimental investigation. Many approaches have been described, but the use of replication-incompetent vectors has considerable promise. Promising progress has been made, but many problems, including low levels of expression and fears of recombination with endogenous retroviruses leading to novel pathogens, continue to be complications.

The actual technology is quite complex, since oncornaviruses have very narrow host ranges determined in large part by the properties of their *env* genes. There are a large number of specific approaches. For a replication-incompetent vector, a packaging cell line is first constructed by transfection of a modified proviral genome containing the *gag* and *env* genes but lacking the packaging signal normally present in the long terminal repeat (LTR; Chapter 19). The proviral DNA is randomly integrated into the host chromosome, and a cell line that expresses gene products from the integrated DNA is obtained by brute force screening of many colonies of cells derived from the original transfection. This cell line is then transfected with a plasmid containing a retrovirus LTR containing a functional packaging signal, and the gene to be expressed under the control of this LTR. A drug resistance marker is also included. The presence of the packaging signal ensures that the mRNA expressed from this vector is efficiently packaged into retrovirus particles that can later be used to efficiently infect the cell of interest for stable expression of the protein.

The lentivirus HIV is also being actively used as a cloning/expression vector. Besides its specific cell tropism, it offers several potential advantages over oncornavirus vectors, including the ability to infect and express its cDNA from non-replicating cells and the lack of requirement for integration of the cDNA for efficient expression of the RNA of interest. These advantages are offset by the need to delete or permanently inactivate the large number of small regulatory genes that make this virus so toxic to immune cells (Chapter 20).

A togavirus vector

A different strategy is used for expression of proteins (often membrane associated) using a vector based upon the Semliki Forest virus – a togavirus similar to Sindbis virus described in Chapter 14. Like other RNA viral genomes, manipulation is carried out using reverse genetics and cDNA copies, but in order to increase the efficiency of expression of the cloned gene, RNA is expressed in bacteria from plasmids bearing the appropriate viral and cloned foreign genes using a phage promoter and polymerase, as described for phage T7 in Chapter 18.

The system takes advantage of the fact that togaviruses express their structural proteins from a subgenomic 26S mRNA that is generated during viral replication in the cell, but is not packaged (as illustrated in Figure 22.7). A helper plasmid expresses the 26S transcript. Full-length recombinant RNA containing genes for the viral replicase complex and a second, cryptic gene encoding the exogenous protein to be expressed is also generated, by transcription of a recombinant plasmid. A suitable animal cell is then transfected with these two RNAs, after which the recombinant RNA becomes amplified and packaged into virus-like particles, which can be used for subsequent infection. Upon such infection, subgenomic RNA is expressed and large quantities of the cloned recombinant protein are expressed, but no virus is produced. Such cells may express the recombinant protein for extended periods.

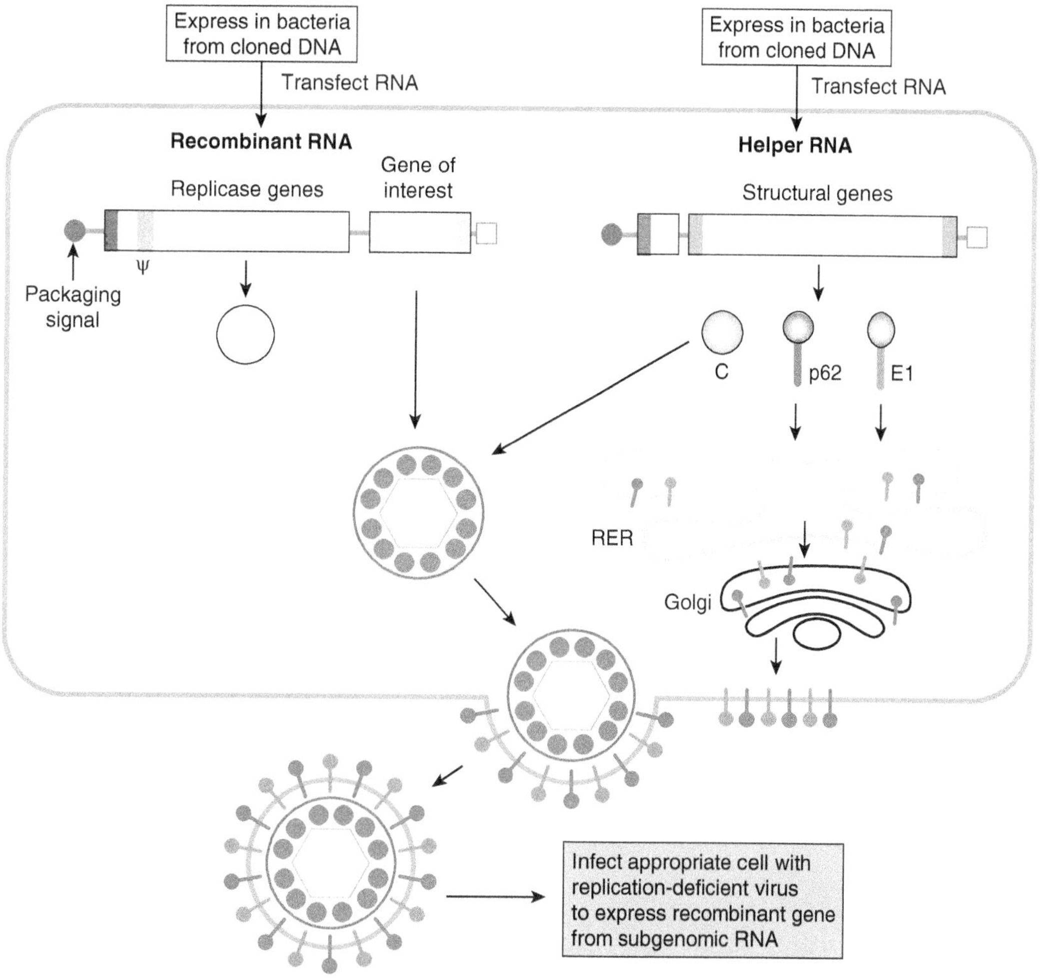

Figure 22.7 A togavirus expression vector. Semliki Forest virus (a togavirus) has been adapted for use as a protein expression vector. The replication cycle involves the translation of viral replication proteins from the genomic 49S (+) strand and the translation of viral structural proteins from the subgenomic 26S (+) RNA transcribed during replication (see Figure 14.8a). For protein expression, two cloned DNA fragments are generated by reverse transcription of the viral (+) RNAs, followed by insertion into expression vectors containing T7 or Sp6 phage promoters (see Chapter 18). The recombinant clone containing the DNA complement of the 49S RNA is engineered to contain the gene of interest in the position of the structural proteins, while the helper clone contains the 26S RNA complement encoding structural proteins. These cloned fragments are expressed as RNA by addition of phage transcriptase and nucleoside triphosphates, and the RNA is transfected into a packaging cell. This transfected cell then generates replication-incompetent virions that, when infected into a suitable cell line, will express the protein of interest from subgenomic RNA generated from replicated input virion RNA.

Defective virus particles

The process of viral genome replication and packaging is not tightly controlled during productive infection by many viruses. If a viral genome produced contains a replication origin and a packaging signal, but lacks one or a number of essential genes, this *defective* genome can still be packaged and released from the infected cell. The virion produced will be a defective particle in that while it can initiate infection in a cell, its genome does not contain all the required genes for replication, and thus the infection will be abortive.

If, however, a cell is infected with both the defective particle and an infectious virus particle, both genomes will replicate in the infected cell, and both can be packaged and released. In such a mixed infection, the normal virus serves as a helper for the defective particle.

During subsequent rounds of high-multiplicity infections of cells with stocks of virus recovered from such mixed infections, propagation of the defective particles is favored, since the defective virions contain genomes that are smaller and less genetically complex than those of the infectious virus, and therefore replicate a bit faster than the full-length ones. This gives them a replication advantage. Eventually, the proportion of defective particles will dominate the mixture, and infectivity will be lost.

Some defective viral genomes, notably vesicular stomatitis virus (VSV), have such a replication advantage that they compete for the virus-encoded replication machinery in the infected cell and exacerbate the loss of infectivity. Such particles are called defective interfering particles (DI particles).

The process of generation of DI particles may have a role in naturally limiting virus infection in the host, although this is not certain. What is important about the phenomenon, however, is that the defective genomes can serve as vectors for carrying other genes. Such defective viral vectors require a helper virus for replication, but in principle at least, they can serve to introduce a gene into cells without the attendant cytopathology and virus-induced cell death that are caused by infection with the non-defective virus.

DIRECTED MUTAGENESIS OF VIRAL GENES

There are a vast number of methods for specifically altering the sequence of a cloned fragment of DNA. Sequence changes could be used to introduce a novel restriction site or to specifically change the sequence of a gene or genetic element. Such changes are extremely useful in generating genetically altered viruses where only a very defined function of a protein under study is changed. A good overall strategy for engineering defined mutations in a viral gene is as follows:

1 Clone the gene of interest in a suitable vector, and then verify by sequence analysis that the gene has been undamaged.
2 Introduce the mutation in the site of interest, and verify by sequence analysis that the alteration is as planned.
3 Express the mutated protein, and check it for altered function or properties.
4 Incorporate the altered gene into a recombinant virus, verify the change, and proceed with the planned investigations.

Site-directed mutagenesis

A common method for introducing site-specific mutations into a cloned DNA fragment is to utilize a synthetic oligonucleotide incorporating the desired change as a primer for synthesis of an otherwise complete copy of the gene. Single-strand DNA clones generated by phage M13 have historically been a convenient vehicle for carrying this out, as shown in Figure 22.8a. There are many other variations of this general scheme, such as incorporating a restriction site at the 5′-end of the primer for ease of cloning, using mutated oligonucleotides as PCR primers to synthesize a large amount of the DNA of interest for subsequent cloning, and so on. Arguably the most direct method of generating a gene containing one or more specific changes is to synthesize the gene in total; and, after checking its sequence, incorporating it into a virus. For example, several synthetic oligonucleotides with matching sequences at their ends are annealed, then used as templates for DNA repair to a larger dsDNA fragment. This process is shown in Figure 22.8b.

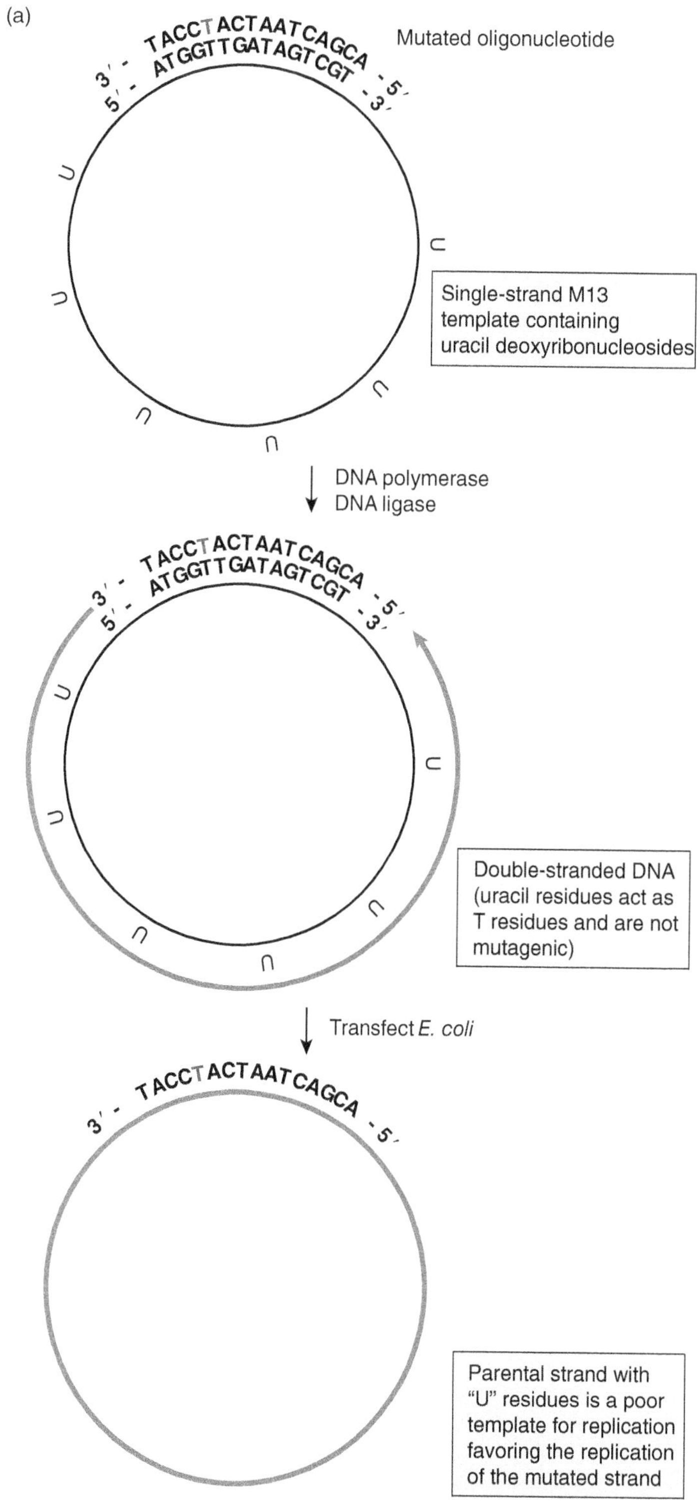

Figure 22.8 Directed mutagenesis of viral DNA. (a) Single-stranded DNA containing the gene of interest inserted into phage M13 is isolated from mutant *E. coli* cells, which incorporated deoxyuridine into DNA in place of thymidine. This DNA is used as a template for synthesis of dsDNA using an oligonucleotide primer in which a single base has been changed (in the example shown, a "T" is substituted for a "C"). This dsDNA is used to transfect *wt E. coli*, and the presence of "U" residues in the template strand results in the preferred use of the mutated strand as a replication template. Newly replicated DNA is then isolated, amplified by transfection and replication, sequence confirmed, and then utilized for recombination back into the virus under study. (b) Synthesis of an "artificial" gene containing a mutation or mutations of choice. A series of single-stranded synthetic oligonucleotides is synthesized. These are constructed so as to have short regions on their ends, which are complementary to other synthetic oligonucleotides containing more of the gene in question. The oligonucleotides at the ends of the construct are often engineered to contain a restriction site for cloning and sequences complementary to viral sequences flanking the insertion site for the modified gene to ease recombination. A whole series of such short, synthetic DNA sequences can be ligated together and dsDNA generated by incubation with exonuclease free DNA polymerase and DNA ligase. If a restriction site is incorporated in the terminal ends, a long synthetic gene can be generated, which can be cloned into a suitable vector, amplified, and used for generation of recombinant virus.

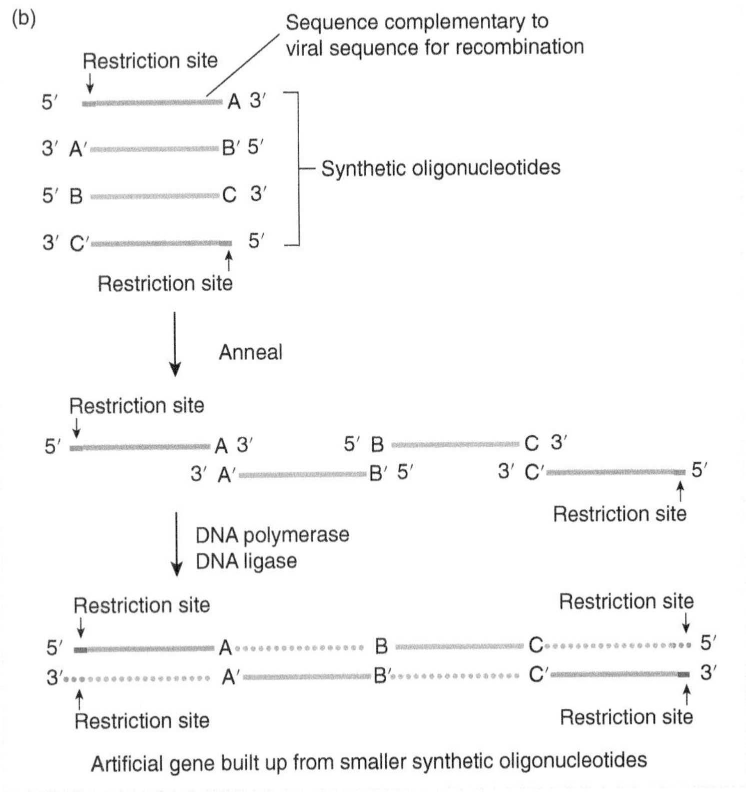

Figure 22.8 *Continued*

When studying the function of a specific protein by defined mutagenesis, it is often of value to generate a series of mutations spaced more or less regularly throughout the gene, followed by comparative analysis of the function of the altered proteins. An example of this approach where a short DNA oligonucleotide encoding, say, an alanine residue (a linker) is introduced at various sites is termed **linker-scanning mutagenesis**, since the mutation "scans" the whole gene. Often such linkers can be introduced into unique or rare restriction sites within the gene – in the latter case, limited partial digestion with the restriction enzyme can result in only a single site of cleavage. Linker-scanning mutagenesis can also be accomplished by generating partial deletions of the gene at a specific site with an exonuclease, followed by partial repair or by using PCR.

GENERATION OF RECOMBINANT VIRUSES

Many techniques of molecular biology allow us to make directed mutations in specific viral genes, and to generate recombinant viruses bearing foreign or specifically altered genes. Recombinant viruses can be used to study the specific blockage of replication caused by interruption of a particular viral gene. If a gene encoding an "indicator" protein such as bacterial β-galactosidase is recombined into a virus, the presence of the virus in specific histological sections of tissue could be determined by localization of a diagnostic enzymatic reaction.

The generation of recombinant viruses involves first and foremost the ability to isolate and separately manipulate specific portions of the viral DNA using the cloning and restriction analyses outlined in earlier sections of this chapter. This DNA can be modified to introduce mutations into the genes being studied. When a desired foreign or modified viral gene has been altered in a desired way, the gene can be cloned into a fragment of viral DNA so that it will be bounded by regions of DNA that are homologous to the viral genome.

Homologous recombination

The gene can then be introduced into the virus by transfection of the DNA containing it along with full-length viral DNA into cells under conditions where the cell is encouraged to use endocytosis to incorporate large amounts of DNA. If one or a few copies of the viral DNA remain intact in the recipient cell, transcription of the genome can lead to initiation of a productive replication cycle.

The recombinant virus is formed by random homologous recombination between the piece of viral DNA carrying the modified viral gene and the full-length infectious DNA. Once the replication cycle is begun, the generated virus will be normal in its ability to spread efficiently to neighboring cells. The recombinant virus can then be isolated either by selection, if possible, or by screening if the new gene does not provide a growth advantage to the virus.

The isolation of a recombinant HSV containing the bacterial β-galactosidase gene inserted into a specific region of the genome is shown in Figure 22.9. In the method shown, a cell culture was infected with virus isolated from plaques formed by virus produced in a transfected cell culture. Some virus can be isolated from individual plaques by probing them with a sterile toothpick or disposable pipette. This virus was used to infect small cultures of cells, and following development of cytopathology and progeny virus, some viral DNA was spotted on a membrane filter and hybridized with a probe for the gene of interest. A positive signal indicates the presence of recombinant virus, and the process is repeated until the virus is pure. Identity of the recombinant can then be confirmed by Southern blotting (shown) or PCR analysis.

Bacterial artificial chromosomes

While straightforward, isolation, purification, and amplification of pure recombinant viruses can be very time-consuming, requiring long periods for plaque development for picking, and multiple rounds of purification. Enveloped viruses such as herpesviruses are notoriously "sticky," and isolation of individual infectious virions free of contamination with particles bearing partial genomes and the like is not easy. Further, some viral mutations are nearly impossible to work with – for example, a mutation in a required gene that is not readily expressed in a complementing cell.

The Human Genome Project and other large sequencing projects have fostered the development of cloning vehicles that can accommodate very large pieces of foreign DNA, and one of these, the bacterial artificial chromosome system, is able to incorporate 100–300 kbp of DNA, which is ideal for cloning herpesviruses and poxviruses. The BAC is based upon the *E. coli* sex or **fertility factor (F′ factor),** which mediates replicative DNA transfer during conjugation. This factor – essentially a plasmid – encodes genes mediating its self-replication and controls its copy number, and exists as an extra-chromosomal element. For BAC cloning, vectors based upon the F′ factor have been engineered to contain multiple cloning sites and a drug resistance gene. When DNA containing a replication-competent herpesvirus genome is grown in such a vector and transfected into an animal host cell, the result is identical to transfection of the viral DNA itself – infectious virus is produced. With vaccinia clones, provision must be made to provide the requisite early gene products required for genome replication, but these can be provided by infection of the transfected cells with fowlpox virus, which does not recombine with vaccinia.

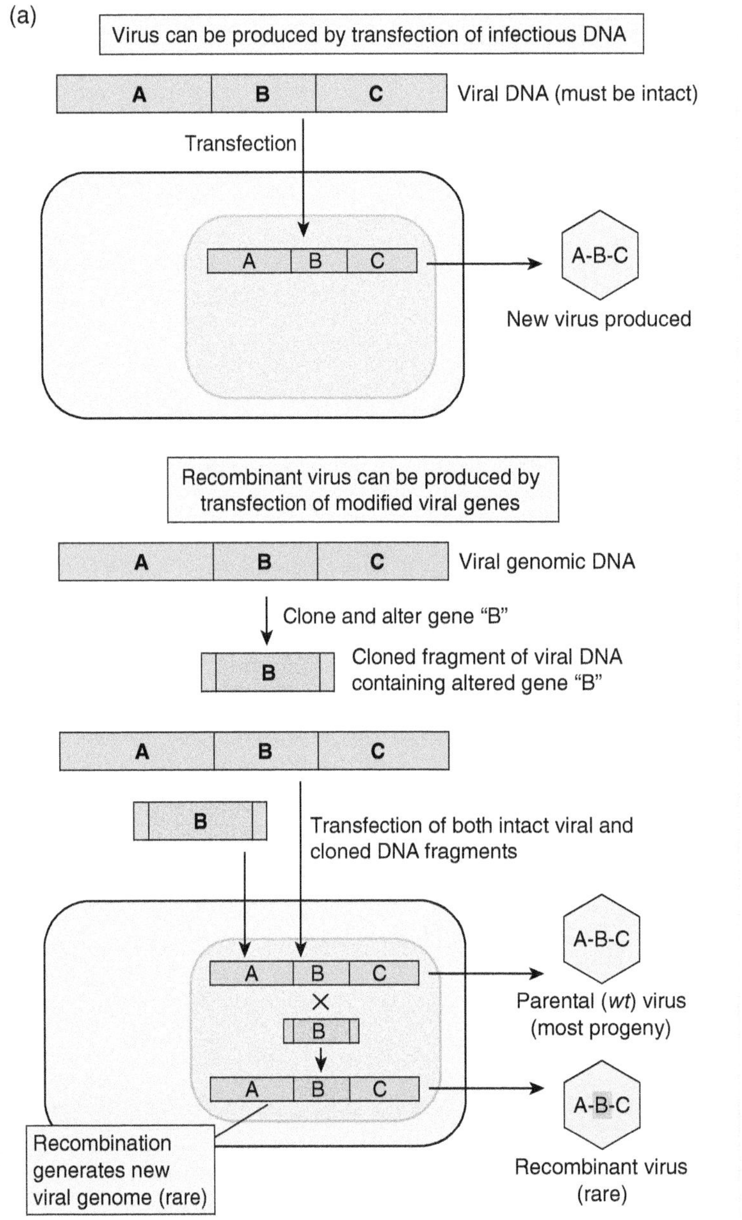

Figure 22.9 Generating and isolating recombinant viruses. (a) As outlined in Chapter 6, Part II, transfection of infectious viral DNA into a permissive cell leads to gene expression. If a full-length viral genome is transfected into a cell, production of infectious virus will ensue. If a fragment of homologous DNA containing a modified or foreign gene is included in the transfection, recombination can occur. While this is a rare event, the appropriate combination of selection and screening for recombinant virus can result in isolation of pure stocks. (b) One approach toward screening for a recombinant virus. In this example, hybridization was used to detect the presence of virus containing the bacterial β-galactosidase gene. This requires physically picking plaques from an infected dish and testing the viral DNA present. The presence of the desired gene was confirmed by the fact that insertion of the 4-kbp β-galactosidase gene results in formation of an altered restriction fragment that can be identified by Southern blot hybridization. (c) The DNA fragment was inserted into a 3.5-kb *Sal*I fragment of HSV-1. The altered fragment size can be seen by hybridization of blots of electrophoretically separated *Sal*I-digested DNA isolated from recombinant viruses. Hybridization was either with the DNA sequences specific for the region of viral DNA used for the homologous recombination, or with a probe specific for the inserted β-galactosidase gene. Hybridization of the blot with the latter probe, however, does not produce a signal with the *wt* fragment into which the gene was inserted.

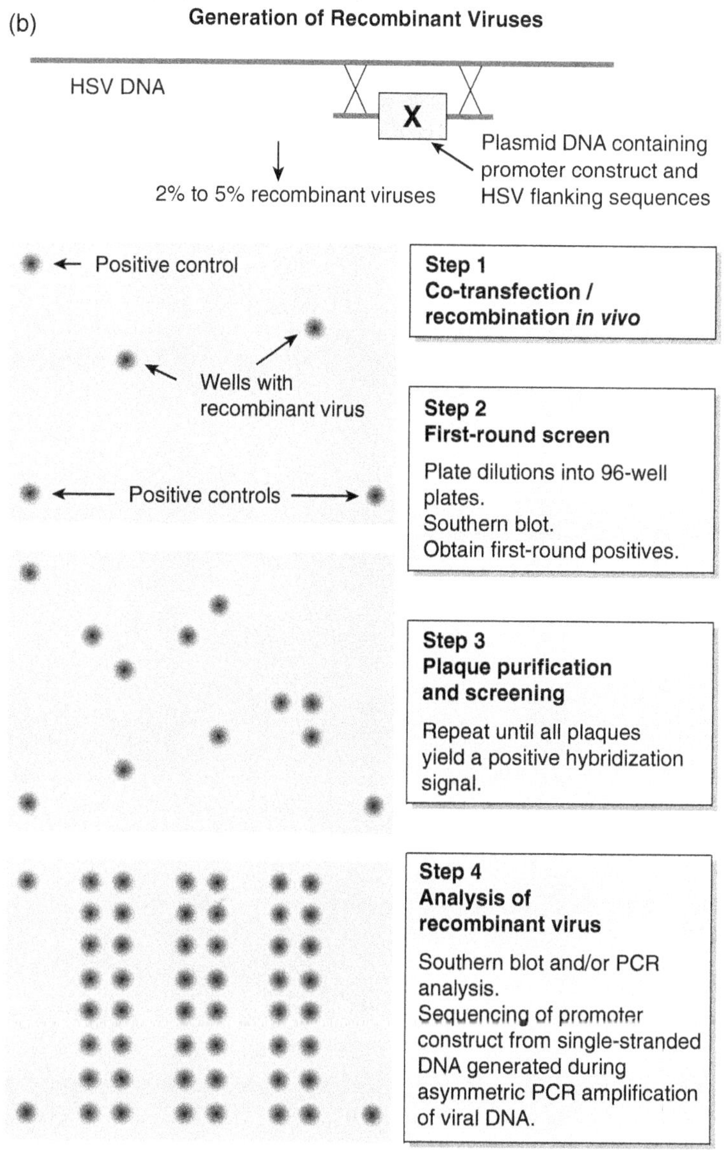

Figure 22.9 *Continued*

The process of cloning the viral genome into the BAC vector generally requires recombinational insertion of the vector into a viral gene, and the recombinant genome is then purified by screening transfected bacterial colonies, alleviating the need for laborious plaque purification. Obviously, however, a virus cloned and generated in such a way will have an interrupted gene or, at least, extra bacterial DNA elements. More recent approaches have been described that utilize recombination to eliminate the vector sequences. Here, the inserted vector element contains a short viral sequence repeated at both ends, and during growth in mammalian cells, the vector sequences are eliminated by internal homologous recombination.

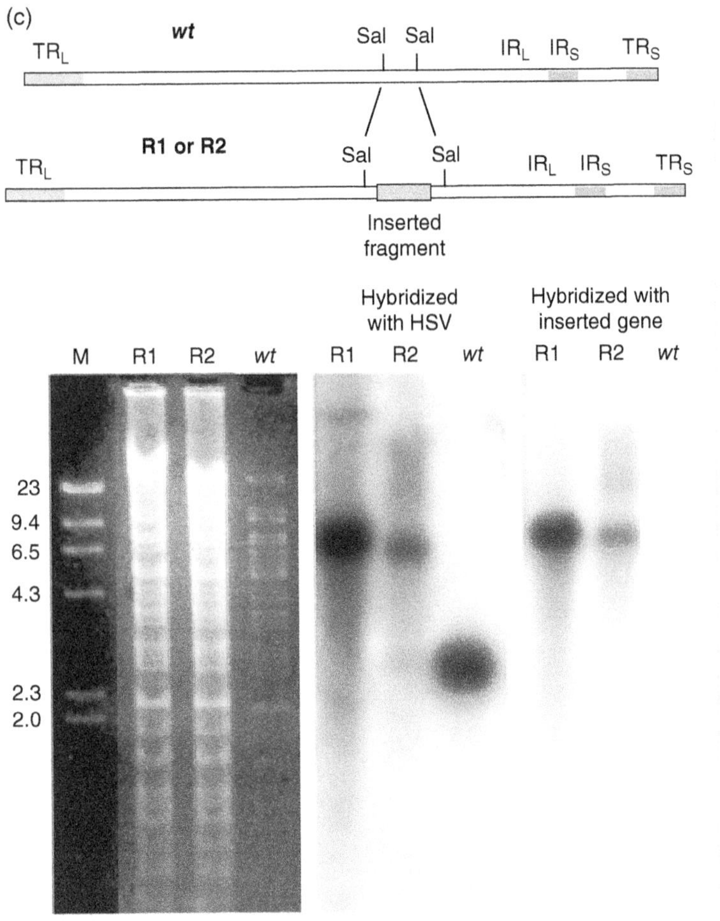

Figure 22.9 *Continued*

CRISPR-cas

CRISPR (clustered regularly interspersed short palindromic repeats) is a family of DNA sequences that are found in the genomes of most bacteria. They are believed to have been originally acquired from bacteriophage that infected ancestral bacteria many years ago. They are known to be part of an antiviral defense of bacteria: These CRISPR DNAs are transcribed as small RNAs that can recognize invading bacteriophage DNA. The second part of this defense system is a protein called **Cas9** (CRISPR-associated protein 9). This protein is an enzyme that recognizes the CRISPR RNAs as guides (guide RNAs) to the phage DNA that they are bound to. Cas9 then binds to these RNAs and the phage DNA, and makes a cut at the RNA–DNA duplexes. This inactivates the phage, thus protecting the bacterium.

This novel mechanism of CRISPR-Cas, removing a piece of DNA, has been exploited as a powerful technique for **genome editing**. By designing guide RNAs that flank a region of DNA that one wants to delete, one can transfect the guide RNAs and the Cas9 into eukaryotic cells and edit out a region or gene of interest. After the Cas9 cuts the DNA, the host cell DNA ligases (part of the non-homologous end-joining machinery) join the cut ends back together, but in the process chew back several nucleotides, making a deletion. Figure 22.10 shows how this editing technique can be used to inactivate a gene of a DNA virus (HSV in this case). There are adaptations of this technology that allow one to make larger directed deletions, as well as gene insertions. This technique is very efficient and often faster than other methods used to make viral recombinants. It is somewhat ironic that an antivirus defense system of phage has been exploited to study the molecular genetics of mammalian viruses!

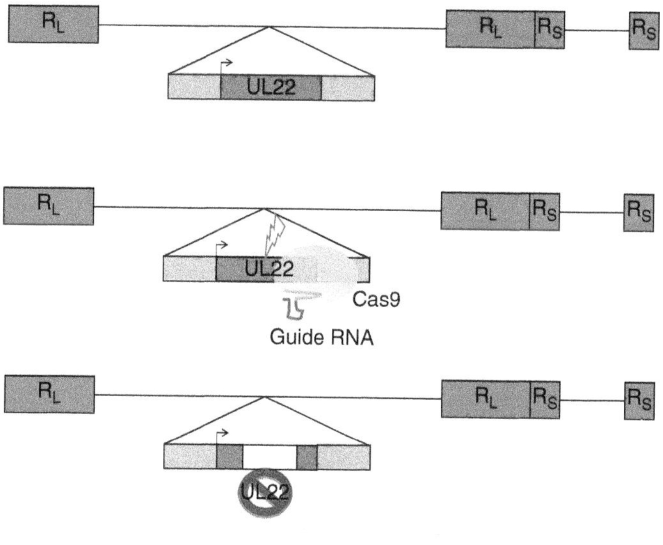

Figure 22.10 Generating a recombinant virus using CRISPR-Cas gene editing. In order to make a deletion in the HSV-1 gene UL22, a guide RNA is designed that has two features: (1) A portion of the RNA is complementary to the region in the UL22 gene where one wants a deletion, and (2) the remaining portion of the guide RNA is a conserved CRISPR sequence that is recognized by the Cas9 protein. The HSV-1 viral DNA is co-transfected into cells, along with a plasmid that expresses the Cas9 protein as well as the guide RNA. This results in the generation of HSV-1 recombinants at high efficiency that have a deletion in the UL22 gene.

QUESTIONS FOR CHAPTER 22

1 A virus such as bacteriophage T7 has a linear, dsDNA genome. Why is it correct to say that a schematic description of this DNA represents both a physical and a genetic map of the genome?

2 Describe the differences between complementation and recombination.

3 You wish to clone a specific fragment of the SV40 dsDNA genome. Using one of the cloning plasmids shown in Figure 22.5, design an experiment to do this.

4 You wish to prepare temperature-sensitive mutants of bacteriophage T4 that are defective in their ability to attach and enter their host cells at high (nonpermissive) temperature. Describe an experimental protocol and selection method you might employ to do this.

5 You have isolated a circular, dsDNA genome of 5000 base pairs (5 kbp) from a virus. You digest this DNA with one of three restriction enzymes or with combinations of the three. The fragments are separated by agarose gel electrophoresis, and their sizes determined. The results are shown in the table below:

Restriction Endonuclease	Fragments and Lengths (kpb)
*Eco*RI	5.0
*Pst*I	3.5, 1.5
*Hinf*1	1.8, 1.75, 1.45
*Eco*RI + *Pst*I	3.5, 1.0, 0.5
*Eco*RI + *Hinf*1	1.8, 1.75, 1.2, 0.25
*Pst*I + *Hinf*1	1.8, 1.0, 0.75 (double-intensity band), 0.7
*Eco*RI + *Pst*I + *Hinf*1	1.8, 1.0, 0.75, 0.7, 0.5, 0.25

From these data, draw the circular restriction map of this viral genome.

Molecular Pathogenesis

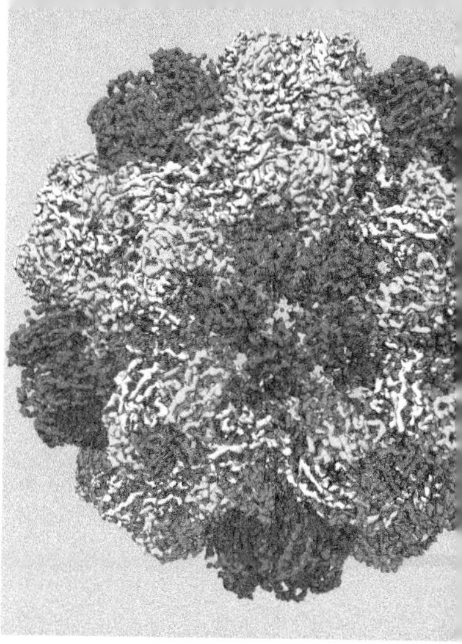

CHAPTER

23

AN INTRODUCTION TO THE STUDY OF VIRAL PATHOGENESIS

Pathogenesis refers to the process of an agent, such as a virus, causing tissue destruction or disease within a plant or animal host. By definition, this process involves tissues that are composed of multicellular structures and often involves multiple tissues and organ systems. Chapter 4 discussed differences in the patterns of disease caused by different viruses.

The study of viral pathogenesis considers viral replication in the broadest sense: the ability of a virus to productively infect an organism; its ability to replicate within a myriad of cell types; its ability to spread from one organ system to another; and, finally, the pathology caused by this entire process. Adding to the complexity of this analysis are the numerous host defenses discussed in Chapter 7 that help limit the spread of a viral infection. Ultimately, the question comes down to: Why is it that some

Basic Virology, Fourth Edition. Martinez "Marty" Hewlett, David Camerini, and David C. Bloom.

viral infections are self-limiting while others can cause severe disease? It is clear that the answer to this question encompasses genetic components of both the host and virus.

The primary tool for studying viral pathogenesis therefore relies upon the use of animal models that allow the analysis of viral interactions with the complexity of cell types, tissue organization, and antiviral arms of the immune response that are present in the intact and living host. Recently the tools of molecular biology have greatly enhanced our ability to follow viral infections *in vivo* and to genetically separate contributions of virus and host that ultimately dictate the outcome of infection. This chapter will explore the development of animal models and how the tools of molecular biology and the postgenomic era have provided new ways of studying virus–host interactions.

ANIMAL MODELS

Choosing a model: natural host versus surrogate models

While the largest body of research into viral pathogenesis focuses on viral pathogens of humans, experimental analyses of human viruses have largely relied on finding surrogates for the natural host. In the best-case scenario, an animal species can be found in which the pathobiology of the infection of the animal with a human virus shares many of the same clinical characteristics that are observed in human infection. Examples of this are the rabbit and guinea pig models for herpes simplex virus (HSV)-1 and HSV-2, respectively, as discussed in Chapter 3. However, this is not always achievable because some human viruses have a limited host range and will not replicate in animals. In addition, there are examples of human viruses that infect animals, but the disease produced is so different from that seen in humans that its study would be of little value. In these cases, a *surrogate virus* is often used to gain insight into the systemic process of infection of the human virus. An example is the use of the mousepox virus, ectromelia, to study the pathogenesis of smallpox. While this has been a very useful surrogate model, it suffers as almost all animal models do in only approximating the human infection, and differences in pathogenesis often exist. Nonetheless, surrogate models have been useful in studying the replication, spread, and immune response that result from a viral infection.

Development of new models: transgenic animals

An example of a human virus without a suitable model for the study of human disease is poliovirus. It was determined that poliovirus cannot infect mice because mice lack the human form of the poliovirus receptor, a protein called CD155. This discovery led researchers to construct a **transgenic mouse** (Figure 23.1) containing the gene for the human CD155 receptor, resulting in a mouse that could now be infected by poliovirus! This allowed the process of **neuroinvasion** of poliovirus to be studied experimentally for the first time, and has provided insight into the basis of the restricted tropism of the vaccine strains of the virus. This groundbreaking transgenic mouse demonstrated that a mouse model for human viruses can be generated if the defect in the ability of the virus to infect the mouse is known. Subsequently similar transgenic mouse models were generated for measles virus, echovirus, and hepatitis B and C viruses, among many others.

Chimeric models: the SCID-hu mouse

There are a number of human viral infections for which no animal models exist and for which the replication restriction in animals either is not known or is multifactorial. Examples include human cytomegalovirus (CMV), HIV, and varicella zoster virus (VZV). One approach that has

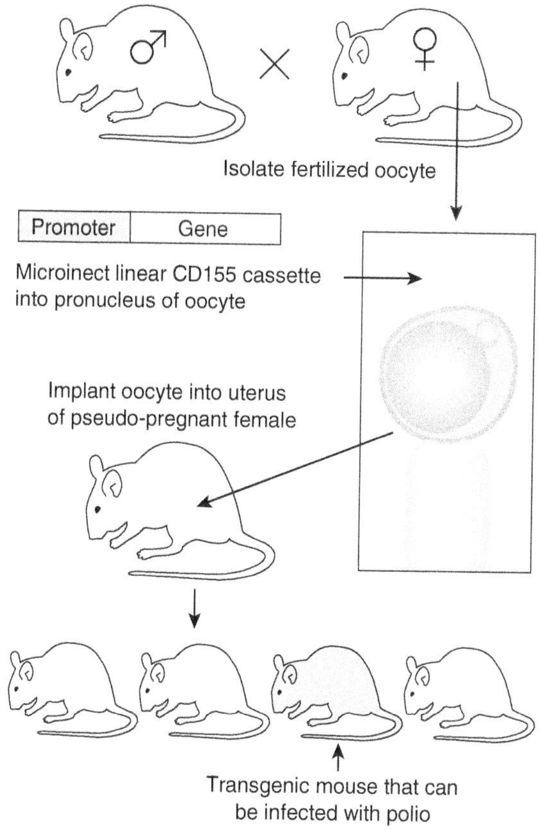

Figure 23.1 Construction of a transgenic mouse. The desired DNA (here, the human poliovirus receptor CD155) is injected into the pronucleus of fertilized mouse eggs, and the resulting embryos are implanted into a number of **pseudo-pregnant** females. A certain proportion of such implanted embryos will result in offspring that contain the desired transgene inserted into the mouse's genome. The mice are screened for these offspring by PCR analysis of tail clippings or blood for the presence of the transgene. Once this first-generation transgenic mouse (founder) is identified, it is then mated with sibling transgenic mice until a pure or homozygous line is produced that now can be infected with polio.

been successful is the construction of a "chimeric mouse" in which a piece of human tissue is implanted into a mouse. In this system, severe combined immunodeficient (SCID) mice are used because they are not able to mount a **xenograft response**, resulting in rejection of the implant. Typically, the human tissue is surgically implanted under the capsule of the kidney, and can be maintained intact and in **homeostasis** for long periods of time. These mice are essentially used as "incubators" for these pieces of human tissue, which are then infected and allow one to study the spread of a human virus through a relevant human organ/tissue. While this technique does not allow analysis of the virus's ability to spread throughout the entire animal, or analysis of the immune response, it has been useful in looking at aspects of replication, spread, and persistence within the more complex cellular environments of a tissue or organ than is possible *in vitro*.

Considerations regarding the humane use of animals

While there is no adequate substitute for animal models in order to study viral pathogenesis, it is important that any study that involves the use of animals be performed in a manner that is considerate of the fact that animals are living entities. It is the obligation of all researchers that great care is taken to ensure that all animals are treated humanely and in a way that minimizes pain or discomfort resulting from the procedures. Research involving animals at universities and research centers in the United States is regulated by National Institutes of Health (NIH) and US Department of Agriculture (USDA) policy guidelines. In addition, there are a variety of independent review groups, such as the Association for Assessment and Accreditation of Laboratory Animal Care (AAALAC), that inspect and accredit animal research programs. Each university is mandated to have an Institutional Animal Care and Use Committee (IACUC) that oversees all animal use. It is responsible for approving all procedures and protocols that will be used, to ensure that proper veterinary care is provided, and to make sure that all personnel that will be working with animals are well-trained in the proper care and humane treatment of animals. One of the doctrines of all IACUCs is to promote the *3 Rs*: refinement, reduction, and replacement. All investigators are asked to justify animal use and search for ways to refine scientific techniques so that there can be a reduction in the number of animals used. Ultimately, replacement of animal procedures with non-animal procedures is a goal if surrogate *in vitro* systems can be developed.

METHODS FOR THE STUDY OF PATHOGENESIS

Once an animal model system has been established, there are a number of established approaches to investigate the process of infection and the progression of disease. Some of these techniques are based on standard virological assays, while others draw on techniques used in pathology. More recently, sensitive molecular assays have augmented our ability to detect both the presence of even small amounts of virus and changes in cellular gene expression in various tissues. Taken together, these techniques allow a comprehensive picture of the pattern of infection and process of disease to be followed longitudinally as the infection spreads through the animal.

Assays of virulence

One of the oldest and still most basic assays of pathogenesis is the determination of the relative virulence, or the ability of a viral strain to cause disease, in a particular animal model. Virulence, which can vary greatly from strain to strain of the same virus, is measured by inoculating groups of animals with serial dilutions of a virus and assaying the resulting disease over time using either a physiological parameter (such as change in body temperature or weight loss) or an endpoint (such as death). The most common way of expressing virulence is as the amount of virus required to kill 50% of the animals (LD_{50}). A full discussion of how LD_{50} endpoints are calculated is presented in Chapter 10.

An important consideration in the evaluation of pathogenesis or virulence of a particular virus is the *route of inoculation*. Not surprisingly, inoculating an animal by injecting virus through an invasive route such as injection into a vein (**intravenous, or i.v.**) or into the brain (**intracranial, or i.c.**) provides a much more sensitive measure of virulence and, hence, a lower LD_{50}. In contrast, inoculating the virus at the periphery such as through intranasal administration or through application at the surface of the skin usually requires more virus to infect the animal and results in a much higher LD_{50} value. Table 23.1 illustrates a determination of the LD_{50}s for HSV-1 following inoculation of the virus on the skin of the foot versus i.c. injection. More virus is needed to infect at the skin because this route of inoculation requires that the virus first replicate in the skin, gain

Table 23.1 Differences in neuroinvasiveness of HSV-1 strains.

	LD_{50} (PFU)	
Route of Inoculation[a]	Strain 17+	Strain KOS
Intracranial (i.c.)	0.1	1
Foot pad (f.p.)	5×10^2	$>1 \times 10^7$ PFU

[a] Six- to eight-week-old Swiss-Webster mice were inoculated with HSV by injecting 10-fold serial dilutions of virus (in 0.01 ml diluent) into the left cerebral hemisphere (i.c.), or applied to the abraded rear footpads (f.p.).

access to the nervous system, replicate there, and ultimately spread to the brain. Replication in the brain and encephalitis ultimately kill mice infected with HSV-1. For HSV-1, as well as most viruses, aspects of innate and local immunity provide a barrier to the virus's entry into critical organs of the host (the nervous system in the case of HSV). As Table 23.1 highlights, doses of less than 500 PFU (plaque-forming units) of HSV-1 strain 17+ on the foot are not always lethal, while larger doses of HSV-1 can overcome the barrier provided by the skin. This highlights an important point in the study of pathogenesis: The route of inoculation is an important way to evaluate not only virulence, but also the relative ability of a virus to spread through the host.

The concept of ability to spread in the host is further illustrated by the comparison of two different strains of HSV-1. As the data in Table 23.1 illustrate, while the i.c. LD_{50}s for strains 17+ and KOS are similar, they vary greatly in their relative LD_{50}s following footpad infection. In fact, strain KOS is essentially *avirulent* when administered at the foot. Therefore, while both strains are virulent when the virus is inoculated into the brain (**neurovirulent**), only strain 17+ is able to invade the nervous system efficiently (**neuroinvasive**). This highlights the fact that very closely related strains of a virus can differ greatly in their pathogenic properties. Similar types of analyses and comparisons can be made for viruses that infect other organ systems, as well.

A final point that should be made relates to the contribution of host genetics and the composition of the immune repertoire to virus susceptibility. It has clearly been demonstrated that there are dramatic variations in the susceptibility of different inbred strains of mice to various viral infections. Often these differences have been mapped, using tools of mouse genetics, to differences in major histocompatibility complex (MHC) composition. Table 23.2 illustrates that while HSV-1 strain 17+ is exquisitely virulent (and neuroinvasive) in the Swiss-Webster strain of mice, C57BL/6 mice are essentially resistant to the spread of the virus following skin inoculation.

In summary, comparison of the relative virulence and potential of a virus to spread following various routes of inoculation can provide important information regarding its ability to spread and cause disease within the host. Virulence is a relative term, however, and is defined by genetic determinants of the virus, by the host, and by the route of infection. As we will see later in this chapter, tools of molecular genetics can be used to tease out the viral and host processes responsible for these differences.

SCID mice with fragments of human thymus and liver implanted under the kidney capsule (SCID-hu thy/liv mouse) can be used to characterize the virulence or pathogenesis of HIV isolates. In this animal model, the implanted human tissue is allowed to develop into a thymus-like organ for several months prior to infection with HIV. Some isolates of HIV rapidly kill human CD4+ thymocytes as they develop in the SCID-hu thy/liv graft, while other isolates are much less pathogenic in this chimeric animal model. Mouse cells are not infected, so the infection is confined to human cells in the chimeric animals.

Figure 23.2 shows the result of infection of human thy/liv grafts in SCID mice with two CCR5-tropic (R5) isolates of HIV-1 derived from the same patient, isolate 32D2 from the

Table 23.2 Differences in susceptibility of mouse strains to HSV-1 infection.

	LD_{50} (PFU)	
Route of Inoculation[a]	Swiss-Webster	C57BL/6
Intracranial (i.c.)	0.1	0.5
Foot pad (f.p.)	5×10^2	$>8 \times 10^6$ PFU

[a] Six- to eight-week-old mice were inoculated with HSV-1 strain 17+ by injecting 10-fold serial dilutions of virus (in 0.05 ml diluent) into the left cerebral hemisphere (i.c.), or applied to the abraded rear footpads (f.p.).

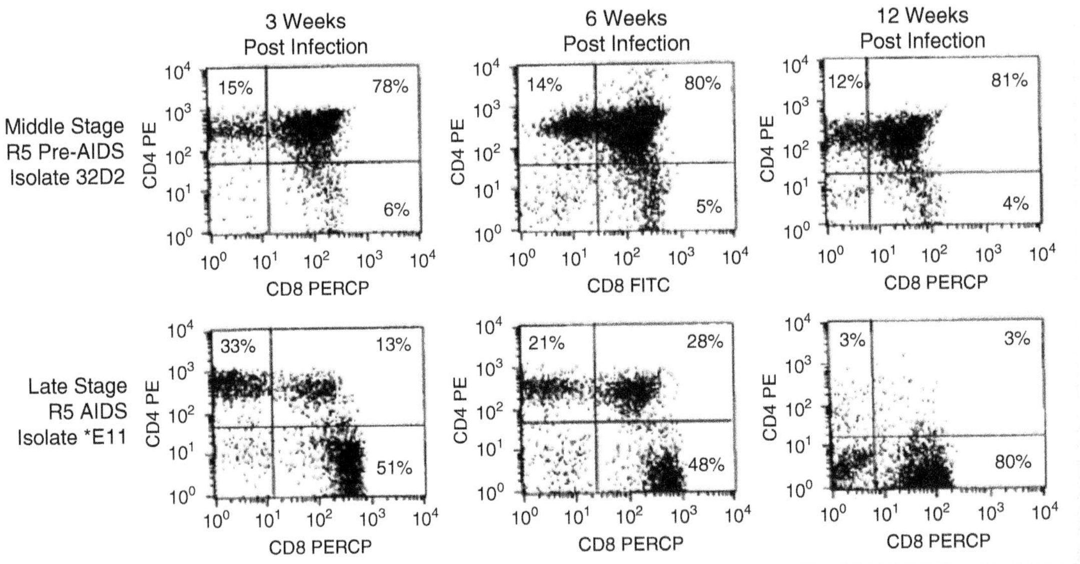

Figure 23.2 Use of SCID-hu mice to study HIV pathogenesis. Human thymus–liver grafts in SCID mice were infected with two isolates of HIV-1 derived from different stages of infection of the same patient. Thymus–liver grafts were biopsied at 3, 6, and 12 weeks post infection, and cells were incubated with CD4 and CD8 monoclonal antibodies conjugated to different fluorochromes for simultaneous two-color analysis of thymocytes by flow cytometry.

mid-asymptomatic stage of infection and *E11 from after the patient developed AIDS. Thymus–liver grafts were biopsied, incubated with fluorochrome-conjugated CD4 and CD8 monoclonal antibodies, and analyzed by flow cytometry. It is readily apparent that over the 12-week time course of the experiment, HIV-1-*E11 depleted nearly all CD4+ thymocytes from the human thymic graft, while HIV-1-32D2 had little effect on the fraction of CD4+ cells in the human thymus–liver grafts.

Analysis of viral spread within the host

Traditional means of assessing the spread of viral infections through a host involve rather tedious "tracer" experiments, where animals are infected at a peripheral location (for example, skin or nose) and then euthanized and dissected at daily intervals after infection. Tissue samples are assayed for infectious virus, viral DNA, or viral RNA transcripts as a means of identifying sites and magnitude of active infection. These studies often require large numbers of animals, and the analysis of virus in the tissues is expensive and tedious. A significant improvement for tracing viral infections has involved the engineering of viruses to express reporter proteins such as green fluorescent protein (GFP) or ß-galactosidase (ß-gal), as shown in Figure 23.3a. In this manner, complete organs or even whole animals can be stained at various times post infection, and the viral spread visualized by the presence of the reporter (Figure 23.3b). An additional advantage of this technique is that one can thinly section the organs and directly visualize which cell types within an organ structure harbor the infected (blue) cells (Figure 23.3c).

While these reporter viruses represented a significant advance over traditional "grind and titer" tracer studies, they still require that animals be euthanized and examined at time intervals since reporter visualization is a terminal assay. Two recent advances in *in vivo* imaging technology have been applied to the tracing of viral infections and are proving to be useful tools. Both

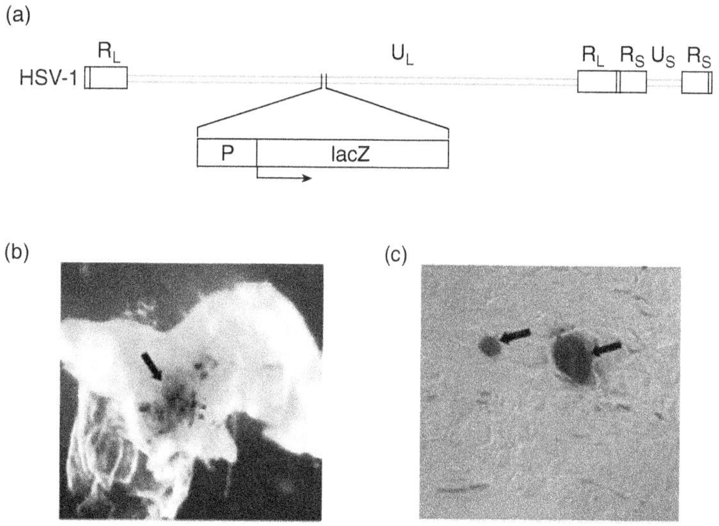

Figure 23.3 Use of ß-galactosidase (ß-gal) as a reporter to trace viral infections. (a) An HSV-1 recombinant has been engineered that contains the *Escherichia coli lacZ* gene inserted into its genome. When the virus infects cells, it will express the *lacZ* gene and produce ß-gal. (b) When the HSV-1 *lacZ* recombinant is used to infect mice on the nose or eye, the virus will spread to the trigeminal ganglion (shown). Use of the chromogenic substrate x-gal stains cells (indicated by arrow) that are expressing ß-gal blue. (c) A thin section (6 μm) of the ganglion allows individual infected cells (indicated by arrows) to be visualized.

of these techniques allow imaging of viral infections within living (anesthetized) animals, and therefore allow longitudinal assessment of changes in the spread of a viral infection on a day-to-day basis. This has the advantages of not only greatly reducing the number of animals that need to be included in a study, but also providing a way of assessing spread within the same animal.

The first of these techniques involves whole-body *magnetic resonance imaging* (MRI).

Larger MRI magnets with very high resolution are now available, and allow viral infections to be monitored largely as a function of changes in the water density of tissue as a result of inflammation (Figure 23.4a). In some cases, enhancing compounds containing paramagnetic atoms (such as fluorine) that improve contrast and sensitivity can be used. One example of this is the treatment of a mouse with a fluorinated form of acycloguanosine.

A second technique for whole-animal *in vivo* imaging involves the detection of bioluminescence. *Bioluminescent imaging* (BLI) makes use of either a virus recombinant expressing a firefly luciferase reporter or a transgenic mouse that contains the luciferase reporter behind a promoter that will be induced by the infection. Luciferase is the enzymatic activity, found in fireflies for instance, that uses the energy of adenosine triphosphate (ATP) and produces luminescence as visible light. The BLI is then performed using a camera that can detect this luminescence, and the location and relative amount of luciferase expression can be assessed (Figure 23.4b). One of the caveats of engineering a reporter into a virus is that one has to consider the possibility that the insertion site might have an attenuating effect on the pathogenesis of the virus. This possibility dictates that the pathogenic properties of the reporter virus be carefully compared to the wild-type (WT) parent virus, typically by using conventional assessments of virulence and replication. A recent advance in the area of bioimaging has been construction of transgenic mice with luciferase driven by a viral promoter that is only activated by a viral infection. In this manner, native viruses can be assessed for relative pathogenesis without having to engineer reporters into the virus.

Resolving the infection to the level of single cells

The above techniques are useful in determining a gross pattern of the sites of viral replication within an animal host. Ultimately, one often needs to determine the specific cell types that a virus is replicating in within a particular organ. For example, if virus is found within the brain,

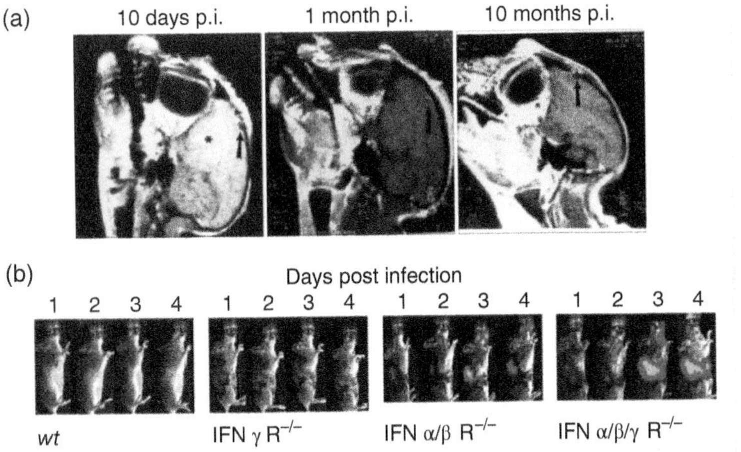

Figure 23.4 (a) MRI images of a mouse infected with HSV-1 10 days post infection (p.i.), 1 month p.i., and 10 months p.i. Arrows represent the injection sites. The light regions 10 days p.i. represent the inflammation resulting from the initial infection, which resolves with time. *Source:* Courtesy of Hunter, W.D., Martuza, R.L., Feigenbaum, F., et al. (1999). Attenuated, replication-competent herpes simplex virus type 1 mutant G207: Safety evaluation of intracerebral injection of nonhuman primates. *J. Virol.* 73(8): 6319–6326. (b) Use of real-time bioluminescence to monitor the spread of HSV-1 in wild-type (*wt*) and interferon-deficient mice. An HSV-1 strain engineered to express firefly luciferase was used to infect the right eye of mice, which were either *wt* or three different strains of interferon receptor (R) KO mice (IFNγ $R^{-/-}$, IFNα/β $R^{-/-}$, or IFNα/β/γ $R^{-/-}$). The mice were imaged for bioluminescence at one, two, three, and four days post infection. The intensity of bioluminescence, reflecting the spread of the infection, is represented by colors, with dark blue being the least intense and red being the most intense. In the *wt* virus and the IFNγ $R^{-/-}$ mice, the luminescence indicates that the active infection is largely confined to the area surrounding the eye. In contrast, the infection of the IFNα/β $R^{-/-}$ or IFNα/β/γ $R^{-/-}$ mice is not controlled efficiently, and the virus disseminates to the brain as well as to the spinal cord and abdominal cavity, as indicated by the enhanced luminescence. *Source:* Courtesy of Luker, G.D., Prior, J.L., Song, J., et al. (2003). Bioluminescence imaging reveals systemic dissemination of herpes simplex virus type 1 in the absence of interferon receptors. *J. Virol.* 77(20): 11082–11093.

is it found within neurons *and* glial cells, or is it localized within certain anatomic regions of the brain, such as the hippocampus or cerebellum? Determining these answers requires analysis of the tissue at higher resolution with microscopy. To visualize the virus, several techniques are employed: (1) immunohistochemistry, which uses labeled antibodies that are specific for the particular virus; (2) *in situ* hybridization (ISH) or fluorescent *in situ* hybridization (FISH) that utilizes DNA or RNA probes to detect the viral nucleic acids; or (3) use of recombinant viruses containing reporters (see above) that allow detection through direct fluorescence (GFP) or enzyme assay (ß-gal).

An exciting recent advance in this area is a technique called *laser capture microdissection* (LCM). This technique allows one to focus on a particular cell that is expressing a reporter indicative of viral infection, and to use a laser to isolate and transfer the individual cell to a piece of plastic film. The cell and others like it can then be analyzed in isolation for the specific viral or cellular genes that are expressed.

CHARACTERIZATION OF THE HOST RESPONSE

Viral infections *in vivo* represent a dynamic interplay between virus replication within the cells and organs of the host and attempts by the host immune response to ultimately contain and eliminate the infecting virus. As mentioned in Chapter 7, host defenses are multifaceted and involve aspects of both innate and acquired immunity. The challenge for the study of viral pathogenesis and its application to the design of new therapeutics such as vaccines is that the immune response to each virus is different, and often there exists a dominant component of the immune response that plays a more critical role in the ability of the host to resolve a particular viral infection. For example, antibody is effective in blocking and helping the host resolve infections by some viruses such as influenza virus; however, it does little against others like HIV. Therefore, determining the components of the host immune response that are dominant for a particular viral infection is a central and inseparable part of the study of viral pathogenesis.

Immunological assays

Classic immunological assays are still used to detect the presence of a humoral or cellular response to a virus infection. Blood can be drawn at various time points after infection and analyzed for the presence of antibody that is specific for a given virus (or specific viral antigen) using serum neutralization assays or enzyme-linked immunosorbent assays (ELISAs), as described in Chapter 7. Other humoral components of the immune response, such as the presence of specific cytokines that have been produced in response to the viral infection, can also be assessed by ELISA-based assays. Knowledge concerning specific cytokine production can be useful in determining whether the immune response to a particular virus tends to be more Th1 or Th2 (Chapter 7).

Recent refinements on these basic assays brought about by the application of biotechnological advances include the application of sensitive reverse transcription polymerase chain reaction (**RT-PCR**) analyses or enzyme-linked immune-absorbent spot (**ELISPOT**) assays to probe small tissue or fluid samples from mice for changes in cellular gene expression in response to viral infection. These techniques allow the detection of local responses to the virus infection (such as at the site of initial infection) long before detectable systemic levels of antibody or cytokines have accumulated.

Protein microarrays that display all the proteins and/or protein fragments and epitopes from one or more virus can be used to characterize the antibody response to viral infection. Figure 23.5 shows the reactivity of serum IgG, saliva IgG, and saliva IgA from 100 donors with all the proteins of HSV-1, HSV-2, HPV-16, and HPV-18. Each column in Figure 23.5 shows the reactivity of a single donor's serum or saliva, and each row of the figure shows reaction with a single protein from the indicated virus. This type of figure is called a *heat map* because colors are used to represent the strength of antibody reactivity. Light green indicates no reaction, dark green and black show progressively stronger reactions, and red indicates the strongest reactions. The protein microarray heat map allows rapid visualization of a large number of antibody interactions with viral proteins. For example, the three upper squares in Figure 23.5 show the reactivity of 100 patient samples with 88 HSV-1 proteins or 8800 individual antibody interactions with viral antigens!

Use of transgenic mice to dissect critical components of the host immune response that modulate the viral infection

The techniques discussed here are invaluable in describing the nature of the immune responses that are mounted against a given pathogen, but it is ultimately most useful to determine the *key players* in the immune response that are responsible for clearing a viral infection. A powerful tool available to address this is to use immunological **knock-out (KO) mice**. These are transgenic mice in which specific effectors of the innate or acquired immune response have been deleted by

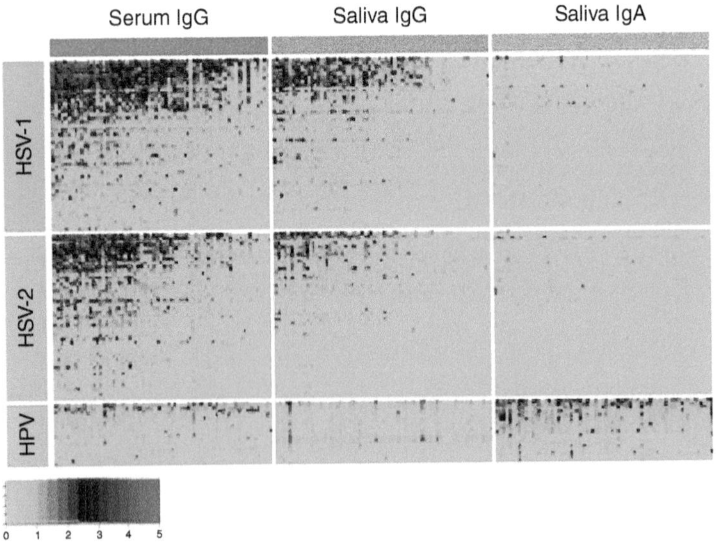

Figure 23.5 Use of a protein microarray to assay patient antibody reactivity with viral proteins. All proteins from HSV-1, HSV-2, HPV-16, and HPV-18 were spotted on nitrocellulose-coated slides. The slides were incubated with patient serum or saliva as indicated and then with a fluorochrome-conjugated antihuman IgG or antihuman IgA antibody as indicated. The intensity of fluorescence, indicating antibody binding to each viral protein, was measured with a scanner equipped with a laser and analyzed with software designed for this purpose. Data from over 19 000 antibody–antigen binding assays carried out on the protein microarray are shown. Each row of the heat map shows reactivity with a single viral protein, and each column indicates the antibody reaction from a single patient sample. The intensity of antibody binding to each viral antigen is depicted by the color of each square; light green indicates no reaction, dark green and black show intermediate reactivity, and red indicates a strong reaction. Figure courtesy of Antigen Discovery Inc., Irvine, CA, USA.

targeted gene disruption. Examples are interferon and other cytokine KO mice, as well as mice deficient in various aspects of **CTL** or Th responses. By infecting such mice with a given virus, one can assess the contribution of individual components of the immune system on the outcome of the viral infection. Many such mice are now commercially available and have greatly advanced our understanding of the specific aspects of the immune response that are especially important in resolving different viral infections. Such information can be useful in designing vaccine strategies to combat both old and new viral pathogens for which there is no existing vaccine or effective treatment. It should be noted that these approaches using engineered mice are only applicable to diseases that have mouse models.

QUESTIONS FOR CHAPTER 23

A new strain of HSV has been isolated that is severely attenuated with respect to virulence. This strain is called XC-50. Your lab has sought to characterize the basis of this restriction, and to determine whether the primary defect is in *neurovirulence* or *neuroinvasion*. Below are the results of several experiments involving this virus. Analyze and interpret the data as directed.

1 Serial dilutions (10-fold) of the XC-50 virus were inoculated into mice i.c. (intracranially) to determine the relative lethality by this route of inoculation (relative to a virulent strain termed 17+). The inoculi ranged from 0.1 PFU to 10 000 PFU per mouse. The following represent the **number of surviving mice/mice inoculated** for each dose of virus 10 days after inoculation.

	XC-50	17+
0.1 PFU	5/5	4/5
1 PFU	5/5	3/5
10 PFU	5/5	0/5
100 PFU	5/5	0/5
1000 PFU	4/5	0/5
10 000 PFU	2/5	0/5

2 What is the **approximate** $\mathbf{LD_{50}}$ for each virus?

3 Based on these results, what can you say about the relative **neurovirulence** of **XC-50?**

Viral Bioinformatics

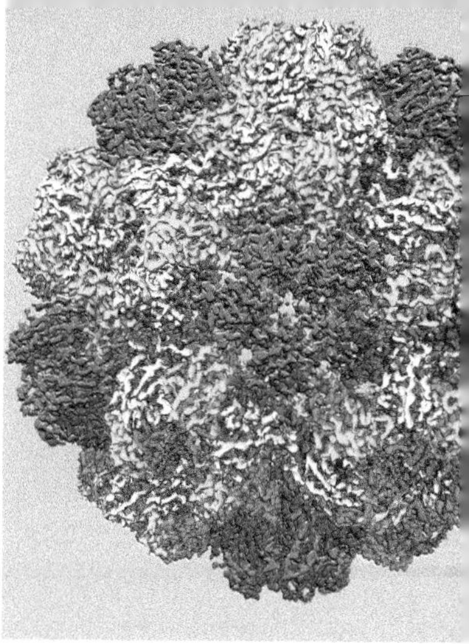

CHAPTER

24

BIOINFORMATICS

The past few years have seen a huge increase in the quantity of biological data due to advances in **high-throughput (HT) technology**, such as DNA sequencing (see Part III, Chapter 11), oligonucleotide microarrays (see Part III, Chapter 12), and proteomics. To cope with this deluge of information, the scientific discipline of bioinformatics has emerged. **Bioinformatics**, or computational biology, is the application of computer science to the management and analysis of biological data. The goal of bioinformatics is to mine the wealth of biological information from various types of data, gain new knowledge, and suggest new pathways of exploration.

Basic Virology, Fourth Edition. Martinez "Marty" Hewlett, David Camerini, and David C. Bloom.

Bioinformatics and virology

Bioinformatics research is rapidly changing all areas of biology, including virology. Traditional laboratory-based work is now being complemented by *in silico* analysis. *In silico* is not really a Latin phrase, but one coined to resemble *in vivo* and *in vitro*, and to mean "carried out by computer simulation," or "in silicon" (computer chips). For instance, one can design an experiment to search a particular database, such as the mouse genome, for all instances of a certain transcription promoter sequence of interest. This kind of computer analysis is then used to predict expression patterns that might ultimately be observed using standard methods.

Virology was the first biological discipline to enter the postgenomic era when, in 1977, the genome of bacteriophage ΦX174 was sequenced. Now, over 40 years later, there are more than 8000 completely sequenced viral genomes.

These sequence data are being used alongside bioinformatics techniques to determine the functional genetic units that make up an organism. In addition to the sequencing of genomes, the response of every gene (all the transcripts from the genes, or the **transcriptome**) and all the proteins (the **proteome**) of an organism can now be analyzed using new HT technologies. The output from these HT technologies is enormous, with some experiments yielding over one million measurements per sample, presenting an impossible task to interpret manually. Bioinformatics research provides the computer software tools that are required to analyze these data and find significant results. In addition to analysis, biological data need to be stored in a consistent, retrievable form, and computer software called database management systems (databases) provides an ideal tool for this purpose.

BIOLOGICAL DATABASES

A biological database is a large, organized body of persistent data, usually associated with computerized software designed to update, query, and retrieve components of the data stored within the system. These databases store a wide variety of biological information ranging from primary sequence repositories (GenBank, http://www.ncbi.nlm.nih.gov) to complex databases that store spatially mapped data such as *in situ* gene expression and cell lineage (Edinburgh Mouse Atlas, http://genex.hgu.mrc.ac.uk). The growth of biological databases over the last two decades has been staggering, as shown in Figure 24.1. These biological databases can be broadly classified into three categories: primary, secondary, and composite databases.

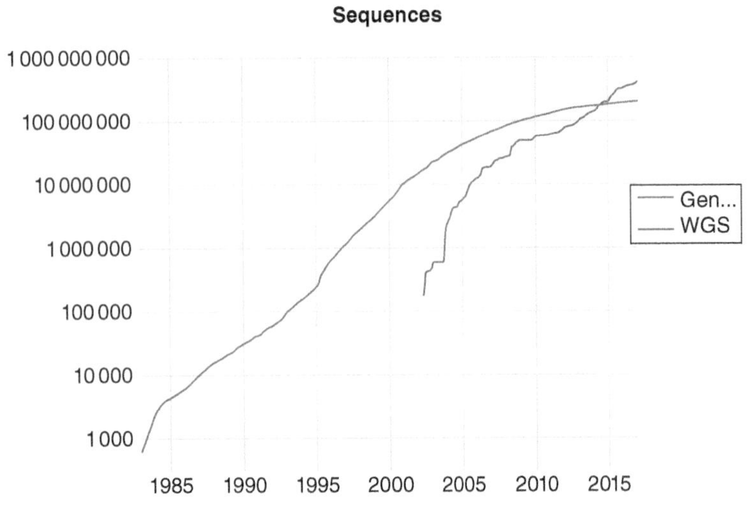

Figure 24.1 Growth of the genomic databases. Data displayed here are taken from the National Center for Biotechnology Information (NCBI) at the National Library of Medicine (NLM), National Institutes of Health (NIH). From 1982 to 1993, very little expansion of the database is seen. However, with the advent of rapid sequencing technologies, dramatic increases in the sequences deposited in the databases took place. Notice that the y-axis is a log scale. In the figure, the Genbank data result from sequencing of specific genomic segments. The WGS data refer to "whole genome shotgun" sequencing results deposited in the database. *Source:* Data courtesy NCBI (http://www.ncbi.nlm.nih.gov/Genbank/genbankstats.html).

Primary databases

Primary databases store raw data, including nucleotide and protein sequence information. They serve as public repositories for researchers to deposit sequence and structural information. The biggest primary databases are the nucleotide sequence repositories. The International Nucleotide Sequence Database Collaboration (INSDC) now operates to cover nucleotide sequence repositories from the National Center for Biotechnology Information (NCBI), the European Molecular Biology Laboratory European Bioinformatics Institute (EMBL-EBI), and the DNA DataBank of Japan (DDBJ). As of the end of 2018, GenBank (NCBI) contains about 2.8×10^{11} nucleotides, representing about 2×10^{8} genomes. The European and Japanese databases contain similar numbers. The INSDC initiative works to allow consistent submission requirements for all deposits into the three databases that make up the collaborative.

The second largest primary databases are the protein sequence databases. UniProt, the central repository for all protein sequence information, contains about 560 000 protein sequence entries. However, only about 100 000 of these protein sequences have been experimentally determined, with the rest having been computationally predicted by analysis of nucleotide sequences in GenBank or inferred by homology to existing proteins. In contrast, the protein structure database (PDB) managed by the Research Collaboratory for Structural Bioinformatics (RCSB) contains only a fraction of the nucleotide and protein sequence databases, with only about 150 000 known structures. These primary databases provide a valuable source of information for further analysis.

Secondary databases

Secondary databases store information derived from primary databases. Secondary databases contain information like the conserved sequence patterns or motifs of nucleic acids or proteins, genomic variants and mutations, and evolutionary relationships. For example, the viral database VIDA contains a complete collection of homologous protein families derived from the raw sequence data of complete and partial viral genomes. The Viral Pathogen Resource (ViPR) contains searchable data for 18 viral families, with about 5800 species. There are many other secondary databases, such as Structural Classification of Proteins (SCOP), that provide a detailed and comprehensive description of the structural and evolutionary relationships between all proteins whose structure is known.

Composite databases

With the numerous primary and secondary databases available, it is necessary to collate this information in a central resource. Composite databases combine information from different primary and secondary resources and make them available in an integrated and more convenient form. Entrez Gene is a composite database hosted at the NCBI (https://www.ncbi.nlm.nih.gov) that compiles information about a gene, such as genomic map, sequence, expression, protein structure, function, and homology data, in a single database record. Combing resources in this manner eliminates the need to search multiple resources and can give the biologist a rapid picture of the total combined knowledge of a protein or gene. In virology, there are a number of these composite resources, such as the HIV drug resistance database (http://hivdb.stanford.edu), which compiles information from published HIV-RT and protease sequences and sequences of HIV isolates from persons participating in clinical trials.

Other databases

There are many other types of specialized databases, such as PubMed (https://pubmed.ncbi.nlm.nih.gov), which contains nearly 30 million literature references in the field of life sciences. If you are interested in finding out about new and existing databases, the journal *Nucleic Acid Research* publishes a freely available database issue in January of each year. The most current version of this issue can be found at https://academic.oup.com/nar/issue/48/D1.

BIOLOGICAL APPLICATIONS

The large volume of data that is stored in these biological databases is difficult to manually interpret, and bioinformatics tools are required to extract meaningful information and knowledge from them. The five most common types of bioinformatics applications are similarity-searching tools, protein function analysis, structural modeling, structural analysis, and sequence analysis.

Similarity-searching tools

Similarity-searching tools enable the biologist to find sequences closely related to their sequence of interest. Finding a closely related sequence may provide information on its molecular function or biological properties. It is possible to annotate a whole viral genome by this method. For instance, the functional annotation for many genes within the murine cytomegalovirus genome has been derived from sequence comparison with human herpesvirus 1 and the human genome.

Similarity searching uses a number of algorithms that employ a linear comparison between a sequence of interest and sequences found within the databases. One of the most powerful tools of this kind is located at NCBI and is available to any user at their website. The tool is called BLAST (Basic Local Alignment Search Tool) and can be found at https://blast.ncbi.nlm.nih.gov. For a BLAST search, you need to simply have a sequence of nucleotides or amino acids that is of interest.

The BLAST search algorithm compares the input sequence with a selected database by attempting structural alignments. This is accomplished by local rather than global protocols. This means that local structural similarities are searched, especially in the case of proteins that may have modular regions of sequence. The output of the search is presented as a table, ordered by statistical measurements. These statistics relate to the likelihood that the alignment could have happened by chance. There are two parameters reported: the bit score and the E-value. The higher the bit score and the lower the E-value, the more likely it is that the search "hit" is an authentic match.

A typical example of a BLAST search is shown in Figure 24.2. A test "unknown" sequence of 50 bases was used for the search. This was input into the initial BLAST page (Figure 24.2a), and the selection filter was set to "Viruses." The search required about 10 seconds, and the results page is shown in Figure 24.2b. The "unknown" sequence was actually harvested from the human herpesvirus 6 p41 gene. You can see that the top matches in the output are all from this virus. Notice, however, the matches with other, more distantly related herpesviruses. If this were truly an unknown sequence, statistical methods would be used to decide upon the legitimacy of a presumed match.

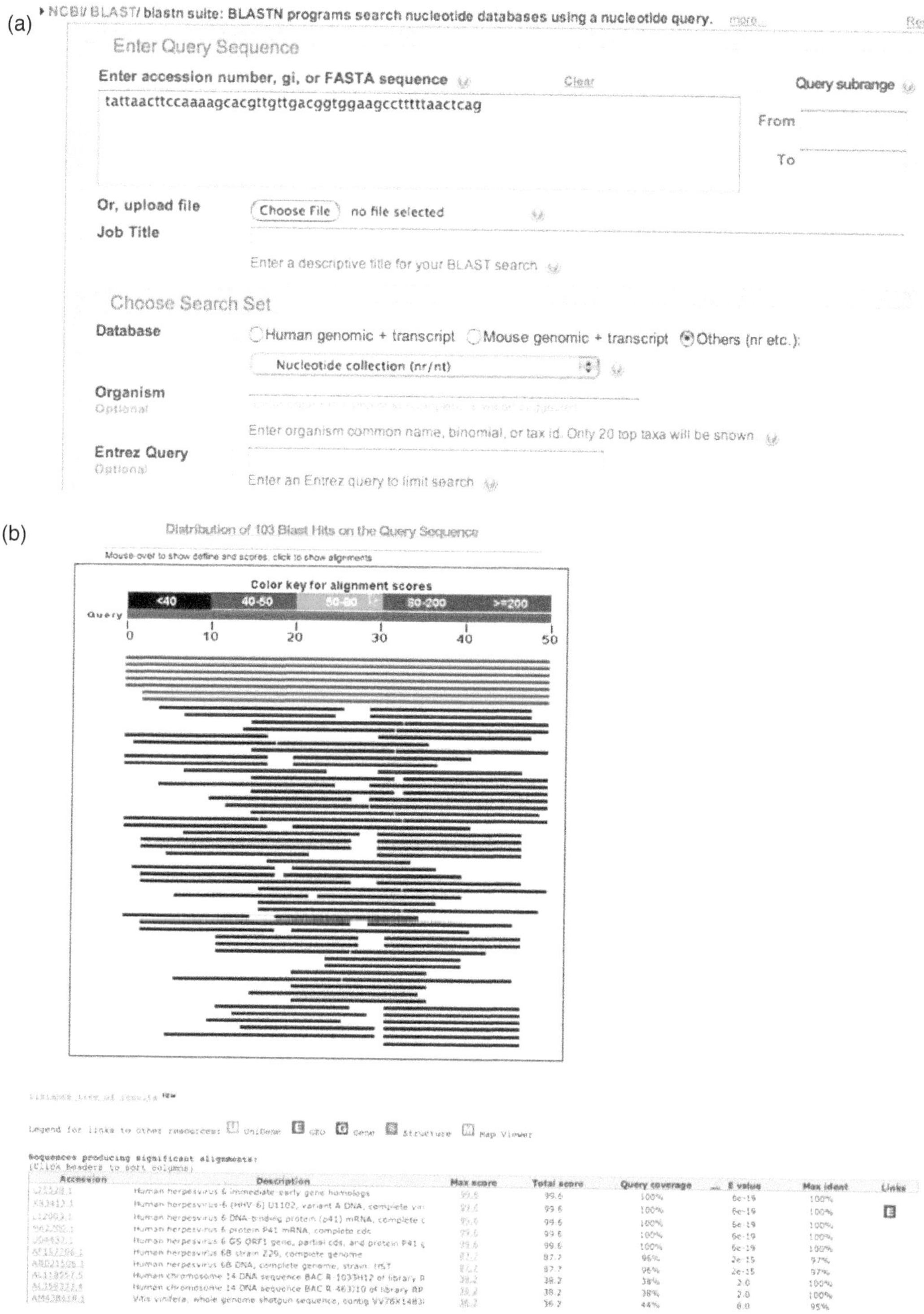

Figure 24.2 Results of a BLAST Search for a Nucleotide Sequence. (a) Initial BLAST input page, with a 50-nucleotide "unknown" sequence entered. Notice that the database filter has been set to "Viruses." (b) Result of the search is shown in graphic format at the top, along with the first 10 "hits" from the database. The search algorithm calculates a score for each hit, along with a significance parameter (E-value). The lower the E-value, the more significant is the match. In this example, the "unknown" sequence is actually 50 bases from the human herpesvirus 6 p41 gene.

Protein functional analysis

Protein functional analysis tools enable the biologist to determine the function of proteins. Protein sequences are compared against secondary or derived databases that contain structural signatures, motifs, and protein domains. By finding a match, functional information can be induced from a protein family. This technique was used by the VIDA, maintained at University College London, to classify 14484 viral proteins into 1156 homologous protein families. The system extracted data from GenBank at the National Institutes of Health (NIH) and used protein sequence alignments to search for homologies. Such matches, along with annotations regarding protein function, were then included in the display of homologies. Unfortunately, as of this writing, the system is no longer available. However, a collection of online tools that allow this kind of comparison can be found at the ViPR site (https://www.viprbrc.org/brc/externalTools.spg?decorator=vipr).

Sequence analysis

Sequence analysis tools encompass a wide variety of functions, including the ability to determine the biological function and/or structure of genes and the proteins for which they code. By using information gathered from previous biological experience, rules can be defined that can then be used to determine features of a sequence such as transcriptional start sites, CpG islands, and hydropathy regions. Programs such as Clustal Omega (http://www.clustal.org/omega) allow you to align multiple sequences. The output is an alignment of sequences so that the user can identify identical and different regions of protein or nucleic acid sequences. This technique can be used to examine possible evolutionary relationships between species and strains of organisms. Output of the program includes displays of relationships as phylograms or cladograms.

Structural modeling

The number of known protein structures (about 118 000) is only a small percentage of the number of known protein sequences (about two million). However, in the absence of a protein structure that has been determined experimentally by X-ray crystallography or nuclear magnetic resonance (NMR) spectroscopy, researchers can try to predict the three-dimensional structure using structural modeling. This method uses experimentally determined protein structures to predict the structure of another protein that has a similar amino acid sequence. For proteins with sequence homology of greater than 45% identity, this method produces good results. Protein modeling has been used to determine the 3D structure of the dengue virus NS3 protease based on the related hepatitis C virus (HCV) NS3 protease. This predicted structure was later confirmed experimentally, showing the power of this approach.

Structural analysis

Structural analysis tools are among the most powerful tools in bioinformatics. They enable the biologist to compare a given structure against known structures in the databases. Comparing protein structure is more sensitive than sequence comparison because the function of a protein is more directly a consequence of its structure rather than its sequence. In favorable cases, comparing 3D structures may reveal biologically interesting similarities that are not detectable by comparing sequences. A number of these tools are available online. One good source is the European Bioinformatics Institute site (www.ebi.ac.uk/Tools/structural.html).

There are many other types of bioinformatics applications that can be utilized to annotate genes and proteins as well as find regulatory elements or signaling pathways.

SYSTEMS BIOLOGY AND VIRUSES

One consequence of the massive amount of data produced by HT technologies has been a turning point in the philosophical approach of molecular biology. So much sequence data have been derived for organisms that the very definition of the gene has been called into question. The solution to this dilemma was not immediately apparent. On the one hand, molecular biology had been committed to the idea that the sequence of the genome would intrinsically contain all the information necessary to define a particular organism. On the other hand, it has become increasingly obvious that the mere knowledge of the sequence is insufficient to define the biological properties of the organism. Thus, prediction of the expressed sequences (transcriptome), open reading frames (ORFeome), and actual proteins (proteome) became ways in which these massive data accumulations could be viewed. However, it remained to be able to understand these data in terms of the whole organism.

One approach that is being used increasingly involves the methods of systems biology. This utilizes a different conceptual approach to understanding the data. Systems biology takes its key from the idea that living systems involve interactions that can be described in terms of networks. These networks have very specific features that make them unique and that lead to properties of the system that are more than the sum of the parts that make up the system. Such networks are called scale-free or small-world networks. Their properties are well known in the fields of physics and social science, but are just becoming understood in biological systems.

Scale-free networks emphasize the importance of the connections between nodes rather than the nodes themselves. In such networks, a few such nodes are connected to large numbers of other nodes.

Figure 24.3 describes the properties of a scale-free network as compared with a random network. Random networks have nodes that are connected to only 27% of all nodes (Figure 24.3a). However, in scale-free networks, the highly connected nodes are connected to 60% of all nodes. This high connectivity gives the network its unique features, including the fact that changes to these particular nodes have wide-reaching effects throughout the network. A good example of such a network is the World Wide Web. Certain websites, such as Google and Amazon, are so highly connected that changes in access to these sites percolate throughout the web, affecting the behavior of distant sites. A recent example of this occurred on March 2, 2017, when a programmer's typographical error in a command took a part of the Amazon Cloud server offline, resulting in the outage of a significant part of the web.

A further consequence of this connectivity means that the number of steps it takes to get from one place in the network (the web) to any other place is much smaller than one would think. The "degrees of separation" concept is inherent in the understanding of scale-free networks. This harkens back to the original work of the social scientist Stanley Milgram, who discovered that there are only about six degrees or steps of separation

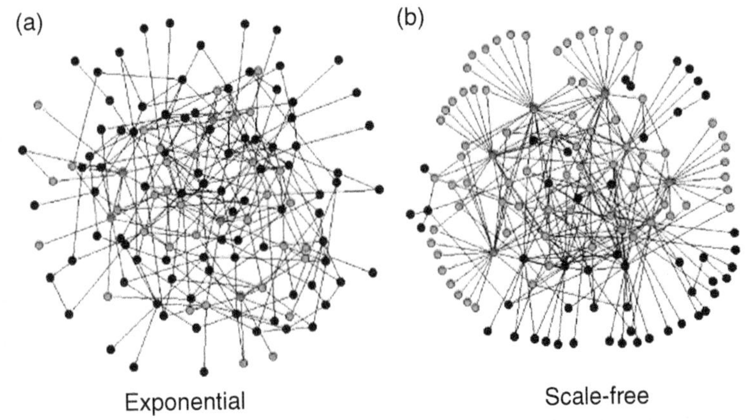

Figure 24.3 Random versus scale-free networks. (a) A random network, in which the five nodes that have the most links are connected to 27% of the nodes. (b) A scale-free network, in which the five nodes that have the most links are connected to 60% of the nodes. *Source:* From Albert, R., Jeong, H., & Barabási, A.L., "Error and attack tolerance of complex networks," *Nature* 406, 378–382 (2000). Reproduced with permission from Springer Nature.

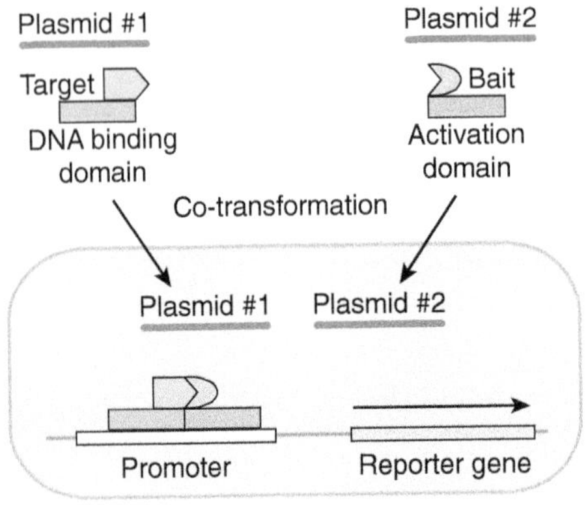

Figure 24.4 Yeast two-hybrid detection system. Transformation vectors are created containing hybrid sequences in which the DNA binding domain and the activation domain of a gene activator are present in separate plasmids (plasmid #1 and plasmid #2). In each case, the sequence of a test protein is inserted such that it is expressed in phase with the transcriptional activator domain. The two plasmids are used to co-transform a cell that also contains a reporter gene downstream from a promoter that is sensitive to control by the transcriptional regulator. When the two test proteins are able to bind together sufficiently well so that the two domains of the transcriptional activator can bind to and enhance transcription of the reporter gene, then a positive interaction is scored. Such assays can be done using microarrays, as described in Chapter 12.

between any two people on earth. This is the origin of the so-called "small-world" feature of these networks.

A major application of this analysis is the study of protein–protein interactions in cells. Again, HT technologies are used. In this case, the yeast two-hybrid system is the key data-gathering tool. This is a very sensitive method for assessing the interactions between two proteins utilizing the activation of transcription from a reporter gene. The system is shown schematically in Figure 24.4.

Pairwise protein interaction assays are done using all known or predicted open reading frames (the ORFeome) in a genomic set. Of course, this generates massive amounts of data that can only be effectively analyzed using computational approaches.

Protein interaction maps (PIMs) are produced as a result of these kinds of analyses. Such maps have now been derived for a number of organisms whose complete ORFeome is known. To date, all such maps have a scale-free topology. That is, there are a few proteins that are highly connected. These hubs are critical nodes in the map, since alterations in these proteins have

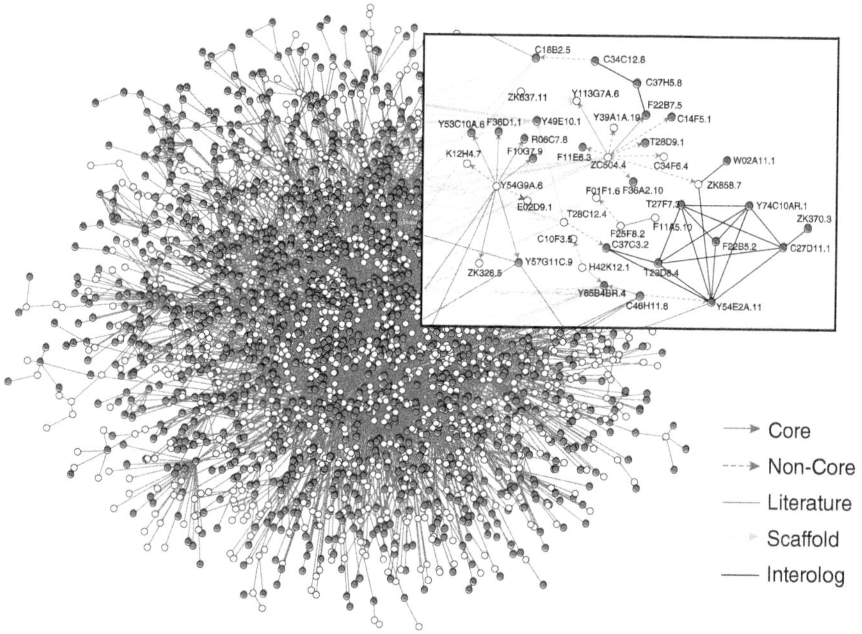

Figure 24.5 Protein interaction map (PIM) for the round worm, *Caenorhabditis elegans*. *Source:* From Li, S., Armstrong, C.M., Bertin, N., et al., "A map of the interactome network of the metazoan *C. elegans*. Science, 303 (2004), 540–543.) Reproduced with permission from American Association for the Advancement of Science.

wide-ranging effects on the network. On the other hand, such networks are resistant to the effects of random changes. Mutational inactivation of random nodes has little effect on the functionality of the network. An example of a PIM is shown in Figure 24.5.

Now that both viral genomes and their hosts' genomes are being sequenced at a rapid rate, the use of HT technology coupled with the yeast two-hybrid system can reveal the interactions between viral proteins and those of their host cells. While this work is just beginning, some interesting preliminary results have been obtained. PIMs for viruses can be derived in two ways: as virus–virus maps or as virus–host maps. An example of such a map is shown in Figure 24.6, where this has been determined for Hepatitis C virus (HCV).

There are at this writing a number of databases with online-accessible tools to examine virus–host protein interactions from a large number of viruses. One of these (VirHostNet 2.0, http://virhostnet.prabi.fr/index.html#) has a robust browsing panel that allows selection of viruses by family and resulting display of interactions between specific viral proteins and a variety of host proteins. Another database (Virhostome Database, http://interactome.dfci.harvard.edu/V_hostome) examines interactions between human proteins and a set of DNA tumor viruses. Finally, VirusMentha (https://virusmentha.uniroma2.it) is a powerful, interactive browser that quickly allows a display of selected viral protein interactions with a large array of host proteins.

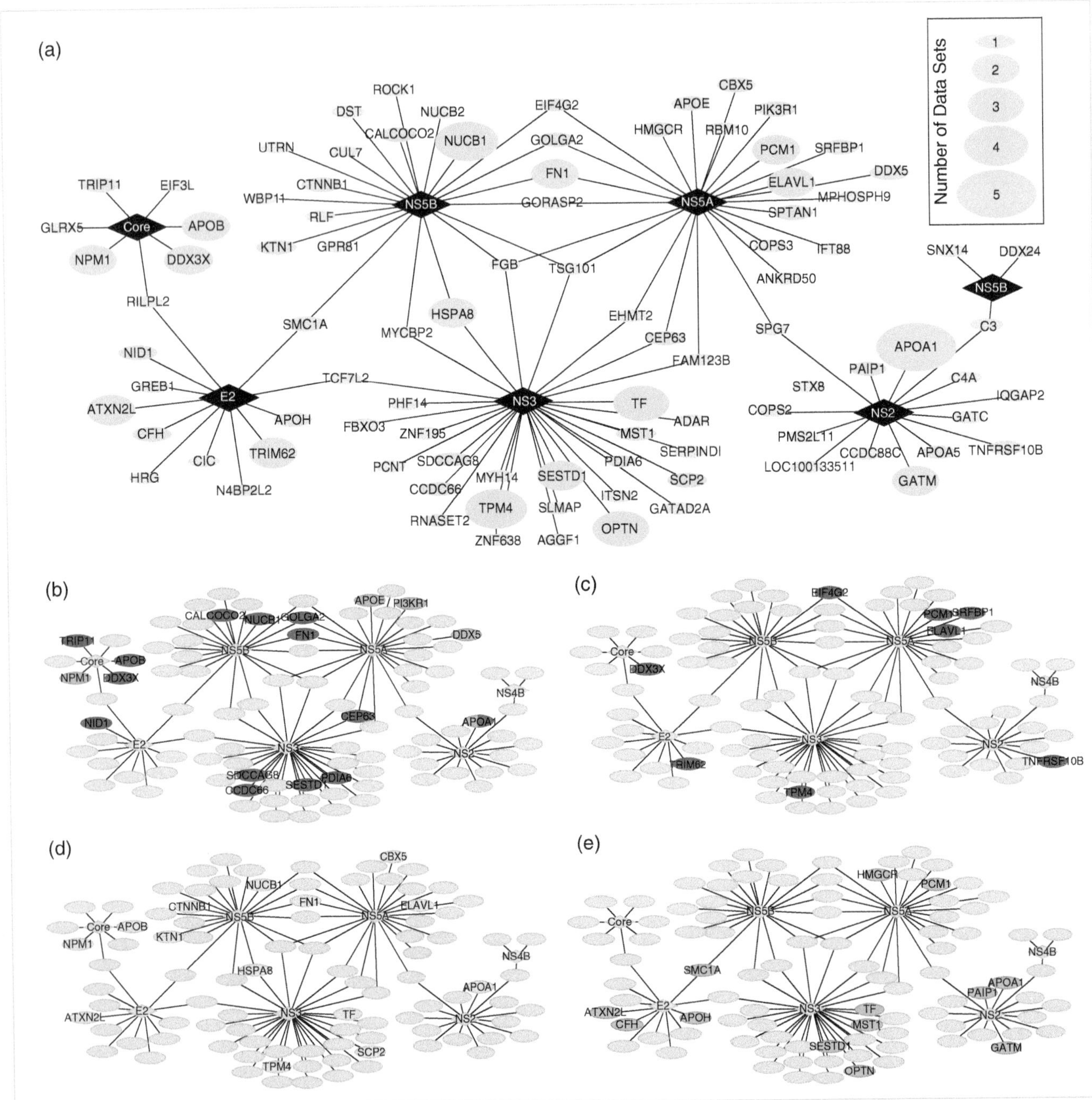

Figure 24.6 Host–virus protein interaction map of hepatitis C virus (HCV): the HCV 2a–human interaction network. (a) Interactions between HCV 2a and human proteins identified in this study. Diamonds indicate HCV proteins; ovals, human proteins; black lines, HCV 2a–human protein interactions. Proteins linked to HCV infection by other *in vitro* HCV–host cell interaction experiments are shown in blue. Node size indicates the number of independent data sets linking the corresponding gene to HCV replication. (b–e) Human proteins previously implicated in HCV infection by experiment type. (b) Red nodes: Physical interactions identified by Y2H; blue nodes: other experimental methods. (c) Purple nodes: siRNA screens. (d) For proteomics, yellow nodes show differentially expressed proteins; rectangles indicate bottleneck proteins as defined in reference below. (e) Green nodes: microarrays. *Source:* From Dolan, P.T., Zhang, C., Khadka, S., et al., Identification and comparative analysis of hepatitis C virus–host cell protein interactions. Mol. BioSyst., (2013), 9, 3199–3209. Copyright 2013 Royal Society of Chemistry.

VIRAL INTERNET RESOURCES

There are many useful websites that provide databases, analysis tools, and links to viral bioinformatics resources:

Name	Description	URL
The Universal Virus Database (ICTVdB)	The International Committee on Taxonomy of Viruses database is a universally available taxonomic research tool for understanding relationships among all viruses. ICTVdB's fundamental goals are to provide researchers with precise virus identification and to link the agreed taxonomy to sequence databases.	http://www.ncbi.nlm.nih.gov/ICTVdb
All the Virology on the WWW	All the Virology on the WWW is a website that collates virology-related websites on the Internet. In addition, it provides an index to virus pictures on the web, The Big Picture Book of Viruses, which also functions as a resource for viral taxonomy. A collection of Online Virology Notes can also be found here.	http://www.virology.net
Viral Bioinformatics Resource Center	This resource provides access to viral genomes and a variety of tools for comparative genomic analyses. At the heart of the system is VOCs (Virus Orthologous Clusters), a database with built-in tools that allows users to retrieve and analyze the genes, gene families, and genomes of 12 different virus families. The database is the source of information for other programs of the workbench for whole-genome alignments, genome display, and gene or protein sequence analysis. Many of these tools can also be used with user-provided sequence data. The workbench tools are Java-based and user-friendly to allow all users, regardless of computer skill level, to access and analyze the data.	http://virology.uvic.cahttp://athena.bioc.uvic.ca
ViPR	The Virus Pathogen Resource is a compendium of tools that allow analysis of a wide variety of viral genomes. Supported by a team of scientific advisors, this database tool is extremely useful.	https://www.viprbrc.org/brc/home.spg?decorator=vipr
Virology Blog	While not a database, this is still a very informative site. The blog is managed by Dr. Vincent Racaniello, who is Professor of Microbiology & Immunology in the College of Physicians and Surgeons of Columbia University. Well worth following for those interested in virology.	http://www.virology.ws
Viral Zone	Maintained by the Swiss Institute of Bioinformatics, this is a web resource for all viral genera and families, providing general molecular and epidemiological information, along with virion and genome figures. Each virus or family page gives an easy access to UniProtKB/Swiss-Prot viral protein entries.	https://viralzone.expasy.org

QUESTIONS FOR CHAPTER 24

1 It has been shown, using yeast two-hybrid systems and microarray analysis, that nonstructural protein 5A of HCV interacts with the interferon-induced protein kinase (PKR). If this interaction represents a real event during the viral life cycle, what might you propose as a possible effect of HCV on the host defense system?

2 The following table presents four "unknown" sequences as single-stranded DNA. Your task is to identify the most probable identity or source of each sequence. Use the BLAST search tool online to carry out this identification.

"Unknown Sequence"	Most Probable Identity
tgtcagatacggatcgtttttaacgaaaaccattaataatcatattcca	
tatctgatttgccaggtttgttaaactatcttgtagttttgactggcaaa	
catattaatcacactagttggtattataatgataaggttatagcgctagc	
tcgtgcggtcttggtaaatgttaccaatttgcacttgggcagggaaccac	

The BLAST search tools can be used by starting at https://blast.ncbi.nlm.nih.gov. Instructions for using the tools are also found at this web address.

3 Suppose that you wish to examine the interaction network for influenza A virus, strain H1N5, with bird host cells. How would you go about this? What kinds of steps would you have to take in order to display all possible interactions of H1N5 proteins with the proteins in the cell during a productive infection?

Viruses and the Future – Problems and Promises

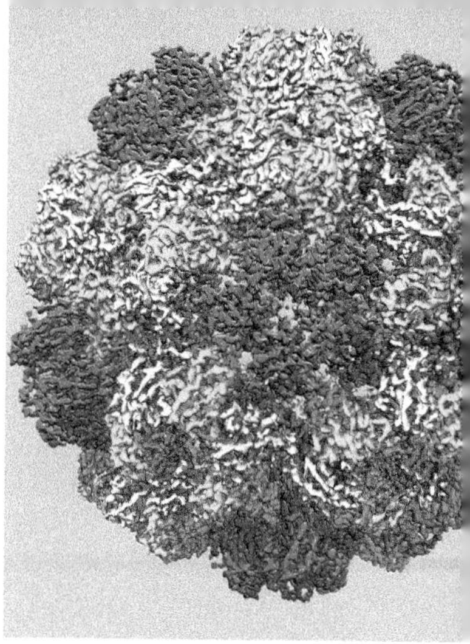

CHAPTER

25

The study of viruses in this century has provided many insights that have been successfully exploited in the development of modern biology and medicine. Virology is now a "mature" discipline, and, while much detail remains to be elucidated, there is no doubt that the basic processes of virus replication, infection, and pathology are understood. The ever-accelerating pace of technology will only serve to facilitate the study of new viral variants and provide new information on old problems.

While technology has provided many benefits to our studies and our lives, especially to those of us in the "technologically developed" (European, American, East Asian) countries, which hold economic sway at this point in the twenty-first century, we should be sophisticated enough to know that exploitation of our understanding of nature is a mixed blessing. Technology has bred many problems. These problems are different in scope and impact from those facing the world at the beginning of the twentieth century, but loom large, especially to those in less economically favored portions of the globe. Viruses are part of the problem and may provide part of the solution.

Basic Virology, Fourth Edition. Martinez "Marty" Hewlett, David Camerini, and David C. Bloom.

CLOUDS ON THE HORIZON – EMERGING DISEASE

The ravages of the "new" disease AIDS, caused by an emergent virus, HIV, has generated much concern, controversy, and public debate. It is not the first virus to cause concern, and will not be the last. Despite its success in garnering media attention, HIV is nowhere close to being near the top among infectious diseases in threats to human welfare. The most formidable viral challenge to our society in the twentieth century was caused by influenza!

The so-called Spanish influenza worldwide pandemic in the last year of World War I killed 20 million people worldwide, and more than 600 000 in this country alone. Many of its victims were young people in the prime of life – those who might be expected to be the least affected by such a disease. The large spike in infectious disease mortality caused by this virus is clearly evident in the statistical display of mortality rates due to infectious disease shown in Figure 25.1.

The lethality of the Spanish influenza (the actual geographic origin of this virus is not known) was in part due to its causing hemorrhaging and extensive cell damage in pulmonary tissue, much like the more recent hantavirus infections seen in the winter of 1994 in the American Southwest. Victims essentially drowned in their own body fluids. Why did this Spanish influenza virus cause so much damage to the lungs while earlier and subsequent strains of flu did not? This question is an important one, and it has been suggested that perhaps the virus was a direct infection of humans of an avian strain instead of first passing through pigs, as is the usual route of emergence of new strains of influenza. As discussed in Chapter 16, such transmission was recently reported for a strain of avian influenza A in Hong Kong. Another possibility is that it encoded an unusual gene or combination of genes responsible for its devastating virulence and unique (for influenza) patterns of pathogenesis.

In an attempt to answer such questions, US Army researchers isolated RNA from preserved lung tissue of victims of the disease and used reverse transcriptase to make complementary DNA (cDNA) copies. This was then subjected to polymerase chain reaction (PCR) using a number of primer sets that are known to be specific for various strains of flu messenger RNAs (mRNAs) present now; and, perhaps surprisingly, the method worked! (The basic methods of carrying out reverse transcription [RT]-PCR were described in Chapter 11, Part III.) The workers then went on to sequence the entire set of eight RNAs that make up the influenza virus genome. Finally, they were able to actually recreate the 1918 strain and then study its virulence properties in mice, a stunning and somewhat controversial achievement. The controversy, of course, surrounds the

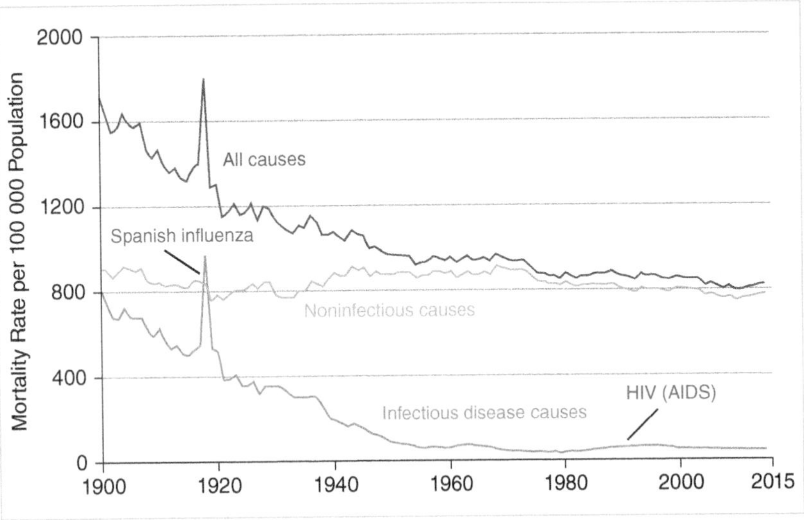

Figure 25.1 Mortality from infectious diseases in the United States between 1900 and 2015. The data show a continuing decline associated with improvements in public health measures and antiseptic methods in hospitals through the end of World War II. The already low rate was then further lowered by the use of antibiotics. The two increases seen are due to the 1918 Spanish influenza worldwide pandemic and the onset of HIV-induced AIDS in the early 1980s. The rate may continue to increase as antibiotic-resistant bacterial strains become established in the population. *Source:* Data are from the US Centers for Disease Control and Prevention.

potential danger that the existence of the virus might pose, especially in our age of concern about potential bioterrorism.

The results of this study revealed that the 1918 influenza virus was, in fact, an avian-like virus that had not been circulating in the human population previously. As a result, the entire population was essentially immunologically naive with respect to this strain. Exactly how it came to infect humans and be transmitted from person to person has not yet been determined from these studies.

Using the reconstructed virus, scientists were able to determine features of the virus that may be related to its high pathogenicity. Three differences were noted between the reconstructed 1918 H1N1 and its less virulent relatives:

1 The virus could gain entry to cells without relying on a cellular protease. The neuraminidase (NA) protein of the virus appears to be responsible for this.

2 The polymerase of the 1918 virus allows it to replicate efficiently in human bronchial cells in culture.

3 The hemagglutinin (HA) of the 1918 strain is associated with extensive pulmonary damage and a high level of virulence in mice.

It is not known whether this 1918 strain of flu is truly "extinct" or whether it is lurking in some influenza reservoir that could be tapped to generate a new, lethal outbreak. If more concern about such a possibility is necessary, the direct transmission of two strains of avian influenza from chickens to humans that we have seen with the H5N1 and H7N9 viruses observed in China and Southeast Asia should provide further impetus, since these viruses target young people and have a relatively high mortality rate compared to other current strains. Thus far, these outbreaks have not been associated with the appearance of a variant of the avian flu that can spread directly between people, but most epidemiologists consider the occurrence as a portent of future risk. Clearly, while we consider flu a mild, inconvenient disease, it can show itself to be a formidable threat!

People tend to respond to the problem most apparent and nearest at hand, and thus, emerging diseases only take center stage when a crisis looms. Hopefully, the medical and scientific community is a bit more discerning in its evaluation of potential risks. The impact of HIV infection and AIDS has been profound and extremely disquieting. If there is anything at all salutatory about the appearance of this virus in the long run, it may be that it has reinforced the certain knowledge that infectious disease is not conquerable; it is only controllable, and such control is expensive. Ironically, it often seems that just as science gets a handle on one deadly virus, new ones emerge to fill the void.

Modern molecular biology has suggested direct approaches toward treatment of HIV-infected individuals; some are outlined in Chapter 8, Part II. These may reverse some or all of the ravages of AIDS. For example, a San Francisco periodical dedicated to tracking AIDS in that city's large and affluent gay community reported in the late 1990s a week without any death from AIDS for the first time since the mid-1980s. Still, as evident in Figure 22.1, HIV is responsible for the first general rise in mortality rates due to infectious disease since the great flu pandemic of 1918, and current therapy strategies will have little effect on the impact of HIV worldwide in the foreseeable future. Even when controlled, HIV and AIDS will be endemic diseases, and only the relatively inefficient mode of passage between individuals is proof against a more major impact.

As ominous as AIDS is – a progressive disease with awful consequences – it may well not be the greatest threat to our well-being from viruses in the next decade or so. The honor of that place may go to a positive-sense strand RNA virus in the Flaviviridae family – hepatitis C virus (HCV). The widespread occurrence of this virus in the general population has only become appreciated near the beginning of this century following development of effective screening tests for its presence in the blood of individuals and of blood stored for transfusion. It is now known that HCV infections account for 40–50% of all chronic liver disease.

Currently, screening for HCV has reduced transmission of the virus through blood supplies to a negligible level, but screening also reveals that there is a high incidence of the virus in healthy individuals, many of whom have never had a transfusion. Similarly, although the virus can be spread by shared needle use among injection drug users, and possibly by venereal routes, many HCV-positive individuals have no history of either drug abuse or high-frequency casual sexual encounters. Clearly, other routes of transmission are effective.

It is not known how important or extensive future problems will be from HCV infections. It is known that most individuals who are HCV positive are healthy and show no signs of liver dysfunction. Further, acute infection with the virus is often cleared with no sequelae. Unfortunately, there is a small but significant proportion of chronically infected individuals who go on to develop cirrhosis and other critical liver damage. In addition, there is a strong statistical correlation between these complications and the development of hepatocarcinoma in some of these afflicted individuals. The fear is that if the unknown mode of spread of the virus is efficient and difficult to control, the reservoir of chronic HCV carriers is already great enough to provide a significant health risk to the entire population.

At present, there are a total of 12 US Food and Drug Administration (FDA)-approved antiviral drugs for the treatment of HCV infections. These drugs range from long-acting interferons (pegylated interferon) to versions of ribavirin to drugs that act in combination with either or both of these agents. The most recently approved treatment uses a two-drug combination called Epclusa, each of which inhibits a portion of the viral replication machinery.

Further study and the development of effective vaccines may obviate any dangers threatened by HCV infections, but the general problem illustrated by this virus is clear. The use of transfusion, the occurrence of reservoirs of unknown pathogens within human populations, and the increasing interaction between populations and individuals within these populations will consistently pose threats to our public health in the form of emerging viral infections. We cannot appreciate the risk engendered by a new virus and its transmission until that virus is identified and screened for, and such identification and screening are only going to be directed at the known threat. It is often impossible to know the dangers posed by a new virus until it is already established in the population.

Sources and causes of emergent virus disease

The Centers for Disease Control and Prevention list many factors that contribute to the emergence of a novel viral disease. They term these "enabling" factors. Of particular importance is the potential evolution of novel infectious strains of virus due to co-infection of the same individual. This is the *sine qua non* of influenza epidemics, but genetic recombination and reassortment have been observed in many mixed infections with viruses in the laboratory, and a rare, nonhomologous recombination could take place at any time. Indeed, this must be the source of so many genes of host origin that are maintained in various viruses and clearly provide a great survival advantage to the virus.

Other enabling factors include breakdown of public health measures due to social and economic disruptions such as war and depressions. Such disruptions occur sporadically throughout the world, and there is no evidence that the rate of occurrence is in decline.

Concentration of people with shared lifestyles can also be an important enabling factor. Certainly, HIV infection and AIDS are not more confined to the subset of the homosexual community habituated to casual sex with large numbers of partners than they are confined to injection drug users, but the concentration of these populations in small urban areas has served an important role in the virus's establishment in American and European populations. In addition, the virus has established itself in female "commercial sex workers," especially in Southeast Asia and in urban areas of Central Africa.

The global economy also has served as an important enabling factor for establishment of viral disease. For example, the Asian Tiger mosquito has moved to California and other parts of the United States in the stagnant water inside worn-out rubber tires that were sent to these places for recycling. The viral diseases spread by this vector are already on the rise and will continue. Another example of a virus that has established a new geographical range due to modern communication is West Nile virus, an agent of arboviral encephalitis. As mentioned in this book, this virus, previously found in Africa and the Middle East, is now well established throughout the United States. The exact details of its arrival are not clear at this time, but the epidemiology of its spread suggests an origin point in the Northeast.

The application of intensive and invasive agricultural methods has led to severe habitat disruption in Africa, South America, and parts of Asia, bringing humans into closer contact with potential vectors of disease transmission. As discussed in Chapter 16, these are important factors in the sporadic outbreak of arenavirus and filovirus (Marburg and Ebola viruses) infections.

One must also be aware that periodic disruptions in typical weather patterns can have an important role in enabling new virus infections. Unusually wet, warm winters led to the emergence of hantavirus in the American Southwest as its natural rodent hosts proliferated and interacted with human populations. It is virtually certain that this has happened before, because both Navajo and Ute indigenous knowledge warns against approaching wild rodents during such periods of unusual weather.

In the larger context of our planet, climate change will have a significant influence on the emergence of viral diseases in locations where they have not yet been a problem. As shifts in climate take place due to atmospheric effects of increasing greenhouse gases, viral vectors such as mosquitoes will be able to establish themselves in ecological niches in which they could not have existed previously. For instance, *Aedes aegypti* and *Aedes albopictus*, the vectors of yellow fever, Dengue, and Zika viruses, are predicted to move into new geographical locations that are further northward than before. We can therefore expect to see public health consequences from this evolving situation.

Obviously, technology in the form of efficient transportation and effective but invasive medical procedures also has a role in fostering the establishment and spread of novel viral diseases. The recent news regarding Zika virus is a case in point regarding modern transportation. The virus, another member of the Flaviviridae family, was first identified in 1947 as a mosquito-borne infection of monkeys in the Zika Forest region of Uganda. Human infections in the region were documented in 1952 by the presence of neutralizing antibodies. By 1983, the virus had spread throughout Equatorial Africa and had made its way to Southeast Asia. From there, between 2007 and 2014, the virus was found in various island nations throughout Oceania, including Micronesia and French Polynesia. Finally, in 2015, the virus made its way to Brazil. At this point in the story, the pathogenesis began to include, in addition to a mild febrile illness, two disturbing features. Infection can lead to neurological sequelae, notably Guillain–Barré syndrome. In addition, pregnant women who are infected have an increased likelihood of delivering a child with microcephaly, with abnormal brain development and a small head. Most recently the virus has tracked north along with the range of the vector, *A. aegypti*, and can now be found in regions of the southeastern United States.

The threat of bioterrorism

On September 11, 2001, our world changed irreversibly after the attacks on the World Trade Center and the Pentagon. In the wake of these events, it has become all too clear that biological weapons also have to be considered among the choices of those wishing to attack this or any other nation.

The Centers for Disease Control and Prevention lists three classes of agents, all of which include viruses. Of most concern is Class A, which includes agents of viral hemorrhagic fever (Ebola/Marburg, Lassa, and Machupo) and smallpox virus (variola major). Class B includes agents of viral encephalitides (the various equine encephalitis viruses, for example). Class C includes emerging viruses such as Sin Nombre virus.

The most immediate worry for many public health agencies is smallpox virus. As discussed in this book, with the eradication of wild virus by the massively successful World Health Organization (WHO) program, few people receive vaccination against this agent. As a result, stocks of vaccine have been built up and plans exist for necessary measures to take in case of dispersion of smallpox within the population. The existing stocks of the virus, scheduled to be destroyed, have been pressed into service to study the efficacy of vaccination protocols as well as potential antiviral treatment strategies.

It is sadly true that biotechnology, just like chemical technology, can be directly applied to human destruction, and stockpiles of biological weapons can be readily made. Although such weapons might be considered to be the last resorts of "rogue states" or terrorists, it is well to remember that the US CIA released the bacteria *Serratia marcescens* from a submarine outside of San Francisco in the 1950s to test methods for spread of biological weapons. While this bacterium is usually considered a harmless one, a high incidence of infections among the elderly and infirm was noted by public health officials at the time. Only after the Freedom of Information Act was enacted was the reason for this unusual outbreak determined.

This is not to argue that technology is evil. Technology in the form of efficient screening, effective countermeasures, and development of new medical products and procedures such as blood substitutes for transfusion is vital in controlling the threats posed by pathogenic microorganisms, whether from the environment or as the result of a deliberate release. Still, there is little likelihood that all or even most can be eliminated by technology. Certainly, HIV, HCV, Zika virus, SARS-CoV-2, and other as-yet-unknown viruses will still exist in the human ecosystem, even in the setting of effective and inexpensive modes of treatment and control. And all our technology cannot guard against the inhuman (and arguably psychotic) decision of any one group to rain destruction on another through the use of biological agents.

WHAT ARE THE PROSPECTS OF USING MEDICAL TECHNOLOGY TO ELIMINATE SPECIFIC VIRAL AND OTHER INFECTIOUS DISEASES?

The discovery and exploitation of antibiotics following World War II led to unreasonable euphoria. Experts widely claimed, and the lay population believed, that infectious bacterial diseases were no longer threats, and that viral diseases would soon be conquered and eliminated.

This has not happened and, with some notable exceptions, will not happen. Misuse of antibiotics, coupled with the ability of pathogenic bacteria to transmit drug resistance horizontally via plasmids, has led to an increasing incidence of multiple-drug-resistant strains of bacteria being detected in hospital laboratories. If there ever was the chance of eliminating bacterial infections, it is long past.

Most viral infections will not be eliminated for different reasons. In order for a viral disease to be eliminated, the virus must have a human or readily controllable reservoir, it must be able to induce an effective and lasting immune response, and there must be an effective vaccine against it. The success of eliminating smallpox and rinderpest shows that, given these conditions, eradication of a disease is possible. Presently, workers with WHO have targeted poliovirus and measles virus for future eradication. Some were predicting that this could be accomplished during the first decade of the twenty-first century, and it was certainly within the realm of possibility. Unfortunately, the most recent reports from WHO indicate that hopes for complete eradication of poliovirus are dimming in light of the ineffectiveness of the program in certain countries undergoing severe internal turmoil. The same is true for measles.

The promise is notable, but what of other viral diseases? What of HIV, cytomegalovirus, and the panoply of viruses that either establish persistent infections, effectively counter the immune response, have a nonhuman reservoir, or cause diseases that are more expensive to control than to live with? The answer is obvious. Viruses, like bacteria and parasites, will continue to be our life partners.

SILVER LININGS - VIRUSES AS THERAPEUTIC AGENTS

We have briefly described laboratory uses of viruses and viral genes. Further, we have identified some values of recombinant viruses in biotechnology such as the development of recombinant virus vaccines (see Chapter 8). These uses will likely increase in the future as the methods to exploit them increase in sophistication.

One of the most promising uses of current knowledge of virology and viral replication is in the area of gene transfer. The idea is simple: to construct (engineer) viruses that have no pathological properties, but retain their ability to selectively interact with and introduce their genes into specific cells and tissues. Another important area of study is the development of viruses specifically engineered to eliminate diseased or malfunctioning cells in the host.

Viruses for gene delivery

A number of laboratories are actively investigating the potential uses of adenoviruses, adeno-associated viruses (AAVs), several different retroviruses, and HSV as agents for gene delivery to specific tissues. Currently promising candidates are listed in Table 22.1 (see Chapter 22 for a full discussion).

Two basic approaches are currently under intense study and development. The first approach is to engineer a virus from which all pathogenic genes or elements are eliminated, but that can effectively express a therapeutic gene. The second is to produce a virus that expresses the appropriate gene but is unable to replicate by itself (a replication-deficient virus). A replication-deficient virus must be grown with the aid of a helper virus, or in a complementing cell. In principle, the methods for the generation and production of such viruses follow those outlined in Chapter 22.

As an example of the potential use of such a therapeutic virus, consider an individual with a genetic disease caused by the mutation of a gene whose function is required for normal functioning of a specific cell or tissue. Such a function might be the ability to produce insulin in cells of the pancreas, or it might involve a factor necessary for the maintenance of healthy neurons in an aging individual. Infection of the appropriate cells with a retrovirus containing this gene could lead to its integration into the host genome and continued expression of the gene. Alternatively, injection of a replication-deficient HSV containing a therapeutic gene into the brain of a person susceptible to early-onset Alzheimer's disease might lead to expression of a remedial protein from a latently maintained genome.

There is, however, one danger to the use of viruses for gene therapy that cannot be fully controlled. This danger is the possibility of inadvertently transferring a contaminating gene or genes along with the therapeutic ones. Of course, appropriate standards of purity and safety will go a long way to reducing such a danger, but it can never be eliminated. A survey of the relationships between viral genes clearly demonstrates that viruses can borrow genes from cells or from other viruses by rare nonhomologous recombination events. If the resulting virus has a strong replication advantage and is pathogenic, it could lead to a novel disease of unpredictable severity.

This problem could be eliminated if the genes mediating the specific tissue tropism of the virus were engineered into another delivery vehicle. One possibility might be a manufactured enveloped particle with viral glycoproteins interactive with a desired cellular receptor on the surface and the desired gene or genes within. Such artificially engineered delivery particles would have many of the advantages of viral vectors, without the drawbacks. Engineering these vehicles is currently a major focus of the emerging field of *nanotechnology*, which has formed through the marriage of the disciplines of biology and engineering. The exciting prospects of this new discipline are discussed further in the "Viruses and Nanotechnology" section.

Using viruses to destroy other viruses

The latent stage of infection with AAV is briefly discussed in Chapter 17. This virus requires a helper for replication, but can integrate and maintain a latent state in cells that it infects without the helper virus. Then, when these cells are infected with an adenovirus, the reactivation of AAV can lead to reduced yields of the infecting pathogen. There has been some semi-serious discussion of adapting this feature of parvoviruses to the development of antiviral strains. Of course, it is not clear how the host immune response to this antivirus defense will be overcome.

Another approach toward using viruses to combat other viruses is to take advantage of the fact that HIV requires a biochemical "handshake" between the CD4 receptor on the surface of the cell it infects and a second co-receptor such as CXCR4 (see Chapters 6 and 20). An HIV "missile" was developed by several laboratories here and in Europe, and was based on a recombinant vesicular stomatitis virus (VSV) that contains genes for both the CD4 and CXCR4 proteins but lacks its own surface protein that interacts with VSV receptors. Thus, this engineered virus can attach to any membrane that contains the HIV envelope glycoprotein, since it interacts with CD4 and CXCR4. This will include HIV but also, significantly, cells infected with HIV that express the envelope glycoprotein (gp120) on their surface.

Preliminary tests demonstrated the ability of the engineered virus to kill HIV-infected cells in culture. However, like all such approaches, this one may be obviated by the host generating an immune response to the therapeutic virus. The fact that this virus only expresses cellular glycoproteins on its membrane, however, suggests that an immune response may not be generated.

Viruses and nanotechnology

As we have seen, many viruses have capsids that can be assembled *in vitro*, using the structural and catalytic proteins that are either isolated from the viruses themselves or produced by gene expression from cloned versions of the viral sequences. Examples of this include the classic work with tobacco mosaic virus, as well as more recent work with cowpea chlorotic mottle virus. In these and other cases, empty capsids, devoid of nucleic acid, can be produced.

Such viral structures have been used in the cutting-edge field of nanotechnology. In effect, the empty structures can be used as containers for a variety of "cargos." Further, by changing the properties of the capsid proteins in ways that do not prevent assembly of the capsid itself, the empty virus particles can be modified with respect to size and both external and internal composition.

Potential applications for viral-based nanoparticles are great. Including materials inside of the capsid, or attaching materials as ligands to the outside of the particles, provides a way of customizing these particles in a wide variety of settings, both medical and industrial. For instance, the nanoparticles might contain charged material, making them useful as minute biosensors. In the case of the filamentous capsid of tobacco mosaic virus, the attachment of charged groups to the surface has allowed the creation of nanowires. This new and rapidly expanding utilization of the viral world truly stretches the imagination.

The place of viruses in the biosphere

As we suggested in Chapter 1, the evolutionary story of viruses is difficult to tell, given that fossil remains are not available to study. However, a number of theoretical models have been used to suggest the place of viruses in the biosphere. It has been proposed that viruses were key to some of the major transitions in living systems over evolutionary time, including the change from RNA to DNA as the genomic material. The idea is that all living systems originated from a last universal common cellular ancestor whose genetic information was maintained as RNA. Viruses that derived from this ancestor initially used RNA but later evolved DNA as a way of protecting the genetic information from attack. Much later, the use of DNA as genetic information returned to the cellular world, and the end result is the chromosomal forms we see at present.

While this proposal is, no doubt, very theoretical, it does highlight the increasing interest that viruses present in questions about the nature of the living systems we observe in the complex sets of interactions on our planet.

WHY STUDY VIROLOGY?

We finish with the same question that was posed in Chapter 1: Why study this complex subject, and how much of this detail can be used in everyday life? For students who use this text as a foundation for further study in medicine or biology, important concepts and details will be reinforced many times and the question is a nonstarter. But we submit that students who never touch this subject again can still profitably remember a few things.

The first is that viruses and all microorganisms, whether pathogenic or benign, are important members of the biosphere and have an important impact on our daily and future activities. This impact goes both ways.

Second, virology is biology "writ small." The principles studied here apply to all biological sciences. Remembering that the field *is* complex, even if the complexities themselves are forgotten, will go a long way to maintaining that healthy skepticism required of a citizen in a technologically complex world to inflated claims and counterclaims.

QUESTIONS FOR CHAPTER 25

1 What factors may account for the sporadic emergence or reemergence of human viral diseases?

2 What are the essential differences between the following strategies designed to deal with viral diseases:
(a) Development of a vaccine
(b) Development of an antiviral drug

3 Viruses are used as gene delivery systems in attempts to modify the genetic information of cells or tissues. What features of viruses make them good candidates for this technology? What features of viruses make this a difficult approach?

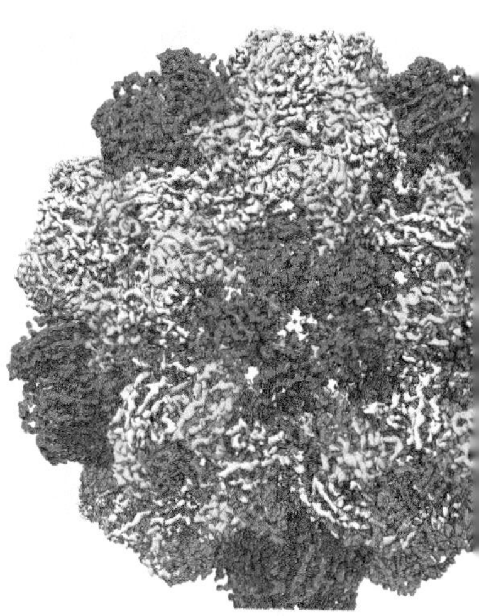

Problems

PART V

1 Varicella zoster virus (VZV) is an example of a virus that causes disease in humans but for which there is no animal model to study pathogenesis. Describe possible approaches that could be used to develop a model to study replication of this virus *in vivo*.

2 Mousepox (ectromelia) virus is extremely lethal for the Balb/c strain of mouse, but is essentially avirulent in the C57BL/6 strain of mouse. The lack of virulence in C57BL/6 mice is not due to an inability of mousepox to productively infect the C57BL/6 mice, but instead due to differences in the immune response. What specific immunological determinants could mediate viral resistance? What experimental approaches could you use to prove what specific arm of the immune system is responsible for the resistance of C57BL/6 mice to mousepox?

3 You are studying the pathogenesis of a recently discovered DNA virus that causes encephalitis in humans. You have established that this virus also infects mice, and that it kills the mice 7 days after the virus is inoculated intranasally. Describe what experimental procedures you could use to do the following: Trace the spread of the virus through the mouse, determine what tissues and cell types it replicates in, and find out if the virus reaches the brain.

4 The figure below shows a short stretch of DNA containing a coding region, along with changes in this DNA brought about by four mutagens (A through D). In each case, determine the nature of the mutational change at the base pair level and the result of the mutation in terms of coding of the sequence.

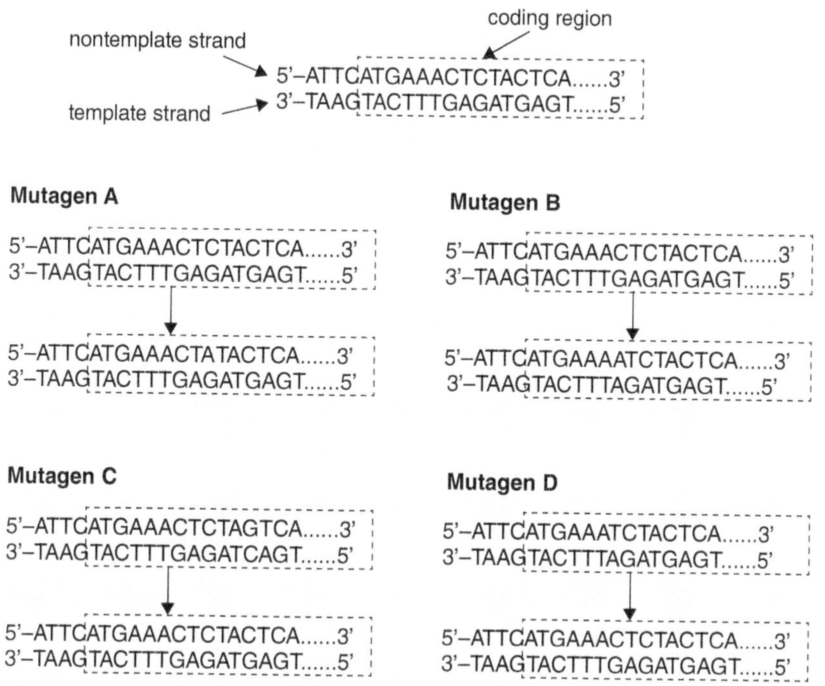

5 The table below contains an additional set of "unknown" nucleotide sequences, as we saw for Question 2 in Chapter 24. Using these online tools:

http://blast.ncbi.nlm.nih.gov
http://bioinformatics.org/sms/orf_find.html
http://motif.genome.jp/

Identify the source of these sequences, the protein encoded by the complete sequence, the function of the protein, and any noted structural features:

Unknown	Sequence
1	ccccatcgccgtggacctgtggaacgtcatgtacacgttggtggtcaaat
2	aagtgacaaaccacatacagacgatcccatactctgcctgtcactctctc
3	attgcttctgtgcctgctccacgctatctaactgacatgactcttgaaga
4	aaagccttaggcatctcctatggcaggaagaagcggagacagcgacgaag

Additional Reading for Part V

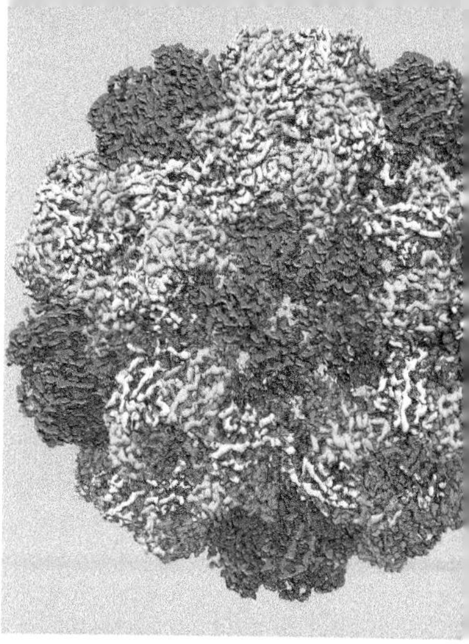

Barabási, A.-L. (2016). *Network Science*. Cambridge: Cambridge University Press.

Campbell, A. and Heyer, L. (2007). *Discovering Genomics, Proteomics, & Bioinformatics*, 2e. Cold Spring Harbor, NY: Cold Spring Harbor Press.

Craighead, J. (2000). *Pathology and Pathogenesis of Human Viral Disease*. New York: Academic Press.

Friedmann, T. (ed.) (2007). *Delivery and Expression of DNA and RNA: A Laboratory Manual*. Cold Spring Harbor, NY: Cold Spring Harbor Press.

Katze, M., Korth, M., Law, L., and Nathanson, N. (2016). *Viral Pathogenesis*, 3e. New York: Academic Press.

Pevsner, J. (2015). *Bioinformatics and Functional Genomics*, 3e. Hoboken, NJ: Wiley.

Sompayrac, L. (2013). *How Pathogenic Viruses Think*. Burlington, MA: Jones and Bartlett.

Watson, J., Witkowski, J., Myers, R., and Caudy, A. (2006). *Recombinant DNA: Genes and Genomics: A Short Course*, 3e. New York: W.H. Freeman.

Watts, D. (2003). *Six Degrees: The Science of a Connected Age*. New York: W.W. Norton.

Appendix – Resource Center

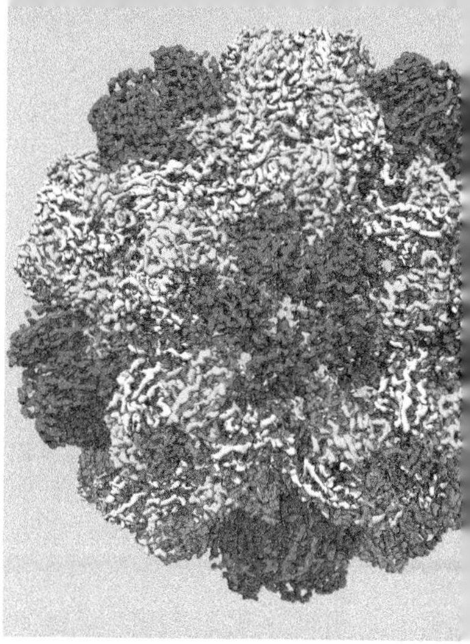

A number of sources are available for more detailed investigation of the topics introduced in this text. The level of coverage varies from basic (like this book) to highly advanced investigations of detailed experimental questions found in primary journal articles. The following should provide a useful set of sources for beginning follow-up investigations.

BOOKS OF HISTORICAL AND BASIC VALUE

Although early texts are generally so out-of-date as to be unusable, three basic "landmark" texts still provide useful information and an appealing richness of coverage:

- Stent, G.S. (1963). *Molecular Biology of Bacterial Viruses*. San Francisco: WH Freeman.
- Luria, S.E. and Darnell, J.E. (1967). *General Virology*, 2e. New York: Wiley.
- Fenner, F., McAuslan, B., Mims, C., Sambrook, J., and White, D.O. (1974). *The Biology of Animal Viruses*, 2e. New York: Academic Press.

The first of these books is a classic. Not only does it provide basic technical information that is invaluable; it also provides a wonderful description of the origins of molecular biology in the study of bacterial viruses. The author, Gunther Stent, along with J. Cairns and J.D. Watson, subsequently edited a collection of reminiscences by many of the original contributors to what we now know as molecular biology and molecular genetics. It was originally published in 1966, and then republished in an expanded version in 2000:

- Cairns, J., Stent, G.S., and Watson, J.D. (eds.) (2000). *Phage and the Origins of Molecular Biology*. Cold Spring Harbor, NY: Cold Spring Harbor Press.

In our opinion, for beginners, this more complete historical source does not significantly improve on the simpler descriptions in the first book.

The second text is also full of historical interest. It was written at a time when the field was just beginning to "explode" from the infusion of what is now modern molecular biology. Its style and organization are a model for almost all subsequent texts.

The third book is still more than a little useful for reading about the interaction between viruses and human populations as well as pathogenesis. The overall style and level of coverage provide another milestone in development of the field. Portions covering pathogenesis and immunology were updated in the following text:

- Mims, C.A. and White, D.O. (1984). *Viral Pathogenesis and Immunology*. Boston: Blackwell Science.

Basic Virology, Fourth Edition. Martinez "Marty" Hewlett, David Camerini, and David C. Bloom.

BOOKS ON VIROLOGY

• Watson, J.D., Baker, T.A., Bell, S.P., et al. (2003). *The Molecular Biology of the Gene*, 5e. Menlo Park, CA: Benjamin/Cummings.
This comprehensive text describes most aspects of modern molecular biology at a level appropriate for advanced undergraduates. There are some excellent sections on gene regulation, and some important bacterial and animal viruses are well covered.

The most comprehensive modern text devoted to virology that contains a wealth of detail concerning individual viruses infecting humans, as well as some detail on the general principles of virology, is the extensive compendium originally conceived by the late Bernard Fields. The set is now in its sixth edition and is called *Field's Virology*:
• Knipe, D.M., Howley, P.M., Griffin, D.E., et al. (eds.) (2015). *Field's Virology*, 6e. New York: Lippincott, Williams and Wilkins.

The chapters in the following book are essentially reviews written by various experts in the field, and as such, the book (of necessity) suffers a bit from unevenness in style and depth of coverage. It is intended for medical and professional students as well as working scientists. The third edition of the book was extracted to make it more manageable in size and cost as a medical text. It is currently in its fourth edition and published as
• Knipe, D.M., Howley, P.M., Griffin, D.E., et al. (eds.) (2001). *Fundamental Virology*, 4e. New York: Raven Press.

A slightly less detailed but very useful general coverage of viruses *in toto* is
• Mahy, B.W.J. and van Regenmortel, M. (eds.) (2008). *Encyclopedia of Virology*, 3e. New York: Academic Press.

The organization is by subject matter, and its effective use requires some basic background knowledge (like that offered in this book).

Short definitions of terms used in virology can often be found in
• Mahy, B.W.J. (2008). *The Dictionary of Virology*, 4e. New York: Academic Press.

Detailed aspects of the pathogenesis of virus infections, completely updated and revised, are covered in the latest edition of
• Katze, M., Korth, M., Law, G.L., and Neal Nathanson, N. (2016). *Viral Pathogenesis*, 3e. Amsterdam: Elsevier.

Recently an update of a book on viral diseases of humans has been published:
• Strauss, E. and Strauss, J. (2007). *Viruses and Human Diseases*, 2e. San Diego: Academic Press.

Another useful reference is a medical source:
• Gorbach, S.L., Bartlett, J.G., and Blacklow, N.R. (eds.) (2004). *Infectious Diseases*, 3e. Philadelphia: WB Saunders.

This encyclopedia is very detailed and intended for medical students and physicians. Nevertheless, it contains a lot of basic information concerning the symptoms and course of viral diseases and is worth a look when a specific subject is of interest.

A myriad of general texts on aspects of virology are available. A book that has coverage perhaps slightly broader than this one is
• Voyles, B.A. (2001). *The Biology of Viruses*, 2e. St. Louis: Mosby.

However, it has not been updated in many years.

Other books of a relatively equivalent level include
• Cann, A.J. (2015). *Principles of Molecular Virology*, 6e. Amsterdam: Elsevier.
• Dimmock, N.J., Easton, A.J., and Leppard, K.N. (2016). *Introduction to Modern Virology*, 7e. Cambridge, MA: Wiley-Blackwell.

Finally, a text at a slightly more advanced level is
• Flint, S.J., Enquist, L.W., Krug, R.M., Racaniello, V.R., and Skalka, A.M. (2008). *Principles of Virology: Molecular Biology, Pathogenesis, and Control*, 3e. Washington, DC: ASM Press.

MOLECULAR BIOLOGY AND BIOCHEMISTRY TEXTS

Virology is intimately linked with molecular biology and biochemistry. A number of excellent and detailed texts covering these topics are currently available. Many, like the Watson text mentioned in the "Books on Virology" section, have some coverage of viruses. A (partial) listing of some of the best would include the following:

- Alberts, B., Johnson, A., and Lewis, J. (2014). *Molecular Biology of the Cell*, 6e. New York: W. W. Norton.

The first edition of this book set the standard for comprehensive, molecular biology–based texts that span microorganisms to humans. It is still a fine source.

Others include

- Lodish, H., Berk, A., Kaiser, C.A., et al. (2007). *Molecular Cell Biology*, 6e. New York: Freeman.
- Krebs, J.E., Goldstein, E.S., and Kilpatrick, S.T. (2018). *Lewin's Genes XII*. New York: Jones and Bartlett.
- Mathews, C.K., Van Holde, K.E., Appling, D.R., and Anthony-Cahill, S.J. (2012). *Biochemistry*, 4e. Menlo Park, CA: Pearson.
- Berg, J.M., Tymoczko, J.L., Gatto, G.J., and Stryer, L. (2015). *Biochemistry*, 8e. New York: Freeman.
- Voet, D. and Voet, J.G. (2010). *Biochemistry*, 4e. New York: Wiley.

DETAILED SOURCES

Many other serials, periodicals, and occasional reviews are available in any good university or medical school library. The primary journals contain detailed, complex, and often opaquely written descriptions of specific experimental studies on one or another aspect of a virus or virus–host interaction. These articles require a lot of background before they make much sense (even to an expert), but they often have valuable figures, schematics, and other *bons mots* that could be of use to a beginning student. A typical source might be something like:

- Devi-Rao, G.B., Aguilar, J.S., Rice, M.K., et al. (1997). HSV genome replication and transcription during induced reactivation in the rabbit eye. *Journal of Virology* 71: 7039–7047.

Major virology journals include *Journal of Virology*, published bimonthly by the American Society of Microbiology (ASM); the bimonthly journal *Virology*, published by Academic Press; and the *Journal of General Virology*, published by (England's) Society for Microbiology. The ASM also publishes journals entitled *Molecular and Cell Biology*, *Journal of Bacteriology*, *Clinical Microbiology Reviews*, as well as many others that cover detailed subject matter. Secondary journals containing material of less general interest include *Virus Research*, *Virus Genes*, and *Intervirology*. These may not be available in all university libraries.

While these journals are probably too detailed to be of much interest to the beginning student, a monthly periodical published by the Centers for Disease Control and Prevention (CDC), *Emerging Infectious Diseases*, provides up-to-date articles on emerging and re-emerging viral infections such as SARS and avian influenza. Though not strictly limited to viruses, these articles are of general interest and are written at a level of complexity about equivalent to *Field's Virology*.

There are also numerous articles concerning viruses and aspects of virology written at a reasonable level of detail that appear periodically in general interest science magazines. The most widely read one is *Scientific American*.

SOURCES FOR EXPERIMENTAL PROTOCOLS

Individual laboratories have long had "recipe books" in which basic procedures and reagents are outlined. The applicability of molecular biology and DNA-cloning techniques is so varied and so general to biological studies that no one person or laboratory can keep in touch with all the methods. To resolve this problem, T. Maniatis at Harvard University compiled a general laboratory manual for such techniques. This rapidly became a world standard. The most current edition is

- Green, M.R., Sambrook, J., and MacCallum, P. (2012). *Molecular Cloning – A Laboratory Manual*, 4e. Cold Spring Harbor, NY: Cold Spring Harbor Laboratory Press.

This has not been the final word, however. Since techniques constantly are updated and improved, and new methods are developed, no book stays current for long. This problem has been met by the publication of

- Ausubel, F.M., Brent, R., Kingston, R.E., et al. (eds.) (1994–present). *Current Protocols in Molecular Biology*. New York: Wiley.

This manual is published in loose leaf and is updated two to four times per year. Updates include revisions, corrections, and new methods. The current compendium runs to several thousand pages and covers everything from cloning to the use of computers for information on genes. While the methods are only of interest for specific use, often there are short explanatory passages outlining general approaches that are useful to even beginning students. All active research laboratories should have access to this series.

Other specialized sets of technique-oriented references are available. One further excellent four-volume source of general methods for dealing with cell culture and other techniques oriented toward the cell is

- Celis, J. (ed.) (2008). *Cell Biology: A Laboratory Handbook*, 3e. Amsterdam: Elsevier.

A bit less detailed reference is

- Feshney, R.I. (2005). *Culture of Animal Cells: A Manual of Basic Techniques*, 5e. New York: Wiley.

Finally, a good scientific dictionary can be helpful in clarifying a term, and medical and biological encyclopedias have value. One recent source that provides rather succinct but generally well-organized definitions and descriptions is

- Cammack, R., Atwood, T., Campbell, P., et al. (eds.) (2006). *Oxford Dictionary of Biochemistry and Molecular Biology*. Oxford: Oxford University Press. Published online in 2008.

THE INTERNET

The internet has, of course, proven itself as an increasingly useful and important source of basic information. Although addresses change and the internet continues to develop rapidly, any good search engine will pull out topical information on a number of viruses, viral diseases, and therapies.

Of special interest are websites maintained by the CDC, the National Institutes of Health, the ASM, and other such organizations. The following URLs were active as of this writing (July 2020).

Virology Sites

In addition to the internet sites listed in Chapter 24, the following might also provide informative reading for students interested in virology.

Some websites that you might want to start with:

For the time being, Ed Wagner's research pages and his virology course page at University of California, Irvine (UCI) are still active. We leave reference here to them as part of our memorial to Ed. We hope they will be useful to you for as long as they are functional.

- E. Wagner: Research page: http://darwin.bio.uci.edu/~faculty/wagner
- E. Wagner: Virology course page: http://eee.uci.edu/98f/07426

Additional virology web pages are personally organized by various faculty and scientists to ease searches for specific topics in virology. As of July 2020, the sites listed here seem useful for general background information. Some have self-study questions and sample examinations. It is important to be aware, however, that there is no guarantee that they will survive or be updated regularly.

The Viral Bioinformatics Resource Centre maintains a useful website (https://4virology.net) that provides access to the sequences of viral genomes and tools for genome analysis. A particularly valuable resource is the site's taxonomic database that allows the student to navigate through the indices of virus families for details about virion structures, summaries of life cycles, and links to additional resources.

A PBS site about vaccines and bioterrorism: http://www.pbs.org/wgbh/nova/bioterror/vaccines.html

An informative website to accompany the book *The Bacteriophages* by Richard Calendar: http://www.thebacteriophages.org

Important Websites for Organizations and Facilities of Interest

- American Society for Virology (ASV): http://www.asv.org
- American Society for Microbiology (ASM): http://www.asm.org
- Melvyl (University of California Library database): https://melvyl.worldcat.org
- Centers for Disease Control and Prevention (CDC): http://www.cdc.gov
- National Library of Medicine: http://www.nlm.nih.gov
- National Center for Biotechnology Information: http://www.ncbi.nlm.nih.gov
- American Association for the Advancement of Science (publisher of the weekly periodical *Science*): http://aaas.org

Technical Glossary

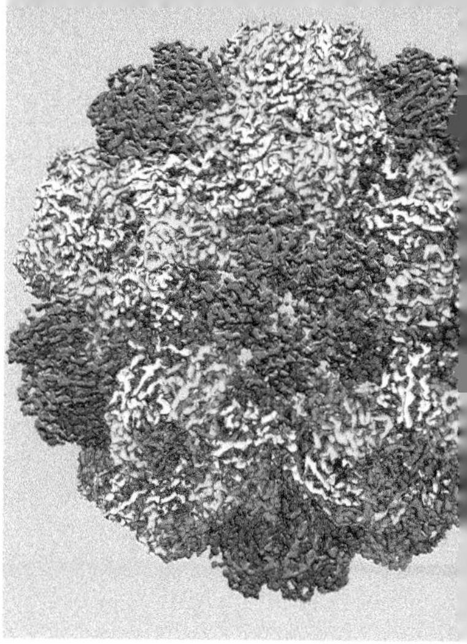

2′,5′-oligoadenylate synthetase (2′,5′-AS) The enzyme that polymerizes ATP into 2′,5′-oligoadenylate; activated by double-stranded RNA and induced by interferon activation of a target cell.

abortive infection Infection of a cell where there is no net increase in the production of infectious virus.

abortive transformation See **transitory (transient or abortive) transformation**.

acid blob activator A regulatory protein that acts in *trans* to alter gene expression and whose activity depends on a region of an amino acid sequence containing acidic or phosphorylated residues.

acquired immune deficiency syndrome (AIDS) A disease characterized by loss of cell-mediated and humoral immunity as the result of infection with human immunodeficiency virus (HIV).

acute infection An infection marked by a sudden onset of detectable symptoms usually followed by complete or apparent recovery.

adaptive immunity (acquired immunity) See **immunity**.

adjuvant Something added to a drug to increase the effectiveness of that drug. With respect to the immune system, an adjuvant increases the response of the system to a particular antigen.

agnogene A region of a genome that contains an open reading frame of unknown function; originally used to describe a 67- to 71-amino-acid product from the late region of SV40.

AIDS See **acquired immune deficiency syndrome**.

aliquot One of a number of replicate samples of known size.

α-TIF The alpha *trans*-inducing factor protein of HSV; a structural (virion) protein that functions as an acid blob transcriptional activator. Its specificity requires interaction with certain host cellular proteins (such as Oct1) that bind to immediate-early promoter enhancers.

ambisense genome An RNA genome that contains sequence information in both the positive and negative senses. The S genomic segment of the Arenaviridae and of certain genera of the Bunyaviridae have this characteristic.

amorphous Without definite shape or form.

aneuploid A eukaryotic cell with an ill-defined number of fragmented chromosomes, as a result of long periods of continuous passage in culture.

animal model for a (specific) disease An experimental system using a specific laboratory animal to investigate aspects of pathogenesis of an infectious disease not usually occurring in that animal.

Basic Virology, Fourth Edition. Martinez "Marty" Hewlett, David Camerini, and David C. Bloom.

antibodies Glycoprotein molecules secreted by B lymphocytes with a defined structure consisting of an N-terminal region of variable amino acid sequence (the Fab region), which combines with specific antigenic determinants, and a C-terminal region of constant amino acid sequence (the Fc region), which serves as a biological marker identifying the molecule as part of the immune response.

antigen Usually a macromolecule that induces an immune response by virtue of small regions that combine with antigen-combining sites on immune cells.

antigen-presenting cell (APC) A cell in which an antigen is processed, followed by expression of epitopes on the surface in conjunction with major histocompatibility antigens. See also **dendritic cells**.

antigen processing Partial degradation of an antigen within an antigen-presenting cell (APC) followed by its expression at the surface of the APC in the presence of a major histocompatibility protein.

antigenic drift The slow change in structure of an antigen over time due to accumulated mutational changes in the sequence of the gene encoding the antigen.

antigenic shift An abrupt change in an antigen associated with a pathogen due to the acquisition of a novel gene substituting for the original one.

antisense oligonucleotide A short oligonucleotide with a sequence complementary to a specific sequence of nucleotides on a nucleic acid molecule; under investigation as a potential high-specificity target for binding and, thus, inactivating viral genes or mRNA.

antiserum The plasma fraction containing antibody molecules of the blood of a subject mounting an immune response to the protein or other molecule against which the antibody is directed.

antiviral drug Any drug that specifically inhibits some process in the replication of a virus without undue toxicity to the host in which the virus is replicating.

antiviral effector molecule (AVEM) One of a number of cellular proteins that function to limit virus replication when activated by the presence of double-stranded RNA (dsRNA) in a cell activated by interferon.

antiviral state A state induced by interferon in a susceptible cell. The cell is partially refractory to virus replication in this state.

apathogenic Not pathogenic.

APC See **antigen-presenting cell**.

apical surface The surface of epithelial cells that face the exterior or an extracellular compartment of the organism. This surface contains specific membrane-associated proteins mediating specific functions; this surface is distinct from the basolateral regions that communicate with adjoining cells.

APOBEC Apolipoprotein B mRNA-editing enzyme complex. This RNA-editing enzymatic activity catalyzes the deamination of specific C residues to U in mRNA.

apoptosis Programmed cell death; a specific linked set of cellular responses to specific stimuli, such as loss of replication control, that leads to a specific course of changes resulting in cell death.

arboviruses Arthropod-borne viruses.

archaebacteria (Archaea) One of the two prokaryotic domains of the biosphere. The Archaea are not really bacteria, but the term archaebacteria remains in use.

artificial chromosome A DNA cloning vector created *in vitro* such that it has the necessary replication structures corresponding to the chromosome of a particular cell. See **bacterial artificial chromosome** (**BAC**) and **yeast artificial chromosome** (**YAC**).

ascites An accumulation of fluid in the peritoneal cavity of the body.

aseptic meningitis An infection of the CNS's surface tissue from which no bacteria or metazoan pathogen can be cultured; therefore, by elimination, a viral infection.

assay A measurement or test.

asymptomatic Without detectable symptoms.

atomic force microscopy (AFM) A technique for displaying surface features of a sample by recording the deflection of a microscopic probe as it passes over the sample. Resolution with this technique can be on the order of the diameter of a DNA molecule (2.0 nm).

attenuate Losing virulence due to an accumulation of mutations in virus-encoded proteins mediating pathogenesis during continued passage of a virus in a natural population or in the laboratory.

autoimmune disease A disease characterized by the subject's immune system attacking and destroying ostensibly normal host tissue.

AVEM See **antiviral effector molecule**.

avirulent A genetic variant of a virulent pathogen that does not cause the disease usually associated with the agent.

B lymphocytes The immune cells that secrete soluble antibodies.

back mutation A change in the genome of an organism or virus that returns the genotype to the wild type or original strain.

bacterial artificial chromosomes (BACs) An *in vitro* constructed cloning vector that utilizes parts of the *Escherichia coli* **F-plasmid** to create a vehicle for inserting foreign DNA into the bacterial chromosome. BACs can be used for the cloning of up to 300 kbp. The cloned DNA is stable, since it becomes part of the bacterial chromosome.

bacterial plasmid An extrachromosomal circular genetic element with the capacity to replicate as an episome in bacteria. This plasmid often may confer antibiotic resistance on its host. It frequently is utilized for maintaining cloned fragments of DNA.

bacterial restriction A set of endonuclease-mediated responses to foreign DNA sequences encoded by bacterial genes designed to destroy the genomes of invading bacteriophages and plasmids.

bacteriophage One of a large group of viruses infecting bacteria.

Baltimore scheme A scheme of virus classification stressing the mechanism for expression of viral mRNA and the way that information is encoded in the viral genome as primary criteria; formalized by David Baltimore.

base plate Complex portion of a tailed bacteriophage's capsid that contains projecting protein "pins" mediating the noncovalent interactions between the phage and host cell that lead to injection of the viral genome.

basolateral surface The "interior" surface of epithelial cells that face other epithelial cells of this and adjoining tissue. The region is distinct from the apical surface that contacts the exterior.

benign tumor A group of cells forming a discrete mass resulting from limited alterations in cellular growth properties; distinguished from a malignant tumor or cancer in that the cells do not metastasize throughout the body.

bioinformatics The new field of computational biology that involves the analysis of complex data sets generated by large-scale sequencing and **high-throughput technology**.

bioterrorism The threat to use or the actual deployment of an infectious agent (fungal, bacterial, or viral) to inflict harm and/or engender fear, especially in a civilian population.

blood–brain barrier The physical and biochemical parameters creating the relative physical and physiological isolation and physical separation of the CNS from the circulatory system and any pathogens that may be present in it.

bp Abbreviation for a base pair.

broad host range Interaction between a virus and host such that the virus can infect a large and diverse number of host organisms. A broad host range for a virus may be associated with the use of a common cellular structure to gain entry.

buoyant density The density at which a macromolecule or virus "floats" in a gradient under conditions of equilibrium in a centrifugal field.

burst size Amount of (usually infectious) virus produced from a single infected cell.

cAMP 3′,5′-cyclic adenosine monophosphate.

cancer A disease of multicellular organisms characterized by genetic modifications to specific cells that result in the formation of malignant tumors composed of cells without the ability to respond to normal growth-control signals limiting their replication and spread in the body.

cancer cell A cell isolated from a malignant tumor and displaying the altered growth and other properties associated with the tumor.

cap The methylated guanosine residue at the 5′ end of eukaryotic mRNA molecules that is added posttranscriptionally as a 5′-5′-phosphodiester.

cap site The location of initiation of transcription of a specific eukaryotic mRNA.

cap snatching or stealing The endonucleolytic removal of the 5′-methylated cap region of nascent eukaryotic mRNAs and transfer to a nascent viral mRNA. The process occurs during the replication of orthomyxoviruses and bunyaviruses.

capillary electrophoresis The separation of charged particles in an applied electric field (10–30 kV) across a capillary tube of small (20–100 mm) diameter. The technique has the advantage over conventional electrophoresis in that it produces quantitatively reproducible separation in a very short (less than one hour) time, using small sample sizes.

capsid The viral protein shell surrounding the virion core and its nucleic acid genome.

capsomer (capsomere) One of a group of identical protein subunits making up a viral capsid.

carcinogen A chemical substance that induces a cancer.

CAT Chloramphenicol acetyl transferase. This bacterial enzyme transfers one or two acetyl groups to the structure of the drug chloramphenicol, making it refractive to uptake by the cell. CAT is a prokaryotic gene and is therefore well-suited for use as a reporter gene in eukaryotic cells (see also **reporter gene**).

catabolite repression A control mechanism for metabolite use in bacteria mediated by levels of cAMP. It ensures that glucose or other energy sources not requiring expression of inducible metabolic enzymes are utilized first.

caveolae-mediated endocytosis A process of cellular entry using assemblies of glycosphingolipids and cholesterol in the membrane, along with the protein caveolin. Caveolae-mediated and **lipid raft endocytosis** differ from the classic **clathrin-mediated** entry system involving clathrin-coated pits in that there is no lysosomal fusion step.

CBP1 (cap-binding protein) A eukaryotic ribosome-associated protein involved in the initiation of translation. It functions by binding the 5′-cap structure of eukaryotic mRNA.

CCR5 A cellular chemokine (growth factor) receptor that is used for attachment during HIV infection of macrophage cells.

CD4, CD8 Specific protein antigenic markers found on the surface of different functional classes of T lymphocytes.

cDNA Complementary DNA; produced by the reverse transcription of an RNA molecule.

cell-mediated immunity (CMI) The portion of the immune response that requires specific recognition between receptors on individual T lymphocytes and antigenic determinants on the surface of antigen-presenting cells.

centrifugation Subjection of a sample to an artificial gravity field resulting from rapid rotation.

chemokine An extracellular signaling molecule capable of activating a target cell with appropriate receptors for its recognition.

chickenpox A childhood disease marked by a distinctive rash and caused by the herpesvirus varicella zoster virus.

***Chlorella* virus** A DNA virus of the green algae *Chlorella*.
chorioallantoic membrane The membrane located between the shell and chicken embryo in an embryonated egg; used as a location for growth of certain viruses such as poxvirus.
cirrhosis Liver damage characterized by tissue hardening and loss of function and circulation.
***cis*-acting genetic element** A genetic element that functions only in the contiguous piece of DNA or RNA in which it is present.
CJD See **Creutzfeldt–Jakob disease**.
clathrin-mediated endocytosis The process by which ligand molecules are taken into the cell after binding to specific receptors on the cell surface. The ligand–receptor complex then invaginates into an endocytotic vesicle that is surrounded by a clathrin coat. Clathrin is a protein that forms a cage-like structure around the vesicle. Also called receptor-mediated endocytosis.
clinical trial A formal process for testing the effectiveness of a drug or vaccine on human subjects in a clinical setting.
clonal selection The selective stimulation of subsets of immune cells reactive with a single antigenic determinant that results in the proliferation of cells specifically responding to that antigen.
clones Genetically identical biological entities derived from a single parental entity.
cloning The word has several meanings. First, it means to produce **clones** from a single parent. This might be done in the normal course of replication, such as with bacteria, or by artificial means, such as with nuclear transplant. Second, when the clone organisms (bacteria, for instance) contain a recombinant DNA molecule, it is said that the DNA has been "cloned." In this sense, cloning then refers to the production of cells containing that recombinant DNA.
CMI See **cell-mediated immunity**.
CNS Central nervous system.
cold virus One of a number of RNA genome–containing viruses known to be causative agents of mild infections of the upper respiratory tract and nasopharynx.
complement A group of serum proteins that bind to an antibody–antigen complex on the surface of a cell, and sequentially undergo a complex maturation process resulting in cell lysis.
complementary strands of nucleic acid Strands of single-stranded nucleic acid whose sequence is determined by the Watson–Crick base-pairing rules relative to a reference strand.
complementation The growth of two replication-deficient mutant organisms or viruses that have mutations in different *trans*-acting genes so that each can supply the deficient gene product of the other.
complementing cell line A cell line that produces those proteins necessary for the growth of defective viruses, especially viruses designed for gene transfer applications.
concatamers A number of genome-sized DNA regions joined together in a linear array as an intermediate in viral replication, such as during the bacteriophage T4 life cycle.
conditional lethal mutations Genetic alterations in a specific protein or genetic element that lead to a replication deficiency under controlled laboratory conditions such as high temperature.
confocal microscopy Technique using a computer-enhanced microscope equipped with laser illumination optics that allow a very small focal plane to be visualized with various wavelengths of light.
conformational epitopes Antigenic determinants that are only present when the antigen is in a specific (usually native) conformation.
contact inhibition The cessation of cell replication or movement when the cell is in contact with other cells of the same type.

continuous cell line A clonal cell line that has been maintained in culture for a large number of passages and is essentially immortal.
copy number The number of molecules of a particular type (protein, mRNA, gene, plasmid) per cell.
cosmid An artificial cloning vector that contains the *cos* sites from bacteriophage λ, allowing the packaging into phage heads of relative large (up to 45 kbp) DNA inserts.
coupled transcription/translation The linked process (often in prokaryotic cells) where synthesis of protein encoded by an mRNA molecule commences during ongoing transcription of the mRNA.
CPE See **cytopathic effect** and **cytopathology**.
Creutzfeldt–Jakob disease (CJD) A slow, noninflammatory infection of the human CNS caused by a prion.
CRISPR/Cas CRISPR (clustered regularly interspersed short palindromic repeats) is a family of DNA sequences, found in the genomes of most bacteria, that are transcribed as small RNAs that recognize invading bacteriophage DNA. The protein Cas9 then binds to these RNAs and makes a cut at RNA–DNA duplexes.
cryoelectron microscopy The preparation of a specimen by embedding it in vitreous (noncrystalline) ice in the absence of stain and its visualization using an electron beam of low intensity. This method preserves much structural integrity of the virion, especially in enveloped viruses.
cryptic ORFs Open translational reading frames in a eukaryotic mRNA downstream of an efficiently translated one and, thus, one usually not recognized by eukaryotic ribosomes.
CTL See **cytotoxic T lymphocyte**.
CXCR4 A cellular chemokine (growth factor) receptor used by HIV for attachment during infection of T cells.
cytochalasin B One of a number of compounds that interfere with formation of the actin-fiber cytoskeleton that anchors the nucleus inside the cell.
cytokine One of a group of proteins (usually glycosylated) secreted by cells that have a specific effect on the growth and behavior of target cells. Examples include the interferons.
cytolysis Cell lysis.
cytopathic effect (CPE) See **cytopathology**.
cytopathology Observable changes to the appearance, metabolic processes, growth, and other properties of a cell induced by a virus infection.
cytotoxic T lymphocyte (CTL) A subtype of T lymphocyte (CD8) that has been activated by an antigen and can target and kill a cell that presents that antigen.
defective interfering particles Defective virus particles that reduce the efficiency of infection by normal virus particles in the same stock.
defective virus particles Virus particles that are normal or apparently normal in appearance but cannot initiate a productive replication cycle.
defensins Small (35–40 amino acids), positively charged, antimicrobial peptides found in both animal and plant cells. Defensins are effective against bacteria, fungi, and enveloped viruses. They function by disrupting the cellular or viral membrane.
denaturation temperature The temperature at which a biological macromolecule loses its functional higher-order structure by virtue of thermal disruption of hydrogen bonding and other forces contributing to its stability. In the case of double-stranded nucleic acids, this is the temperature where the two complementary strands can no longer associate (also the melting temperature, or Tm).
denature To disrupt the higher-order structure of a protein or nucleic acid–containing macromolecule.
dendritic cells Cells derived from bone marrow that play a major role in presenting processed antigen to T cells.

desiccation The act of drying out.

dicer A ribonuclease of the RNase III family that cuts long dsRNA into short interfering RNA (siRNA) fragments as part of the RNA interference (RNAi) gene-silencing system.

differential display analysis A method for comparing the population of cellular transcripts in a population of cells before and after an induced metabolic change, such as infection with a virus or stimulation with a cytokine.

differential polyadenylation site usage The termination of eukaryotic transcription employing alternate polyadenylation signals. The process generates overlapping mRNAs with different 3′ terminal regions.

dilution endpoint The dilution in a quantal assay at which the agent being tested cannot evoke a positive response.

disease-based classification scheme One of a number of classification schemes for viruses based on the disease caused.

DNA end problem A dilemma in the replication of a linear DNA molecule that results from the constraint that the polymerase must synthesize in the 5′ to 3′ direction from a primer. The result is the loss of bases from the end of such a linear molecule, such as occurs in replication of the telomeric regions of eukaryotic DNA.

DNA ligase One of a class of enzymes involved in DNA replication and repair that join two fragments of DNA together by forming a phosphodiester bond.

DNA polymerase One of a class of enzymes of complex structure that catalyze the synthesis of a new strand of DNA complementary to the template strand in a primer-dependent reaction.

DNA vaccine A vaccine comprising a DNA molecule containing the gene for one or more antigenic proteins that can be expressed to elicit immunity when the DNA is injected into a test subject.

dsRNA Double-stranded RNA.

dsRNA-dependent protein kinase (PKR) The enzyme induced by interferon action on a target cell that phosphorylates eIF-2 in the presence of dsRNA, thus resulting in inhibition of protein synthesis; part of the interferon-induced antiviral state.

EBV See **Epstein–Barr virus.**

eclipse period The time during the replication cycle of a virus when no infectious virus can be isolated; in other words, the time between virus adsorption and genome penetration and the appearance of newly synthesized infectious virus.

ED_{50} Median effective dose; in a quantal assay, the dilution of a pathogen sufficient to ensure that 50% of standard aliquots of that dilution will contain the infectious agent.

effector T cells Cells of the immune system that act on antigen-bearing cells as part of the immune response. Two major classes include the helper T cells (T_H cells), which interact with antigen-presenting cells in the stimulation of reactive B lymphocytes, and cytotoxic or killer T cells (T_C), which act on antigen-bearing cells to destroy them.

electron microscope An instrument for viewing biological specimens at a resolution greater than the wavelength of light, using electrons accelerated to a high energy and, thus, short wavelength.

ELISA See **enzyme-linked immunosorbent assay**.

ELISPOT A procedure typically used to detect small amounts of cytokines present in tissues or fluids. Similar to an **ELISA**, the primary difference is that the fluid or tissue extract to be tested is placed in a well that has been coated with cytokine-specific antibodies. In an ELISA, the antibodies are often added to wells that were previously coated with the tissue or fluid.

encephalitis An inflammation of the brain or tissues of the upper CNS.

encephalopathy A noninflammatory disease of the brain.

endocytosis The process of incorporation of viruses or large molecules into a cell by formation of a specific vesicle at the cell surface that engulfs the material and transports it into the cell's interior.

endonuclease A nuclease that initiates hydrolysis of a nucleic acid by attack at an interior phosphodiester bond.

endoplasmic reticulum The complex membrane system in eukaryotic cells that is continuous with the nucleus. It is the site of lipid and membrane synthesis as well as the synthesis of proteins destined to be secreted or remain membrane-associated in the cell.

enhancers *Cis*-acting control sequences in eukaryotic DNA that facilitate transcription from a promoter located a relatively long distance from them.

enteroviruses A group of RNA viruses replicating in the vertebrate gut and causing mild to severe enteric disease.

env The gene encoding the polyprotein translation product that contains the envelope glycoproteins of a retrovirus.

envelope (membrane) The lipid bilayer with associated proteins that encompasses a cell or a virus.

enzyme-linked immunosorbent assay (ELISA) An enzymatic method for measuring an immune reaction by binding an enzyme to the Fc region of an antibody molecule and using the activity of this enzyme to indicate the presence of the antibody to which it is bound.

epidemiology The study of the spread and control of infectious disease in human populations.

epidermis The outer surface of the skin.

episome An extrachromosomal (usually circular) genetic element able to replicate in concert with chromosomes of the cell in which it resides.

epitopes Small regions of (usually) hydrophilic amino acids making up specific antigenic determinants in protein antigens.

epizoology The study of the spread and control of infectious disease in nonhuman populations.

Epstein–Barr virus (EBV) A human herpesvirus.

error frequency The rate of introduction of errors during the replication of a DNA or RNA genome.

etiology The cause of a disease or pathologic condition.

Eubacteria One of two prokaryotic domains of the living world, the other being the Archaea.

eukaryote One of two kinds of cells that constitute living systems. Eukaryotes have internal membrane-surrounded structures, especially a membrane-defined nucleus containing the genetic material.

eukaryotic translation initiation factors (CBP1, eIF-2, eIF-3, eIF-4A, eIF-4B, eIF-4C, eIF-4F, eIF-5, eIF-6) Ribosome-associated proteins that function in the initiation of translation of an open reading frame on eukaryotic mRNA.

exocytotic (exocytic) vesicles Membrane vesicles within a cell that carry macromolecules or viruses for release at the cell surface.

exons The portions of RNA that remain as mature mRNA after the removal of introns by splicing.

exonuclease A progressive nuclease that attacks its polynucleotide substrate only from a free end.

explant The removal of intact tissue from an organism, followed by maintenance in culture medium.

extremophile An organism that can grow at extreme limits of environmental conditions, such as high temperature (thermophile) or high salt (halophile).

F pilus See **sex pilus**.

Fab region The N-terminal half of an antibody molecule that has both regions of constant and variable sequences. The antigen-combining sites are in the regions of variable sequence.

Fc region The C-terminal half of an antibody molecule that shares the same sequence as all other antibodies of that class.

feeder layer A layer of cultured cells included with explanted tissue to provide optimal conditions for survival of the tissue.

feline panleukopenia A (generally) fatal viral disease of cats, characterized by extreme reduction in circulating leukocytes (white blood cells).

fertility factor (F′ factor) A double-stranded DNA plasmid found in certain bacteria that confers the ability to transfer DNA from one cell to another by conjugation. The F′ factor or plasmid encodes genes that carry out this process, including genes for the F pilus, the bacterial structure that links two cells during transfer.

flu See **influenza**.

fluor A substance that absorbs light of a particular wavelength and subsequently emits light, usually of a lower wavelength. The process is called fluorescence.

focus of infection Identification of areas of cells that have been infected with virus on a tissue culture plate. The areas are recognized by the **cytopathology** produced by the virus in question. Such areas may be observable microscopically or, in some cases, macroscopically. Foci of infection may be used quantitatively for the enumeration of biologically active virus particles.

frame shift A mutation that affects the sequence of the encoded protein by altering the translational reading frame.

fulminant infection A severe, sudden, often fatal infection characterized by rapid invasive spread of the infectious agent; often refers to a particularly severe form of hepatitis.

fusion (membrane fusion) The process in which the membrane envelope of a virus combines with a cell membrane or vesicle in the process of virus entry.

gag The gene encoding the polyprotein translation product that contains the capsid proteins of a retrovirus; the group-specific antigens of a retrovirus.

gene therapy The treatment of a genetically based disease by the artificial insertion of a corrected version of the gene in question into the patient by one of several mechanisms.

genetic marker A genetic characteristic that can be screened or selected for.

genetic recombination The creation of new genotypic arrangements by the breakage and rejoining of chromosomes.

genome The nucleic acid molecule that encodes the genetic information of an organism.

genomics The study of the sequence, structure, and function of genetic information, especially using computational methods for the analysis of large data sets.

genotype The genetic makeup of an organism.

German measles (rubella) A usually mild rash and fever caused by an RNA virus, characterized by severe neurological damage to a fetus in the first trimester of gestation.

glycoproteins Proteins that have sugar residues covalently linked to specific amino acids. Such proteins are often secreted from cells or associated with membranes in such a way that the major portion projects through the membrane.

glycosylation The process of adding sugar residues onto membrane and excreted glycoproteins in the Golgi apparatus.

Golgi apparatus Vesicles in the eukaryotic cell that receive newly synthesized lipids and proteins from the endoplasmic reticulum and transport them to the correct location in the cell. Specific chemical modifications to the proteins and lipids transported take place in the Golgi apparatus.

growth factors Complex macromolecules that function to signal specific cells to replicate.

HAART Highly active antiretroviral therapy. The use of four or five different antiviral drugs in an attempt to dramatically reduce the viral load of a patient infected with HIV.

HCCs See **hepatocellular carcinomas.**

helicase One of a class of enzymes involved in DNA replication that unwind DNA by catalyzing a local denaturation of the DNA duplex at the point of action.

helix A spiral.

helper virus A virus in a mixed infection (usually in cultured cells) that provides a complementing function so that a co-infecting defective virus can replicate.

hemagglutination (hemadsorption) The ability of a virus membrane or a virus-infected cell to stick to red blood cells, caused by the action of one or several specific viral glycoproteins.

hepatitis virus One of a group of viruses (many unrelated) that target the liver.

hepatocellular carcinomas (HCCs) Malignant tumors derived from cells of the liver.

herd immunity A qualitative state in a population exposed to an infectious disease where the existence of a sufficient number of recovered and immune individuals restricts the spread of the disease.

herpes simplex virus (HSV) One of two closely related neurotropic human viruses containing a DNA genome and characterized by the ability to form latent infections. HSV type 1 (HSV-1) normally infects facial tissue, while HSV-2 infects genital tissue.

herpes zoster virus (HZV) The causative agent of chickenpox.

herpesvirus One of a large group of related viruses containing large DNA genomes and possessing a similar structure whose infection is characterized by establishment of a latent infection.

high-throughput sequencing (HTS) Technology that sequences large segments of DNA and RNA quickly and cheaply.

histology Microscopic study of cellular structure and organization.

HIV See **human immunodeficiency virus.**

homeostasis A system that is maintained in equilibrium in a stable physiological state is said to be in homeostasis.

host range The organism or group of organisms that a virus can infect.

HSV See **herpes simplex virus.**

HTLV See **human T-cell leukemia virus.**

human immunodeficiency virus (HIV) A human retrovirus; the causative agent of AIDS.

human T-cell leukemia virus (HTLV) A human retrovirus that is a causative agent of some forms of leukemia.

humoral immunity Immunity due to antibody molecules circulating in the blood and lymphatic system.

hybrid (nucleic acid hybrid) A double-stranded nucleic acid molecule formed by Watson–Crick base pairing of a given sequence with its complementary sequence, which can be either RNA or DNA.

hybridoma cell A cell derived by the fusion of an antibody-secreting B cell and a myeloma cell. Hybridomas are clonal and immortal, and secrete the antibody that was secreted by the parental B cell.

hydrophilic Used to describe a molecule or portion thereof that is hydrated in solution due to energetically favorable interactions with water molecules.

hydrophobic Used to describe a molecule or portion thereof whose interaction with water molecules is energetically unfavorable.

HZV See **herpes zoster virus.**

i.c. See **intracranial** or **intracerebral.**

ICAM See **intercellular adhesion molecule.**

icosahedron A regular solid polygon made up of 12 vertices and 20 faces.

ID_{50} Median infectious dose; in a quantal assay, the dilution at which half the tested aliquots are able to initiate an infection (i.e., contain infectious virus).

IFN See **interferon.**

immunity The ability to resist or defend against a pathogen by innate immune responses or acquired immune responses to particular antigens by clonal selection of reactive lymphocytes.

immunofluorescence A method of detecting and localizing an antibody bound to its cognate antigen by use of a fluorescent dye attached to the Fc region of the antibody molecule. This allows microscopic observation using ultraviolet illumination.

immunologically naive Refers to a subject that has never been infected with the infectious agent in question.

IN The retrovirus gene encoding the enzymatic function that catalyzes the integration of proviral cDNA into the host chromosome.

***in situ* hybridization** One of a number of methods for localizing a specific nucleic acid sequence or species within a cell, accomplished by fixing the cell, making it permeable, and hybridizing an appropriate probe under conditions where cellular structure is maintained.

inapparent infection An infection not characterized by overt symptoms of disease, but in which there is active replication of the pathogen.

inbred Characteristic of offspring between two genetically closely related parents.

incubation period The time between initial infection and the onset of notable symptoms of a disease.

index case The first documented case in an epidemiological investigation of a disease outbreak.

inducible genes Genes whose expression can be induced under appropriate conditions.

influenza (flu) A generally mild infectious disease of the upper respiratory tract caused by a group of viruses with segmented RNA genomes and characterized by rapid genetic change.

informed consent Permission given for an experimental medical or research procedure only after a full disclosure of the possible dangers to the individual and the benefit to medical knowledge.

initiator tRNA *N*-formylmethionine-tRNA (fMet tRNA), which initiates the first amino acid in translation of bacterial proteins. In eukaryotic cells, the initiator tRNA is Met tRNA with no formylation.

innate immunity Also called nonspecific immunity, this is the collection of structures and responses to an invading pathogen, including anatomical barriers, the complement system, neutrophils, macrophages, and the interferon response, that are induced by the cell after it detects a virus.

inoculation Process of introducing a substance into an organism.

integral membrane protein A cell membrane-associated protein within but not extending appreciably beyond the membrane envelope.

intercellular adhesion molecule (ICAM) One of a large family of glycoproteins projecting through the cellular envelope that mediate the association between cells and between cells and surfaces; used by some viruses (notably poliovirus) as a receptor for entry.

interference In a mixed-virus infection, a phenomenon where some function encoded by the interfering virus reduces the efficiency of replication of the wild-type virus. See **defective interfering particles.**

interferon (IFN) A group of proteins (cytokines) secreted by virus-infected and certain other cells that act to induce a specific set of cellular antiviral and antitumor responses in other cells.

interleukin A cytokine secreted by an effector cell of the immune system that functions to stimulate other immune cells.

internal ribosome entry site (IRES) A feature in the secondary structure near the 5′ end of a picornaviral RNA genome that allows eukaryotic ribosomes to bind and begin translation without binding to a 5′ capped end.

intracellular trafficking proteins Intracellular proteins whose main function is to recognize specific molecules and guide them to the appropriate subcellular location.

intracerebral (i.c.) Literally, in the cerebrum. However, commonly used to mean in the brain, such as an injection route.

intracranial (i.c.) General term for injecting virus into the brain of an animal to assess viral replication or virulence.

intravenous (i.v.) Injection of a substance (or virus) into the vein. In the mouse, this typically means injecting into the tail vein.

intrinsic immune response The portion of the nonspecific immune response that is composed of proteins present in all cells, even uninfected ones, whose job is to detect and defend against pathogens. This includes apoptosis and autophagy pathway proteins as well as antiviral miRNAs.

introns The portions of a eukaryotic RNA removed by splicing.

iontophoresis Movement of a positively charged compound in an electric field.

IRES See **internal ribosomal entry site.**

isoform One of different structural forms of a protein. Isoforms may differ in their activity.

i.v. See **intravenous.**

Jennerian vaccine A live-virus vaccine that elicits an immune response to a related pathogenic virus infecting another species.

kb Abbreviation for kilobase (one thousand bases); used in designating the size of DNA and RNA. If the molecules are double stranded, the appropriate unit is kilobase pairs (kbp).

keratinized tissue Tissue marked by a large amount of keratin, such as at the surface of the skin, the cornea, and hair.

killed-virus vaccine A vaccine made up of a virus suspension that has been chemically treated so that it is no longer able to cause a productive infection.

knock-out (KO) mouse A type of transgenic mouse that has been engineered in such a way that a gene is interrupted or deleted such that the function of that gene is altered. KO mice that have deletions in components of immune effectors (such as cytokines) are useful in the study of viral pathogenesis.

Koch's rules A set of criteria that must be met to demonstrate that a specific microorganism is the causative agent of an infectious disease; named for Robert Koch, the nineteenth-century German microbiologist who first formulated them.

Kozak sequence The sequence ANNAUGG, which was identified by Marilyn Kozak as being a favored sequence for the initiation of protein translation on eukaryotic mRNA.

kuru A human encephalopathy caused by a prion and associated with ritualized funeral cannibalism.

lagging strand The strand of DNA being replicated in which synthesis is discontinuous.

λ-arms Distal ends of the linear lambda genome, containing information for the replication of the viral DNA. The lambda arms are used as parts of cloning vectors designed to take larger inserted DNA sequences.

Last Universal Common Ancestor (LUCA) In an evolutionary cladogram, the (usually theoretical) common ancestral form or forms that precede the development of further diverging forms. For instance, the ancestral cell or cells that led to the development of both prokaryotes and eukaryotes.

latency-associated transcripts (LATs) Transcripts expressed by many neurotropic herpesviruses during the latent phase of infection.

latent infection Usually refers to a period following acute herpesvirus infection in which the viral genome is present in specific cells, but in which genes encoding the replication genes are not expressed and viral replication does not take place.

LATs See **latency-associated transcripts.**

LD_{50} Median lethal dose; in a quantal assay, the dilution at which half the tested aliquots are able to initiate a lethal infection (i.e., contain sufficient infectious virus to kill half the test subjects).

lentivirus A group of retroviruses, such as HIV and visna virus, characterized by a slow, progressive pathogenic course.

leucine zipper A protein motif that involves the hydrophobic interaction between two amphipathic helices in which one side of each helix contains an alignment of leucine residues.

Leviviridae A family of positive-sense, single-stranded RNA bacteriophages, including MS2.

linker-scanning mutagenesis A deletion and replacement strategy in which specific base pairs of a sequence are altered *in vitro* without affecting the relative spatial arrangements and reading frames.

lipid raft endocytosis Entry into cells mediated by domains of lipids on the outer surface of the membrane, featuring glycosphingolipids and cholesterol. Similar to **caveolae-mediated endocytosis**, but without the presence of the protein caveolin.

live-virus vaccine A vaccine made up of a virus that has been specifically attenuated, usually by serial passage in a nonhuman host cell.

long terminal repeat (LTR) The 5′ and 3′ terminal regions of the proviral-integrated DNA of a retrovirus that contains control regions for viral RNA transcription.

LTR See **long terminal repeat.**

lymph nodes Small bodies of lymphatic tissue within the lymphatic system to which antigenic material is transported by antigen-presenting cells.

lyophilize Freeze-dry.

lysogeny Ability of certain bacteriophages (notably, bacteriophage l) to integrate its genome into that of the host bacteria and remain associated as a genetic passenger as the bacteria replicate.

mAbs See **monoclonal antibodies.**

macrophage The primary antigen-presenting cell of the lymphatic system.

macropinocytosis The entry into the cell of large quantities of fluids from the extracellular space.

major histocompatibility antigen A complex set of membrane glycoproteins encoded by the major histocompatibility complex (MHC). These antigens are on the surface of an antigen-presenting cell and determine whether or not an immune cell recognizes the presenting cell as "self." This is a necessary step in mounting an immune response.

male-specific phage A bacteriophage that uses the sex pilus of the host cell as a receptor.

malignant tumor See **cancer.**

MAVS (mitochondrial antiviral signaling) protein A feature of the **innate immune** response. MAVS mediates the action of NF-κ-B and IRF-3, transcription factors involved in the expression of interferon-β, and therefore a stimulation of the antiviral state in the cell.

MCSs See **multiple cloning sites.**

meningitis An infection of the lining of the brain and brain stem.

metastasis The process by which a cancer (malignant) cell breaks away from the tumor in which it originated, spreads to a new location in the body, and establishes a new tumor.

MHC See major histocompatibility antigen.

microarrays An ordered series of small (<200 μm) spots of material (nucleic acid or protein) immobilized on a solid surface (see also **microchip**) such that their interaction with a target molecule in solution can be observed. Microarrays usually contain thousands of such sample spots.

microchip A small glass surface containing thousands of immobilized samples to be used in a microarray analysis. If the samples are DNA, it may be called a DNA chip or a genome chip.
microglial cells Cells of the CNS that function as immune cells.
mixed infection A (viral) infection in which two or more distinct genotypes are able to infect the same cell or individual at the same time.
MOI See **multiplicity of infection**.
molecular mimicry The immunological resemblance between two unrelated proteins, for instance, a viral protein and a cellular protein. The resemblance usually involves one part of the protein structure. Such mimicry can lead to immunological consequences for the host during a viral infection, such as precipitation of autoimmune phenomena.
monoclonal antibodies (mAbs) Antibodies produced by a single clone of identical B cells or hybridoma cells.
monopartite genome A viral genome made up of a single segment.
monospecific An antibody or antiserum preparation that reacts only with the antigen of interest.
mosaicism Refers to organized tissue in which there is more than one distinct genotype intermixed with another.
mucosa The epithelial layer lining the digestive, respiratory, or urogenital tract.
multipartite genome A viral genome comprising two or more fragments.
multiple cloning sites (MCSs) Sequences of closely spaced restriction enzyme cleavage sites constructed into a cloning vector to make available several possible points of insertion for DNA.
multiple sclerosis A neurodegenerative, autoimmune disease of the CNS. There is good evidence that at least some forms result from a complication caused by the persistence of an infectious agent (probably a virus) from an acute infection that occurred many years previously.
multiplicity of infection (MOI) Average ratio of infectious virus particles to target cells in a given infection.
mutation An inheritable change in the base sequence of the nucleic acid genome of an organism.
Mx One of a family of proteins induced by interferon action on a target cell. Many of these proteins have unknown functions, but MxA specifically interferes with the initial infection of cells by influenza and vesicular stomatitis virus.
myelitis Inflammation of the spinal cord.
myeloma cells Immortal tumor cells derived from lymphocytes. Such cells are useful in creating hybridomas to produce monoclonal antibodies, but they do not produce normal antibodies themselves.
myristoylation Modification of a protein by the covalent addition of myrstic acid to specific glycine residues. Myristoylation gives the protein a hydrophobic (fatty acid) anchor for membrane insertion.
myxoma virus A poxvirus that normally infects South American hares. It was introduced (with mixed success) into Australia in an attempt to control the devastating increase in the population of European rabbits introduced by English settlers.
narrow host range Interaction between virus and host such that only a very limited kind of host cells are susceptible to infection. HIV has a narrow host range in that if infects only human cells with certain specific cell surface receptors.
negative-sense RNA An RNA molecule whose sense is opposite that of mRNA.
negative-sense RNA virus A single-stranded RNA virus whose genome is the opposite sense of mRNA. The viral genome must be transcribed into mRNA by a virion-associated enzyme as the first step in virus gene expression.

negative strand See **negative-sense RNA**.
neoplasm A tumor or localized group of proliferating cells that have become independent of the normal control of cell replication.
neuroinvasive Refers to the specific ability of a virus to gain access to or "invade" the nervous system from the periphery. Ultimately this involves breaching the blood–brain barrier.
neurotropic virus A virus targeting cells of the nervous system.
neurovirulent Refers to the ability of a virus to cause death by replication in the nervous system. Often refers to replication specifically within neurons of the CNS.
nonpermissive cell A cell that will not support the (efficient) replication of a specific virus.
nonpermissive temperature The temperature at which a conditionally lethal, temperature-sensitive mutant will be nonfunctional.
nonproductive infection An infection in which the virus interacts with the host cells so that its infectivity is lost, but no progeny virus are produced.
nonstructural protein In a virus infection, a protein expressed by the virus that does not function or *is not* found in the infectious virus particle. See **structural protein**.
nosocomial An infection associated with or acquired during a stay in a hospital or health care facility such as a nursing home.
nuclear location signal (NLS) A short sequence of amino acids, rich in lysine, at the amino terminus of proteins destined to be transported into the nucleus of the cell.
nucleoprotein A protein–nucleic acid complex.
Oct1 The cellular octamer-binding protein that binds to eight nucleotides in double-stranded DNA with the nominal sequence TATGARAT (R is any purine). This protein serves as an "adapter" for the binding of the HSV α-TIF transcriptional activator to enhancers of immediate-early promoters.
Okazaki fragments Short fragments of nascent DNA synthesized using an RNA primer that is an early intermediate of DNA chain growth on the discontinuous (lagging) strand.
oncogenes (c-onc, v-onc) Genes encoding the proteins originally identified as the transforming agents of oncogenic viruses, some of which were shown to be normal components of cells; v-onc is the viral version of an oncogene, while c-onc is the cellular version of the same gene.
open translational reading frame (ORF) A sequence of bases read three at a time between a translation initiator signal (AUG) and a translational termination signal (UAA, UAG, or UGA) in an mRNA or in the DNA encoding that mRNA. Each triplet of bases specifies a specific amino acid in the protein encoded by the ORF.
operator The region of a regulated bacterial gene to which the product of a regulatory gene binds to modulate transcription.
operon A set of regulated genes in bacteria expressed as a single transcript modulated by the activity of a nearby regulatory gene.
opportunistic infection An infection that takes advantage of a depleted or deficient immune system to establish itself in a host. AIDS has, as one of its sequelae, opportunistic infections by otherwise innocuous pathogens.
opposite polarity The orientation of a strand of nucleic acid whose phosphodiester backbone is in the opposite 5′ to 3′ direction relative to another strand.
ORF See **open translational reading frame**.
ORFeome The map of all possible **open translational reading frame**s (ORFs) with a genome or set of genomic sequences.
ori See **origin of replication**.
origin-binding protein A protein involved in initiating rounds of DNA (or RNA) replication by binding to the specific origin sequence.
origin of replication (ori) Specific site in a DNA or RNA genome at which a round of replication is initiated.

outbred Results of a genetic cross between two unrelated organisms.

packaging signal A specific sequence of bases within the genome of a virus that functions in the association and insertion of the genome into the procapsid.

palindromic sequence A sequence of nucleotides in a double-stranded molecule that is self-complementary and, thus, has the same 5′ to 3′ sequence on both complementary strands; for example, GATATC.

palliative treatment A treatment of a disease or condition designed to minimize discomfort.

pandemic An epidemic infection that involves a large percentage of the world population during a single period of time.

papillomaviruses A large group of viruses with DNA genomes that are classified as papovaviruses and cause warts.

parameters (of an experiment) Measurable characteristics of an experiment.

particle-to-PFU ratio Ratio of total virus particles to infectious particles in a specific virus stock.

parvovirus One of a group of viruses with small, single-stranded DNA genomes.

passage (serial passage) In virology, the sequential infection, harvest, and reinfection of a virus into a host or cell culture.

pathogen A disease-causing organism or entity.

pathogenesis The mechanism of causing a disease.

PCNA (proliferating cell nuclear antigen) A protein involved in DNA replication in eukaryotic cells. One of a group of cyclins, PCNA is an accessory protein of DNA polymerase delta, and is involved in leading strand elongation during replication.

PCR See **polymerase chain reaction.**

Peyer's patches (gut-associated lymphatic tissue) Lymphatic tissues in the gut that allow antigenic proteins and pathogens to interact directly with the immune system.

phagemids Cloning vectors constructed with a combination of single-stranded DNA phage (e.g., f1) and plasmid genes. Phagemids are able to replicate as single-stranded or double-stranded molecules.

phenotype The observable characteristics of an organism that are determined by its genetic makeup.

picornavirus Small viruses with RNA genomes. Poliovirus is an example.

pilot protein A protein associated with the genome of tailed bacteriophages that functions to begin the process of genome injection into the host cell once the cell wall has been breached through the association with the base plate.

placebo An inert or innocuous substance used as a negative control for testing a drug in clinical trials.

plaque-forming unit (PFU) A unit of infectious virus determined by the ability of the virus to form a plaque or area of lysed cells on a "lawn" of susceptible cells.

plasmid-like replication The replication of an extrachromosomal DNA element, especially a circular molecule such as the genome of SV40 or polyomavirus.

plasmodesmata The cytoplasmic connections between plant cells.

Poisson analysis The statistical analysis of the probability of a given event happening after a small number of trials.

polarity For a nucleic acid, the direction in which the sequence is read (i.e., 5′ to 3′ or vice versa).

polyA polymerase The enzyme that adds the polyA tail at the 3′ end of eukaryotic mRNA molecules.

polyadenylation signals A specific sequence (AAUAAA) in a nascent eukaryotic mRNA that specifies the site for endonucleolytic cleavage of the transcript and addition of the polyA tail.

polycistronic mRNA An mRNA molecule that contains multiple open reading frames, usually found in prokaryotic mRNAs.

polyclonal Refers to an antiserum containing a number of different types of antibody molecules directed against various determinants on an antigen. These antibodies are secreted by various B cells, each derived from a distinct precursor by clonal selection.

polydnavirus A DNA virus replicating in the ovaries of certain parasitic wasps that can suppress the immune response of the caterpillar prey to the developing wasp embryo.

polymerase chain reaction (PCR) A linked set of reactions using sequence-specific primers and a high-temperature DNA-dependent DNA polymerase; used to amplify a specific DNA sequence by multiple rounds of primer-directed DNA synthesis.

polythetic Defining the relationships between members of the groups based on several similarities rather than a single, common feature.

positive-sense RNA RNA whose sequence is the sense of mRNA.

positive-sense RNA virus A single-stranded RNA virus whose genome is the same sense as mRNA.

positive strand See **positive-sense RNA**.

posttranscriptional modifications Cellular enzymatic modifications to the primary structure of a transcript following synthesis from the DNA (or RNA) template; include polyadenylation, capping, and splicing.

pre-biotic Referring to the time in the history of earth before the existence of cellular forms of life, or before the existence of living structures, such as self-replicating molecules.

pre-initiation complex The assembly of transcription factors and RNA polymerase II at the TATA box and cap site of a eukaryotic transcript that forms just prior to the initiation of transcription.

Pribnow box An AT-rich sequence 10–12 base pairs upstream of the start site of prokaryotic transcription that serves as an association site for RNA polymerase; analogous to the TATA box of eukaryotic mRNA.

primary cells Cells isolated directly from the tissue of origin that display all the histological and growth properties of cells in the tissue of origin.

primase The enzymatic activity that catalyzes the synthesis of the RNA primer that begins DNA replication.

primosome The complex of primase and DNA helicase that is involved in the synthesis of the RNA primer for DNA replication.

prion An infectious agent spread by ingestion that does not appear to contain any genetic material. It is thought to be a host cell protein that is folded in such a way as to lead to neurological degeneration and can induce similar conformational changes in identical proteins originally folded in a benign manner in the infected individual.

procapsid A precursor to a mature viral capsid.

prodromal period A time prior to the onset of full symptoms of a disease when specific physiological responses resulting from it can be discerned by a practiced observer.

productive infection A virus infection of cells in which more infectious virus is produced than was present to initiate the infection.

professional antigen-presenting cells Cells of the immune system whose function is to take up antigens by endocytosis, degrade them into fragments, and display the fragments on their surface in complex with class II MHC. Such cells include macrophages and B lymphocytes.

programmed cell death See **apoptosis**.

promoter A region of DNA proximal to the transcript start site of a gene that controls the formation of the pre-initiation complex and transcription.

propagate Spread by replication.

prophage The form of the genome of a lysogenic bacteriophage integrated into the host cell chromosome.

prophylactic Preventative.

***prot* (protease)** The gene expressed by a number of viruses, especially retroviruses, that catalyzes the proteolytic maturation of polyproteins into specific virus proteins.

proteome The data set of all proteins encoded by and expressed from a genome or set of sequences.

proteosome A complex structure within eukaryotic cells that is the site of protein degradation. Proteins destined for turnover at the proteosome have been tagged by the addition of **ubiquitin**.

prototrophy The ability of bacteria to grow and replicate with only an energy and carbon source (usually a sugar) and inorganic sources of nitrogen, sulfur, and phosphorus.

provirus The double-stranded cDNA produced by reverse transcriptase as the first step in infection by a retrovirus.

pseudo-pregnant female In the production of a **transgenic mouse**, a female mouse of breeder age that has been hormonally treated to make her receptive to receive an embryo.

pseudovirion A virus-like particle that has the external structure of a virus but does not contain the viral genome.

pulse As applied to a virus infection or other molecular and biochemical applications, a short time interval in which an experimental modification of conditions such as the addition of a radioactive precursor is carried out.

pulse-chase experiment An experimental protocol in which a radioactive precursor is provided to a system for a defined, usually short, amount of time, which is then followed by the addition of a large excess of unlabeled precursor (chase) to dilute the pool of material; used to follow the fate of material synthesized during the pulse period.

quantal assay A statistical assay of infectious virus based on dilution endpoints.

quasi-species swarm A population of RNA viruses in which, by random errors made during genome replication, a large number of possible variants are represented, for instance, any population of HIV particles.

R-loop mapping A method of visualizing genes using electron microscopy to detect the looping out of DNA from an RNA–DNA duplex formed by the hybridization of mRNA and the double-stranded DNA gene encoding it.

random reassortment The random mixing of genomic segments following mixed infection with a virus with a segmented genome; an important mechanism for generating genetic diversity in reoviruses and influenza viruses.

rate zonal centrifugation A technique of subjecting a macromolecule, organelle, or virus particle to a high centrifugal field so that its rate of sedimentation allows it to be separated from other materials that have different sedimentation rates.

reactivation (recrudescence) Periodic reappearance of an infectious agent following a period of latency; a hallmark of herpesvirus infections.

real-time PCR A method for the quantitative detection of amplified products of **PCR**. Real-time PCR measures the increase in copy number during early stages of the amplification process, using fluorescent reaction components. This is in contrast to traditional PCR, which is an endpoint analysis.

reanneal To allow the two complementary strands of a double-stranded nucleic acid that have been separated (denatured) by chemical or thermal denaturation to re-form the original duplex.

receptor A specific macromolecule (usually a protein) on the surface of a cell that interacts with one or several specific proteins present in the exterior medium. In the case of a virus infection, the cellular receptor interacts with a specific viral structural protein or proteins to initiate infection.

recombinant See **genetic recombination**.

recrudescence A new outbreak of symptoms after a time during which symptoms were absent or greatly reduced.

reovirus Viruses with a segmented, dsRNA genome.

replica plate A method of replicating an organism or virus on a solid or semisolid surface so that the spatial relationship between individual clones is maintained for screening or selection.

replication fork The growing point in the replication of a DNA duplex molecule.

replicative intermediate (RI) The replicating structure, consisting of a template strand and multiple complementary progeny strands, found in cells infected with a single-stranded RNA virus. There are two forms: RI-1 is the complex synthesizing complementary strands using virion (genomic) or virion-sense RNA as a template; and RI-2 contains RNA complementary to genomic RNA as the template and genomic-sense RNA as the product strands.

replicon A region of DNA or RNA genome whose replication is controlled by an **origin of replication (ori)**.

reporter gene A gene not normally found within a eukaryotic cell whose expression can be used as a measure of gene expression in that cell. Chloramphenicol acetyl transferase (CAT) is a bacterial enzyme that can be used in this way for the study of gene expression in eukaryotes.

repressor A regulatory protein that blocks the expression of a gene by binding to a specific regulatory sequence in the DNA encoding it.

reservoir The source of an infectious agent.

resolvase An enzyme that can convert DNA recombination intermediates, such as the Holliday structure, to separate molecules by endonucleolytic scission and rejoining.

restriction enzyme A DNA endonuclease that is expressed in bacterial cells in order to degrade foreign DNA. The recognition sites are often palindromic, and various specific restriction endonucleases recognize exact double-stranded DNA sequences of 4–8 bases. The ability to cleave large pieces of DNA into specific fragments with such enzymes is an important basic tool in molecular cloning methods.

retroelements Regions of genomic DNA that contain structures such as long terminal repeats and genes for expressing reverse transcriptase and other enzymatic functions seen in retroviruses. Retrotransposons are one type of retroelement.

reverse genetics The reverse of traditional genetics. Rather than starting with a phenotype or gene product and searching for a gene, the experimenter begins with a region of known DNA sequence and then seeks to discover the phenotype or product of the presumed gene.

reverse transcriptase (Pol) An enzyme, originally discovered in retroviruses, that uses single-stranded RNA as a template for the synthesis of a cDNA sequence; can also use DNA as a template, and can degrade RNA from a DNA–RNA hybrid. This latter activity is termed RNase H activity.

reverse transcription–polymerase chain reaction (RT-PCR) A PCR reaction performed in conjunction with or following a reverse transcription reaction for detecting and quantifying small amounts of RNA. The reverse transcription reaction (catalyzed by the enzyme reverse transcriptase) synthesizes a cDNA copy of the RNA. The PCR reaction then amplifies specific cDNA sequences, depending on the primers used.

RI-1 (RI-2) See **replicative intermediate**.

ribonucleoprotein A complex of RNA and protein; see **nucleoprotein**.

ribosomal RNA (rRNA) The RNA species that make up structural and, in one case, enzymatic portions of both prokaryotic and eukaryotic ribosomes.

ribozyme An enzymatic activity of certain RNA molecules, such as the self-splicing of plant viroid RNA.

RIG-1 Retinoid-inducible gene 1. RIG-1 is a negative regulator of cellular growth that can induce differentiation or apoptosis.

RNA-dependent transcriptase An enzyme synthesizing RNA (usually mRNA) using Watson–Crick base-pairing rules and RNA as a template.

RNA editing An enzymatic change in the sequence of RNA by directly changing one of the bases. The process occurs in the replication of hepatitis delta virus as well as in the biogenesis of plant mitochondrial mRNA.

RNase H A ribonucleolytic activity of the reverse transcriptase enzyme that degrades the RNA portion of an RNA–DNA hybrid.

RNase L A ribonucleolytic activity that degrades mRNA when induced by 2′,5′-oligoadenylate in the presence of dsRNA; part of the antiviral state induced by interferon.

RT-PCR See **reverse transcription–polymerase chain reaction.**

R:U_5:(PB):leader region The 5′ terminal region of a retrovirus genome, containing a sequence also present on the 3′ end (R), a sequence unique to the 5′ end (U5), a binding site for the tRNA primer of replication (PB), and an untranslated region between PB and the first open reading frame (leader).

rubella German measles.

s value A numerical measure of the sedimentation rate of a macromolecule, organelle, or virus when the material is subjected to high centrifugal fields under defined conditions.

sarcoma A malignant solid tumor of mesodermal cells, such as muscle.

scaffolding proteins Proteins that are involved in the assembly of a viral capsid but are not part of the mature virion.

scale-free network A network in which a small percentage of nodes are connected to large numbers of other nodes. Such networks are called "scale-free" because their topography appears the same at any level of resolution (scale).

SCID mouse Severe combined immune-deficient mouse. These animals are useful in experimental models and have been specifically bred so that they lack both T-cell and B-cell immunity.

scrapie A slow, progressive, prion-caused neurological disease of sheep that leads to paralysis and death.

screen To isolate manually or automatically a desired mutant or cell line using observable differences between the desired phenotype and the background.

segmented genome A viral genome existing as two or more separate segments of nucleic acid; for example, the genome of influenza viruses.

select To isolate a desired genotype by incubating it and other genotypes under conditions where only the desired entity can replicate efficiently.

selfish genes Genes whose only function is to replicate themselves, providing no advantage to the entity carrying them; originally defined by Francis Crick to explain the existence of certain self-replicating genetic elements.

senesce To age to infirmity; in the case of normal cells in culture, the gradual loss of the ability to divide after multiple serial passages.

sequela (plural: sequelae) The long-term consequence(s) of a disease.

sequencing libraries Libraries containing genetic sequences of organisms.

serotype An organism or microbe with a distinct immunological signature.

severe acute respiratory syndrome (SARS) An emerging viral disease that presented first in Southeast Asia and China, followed by an outbreak in Canada. The disease, characterized by acute pulmonary edema, is caused by a coronavirus.

sex pilus One of the projecting portions of the bacterial cell wall that can instigate mating with a bacterial cell of another sex; used as a receptor by male-specific bacteriophages.

Shine-Dalgarno sequence A sequence element in prokaryotic mRNA, discovered by John Shine and Lynn Dalgarno, that binds to 16S rRNA and signals the site of initiation of translation at the nearest AUG downstream.

signal transduction cascade A series of reactions, precipitated by the binding of a factor to a cellular receptor. The series, which may involve phosphorylations, as an example, results in the activation of a final regulatory molecule, such as a transcription factor.

signaling cascade See **signal transduction cascade**.

small interfering RNAs (siRNAs) Small (21 nt) double-stranded RNAs produced from larger dsRNAs by the action of **dicer**. siRNA unwinds and interacts with RISC (RNA-induced silencing complex) to bind to specific mRNAs and degrade them, thus silencing the production of protein from that gene.

smallpox (variola) The first human viral disease to be eradicated from the population; caused by a cytoplasmic DNA virus. The disease had two forms caused by different viral serotypes: Variola major was the most severe form, while variola minor was a form characterized by a lower death rate and generally milder symptoms.

snRNA One of a class of small nuclear RNAs of eukaryotic cells involved in splicing; functions with cognate proteins to form a nuclear organelle termed a *small nuclear ribonucleoprotein particle* or snRNP.

spliceosomes The nuclear organelles made up of snRNA, unspliced pre-mRNA, and specific proteins that are formed as the first step in the splicing reaction of eukaryotic mRNA.

splicing The process of removing interior segments (introns) of eukaryotic pre-mRNA as part of posttranscriptional modification.

ssDNA-binding protein A protein involved in DNA replication that prevents reannealing of the two denatured strands at the replication point by binding to the single-stranded DNA in a sequence-independent manner.

SSPE See **subacute sclerosing panencephalitis**.

stimulation index A quantitative measure of T-cell proliferation in response to stimulation by a recognized antigen.

stochastic Random.

strong stop A step in the transcription of a retroviral genome into cDNA by reverse transcriptase; occurs after synthesis of cDNA from the tRNA primer to the 5′ end of the genome, and is a result of translocation of the newly synthesized cDNA strand to the other end of the RNA template.

structural protein A viral protein that is normally found specifically associated with the infectious virus particle.

subacute sclerosing panencephalitis (SSPE) A rare autoimmune disease caused by the host immune system destroying neural tissue in which noninfectious measles virus is maintained following recovery from the acute phase of infection; usually occurs within five years of the initial infection.

subcutaneous Under the skin.

subgenomic mRNA An RNA transcript of an RNA viral genome that contains only part of the sequence present in the entire genome.

subunit vaccine A vaccine preparation that contains only a viral antigenic protein.

suppression Change in the phenotype of a mutant by the effect of a mutation in a different gene, negating the effect of the original mutant.

symmetry Arrangement of repeating subunits in a structure; refers to the arrangement of virus capsomers within the virion.

symptoms Diagnostic features of a disease.

syncytia Cells whose cytoplasm has fused; one type of cytopathology induced by infection with some viruses.

syndrome A set of symptoms characteristic of a specific disease. This term is often used to describe a long-lasting condition.

T (large T) antigen The autoregulatory multifunctional early protein expressed by papovaviruses that functions to initiate rounds of viral DNA replication, inactivate cellular tumor suppressor genes, and activate late transcription.

t (small t) antigen The early protein co-linear with the N-terminal portion of T antigen expressed during infection with polyomaviruses. Its function is dispensable for virus replication in cultured cells.

T helper 1 cells (Th1) A subset of T-helper cells. Th1 cells suppress the activity of **Th2** cells and produce IL-1, IL-2, interferon-γ, and TFN-β. Th1 cells activate macrophages, assist B cells, and are involved in delayed hypersensitivity.

T helper 2 cells (Th2) A subset of T-helper cells. Th2 cells secrete a variety of interleukins, as well as being involved in a number of features of cell-mediated immunity.

T lymphocytes Immune cells that react with other cells bearing foreign antigens on their surface due to viral infection or genetic alteration.

TAP Transporter associated with antigen processing. Transport proteins responsible for moving fragments of antigens into the Golgi prior to presentation in complex with MHC I.

target tissue (organ) The tissue or organ of an individual infected with a pathogen that, when infected, is responsible for the appearance of the characteristic symptoms of the disease.

TATA box The AT-rich region (canonical sequence TATAA) about 25 base pairs upstream of the cap site of eukaryotic transcripts that serves as the nucleation point for the assembly of the pre-initiation transcription complex.

TATGARAT sequence The nominal eight-base sequence in double-stranded DNA to which the Oct1 and related proteins bind. In HSV-1 these elements occur in the enhancers for the immediate-early gene promoters, and the binding of Oct1 leads to subsequent association with the α-TIF protein and transcriptional activation.

$TCID_{50}$ Median tissue culture infectious dose; in a quantal assay, the dilution at which half the tested aliquots are able to initiate an infection in a test culture of cells (i.e., contain an infectious virus particle or PFU).

tegument The matrix of proteins and other material between the capsid and envelope of a herpesvirus virion.

telomere A structural element at the end of a eukaryotic chromosome that contains multiple repeated DNA sequences and a closed end.

temperature-sensitive (ts) mutation A conditional lethal mutation where an altered protein cannot assume its correct folding and structure at a high (nonpermissive) temperature and, thus, cannot function. At a lower (permissive) temperature, the protein can assume its correct structure and function normally.

termination factor (ρ factor) A protein that is involved in the termination of transcription of one class of prokaryotic mRNA molecules.

tetramer assay A synthetically produced complex consisting of four MHC molecules that contain a viral peptide fragment that is used to detect and quantify T cells that specifically recognize a virus protein.

therapeutic index A numerical measure of the effectiveness of a drug or therapeutic method. It is most simply the ratio of some numerical measure of the effectiveness against the disease or infectious agent to a measure of undesirable side effects.

thymidine kinase (TK) An enzyme involved in the salvage of pyrimidines within the cell; encoded by a number of herpesviruses and some other large DNA viruses.

tissue tropism The property of a virus that allows it to grow in a specific tissue.

titer Quantitative measure of the amount of a medically important substance or entity.

"toll-like" receptor (TLR) Transmembrane proteins that are involved in pathogen recognition and activation of both the innate and adaptive immune responses. TLRs are so named because of homology to the *Drosophila Toll* gene product.

topoisomerase An enzyme that can change the superhelicity of a double-stranded DNA molecule.

***trans*-acting elements** Genetic elements, transcripts, or proteins functioning throughout the cell in which it is expressed; the converse of a *cis*-acting element, which only functions on elements within the contiguous genome in which it occurs.

transfer vector Also called gene transfer vector, a viral vector designed to transfer a gene to a cell or tissue for **gene therapy** purposes.

transcription The enzymatic synthesis of RNA from either a complementary DNA or RNA template.

transcription factors A group of nuclear proteins of eukaryotes involved with RNA polymerase in the transcription of RNA from a DNA template. Many transcription factors are able to bind to specific sequences within the promoters and other regulatory regions of transcripts.

transcription-termination/polyadenylation signal (cleavage/polyadenylation signal) A sequence of bases that occur over 25–100 base pairs at the 3′ end of the gene, encoding a specific transcript that signals the point at which the RNA polymerase disassociates from the template. A major feature of this region is the presence of one or more sequences that are templates for polyadenylation signals in the transcript.

transfection The process of introduction of nucleic acid into a cell by nonspecific chemical means.

transformation The alteration of a cell by insertion of one or more foreign or mutant genes.

transgenic mouse A mouse that has been engineered to express a foreign **transgene.** This gene can be from a virus, another organism, or a reporter gene.

transient expression The temporary expression of genetic information in a cell after the insertion of genome sequences into that cell by some artificial means (e.g., **transfection**). In this case, the new information is not stably incorporated into the genetic material of the cell, but is expressed temporarily during the lifetime of the cell.

transitory (transient or abortive) transformation The change of a cell's growth characteristics by the temporary expression of a transforming gene product without a genotypic change in that cell; a temporary state of transformation, such as with a nonintegrating plasmid.

translocation In virus infection, the process of the genome-containing portion of the virion being biochemically transported across the cell or nuclear membrane.

transposase An enzyme mediating transposition of a transposable genetic element.

transposon A genetic element that can move by recombination from one location to another in a genome; often encodes genes that catalyze transposition.

tropism The tendency of a virus or other pathogen to favor replication in a specific set of tissues or site in the body.

tumor antigen (T antigen) Generally, an antigen found on or in a tumor but not a normal cell; specifically, an early multifunctional protein expressed by polyomavirus and SV40.

tumor suppressor genes Cellular genes whose function is to block uncontrolled cell replication.

ubiquitin A small (76 amino acids) protein in eukaryotic cells that is covalently linked to proteins destined for degradation at the **proteosome**.

VA RNA One of two small (about 160 nucleotides) transcripts transcribed early from the adenovirus genome by host RNA polymerase III; responsible for inhibiting the effects of interferon-α and interferon-β.

vaccination The process of using an inactivated or attenuated pathogen (or portion thereof) to induce an immune response in an individual prior to his or her exposure to the pathogen.

variola See **smallpox**.

variolation The practice of injecting dried exudate from recovering smallpox patients into an immunologically naive individual in order to generate protective immunity.

vector The agent or means by which an infectious agent is spread from one individual to another; also refers to an engineered plasmid or virus designed to transfer genes into an organism.

vegetative DNA replication Exponential viral genome replication.

vesicle A membrane-bound cellular compartment; also a fluid-filled blister or pouch that contains the infectious agent during an infectious disease.

viral cloning vector A cloning vector based upon the genome and replication strategy of a particular virus.

viremia The presence of a virus in the blood and circulatory system.

virion A virus particle that appears structurally complete when viewed in the electron microscope.

viroid A plant pathogen that is the smallest known nucleic acid–based agent of infectious disease. Each is a 250- to 350-base circular RNA molecule that encodes no protein and is not encapsidated. Hepatitis delta virus shares some features of viroids, including a highly structured circular RNA genome.

virosphere The complete array of viruses present in the biosphere.

virulence A measure of the severity of the disease-causing potential of a pathogen.

virus passage Transmission and generation of virus stocks by multiple rounds of virus replication, usually in the laboratory.

Watson–Crick base-pairing rules The basic rules that describe how one strand of a double-stranded nucleic acid can specify the sequence of the complementary strand; named for the two scientists who formally proposed them based on chemical and X-ray crystallographic data of double-stranded DNA. Most simply stated: (i) A pairs with T (or U in RNA), G pairs with C; and (ii) the two strands are antiparallel.

xenograft A piece of tissue or organ from a different species that is implanted or transplanted into an animal.

X-ray crystallography The determination of structure at the atomic level by the analysis of X-rays that are reflected from the planes of a crystal prepared from a macromolecule or virus particle.

yeast artificial chromosome (YAC) A cloning vector constructed to be a self-replicating element in yeast cells. The YAC contains the centromeric region, origin of replication, and telomeres from a yeast chromosome. YACs have very large (up to 1000 kbp) cloning capacities.

zoonosis A virus disease of another animal species that can cause a human disease.

Index

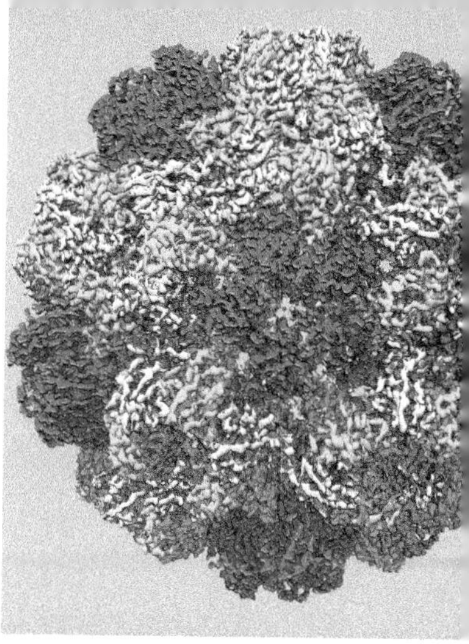

Basic Virology, Fourth Edition. Martinez "Marty" Hewlett, David Camerini, and David C. Bloom.